Handbook of Experimental Pharmacology

Volume 111

AF768228

Pharmacology of Smooth Muscle

Contributors

M.L. Cohen, K.E. Creed, P. D'Orléans-Juste, G. Edwards
M. Fujiwara, G. Gabella, S. Iino, T. Itoh, A.L. Killam
K. Kitamura, E. Klinge, J. Knoll, A.J. Knox, H. Kuriyama
P.F. Mannaioni, E. Masini, J. Mironneau, I. Muramatsu, H. Ozaki
J.Gy. Papp, G. Pogátsa, D. Regoli, N.-E. Rhaleb, N. Rouissi
K.M. Sanders, H. Schröder, K. Schrör, N.O. Sjöstrand, L. Szekeres
A.E. Tattersfield, T. Tomita, A.H. Weston

Editors

Laszló Szekeres and Julius Gy. Papp

Springer-Verlag
Berlin Heidelberg New York London Paris
Tokyo Hong Kong Barcelona Budapest

Professor L. SZEKERES, M.D., Ph.D., D.Sc.
Department of Pharmacology
Albert-Szent-Györgyi Medical University
Dóm tér 12, P.O. Box 115
H-6720 Szeged, Hungary

Professor JULIUS GY. PAPP, M.D., Ph.D., D.Sc.
Department of Pharmacology
Albert-Szent-Györgyi Medical University
Dóm tér 12, P.O. Box 115
H-6720 Szeged, Hungary

With 78 Figures and 36 Tables

ISBN-13: 978-3-642-78922-9 e-ISBN-13: 978-3-642-78920-5
DOI: 10.1007/978-3-642-78920-5

Library of Congress Cataloging-in-Publication Data. Pharmacology of smooth muscle/contributors, M.L. Cohen . . . [et al.]; editors, Julius Gy. Papp and Laszló Szekeres. p. cm. – (Handbook of experimental pharmacology; v. 111) Includes bibliographical references and index.
1. Smooth muscle – Physiology. 2. Smooth muscle – Effect of drugs on. I. Cohen, M.L. II. Papp, J.Gy. III. Szekeres, L. IV. Series. [DNLM: 1. Muscle, Smooth – physiology. 2. Muscle, Smooth – drug affects. W1 HA51L v. 111 1994/WE 500 p536 1994] QP905.H3 vol. 111 [QP321.5] 615'.1 s – dc20 [612.7'4] DNLM/DLC 94-17635

Softcover reprint of the hardcover 1st edition 1994

Typesetting: Best-set Typesetter Ltd., Hong Kong

SPIN: 10077685 27/3130/SPS – 5 4 3 2 1 0 – Printed on acid-free paper

Preface

Eighty years have passed since Arthur Heffter, the founder of this handbook series, invited in 1913 eminent scientists from different parts of the world to contribute. At that time 6–10 years were needed to publish the first two volumes, which appeared between 1919 and 1923.

During these 80 years, pharmacology as an independent science has undergone tremendouus development, which is reflected truly and comprehensively by the ever-growing number of volumes in the now "classic" series of *Heffter-Heubner's Handbook of Experimental Pharmacology*. The Editorial Board of distinguished, world wide known, experienced pharmacologists assumed responsibility for finding and editing the most current and most interesting topics, keeping in mind that some "evergreen topics" should be brushed up from time to time when sufficient new knowledge has accumulated.

In this sense it is surprising that the highly popular topic of ever-growing importance, namely "pharmacology of smooth muscle" has, in the knowledge of the editors of this volume, never been treated as such. Even the classic volume on the structure and function of smooth muscle edited by Bülbring, Brading, Jones and Tomita (*Smooth Muscle*, 1981) is more than 12 years old. So we think it is justified to say that the present volume really fills a gap. We were lucky to be able to invite eminent scientists working in this field and persuade them of the importance of their contributions, which cover the most important aspects of this wide-ranging topic.

The guiding concept of the book is a systematic survey of our present knowledge of the structure and physiological functions of smooth muscle and its response to endogenous substances and pharmacological agents. The latter includes findings on different organ systems containing smooth muscle, with the exception of the vascular system, which is treated in a special chapter owing to the great amount of new knowledge in this area which has accumulated in recent years. In several chapters an attempt is made to interpret the possible molecular mechanism of action of agents acting on smooth muscle.

Although 6–10 years were not needed to bring this volume to publication, as at the beginning of this series, nevertheless it has taken several years to collect and edit the material. So we are fully aware that, as with every large survey of current knowledge, this volume will be outdated to some extent at the moment of publication; new results which emerge every day can render firmly looking statements obsolete. We also know that even the most comprehensive handbook can never be complete. In this respect invitation of numerous authors is of some help. This is, unfortunately, associated with the disadvantage of some overlap between certain chapters, although we have tried our best with careful planning to avoid repetitions.

In spite of all these deficiencies the editors strongly hope that the present volume will be useful for those who wish to gain a deeper insight into smooth muscle pharmacology.

We would like to thank Professor Hans Herken (Berlin-Dahlem) for initiating this work and for his continued interest in its progress. We also feel greatly indebted to Mrs. Doris M. Walker (Biomedicine Editorial, Springer-Verlag) for keeping up our contact with the contributors, the publishing house and the Editorial Board. Without her active, always timely interventions and excellent cooperation we would never have succeeded in accomplishing our editorial task.

Szeged, Hungary
September 1994

L. SZEKERES
J.GY. PAPP

List of Contributors

COHEN, M.L., Lilly Research Laboratories, Eli Lilly Company, Lilly Corporate Center, Indianapolis, IN 46285, USA

CREED, K.E., School of Veterinary Studies, Murdoch University, Murdoch, Western Australia 6150, Australia

D'ORLÉANS-JUSTE, P., Department of Pharmacology, Faculty of Medicine, University of Sherbrooke, 3001, 12° Avenue North, Sherbrooke, Quebec, J1H 5N4, Canada

EDWARDS, G., Smooth Muscle Pharmacology Group, School of Biological Sciences, G 38 Stopford Building, University of Manchester, Manchester M13 9PT, Great Britain

FUJIWARA, M., Department of Pharmacology, Faculty of Pharmaceutical Sciences, Mukogawa Women's University, Nishinomiya, Hyogo 663, Japan

GABELLA, G., Department of Anatomy and Developmental Biology, University College London, Gower Street, London WC1E 6BT, Great Britain

IINO, S., Department of Physiology, School of Medicine, Nagoya University, Nagoya 466, Japan

ITOH, T., Department of Pharmacology, Nagoya City University Medical School, Nagoya 467, Japan

KILLAM, A.L., Department of Pharmacology and Toxicology, B440 Life Sciences Building, Michigan State University, East Lansing, MI 48824, USA

KITAMURA, K., Department of Pharmacology, Faculty of Medicine, Kyushu University, Fukuoka 812, Japan

KLINGE, E., University of Helsinki, Department of Pharmacy, Division of Pharmacology and Toxicology, P.O. Box 15 (Kirkkokatu 20), University of Helsinki, SF-00014 Helsinki, Finland

KNOLL, J., Department of Pharmacology, Semmelweis University of Medicine, Nagyvarad tér 4, P.O.B. 370, H-1445 Budapest, Hungary

KNOX, A.J., Respiratory Medicine Unit, City Hospital, Hucknall Road, Nottingham NG5 1PB, Great Britain

KURIYAMA, H., Department of Pharmacology, Faculty of Medicine, Kyushu University, Fukuoka 812, Japan

MANNAIONI, P.F., Dipartimento di Farmacologia, Preclinica e Clinica "Mario Aiazzi Mancini", Università degli Studi di Firenze, Viale G.B. Morgagni, 65, I-50134 Firenze, Italia

MASINI, E., Dipartimento di Farmacologia, Preclinica e Clinica "Mario Aiazzi Mancini", Università degli Studi di Firenze, Viale G.B. Morgagni, 65, I-50134 Firenze, Italia

MIRONNEAU, J., Laboratoire de Physiologie Cellulaire et Pharmacologie Moléculaire, CNRS-URA 1489, Université de Bordeaux II, 3, Place de la Victoire, F-33076 Bordeaux, France

MURAMATSU, I., Department of Pharmacology, Fukui Medical School, Matsuoka, Fukui 910-11, Japan

OZAKI, H., Department of Veterinary Pharmacology, Faculty of Agriculture, University of Tokyo, Bunkyo-ku, Yayoi 1-1-1, Tokyo 113, Japan

PAPP, J.GY., Department of Pharmacology, Albert-Szent-Györgyi Medical University, Dóm tér 12, P.O. Box 115, H-6720 Szeged, Hungary

POGÁTSA, G., Research Department, National Institute of Cardiology, Haller ùt 29, H-1096 Budapest, Hungary

REGOLI, D., Department of Pharmacology, Faculty of Medicine, University of Sherbrooke, Sherbrooke, Quebec, J1H 5N4, Canada

RHALEB, N.-E., Department of Pharmacology, Faculty of Medicine, University of Sherbrooke, Sherbrooke, Quebec, J1H 5N4, Canada

ROUISSI, N., Department of Pharmacology, Faculty of Medicine, University of Sherbrooke, Sherbrooke, Quebec, J1H 4N4, Canada

SANDERS, K.M., Department of Physiology, University of Nevada School of Medicine, Anderson Building, Reno, NV 89557, USA

SCHRÖDER, H., Institut für Pharmakologie, Heinrich-Heine-Universität Düsseldorf, Moorenstr. 5, D-40225 Düsseldorf, Germany

SCHRÖR, K., Institut für Pharmakologie, Heinrich-Heine-Universität Düsseldorf, Moorenstr. 5, D-40225 Düsseldorf, Germany

SJÖSTRAND, N.O., Department of Physiology, Karolinska Institutet, Solnavägen 1, S-104 01 Stockholm, Sweden

SZEKERES, L., Department of Pharmacology, Albert-Szent-Györgyi Medical University, Dóm tér 12, P.O. Box 115, H-6720 Szeged, Hungary

TATTERSFIELD, A.E., Respiratory Medicine Unit, City Hospital, Hucknall Road, Nottingham NG5 1PB, Great Britain

TOMITA, T., Department of Physiology, School of Medicine, Nagoya University, Nagoya 466, Japan

WESTON, A.H., Smooth Muscle Pharmacology Group, School of Biological Sciences, G 38 Stopford Building, University of Manchester, Manchester M13 9PT, Great Britain

Contents

Section I: General Aspects

Section II: Endogenous Substances and Smooth Muscle

CHAPTER 4

Eicosanoids and Smooth Muscle Function

CHAPTER 6

Effect of Histamine on Smooth Muscle

CHAPTER 7

Angiohypotensin

CHAPTER 9

Serotonin and Smooth Muscle

Section III: Pharmacological Agents and Smooth Muscle

CHAPTER 10

Excitation-Contraction Coupling in Gastrointestinal Smooth Muscles

CHAPTER 11

Airway Smooth Muscle

CHAPTER 19

Altered Responsiveness of Vascular Smooth Muscle to Drugs in Diabetes
G. Pogátsa. With 9 Figures 693

Section I
General Aspects

CHAPTER 1
Structure of Smooth Muscles

G. GABELLA

A. Introduction

Smooth muscle is widespread; it is found in all viscera and vessels, and is a major component in the wall of all tubular organs (one of the few exceptions is the bile duct of some species, such as the rat, which has no muscle). The appearance of many smooth muscles under a microscope varies to an extent which is surprisingly small by comparison with the wide range of functional properties found in different muscles. Yet, the structure of the muscle, which comes to life only at the level of resolution that can be achieved by electron microscopy, has features which account for its different mechanical properties.

Some structural properties are common to all smooth muscle cells: they are single (uninuclear) cells; they have no transverse striation; they have no T-tubules but abundant caveolae; they have numerous cell-to-cell and cell-to-stroma junctions. Other aspects of the muscle structure vary from organ to organ and are related to the characteristic functions of each muscle. Specific features are found at all levels of the muscle structure: (a) in the three-dimensional arrangement of muscle cells, layers, bundles, cords (Figs. 1, 2); (b) in the packing density of muscle cells, the amount of intercellular material, the occurrence and distribution of non-muscle cells (Fig. 3); (c) in the cytological characteristics of muscle cells; and (d) in the nature and distribution of proteins associated with the contractile apparatus, the cytoskeleton, the cell membrane and the stroma.

B. Smooth Muscle Cells

I. Cell Size

Smooth muscle cells are elongated, 20–100 times as long as they are wide at their widest point (Fig. 1). In contrast to the cylindrical shape of skeletal muscle fibres and cardiac muscle cells, smooth muscle cells are spindle shaped, or at least they taper from the nucleus-containing central portion to their two ends. This shape (schematically two cones joined by their bases) is dictated by the insertion of the filaments over the entire extent of the cell surface and is essential for the mechanical properties of this type of muscle.

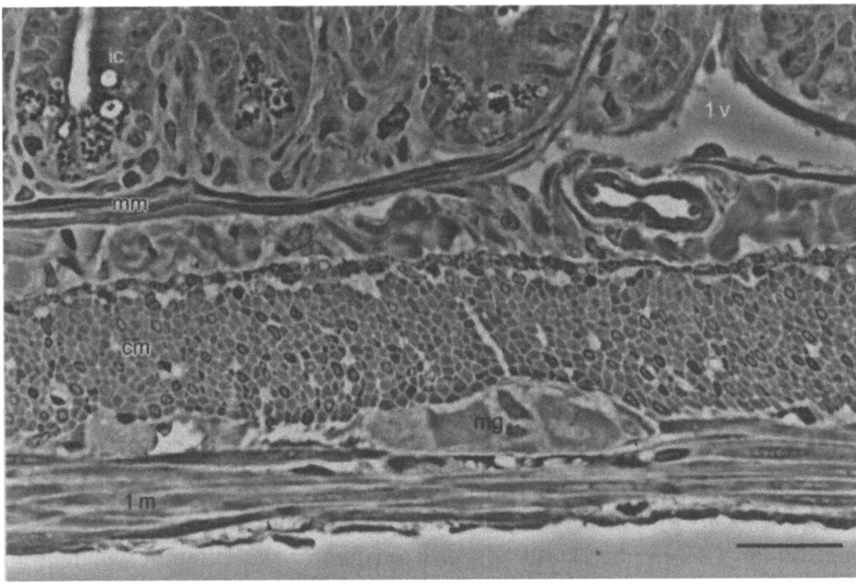

Fig. 1. Transverse section of the wall of the rat ileum. *cm*, circular muscle layer (in transverse section); *ic*, intestinal crypts; *lm*, longitudinal muscle layer; *lv*, lymphatic vessel (lacteal); *mg*, ganglion of the myenteric plexus; *mm*, muscularis mucosae. *Calibration bar*, 25 μm

The cells are mononucleated, the nucleus measuring 10–25 μm in length and occupying 20%–50% of the cell transverse profile. Binucleate cells are very rare.

Visceral muscle cells measure up to 1 mm in length and 2500–3500 μm^3 in volume (the volume of a sphere of 17–19 μm in diameter), with only limited variation between different muscles and between animal species. However, they are markedly larger in hypertrophic muscles. Very large muscle cells are consistently found in the muscle coat of the stomach of amphibia (HEIDLAGE and ANDERSON 1984). In contrast, vascular muscle cells are shorter and smaller, especially in the vessels with high intraluminar pressure (arteries).

An important parameter of smooth muscle cells is the ratio of surface to volume. Visceral muscle cells, for example in the taenia coli of the guinea pig, with a volume of 3500 μm^3 and a surface area (without taking the caveolae into account) of about 5300 μm^2, have a surface to volume ratio of 1.5, i.e. there is 1.5 μm^2 of cell surface for every μm^3 of cell volume (this value is the same as that found in human erythocytes (LINDERKAMP and MEISELMAN 1982)). Other muscle cells, for example those of elastic arteries, which have a smaller volume and a more irregular contour, also have a higher surface to volume ratio, for example 2.7 in the rat aorta (OSBORNE-

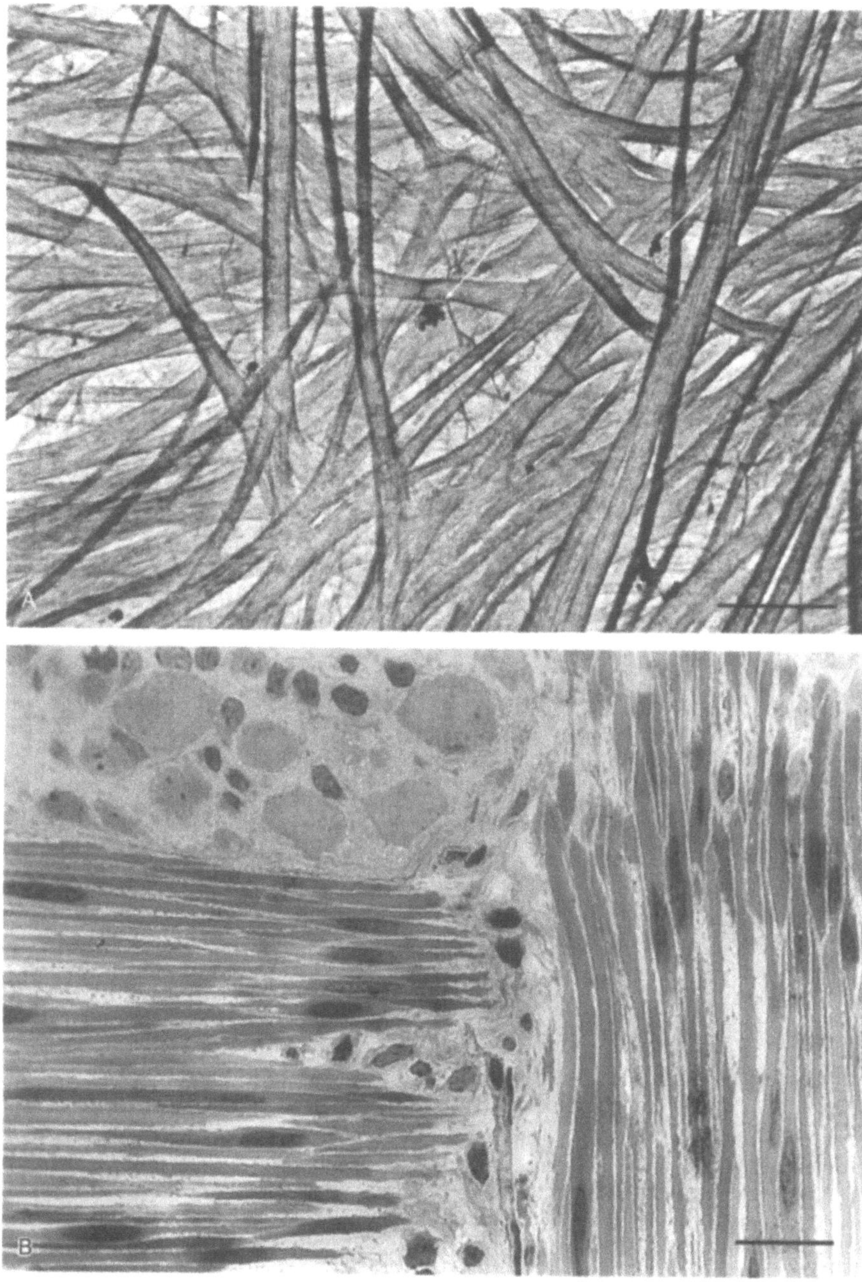

Fig. 2. A Muscle coat of the guinea pig urinary bladder, in a whole-mount preparation. Note the variability in size of the muscle bundles and their irregular orientation. The *arrows* point to small intramural nerve ganglia. *Calibration bar*, 1 mm. **B** Muscle coat of the guinea pig ileum, in a section near parallel to the serosal surface. Note the regular arrangement of the muscle cells in each layer, and the orthogonal arrangement of the two layers. *cm*, circular muscle; *lm*, longitudinal muscle layer; *n*, neurons in a ganglion of the myenteric plexus; *g*, glial cells. *Calibration bar*, 25 μm

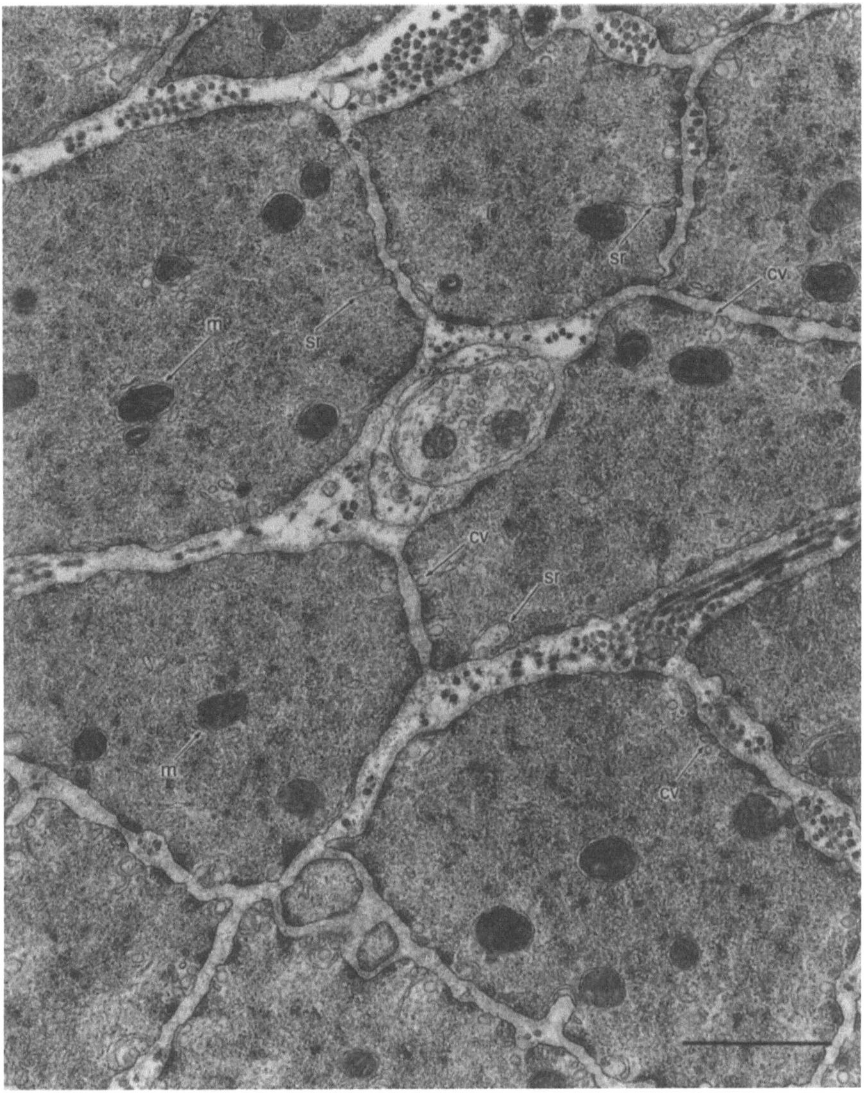

Fig. 3. Muscle cells of the rat urinary bladder in transverse section. The muscle cell profiles are occupied by myofilaments and by mitochondria. Collagen fibrils, mainly in cross-section, are abundant in the extracellular space. In the centre is a varicose nerve ending, packed with clear axonal vesicles and with two mitochondria. The nerve ending is partly coated by a slender Schwann cell process and forms an intimate junction with a muscle cell. *cv*, caveolae; *m*, mitochondria; *sr*, sarcoplasmic reticulum. *Calibration bar*, 1 μm

PELLEGRIN 1978), whereas in large muscle cells of the amphibian stomach or of hypertrophic viscera in general, the ratio is less than 1.

The packing density of muscle cells varies between muscles, and it depends not only on the size of the cells and the arrangement of the bundles but also on the amount of extracellular materials (see p. 19). High densities are found in the muscle coats of large viscera (Figs. 1, 3). For example in the guinea pig taenia coli the cells lie parallel to each other and are packed to a density of about 190000/mm^3, and 1 g tissue contains therefore about 1 m^2 cell membrane (and 1.7 m^2 if the caveolae are taken into account).

II. Cell Membrane

Ultrastructurally, there is an obvious distinction between the expanses of the cell membrane (plasmalemma) occupied by caveolae and those occupied by dense bands, and they probably represent different "domains" of the cell membrane (Figs. 3, 4). In freeze-fracture preparations the cell membrane displays about 450 intramembrane particles on the P-face and about 300 on the E-face (DEVINE and RAYNS 1975), mainly concentrated in the regions occupied by caveolae. Chemical differences between the two domains include the presence of dystrophin (BYERS et al. 1991; NORTH et al. 1993) and of a calcium pump (FUJIMOTO 1993) specifically in the region of the caveolae, whereas the regions of the dense bands is specifically occupied by vinculin and other cytoskeletal molecules (see below). Strictly related with the cell membrane are caveolae, dense bands, the basal lamina and various cell junctions.

III. Caveolae

Caveolae are inpocketings of the cell membrane which are flask shaped and uniform in size and measure about 70 nm across and 120 nm in length, their long axis being orthogonal to the cell surface (Fig. 4). Their neck, clearly visible in freeze-fracture preparations, outlines a circular space only 35 nm across which opens in the extracellular space and is covered but not penetrated by the basal lamina. Small particles of dense material line the cytoplasmic side of the caveolar neck, while a ring of intramembrane particles lies around the opening of the caveola. The membrane of caveolae shows fewer intramembrane particles than the cell membrane proper. In visceral muscle cells caveolae tend to be arranged in longitudinal rows, interposed between dense bands, and occupying almost half of the cell surface over the entire cell length. In the guinea pig taenia coli there are 32–35 caveolae/μm^2 and 170000/cell (GABELLA and BLUNDELL 1978). On the whole, caveolae add about 70% to the amount of plasma membrane present at the cell surface. In different words, more than one-third of the plasma membrane at the cell surface is in the form of caveolae, while the rest constitutes the cell surface proper. This value varies between different

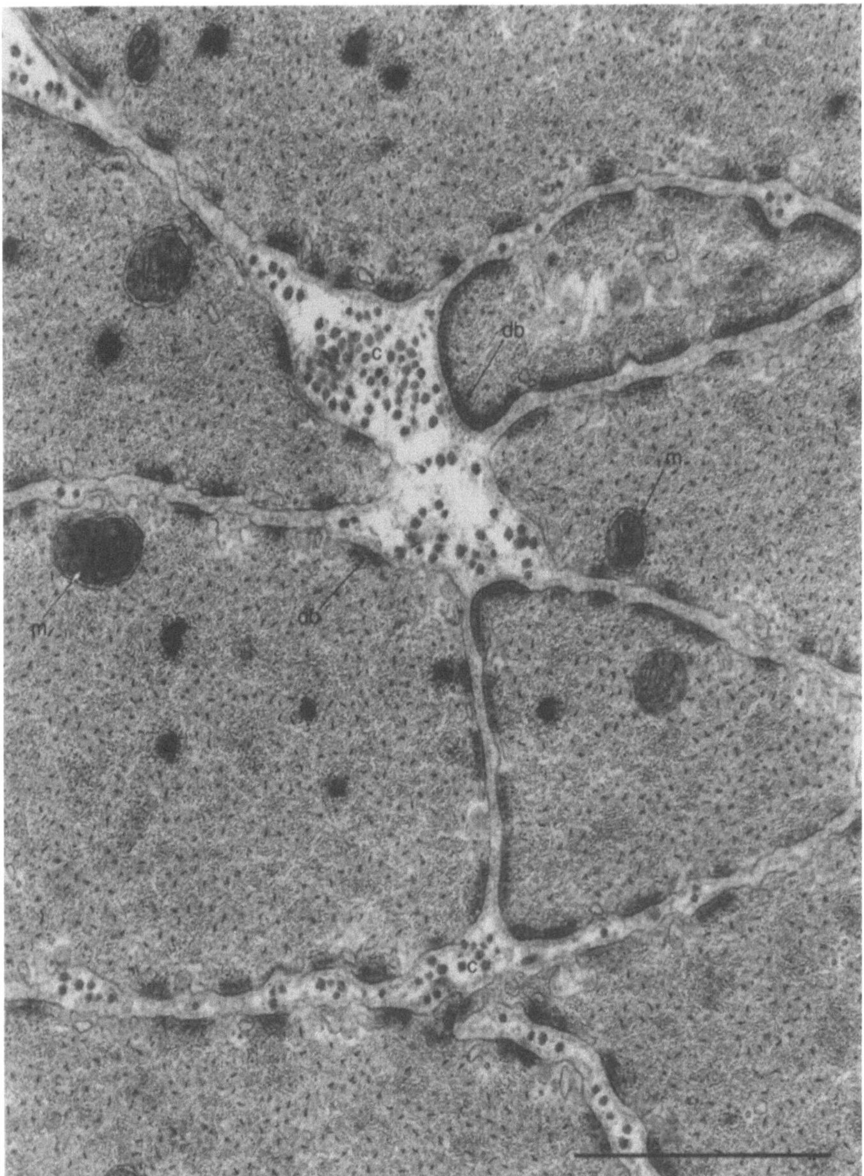

Fig. 4. Muscle cells of the circular muscle of the chicken ileum, in transverse section. The *dark dots* in the extracellular space are profiles of collagen fibrils. Apart from a few mitochondria, the muscle cell profiles are mainly occupied by myofilaments and by dense bodies. Note the dense encrustation of the cell membrane, or dense bands; these are particularly well developed in the smallest of the cell profiles. *c*, collagen fibrils; *db*, dense bands; *m*, mitochondria. *Calibration bar*, 1 μm

muscles, but none are devoid of caveolae. Clusters of caveolae opening into a common cavity which reaches the cell surface are not uncommon in certain muscles. Caveolae often lie close to cisternae and tubules of sarcoplasmic reticulum.

The number of caveolae in muscle cells is stable and is not affected by contraction, relaxation and stretch of the muscle. Although the distribution of caveolae is affected by the mechanical activity of the tissue, the number, size and diameter of the neck appear constant (GABELLA and BLUNDELL 1979). Extracellular space tracers, such as colloidal lanthanum, ferritin or peroxidase, gain access to the inside of caveolae: neither the basal lamina nor the narrow caveolar neck is a barrier to the diffusion of these tracers. The ferritin experiments also show that caveolae do not display pinocytotic activity: over a period of 90 min caveolae do not separate from the cell membrane and do not form truly intracellular vesicles (DEVINE et al. 1972). The plasmalemmal calcium pump ATPase, the enzyme that extrudes calcium from the sarcoplasm and thus maintains calcium homeostasis (CARAFOLI 1991), is localized in the membrane of caveolae in immunocytochemical preparations of smooth muscle cells and other cell types (FUJIMOTO 1993), as already suggested by earlier immunohistochemical studies (OGAWA et al. 1986; NASU and INOMATA 1990). High concentrations of calcium are detected by electron probe inside caveolae of smooth muscle (POPESCU and DICULESCU 1975). A protein probably involved in calcium entry through the plasma membrane, an inositol 1,4,5-trisphosphate receptor, is localized in the caveolae (FUJIMOTO et al. 1992). Both these histochemical observations apply to the caveolae not only of smooth muscle cells but also of other cells, especially endothelial cells, and therefore it has been suggested or assumed that the caveolae are basically similar in all cell types. This point remains dubious and conflicts with the evidence of features unique to smooth muscle caveolae.

Caveolae are found in a variety of cell types, especially those which are exposed to extensive mechanical deformation (endothelial cells, pneumocytes, muscle cells). Those of smooth muscle cells are characterized by the distribution in rows, possibly due to spatial constraints related to the presence of dense bands, by the gathering of intramembrane particles around the neck, and by the association with sarcoplasmic reticulum. In addition to the proposal that caveolae are involved in calcium transport across the cell membrane (POPESCU et al. 1974; CRONE 1986; FUJIMOTO 1993), it has been suggested that they participate in the control of cell volume (GARFIELD and DANIEL 1976), or act as miniature stretch receptors (PRESCOTT and BRIGHTMAN 1976), or serve the role of creating a specialized compartment of the extracellular space. The idea that they may provide additional membrane that is brought to the surface of the cell, by flattening of the caveolae, when the muscle is stretched, has been proven for the muscle cells of *Aplysia* (PRESCOTT and BRIGHTMAN 1976), but it does not seem to be applicable to vertebrate smooth muscle cells.

IV. Cell Junctions

Cell-to-cell junctions are very abundant in smooth muscles, and they serve at least two fundamental functions: mechanical coupling and excitation coupling. Mechanical coupling allows the contraction of each muscle cell to be added to that of the adjacent muscle cells and to be transmitted along the muscle. Equally important in this respect is the junction between muscle cells and stroma. Excitation coupling allows a virtually simultaneous activation of many muscle cells to take place, by spreading the excitation arising from nerve endings or from pacemaker cells to large assemblies of muscle cells. The junctions are structural specializations of the cell membrane, with the involvement of additional materials; they are symmetrical in appearance, and rather variable in extent and number in different muscles and species and in different preparations.

Gap junctions are found in smooth muscles and in many other tissues and they allow the direct and non-selective exchange of ions and small molecules between cells (ionic coupling, hence electrical coupling). Their structural and physiological features are the same as found in other tissues, such as liver and cardiac muscle. Gap junctions are recognized in thin sections because the intercellular space is reduced to less than 3 nm, the membranes are strictly parallel to each other, and there is no incrustation of dense material onto the cytoplasmic side of the membranes (Fig. 5A). In freeze-fracture preparations gap junctions are identified as distinct clusters of intramembrane particles (connexons), 8–10 nm in diameter, or of small intramembrane pits. The spatial density of connexons is about $7000/\mu m^2$, but the gap junctions of smooth muscles are small by comparison with those of other tissues, such as the heart, and they rarely exceed $0.2\,\mu m^2$. The clusters of particles are clearly outlined and they are usually elongated along the cell length; the membrane is generally flat and not markedly incurvated even with extensive changes in cell shape, for example in isotonic contraction. The largest junctions may contain about 1400 connexons from each of the two membranes, while the smallest ones are made of only 3–6 connexons. The main component of a connexon is the protein connexin. Connexin-43 from myometrium and from colonic musculature is immunologically identical to the connexin isolated from heart muscle (BEYER et al. 1989; RISEK et al. 1990). So, the main component of gap junctions in cardiac and smooth muscle is produced by the same gene (α_1), unlike gap junctions in liver and other epithelial tissues (RISEK et al. 1990).

The abundance of gap junctions varies between different smooth muscles. High values are found in the circular muscle of the small intestine. In the guinea pig duodenum and ileum, 0.5% and 0.2% of the circular muscle cell surface is occupied by gap junctions: this amounts to a total of 25 and $11\,\mu m^2$/cell, or 175 000 and 77 000 connexons/cell, respectively (GABELLA and BLUNDELL 1979). In contrast, in the adjacent longitudinal muscle, gap junctions are either absent or very small and few in number. There is,

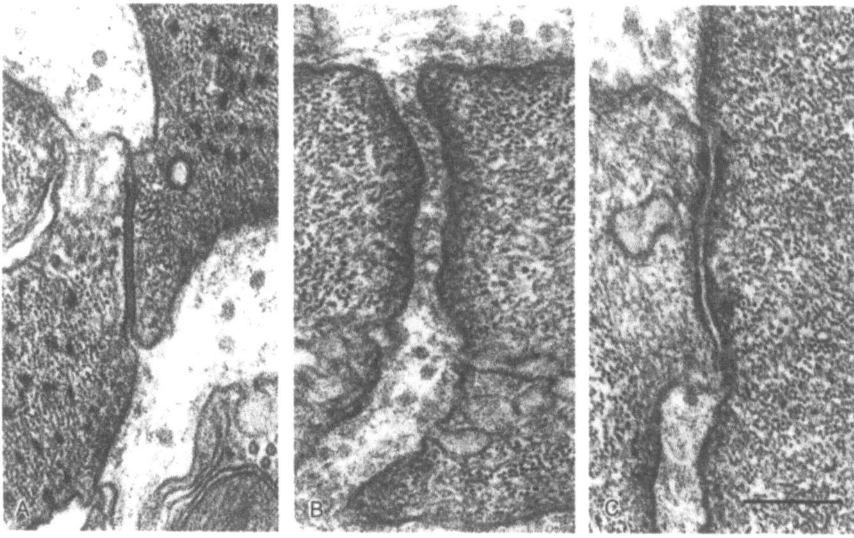

Fig. 5A–C. Three pairs of intestinal muscle cells, showing a gap junction (**A**), an intermediate junction (**B**) and a third type of junction (**C**). (From Gabella 1989). *Calibration bar*, 0.2 μm

however, evidence of dye coupling (transfer of intracellularly injected Lucifer yellow), in muscle cells of the guinea pig ileum longitudinal muscle in vitro (Zamir and Hanani 1990), clearly suggesting that in those experimental conditions the muscle cells are joined by gap junctions. In the caecum of the guinea pig too gap junctions are rare in the taenia but common in the adjacent circular muscle. The highest density of gap junctions is found in the sphincter pupillae of mammals, a muscle which is also very densely innervated. Many other smooth muscles are devoid of gap junctions, for example the detrusor muscle of the rat bladder. The problem of the extent of ionic coupling in muscles with very few or no gap junctions remains to be investigated. Alternative mechanisms to provide cell-to-cell propagation without gap junctions in cardiac and smooth muscle have been proposed (Sperelakis 1991).

The dynamic properties of smooth muscle gap junctions are not well understood. The junctions are structurally stable, and are not disrupted by contraction or stretch, but the rate of turnover of connexons is unknown. New gap junctions form rapidly in vitro in the tracheal muscle treated with potassium channel blockers (Kannan and Daniel 1978). Gap junctions are formed rapidly and in large numbers in myometrial muscle cells a few hours before parturition, so that they come to occupy 0.2%–0.4% of the cell surface (Garfield et al. 1977). Further studies have shown that progesterone suppresses the formation of myometrial gap junctions, while progesterone promotes it (Dahl and Berger 1978; Mackenzie and Garfield 1985).

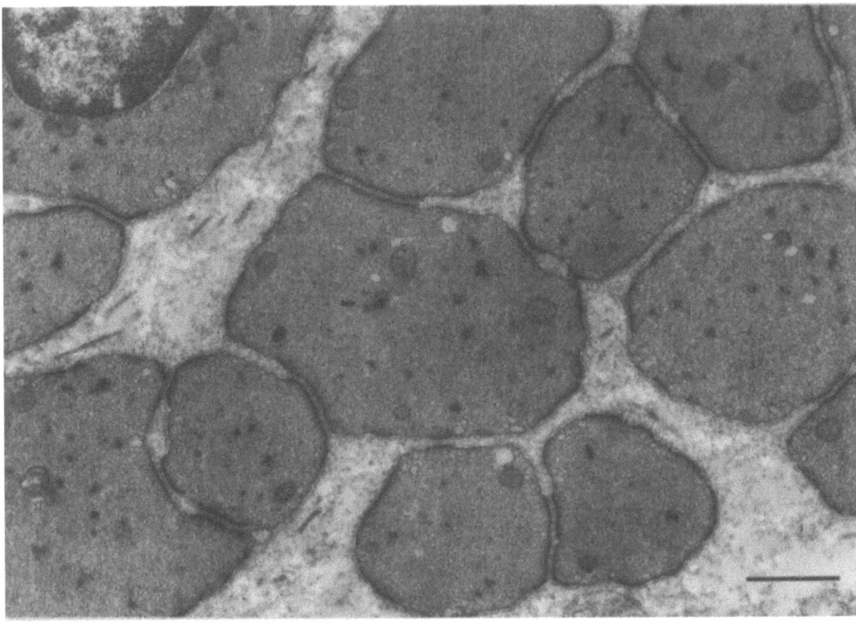

Fig. 6. Muscle cells of the detrusor muscle of the bladder of a baboon. Note the conspicuous development of intermediate junctions. *Calibration bar*, 1 μm

Adherens type junctions of smooth muscles are generally called intermediate junctions and they provide strong adhesion of the two membranes across a gap of up to 60 nm occupied by special extracellular materials (Fig. 5B). The junction is formed by two dense bands, one from each cell, and is thus linked to bundles of actin filaments (Fig. 6).

There are other types of junctions in smooth muscles, of uncertain functional significance. They include a patch of apposition of the two membranes, with an intercellular gap of up to 60 nm and incrustation of dense material on the cytoplasmic side but no association with actin filament bundles (Fig. 5C). There are also large areas where two muscle cells lie within about 30 nm from each other, with exclusion of basal lamina and absence of dense material, and not uncommonly there are laminar or finger-like processes penetrating into an adjacent cell. There are also some heterologous junctions (junctions between cells of different type), including the junctions between nerve endings and muscle cells. There are no tight junctions in smooth muscles.

V. Dense Bands

Dense bands are structures associated with the cell membrane where contractile apparatus and cytoskeleton are anchored to the cell surface (Fig. 4). When a band is paired with a similar structure in an adjacent cell a cell-to-

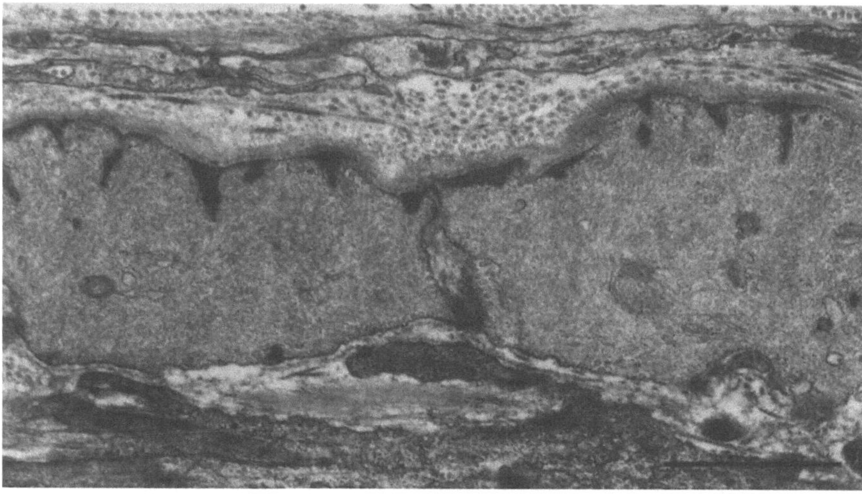

Fig. 7. Longitudinal section through an arteriole of ileal submucosa of a pig. Endothelium *at bottom*, adventitia *at top*. Between the two the media shows profiles of a muscle cell transversely sectioned (the cell makes several turns around the lumen). Note that on the muscle cell membrane facing the adventitia the dense bands are more numerous and larger and project more deeply into the sarcoplasm than those in the membrane facing the endothelium. *Calibration bar*, 2 μm

cell junction of the adherens type is formed (intermediate junction); in the other cases dense bands are coupled with collagen or elastic fibres and can be regarded as part of cell-to-stroma junctions. Because they have structural features in common with the cytoplasmic dense bodies, dense bands are also called membrane-associated dense bodies. They are about 30 nm thick and 0.2–0.2 μm wide; their length exceeds 1 μm, but it is more difficult to establish. Dense bands are distributed over the entire cell surface and occupy 30%–50% of the cell profile, and an even greater percentage at the tapering ends of the cell (Fig. 4). In muscle cells of large arteries some dense bands are much thicker and project into the sarcoplasm in the shape of a wedge (Büssow and Wulfhekel 1972). In arterioles, where the muscle cells are tightly coiled and make several turns around the lumen, the dense bands are more numerous, wider and deeper on the side of the cell facing the adventitia than on the side facing the endothelium (Fig. 7) (Gabella 1983). The linkage with actin filaments is very evident both in dense bands and dense bodies. The similarity with Z-lines is stressed by the fact that actin filaments are inserted with the same polarity: the arrowheads formed by decoration of the actin filaments with the subfragment S1 are directed away from the point of insertion in the dense material (Bond and Somlyo 1982). The actin filaments involved are always grouped in a tight bundle (cable of actin) and they approach the membrane at a very small angle. In many, but not all, dense bands there are also intermediate filaments; they approach a dense band at a variable angle and their spatial arrangement appears less regular than that of actin bundles.

The association of the actin filaments and the cell membrane is provided by a layer of electron dense material, usually 30–50 nm thick (Fig. 8). The proteins vinculin (GEIGER 1979) and talin (BURRIDGE and CONNELL 1983) are present at these sites (they are absent in dense bodies) together with alpha-actinin (GEIGER et al. 1981) and they provide the molecular link between actin filaments and transmembrane proteins. Vinculin lies closer to the cell membrane than alpha-actinin (GEIGER et al. 1981), while talin is closer to the cytoplasm (VOLBERG et al. 1986), although it is the former that has specific binding affinity for actin (BURRIDGE amd MANGEAT 1984). Talin, however, is characteristic of those dense bands which form a cell-to-stroma junction, and it is not found in those which form cell-to-cell junctions (DRENCKHAHN et al. 1988). In isolated muscle cells from the chicken gizzard the distribution of dense bands has been studied by labelling them with antibodies for talin and vinculin (DRAEGER et al. 1989); it was found that dense bands, totalling on average 67/cell, are arranged in periodic, mainly transverse bands along the cell surface. This distribution is at variance with that observed in thin sections of the same muscle in situ (GABELLA 1985). A longitudinal rib-like geometry of dense bands has been described in isolated muscle cells of guinea pig taenia coli and vas deferens (SMALL 1985). These "ribs" run parallel to each other along the cell length, and they remain parallel even when the cell shortens; many of them seem to extend for the full length of the cell (SMALL 1985).

VI. Sarcoplasmic Reticulum

Sarcoplasmic reticulum is well developed in smooth muscle cells. It consists of tubules and flattened sacs, studded or not with ribosomes (rough and smooth sarcoplasmic reticulum). Its shape, distribution and amount are very variable. The individual cisternae are probably very mobile organelles and it is fair to assume that the communication between cisternae in all parts of the cell is more extensive than can be appreciated on ordinary sections (~100 nm in thickness) used for electron microscopy. The complex extent of the sarcoplasmic reticulum can best be appreciated when it is stained in toto with osmium ferrocyanide (FORBES et al. 1979).

The extent of the sarcoplasmic reticulum has been estimated morphometrically in several muscles, in an attempt to correlate calcium storage capacity of a muscle with the extent of its sarcoplasmic reticulum, although the reticulum must be serving other roles too. Values range from over 5% in the aorta and pulmonary artery to about 2% in rabbit taenia coli (DEVINE et al. 1972) or to 1.5% in guinea pig taenia coli and vas deferens (MCGUFFEE and BAGBY 1976). Quantitative studies by electron probe on the guinea pig portal vein have shown that the calcium concentration within the sarcoplasmic reticulum (28 mmol/kg dry weight) is sufficiently high to activate

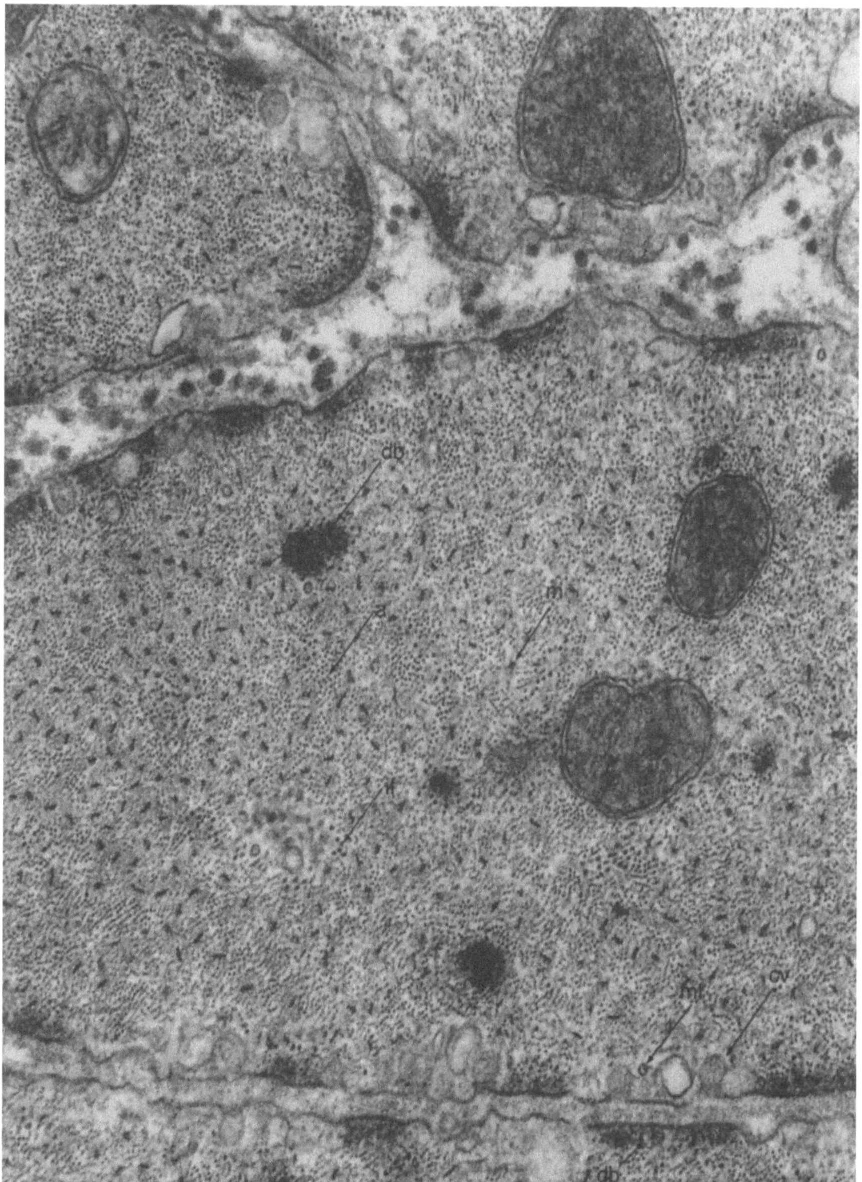

Fig. 8. Muscle cells of the chicken ileum. *a*, bundles of actin filaments; *cv*, caveolae; *db*, dense bands and dense bodies; *if*, intermediate filaments; *m*, myosin filaments; *mt*, mitochondria; *t*, microtubules. *Calibration bar*, 0.5 μm

full contraction of the muscle in the absence of extracellular calcium (BOND et al. 1985). In muscles rich in sarcoplasmic reticulum, the overall amount of membrane deployed in the reticulum is greater than that deployed at the cell surface.

Some of the elements of sarcoplasmic reticulum lie parallel to and close to the plasma membrane: bridging the gap of 12–20 nm there are ill-defined electron-dense structures with a periodic distribution (DEVINE et al. 1972). These bridges between sarcoplasmic reticulum and plasma membrane resemble the peripheral or surface couplings of cardiac muscle cells (SOMMER and JOHNSON 1968; FAWCETT and MCNUTT 1969). Tubules and cisternae of reticulum lie beneath or between rows of caveolae; the membrane-to-membrane separation is of the order of 15–20 nm, but bridging structures are not found.

VII. Filaments

The three classes of filaments found in smooth muscle cells are actin filaments, myosin filaments and intermediate filaments (Fig. 8). Associated with actin filaments are the dense bodies (and the dense bands described above). In addition to intermediate filaments, the cytoskeleton always comprises a small number of microtubules.

1. Myosin Filaments

Myosin (thick) filaments, approximately cylindrical but with a slightly irregular profile and tapering ends, measure 2.2 μm (ASHTON et al. 1975) or less in length (SMALL et al. 1990). The arrangement of the myosin molecules in thick filaments remains uncertain and the orientation and stability of the filaments during contraction is controversial. Myosin is present in polymeric, filamentous form both in the contracted and in the relaxed non-phosphorylated state (SOMLYO et al. 1981). The filaments may vary in length or number during contraction (GILLIS et al. 1988). The myosin molecules may be arranged as in myosin filaments of striated muscles, with bipolar symmetry and a bare area in the middle. An alternative, and more convincing, hypothesis, is that there is side polarity, i.e. that each filament has two sides which are parallel and span the full length of the filament: the myosin heads are oriented in the same direction on one side, and in the opposite direction on the other side, and there is no bare area (CRAIG and MEGERMAN 1977). This is the "side-polar" model of myosin filament (CRAIG and MEGERMAN 1977; COOKE et al. 1989), or the closely related "mixed-polarity" model (SMALL 1977; HINSSEN et al. 1978). By rotating isolated myosin filaments around their length, COOKE et al. (1989) found that cross bridges occurred only along two sides of the filaments, as predicted by the side polar model, rather than along the entire circumference, as predicted by the mixed polarity model. A side polar structure would also allow or favour the formation of ribbons of myosin, which are observed in some preparations but are now regarded as artifacts. With a side-polar arrangement, actin filaments could slide along the entire myosin filament and they could slide in opposite directions along the same filament.

2. Actin Filaments

Actin (thin) filaments, ~7 nm in diameter and of considerable but undetermined length, are preserved in most preparations of smooth muscle (Fig. 8). They are usually arranged not in rosettes around the myosin filaments but in bundles or cables of 10–20 filaments with a square lattice. Actin cables divide and merge along producing a pattern which has not yet been well worked out. Many actin filaments do not interact with a myosin filament, at least over part of their length. The ratio between actin and myosin filaments in smooth muscle cells is unknown (it is usually 2 to 1 in striated muscles) because the length of actin filaments is unknown (on the basis of estimates on cell fragments from isolated cells in supercontracted state of the chicken gizzard it has been reported that actin filaments average 4.5 μm in length (SMALL et al. 1990)). Even the ratio of filament intersections in smooth muscles is uncertain because the preservation of filaments for electron microscopy is difficult and inconsistent. In visceral muscle cells there may be as many as 12 actin filament intersections per myosin filament intersection (BOIS 1973), but there are probably large variations between different muscles. The weight ratio of actin to myosin is considerably higher in smooth muscles than in skeletal muscles, and it is higher in vascular than in visceral muscles (MURPHY et al. 1977).

Actin filaments penetrate into dense bodies and dense bands, with which they are intimately associated, the dense bodies usually receiving a bundle of actin filaments at each end (ASHTON et al. 1975; COOKE 1976). The polarity of the inserted filaments is similar to that found in actin filament inserted to Z-bands of striated muscle fibres: when the filaments are decorated with myosin subfragment 1 (S1) the arrowheads thus formed point away from the point of insertion (BOND and SOMLYO 1982; TSUKITA et al. 1983).

3. Dense Bodies

Dense bodies are scattered through the sarcoplasm, without an apparent pattern (Fig. 4). Up to 0.35 μm wide and 0.6–1.2 μm long, they are elongated in the direction of the myofilaments, heterogeneous in size and somewhat spindle-shaped; the size variation is marked even within a single muscle cells, and more so when different muscles and animal species are compared. The insertion of actin filaments into the dense bodies is well established, hence an analogy with the Z-lines of striated muscles (PEASE and MOLINARI 1960). In both cases, and in the case of dense bands, the polarity of actin filaments is constant and identical. Two bundles of actin filaments are inserted on either end of a dense body, and they have opposite polarities. Dense bodies are rich in alpha-actinin (SCHOLLMEYER et al. 1976; FAY et al. 1983), but, unlike dense bands, they do not contain vinculin. Immunohistochemical staining of dense bodies for desmin and filamin (SMALL et al. 1986) is due to the intermediate filaments that are associated with the periphery of dense bodies (Fig. 8). Dense bodies can be observed in living isolated muscle cells from the toad stomach, by virtue of their optical density in phase contrast microscopy (after saponin permeabilization) or by their staining with antibodies against alpha-actinin (FAY et al. 1983; KARGACIN et al. 1989), and their movement can be followed during muscle contraction (see p. 22). With these approaches a certain regularity in the distribution of dense bodies has been noted: they are arranged in string-like arrays, the strings running parallel to each other with some degree of lateral register of dense bodies in adjacent strings (FAY et al. 1983).

4. Intermediate Filaments

Intermediate filaments – a cytoskeletal component present in most cell types – are abundant in smooth muscle cells, and far more abundant than in cardiac or skeletal muscle (HUIATT et al. 1980): in the chicken gizzard, intermediate filaments contribute about 8% of the total protein content of the tissue (HUIATT et al. 1980). They measure 10 nm in diameter, have sharp profiles and are of indeterminate length. Unlike actin and myosin, they are insoluble in solutions of high ionic strength (COOKE and CHASE 1971) and their extraction from a muscle requires denaturing conditions such as urea or low pH (HUIATT et al. 1980). Incubation of muscle strips in these conditions produces cells in which the only filaments visible are intermediate filaments (COOKE 1976).

The main component of intermediate filaments in striated muscles and in visceral smooth muscles is the protein desmin; occasionally small amounts of vimentin are also present. In contrast, in vascular muscle cells intermediate filaments are made of vimentin with little or no desmin (GABBIANI et al. 1981). In myoepithelial cells of exocrine glands, such as mammary, sweat and salivary glands, the intermediate filaments are of the keratin type (FRANKE et al. 1980).

Intermediate filaments are located in an axial bundle that opens up to surround the nucleus (STROMER and BENDAYAN 1988); this arrangement, however, is not apparent in all muscle cells. Intermediate filaments are also found surrounding, or looping around, dense bodies, and beneath several of the dense bands.

A separation between myofilament domain and cytoskeleton domain proposed by SMALL and SOBIESZEK (1980) has not been confirmed by more recent studies (STROMER and BENDAYAN 1988). At least at the level of the dense bodies there is clear interaction between the two components.

C. Smooth Muscle as a Tissue

I. Non-muscle Cells

While the impression one gets by looking at a section of a muscle in micrographs is that of an assembly of muscle cells, one is actually dealing with a tissue in which several other cell types are present and in which the extracellular material, or stroma, is of paramount importance. Among non-muscle cells are vascular cells, mainly endothelial cells of capillaries and pericytes, fibroblasts, interstitial cells, mast cells, Schwann cells and nerve fibres. These cell types are not found in every muscle, and the pattern of these cells is often characteristic. Interstitial cells of Cajal are found in the musculature of the alimentary tract, and it is doubtful that non-muscle cells found in the interstitium of other muscles are equivalent cells. In muscles with a distinct bundle structure fibroblasts rarely penetrate into muscle cell bundles; they are absent in the media of blood vessels, with the important exception of the arteries of birds where layers of fibroblasts alternate with layers of muscle cells. Capillaries are absent or very rare in some muscles, such as the media of arteries or certain intestinal muscles (longitudinal muscle of the ileum in most species, taenia coli of the rabbit), whereas other muscles are well vascularized. There are about 500 capillaries/mm^2 in a transverse section of the guinea pig taenia coli (GABELLA 1988). The largest variability is found in the extent of the innervation. Nerves are absent in some smooth muscles such as the avian amnion or the largest arteries, while other muscles such as the bladder detrusor or the sphincter pupillae have innumerable nerve endings, or varicosities, expanded and packed with axonal vesicles, and making a close contact with each muscle cell.

Smooth muscles have no perimysium or epimysium and most of them have no tendons at all, although muscle cords and layers are usually very sharply outlined. Among the few smooth muscles with tendinous insertions are the avian gizzard, the anococcygeus muscle, and the costo-uterine muscle. At the level of the insertion onto the tendon the muscle cells develop deep longitudinal invaginations of the cell membrane, with conspicuous dense bands and an extensive contact with collagen fibrils. The tendons themselves are similar to those found in striated muscles.

II. Extracellular Materials

The major components of the stroma are collagen fibrils, elastic fibres and lamellae and the so-called ground substance.

Collagen concentration in smooth muscles, measured via the hydroxyproline content, is invariably greater than in striated muscles. In the guinea pig taenia coli there is 32 mg collagen/g tissue, in the myometrium 36, in the aorta 78, in the myocardium 6, in skeletal muscle 9 and in tendon

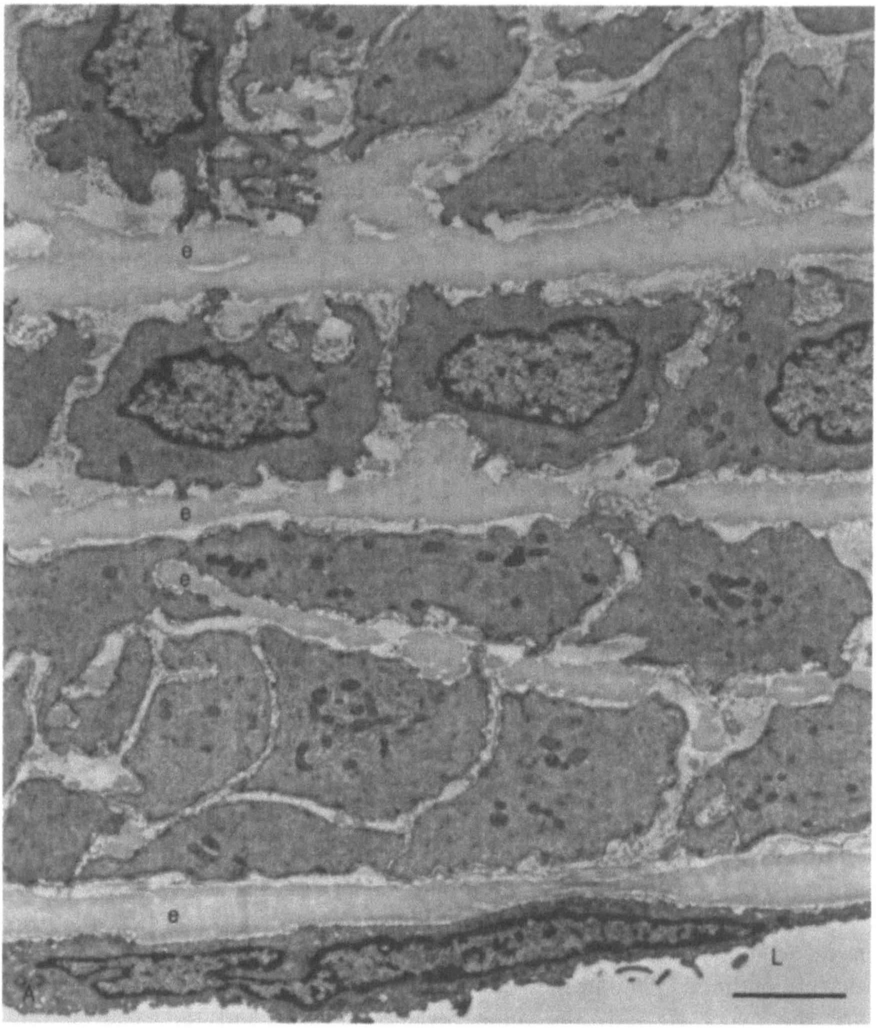

Fig. 9A,B. Longitudinal sections through the mesenteric artery of a rat, near its emergence from the aorta (**A**) and near the intestine (**B**), shown at the same magnification. *c*, bundles of collagen fibrils; *e*, elastic fibres and elastic lamellae, including the inner elastic lamina; *L*, lumen, lined by an endothelial cell showing its nucleus and cut longitudinally. The muscle cells are in transverse section. In **A** only part of the thickness of the media is shown. Note the large development of elastic material and the extent of the intercellular spaces. In **B** the muscle cells are closer to each other and the spaces are occupied by collagen bundles and there are numerous cell-to-cell contacts. *At the top* is the adventitia with fibroblasts and bundles of collagen. *Calibration bars*, 2.5 μm

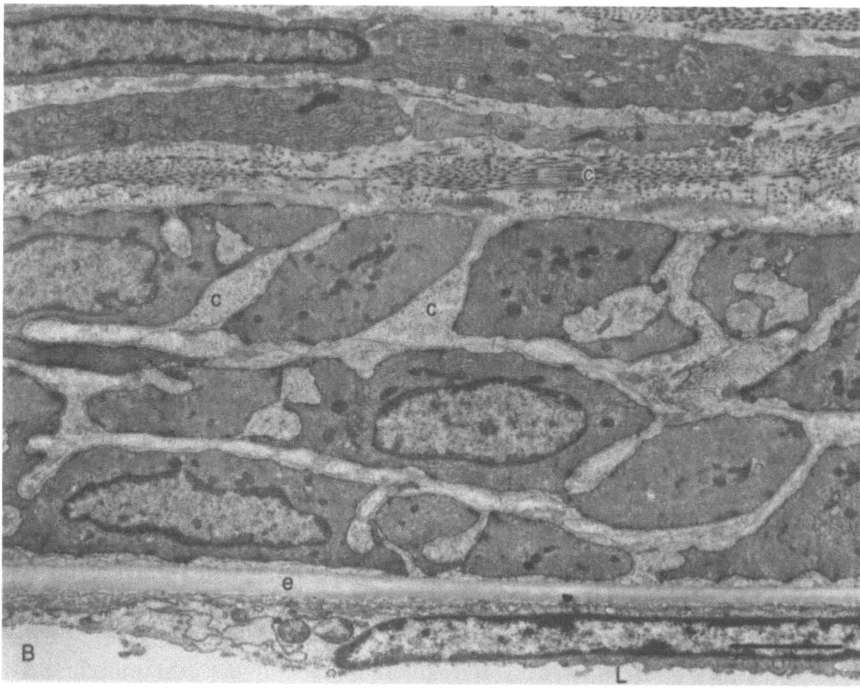

Fig. 9B

197 (GABELLA 1984). Most of the collagen content is accounted for by the collagen fibrils, cable-like structures of 30–35 nm diameter and of indeterminate but extremely long length. Some of the collagen, however, for example that of the basal lamina, which is type IV collagen is in non-fibrillar form. Elastic fibres and lamellae are very abundant in large (elastic) arteries, where the extracellular space amounts to more than 50% of the volume of the media. The elastic material is progressively reduced along the arterial tree (Fig. 9).

D. Changeable Structure of Smooth Muscles

In spite of the fact that microscopy mostly produces static views of cells and tissues, including smooth muscles, these are very dynamic structures. During mechanical activity, of course, they change by shortening, elongating, fattening, hardening and they do so in a reversible manner. In addition, muscles change shape in development while growing and differentiating, and even when they have reached the adult stage they can grow further by hypertrophy or can undergo atrophy, changes which are all related to adaptability of the muscle to variations in the loads and functional demands imposed. Therefore, the concept of a mature, stable and structurally defined smooth muscle is to some extent an idealized notion.

I. Contraction

In isotonic contraction there is a marked shortening of the muscle, which is in practice limited by structural constraints in the wall of the organ. In vitro, an isolated muscle strip can shorten, reversibly, to less than 20% its resting length, a shortening which would only rarely occur in a muscle in situ, and isolated muscle cells in vitro can contract until they acquire an almost spherical shape, although usually these extreme changes in shape are irreversible.

Contraction is very likely to be due to sliding of the two sets of myofilaments, as in striated muscle, although there is not yet direct experimental evidence for this. While there is agreement on the fact that in distended muscle cells the myofilaments, and any "unit" that may form, lie approximately parallel to each other, there is controversy as to their orientation in isotonically contracted cells. When the muscle cells are fully shortened, the "units" of myofilaments are described as forming with each other angles of 25° (SMALL 1974) or 45° (FISHER and BAGBY 1977), and on these observations (and on earlier studies on changes in birefringence of contracting muscle cells (FISCHER 1944)) is based one of the best known models of smooth muscle contraction (SMALL and SOBIESZEK 1980). In the opinion of the present writer, the myofilaments in visceral muscle cells contracting in situ remain approximately parallel to each other and approximately parallel to the cell length (GABELLA 1984). However, the lateral expansion of the shortening muscle cells is accompanied by some torsion of the superficial parts of the cell and the surrounding stroma, as seen in thin sections of muscles contracting in situ (GABELLA 1984). The process of torsion is also clearly documented in isolated muscle cells (WARSHAW et al. 1987), where it is probably much enhanced by the lack of constraints from surrounding cells and collagen network. If this element of torsion really exists, as the evidence suggests, it has to be reconciled with the observation that there is no substantial reorientation of filaments and dense bands in shortened muscle cells (GABELLA 1984).

The shortening of a muscle is accompanied by its lateral expansion, as there is no significant change in the volume of the muscle. Similarly, the individual cells shorten and increase substantially their cross sectional area; their surface becomes corrugated due to the formation of deep laminar folds and other excrescences of the cell membrane. Many collagen fibrils close to the cell surface, which run almost longitudinally when the cells are elongated, run almost transversely when the cells are maximally shortened. These structural changes are important for the mechanics of smooth muscle. In this tissue, contraction is more satisfactorily explained as an active change in shape rather than as a simple cell shortening.

In contrast to the isotonic contraction, during isometric contraction a muscle does not change in length; there is nevertheless some shortening of the muscle cells to take up any slack in the muscle. There is, therefore, in isometric contraction some longitudinal displacement within the contractile apparatus, which is, however, very small by comparison with that occurring in isotonic contraction. Under the electron microscope, isometrically contracted muscle cells of viscera display a chequered appearance due to the presence of several distinct areas which are devoid of myofilaments and are therefore of less dense appearance. These areas or domains are mainly found beneath the cell membrane, although the two largest ones are found at the poles of the nucleus. The myofilaments are exactly longitudinally arranged. GILLIS et al. (1988) have noted that the packing density of myofilaments in isometrically contracted cells of the rat anococcygeus musce is increased by a factor of two by comparison with the muscle at rest, and the cells appear of a smaller volume.

An additional type of contraction (passive contraction) is the one occurring in inactive muscle cells or muscle bundles which shorten by the effect of contraction in adjacent cells or bundles.

II. Development

Differentiation of smooth muscles begins early in embryonic life. In the chicken gizzard, myofibrillar proteins (myosin heavy chain and actin) and desmin can be detected at the end of the 1st week in ovo, first by biochemical methods (HIRAI and HIRABAYASHI 1983; STUEWER and GRÖSCHEL-STEWART 1985), then by immunohistochemistry (STUEWER and GRÖSCHEL-STEWART 1985; HIRAI and HIRABAYASHI 1986). Spontaneous contractile activity is recorded from day 8 (DONAHOE and BOWEN 1972) and myofilaments are seen by electron microscopy by day 9 (BENNET and COBB 1969). The precursors are mesenchymal cells which through the expression of myofibrillar proteins acquire the main feature of muscle cells.

The main increase of the muscle volume occurs after the differentiation of the precursors into muscle cells. In the chicken gizzard the musculature increases almost 1000-fold from the 10th day in ovo to 6 months after hatching. The increase is accounted for partly by an increase in volume of the muscle cells (from about 600 μm^3 to about 3000 μm^3, both in mammalian and avian visceral muscles) and partly by a substantial increase in cell number (GABELLA 1989). Increase in cell number is by mitotic division of the muscle cells (Fig. 10). The smooth muscle cells undergo mitosis even when they have developed their contractile apparatus and their cell membrane specializations, that is, terminal differentiation is compatible with cell proliferation. In this respect smooth muscle cells resemble cardiac muscle cells (RUMYANTSEV and SNIGIREVSKAYA 1968; ZAK 1973), whereas in skeletal muscles the myoblasts stop dividing at the onset of myosin expression (MOSS and LEBLOND 1970). In the chicken gizzard the peak of mitosis occurs between day 17 and 19 in ovo, when the cells have reached an advanced degree of differentiation. The central part of the cell swells, displays the mitotic apparatus, including characteristic paired cisternae of sarcoplasmic

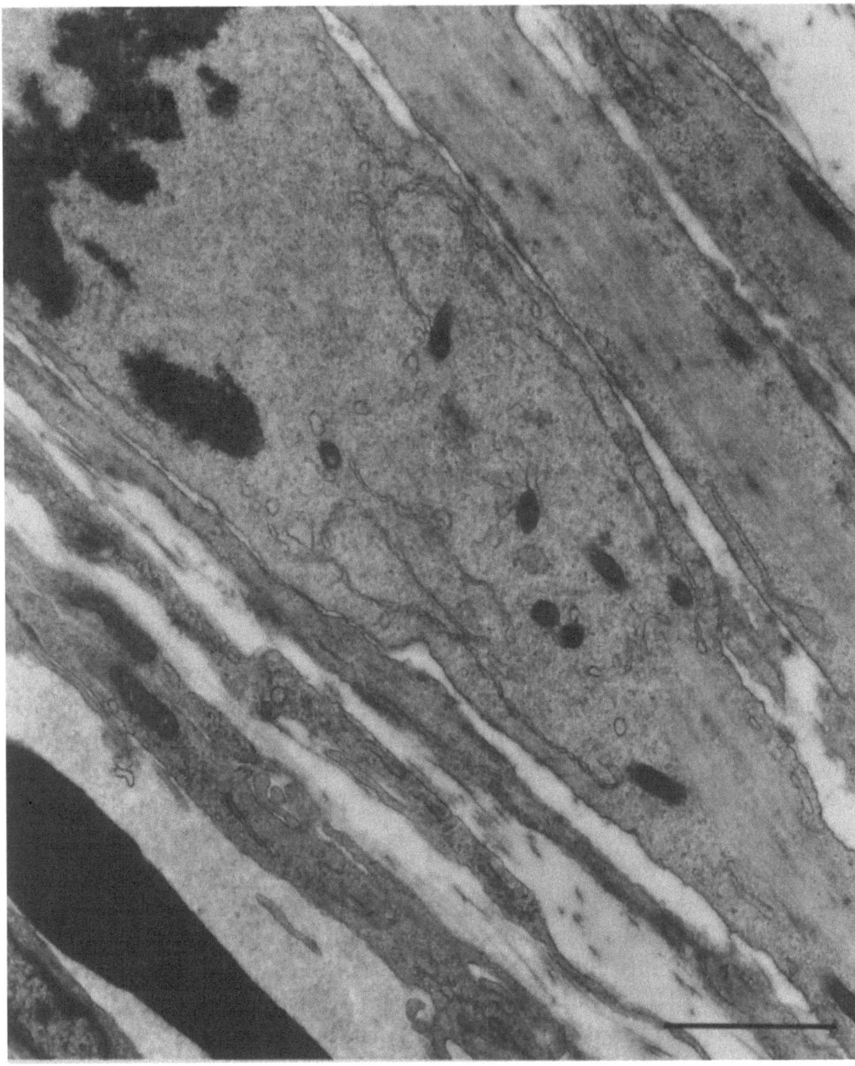

Fig. 10. Gizzard of a 19-day-old chick embryo, showing a muscle cell undergoing mitosis. Contractile elements and mitochondria are visible *at the bottom right corner*, chromosomes *at the top left corner*; granular material and large cisternae of cytoplasmic reticulum are visible in the middle portion of the cell. *At the bottom left corner* is a blood vessel with a red blood cell, *at the opposite corner* are two muscle cells in longitudinal section. *Calibration bar*, 2 μm

reticulum associated with the chromosomes and possibly originating from the breakdown of the nuclear envelop (KAMIO et al. 1977). The remaining part of the dividing cell is indistinguishable from its neighbours, in its organelles, filaments and membrane features (Fig. 10). The equatorial plate of chromosomes is invariably oriented transversely to the long axis of the cell, and the two daughter cells are initially lined up end-to-end: they also remain in communication with each other for some time after division by a narrow cytoplasmic bridge occupied by a bundle of microtubules. In the large muscles of viscera mitoses are distributed without an apparent pattern and they are found in any part of the muscle. The addition of new cells does not occur preferentially at the surface of a muscle layer or at the end of a muscle cord: it occurs within the texture of the existing muscle (intussusceptive growth), and this requires a rearrangement of the older cells and their connections (GABELLA 1989). In certain smooth muscles, such as the ureter and vas deferens, muscle cell mitoses have not been observed (LEESON and LEESON 1965; YAMAUCHI and BURNSTOCK 1969) or they have been seen after muscle cell de-differentiation, as happens in cell cultures (CAMPBELL et al. 1974). In the musculature of the mammalian intestine the dividing cells are differentiated muscle cells, as in the chicken gizzard, and, in the guinea pig, the peak of mitosis occurs at the time of birth or soon afterwards. Mitoses, however, do not stop at this stage, as indicated also by the uptake of tritiated thymidine, but they continue throughout life, although at a progressively reduced rate (Fig. 11). Mitoses are occasionally found in smooth muscles of adult or ageing subjects, and they can resume on a large scale when the muscle is stimulated to hypertrophy (see next section). In vascular muscles, cell divisions are abundant during the fetal life but they are still very common after birth. In the rat aorta, for example, the number of muscle cells increases threefold during the first 2 weeks after birth (OLIVETTI et al. 1980).

The timing of development shows considerable variation between different muscles. Vascular muscles generally differentiate earlier than visceral muscles. The intestinal musculature precedes genital muscles, and the iris musculature differentiates earlier than both. There can be differences even within an organ, for example in the intestine, where the circular musculature differentiates a little earlier than the longitudinal musculature. A special case is the smooth muscle in the amnion of birds which in the chicken is already fully differentiated around the 11th day in ovo (EVANS and EVANS 1964) and regresses afterwards.

Within a given developing smooth muscle the cell composition is uniform and all muscle cells appear to be at the same stage of development (Fig. 12). There are no undifferentiated cells in the tissue which remain available for development at a later stage. Mitochondria are more abundant in developing than in mature muscle cells: in the chicken gizzard they occupy about 8% of the cytoplasm at 14 day in ovo, about 7% at day 19 and about 5% in the adult (GABELLA 1989). The sarcoplasmic reticulum, especially the smooth type, is well developed from the earliest stages, whereas caveolae appear relatively late. The early appearance of myofilaments coincides with the marked elongation of the cell. Dense bands appear later than the dense bodies, so that initially the contractile apparatus has only a limited insertion to the cell membrane (GABELLA 1989). The proteins talin and vinculin are

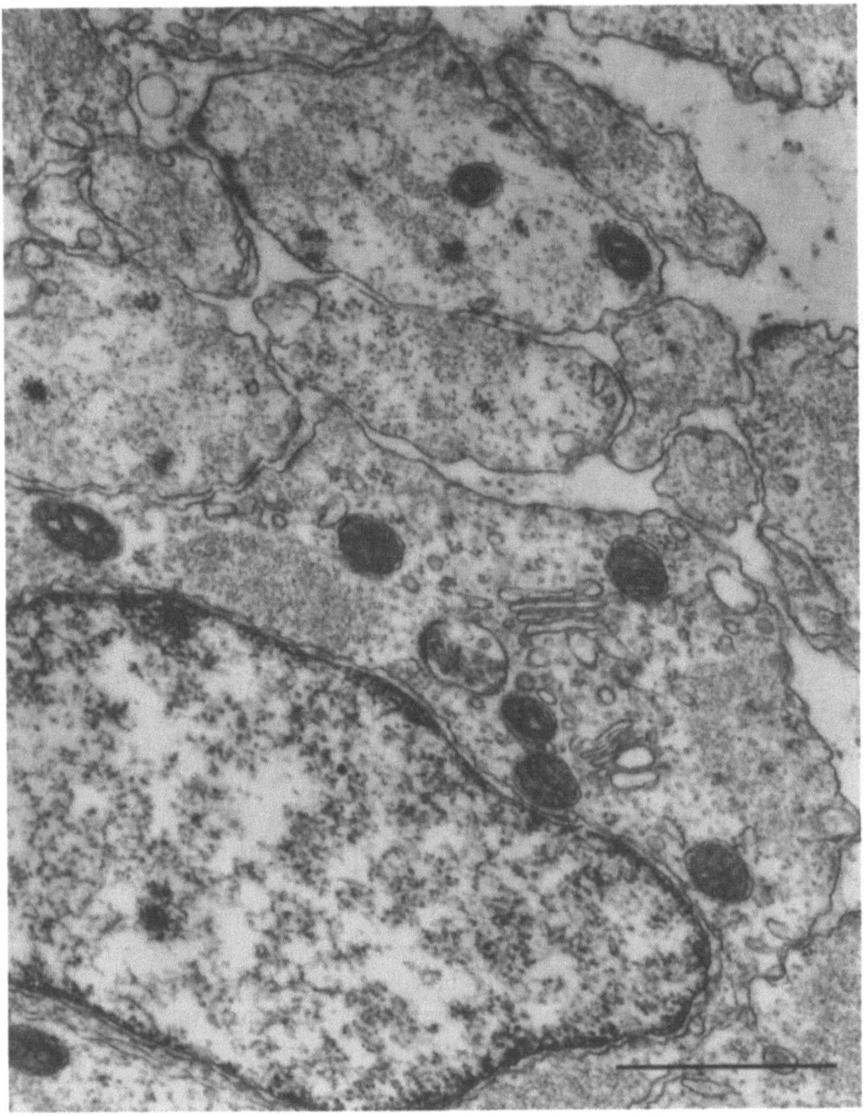

Fig. 11. Oblique section through the wall of the ileum of a 4-day-old guinea pig injected with tritiated thymidine. Labelled nuclei (some indicated with an *arrow*) are seen in muscle cells of both layers, in endothelial cells, and in a glial cell of a myenteric ganglion. There is no labelling of the ganglion neurons. *Calibration bar*, 100 μm

initially present in the cytoplasm (VOLBERG et al. 1986). Talin becomes associated with the cell membrane at the time of formation of dense bands, around the 16–18th day in the chick embryo, while vinculin does not bind to dense bands until after hatching (VOLBERG et al. 1986). Gap junctions are

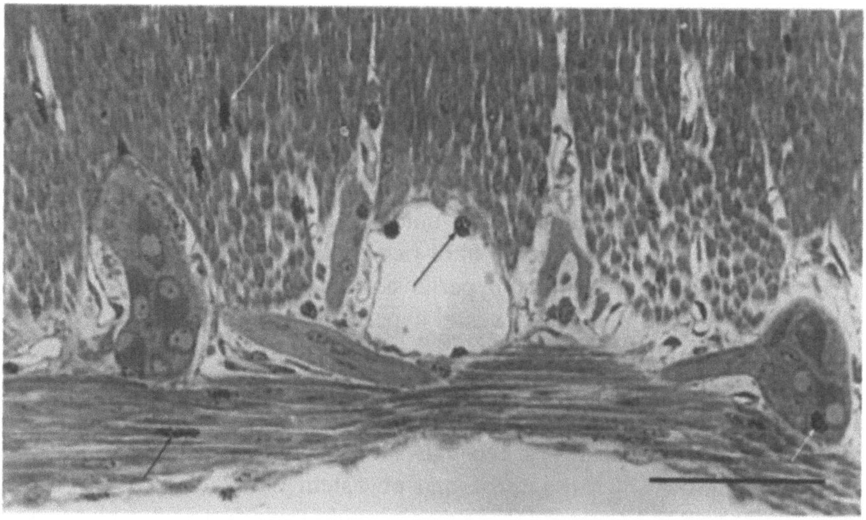

Fig. 12. Muscle cells of the gizzard of a 14-day-old chick embryo. Note the loose arrangement of the contractile apparatus and the scarcity of dense bands. Mitochondria, Golgi apparatus and sarcoplasmic reticulum are well developed. Caveolae, basal lamina and extracellular material are poorly developed at this stage. *Calibration bar*, 1 μm

initially absent in embryonic smooth muscle. In the chicken gizzard they appear at day 16 in ovo (LA MANTIA and SHAFIQ 1982; GABELLA 1989) and in the guinea pig ileum around the time of birth. Initially they are made of only five to ten intramembrane particles, and they grow progressively by addition of new particles. The stroma is initially very sparse and it becomes substantial only towards the end of the embryonic life, during a time when the rough sarcoplasmic reticulum is well in evidence. Strong indirect evidence suggests that the muscle cells are involved in the synthesis and secretion of most of the extracellular materials.

In intestinal smooth muscles the density of innervation (for example, the ratio of number of axon profiles and number of muscle cell profiles in a transversely sectioned muscle) is higher than in the adult. The axons are mainly gathered into large bundles, but many of these lie very close to muscle cells: many of their axons, some of which are packed with vesicles, are separated by a gap of less than 20 nm from the muscle cell membrane.

III. Hypertrophy

The potential for growth is not exhausted in the adult subject. The growth in excess to the physiological size achieved in the adult is hypertrophic growth, and it takes place when new ("abnormal") functional demands are imposed on the muscle. In most cases the process is an overload hypertrophy, i.e. it is due to an overdistension of the muscle. In hollow organs the overdistension is usually related to an increase in the contents for example because of an outlet obstruction. The process can occur in physiological conditions (in the myometrium during gestation) and is found in many pathological conditions: urinary bladder hypertrophy in the presence of prostate enlargement or other impairments of urine outlet; intestinal hypertrophy oral to an obstruction of the lumen by a tumour or a malformation; hypertrophy of the caeum with a fibre-rich diet (JACOBE 1985); blood vessel hypertrophy in certain forms of hypertension or in arterial stenosis (FOLKOW 1982; OWENS 1989); pulmonary vessel hypertrophy in conditions of hypoxia; bladder hypertrophy with polyuria; hypertrophy of ureter or biliary pathways in the presence of obstructions. In all these examples, hypertrophy is not an acute process: a complete obstruction produces overdistension of the organ wall, which then perforates or becomes irreversibly damaged. In contrast, a chronic condition which is intense and persistent but not such as to cause organ failure allows the growth response of the muscle to take place, and the distension is then accompanied by increase in muscle volume. The mechanical performance of the muscle is, at least initially, increased and the organ can cope with the increased functional demands, for example by overcoming the increased outlet resistance.

The increase in muscle volume in hypertrophic viscera is substantial. In rats with urethral obstruction-induced bladder hypertrophy the weight of the bladder increases up to 15-fold (from 70 to 1000 mg) in about 8 weeks, and similar increases are observed in the small intestine oral to a stenosis. In the bladder, in spite of the change in size and weight, the shape of the bladder (the ratio between the three main axes) is unchanged, while in the intestine the increase in muscle volume is due to an increase in thickness of the muscle coat (both muscle layers being similarly affected) and in diameter of the intestine, with no change in the length of the intestine.

In hypertrophic smooth muscles, muscle cells are much increased in size, up to 13 500 μm^3 in the intestine and more in the bladder. In some cases, for example in the intestine, this increase accounts for only part of the increase in muscle volume, the rest being accounted for by hyperplasia, i.e. an increase in the number of muscle cells. This is due to mitosis of muscle cells, both in the circular and the longitudinal layer. As is found in normal development, cell division is not preceded by dedifferentiation: it occurs in muscle cells with their full complement of organelles, filaments and membrane specializations. Whether cell proliferation is accompanied by a suppression of the expression of contractile proteins, as is the case in muscle cell dividing in vitro (BLANK et al. 1988), remains to be proven. In the hypertrophic bladder muscle there is a substantial increase in total DNA content (UVELIUS et al. 1984), suggesting the occurrence of mitoses, although these have not been seen at the stages that have been studied by microscopy. In the muscle cells of the hypertrophic aorta of hypertensive rats, the increase in DNA content is accounted for by polyploidy and there is little or no cellular proliferation (OWENS and SCHWARTZ 1983). In contrast, in the rat hypertrophic mesenteric artery, the enlargement is mainly due to muscle cell division (MULVANY et al. 1985). The same is the case in rat and rabbit aorta in obstruction-induced hypertrophy (BEVAN et al. 1976; OWENS and REIDY 1985). Even when polyploid, the muscle cells remain mononucleated (OWENS 1989).

The hypertrophic muscle cells are much larger in sectional profile, the increase in cell size being mainly accounted for by an increase in girth rather than in length. Even in relaxed conditions, the cell profiles are corrugated with deep, tunnel-like invaginations of the cell membrane. These changes of the cell surface partly counteract the fall in the surface-to-volume ratio that arises from the fact that the surface grows less than the volume. Even so, the surface to volume ratio is markedly smaller than in control tissues, 0.8:1 as opposed to 1.4:1 in the guinea pig intestine. Caveolae are present at the same spatial density and with the same appearance as in control muscle cells. In hypertrophic intestinal muscle cells, gap junctions are increased in size while their number per unit surface of cell membrane is unchanged; therefore, the total number of gap junctions per cell and the percentage area occupied by gap junctions are significantly increased (GABELLA 1990).

There are small changes in the spatial density of organelles in the hypertrophic muscle cells. The percentage volume occupied by mitochondria is slightly decreased, but, on account of the large increase in cell size, the total number of mitochondria is much higher than in controls. The percentage volume of sarcoplasmic reticulum is increased. Some cells contain large cisternae of rough sarcoplasmic reticulum, an observation suggesting an activation of the synthetic and secretory activity of the muscle cell. In most cells the smooth sarcoplasmic reticulum is well developed, with fenestrated cisternae and tubules very common near the cell surface.

The force generated by a hypertrophic smooth muscle is increased, on account of its larger volume. However, the force generated per unit transverse area of muscle is either unchanged (in the ureter (HAUSSMAN et al. 1979)) or somewhat reduced (in the portal vein (JOHANSSON 1976) and in the intestine (GABELLA 1990)). The amount of contractile and cytoskeletal filaments is much increased in the hypertrophic muscle cells. In hypertrophic intestine and blood vessels the intermediate filaments increase more than other filaments and large bundles of them become apparent in the cell (GABELLA 1990; BERNER et al. 1981). New filaments of myosin are not formed (BERNER et al. 1981, portal vein) or they are formed to a lesser extent than new filaments of actin (GABELLA 1990, intestine). There is also an increase in the total collagen content of the muscle, most of it probably produced by the muscle cells themselves, and in the hypertrophic intestine even an increase in its concentration. Many new blood vessels (capillaries) are formed, so that the hypertrophic muscle remain almost as well vascularized as the control tissue.

References

Ashton FT, Somlyo AV, Somlyo AP (1975) The contractile apparatus of vascular smooth muscle: intermediate high voltage stereo electron microscopy. J Mol Biol 98:17–29
Bennett T, Cobb JLS (1969) Studies on the avian gizzard: the development of the gizzard and its innervation. Z Zellforsch 98:599–621
Berner PF, Somlyo AV, Somlyo AP (1981) Hypertrophy-induced increase of intermediate filaments in vascular smooth muscle. J Cell Biol 88:96–101
Bevan RD, van Marthens E, Bevan J (1976) Hyperplasia of vascular smooth muscle in experimental hypertension in the rabbit. Circ Res 38 [Suppl II]:58–62
Beyer EC, Kistler J, Paul DL, Goodenough DA (1989) Antisera directed against connexin 43 peptides react with a 43-kD protein localized to gap junctions in myocardium and other tissues. J Cell Biol 108:595–605
Blank R, Thompson M, Owens G (1988) Cell cycle versus density dependence of smooth muscle alpha actin expression in cultured rat aortic muscle cells. J Cell Biol 107:299–306
Bois RM (1973) The organization of the contractile apparatus of vertebrate smooth muscle. Anat Rec 93:138–149
Bond M, Somlyo AV (1982) Dense bodies and actin polarity in vertebrate smooth muscle. J Cell Biol 95:403–413
Bond M, Kitizawa T, Somlyo AP, Somlyo AV (1985) Release and recycling of calcium by the sarcoplasmic reticulum in guinea pig portal vein smooth muscle. J Physiol (Lond) 355:677–695
Burridge K, Connell L (1983) Talin: a cytoskeletal component concentrated in adhesion plaques and other sites of actin-membrane interaction. Cell Motil 3:405–417
Burridge K, Mangeat P (1984) An interaction between vinculin and talin. Nature 308:744–745
Büssow H, Wulfhekel U (1972) Die Feinstruktur der glatten Muskelzellen in den grossen muskulären Arterien der Vögel. Z Zellforsch Mikrosk Anat 125:339–352
Byers TJ, Kunkel LM, Watkins SC (1991) The subcellular distribution of dystrophin in mouse skeletal, cardiac, and smooth muscle. J Cell Biol 115:411–421
Campbell GR, Chamley JH, Burnstock G (1974) Development of smooth muscle cells in tissue culture. J Anat 117:295–312
Carafoli E (1991) Calcium pump of the plasma membrane. Physiol Rev 71:129–153
Cooke P (1976) A filamentous cytoskeleton in vertebrate smooth muscle fibers. J Cell Biol 68:539–556
Cooke PH, Chase RH (1971) Potassium chloride-insoluble myofilaments in vertebrate smooth muscle cells. Exp Cell Res 66:417–425
Cooke PH, Fay FS, Craig R (1989) Myosin filaments isolated from skinned amphibian smooth muscle cells are side-polar. J Muscle Res Cell Motil 10:206–220
Craig R, Megerman J (1977) Assembly of smooth muscle myosin into side-polar filaments. J Cell Biol 75:990–996
Crone C (1986) Modulation of solute permeability in microvascular endothelium. Fed Proc 45:77–83
Dahl GP, Berger W (1978) Nexus formation in the myometrium during parturition and induced by estrogen. Cell Biol Int Rep 2:381–387
Devine CE, Rayns DG (1975) Freeze-fracture studies of membrane systems in vertebrate muscle. II. Smooth muscle. J Ultrastruct Res 51:293–306
Devine CE, Somlyo AV, Somlyo AP (1972) Sarcoplasmic reticulum and excitation-contraction coupling in mammalian smooth muscles. J Cell Biol 52:690–718
Donahoe JR, Bowen JM (1972) Analysis of the spontaneous motility of the avian embryonic gizzard. Am J Vet Res 33:1835–1848

Draeger A, Stelzer EHK, Herzog M, Small JV (1989) Unique geometry of actin-membrane anchorage sites in avian gizzard smooth muscle cells. J Cell Sci 94:703–711
Drenckhahn D, Beckerle M, Burridge K, Otto J (1988) Identification and subcellular location of talin in various cell types and tissues by means of [^{125}I]vinculin overlay, immunoblotting and immunocytochemistry. Eur J Cell Biol 46:513–522
Evans DHL, Evans EM (1964) The membrane relationships of smooth muscles: an electron microscope study. J Anat 98:37–46
Fawcett DW, McNutt NS (1969) The ultrastructure of the cat myocardium. I. Ventricular papillary muscle. J Cell Biol 42:1–45
Fay FS, Fujiwara K, Rees DD, Fogarty KE (1983) Distribution of α-actinin in single isolated smooth muscle cells. J Cell Biol 96:783–795
Fischer E (1944) The birefringence of striated and smooth mammalian muscles. J Cell Comp Physiol 23:1130–130
Fisher BA, Bagby RM (1977) Reorientation of myofilaments during contraction of a vertebrate smooth muscle. Am J Physiol 232:C5–C14
Folkow B (1982) Physiological aspects of primary hypertension. Physiol Rev 62:347–504
Forbes MS, Rennels ML, Nelson E (1979) Caveolar systems and sarcoplasmic reticulum in coronary smooth muscle cells of the mouse. J Ultrastruct Res 67:325–339
Franke WW, Schmid E, Freudenstein C, Appelhans B, Osborn M, Weber K, Keenan TW (1980) Intermediate-sized filaments of the prekeratin type in myoepithelial cells. J Cell Biol 84:633–654
Fujimoto T (1993) Calcium pump of the plasma membrane is localized in caveolae. J Cell Biol 120:1147–1157
Fujimoto T, Nakade S, Miyawaki A, Mikoshiba K, Ogawa K (1992) Localization of inositol 1,4,5-trisphosphate receptor-like protein in plasmalemmal caveolae. J Cell Biol 119:1507–1513
Gabbiani G, Schmid E, Winter S, Chaponnier C, de Chastonay C, Vandekerckhove J, Weber K, Franke WW (1981) Vascular smooth muscle cells differ from other smooth muscle cells: Predominance of vimentin filaments and a specific α-type actin. Proc Nat Acad Sci USA 78:298–302
Gabella G (1983) Asymmetric distribution of dense bands in muscle cells of mammalian arterioles. J Ultrastruct Res 84:24–33
Gabella G (1984) Structural apparatus for force transmission in smooth muscles. Physiol Rev 64:455–477
Gabella G (1985) Chicken gizzard. The muscle, the tendon and their attachment. Anat Embryol 171:151–162
Gabella G (1988) Structure of the intestinal musculature. In: Wood JD (ed) Motility and circulation. American Physiological Society, Bethesda, pp 103–139 (Handbook of physiology, vol 4)
Gabella G (1989) Development of smooth muscle: ultrastructural study of the chick embryo gizzard. Anat Embryol 180:213–226
Gabella G (1990) Hypertrophy of visceral smooth muscle. Anat Embryol 182: 409–24
Gabella G, Blundell D (1979) Nexuses between the smooth muscle cells of the guinea-pig ileum. J Cell Biol 82:239–247
Garfield RE, Daniel EE (1976) Relation to membrane vesicles to volume control and Na^+-transport in smooth muscle: studies on Na^+ rich tissues. J Mechanochem Cell Motil 4:157–176
Garfield RE, Sims SM, Daniel EE (1977) Gap junctions: their presence and necessity in myometrium during parturition. Science 198:958–959
Geiger B (1979) A 130K protein from chcken gizzard: its localization at the termini of microfilament bundles in cultured chicken cells. Cell 18:193–205

Geiger B, Dutton AH, Tokayasu KT, Singer SJ (1981) Immunoelectron microscope studies of membrane-microfilament interaction. The distribution of alpha-actinin, tropomyosin and vinculin intestinal epithelial brush border and chicken gizzard smooth muscle cells. J Cell Biol 91:614–628
Gillis JM, Cao ML, Godfraind-De Becker A (1988) Density of myosin filaments in the rat anococcygeus muscle, at rest and in contraction. II. J Muscle Res Cell Motil 9:18–28
Haussman M, Biancani P, Weiss RM (1979) Obstruction-induced changes in longitudinal force-length relations of rabbit ureter. Invest Urol 17:223–226
Heidlage JF, Anderson NC Jr (1984) Ultrastructure and morphometry of the stomach muscle of Amphiuma tridactylum. Cell Tissue Res 236:393–397
Hinssen H, D'Haese J, Small JV, Sobieszek A (1978) Mode of filament assembly of myosins from muscle and non-muscle cells. J Ultrastruct Res 64:282–302
Hirai SI, Hirabayashi T (1983) Developmental changes of proein constituents in chicken gizzards. Dev Biol 97:483–493
Hirai SI, Hirabayashi T (1986) Development of myofibrils in the gizzard of chicken embryos. Intracellular distribution of structural proteins and development of contractility. Cell Tissue Res 243:487–493
Huiatt TW, Robson RM, Arakawa N, Stromer MH (1980) Desmin from avian smooth muscle. Purification and partial characterization. J Biol Chem 255:6981–6989
Jacobs LR (1985) Differential effects of diatary fibers on rat intestinal circular muscle cell size. Dig Dis Sci 30:247–252
Johansson B (1976) Structural and functional changes in rat portal vein after experimental portal hypertension. Acta Physiol Scand 98:381–383
Kamio A, Huang WY, Imai H, Kummerow FA (1977) Mitotic structures of aortic smooth muscle cells in swine and in culture: paired cisternae. J Electron Microsc (Tokyo) 26:29–40
Kannan MS, Daniel EE (1978) Formation of gap junctions by treatment in vitro with potassium conductance blockers. J Cell Biol 78:338–348
Kargacin GJ, Cooke PH, Abramsom SB, Fay FS (1989) Periodic organization of the contractile apparatus in smooth muscle revealed by the motion of dense bodies in single cells. J Cell Biol 108:1465–1475
La Mantia J, Shafiq SA (1982) Developmental changes in the plasma membrane of gizzard smooth muscle of the chicken. A freeze-fracture study. J Anat 134: 243–253
Leeson TS, Leeson GR (1965) The rat ureter. Fine structural changes during development. Acta Anat 62:60–79
Linderkamp O, Meiselman HJ (1982) Geometric, osmotic, and membrane mechanical properties of density-separated human red cells. Blood 59:1121–1127
Mackenzie LW, Garfield RE (1985) Hormonal control of gap junctions in the myometrium. Am J Physiol 248:C296–C302
McGuffeee LJ, Bagby RM (1976) Ultrastructure, calcium accumulation, and contractile response in smooth muscle. Am J Physiol 230:1217–1224
Moss F, Leblond C (1970) Nature of dividing nuclei in skeletal muscle of growing rats. J Cell Biol 44:459–462
Mulvany M, Baandrup U, Gundersen H (1985) Evidence for hyperplasia in mesenteric resistance vessels of spontaneously hypertensive rats using a three-dimensional disector. Circ Res 57:794–800
Murphy RA, Driska SP, Cohen DM (1977) Variations in actin to myosin ratios and cellular force generation in vertebrate smooth muscles. In: Casteels R (ed) Excitation-contraction coupling in smooth muscle. Elsevier/North-Holland, Amsterham, pp 417–424
Nasu F, Inomata K (1990) Ultracytochemical demonstration of Ca^{1+}-ATPase activity in the rat saphenous artery and its innervated nerve terminal. J Electron Micr 39:487–491

North AJ, Galazkiewicz B, Byers TJ, Glenney JR, Small JV (1993) Complemetary distribution of vinculin and dystrophin define two distinct sarcolemma domains in smooth muscle. J Cell Biol 120:1159–1167
Ogawa KS, Fujimoto K, Ogawa K (1986) Ultracytochemical studies of adenosine nucleotidases in aortic endothelial and smooth muscle cells – Ca^{2+}-ATPase and Na^{+},K^{+}-ATPase. Acta Histochem Cytochem 19:601–620
Olivetti G, Anversa P, Melissari M, Loud AV (1980) Morphometric study of early postnatal development of the thoracic aorta in the rat. Circ Res 47:417–424
Osborne-Pellegrin MJ (1978) Some ultrastructural characteristics of the renal artery and abdominal aorta of the rat. J Anat 125:641–652
O'Shea JM, Robson RM, Huiatt TW, Hartzer MK, Stromer MH (1979)
Owens GK (1989) Control of hypertrophic versus hyperplastic growth of vascular smooth muscle cells. Am J Physiol 257:H1755–H1765
Owens G, Reidy M (1985) Hyperplastic growth response of vascular muscle cells following induction of acute hypertension in rats by aortic coarctation. Circ Res 57:695–705
Owens GK, Schwartz SM (1983) Vascular smooth muscle cell hypertrophy and hyperploidy in the Goldblatt hypertensive rat. Circ Res 53:491–501
Pease DC, Molinari S (1960) Electron microscopy of muscular arteries: pial vessels of the cat and monkey. J Ultrastruct Res 3:447–468
Popescu LM, Diculescu I (1975) Calcium in smooth muscle sarcoplasmic reticulum in situ. Conventional and X-ray analytical electron microscopy. J Cell Biol 67:911–918
Popescu LM, Diculescu I, Zelck V, Ionescu N (1974) Ultrastructural distribution of calcium in smooth muscle cells of guinea-pig taenia coli. A correlated electron microscopic and quantitative study. Cell Tissue Res 154:357–378
Prescott L, Brightman MW (1976) The sarcolemma of Aplysia smooth muscle in freeze-fracture preparations. Tissue Cell 8:241–258
Risek B, Guthrie S, Kumar N, Giula NJ (1990) Modulation of gap junction transcript and protein expression during pregnancy in the rat. J Cell Biol 110:269–282
Rumyantsev PP, Snigirevskaya E (1968) Ultrastructure of differentiating cells of the heart muscle in the state of mitotic division. Acta Morphol Acad Sci Hung 16:271–283
Schollmeyer JE, Furcht LJ, Goll DE, Robson RM, Stromer MH (1976) Localization of contractile proteins in smooth muscle cells and in normal and transformed fibroblasts. In: Goldman AR, Pollard T, Rosenbaum J (eds) Cell motility. Cold Spring Harbor Laboratory, Cold Spring Harbor, pp 361–388
Small JV (1974) Contractile units in vertebrate smooth muscle cells. Nature 249: 324–327
Small JV (1977) Studies on isolated smooth muscle cells: the contractile apparatus. J Cell Sci 24:327–349
Small JV (1985) Geometry of actin-membrane attachments in the smooth muscle cell: the localizations of vinculin and α-actinin. EMBO J 4:45–49
Small JV, Sobieszek A (1980) The contractile apparatus of smooth muscle. Int Rev Cytol 64:241–306
Small JV, Furst DO, DeMay J (1986) Localization of filamin in smooth muscle. J Cell Biol 102:210–220
Small JV, Herzog M, Barth M, Draeger A (1990) Supercontracted state of verte brate smooth muscle cell fragments reveals myofilament lengths. J Cell Biol 111:2451–2461
Somlyo AV, Butler TM, Bond M, Somlyo AP (1981) Myosin filaments have non-phosphorylated light chains in relaxed smooth muscle. Nature 294:567–569
Sommer JR, Johnson EA (1968) Cardiac muscle. A comparative study of Pukinje fibers and ventricular fibers. J Cell Biol 36:497–526

Sperelakis NJ (1991) Electric field model: an alternative mechanism for cell-to-cell propagation in cardiac muscle and smooth muscle. J Gastrointest Motil 3:64–83
Stuewer D, Gröschel-Stewart U (1985) Expression of immunoreactive myosin and myoglobin in the developing chicken gizzard. Roux's Arch Dev Biol 194:417–424
Stromer MH, Bendayan M (1988) Arrangement of desmin intermediate filaments in smooth muscle cells as shown by high-resolution immunocytochemistry. Cell Motil Cytoskel 11:117–125
Tsukita S, Tsukita S, Ishikawa H (1983) Association of actin and 10 nm filaments with the dense body in smooth muscle cells of the chicken gizzard. Cell Tissue Res 229:233–242
Uvelius B, Persson L, Mattiasson A (1984) Smooth muscle cell hypertrophy and hyperplasia in the rat detrusor after short term infravesical outflow obstruction. J Urol 131:173–176
Volberg T, Sabanay H, Geiger B (1986) Spatial and temporal relationships between vinculin and talin in the developing chicken gizzard smooth muscle. Differentiation 32:34–43
Warshaw DM, McBride WJ, Work SS (1987) Corkscrew-like shortening in single smooth muscle cells. Science 236:1457–1459
Yamauchi A, Burnstock G (1969) Post-natal development of smooth muscle cells in the mouse vas deferens. J Anat 104:1–15
Zak R (1973) Cell proliferation during cardiac growth. Am J Cardiol 3:211–219
Zamir O, Hanani M (1990) Intercellular dye-coupling in intestinal smooth muscle. Are gap junctions required for intercellular coupling? Experientia 46:1002–1005

CHAPTER 2

Ionic Channels in Smooth Muscle

T. TOMITA and S. IINO

A. Introduction

Our fundamental knowledge of membrane currents and ionic channels in smooth muscle has advanced considerably following the introduction of the patch-clamp method about 10 years ago, and it is a field that is still rapidly expanding. Since the patch-clamp method uses single cells or membrane patches, however, there are limits to the precision of the correlation that can be obtained between the ionic currents analysed and their functional role, because smooth muscle functionally is a multicellular system. A combination of patch-clamp experiments and intracellular recording of tissues may be the best way to further our understanding of membrane currents. More information is also required about functional cell-to-cell coupling.

In general, Ca^{2+} influx is important for the initiation of spontaneous and agonist-mediated contractions, K^+ contracture or Na^+-deficient contracture. The Ca^{2+} pathway in the plasma membrane is often classified into voltage-operated (or -gated) channels and receptor-operated channels (BOLTON 1979; MEISHERI et al. 1981). The receptor-operated channels may have poor ionic selectivity and are much less susceptible to organic Ca^{2+}-channel blockers than voltage-operated channels. Receptor-operated Ca^{2+} pathways need further analysis, but the noradrenaline-induced contraction has been explained at least in some vascular muscles by the effect of the voltage-operated Ca^{2+} channels alone, without assuming the involvement of receptor-operated pathways (NELSON et al. 1990).

In addition to these Ca^{2+} influx pathways, stretch-activated channels and carrier-mediated transport systems, such as Na^+–Ca^{2+} exchange, may also play an important role in contractions that are less dependent on membrane potential or are caused by a decrease in the Na^+ concentration gradient across the plasma membrane. In this chapter, only voltage-operated Ca^{2+} channels are reviewed as the subject of Ca^{2+} influx.

The conductance of K^+ is important for determining membrane potential and excitability. Many different K^+ channels have been found in smooth muscle, and the functional roles of each channel are gradually becoming clearer. Na^+ and Cl^- channels have been rather neglected, but recently have begun to be clarified. Further experiments are necessary to explain the diversity of smooth muscle function in terms of ionic channels.

B. Voltage-Operated Ca^{2+} Channel

I. Type

In several smooth muscles, producing both phasic contraction with action potential and tonic contraction without action potential (or quiescent), the presence of two different types of Ca^{2+} channel has been reported based on the demonstration of two different single-channel currents (Table 1) and also on the property of inward currents (Table 2). Typically, the activity of one channel (small conductance) is transient (T-type) and that of the other channel (large conductance) is long-lasting (L-type); the T-type channel is activated at a more negative membrane potential (lower threshold) and is less susceptible to dihydropyridine Ca^{2+}-channel blockers than the L-type.

Table 1. Distribution of different types of Ca^{2+} channel (studied with whole-cell clamp method)

	Reference
L-type	
Rabbit jejunum (longitudinal) (1.5 m*M* Ca^{2+}, Ba^{2+})	Aaronson and Russels 1991
Guinea pig taenia caeci (2.5 m*M* Ca^{2+}, 10 m*M* Ba^{2+})	Ganitkevich et al. 1991
Guinea pig coronary artery (2.5 m*M* Ca^{2+}, 10 m*M* Ba^{2+})	Ganitkevich et al. 1991
Rabbit ileum (longitudinal) (100 m*M* Ba^{2+})	Inoue et al. 1989
Guinea pig stomach (corpus) (5 m*M* Ca^{2+})	Katzka and Morad 1989
Canine colon (circular) (1.8 m*M* Ca^{2+})	Langton et al. 1989
Rabbit coronary artery (2.5 m*M* Ca^{2+}, 110 m*M* Ba^{2+})	Matsuda et al. 1990
Guinea pig portal vein (2.4 m*M* Ca^{2+})	Okashiro et al. 1992
Canine stomach (corpus, circular) (2.5 m*M* Ca^{2+})	Sims 1992
Rat bronchus (80 m*M* Ba^{2+})	Worley and Kotlikoff 1990
Human tracheobronchus (80 m*M* Ba^{2+})	Worley and Kotlikoff 1990
T-type	
Canine trachea (5, 20 m*M* Ca^{2+})	Kotlikoff 1988
L- and T-type	
Rat aorta (cultured) (20 m*M* Ca^{2+})	Akaike et al. 1989
Rat mesenteric artery (110 m*M* Ba^{2+})	Bean et al. 1986
Rabbit ear artery (110 m*M* Ba^{2+})	Benham et al. 1987b
Guinea pig coronary artery (10 m*M* Ba^{2+})	Ganitkevich and Isenberg 1990
Rat portal vein (cultured) (5 m*M* Ca^{2+}, 10 m*M* Ba^{2+})	Loirand et al. 1989
Guinea pig mesentric artery (100 m*M* Ba^{2+})	Ohya and Sperelakis 1989a
Rat azygos vein (cultured) (10, 20 m*M* Ca^{2+})	Sturek and Hermsmeyer 1986
Rat tail artery (20 m*M* Ba^{2+})	Wang et al. 1989
Canine saphenous vein (20 m*M* Ca^{2+}, Ba^{2+})	Yatani et al. 1987

Table 2. Single-channel conductances of two different channels found in the same cell

Smooth muscle	Conductance (pS)	$[Ba^{2+}]_o$ (m*M*)	Reference
Rabbit ear artery	8 and 25 (21°C)	110	BENHAM et al. 1987b
Guinea pig coronary artery	8 and 28 (22°–25°C)	110	GANITKEVICH and ISENBERG 1990
Guinea pig portal vein	12 and 22 (room temp.)	50	INOUE et al. 1990
Rabbit mesenteric artery	8 and 15	80	WORLEY et al. 1986
Rabbit basilar artery	12 and 25 (20°–24°C)	80	WORLEY et al. 1991
Canine saphenous vein	8 and 24 (20°–22°C)	90	YATANI et al. 1987
Guinea pig taenia caeci	7 and 30 (25°–35°C)	100	YOSHINO et al. 1989

The difference in susceptibility to Ca^{2+} channel blockers between the two types in smooth muscle is often not as great as that found in other excitable tissues. Both types of single channel in rabbit mesenteric artery (WORLEY et al. 1986) and rabbit basilar artery (WORLEY et al. 1991) have been shown to be susceptible to nisoldipine (50 n*M*) and nimodipine (2 n*M*), respectively. In cultured cells from rat aorta, the IC_{50} of flunarizine, nicardipine and verapamil is 0.1 μM, 0.2 μM and 0.9 μM for the L-type, and 0.2 μM, 0.8 μM and 70 μM for the T-type channel, respectively, when the external Ca^{2+} concentration is 20 m*M* (AKAIKE et al. 1989).

In order to record single-channel currents, Ca^{2+} is usually replaced with a high concentration of Ba^{2+} and a channel activator, Bay K8644, is included in patch pipettes. The channel with a small conductance (7–12 pS) probably corresponds to the T-type and the one with a large conductance (15–30 pS) to the L-type (Table 2). The conductance of each type is similar in different smooth muscles. The small difference could be partly due to experimental conditions such as difference in Ba^{2+} concentration, temperature and membrane potential.

In smooth muscle, L-type channels are generally preponderant, but canine trachea is considered to contain mainly T-type channel (KOTLIKOFF 1988). In this muscle, the threshold voltage for I_{Ca} is −55 mV with 5 m*M* external Ca^{2+}, the half-inactivation potential is about −35 mV and the apparent K_i of nifedipine is about 1 μM with a holding potential of −70 mV.

It is often difficult to distinguish the two types of channel clearly with the whole-cell clamp method. Rabbit jejunum (longitudinal muscle) has a rapidly inactivating inward current at a holding potential of −90 mV, but this component disappears when the holding potential is reduced to −60 mV. Although this suggests the presence of T-type channels, the shape of the I–V curve and nifedipine sensitivity do not support this theory (AARONSON and RUSSELL 1991). Similarly, in rat vas deferens, there are fast and slow components in inward currents (10.8 m*M* Ba^{2+}), the former being more resistant to nicardipine, but they have similar potential dependency (NAKAZAWA et al. 1988). The T- and L-type channel currents (1.8 m*M* Ca^{2+}) in guinea pig taenia caeci have similar potential dependency for activation (YOSHINO et al. 1989). It is possible that a current component which is apparently T-type is produced by modification of inactivation kinetics of L-type channels at least in some smooth muscles. Although two different types of Ca^{2+} channel have been demonstrated in single-channel studies on some smooth muscles, direct correlation between these channels and membrane currents recorded with the whole-cell clamp method does not follow.

The presence of a typical T-type channel has not been reported in visceral smooth muscles, but it has recently been observed that the longitudinal muscle of newborn rat ileum has typical T-type channel currents which are not affected by 30 μM nifedipine (SMIRNOV et al. 1992a). In this muscle, however, the high-threshold channel current which appears to be L-type channel current is also fairly resistant to nifedipine. N-type Ca^{2+} channels may be responsible for this component of inward currents, but further confirmation is necessary.

In guinea pig basilar artery, two different components in inward currents have been identified, one a typical L-type channel current and the other an atypical T-type channel current (SIMARD 1991; WEST et al. 1992). The latter current is less sensitive to nifedipine blockade and shows little inactivation near the threshold. This is termed a "B-type" channel current. It is, however, possible that this channel is fundamentally an L-type current, but that it is kinetically modified, as discussed above. Further clarification of the subclass of Ca^{2+} channels in smooth muscle is necessary.

II. Ionic Selectivity

Ba^{2+} and Sr^{2+}, but not Mg^{2+}, can pass Ca^{2+} channels. In guinea pig taenia caeci, currents show saturation with the concentration of the divalent cations. The relationship between the maximum current amplitude and the ion concentration fits a Langmuir curve with apparent dissociation constants of 1.2 m*M* for Ca^{2+}, 1.8 m*M* for Sr^{2+} and 9.6 m*M* for Ba^{2+} (22°–24°C) (GANITKEVICH et al. 1988). When Ca^{2+} is replaced with Ba^{2+}, inward currents become larger and the current-voltage relationship (I–V curve) shifts toward a more negative potential (the threshold potential becomes more negative). This can be explained by the weaker binding of Ba^{2+} (weaker reduction of surface negative charges by Ba^{2+}) (GANITKEVICH et al. 1988).

Ionic selectivity of T-type channels is similar to Ca^{2+} and Ba^{2+}, but Ba^{2+} is generally more permeable than Ca^{2+} in L-type channels. For example, in rat aorta (cultured), the ratio of peak inward currents of Ba^{2+} and Ca^{2+} was 1.0 for T- and 1.6 for L-type currents when compared at a concentration of 20 m*M* (AKAIKE et al. 1989). Micromolar Ca^{2+} contamination in Ba^{2+}-containing medium may have a significant inhibitory effect on Ba^{2+} currents, as found in rabbit portal vein (KATZKA et al. 1992), so that the actual difference in permeation of Ca^{2+} and Ba^{2+} may be greater than the previously reported values.

Removal of the external divalent cations increases the membrane permeability to Na^+. This is known to result from Na^+ flow through Ca^{2+} channels (rat myometrium: JMARI et al. 1987; guinea pig stomach: KATZKA and MORAD 1989; rabbit ileum: OHYA et al. 1986; canine stomach: VOGALIS et al. 1992). Thus, the currents recorded in a solution containing a very low concentration of Ca^{2+} and no Mg^{2+} are not influenced by tetrodotoxin, but blocked by Ca^{2+} channel blockers.

III. Inactivation

There are at least three states for Ca^{2+} channels: resting, open and inactivated. Channels can be opened by depolarization from the resting state, but are closed at various rates during maintained depolarization by inactivating processes. Once inactivated by depolarization, they can reopen only when the membrane is repolarized to remove inactivation. The amplitude and the time course of inward currents are determined by the availability of channels in the open state.

The rate of decrease in inward Ca^{2+} currents (I_{Ca}) varies considerably under different experimental conditions, with different types of smooth muscle or even in different cells of the same smooth muscle. The time course of I_{Ca} is likely to be controlled by many factors, including membrane potential and intracellular Ca^{2+} concentration.

It has been observed in many smooth muscles that inactivation of I_{Ca} slows as I_{Ca} peak amplitude decreases with larger depolarization and that inactivation of the inward current is much slower when charge carriers are Ba^{2+} (guinea pig taenia caeci: GANITKEVICH et al. 1987, 1991; rat myometrium: JMARI et al. 1986; canine stomach: VOGALIS et al. 1992). Similarly, inactivation of Ca^{2+} channels is much slower when Na^+ is the charge carrier than Ca^{2+} in guinea pig taenia (GANITKEVICH et al. 1991), rat myometrium (JMARI et al. 1987), rabbit ileum (OHYA et al. 1986) and canine stomach (VOGALIS et al. 1992).

In canine gastric circular muscle, the inward current recovers with a similar time course to the decrease in $[Ca^{2+}]_i$ following a depolarizing pulse, suggesting that $[Ca^{2+}]_i$ regulates the basal availability of Ca^{2+} channels (VOGALIS et al. 1992). Since ethylene glycol tetraacetic acid (EGTA) in patch pipettes has little effect, it is considered that the "submembrane" compartment of $[Ca^{2+}]_i$ regulates the development of inactivation.

In guinea pig coronary artery, an increase in the external Ca^{2+} concentration does not affect the rate of inactivation. No significant difference is found in the rate of inactivation when Na^{+} is charge carrier in Ca^{2+} channels, in contrast to slower inactivation in the taenia. As suggested by GANITKEVICH et al. (1991), it is possible that inactivation is fundamentally determined by the property of the potential-sensitive gate of the channel, but that this is modified by Ca^{2+} and other factors. Therefore, the gating property and also the degree of modification by Ca^{2+} may determine the rate of inactivation differently in different types of smooth muscle.

Inactivation of T-type current is considered to be only potential dependent (cultured rat aorta: AKAIKE et al. 1989; rat portal vein: LOIRAND et al. 1989).

IV. Blocking Agents

The inhibitory effects of verapamil are dependent on the holding potential level and also on the frequency of stimulation, i.e. verapamil can block the channel in an open state (use-dependent block). When the holding potential was -50 mV and depolarizing pulses to 0 mV were applied at a frequency of 4/min, 10 μM verapamil was very effective, but when the holding potential was -90 mV it had little effect in the guinea pig taenia (GANITKEVICH et al. 1986).

On the other hand, dihydropyridines have a high affinity to the channel in the inactivated state, so that the degree of inhibition increases with membrane depolarization [rat mesenteric artery: BEAN et al. 1986; rabbit mesenteric artery: NELSON and WORLEY 1989; rat bronchus and human tracheobronchus (cultured): WORLEY and KOTLIKOFF 1990]. For example, in the rabbit mesenteric artery, the IC_{50} of nisoldpine for the inhibition of single Ca^{2+} channel currents decreased from 12.1 to 1.9 nM by changing the holding potential from -100 to -50 mV in the presence of 1 μM Bay K8644 (NELSON and WORLEY 1989).

L-Type channels are considered to be more susceptible to dihydropyridines. Thus, if T- and L-type channels are both involved, the time course of the inward current would be expected to become faster in the presence of the channel blocker, because of the removal of slower L-type currents. On the other hand, since the blockers may accelerate inactivation of L-type currents, and may potentiate an early phase of L-type currents, the change in current decay cannot simply be used for evidence in support of the presence of T-type channels (AARONSON et al. 1988; AARONSON and RUSSELL 1991).

Sensitivity to Ca^{2+} channel blockers seems to differ to some degree in different smooth muscles (KATZKA and MORAD 1989). This could be partly due to unhomogeneous distribution of different types of channels, but could also be due to the different affinity to the blockers even in the same type of channels, as the rate of inactivation varies in different smooth muscles.

V. Intracellular ATP

Direct modulation of L-type channels by intracellular ATP has been observed in single smooth muscle cells isolated from intestine, mesenteric artery and portal vein of rabbit and it has also been shown that intrapipette perfusion of 0.3 m*M* ATP prevents "rundown" of inward currents (rabbit ileum, longitudinal muscle: OHYA et al. 1987b; rabbit small intestine and portal vein: OHYA and SPERELAKIS 1988; guinea pig jejunum and mesenteric artery: OHYA and SPERELAKIS 1989b). T-type channels are apparently not affected by intracellular ATP (guinea pig portal vein: OHYA and SPERELAKIS 1989b).

In guinea pig portal vein, which seems to have a dominant L-type channel, OKASHIRO et al. (1992) failed to obtain a clear correlation between the current size (or configuration) and ATP concentration in pipettes. Cyanide (0.5–1 m*M*) inhibited inward currents, but the recovery on cyanide washout was often very poor, although mechanical inhibition in muscle strips always recovered perfectly. This may suggest that metabolic impairment facilitates some process responsible for channel "rundown" under the whole-cell clamp condition. A decrease in ATP supply may inhibit Ca^{2+} channel indirectly through poor regulation of intracellular Ca^{2+} concentration ($[Ca^{2+}]_i$). We have obtained data which show that cyanide treatment increases $[Ca^{2+}]_i$ during mechanical inhibition in guinea pig gastric muscle (HUANG et al. 1993).

VI. pH Effect

Hydrogen ion (H^+) concentration is physiologically 40 n*M* outside and about 100 n*M* inside the cell. This concentration may affect the excitability significantly, but this effect has not been thoroughly investigated. In guinea pig gastric muscle, changes in the extracellular pH (pH_o) between 6.0 and 8.0 altered the amplitude of I_{Ca} (L-type) (KATZKA and MORAD 1989). The peak I_{Ca} increased linearly with increasing pH. Alkalinization did not affect the inactivation kinetics, but acidification increased the rate of inactivation. Similar results have been obtained in guinea pig basilar artery (WEST et al. 1992). Since the I–V curve is not shifted along the V axis and the holding current is also not affected by altering the external pH, Ca^{2+} channels are likely to be affected directly. T-type channel currents are less affected by pH than L-type channel currents.

In rabbit portal vein, the I_{Ca} (2.4 m*M* Ca^{2+}) is markedly potentiated by ammonium chloride (NH_4Cl, 20 m*M*) and inhibited by butyrate applied externally without shifting the I–V curve along the voltage axis and the rate of inactivation (TOMITA et al. 1991). Since NH_4Cl alkalinizes and butyrate acidifies the inside of cells, the effect of changes in intracellular pH (pH_i) seems similar to the change in pH_o.

C. K^+ Channel

The properties of smooth muscle differ greatly depending on the organ to which it belongs. The diversity of the excitability of the plasma membrane is considered to be closely related to the difference in outward K^+ currents. Several different types of K^+ channel have been found in smooth muscle. It is very important to know the properties and density of these K^+ channels in each type of smooth muscle in relation to the properties of the smooth muscle. It is likely that alterations in K^+ conductance are involved as a major mechanism in many pharmacological responses. Furthermore, we still do not know the exact mechanism underlying the resting membrane potential, although the contribution of K^+ conductance is certainly a major factor. We can expect further advances to clarify the functional roles of various K^+ channels.

K^+ channels may be divided into five main types (Latorre 1991): (a) voltage-dependent channels which can be subdivided into fast inactivating type and delayed rectifier type; (b) Ca^{2+}-activated (K_{Ca}) channels with two subclasses of large (maxi, LK_{Ca}) and small (SK_{Ca}) conductance; (c) inward rectifying channels; (d) receptor-coupled channels; and (e) ATP-regulated channels (K_{ATP}). These are all found in smooth muscle, but K_{Ca} and K_{ATP} channels have been most extensively examined to date.

I. Ca^{2+}-Activated K^+ Channel

The dominant K^+ current in smooth muscles flows through LK_{Ca} channels that are activated by intracellular Ca^{2+} and depolarization. This channel has been most extensively investigated (rabbit jejunum, guinea pig mesenteric artery: BENHAM et al. 1985, 1986; canine colon: CARL and SANDERS 1989; rabbit portal vein: HUME and LEBLANC 1989; INOUE et al. 1985; guinea pig urinary bladder: KLÖCKNER and ISENBERG 1992; guinea pig stomach: MITRA and MORAD 1985; rabbit ileum: OHYA et al. 1987a). Single-channel conductance is about 100 pS at a physiological K^+ gradient. This channel can be blocked by tetraethylammonium (TEA) (INOUE et al. 1985; BEECH and BOLTON 1989b). The dissociation constant (K_D) of externally applied TEA is 159–196 μM at 0 mV, 5/124 (mM) K^+ for mesenteric artery (LANGTON et al. 1990) and 300 μM for portal vein (BEECH and BOLTON 1989). 4-Aminopyridine (4-AP) is ineffective (CARL and SANDERS 1989).

SK_{Ca} channels with smaller unitary conductance have been reported in several smooth muscles (e.g., rabbit portal vein; INOUE et al. 1985, 1986). The conductance in rabbit portal vein is about 30–50 pS at a physiological K^+ gradient, and this is not susceptible to TEA (10 mM), in contrast to the LK_{Ca} channel. Another type of K^+ channel which is activated by external Ca^{2+} has also been reported for rabbit portal vein (INOUE et al. 1986). This channel has a unitary conductance between those of LK_{Ca} and SK_{Ca}. TEA applied to the cytosolic side is much more potent than that applied to the outside.

Although the properties of the Ca^{2+}-binding sites of LK_{Ca} channels are similar to those of calmodulin and troponin C (LATORRE et al. 1989), calmodulin is not considered to be involved in the channel activation in pregnant rat myometrium (KIHIRA et al. 1990; SEREBRYAKOV et al. 1989). When channel activity is low at $0.1\,\mu M$ cytosol Ca^{2+}, calmodulin antagonists W-7 and trifluoperazine applied to the cytoplasmic face increased channel open probability.

K_{Ca} is likely to be involved in spontaneous transient outward current (STOC: rabbit jejunum, ear artery: BENHAM and BOLTON 1986; rabbit jejunum: BOLTON and LIM 1989; hog carotid artery: DÉSILETS et al. 1989; rabbit portal vein: HUME and LEBLANC 1989; rabbit ileum: OHYA et al. 1987a; rabbit ileum: SAKAI et al. 1988). STOCs are inhibited by caffeine ($3\,mM$) or ryanodine ($30\,\mu M$) (XIONG et al. 1991). Ca^{2+} released spontaneously from intracellular stores is thought to be responsible for the current. In rabbit portal vein, the occurrence of this current significantly decreased when patch electrodes contained no ATP (HUME and LEBLANC 1989). We have observed that STOC in guinea pig portal vein was inhibited by cyanide application (H. TOKUNO, S. IINO, T. TOMOTA, unpublished observations). A sufficient supply of ATP seems necessary for spontaneous release of Ca^{2+}.

Cyclic AMP is considered to increase K_{Ca} channel activity through phosphorylation by cyclic AMP-dependent protein kinase (guinea pig trachea: KUME et al. 1989; cultured muscle cells of rat aorta: SADOSHIMA et al. 1988; canine colon: CARL et al. 1991). This has been confirmed in chemically skinned muscle cells with β-escin obtained from porcine trachea and guinea pig urinary bladder using the cell-attached patch-clamp method (MURAKI et al. 1992).

In canine colonic circular muscle, acetylcholine reduces outward currents through K_{Ca} channels and this is shown to be mediated by pertussis toxin-sensitive G proteins (COLE and SANDERS 1989b).

II. ATP-Regulated Channel

In several smooth muscles the presence of ATP-regulated (or ATP-dependent) K^+ (K_{ATP}) channels has been reported, particularly in relation to a possible contribution to vasodilator actions, as will be described. This is a class of K^+ channel that is activated when intracellular ATP concentration ($[ATP]_i$) is lowered, first reported in cardiac muscle by NOMA (1983). K_{ATP} channels are essentially time independent and only very weakly voltage dependent.

The single-channel conductance of the K_{ATP} channel in rabbit mesenteric artery is reported to be 20 pS at 0 mV (20°–24°C), 6/120 (mM) K^+ (NELSON et al. 1990), and also 135 pS at a 60/120 (mM) K^+ gradient (STANDEN et al. 1989). K_{ATP} channels in porcine coronary artery have a conductance of 148 pS at a 5.4/140 (mM) K^+ gradient, but they are also activated by intracellular Ca^{2+} (0.1–10 μM) (SILBERBERG and VAN BREEMEN 1990).

In the smooth muscle cells of rabbit and rat mesenteric arteries, an ATP concentration for half-maximal inhibition of K_{ATP} channel in the inside-out patch is considered to be nearly the same (STANDEN et al. 1989) as values of 10–200 μM reported for this channel in other tissues (ASHCROFT and ASHCROFT 1990). $[ATP]_i$ is likely to be much higher than these values and it is maintained at a relatively constant level even during metabolic inhibition, so that it is unlikely to be reduced low enough to activate the K_{ATP} channel. A possible contribution of the K_{ATP} channel has, however, been suggested in hypoxic dilatation of coronary artery (DAUT et al. 1990). It may be that the conditions used for the patch-clamp experiments have modified the property of the channel. Under more physiological conditions a small decrease in $[ATP]_i$ may activate the channel. When the cardiac cells were permeabilized by saponin ("open-cell attached" patch), the half-maximal inhibition occurred at 500 μM (KAKEI et al. 1985). The presence of physiological Mg^{2+} and ADP may increase the $[ATP]_i$ necessary for inhibition (NICHOLS and LEDERER 1991; NICHOLS et al. 1991).

Regulation of this channel has not been fully investigated in smooth muscles, but it seems that hydrosis of ATP is not necessary to close the channel and that AMP, guanosine diphosphate, and ADP in the absence of Mg^{2+} are also effective (NICHOLS and LEDERER 1991). On the other hand, in the guinea pig portal vein, even in the presence of 5 mM ATP in the patch pipette, ADP (0.1 μM) with 0.7 mM Mg^{2+} has been considered to activate K_{ATP} channels (PFRÜNDER et al. 1993).

There are several compounds which are considered to activate K_{ATP} channels. These include cromakalim (BRL 34915), lemakalim (BRL 38227, the enantiomer of cromakalim), nicorandil, pinacidil, diazoxide, minoxidil, and RP 49356; and they are generally called K^+ channel openers (ROBERTSON and STEINBERG 1990; WESTON and ABBOTT 1987; WESTON 1989). These agents probably act by decreasing K_{ATP} channel sensitivity to ATP. Activation by 1 μM cromakalim has been directly shown at the single-channel level (STANDEN et al. 1989).

The single-channel conductance of the channel activated by lemakalim in rat portal vein is estimated to be 17 pS [5/120 (mM) K^+, 26°C] based on noise analysis (NOACK et al. 1992). This channel may be similar to that reported for rabbit portal vein (BEECH and BOLTON 1989b), but differs from typical K_{ATP} channels which are found in isolated patches and have much larger unitary conductance, as described above. Similarly, activation of K^+ channels having 7.5 pS by K^+ channel opener has also been reported (NAKAO and BOLTON 1991).

It is generally thought that Ca^{2+}-activated K (LK_{Ca}) channels are insensitive to K^+ channel openers (BEECH and BOLTON 1989b; BERRY et al. 1991; LANGTON et al. 1991; OKABE et al. 1990; PAVENSTÄDT et al. 1991), but the actual situation is not as simple. There are also several reports that K^+

channel openers activate K^+ channel sensitive to Ca^{2+} (canine colon: CARL et al. 1992; rabbit aorta, trachea, and pig coronary artery: GELBAND et al. 1989; GROSCHNER et al. 1991; HU et al. 1990; rat portal vein: KAJIOKA et al. 1990; OKABE et al. 1990; KLÖCKNER et al. 1989; NAKAO and BOLTON 1991). K_{Ca} channels activated by K^+ channel openers in canine colon are insensitive to intracellular ATP (CARL et al. 1992). The other type of K^+ channels also activated is the channel (30 pS) sensitive to extracellular Ca^{2+} (INOUE et al. 1989) and the channel sensitive to both intracellular Ca^{2+} and ATP (KAJIOKA et al. 1990). Thus, it is difficult to relate the effect of K^+ channel openers to a typical K_{ATP} channel directly. Inhibition of K_{Ca} channels caused by ATP is likely to be at least partly due to Ca^{2+} chelating action of ATP since, when MgATP was used instead of Na_2ATP, the ATP effect was absent (porcine artery and guinea pig urinary bladder, KLÖCKNER and ISENBERG 1992).

Sulphonylureas, such as glibenclamide and tolbutamide, are known to block K_{ATP} channels, as first demonstrated in pancreatic B cells (STURGESS et al. 1985; TRUBE et al. 1986). Blockade of single K_{ATP} channels with $20\,\mu M$ glibenclamide is shown in arterial muscle (STANDEN et al. 1989). Glibenclamide may also inhibit the other types of K^+ channel, as shown in portal vein (BEECH and BOLTON 1989b). In mesenteric artery, glibenclamide ($10-20\,\mu M$) is ineffective on K_{Ca} channels (LANGTON et al. 1990).

Lemakalim relaxes guinea pig and rabbit coronary arteries contracted by the H_1-receptor agonist 2-(2-aminoethyl)pyridine accompanied by hyperpolarization. These effects are antagonized by glibenclamide ($1-35\,\mu M$), but not by TEA ($10\,mM$) (ECKMAN et al. 1992). In rat portal vein, inhibition of the spontaneous myogenic contractions by K^+ channel openers (cromakalim, minoxidil, diazoxide) is antagonized by TEA ($0.3-10\,mM$) and 4-AP ($1-10\,mM$), but not by charybdotoxin and apamin, K_{Ca} channel blockers (WINQUIST et al. 1989). These results suggest that K_{Ca} channel is not involved in the inhibitory action of K^+ channel openers at least in these smooth muscles.

Coronary vasodilation induced by hypoxia is blocked by glibenclamide, suggesting involvement of K_{ATP} channels in the vasodilation (DAUT et al. 1990). We have also obtained evidence that cyanide treatment activates K_{ATP} channels in the rabbit portal vein, because the cyanide-induced increase in outward currents is partially inhibited by glibenclamide. In the guinea pig gastric muscle, however, an increase in outward currents produced by cyanide was not reduced by glibenclamide, although the cromakalim-induced increase was readily blocked (H. TOKUNO, T. TOMITA, unpublished observations). The difference may not be due to the low density of K_{ATP}, because cromakalim produced a large glibenclamide-sensitive hyperpolarization. It might be that the contribution of glycolysis in ATP supply to the plasma membrane is much greater in gastric than vascular muscle. The functional roles of K_{ATP} channels in smooth muscles should be further investigated.

III. Fast-Inactivating K^+ (K_{FI}) Channel

When a depolarizing voltage pulse is applied, a transient outward current appears in some smooth muscle cells. This current is considered to flow through the K_{FI} channels (rabbit portal vein: BEECH and BOLTON 1989a; rabbit pulmonary artery: CLAPP and GURNEY 1991, OKABE et al. 1987a; rabbit portal vein: HUME and LEBLANC 1989; guinea pig ureter; IMAIZUMI et al. 1990, LANG 1989; rat ileum: SMIRNOV et al. 1992b). It has properties similar to the A current found in neurons and can be blocked by 4-AP, but not by TEA, apamin or charybdotoxin. Since this current is expected to oppose the generation of action potential, the high population of this channel may explain the inexcitability of pulmonary artery under physiological conditions (CLAPP and GURNEY 1991). The conductance of the single channel responsible for the K_{FI} current is estimated to be 14 pS in guinea pig ureter in a physiological ionic gradient (IMAIZUMI et al. 1990).

Another class of transient outward current which is Ca^{2+} dependent, in rabbit portal vein, is significantly reduced by removal of the external Ca^{2+} and not observed with patch pipettes containing 5 m*M* EGTA, suggesting a contribution of the K_{Ca} channel (HUME and LEBLANC 1989). The transient outward current observed in rabbit ileum (longitudinal muscle) seems dependent on the inward current (OHYA et al. 1987a) and is likely to be Ca^{2+} dependent, but it is still not clear whether activation of this channel is largely mediated by Ca^{2+} influx on depolarization.

IV. Delayed Rectifying K^+ (K_{DR}) Channel

This channel, which is activated by membrane depolarization and is probably responsible for the repolarization of action potential together with K_{FI} channel, is considered to be Ca^{2+} independent (canine colon: COLE and SANDERS 1989a; rabbit portal vein: HUME and LEBLANC 1989). K_{DR} channels in rabbit portal vein can be inhibited by phencyclidine, quinidine, 4-AP and TEA, and their potency is in this order (BEECH and BOLTON 1989a), so that they are inhibited by 4-AP but relatively insensitive to TEA. Similar results have been obtained in the rabbit pulmonary artery (CLAPP and GURNEY 1991; OKABE et al. 1987a). On the other hand, K_{DR} channels are not susceptible to charybdotoxin (100 μM) and apamin (100 n*M*) (BEECH and BOLTON 1989a). Similarly, K_{DR} in canine trachea (KOTLIKOFF 1990) and in rat basilar artery (STOCKBRIDGE et al. 1992) is not inhibited by charybdotoxin, but is reduced by procaine (1 m*M*) and strychnine (1 m*M*) in the basilar artery.

V. Time-Independent K^+ (K_{TI}) Channel

This channel provides the instantaneous (background) current which shows outward rectification and is probably mainly responsible for maintaining the resting potential. The properties of this current have been studied in rabbit pulmonary artery (Clapp and Gurney 1991), canine colon (Cole and Sanders 1989a), rabbit portal vein (Hume and Leblanc 1989) and dog coronary artery (Wilde and Lee 1989). This channel is significantly reduced by removal of the external Ca^{2+} and inhibited by TEA (0.5–20 mM), Ba^{2+} (2 mM) and 4-AP (2 mM).

VI. Inward-Rectifying K^+ (K_{IR}) Channel

Hyperpolarization beyond −60 to −110 mV activates inward currents in some cells (rabbit jejunum: Benham et al. 1987a; rat basilar artery: Stockbridge et al. 1992). The current in rabbit jejunum may be carried not only by K^+ but also by Na^+ and it is blocked by external caesium (1 mM), but is less affected by Ba^{2+} (10 mM). In circular muscle of pregnant rat myometrium, a similar current was recorded (K. Okabe, H. Inoue, T. Osa, personal communication). We have also observed K_{IR} channel current in rat urinary bladder. This current was not significantly affected by TEA (10–30 mM), nicardipine (3 μM) or cadmium (100 μM), but was markedly inhibited by 4-AP (5 mM) (H. Tokuno, T. Tomita, unpublished observation).

D. Na^+ Channel

Since action potential in smooth muscle is not abolished by tetrodotoxin (TTX), it has been generally assumed that Na^+ current (I_{Na}) does not significantly contribute to excitation. However, in patch-clamp experiments I_{Na} has been found in several types of smooth muscle. The current has a rapid time course, being activated in 1–2 ms and fully inactivated within 10–30 ms. Sensitivity to TTX seems to vary to a large extent in different smooth muscles. The previous finding of the inability of TTX to block the action potential is due to the fact that the Ca^{2+} channels play a major role in the generation of action potential. The functional roles of I_{Na} have not been fully investigated, but may be important for the spread of excitation by increasing the rate of rise of action potential at least in some smooth muscles. In this sense, the finding that the current density of I_{Na} in rat myometrium (longitudinal) increases during gestation (Inoue and Sperelakis 1991) is interesting.

In rat stomach and guinea pig ureter (10°C), I_{Na} appeared at about −40 mV in stomach and −30 mV in ureter, when depolarized from a holding potential of −80 mV, and reached a peak at −10 mV (stomach) or 0 mV (ureter) (Muraki et al. 1991). The half-inactivation potential was −63 mV for fundus and −39 mV for ureter at 21°C. In rat portal vein, the threshold for initiation of Na^+ inward current was −30 mV at a holding potential of −90 mV, and the maximum current was obtained at 0 mV at 30°C (Mironneau et al. 1990). In rat and human colon (circular muscle), the threshold voltage was −50 mV for rat and −40 mV for human, and peak current was obtained at −20 mV for rat and 0 mV for human (XIONG et al. 1993). The cell having I_{Na} was more frequently obtained from the proximal (ascending) than the distal (descending) colon.

Most of the Na^+ channels are highly sensitive to TTX. The IC_{50} of TTX is reported to be 3 n*M* in rat portal vein (MIRONNEAU et al. 1990), 4.5 n*M* in rat ileum (longitudinal) (SMIRNOV et al. 1992a), 8.7 n*M* in rabbit pulmonary artery (OKABE et al. 1988), 8.8 n*M* in pregnant rat myometrium (MARTIN et al. 1990), 10 n*M* in rat portal vein (OKABE et al. 1990), 11 n*M* in guinea pig ureter (MURAKI et al. 1991), 14 n*M* in human colon (XIONG et al. 1993), 27 n*M* in pregnant rat myometrium (longitudinal) (OHYA and SPERELAKIS 1989a), and 130 n*M* in rat colon (XIONG et al. 1993). On the other hand, in rat stomach fundus the IC_{50} is found to be 870 n*M* (MURAKI et al. 1991), and it is 30 μM in azygos vein (cultured) (STUREK and HERMSMEYER 1986). In rabbit pulmonary artery, 30 μM TTX was used to block I_{Na}, but lower concentrations were not examined (OKABE et al. 1987b).

Na^+ channels may be classified into two groups depending on their sensitivity to TTX. It is, however, possible that the cell dispersion procedure might alter the channel property and this might partly introduce differences in TTX sensitivity. The density of Na^+ channel certainly varies in different types of smooth muscle, but it also varies in different cells within the same tissue. It may change with age; portal vein of younger rats (3 weeks) is considered to show larger I_{Na} than that of adult rats (12 weeks) (OKABE et al. 1990). In rat stomach fundus, I_{Na} could be recorded from all cells examined, whereas in guinea pig ureter it was observed only in 10%–20% of the cells examined (MURAKI et al. 1991). We can expect further studies to clarify the properties and population of Na^{2+} channels in relation to their physiological function.

E. Cl^- Channel

Intracellular Cl^- concentration is considered to be much higher than that predicted from a passive distribution (AICKIN and BRADING 1982). The Cl^- equilibrium potential (E_{Cl}) in guinea pig vas deferens is estimated to be −24 mV. Similarly, the E_{Cl} in many smooth muscles is likely to be much less negative than the resting potential. Therefore, an increase in Cl^- conductance of the plasma membrane, for example, in response to receptor activation, would cause depolarization. In smooth muscle, two different Cl^- channels have been reported: Ca^{2+}-activated and voltage-dependent channels.

I. Ca-Activated Cl^- Channel

In receptor-mediated contractions, intracellular Ca^{2+} released from the stores may activate these Cl^- channels. This possibility has been considered for the response to noradrenaline in rabbit ear artery (AMÉDÉE et al. 1990), rabbit portal vein (BYRNE and LARGE 1988), rat portal vein (PACAUD et al. 1989, 1991), and guinea pig mesenteric vein (VAN HELDEN 1988), acetylcholine in canine and guinea pig trachea (JANSSEN and SIMS 1992), endothelin in porcine coronary and human mesenteric artery (KLÖCKENER and ISENBERG 1991), and neurokinin in rabbit colon (longitudinal) (MAYER et al. 1991). Activation of this channel would depolarize the membrane by Cl^- efflux (inward currents) and secondarily activate voltage-dependent Ca^{2+} channels to cause contraction. As predicted by this theory, depletion of intracellular Ca^{2+} stores, for example by ryanodine, blocked noradrenaline-induced contractions in rat portal vein (PACAUD et al. 1991).

Single Cl^- channels were recorded in response to caffeine (20 m*M*) from myocytes of human mesenteric artery, and their single channel conductance was estimated to be 2.8 pS (KLÖCKNER 1993). In cultured cells from rat portal vein, Cl^- currents evoked by caffeine (10 m*M*) at −70 mV (or by depolarizing pulses to 0 mV) were inhibited by DIDS (4,4′-diisothiocyanostilbene-2,2′-disulphonic acid), A-9-C (anthracene-9-carboxylic acid) or DPC (diphenylamine-2,2′-dicarboxylic acid) with IC_{50}s of 16.5, 117 and 306 μM, respectively (BARON et al. 1991). Similarly, acetylcholine response in canine and guinea pig trachea is strongly inhibited by 100 μM niflumic acid and 1 m*M* SITS (4-acetamido-4′-isothiocyanatostilbene-2,2′-disulphonic acid) in the absence of Na^+ (JANSSEN and SIMS 1992).

In K^+-free solution spontaneous transient inward currents (STICs) have been observed in rabbit portal vein (WANG et al. 1992). This current is considered to be generated by an increase in Cl^- conductance caused by intracellular Ca^{2+} release, as for STOCs described above. The decay of STICs is, however, much slower than that of STOCs due to a difference in channel property (HOGG et al. 1993). STICs can be blocked by 1 m*M* A-9-C and 2 m*M* SITS (WANG et al. 1992), but the effect was voltage dependent, the amplitude being much reduced and the decay made much slower by A-9-C at stronger depolarization (HOGG et al. 1993).

II. Voltage-Dependent Channel

Voltage-dependent channels are generally categorized into three groups according to single-channel conductance: small (less than 15 pS), middle (several tens pS) and large conductance (larger than 300 pS). Only large Cl^- channels have been studied in smooth muscle. In rabbit colon (longitudinal muscle) the single-channel conductance is 309 pS at 130/130 (m*M*) Cl^- gradient and the channel activity is not affected by cytosolic Ca^{2+} up to 1 m*M* (SUN et al. 1992). Channel inhibition by 5-nitro-2-(3-phenylpropylamino)-benzoate (NPPB, 70 μM) and DIDS (10 μM) has been observed in some cells when applied to the cytosolic side (in three of seven with 70 μM NPPB and in three of five with 10 μM DIDS). Similarly, in cultured cells from fetal rat aorta, DIDS (5–25 μM) applied to the cytosolic side inhibited large Cl^-

channels in inside-out patches (Kokubun et al. 1991). A-9-C and furosemide were ineffective.

The channel remained closed at resting membrane potential, but could be activated by substance P methylester (1 p*M*) when examined with outside-out patches using pipettes containing 1 m*M* GTP (Sun et al. 1992).

Plasma membrane obtained from guinea pig myometrium without treating with enzymes contained large Cl^- channels (conductance of about 430 pS when recorded with the inside-out configuration at 130/145 (m*M*) Cl^-) (Coleman and Parkington 1987). This was not affected by TEA (10 m*M*).

References

Aaronson PI, Russell SN (1991) Rabbit intestinal smooth muscle calcium current in solutions with physiological calcium concentration. Exp Physiol 76:539–551

Aaronson PI, Bolton TB, Lang RJ, MacKenzie I (1988) Calcium currents in single isolated smooth muscle cells from the rabbit ear artery in normal-calcium and high-barium solutions. J Physiol (Lond) 405:57–75

Aickin CC, Brading AF (1982) Measurement of intracellular chloride in guinea-pig vas deferens by ion analysis, ^{36}chloride efflux and micro-electrodes. J Physiol (Lond) 326:139–154

Akaike N, Kanaide H, Kuga T, Nakamura M, Sadoshima J, Tomoike H (1989) Low-voltage-activated calcium current in rat aorta smooth muscle cells in primary culture. J Physiol (Lond) 416:141–160

Amédée T, Benham CD, Bolton TB, Byrne NG, Large WA (1990) Potassium, chloride and non-selective cation conductances opened by noradrenaline in rabbit ear artery cells. J Physiol (Lond) 426:551–568

Ashcroft SJH, Ashcroft FM (1990) Properties and functions of ATP-sensitive K-channels. Cell Signal 2:197–214

Baron A, Pacaud P, Loirand G, Mironneau C, Mironneau J (1991) Pharmacological block of Ca^{2+}-activated Cl^- current in rat vascular smooth muscle cells in short-term primary culture. Pflugers Arch 419:553–558

Bean BP, Sturek M, Puga A, Hermsmeyer K (1986) Calcium channels in muscle cells isolated from rat mesenteric arteries: modulation by dihydropyridine drugs. Circ Res 59:229–235

Beech DJ, Bolton TB (1989a) A voltage-dependent outward current with fast kinetics in single smooth muscle cells isolated from rabbit portal vein. J Physiol (Lond) 412:397–414

Beech DJ, Bolton TB (1989b) Properties of the cromakalim-induced potassium conductance in smooth muscle cells isolated from the rabbit portal vein. Br J Pharmacol 98:851–864

Benham CD, Bolton TB (1986) Spontaneous transient outward currents in single visceral and vascular smooth muscle cells of rabbit. J Physiol (Lond) 381: 385–406

Benham CD, Bolton TB, Lang KJ, Takewaki T (1985) The mechanisms of action of Ba^{2+} and TEA on single Ca^{2+}-activated K^+ channels in arterial and intestinal smooth muscle cell membrane. Pflugers Arch 403:120–127

Benham CD, Bolton TB, Lang RJ, Takewaki T (1986) Calcium-activated potassium channels in single smooth muscle cells of rabbit jejunum and guinea-pig mesenteric artery. J Physiol (Lond) 371:45–67

Benham CD, Bolton TB, Denbigh JS, Lang RJ (1987a) Inward rectification in freshly isolated single smooth muscle cells of the rabbit jejunum. J Physiol (Lond) 383:461–476

Benham CD, Hess P, Tsien RW (1987b) Two types of calcium channels in single smooth muscle cells from rabbit ear artery studied with whole-cell and single-channel recordings. Circ Res 61 [Suppl I]:10–16
Berry JL, Elliot KRF, Foster RW, Green KA, Murry MA, Small RC (1991) Mechanical, biochemical and electrophysiological studies of RP 49356 and cromakalim in guinea-pig and bovine trachealis muscle. Pulm Pharmacol 4:91–98
Bolton TB (1979) Mechanisms of action of transmitters and other substances on smooth muscle. Physiol Rev 59:606–718
Bolton TB, Lim SP (1989) Properties of calcium stores and transient outward currents in single smooth muscle cells of rabbit intestine. J Physiol (Lond) 409:385–401
Byrne NG, Large WA (1988) Membrane ionic mechanisms activated by noradrenaline in cells in isolated from the rabbit portal vein. J Physiol (Lond) 404:557–573
Carl A, Sanders KM (1989) Ca^{2+}-activated K channels of canine colonic myocytes. Am J Physiol 257:C470–C480
Carl A, Kenyon JL, Uemura D, Fusetani N, Sanders KM (1991) Regulation of Ca^{2+}-activated K^+ channels by protein kinase A and phosphatase inhibitors. Am J Physiol 261:C387–C392
Carl A, Bowen S, Gelband CH, Sanders KM, Hume JR (1992) Cromakalim and lemakalim activate Ca^{2+}-dependent K^+ channels in canine colon. Pflugers Arch 421:67–76
Clapp LH, Gurney AM (1991) Outward currents in rabbit pulmonary artery cells dissociated with a new technique. Exp Physiol 76:677–693
Cole WC, Sanders KM (1989a) Characterization of macroscopic outward currents of canine colonic myocytes. Am J Physiol 257:C461–C469
Cole WC, Sanders KM (1989b) G protein mediated suppression of Ca^{2+}-activated K current by acetylcholine in smooth muscle cells. Am J Physiol 257:C596–C600
Coleman HA, Parkington HC (1987) Single channel Cl^- and K^+ currents from cells of uterus not treated with enzymes. Pflugers Arch 410:560–562
Daut J, Maier-Rudolph W, von Beckerath N, Mehrke G, Günther K, Goedel-Meinen L (1990) Hypoxic dilatation of coronary arteries is mediated by ATP-sensitive potassium channels. Science 247:1341–1344
Désilets M, Driska SP, Baumgarten CM (1989) Current fluctuations and oscillations in smooth muscle cells from hog carotid artery. Circ Res 65:708–722
Eckman DM, Frankovich JD, Keef KD (1992) Comparison of the actions of acetylcholine and BRL 38227 in the guinea-pig coronary artery. Br J Pharmacol 106:9–16
Ganitkevich VY, Isenberg G (1990) Contribution of two types of calcium channels to membrane conductance of single myocytes from guinea-pig coronary artery. J Physiol (Lond) 426:19–42
Ganitkevich VY, Shuba MF, Smirnov SV (1986) Potential-dependent calcium inward current in a single isolated smooth muscle cell of the guinea-pig taenia caeci. J Physiol (Lond) 380:1–16
Ganitkevich VY, Shuba MF, Smirnov SV (1987) Calcium-dependent inactivation of potential-dependent calcium inward current in an isolated guinea-pig smooth muscle cell. J Physiol (Lond) 392:431–449
Ganitkevich VY, Shuba MF, Smirnov SV (1988) Saturation of calcium channels in single isolated smooth muscle cells of guinea-pig taenia caeci. J Physiol (Lond) 399:419–436
Ganitkevich VY, Shuba MF, Smirnov SV (1991) Inactivation of calcium channels in single vascular and visceral smooth muscle cells of the guinea-pig. Gen Physiol Biophys 10:137–161
Gelband CH, Lodge NJ, van Breemen C (1989) A Ca^{2+}-activated K^+ channel from rabbit aorta: modulation by cromakalim. Eur J Pharmacol 167:201–210

Groschner K, Silberberg SD, Gelband CH, van Breemen C (1991) Ca^{2+}-activated K^+ channels in airway smooth muscle are inhibited by cytosolic ATP. Pflugers Arch 417:517–522
Hogg RC, Wang Q, Large WA (1993) Time course of spontaneous calcium-activated chloride currents in smooth muscle cells from the rabbit portal vein. J Physiol (Lond) 464:15–31
Hu S, Kim HS, Okokke P, Weiss GB (1990) Alterations by glyburide of effects of BRL 34915 and P1060 on contraction, $^{86}Rb^+$ efflux and maxi-K^+ channel in rat portal vein. J Pharmacol Exp Ther 253:771–777
Huang S-M, Chowdhury JU, Kobayashi K, Tomita T (1993) Inhibitory effects of cyanide on mechanical and electrical activities in the circular muscle of gastric antrum of guinea-pig stomach. Jpn J Physiol 43:229–238
Hume JR, Leblanc N (1989) Macroscopic K^+ currents in single smooth muscle cells of the rabbit portal vein. J Physiol (Lond) 413:49–73
Imaizumi Y, Muraki K, Watanabe M (1990) Characteristics of transient outward currents in single smooth muscle cells from the ureter of the guinea-pig. J Physiol (Lond) 427:301–324
Inoue I, Nakaya Y, Mori H (1989) Extracellular Ca^{2+}-activated K channel in coronary artery smooth muscle cells and its role in vasodilation. FEBS Lett 255:281–284
Inoue R, Kitamura K, Kuriyama H (1985) Two Ca-dependent K-channels classified by the application of tetraethylammonium distribute to smooth muscle membranes of the rabbit portal vein. Pflugers Arch 405:173–179
Inoue R, Okabe K, Kitamura K, Kuriyama H (1986) A newly identified Ca^{2+} dependent K^+ channel in the smooth muscle membrane of single cells dispersed from the rabbit portal vein. Pflugers Arch 406:138–143
Inoue Y, Sperelakis N (1991) Gestational change in Na^+ and Ca^{2+} channel current densities in rat myometrial smooth muscle cells. Am J Physiol 260:C658–C663
Inoue Y, Oike M, Nakao K, Kitamura K, Kuriyama H (1990) Endothelin augments unitary calcium channel currents on the smooth muscle cell membrane of guinea-pig portal vein. J Physiol (Lond) 423:171–191
Janssen L, Sims SM (1992) Acetylcholine activates non-selective cation and chloride conductances in canine and guinea-pig tracheal myocytes. J Physiol (Lond) 453:197–218
Jmari K, Mironneau C, Mironneau J (1986) Inactivation of calcium channel current in rat uterine smooth muscle: evidence for calcium- and voltage-mediated mechanisms. J Physiol (Lond) 380:111–126
Jmari K, Mironneau C, Mironneau J (1987) Selectivity of calcium channels in rat uterine smooth muscle: interactions between sodium, calcium and barium ions. J Physiol (Lond) 384:247–261
Kajioka S, Oike M, Kitamura K (1990) Nicorandil opens a calcium-dependent potassium channel in smooth muscle cells of the rat portal vein. J Pharmacol Exp Ther 254:905–913
Kakei M, Noma A, Shibasaki T (1985) Properties of adenosine-triphosphate-regulated potassium channels in guinea-pig ventricular cells. J Physiol (Lond) 363:441–462
Katzka DA, Morad M (1989) Properties of calcium channels in guinea-pig gastric myocytes. J Physiol (Lond) 413:175–197
Katzka DA, Cox R, Davidoff AJ, Morad M (1992) Permeation of divalent cations through the Ca^{2+} channel of rabbit portal vein myocytes. Am J Physiol 262: H326–H330
Kihira M, Matsuzawa K, Tokuno H, Tomita T (1990) Effects of calmodulin antagonists on calcium-activated potassium channels in pregnant rat myometrium. Br J Pharmacol 100:353–359
Klöckner U (1993) Intracellular calcium ions activate a low-conductance chloride channel in smooth-muscle cells isolated from human mesenteric artery. Pflügers Arch 424:231–237

Klöckner U, Isenberg G (1991) Endothelin depolarizes myocytes from porcine coronary and human mesenteric arteries through a Ca-activated chloride current. Pflügers Arch 418:168–175

Klöckner U, Isenberg G (1992) ATP suppresses activity of Ca^{2+}-activated K^{+} channels by Ca^{2+} chelation. Pflügers Arch 420:101–105

Klöckner U, Trieschmann U, Isenberg G (1989) Pharmacological modulation of calcium and potassium channels in isolated vascular smooth muscle cells. Arzneimittelforschung Drug Res 39:120–126

Kokubun S, Saigusa A, Tamura T (1991) Blockade of Cl channels by organic and inorganic blockers in vascular smooth muscle cells. Pflugers Arch 418:204–213

Kotlikoff MI (1988) Calcium currents in isolated canine airway smooth muscle cells. Am J Physiol 254:C793–C801

Kotlikoff MI (1990) Potassium currents in canine airway smooth muscle cells. Am J Physiol 259:L384–L395

Kume H, Takai A, Tokuno H, Tomita T (1989) Regulation of Ca^{2+}-dependent K^{+}-channel activity in tracheal myocytes by phosphorylation. Nature 341:152–154

Lang RJ (1989) Identification of the major membrane currents in freshly dispersed single smooth muscle cells of guinea-pig ureter. J Physiol (Lond) 412:375–395

Langton PD, Burke EP, Sanders KM (1989) Participation of Ca currents in colonic electrical activity. Am J Physiol 257:C451–C460

Langton PD, Nelson MT, Huang Y, Standen NB (1990) The block of Ca^{2+}-activated K^{+} channels by external TEA^{+} and tetrapentylammonium ions in isolated arterial smooth muscle cells from rat and rabbit. J Physiol (Lond) 426:70

Langton PD, Nelson MT, Huang T, Standen NB (1991) Block of calcium-activated potassium channels in mammalian arterial myocytes by tetraethylammonium ions. Am J Physiol 260:H927–934

Latorre R (1991) Metabolic control of K^{+} channels: an overview. J Bioenerg Biomembr 23:493–497

Latorre R, Oberhauser A, Labarca P, Alvarez O (1989) Varieties of calcium-activated potassium channels. Annu Rev Physiol 51:385–399

Loirand G, Mironneau C, Mironneau J, Pacaud P (1989) Two types of calcium currents in single smooth muscle cells from rat portal vein. J Physiol (Lond) 412:333–349

Martin C, Arnaudeau S, Jmari K, Rakotoarisoa L, Sayet I, Dacquet C, Mironneau C, Mironneau J (1990) Identification and properties of voltage-sensitive sodium channels in smooth muscle cells from pregnant rat myometrium. Mol Pharmacol 38:667–673

Matsuda JJ, Volk KA, Shibata EF (1990) Calcium currents in isolated rabbit coronary artery smooth muscle myocytes. J Physiol (Lond) 427:657–680

Mayer EA, Sun XP, Supplison S, Kodner A, Regoli M, Sachs G (1991) Neurokinin receptor-mediated regulation of $[Ca]_i$ and Ca-sensitive ion channels in mammalian colonic muscle. Ann NY Acad Sci 632:439–441

Meisheri KD, Hwang O, van Breemen C (1981) Evidence for two separate Ca^{2+} pathways in smooth muscle plasmalemma. J Membr Biol 59:19–25

Mironneau J, Martin C, Arnaudeau S, Jmari K, Rakotoarisoa L, Sayet I, Mironneau C (1990) High-affinity binding sites for [^{3}H]saxitoxin are associated with voltage-dependent sodium channels in portal vein smooth muscle. Eur J Pharmacol 184:315–319

Mitra R, Morad M (1985) Ca^{2+} and Ca^{2+}-activated K^{+} currents in mammalian gastric smooth muscle cells. Science 229:269–272

Muraki K, Imaizumi Y, Watanabe M (1991) Sodium currents in smooth muscle cells freshly isolated from stomach fundus of the rat and ureter of the guinea-pig. J Physiol (Lond) 442:351–375

Muraki K, Imaizumi Y, Watanabe M (1992) Ca-dependent K channels in smooth muscle cells permeabilized by β-escin recorded using the cell-attached patch-clamp technique. Pflugers Arch 420:461–469

Nakao K, Bolton TB (1991) Cromakalim-induced potassium currents in single dispersed smooth muscle cells of rabbit artery and vein. Br J Pharmacol 102:155p
Nakazawa K, Saito H, Matsuki N (1988) Fast and slowly inactivating components of Ca-channel current and their sensitivity to nicardipine in isolated smooth muscle cells from rat vas deferens. Pflugers Arch 411:289–295
Nelson MT, Worley JF (1989) Dihydropyridine inhibition of single calcium channels and contraction in rabbit mesenteric artery depends on voltage. J Physiol (Lond) 412:65–91
Nelson MT, Huang JE, Brayden JE, Hescheler JK, Standen NB (1990) Arterial dilations in response to calcitonin gene-related peptide involve activation of K^+ channels. Nature 344:770–773
Neveau D, Nargeot J, Richard S (1993) Two high-voltage-activated, dihydropyridine-sensitive Ca^{2+} channel currents with distinct electrophysiological and pharmacological properties in cultured rat aortic myocytes. Pflugers Arch 424: 45–53
Nichols CG, Lederer WJ (1991) Adenosine triphosphate-sensitive potassium channels in the cardiovascular system. Am J Physiol 261:H1675–H1686
Nichols CG, Ripoll C, Lederrer WJ (1991) K_{ATP} channel modulation of the guinea-pig ventricular action potential and contraction. Circ Res 68:280–287
Noack T, Deitmer P, Edwards G, Weston AH (1992) Characterization of potassium currents modulated by BRL 38227 in rat portal vein. Br J Pharmacol 106:717–726
Noma A (1983) ATP-regulated K^+ channels in cardiac muscle. Nature 305:147–148
Ohya Y, Sperelakis N (1988) Whole-cell voltage clamp and intracellular perfusion technique on single smooth muscle cells. Mol Cell Biochem 80:79–86
Ohya Y, Sperelakis N (1989a) Fast Na^+ and slow Ca^{2+} channels in single uterine muscle cells from pregnant rats. Am J Physiol 257:C408–C412
Ohya Y, Sperelakis N (1989b) ATP regulation of the slow calcium channels in vascular smooth muscle cells of guinea-pig mesenteric artery. Circ Res 64: 145–154
Ohya Y, Terada K, Kitamura K, Kuriyama H (1986) Membrane currents recorded from a fragment of rabbit intestinal smooth muscle cell. Am J Physiol 251: C335–C346
Ohya Y, Kitamura K, Kuriyama H (1987a) Cellular calcium regulates outward currents in rabbit intestinal smooth muscle. Am J Physiol 252:C401–C410
Ohya Y, Kitamura K, Kuriyama H (1987b) Modulation of ionic currents in smooth muscle balls of the rabbit intestine by intracellularly perfused ATP and cAMP. Pflugers Arch 408:465–473
Okabe K, Kitamura K, Kuriyama H (1987a) Features of 4-aminopyridine sensitive outward current observed in single smooth muscle cells from the rabbit pulmonary artery. Pflugers Arch 409:561–568
Okabe K, Terada K, Kitamura K, Kuriyama H (1987b) Selective and long-lasting inhibitory actions of the dihydropyridine derivative, CV-4093, on calcium currents in smooth muscle cells of the rabbit pulmonary artery. J Pharmacol Exp Ther 243:703–710
Okabe K, Kitamura K, Kuriyama H (1988) The existence of a highly tetrodotoxin sensitive Na channel in freshly dispersed smooth muscle cells of the rabbit main pulmonary artery. Pflugers Arch 411:423–428
Okabe K, Kajioka S, Nakao K, Kitamura K, Kuriyama H, Weston AH (1990) Actions of cromakalim on ionic currents recorded from single smooth muscle cells of the rat portal vein. J Pharmacol Exp Ther 252:832–839
Okashiro T, Tokuno H, Fukumitsu T, Hayashi H, Tomita T (1992) Effects of intracellular ATP on calcium current in freshly dispersed single cells of guinea-pig portal vein. Exp Physiol 77:719–731
Pacaud P, Loirand G, Mironneau C, Mironneau J (1989) Noradrenaline activates a calcium-activated chloride conductance and increases the voltage-dependent

calcium current in cultured single cells of rat portal vein. Br J Pharmacol 97:139–146

Pacaud P, Loirand G, Baron A, Mironneau C, Mironneau J (1991) Ca^{2+} channel activation and membrane depolarization mediated by Cl^- channels in response to noradrenaline in vascular myocytes. Br J Pharmacol 104:1000–1006

Pavenstädt H, Lindeman V, Lindeman S, Kunzelmann K, Späth M, Greger R (1991) Effect of depolarizing and hyperpolarizing agents on the membrane potential difference of primary cultures of rabbit aorta vascular smooth muscle cells. Pflugers Arch 419:69–75

Pfründer D, Anghelescu I, Kreye VAW (1993) Intracellular ADP activates ATP-sensitive K^+ channels in vascular smooth muscle cells of the guinea-pig portal vein. Pflugers Arch 423:149–151

Robertson DW, Steinberg ML (1990) Potassium channel modulators: scientific applications and therapeutic promise. J Med Chem 33:1529–1541

Sadoshima J, Akaike N, Kanaide H, Nakamura M (1988) Cyclic AMP modulates Ca-activated K channel in cultured smooth muscle cells of rat aorta. Am J Physiol 255:H754–H759

Sakai T, Terada K, Kitamura K, Kuriyama H (1988) Ryanodine inhibits the Cadependent K current after depletion of Ca store in smooth muscle cells of the rabbit ileal longitudinal muscle. Br J Pharmacol 95:1089–1100

Serebryakov VN, Bregestosvski PD, Zamoyski VL, Stinnakre J (1989) Effects of quinine and of calmodulin inhibitors on calcium-dependent potassium channels of human aorta. Pflugers Arch 414 [Suppl 1]:S185

Silberberg SD, van Breemen C (1990) An ATP, calcium and voltage sensitive potassium channel in porcine coronary artery smooth muscle cells. Biochem Biophys Res Commun 172:517–522

Simard JM (1991) Calcium channel currents in isolated smooth muscle cells from the basilar artery of the guinea pig. Pflugers Arch 417:528–536

Sims SM (1992) Calcium and potassium currents in canine gastric smooth muscle cells. Am J Physiol 262:G859–G867

Smirnov SV, Zholos AV, Shuba MF (1992a). Potential-dependent inward currents in single isolated smooth muscle cells of the rat ileum. J Physiol (Lond) 454: 549–571

Smirnov SV, Zholos AV, Shuba MF (1992b) A potential-dependent fast outward current in single smooth muscle cells isolated from the newborn rat ileum. J Physiol (Lond) 454:573–589

Standen NB, Quayle JM, Davies NW, Brayden JE, Huang Y, Nelson MT (1989) Hyperpolarizing vasodilators activate ATP-sensitive K^+ channels in arterial smooth muscle. Science 245:177–180

Stockbridge N, Zhang H, Weir B (1992) Potassium currents of rat basilar artery smooth muscle cells. Pflugers Arch 421:37–42

Sturek M, Hermsmeyer K (1986) Calcium and sodium channels in spontaneously contracting vascular muscle cells. Science 233:475–478

Sturgess NC, Ashford ML, Cook DL, Hales CN (1985) The sulfonylurea receptor may be an ATP-sensitive potassium channel. Lancet 31:474–475

Sun XP, Supplisson S, Torres R, Sachs G, Mayer E (1992) Characterization of large-conductance chloride channels in rabbit colonic smooth muscle. J Physiol (Lond) 448:355–382

Tomita T, Tokuno H, Ito M, Syed MM, Matsumoto T (1991) Effects of ammonium on vascular smooth muscle. In: Sperelakis N, Kuriyama H (eds) Ion channels of vascular smooth muscle cells and endothelial cells. Elsevier, New York, pp 217–223

Trube G, Rohrsman P, Ohno-Shosaku T (1986) Opposite effects of tolbutamide and diazoxide on the ATP-dependent K^+ channel in mouse pancreatic B-cells. Pflugers Arch 407:493–499

Van Helden DF (1988) An α-adrenoceptor-mediated chloride conductance in mesenteric veins of the guinea-pig. J Physiol (Lond) 401:489–501

Vogalis F, Publicover NG, Sanders KM (1992) Regulation of calcium current by voltage and cytoplasmic calcium in canine gastric smooth muscle. Am J Physiol 262:C691–C700

Wang Q, Hogg RC, Large WA (1992) Properties of spontaneous inward currents recorded in smooth muscle cells isolated from the rabbit portal vein. J Physiol (Lond) 451:525–537

Wang R, Karpinski E, Pang PKT (1989) Two types of calcium channels in isolated smooth muscle cells from rat tail artery. Am J Physiol 256:H1361–H1368

Wellner M-C, Isenberg G (1993) Properties of stretch-activated channels in myocytes from the guinea-pig urinary bladder. J Physiol (Lond) 466:213–227

West GA, Leppla DC, Simard JM (1992) Effects of external pH on ionic currents in smooth muscle cells from the basilar artery of the guinea-pig. Circ Res 71:201–209

Weston AH (1989) Smooth muscle K^+ channel openers; their pharmacological and clinical potential. Pflugers Arch 414 [Suppl 1]:S99–S105

Weston AH, Abbott A (1987) New class of antihypertensive acts by opening K^+ channels. Trends Pharmacol Sci 8:283–284

Wilde DW, Lee KS (1989) Outward potassium currents in freshly isolated smooth muscle cell of dog coronary arteries. Circ Res 65:1718–1734

Winquist RJ, Heaney LA, Wallace AA, Baskin EP, Stein RB, Garcia ML, Kaczorowski GJ (1989) Glyburide blocks the relaxation response to BRL 34915 (cromakalim), minoxidil sulfate and diazoxide in vascular smooth muscle. J Pharmacol Exp Ther 248:149–156

Worley JF, Kotlikoff MI (1990) Dihydropyridine-sensitive single calcium channels in airway smooth muscle cells. Am J Physiol 259:L468–L480

Worley JF, Deitmer JW, Nelson MT (1986) Single nisoldipine-sensitive calcium channels in smooth muscle cells isolated from rabbit mesenteric artery. Proc Natl Acad Sci USA 83:5746–5750

Worley JF, Quayle JH, Standen NB, Nelson MT (1991) Regulation of single calcium channels in cerebral arteries by voltage, serotonin, and dihydropyridines. Am J Physiol 261:H1951–H1960

Xiong Z, Kajioka S, Sakai T, Kitamura K, Kuriyama H (1991) Pinacidil inhibits the ryanodine-sensitive outward current and glibenclamide antagonizes its action in cells from the rabbit portal vein. Br J Pharmacol 102:788–790

Xiong Z, Sperelakis N, Noffsinger A, Fenoglio-Preise C (1993) Fast Na^+ current in circular muscle cells of the large intestine. Pflugers Arch 423:485–491

Yatani A, Seidel CL, Allen J, Brown AM (1987) Whole-cell and single channel calcium currents of isolated smooth muscle cells from saphenous vein. Circ Res 60:523–533

Yoshino M, Someya T, Nishio A, Yazawa K, Usuki T, Yabu H (1989) Multiple types of voltage-dependent Ca channels in mammalian intestinal smooth muscle cells. Pflugers Arch 414:401–409

CHAPTER 3

Excitation-Contraction Coupling Mechanisms in Visceral Smooth Muscle Cells

T. ITOH and H. KURIYAMA

A. Introduction

Nearly 3 decades ago, the mechanisms underlying the propagation of the excitation of the sarcolemmal membrane into the contractile machinery in skeletal muscle were termed the "excitation-contraction coupling" (E-C coupling) process by SANDOW (1965). Advances in morphology, physiology and biochemistry have enabled us to understand and describe the basic mechanisms of E-C coupling in skeletal muscle as follows: excitation of the sarcolemma propagates into the transverse tubular structure (T-system) and induces depolarization of the triad at the junctional area between the T-system and the sarcoplasmic reticulum (SR). This triggers release of Ca from the SR, and the consequently increased cytosolic Ca triggers a contraction through conformational changes in the contractile proteins (actin and myosin) following the formation of a Ca-troponin C complex together with hydrolysis of ATP by myosin ATPase. These processes finally cause the activation of cyclic changes in the actin-myosin interaction (EBASHI and ENDO 1968). Thus, the E-C coupling mechanism is thought crucially to involve the mobilization of Ca from the SR.

Recently, in central neural cells and also in skeletal and cardiac muscle cells, the amino acid sequences of at least two different Ca-releasing channels – binding proteins for inositol 1,4,5-trisphosphate (IP_3) and ryanodine – have been recognized in the endoplasmic reticulum (ER) or SR, thanks to advances in molecular biology (FURUICHI et al. 1989; MIGNERY et al. 1989; TAKESHIMA et al. 1989). In addition, in the triad of skeletal muscle, a Ca sensor together with a Ca receptor has been found and identified as a dihydropyridine-binding site (TANABE et al. 1987).

In visceral smooth muscle, there are morphological differences and biochemical differences in the triggering mechanisms of contraction compared with skeletal and cardiac muscles. Thus, individual cells are much smaller than in skeletal and cardiac muscle (5–15 μm width and 100–250 μm length), but in most smooth muscle tissues each cell is connected to neighbouring cells by tight junctions at several sites, thus forming a functional syncytium as in cardiac muscle. The T-system found in cardiac and skeletal muscle is not present in visceral smooth muscle and thus the triad structure is not either. Furthermore, in skeletal muscle, the Ca required for activation of the contractile proteins acts as a repressor of actin-myosin interaction, but in visceral smooth muscle Ca binds to calmodulin and this Ca-calmodulin binding triggers the phosphorylation of myosin light chain and leads to the interaction between actin and myosin (SOBIESZEK 1977; HARTSHORNE and GORECKA 1980; ADELSTEIN and EISENBERG 1980; HARTSHORNE 1987; ITO and HARTSHORNE 1990; WALSH 1985; RÜGEE 1986; KAMM and STULL 1985). This means that, although the final goal of the E-C coupling mechanism is the same (Ca mobilization), the coupling process between Ca and the contractile proteins occurs differently in skeletal and visceral smooth muscles. Furthermore, troponin is absent in smooth muscle but leiotonin (actin-linked Ca-binding protein, EBASHI et al. 1987), calponin (troponin T-like substance) and caldesmon (SOBUE et al. 1981; WALSH 1990) are present, and these proteins are thought to modulate the contractile process through modulation of actin filaments, these effects being exerted especially on the tonic contraction (TAKAHASHI et al. 1988; WALSH 1990).

In addition, when excitatory agonists are added in the bath, visceral smooth muscle is activated by contraction-inducing mechanisms which are distinctly different from the E-C coupling to activate contractile proteins in skeletal muscle. Thus, the increase in cellular Ca required to produce a contraction is evoked through activation of a "pharmacomechanical" coupling (P-M coupling) process (SOMLYO and SOMLYO 1968). In 1964, SU et al. reported that in vascular tissues (main pulmonary artery; non-spontaneously active tissues) some agonists produce contraction with no change in the membrane potential. SOMLYO and SOMLYO (1968) also observed such responses in several vascular tissues using different agonists. Other investigators confirmed this phenomenon (see review of KURIYAMA et al. 1982). The mechanism underlying the P-M coupling process (contraction occurs with no change in the membrane potential in the sarcolemma) has been partly clarified following the discovery of the role of the second messenger (signal transducer) IP_3 in vascular smooth muscles. Thus, some excitatory agonists activate individual receptors and then hydrolyse phosphatidyl 4,5-bisphosphate ($PI\text{-}P_2$), which is distributed in the sarcolemma, through activation of phospholipase C (PLC) (BERRIDGE 1984, 1988; BERRIDGE and IRVINE 1984; ABDEL-LATIF 1986; SHEARS 1989). This IP_3 is able to release Ca from the SR in visceral smooth muscle (SUEMATSU et al. 1984; SOMLYO et al. 1985; YAMAMOTO and VAN BREENMEN 1985; HASHIMOTO et al. 1986). This process does not directly require a change in the ionic permeability of smooth muscle membranes. Of course, activation of the receptor by some agonists does modify the membrane's ionic permeability with or without activation of the P-M coupling mechanism (receptor-operated cation channels, BOLTON 1979). In addition, agonist-induced increases in the Ca sensitivity of the contractile proteins in smooth muscle have been reported, i.e. without any change in the Ca concentration in the cytosol, agonists enhance the contraction as compared with the contraction associated with the same amount of Ca after application of high K (estimated using α-toxin or β-escin treated skinned muscle tissues; HIMPENS et al. 1989; NISHIMURA et al. 1988; KITAZAWA et al. 1989; KOBAYASHI et al. 1989; SOMLYO et al. 1989; HIMPENS and CASTEELS 1990; ITOH et al. 1991).

Agonist-receptor coupling activates a receptor-operated ion (mainly cation) channel and depolarizes the membrane, thus indirectly causing activation of the voltage-dependent Ca channel. Furthermore, the receptor-operated ion channel may permeate Ca during receptor activation as reported in the rabbit ear artery (BENHAM and TSIEN 1987, 1988; BENHAM 1989). Moreover, such receptor activation may directly modify the voltage-dependent Ca channel in a positive or negative manner (DROOGMANS et al. 1987; NELSON et al. 1988; OIKE et al. 1990; XIONG et al. 1991). Therefore, agonist-induced contraction may occur via IP_3-induced Ca release from the SR, activation of the receptor-operated cation channel with or without subsequent voltage-dependent Ca channel activation and an agonist-induced increase in the Ca sensitivity of the contractile proteins.

Thus, in visceral smooth muscles the E-C and P-M coupling processes are evoked independently or cooperatively by individual agonists and these phenomena may occur in different ways in individual visceral tissues. This variety of triggering mechanisms for contraction may effect tissue specificity in visceral smooth muscle tissues in the maintenance of homeostasis.

In this chapter, we review recent investigations on the E-C and P-M coupling mechanisms in visceral smooth muscle in relation to the actions of drugs.

B. General Features of the Sarcolemmal Membrane in Relation to the Influx of Calcium

Influxes of Ca across the sarcolemmal membrane mainly occur through five different paths: by activation of the voltage-dependent Ca channel, by receptor-operated ion (mostly cation) channels, by reversed Na-Ca exchange diffusion, by stretch-induced influx of Ca and by passive diffusion of Ca (due to a 10^4 times concentration gradient). The first two have been rather better clarified than the others.

The voltage-dependent Ca influx can be detected by the observation of two phenomena: action potentials and transient or sustained membrane depolarization induced by agonists and high K.

I. Action Potentials Recorded by Microelectrode Methods

Here, we briefly introduce some classical observations made on action potentials in visceral smooth muscle using the microelectrode and double sucrose-gap methods. The action potential is a sign of activation of the voltage-dependent Ca channel.

1. Action Potentials

Visceral smooth muscle cells distributed in individual visceral including vascular tissues show resting membrane potential levels over the range −45 to −75 mV (KURIYAMA 1970, 1981; CREED 1979), and some tissues produce spontaneously discharged action potentials (with or without an overshoot potential), but other tissues do not. Action potentials occur as burst discharges between silent periods or as continuous train discharges. As a result of the use of the microelectrode and double sucrose gap methods, the action potential generated in visceral smooth muscle has been postulated to be mainly due to an influx of Ca through activation of the voltage-dependent Ca channel, as estimated from its insensitivity to tetrodotoxin (TTX), resistance to Na-free solution and sensitivity to organic and inorganic Ca antagonists, though there are some exceptions, such as the myometrium (HOLMAN 1958; BÜLBRING and KURIYAMA 1963a,b; KURIYAMA 1963b; HOTTA and TSUKUI 1968; INOMATA and KAO 1976). The amplitudes and frequencies of spontaneously activated or evoked action potentials are irregular. Generation of the overshoot action potential is not consistently observed, except following application of K-channel blockers such as tetraethylammonium (TEA), 4-aminopyridine (4-AP), procaine or Sr (KURIYAMA et al. 1982) or in Na-free solution containing Tris-Cl (BÜLBRING and KURIYAMA 1963a). The localization of pacemaker cells in visceral tissues has not yet been clearly defined, except in the ureter (pyrolus region; SANDERS and PUBLICOVER 1989; SANDERS and VOGALIS 1989) and in the circular muscles of the samll intestine (interstitial cells of Cajal; THUNEBERG 1989). The maximum rate of rise of the action potential in visceral smooth muscle in 10–50 V/s, these values being much smaller than those observed in cardiac and skeletal muscles (250–350 V/s).

2. Nerve-Mediated Responses

Action potentials can be evoked on an excitatory junction potential (EJP) or on repetitive generation of EJPs elicited by sympathetic, cholinergic and non-adrenergic, non-cholinergic (NANC) excitatory nerve stimulation (BURNSTOCK and HOLMAN 1961; HOLMAN 1970, 1981; SPEDEN 1970; BURNSTOCK 1980, 1981a–c; STJÄRNE 1989; STARKE et al. 1989; HIRST and EDWARD 1989). Many experiments have been carried out using the microelectrode method and the results can be summarized as follows. In some visceral tissues, including resistance vascular tissues, spontaneous small depolarizations (from within the noise level to about 10 mV in vas deferens and resistance vessels) occur without any electrical nerve stimulation, and the amplitude histogram of these depolarizations mostly shows a skew distribution curve. These depolarizations are blocked by the application of guanethidine or TTX in vascular tissues and vas deferens. Furthermore, reserpine inhibited the release of noradrenaline (NA) with a minute reduction in the amplitude of the EJPs. Therefore, these potential changes are thought to be due to spontaneous release of transmitter from nerve terminals and are termed miniature EJPs (NEILD 1983). Peripheral nerve stimulation, using field or transmural stimulation, produced small-amplitude EJP depolarization and repetitive stimulation facilitated the amplitude of the EJPs. When the depolarization exceeded the threshold required for the generation of the action potential, single or repetitively generated action potentials occurred on the EJPs. When high frequencies of stimulation were applied, EJPs summated and produced a large sustained depolarization. These EJPs evoked by peripheral nerve stimulation were also blocked by TTX or guanethidine. In some tissues (e.g. intestine), EJPs modulated the spontaneous burst spike discharge. In tracheal and broncheal muscle, cholinergic nerve stimulation evoked EJPs but these potential changes never produced the action potential (ITO and TAJIMA 1981). In intestinal longitudinal and circular muscle tissues, spontaneously generated spikes were inhibited by the generation of an inhibitory junction potential (IJP). Field stimulation of visceral tissues often evoked both the EJP and IJP via peripherally distributed cholinergic, sympathetic or NANC nerve stimulation. The resultant nerve stimulation induced the release of various substances such as noradrenaline (NAd), adenosine 5′ triphosphate (ATP), acetylcholine (ACh), vasoactive intestinal peptide (VIP) and other unidentified putative peptides. In addition, VIP, calcitonin gene related peptide (CGRP), histamine and serotonin are also thought to be released from their own nerve terminals as transmitters themselves or as cotransmitters of known classical transmitters (STARKE et al. 1989; HIRST and EDWARD 1989; STJÄRNE 1989).

The substances NAd and ATP released from sympathetic nerve terminals in vascular tissues evoke a slow depolarization and an EJP, respectively, while in the intestine NAd evokes a slow hyperpolarization (though in the case of the ileocaecal junction NAd produced a depolarization; GILLESPIE 1964) and ATP is thought to produce a transient apamin-sensitive IJP (BURNSTOCK 1980).

In some visceral tissues, spike generation did not occur but nerve stimulation still produced a contraction. This phenomenon has been postulated to be due to a lack of the voltage-dependent Ca channel, the contraction occurring through synthesis of IP_3. For example, in the eye ciliary muscle, electrical stimulation evoked neither EJPs nor action potentials, but on application of a high K solution a contraction occurred which was blocked by atropine (YOSHITOMI and ITO 1985). In tracheal muscle, cholinergic nerve stimulation evoked an EJP, but the action potential was never evoked, as described above. This implies that nerve terminals depolarized by high K released ACh and this transmitter produced a contraction, but depolarization induced by high K itself did not produce a contraction.

These observations indicate that in visceral smooth muscle influx of Ca occurs on generation of the action potential (voltage-dependent Ca channel), but that this phenomenon is not a prerequisite for the generation of contraction.

II. Features of the Voltage-Dependent Ca Channel in Visceral Smooth Muscle Membranes Measured Using the Voltage- and Patch-Clamp Procedures

After the patch-clamp procedure using the cell-attached and cell-free configurations had been introduced to the investigation of visceral smooth muscle, the features of ion channels related to the influx of Ca began to become clear. Three different voltage-dependent Ca channel subtypes are conventionally recognized from their own features in various excitable cells. Thus, mainly from work in neural cells, the voltage-dependent Ca channel was classified into T- (transient, I_{high}, I_{fast}), N- (neither T nor L; I_{middle}) and L- (long-lasting, I_{low}, I_{slow}) subtypes. These subtypes differ in a number of respects. Thus, they differ in: (a) their activation potential level and their inactivation potential level, (b) the values of their unitary current conductances, (c) their ion selectivities to Ca and Ba, (d) their K_d values for Cd and Ni, (e) their sensitivity to ω-conotoxin and (f) their sensitivity to 1,4-dihydropyridine derivatives (DHPs) (TSIEN et al. 1987).

However, such a classification is not strictly applicable to the voltage-dependent Ca channel in visceral smooth muscle. Specifically, the N-subtype of voltage-dependent Ca channel has not been found, and in some tissues (the rabbit ileum and guinea pig aorta) only one type of channel (the L-subtype) is present (WORLEY et al. 1986; CAFFREY et al. 1986; BENHAM et al. 1987b; YATANI et al. 1987; KAWASHIMA and OCHI 1987; YOSHINO et al. 1988; INOUE et al. 1989; OIKE et al. 1990). However in visceral smooth muscle, the unitary current conductance could be classified into three subtypes, tentatively from their different ionic conductances and also from their sensitivities to DHP. It must be mentioned here that the unitary current amplitude depends on the ionic environment. For example, increased concentration of the divalent ion (Ca or Ba) in the bath solution (cell-attached or outside-out patch) or in the pipette solution (inside-out patch) enhanced the amplitude of the unitary current, in a concentration-dependent manner. The concentration of Ca or Ba affected the membrane conductance roughly as the square root of the ion concentration. Within limited concentration ranges, the amplitude and conductance of the unitary current increased in a concentration-dependent manner. The recorded unitary current conductances have been classified into three subtypes (I_{Ca}L, 20–30 pS; I_{Ca}M, 12–15 pS; I_{Ca}S, 7–8 pS). In addition, the I_{Ca}M could be further classified into two subtypes: DHP-sensitive and -insensitive ones.

In dispersed smooth muscle cells of some tissues in experiments using the whole-cell voltage-clamp method, the macroscopic Ca current recorded in K-free solution (K in the pipette solution was replaced with Cs and in the bath K was replaced with TEA) could also be classified into two subtypes

(L- and T-types), from their sensitivity to Ca antagonists and also from the curren-tvoltage relationship measured at holding potentials of −60 mV or −80 mV. These two subtypes roughly corresponded with L-(I_{Ca}) and T-(I_{Ca}S) subtypes, respectively, to judge from their natures as measured using the cell-attached and cell-free patch-clamp procedures. The macroscopic current has been recorded using the whole-cell voltage-clamp procedure in various visceral smooth muscle cells by many investigators. However, in the guinea pig basilar artery, the Ca channel was composed of L- and T-subtypes as estimated from their unitary current conductances, but the macroscopic current showed a monophasic voltage-current relationship, and the small hump at the negative potential level, which is thought to be a sign of the presence of the T-type, was not observed (Inoue et al. 1989; Oike et al. 1990).

Action potentials (inward current) evoked in most vascular smooth muscle cells are thought to be due to activation of the voltage-dependent Ca channel alone. However, recent investigations made using the whole-cell voltage-clamp procedure have indicated that in some vascular smooth muscle tissues (pulmonary artery and portal vein) the inward current is composed of slow Ca and fast Na inward currents at a holding potential of −80 mV (Okabe et al. 1988). However, at a holding potential of −60 mV, only the Ca current was evoked. Furthermore, the time course of this Na current was very fast and it lasted in the order of a few milliseconds. This Na current was not sensitive to Ca antagonists, such as nifedipine and diltiazem, but very sensitive to low concentrations of TTX, the K_d value for TTX being 18.7 n*M* (Okabe et al. 1988). This value corresponded with the value obtained for the Na current in nerve fibres (n*M* order; Benoit et al. 1985), and the Na current recorded in smooth muscle cells was much more sensitive to TTX than those observed in skeletal and cardiac muscles (μM order; Benndorf et al. 1985; Hagiwara and Byerly 1983). However, in cultured vascular smooth muscle cells and myometrium, the presence of a TTX-less sensitive (cardiac and skeletal muscle type) Na current has also been reported (Sturek and Hermesmeyer 1986; Amedee et al. 1986). Presumably, the nature of this Na channel differs from that observed in freshly dispersed smooth muscle cells of the matured rabbit pulmonary artery and guinea pig portal vein. Chloramine-T had an interesting action on both the Na and the Ca current, i.e. this drug accelerated the fast Na current, but blocked the Ca current (Okabe et al. 1988). However, at present the physiological role of the Na current is evoked at a holding potential of −80 mV, but not at −60 mV, which is close to the resting membrane potential level observed in many vascular tissues thus, the physiological role of the Na^+ current is unclear.

III. Factors Modifying the Activity of the Voltage-Dependent Ca Channel

1. Ca Antagonists and Agonists

Detailed investigations of the actions of Ca antagonists on the Ca channel will be described in another chapter in this volume (KITAMURA and KURIYAMA, Chap. 16), and here therefore we only briefly describe the action of Ca antagonists in visceral smooth muscle.

Ca antagonists include many drugs, such as: (a) DHP derivatives (nifedipine, nicardipine, nisoldipine, nitrendipine, nimodipine, manidipine, PNP200-110, benidipine, FRC8653, CV-4093, etc.), (b) phenylalkylamine (papaverine) derivatives (gallopamil, verapamil), (c) benzothiazepine derivatives (diltiazem, TA3090), (d) the piperazine derivative flunarizine and (e) others (FLECKENSTEIN 1985; GODFRAIND et al. 1986). These agents consistently block the action potential through inhibition of the L-type of Ca channel in a selective manner. Pharmacological investigations of Ca antagonists have been mainly carried out using electrophysiological and mechanographic methods and also by binding experiments using the radiolabelled Ca antagonist to the binding proteins. Since many organic and inorganic drugs directly or indirectly act on the voltage-dependent Ca channel, a "Ca antagonist" is defined as an organic agent which selectively and directly acts on the voltage-dependent L-type Ca channel.

The potency with which Ca antagonists act on the voltage-dependent Ca channel differs. Using the whole-cell voltage-clamp procedure, the effects of various classical Ca channel antagonists, namely, nifedipine (DHP derivative), verapamil (phenylalkylamine derivative), diltiazem (benzothiazepine derivative) and flunarizine (piperazine derivative), were systematically investigated using the same cells (TERADA et al. 1987a–c). The results indicated that DHPs show the strongest potency in blocking the Ca channel (K_d value for nifedipine was 24 nM and for diltiazem 1.4 mM), and that some Ca antagonists (nifedipine and flunarizine) markedly shifted to the left the steady state inactivation curve (measured using pre-pulses of various durations and amplitudes and command pulses of 0 mV in intensity and 300 ms in duration). The use dependency of drug action on the amplitude of the inward current, as estimated from different frequency stimulations (0.01–1 Hz), occurred more strongly for verapamil and gallopamil than for nifedipine, diltiazem and flunarizine. Thus, verapamil seems to act mainly at the open state whereas DHPs act at the inactivated and open states of the Ca channel. However, recent detailed investigations indicate that more or less all these agents act on the resting, open and inactivated states but with different potencies (TERADA et al. 1987a–c; INOUE et al. 1989). From the results obtained on the voltage-dependent Ca unitary current, it has been concluded that all Ca antagonists inhibit the open probability and the duration of the open times of the voltage-dependent Ca channel without a reduction in the amplitude of the unitary current (YATANI et al. 1987; INOUE et al. 1989; OIKE et al. 1990).

Some DHP derivatives (Bay K 8644 or YC 170, CGP 28–392, H160/51, Sandoz 202–791) possess the properties of a Ca agonist, i.e., they increase the open probability and prolong the open time of the unitary Ca current from mode 1 (short opening) to mode 2 (long opening) as measured using the patch-clamp procedure. This is reflected in an enhancement of the amplitude of the action potential (HESS et al. 1984; BECHEM and SCHRAMM 1987).

It has been found for DHP (and for some other Ca antagonists such as diltiazem) that their enantiomers possess different actions. For example, the (−)-enantiomer possesses a Ca agonistic action, but the (+)-enantiomer possesses a weak Ca antagonistic action. Furthermore, the enantiomers of some DHP drugs defined as Ca antagonists possessed a Ca agonistic action (PN200-110, nitrendipine, CV-4093, benidipine and FRC8653) (HESS et al. 1984; BROWN et al. 1986; BECHEM and SCHRAMM 1987; TERADA et al. 1987a–c; OIKE et al. 1990). Moreover, some drugs, such as (+)-FRC8653, had an antagonistic action at a holding potential of −80 mV but this was inverted to an agonistic action at a holding potential of −40 mV, in a membrane-potential-dependent manner (OIKE et al. 1990).

To explain these agonistic and antagonistic DHP actions, various underlying mechanisms have been postulated. The action of DHP has a voltage dependency as observed from the current-voltage relationship, a holding potential dependency for channel activation and for steady state inactivation of the channel. Therefore, these properties have been analysed in terms of the modulated receptor hypothesis by analogy with the action of local anaesthetics (HILLE 1977; HONDEGHEIM and KATZUNG 1977; HESS et al. 1984; BROWN et al. 1986; SANGUINETTI and KASS 1984; KASS and KRAFTE 1987; TERADA et al. 1987a–c; HERING et al. 1987; KASS 1987; INOUE et al. 1989). Furthermore, the complex actions of DHP have been discussed using a modified hypothesis derived from results obtained by Ca transient measurements using agonist and antagonists of the Ca channel (GURNEY et al. 1985; SANGUINETTI et al. 1986; HERING et al. 1989).

Inorganic substances such as Cd, Co and Mn also block the voltage-dependent Ca channel. However, the inhibitory action of these agents differed from the actions of organic Ca antagonists on the channel, e.g. DHP reduced the open probability and open time of the channel, while Cd closed the open channel for short periods without affecting the amplitude of the current.

2. Dihydropyridine-Binding Protein (Ca Channel)

Recently, the amino acid sequences of the DHP-sensitive Ca channel protein extracted from the triad region have been elucidated. This protein is composed of $\alpha 1$-, $\alpha 2$-, β- and γ-subunits (TANABE et al. 1987, 1988), and the primary amino acid sequences and cDNA have been clarified in the individual subunits (CATTERALL et al. 1988; ELLIS et al. 1988; TANABE et al. 1988, 1990a,b; MIKAMI et al. 1989; RUTH et al. 1989; JAY et al. 1990; BOSSE et al. 1990; BIEL et al. 1990). The subunit responsible for permeating Ca is the α1-subunit (170-kDa protein, consisting of 1873 amino acids in skeletal muscle; TANABE et al. 1987; TAKAHASHI et al. 1988), whereas the α2-, β- and γ-subunits contribute to the stabilization of the receptor (Ca channel) and also positively modify features of the channel activity (MIKAMI et al. 1989).

In skeletal muscle, the triad structure plays an essential role in the excitation-contraction (E-C) coupling mechanism which enables Ca to be released from the SR. In this muscle tissue, the DHP-sensitive L-subtype is distributed on the triad and acts as a voltage sensor (the voltage-sensor hypothesis of the DHP-sensitive L-subtype of Ca channel; SCHNEIDER and CHANDLER 1973; TANABE et al. 1987). Because of its many structural similarities, this receptor is also thought to have a channel property. Thus, this receptor (channel) possesses two roles: as a charge carrier and as a voltage sensor. This hypothesis has been confirmed using the following most elegant technique.

Using muscle cells of mutant mice with muscular dysgenesis (*mdg*) (genetic absence of the E-C coupling process and L-subtype Ca channel; GLUECKSOHN-WAELSCH 1963; POWELL and FAMBROUGH 1973; BOURNAUD and MALLART 1987; BEAM et al. 1986; RIEGER et al. 1987; TANABE et al. 1988), TANABE et al. (1988, 1990b) transferred rabbit expression plasmid (pCAC 6) of the DHP receptor complementary DNA into nuclei of the primary cultured myotubes of the *mdg* mouse and found that this myotube then possessed the ability to trigger the E-C coupling process and to activate the L-subtype Ca current. Moreover, the myotube of the *mdg* mouse then possessed a voltage sensor on the transverse tubule for the release of Ca from the SR (TANABE et al. 1988). TANABE et al. (1990b) investigated the sites responsible for the activation of the above two processes following the implantation of pCAC 6 of the rabbit expression plasmid (pCARD1) of the DHP receptor complementary DNA into nuclei of the primary cultured myotubes of the *mdg* mouse, and they concluded that, following expression of the cardiac type of DHP receptor in the *mdg* mouse, the myotube of the *mdg* mouse then showed much the same property as that observed in intact cardiac muscle (sabsence of contraction after removal of Ca from the bath). Moreover, they implanted the expression plasmid of various chimeric DHP receptor cDNAs in dyskinesic myotubes (pCSK 7, pCSK 1–4), and found the actual sites responsible for the channel function as the triggering step of the E-C coupling mechanism to be the putative cytoplasmic region between repeats II and III of the skeletal muscle DHP receptor in the α1-subunit (TANABE et al. 1990a).

It is a well-known pharmacological fact that in Ca-free solution depolarization induced by high K or electrical stimulation produces a contraction in skeletal muscle but not in cardiac muscle, in spite of both tissues possessing the triad structure. Therefore, the E-C coupling process, including the functions of the triad, differs between skeletal and cardiac muscles. Recently it has become clear that structural and functional differences in the L-type Ca channel distributed on the triad are present in skeletal and cardiac muscles (TANABE et al. 1990b).

In smooth muscle, it is also thought that the Ca channel is composed of 5 subunits as observed in skeletal and cardiac muscles, and the α1-subunit is composed of a 242.5-kDa protein in airway smooth muscle and a 243.6-kDa protein (2169 amino acids) in aortic smooth muscle (BIEL et al. 1990; BOSSE et al. 1990), these α1-subunits being much larger than those observed in skeletal muscle. These proteins had 65% homology with that of skeletal muscle and 93% homology with that of cardiac muscle cells. Presumably, the sarcolemmal Ca channel may have different features from the triad Ca channel. It is, therefore, postulated that the voltage-dependent Ca channel distributed in the sarcolemma may have a different structure and different properties from those distributed on the triad structure in cardiac and skeletal muscles. This is beccause, in smooth muscle, the triad structure is absent (Ca-handling site), and only a part of the sarcolemma (SL) has a close contact with the SR (by a foot structure between the SL and SR as estimated using the electron microscope; GABELLA 1981) and the sensitivity of the Ca channel to DHP-Ca antagonists is much higher than that of cardiac muscle, as estimated from the action of Ca antagonists measured in voltage-clamp and radioligand experiments (TERADA et al. 1987a,b; SUMIMOTO et al. 1988).

3. K Channel Modifying Drugs

The amplitudes and frequencies of trains of action potentials recorded using the microelectrode or double sucrose gap method are irregular and overshoot potentials are not consistently recorded (KURIYAMA 1970, 1981; KURIYAMA et al. 1982). When TEA, procaine or Ba, inhibitors of K permeability, are added to the Krebs solution, regular action potentials with an overshoot together with a depolarization of the membrane can be recorded. During the depolarization induced by these agents, there is an increase in the membrane resistance, as measured from the electrotonic potentials evoked by alternately applied inward and outward currents. Even electrically quiescent tissues such as the pulmonary artery and aorta produced the abortive action potential in the presence of any of these agents. Thus, Ca channel activation is also modified by K permeability changes via inhibition or delayed onset of K channel activation (ITO et al. 1970; HARA et al. 1980; KITAMURA et al. 1986).

Using the voltage- and patch-clamp procedures, K channels have been classified into various subtypes, (a) from their sensitivity to the cytosolic Ca, into Ca-dependent (maxi-K and small K; SK) and -independent (middle K) K channels, and (b) from their voltage dependency, into a voltage-dependent (delayed rectifier K channel and A current generating channel) and voltage-independent (inward rectifying or anomalous rectifying) K channels (BENHAM et al. 1986). Using various agents, these K channels have also been classified into various subtypes, phenomenologically: thus, there are apamin-sensitive (SK) and apamin-insensitive (maxi-K, large K) Ca-dependent K channels; charybdotoxin (scorpion)-sensitive (maxi-K) and charybdotoxin-insensitive (SK) Ca-dependent K channels. In addition, TEA-sensitive and TEA-less sensitive K channels and 4-AP-sensitive and 4-AP-less sensitive K channels are recognized (BENHAM et al. 1985; OHYA et al. 1986, 1987c; SAKAI et al. 1988; BOLTON and LIM 1989; HUME and LEBLANE 1989; KAJIOKA et al. 1990). The TEA-sensitive K channel and the 4-AP-sensitive K channel observed in the rat and rabbit portal veins seemed to differ in nature. In the portal veins, the maxi-K channel was more sensitive to TEA than the SK channel but the reverse was true for the actions of 4-AP (KAJIOKA et al. 1990). However, these results were not consistent in all visceral smooth muscle (STANDEN et al. 1989; NELSON et al. 1990). The enhancement of the spike amplitude induced by K-channel blockers may be due to a suppression of the early onset of the generation of the K outward current relative to the onset of the Ca inward current, as postulated from an investigation made using the microelectrode method (KURIYAMA et al. 1982).

K-channel openers, such as nicorandil, cromakalim (lemakalim) or pinacidil, inhibit spontaneous and evoked spike generation by lowering the threshold required to evoke the action potential and increase the K or Rb permeability as measured using the voltage-clamp procedure or ion flux procedure (FURUKAWA et al. 1981; ITOH et al. 1981a; KARASHIMA et al. 1982; INOUE et al. 1984; KAJIWARA et al. 1984; YAMANAKA et al. 1985; ALLEN et al. 1986a,b; HAMILTON et al. 1986; SUMIMOTO and KURIYAMA 1986; WEIR and WESTON 1986a,b; CLAPHAM and WILSON 1987; COLDWELL and HOWLETT 1987, 1988; KREYE et al. 1987; QUAST 1987; SOUTHERTON et al. 1988; NAKAO et al. 1988; HAMILTON and WESTON 1989; OKABE et al. 1990). A target channel for these K-channel openers has been found to be an ATP- and glibenclamide (antidiabetic agent)-sensitive K channel (NOMA 1983; ASHCROFT 1988; QUAST and COOK 1989; ESCANDE et al. 1988; STANDEN et al. 1989; HAMILTON et al. 1986; KAJIOKA et al. 1990). Glibenclamide inhibited the ATP-sensitive K channel but this agent had an additional action on smooth muscle cells (XIONG et al. 1991). In smooth muscle, the inhibitory action of glibenclamide was weaker than that observed in pancreatic beta cells and cardiac muscle. Cromakalim, in addition to its action as a K-channel opener, possesses a property as a Ca antagonist (NAKAO et al. 1988; KAJIOKA et al. 1990) and pinacidil inhibits the synthesis of IP_3 through hyperpolarization of the membrane (ITOH et al. 1990). In addition, minoxidil and pinacidil have the effect of causing hair growth (QUAST and COOK 1989).

Although these K-channel openers act on the ATP-sensitive K channel, a consensus concerning the channel's properties has not yet been obtained, i.e.in some vascular tissues, these agents act on the Ca-independent, large-conductance K channel, on the extracellular Ca-sensitive Max-K channel or on the Ca-sensitive SK (HERMSMYER 1988; KUSANO et al. 1987; BEECH and BOLTON 1989; INOUE et al. 1989; NELSON et al. 1990; OKABE et al. 1989; KAJIOKA et al. 1990). Presumably, regional differences in the features of the ATP-sensitive K channel may occur and appear as tissue specificities, and such differences may be specific features of vascular tissues.

Thus, K-channel openers hyperpolarize the membrane and inhibit the activation of the Ca-dependent K channel. In addition, synthesis of IP_3 has a causal relation to the membrane potential of smooth muscle cells, and the K-channel opener-induced hyperpolarization causes a reduction in the synthesis of IP_3 (ITOH et al. 1990; ITO et al. 1991). Thus, both effects tend to relax the vascular tissue. Some of these drugs also act not only on vascular tissues, but also on trachea and urinary bladder as relaxants. In addition, some of these drugs have no effects on the ATP-sensitive K channel in pancreatic beta cells (COOK et al. 1989; QUAST and COOK 1989). Therefore, concerning the actions of K-channel openers, not only regional differences in vascular tissues but tissue specificities should be considered.

4. Endothelial Cell Releasing Factors

Endothelial cells play multiple roles in regulating smooth muscle function. Endothelial cells release contractile substances such as endothelin and thromboxane A_2 (TXA_2 or TXA_2-like substance) and relaxing substances such as endothelial cell derived relaxing factor (EDRF), prostaglandin I_2 (PGI_2 or prostacyclin) and endothelial cell derived hyperpolarizing factor (EDHF; the substance is not yet identified).

a) Prostaglandin I_2

In vascular smooth muscle tissues, PGI_2, a product of arachidonic acid via the action of cycloxygenase, is released from endothelial cells (MONCADA et al. 1977; MONCADA and VANE 1979). This agent hyperpolarizes the membrane through the action of synthesized cyclic AMP on the K channel, i.e. cyclic AMP activates the Ca-dependent K channel (SADOSHIMA et al. 1988; KUME et al. 1989; FUKUMITSU et al. 1990) through okadaic-acid-sensitive phosphorylation (KUME et al. 1989). In addition, cyclic AMP reduces the cytosolic Ca through the actions of its dependent protein kinase (protein kinase A).

b) Endothelium-Derived Relaxing Factor

In 1980, a relaxing factor released from endothelial cells was recognized as EDRF by Furchgott's group (FURCHGOTT and ZAWADZKI 1980; FURCHGOTT 1988; FURCHGOTT et al. 1981, 1988). The final product of this factor is thought to be nitric oxide (NO) derived from L-arginine, and it acts on smooth muscles as an accelerator of cyclic GMP synthesis (PALMER et al.

1987, 1989). In other words, EDRF acts on vascular smooth muscle as an endogenous nitroso compound (IGNARRO et al. 1987a,b; IGNARRO 1990). Subsequently, the release of EDHF was also discovered. This agent hyperpolarizes the membrane independently from the action of EDRF and partly relaxes the smooth muscle. The relaxation induced by EDHF is partly due to a suppression of spike generation in vascular smooth muscle cells (FELETOU and VANHOUTTE 1988; KOMORI and SUZUKI 1987a,b; CHEN and SUZUKI 1989; NISHIYE et al. 1989). This EDHF-induced hyperpolarization was insensitive to methylene blue and to haemoglobin (cyclic GMP synthesis inhibitor). The hyperpolarization was blocked by the application of glibenclamide, an inhibitor of K-channel openers and an antidiabetic agent (STANDEN et al. 1989). On the other hand, NAKASHIMA et al. (1991) reported that glibenclamide did not prevent the hyperpolarization induced by ACh in the intact coronary artery of the rabbit. Furthermore, L-nitroarginine (L-NNA), a synthesis inhibitor of NO from L-arginine, had no effect on this hyperpolarization. However, it has also been reported that NO hyperpolarizes the smooth muscle cell membrane; thus NO may act on the synthesis of cyclic GMP or act directly on smooth muscle membranes and thus cause hyperpolarization of the membrane. These actions of NO are prevented by L-methyl arginine (L-NMA) or L-NNA (BULT et al. 1990; TARE et al. 1990).

c) Endothelium-Derived Hyperpolarizing Factor

In vascular smooth muscle, ACh and oxotremoline both release EDRF and relaxed vascular tissues, while ACh did, but oxotremoline did not, hyperpolarize the smooth muscle membrane (KOMORI and SUZUKI 1987a,b). From the actions of atropine and pirenzepine, it appears that M1 and M2 receptors in endothelial cells may release EDHF and EDRF, respectively, and cause vasorelaxation. Therefore, KOMORI and SUZUKI (1987a) postulated that EDRF and EDHF differ in nature. Whether EDHF is the same substance as NO or not will presumably be clarified by further investigations.

d) Endothelin

In addition to the relaxing factors, contracting factors such as endothelin (21 amino acid residues; YANAGISAWA et al. 1988) and thromboxane A_2 (TXA_2) or TXA_2-like substance (VANHOUTTE et al. 1986; SHIRAHASE et al. 1987) are also thought to be released from endothelial cells. These agents increase the synthesis of IP_3 in the cerebral artery (SHIRAHASE et al. 1987). Endothelin consists of at least three different gene-related subtypes (endothelin 1–3) and has at least two different receptors identified from gene manipulation (MASAKI 1989; VANE 1990; WEBB 1991). Endothelin-1 produces a contraction with depolarization of the membrane (YANAGISAWA et al. 1988; SILBERBERG et al. 1989; GOTO et al. 1989; AUGUET et al. 1988) and this depolarization was postulated to be due to activation of the Na channel (NAKAO et al. 1988). In addition, endothelin acts directly on the voltage-dependent Ca channel (GOTO et al. 1989). Detailed experiments indicate that this agent acts on both L-and T-types of voltage-dependent Ca channel, i.e. endothelin increased the open probability and prolonged the opening time of both channels as measured using the cell-attached patch-clamp method (INOUE et al. 1990). However, these actions of endothelin were unlikely to be due to a direct action on the channel because bath application of endothelin was more potent than intrapipette application (INOUE et al. 1990). Presumably, endothelin acts through unknown second messengers, because IP_3 inhibits the voltage-dependent Ca channel through increased Ca concentration in the cytosol and it has been observed that cyclic AMP and cyclic GMP do not act on the voltage-dependent Ca channel (OHYA et al. 1988). In vascular smooth muscle tissues, endothelin was reported to increase the synthesis of IP_3 (MIASHIRO et al. 1988; KAI et al. 1989; HIRATA et al. 1988; RESNIK et al. 1988; MARSDEN et al. 1989; SUGIURA et al. 1989) and many investigators have postulated that the synthesis of IP_3 may be a main action of endothelin in inducing vasoconstriction. However, the underlying mechanism is still not completely resolved. It has also been postulated that endothelin may act on two different endothelin receptors (ET_A and ET_B). Of these, ET_A shows high specificity for endothelin-1 and is suggested to be on vascular smooth muscle, whereas ET_B accepts all three endothelins equally (non-selective) (MASAKI 1989; VANE 1990; WEBB 1991).

e) Thromboxane A_2

It is generally thought that TXA_2 is mainly synthesized in platelet cells and PGI_2 in endothelial cells (MONCADA and VANE 1979). PGI_2 increases the synthesis of cyclic AMP and causes vasodilation. Concerning the action of TXA_2, however, recent investigations have indicated that endothelial cells could release TXA_2 or TXA_2-like substances and produce contraction in vascular tissues (DEMAY and VANHOUTTE 1983; SHIRAHASE et al. 1987). Depolarizations of the membrane induced by TXA_2 accelerate spike generation and release Ca from the SR via synthesized IP_3, and enhance the amplitude of contraction as observed on application of carbocyclic TXA_2 in some vascular tissues (MAKITA 1983).

IV. Receptor-Operated Ion Channel

1. General Features of the Receptor Structure

Due to recent advances in molecular biology, we are able to visualize the primary structure of the amino acid sequence of some receptor proteins.

Agonist receptors can be classified into three main categories: (a) the receptor-ion channel complex forming systems, such as the receptor-GTP-binding protein-ion channel complex formation, (b) the receptor-GTP-binding protein complex forming systems, such as in the formation of second messengers (the receptor-GTP-binding protein-adenylate cyclase complex forming system, the receptor-GTP-binding protein-phospholipase C complex forming system and the receptor-GTP-binding protein-phosphodiesterase complex forming system) and (c) the receptor-autophosphorylation system (tyrosine kinase activating type).

Typical examples of the receptor-ion channel complex forming system are the nicotinic, glutamic acid, $GABA_A$ and glycine receptors, which are each composed of a few subunits (heterogomers, composed of 2α-, β-, γ- and δ-subunits for the nicotinic receptor; and tetragomers, composed of 2α- and 2β-subunits for the $GABA_A$ receptor). The first two receptors, the nicotinic and glutamic acid receptors, are excitatory ones and mainly permeate the Na ion but are non-selective cation channels, and the last two receptors ($GABA_A$ and glycine receptors) are inhibitory ones and mainly permeate the Cl ion but are non-selective anion channels. The nicotinic receptor has also been differentiated into at least three different subtypes and the GABA receptor has been classified into the $GABA_A$ and $GABA_B$ receptors. Two molecules of acetylcholine bind to two α-subunits, GABA binds to the β-subunit and benzodiazepine binds to the α-subunit. N- and C-terminals (NH_2 and COOH, respectively) of each peptide subunit are located on the extracellular side and these peptides, composed of α-helical structures and hydrophobic structures, penertrate the membrane four times. However, these receptors are not distributed in smooth muscle cells. In addition, the receptor-autophosphorylation complex forming system, insulin and epithelial growth factor receptors are also postulated to be absent from smooth muscle cells. Therefore, the receptor-GTP-binding protein complex forming system may be the one closely related to smooth muscle function.

In visceral smooth muscle, receptors for neurotransmitters and autacoids have been described. These are, e.g. the α-, β-adrenoceptors (α_{1A}-, α_{1B}-, α_{2A}- and α_{2B}-subtypes for the α-adrenoceptor and β_1-, β_2- and β_3-subtypes for the β-adrenoceptor), muscarinic (M_1–M_3) and serotonergic receptors (5-HT_1 and 5-HT_3), histaminergic (H_1–H_3) and purinergic (PI_A and PI_B, and PII_X, PII_Y) receptors and others. They have been classified from classical pharmacological and from genetic-coding procedures and, for some of them, the primary amino acid sequences have been clarified. To explain receptor activation by agonists, the presence of various GTP-binding proteins is required (the receptor-GTP-binding protein complex forming system). These types of the receptor are composed of a unitary monomer and this peptide chain penetrates the cell membrane seven times with the N-terminal located at the extracellular side and the C-terminal at the cytosolic side. To explain the physiological functions following receptor activation requires the active state of GTP-binding protein. The kind of GTP-binding proteins that are related to receptor activity will be reviewed later (see Sect. B.V).

2. Neurotransmitter-Induced Changes in the Membrane Potential

In many visceral smooth muscle tissues, excitation and inhibition occur on activation of peripheral nerves such as the adrenergic, cholinergic and

NANC nerve systems. Following the release of NAD, ACh, ATP, VIP or CGRP from nerve terminals, individual receptors are activated and modify the ionic channels.

In experiments using the microelectrode and double sucrose-gap methods, peripheral nerve stimulation produces the EJP or IJP, slow depolarization or slow hyperpolarization. The different electrical responses may reflect different densities of receptor distribution, the qualities of individual transmitters or the distance between the releasing sites in the nerve terminal (including varicosites), and the postjunctional receptor may produce multiple electrical responses. The EJP is thought to generate mainly an increase in the Na and Ca permeabilities and the IJP an increase in the K permeability. For example, ACh generates the EJP in intestinal tissues, slow hyperpolarization in some vascular tissues and slow depolarization in other vascular tissues. In many resistance vascular tissues, ionophoretic application of ATP generates an EJP-like potential change, but in some of them ATP produces a slow hyperpolarization. ATP is thought to produce the IJP in intestinal smooth muscle. Therefore, in visceral smooth muscle tissues, responses to the neurotransmitter are thought to differ according to the density of distribution of individual receptor type and the features of the receptor-operated ion channel in the particular tissue and species (BOLTON 1979; KURIYAMA et al. 1982). EJPs and IJPs generated by repetitive peripheral nerve stimulation undergo facilitation due to increased Ca mobilization at nerve terminals (HOLMAN 1981; KURIYAMA et al. 1982; HIRST and EDWARD 1989). The Ca channel responsible for the increased cytosolic Ca at nerve terminals seems unlikely to be the L-type Ca channel but rather the N- or T-subtype.

The generation of the EJP can trigger the action potential, which propagates the excitation to neighbouring cells, and also the EJP can produce contraction itself or through the generation of the action potential. On the other hand, IJPs block the generation of the action potential, which can lead to relaxation of the tissue (HIRST and EDWARD 1989; STJÄRNE 1989; STARKE 1987; STARKE et al. 1989). However, innervation patterns differ in different tissues, e.g. in vascular tissues innervation occurs at the adventitial layer and does not penetrate into the cells, whereas in intestinal smooth muscle tissues exogenous nervous systems mainly terminate at the enteric plexus (Auerbach's plexus) and a part of the nerve penetrates into the muscle layers (GABELLA 1981; BURNSTOCK 1972, 1981a–c, 1982; BURNSTOCK and COSTA 1975). They have no distinct nerve terminal region as observed in the end plate region, but possess many varicosites at short intervals at nerve terminal regions and these form functional nerve terminals. Varicosites can release transmitters following nerve excitation (GABELLA 1981; STJÄRNE 1989).

ATP and VIP are thought to be cotransmitters of NAd and ACh, respectively, in perivascular nerves, and histamine and serotonin also act as transmitters. VIP is generally thought to be a substance contained in peripheral sensory nerves; however, this substance is released from motor nerves and acts as a negative feedback controller of transmitter release. There is as yet no consensus as to whether the EJP in vascular smooth muscle cells is generated by NAd or its cotransmitter, ATP (the γ-adrenoceptor hypothesis, HIRST and NEILD 1980, 1981; HIRST et al. 1982; and the purinergic receptor hypothesis, BURNSTOCK 1972; SNEDDON and BURNSTOCK 1984; BURNSTOCK and KENNEDY 1986a,b; SUZUKI 1986; NEILD and KOTAECHA 1986). The published results relevant to the identification of the receptor responsible for producing the electrical events may be summarized as follows: (a) α_1-adrenoceptor blockers do not enhance, but α_2-adrenoceptor blockers do enhance, the amplitude of EJPs. However, α_1- or α_2-blockers do inhibit the slow depolarization of the membrane which occurs on repetitive stimulation (CHEUNG 1982; KURIYAMA and MAKITA 1982, 1984; SUZUKI 1981, 1985). (b) Reserpine markedly reduces NAd release, but the amplitude of the EJP is not markedly inhibited. Guanethidine both blocks NAd release and reduces EJP amplitude (MISHIMA et al. 1984; MIYAHARA and SUZUKI 1987). (c) Ionophoretic application of NAd (in the cerebral artery; HIRST and NEILD 1981) or ATP (mesenteric artery; SUZUKI 1985) produces an EJP-like transient potential change. α-, β-methylene ATP, a desensitizer of the purinergic II receptor (BURNSTOCK and KENNEDY 1986a,b), depolarizes the membrane and inhibits the EJP (ISHIKAWA 1985). Another PII-receptor blocker, arylazidoaminopropyonyl ATP, binds irreversibly to the receptor site and blocks the generation of EJPs (SNEDDON and WESTFALL 1984; SNEDDON and BURNSTOCK 1984; CHEUNG and FUJIOKA 1987).

Vasoactive intestinal polypeptide has also been reported to be an inhibitory cotransmitter substance for muscle tone in visceral smooth muscle (intestine, trachea, vas deferens, penis cavili and others; EPELABUM et al. 1979; LUNDBERG et al. 1982; MATSUZAKI et al. 1980; ITO and TAKEDA 1982; LATINEN et al. 1985; ELLIS and FARMER 1989; HAKODA and ITO 1990), and somatostatin and VIP act as inhibitors at presynaptic sites, inhibiting the transmitter release from the cholinergic innervation of the trachea (STJERNQUIST and OWMAN 1984; SEKIGAWA et al. 1988; HAKODA and ITO 1990). HAKODA and ITO (1990) studied the action of exogenously applied VIP on the dog and cat tracheal muscles and concluded that, in addition to a direct action on the smooth muscle cells, low concentrations of VIP have a vagal prejunctional action which inhibits excitatory neuroeffector transmission, as a NANC inhibitory response. Endogenously released VIP also inhibits the release of ACh from vagal nerves, as estimated from the actions of VIP antiserum and antagonists.

Calcitonin gene related peptide has also been nominated as a NANC transmitter in vascular smooth muscle tissues (KAWASAKI et al. 1988, 1990), in which it has been reported to produce relaxation of precontracted tissue. Release of CGRP is inhibited by TTX, capsaicin and NAd. Yohimbine and neuropeptide Y antagonize the action of CGRP. CGRP does not modify the release of NAd from nerve terminals. NAd and neuropeptide Y additively modify the inhibition of mechanical responses induced by CGRP. Capsaicin enhanced, but CGRP did not modify, the release of NAd from adrenergic nerve terminals. Thus, vascular tissues are innervated by adrenergic and NANC systems. The latter acts directly on smooth muscle cells and also modifies the activity of presynaptic nerve terminals.

Cholinergic nerves are densely distributed in visceral smooth muscles; however, in vascular tissues they are thought to be only sparsely distributed. However, BEVAN and BRAYDEN (1987) reported dense distributions of choline acetylase in vascular tissues and it is now clear that the rabbit lingual artery is innervated by cholinergic nerves (BRAYDEN and LARGE 1986; SUZUKI 1989).

In addition, nerve terminals possess many agonist receptors which positively (isoprenaline and adrenaline) or negatively (NAd, histamine, serotonin, opium compounds) regulate transmitter release, as estimated from the bioassay of transmitter release and also from the amplitude of EJPs. For example, at adrenergic nerve terminals, the β_2-adrenoceptor accelerates and the α_2-, D_2-, PI-, H_3- or opioid (μ-, δ-, κ-, ε-subtypes) receptor inhibits both transmitter release and the amplitude of EJPs (LANGER 1974; STARKE 1987; LANGER and LEHMAN 1988; ILLES 1989; STJÄRNE 1989; STARKE et al. 1989).

Although many receptors distributed on nerve terminals positively or negatively control transmitter release, it is not yet completely understood how the cytosolic Ca in nerve terminals is regulated by such receptor activations. Isoprenaline, NAd and cyclic AMP accelerate the release of NAd through Ca-dependent processes (BEVAN et al. 1980; CUBBEDDU et al. 1975; GÖTHERT and HENTRICH 1984; STJÄRNE and BRUNDIN 1976; WOOTEN et al. 1973) and inhibitors of the type II and III isoforms of cyclic AMP phosphodiesterase and dibutyryl cyclic AMP, a permeable derivative of cyclic AMP, facilitated the release of NAd induced by high frequencies of stimulation (GÖTHERT and HENTRICH 1984). On the other hand, with low-frequency stimulation, these agents did not modify the release of NAd from nerve terminals (HEDQVIST et al. 1978). However, accumulated results indicate that stimulation of the β-adrenoceptor consistently accelerates the release of NAd evoked by low and high frequencies of stimulation and enhances the amplitude of EJPs.

On the other hand, it has been suggested that Ca is the trigger which activates guanylate cyclase indirectly through activation of phospholipase A and C (PURNEY et al. 1986; VILLARROEL et al. 1989). Recently, GREENBERG et al. (1990) reported, after using the radioisotope of NAd ([2-^{14}C]NAd) and recording the mechanogram from canine mesenteric smooth muscle, that cyclic GMP may be an inhibitory modulator of Ca and of depolarization-dependent NAd release from sympathetic nerves, whereas neural cyclic AMP may not be a primary modulator of neurotransmission in vascular smooth muscle. Further experiments would be required to clarify the role of second messengers in Ca mobilization at nerve terminals.

Concerning the opioid receptor, it is thought that prototypes of agonists for the μ-, δ-, κ- and ε-types are morphine, leu-enkephalin, ethylketocyclazocin and β-endorphin, respectively. All these agents are antagonized by naloxone (ILLES 1989). In the rabbit pulmonary artery, the κ-subtype is located at nerve terminals (SEELHORST and STARKE 1986), whereas the δ- and κ-subtypes are distributed in the rabbit ear (ILLES and BETTERMAN 1986), ileocolic and jejunal arteries (VON KÜGELGEN et al. 1985; ILLES et al. 1986) and portal vein (SZABO et al. 1987). The ε-subtype was found to be distributed in the rat tail artery (BUCHER et al. 1987). Therefore, the distribution of the opioid receptor subtypes in nerve terminals differs in different tissues.

As described previously, in peripheral sympathetic (noradrenergic) nerves, there are many varicosities. Individual varicosities distributed in muscle tissues possess the ability to release transmitter spontaneously or on excitation of the nerves. The roles of varicosities were hypothesized by STJÄRNE et al. (1990) in the following way. Nerve impulses induce the release of transmitters from the varicosities in a monoquantal and highly intermittent manner. A nerve impulse invades varicosities in an all-or-none manner and invasion failure is probably rare. The release probability is not controlled by the amplitude and duration of the invading action potential (or resulting Ca influx) but there may be an unidentified permissive factor, which may be actively transported intra-axonally, probably in association with organelles. The permissive factor is controlled upstream of the varicosity, which itself is sensitive to Ca influx through a channel insensitive to DHP (T- or N-channel). A high resting K efflux restricts the availability of the permissive factor, as estimated from the action of TEA and 4-AP. This K efflux may be the main mechanism maintaining the low release probability. Prejunctional agonists do not inhibit transmitter secretion by causing a conduction block or reduction in the influx of Ca, but by depressing the Ca-dependent activation and/or transport of permissive factor. They act, at least in part, via receptors upstream of the varicosity. STJÄRNE et al. (1990) proposed this mechanism as an "upstream" hypothesis to explain the role of varicosites distributed at nerve terminals in visceral smooth muscle. More detailed experiments would be required to clarify the nature of the permissive factor and the action of the prejunctional receptors. Direct measurements are required of the Ca mobilization at nerve terminals in relation to transmitter release.

It is not yet clear whether the NANC transmitter substance is ATP or NAd (FURNESS and COSTA 1989). Recent investigations have suggested that NO derived from L-arginine is a candidate for the NANC transmitter substance. NO has been proposed as the transmitter of the NANC nerves in the rat and mouse anococcygeus muscle (GILLESPIE et al. 1989; RAMAGOPAL and LEIGHTON 1989; GIBSON et al. 1990) and this transmission was blocked by L-NMA (GIBSON et al. 1990). BULT et al. (1990) concluded that, in the canine ileocolonic junction, NO is an inhibitory transmitter released from NANC nerves.

3. Activation of the Receptor-Operated Ion Channel

The effects of transmitter substances such as ACh, NAd, histamine and ATP have been investigated using the voltage- and patch-clamp procedures. In dispersed smooth muscle cells of the guinea pig small intestine, the muscarinic receptor activators ACh, carbachol and oxotremoline produce an inward current, in a concentration-dependent manner. This inward current persisted in Ca-free solution containing 1 m*M* ethyleneglycol tetra-acetic acid (EGTA) and Cl-deficient (substituted with isothianate) solution, but ceased in Na-deficient solution (substituted with choline plus atropine). This means that, in this tissue, activation of the muscarinic receptor is coupled with a cationic channel that is rather selective to Na. However, in the toad stomach, ACh produced an outward current due to inhibition of the K current (M current; SIMS et al. 1985, 1988).

a) ATP

In the rabbit ear artery, ATP activated (purinergic (PII) receptor) activated the Na and Ca permeable channel, as estimated from the cell-attached patch-clamp procedure (BENHAM and TSIEN 1987). The channel openings induced by ATP in Na-free and also in Ca-free solutions were measured and, from the results, the authors concluded that this purinergic-receptor-operated channel is permeable to both Na and Ca, the ratio of permeation between Na and Ca being 3:1. Further investigations made on dispersed cells using the whole-cell voltage-clamp procedure together will Ca transient measurements (Indo I) suggested that ATP induces an increase in the cytosolic Ca large enough to evoke contraction by activations of the PII-receptor-coupled cation channel (BENHAM 1989). The amount of Ca influx during the activation of the receptor-operated cation channel was estimated to be responsible for 10% of the total inward current. In the guinea pig portal vein and vas deferens, ATP produces two distinct inward currents as measured using the voltage-clamp procedure, i.e. an initial transient and a slowly developed and sustained inward current. The former, but not the latter, produces a rapid desensitization and is blocked by application of α-, β-methylene ATP (AMP-CPP), a blocker of the PIIx receptor. The sustained inward current could be recorded in Na-free or Ca-free solution (XIONG et al. 1991). In the case of the Ca-free solution, the sustained inward current was generated by Na influx whereas, in Na-free solution, the charge carrier is mainly Cl, as estimated from the reversal potential of the inward current measured at various membrane potential levels, and also the current measured in the presence of niflumic acid, a Cl current blocker, or in Cl-deficient solution. Activation of Cl permeability in Na-free solution had a Ca dependency and thus a minute amount of Ca influx may activate this Ca-dependent Cl permeability. Thus, under physiological conditions, ATP may activate the PII-receptor-operated cation channel and this activation may lead to permeation of Na via the receptor-operated ion channel with a minor proportion of Ca (less than a few per cent) (XIONG et al. 1991).

b) Histamine

In the saphenous artery, histamine activated the H_1-receptor and produced an inward current of minute amplitude, as measured using the whole-cell voltage-clamp procedure. In addition, this receptor activation increased IP_3 synthesis. Activation of the H_2- and H_3-receptors does not generate the receptor-operated inward current, but H_2-receptor activation accelerates the synthesis of cyclic AMP. The H_1-induced current is enhanced in amplitude by intracellular perfusion with GTP or GTPγS and blocked by pretreatment with the H_1-blocker, mepyramine. In the saphenous artery, NAd and serotonin also modify the voltage-dependent Ca channel and this enhancing action is inhibited by GTPγS, but accelerated by GTP, as observed for the actions of histamine. This modification of the voltage-dependent Ca channel via activation of the H_3 receptor by histamine was inhibited by pretreatment with pertussis toxin (OIKE et al. 1992). A detailed investigation of the action of GTP-binding protein in relation to the function of the receptor-operated cation channel has not yet been systematically made on visceral smooth muscle cells other than those of the saphenous artery. It is of interest to know how the GTP-binding-protein activations triggered by hydrolysable and non-hydrolysable GTP can have different effects on the voltage-dependent Ca channel.

c) Acetylcholine

It is well known that activation of muscarinic receptors by ACh leads to a wide variety of excitatory and inhibitory responses in various smooth muscle tissues. This is probably due to the existence of several subtypes of muscarinic receptor linked to distinct effectors, including ion channels (FUKUDA et al. 1989). Ion channels related to muscarinic receptor activation in smooth muscle are classified into three broad categories from their mode of activation and modulation by ACh. The first category includes ion channels that would be directly activated via muscarinic receptors, e.g., a K conductance suppressed by ACh (designated as M current) and a non-selective cationic conductance elicited by ACh (I_{nsACh}). The second category includes those which require intracellular second messengers for their activation, and the last includes those which are primarily activated by factors other than ACh, but could be modulated by a muscarinic receptor. The M current was discovered in an amphibian sympathetic ganglion (BROWN and ADAMS 1980) and subsequently identified in stomach smooth muscle (SIMS et al. 1985; LAMMEL et al. 1991). The M current is a new class of voltage-dependent, Ca-insensitive K channel that seems endogenously active, bringing the membrane potential close to its activation threshold. When ACh or some tachykinins, substance P or substance K were present, this M current was strongly suppressed (SIMS et al. 1986). This leads to membrane depolarization superimposed with action potentials and subsequent contractions. The M current appears to be regulated by the level of cyclic AMP via the β-adrenergic receptor (SIMS et al. 1988).

In the mammalian gastrointestinal tract, it has long been known that ACh produces contraction of smooth muscle by its depolarizing action, associated mainly with an increased Na permeability. This observation was further supported by recordings of the ACh-induced non-selective cation conductance and channel activities (I_{nsACh}) using single cells dispersed from the rabbit jejunum and the guinea pig ileum (BENHAM et al. 1985; INOUE et al. 1987). The I_{nsACh} channel has a single-channel conductance of 25 pS and is permeable to many monovalent and divalent cations (INOUE and ISENBERG 1990a). The latter property suggests that a significant amount of Ca enters smooth muscle cells through the I_{nsACh} channel. The open probability of the I_{nsACh} channel decreases with hyperpolarization (INOUE et al. 1987; INOUE and ISENBERG 1990a) and its steady state activation curve is at its steepest near the resting membrane potential, with a half maximum activation potential of −50 mV. This means that even minute fluctuations in the membrane potential could greatly alter the probability of the I_{nsACh} channel being open, and provide an important mechanism controlling the ability of ACh to depolarize the membrane efficiently. The I_{nsACh} channel has been found to be very sensitive to cytosolic Ca in the physiological range ($0.1-1\,\mu M$), although Ca would not be a primary activator of the channel (INOUE and ISENBERG 1990b). A Ca influx of 1–2 pC into a cell, which could be carried by a single action potential, appears to be sufficient to elicit a large increment (over four fold) in the activity of the I_{nsACh} channel. This mechanism, together with the voltage-dependent gating, would serve as strong positive feedback to potentiate and prolong the depolarizing action of ACh, and become important particularly when a low concentration of ACh is present in the vicinity of the channel, in which condition only a small fraction of the I_{nsACh} channels are active.

In smooth muscle, charybdotoxin-sensitive and large-conductance Ca-dependent K channels have been shown to be activated in response to the elvation in cytosolic Ca caused by muscarinic agonists or IP_3 (KLÖCKNER and ISENBERG 1985; BENHAM and BOLTON 1986; KOMORI and BOLTON 1991). However, there is a considerable shortage of information concerning other Ca-dependent K channels, such as the Ca-dependent Cl channel, the apamin-sensitive small-conductance Ca-dependent K channel and the Ca-dependent non-selective cation channel. Only recently has it been suggested that ACh and other Ca-mobilizing substances (noradrenaline, caffeine) activate a large-conductance Ca-dependent non-selective cation channel in the rat portal vein (LOIRAND et al. 1991).

V. Mutual Relationship Between the Voltage-Dependent and Receptor-Activated Ion Channels

It is now accepted that the majority of voltage-dependent, Ca-dependent channels undergo modulation by neurotransmitters. Well-characterized examples are found in voltage-gated Ca channels, delayed-rectifying K channels and Ca-dependent K channels, many of which seem to involve biochemical events such as a phosphorylation-dephosphorylation cycle (LEVITAN 1988). DHP-sensitive Ca channels in smooth muscle are known to be up- or down-regulated by various neurotransmitters and autacoids. It has been reported that muscarinic agonists increase DHP-sensitive I_{Ca} in the toad stomach (CLAPP et al. 1987) and in the rabbit coronary artery (MATSUDA et al. 1990), but reduce it in the rabbit jejunum and guinea pig ileum (RUSSEL and AARONSON 1990; INOUE and ISENBERG 1990b). The involvement of protein kinase C (VIVAUDOU et al. 1988) has been suggested for the toad stomach but the detail remains unclear.

As described previously, Ca influx may occur by activation of the voltage-dependent Ca and receptor-operated cation channels. However, the amount of Ca increase via the receptor-operated cation channel was much smaller than that induced by activation of the voltage-dependent Ca channel. Receptor activation modifies the voltage-dependent Ca channel through the actions of GTP-binding protein and, furthermore, receptor-operated synthesis of known and unknown second messengers through the actions of GTP-binding proteins may indirectly modify the voltage-dependent Ca channel through regulation of the cytosolic Ca.

a) GTP-Binding Proteins

The nature and function of the large-molecular GTP-binding protein which is composed of α-, β- and γ-subunits and protein have been reviewed elsewhere (GILMAN 1987; BIRNBAUMER 1990). Briefly, components of signal-transducing pathways are encoded by different gene families. So far nine genes have been identified which code for the α-subunit of GTP-binding proteins in mammalian cells (Gs1α, Gs2α, Gi1α, Gi2α, Gi3α, Goα, Gxα, Gt1α and Gt2α). These GTP-binding proteins were named from their toxin sensitivity and localization (cholera toxin-sensitive, pertussis toxin-sensitive), distribution in the brain and localization in the retinal photoreceptor, respectively. The natures of the β- and γ-subunits seem to be the same for the above nine GTP-binding proteins. The α-, β- and γ-subunits of GTP-binding protein (especially the α-subunit) differed in their molecular weights, modulation by bacterial toxins (pertussis toxin and cholera toxin) and their coupled function. The structures of each subtype of gene families were strongly conserved among different organisms. They cross-talked to each other and formed a functional network. Signals, in many cases, are transmitted by protein-protein interaction, and they are often modulated by a ligand-induced conformational change. GTP-binding protein binds with triphosphate compounds such as ATP and GTP, and is transformed to an active form which can interact with other molecules.

The GTP-binding protein provokes negative and positive signals, and these paths often compete with or neutralize each other. The main action of GTP-binding protein is thought to occur through activation of the α-subunit. However, the actions of the β- and γ-subunits have been elucidated in relation to the actions of the agonist-dependent phosphorylation and the arachidonic cascade map. This means that after agonist is bound to the receptor (investigated mainly in the case of the muscarinic receptor) and α-subunit-GTP complex may act on the K channel or other effector system, whereas the β- and γ-subunit may inactivate receptor activation (BIRNBAUMER 1990). On the other hand, activation of both β- and γ-subunits, in the absence of the α-subunit, accelerated arachidonic acid synthesis in cardiac muscle, as estimated from the activity of the K channel measured using the patch-clamp procedure (KURACHI et al. 1989). Individual neurotransmitters and putative substances activated receptors distributed on the sarcolemma through the action of GTP-binding proteins. Thus, in the presence of GTP, the α-subunit is dissociated from the β- and γ-subunit complex and forms a GTP-α-subunit complex, thus activating adenylate cyclase, guanylate cyclase or phospholipase C and synthesizing second messengers, such as cyclic AMP, cyclic GMP or IP_3 with diacylglycerol (DG), respectively. In addition, small (molecular weight about 21 kDa) cytosolically distributed GTP-binding proteins (more than 20 different subtypes) have been recently identified, and these proteins have been named low-molecular GTP-binding proteins, *ras* p21-like substances. These proteins had no subunit and facilitated ribosylation by pertussis toxin (GILMAN 1987; BIRNBAUMER 1990). The physiological role of these small-molecular GTP-binding proteins in smooth muscle cells is now under investigation.

b) Acetylcholine

In cardiac muscle, activation of the muscarinic-receptor produced activation of the K channel which was correlated with activation of the Giα-subunit (pertussis toxin sensitive). This activation of the Giα-subunit may not directly correlate with synthesis of second messengers, but may contribute to activation of the receptor-operated ion channel. This is because synthesis of cyclic AMP is thought to occur through activation of Gsα, and activation of Giα is thought to act as an inhibitory factor for the synthesis of cyclic AMP. The actions of ACh in relation to the voltage-dependent Ca channel have been described in the previous section.

c) Histamine

As described previously, H_1-receptor activation in smooth muscle cells of the basilar artery produces an inward current, while activation of the H_3-receptor in vascular smooth muscle modifies the voltage-dependent Ca channel, without any generation of the inward current (OIKE et al. 1992). In general, the H_3-receptor in visceral tissues has been thought to be distributed on nerve terminals and to negatively regulate transmitter release (ISHIKAWA and SPERELAKIS 1987). However, the presence of this H_3 receptor in smooth muscle membranes is now becoming apparent. Activation of the H_3 receptor accelerates activation of the voltage-dependent Ca channel. It is interesting that intracellular application of GTP enhanced, but GTPγS (from 3 to $100 \mu M$) consistently inhibited, this accelerating action. However, the H_1-receptor-operated inward current was accelerated by both GTP and GTPγS (OIKE et al. 1992). The underlying mechanisms inducing the different actions of GTP and GTPγS on the H_3-receptor induced voltage-dependent Ca channel have not yet been clarified. On the other hand, in vascular smooth muscle, intracellularly applied ATP itself accelerated the voltage-dependent Ca channel, but extracellularly applied ATP (activation of the purinergic receptor) inhibited the voltage-dependent Ca channel (OHYA et al. 1986; XIONG et al. 1991). In addition, GTP and GTPγS accelerated the inhibitory actions of the voltage-dependent Ca channel in the presence of ATP.

d) Catecholamines

In vascular smooth muscle cells, it has been reported that NAd enhances the voltage-dependent Ca channel by lowering the threshold for its activation (NELSON et al. 1988). On the other hand, it has also been reported that NAd inhibits this channel (DROOGMANS et al. 1987). This controversy may be partly related to the cytosolic Ca, i.e. NAd increases the amount of cytosolic Ca through synthesis of IP_3 and also generates the receptor-operated inward current (permeation of Na and Ca). In fact, increased cytosolic Ca accelerates the inactivation of the voltage-dependent Ca channel (OHYA et al. 1987b). In cardiac muscle cells, cyclic AMP, synthesized by β-adrenoceptor-GTP-binding protein (Gαi)-adenylate cyclase activation processes, activates the cyclic AMP-dependent protein kinase (A kinase). This A kinase activation phosphorylates the channel protein (C-terminal), accelerates the voltage-dependent Ca influx and finally produces a positive inotropic action in the cardiac muscle (KAMEYANA et al. 1986; TRAUTWEIN and HESCHELER 1990). However, such an enhancing action on the voltage-dependent Ca channel by β-adrenoceptor activation was not observed in visceral smooth muscle cells (OHYA et al. 1986, 1987a) though there was an acceleration of the Ca-dependent K channel (OHYA et al. 1986; SADOSHIMA et al. 1988). In the guinea pig taenia coli, it has been reported that forskolin accelerated the voltage-dependent Ca channel (FUKUMITSU et al. 1990). Nitroprusside in smooth muscles of the portal vein (ITO et al. 1978) and nitroglycerin in the guinea pig taenia coli (IMAI and KITAGAWA 1981) slightly hyperpolarized the membrane, but such hyperpolarization was not observed in other tissues. These responses are thought to be related to cyclic GMP synthesis. Atrionatriuretic peptide (ANP) relaxed tissues precontracted by NAd but had no effect on the smooth muscle membranes of the rabbit renal artery (FUJII et al. 1986). However, it has been reported that, in the same cultured smooth muscle cells, ANP and 8-bromocyclic GMP both inhibited the action potential (OUSTERHOUT and SPERELAKIS 1987). In the rabbit aorta, it was reported that cyclic AMP and cyclic GMP inhibited the Ba current recorded using the voltage-clamp procedure (OUSTERHOUT and SPERELAKIS 1987; SPERELAKIS and OHYA 1990). Therefore, the actions of cyclic nucleotides are not consistent in different smooth muscles. Further detailed experiments would be required to clarify the action of cyclic nucleotides on the voltage-dependent Ca channel.

e) Endothelin

In rabbit basilar arteries, endothelin activated both the L- and T-subtype of the voltage-dependent Ca channel in experiments using the cell-attached patch-clamp procedure, and this enhancing action was more potent on bath application than on pipette application. Therefore, endothelin is unlikely to act directly on the L- and T-subtypes of the channel, but it may act through the synthesis of an unknown second messenger (INOUE et al. 1990).

VI. Factors Modifying the Mechanical Responses with No Change in the Ca Concentration

FUJIWARA et al. (1989) reported that GTP and GTPγS both enhance the Ca sensitivity of contractile proteins in saponin-treated skinned muscles, i.e. both substances enhanced the Ca-induced contraction in the presence of certain buffered Ca concentrations, enhanced the phosphorylation of the 20-kDa protein of myosin light chain (MLC) and the cycling rate of actin-myosin interaction as estimated from the shortening velocity of tissues. GTPγS had a more potent enhancing action on the contraction than GTP. However, these substrates have no effect on the contraction evoked in the absence of ATP (rigor contraction) or the Ca-independent contraction under the treatment with ATPγS with Ca. Increase in the Ca sensitivity does not imply any precise underlying mechanisms; however, GTP and GTPγS enhanced the contraction either at some step in the formation of the Ca-calmodulin-myosin light-chain kinase (MLCK)-20-kDa protein of the MLC complex or by acting on MLC phosphatase. STULL et al. (personal communication) postulated that GTP and GTPγS may inhibit the action of myosin phosphatase, thus causing enhancement of mechanical response. Increases in the Ca sensitivity of the contraction induced by agonists on Ca-induced contraction were clearly demonstrated in α-toxin-treated skinned smooth muscle tissues by KITAZAWA et al. (1989).

Regulation of MLC phosphatase may modify the contraction in smooth muscle tissues. In fact, okadaic acid, an inhibitor of phosphatase, accelerates contraction and increases the amount of phosphorylated MLC. At the present moment, we know that there are various subtypes of phosphatase in various cells, such as central nervous system and liver cells, and it is thought that the MLC phosphatase in visceral smooth muscle can be classified as phosphatase I (BIALOJAN et al. 1988; TAKAI et al. 1987, 1989). Inhibition of phosphatase may enhance the contraction with no increase in the Ca concentration. Furthermore, HIMPENS et al. (1988) have reported that differences in the phasic and tonic types of contraction (studied in ileum longitudinal muscles and pulmonary artery, respectively) are characterized by differences in the activity of MLC phosphatase. This is because the cytosolic Ca remained unchanged, but the phosphorylation of the 20-kDa protein of MLC was smaller in the phasic than in the tonic type, and the contraction was shorter lasting in the former than in the latter muscle tissue. On the other hand, Ca sensitivity in the phasic type of tissue is markedly greater than that in the tonic type of muscle tissue (HIMPENS et al. 1989). It is also reported that okadaic acid enhances the contraction in the former (the phasic type; SOMLYO et al. 1989). Therefore, SOMLYO et al. (1989) postulated that the phasic and tonic types of tissue possess different types of MLCK and 20-kDa MLC phosphatase. Concerning this postulation, HIMPENS and CASTEELS (1990) supported the above contention that differences in the MLCK and 20-kDa protein of MLC phosphatase activities may induce the differences in the hysteresis of the Ca-tension relationship, and this may also induce the differences that occurred in the contractions in high K and carbachol in smooth muscle.

As briefly described, some agonists produce larger contractions than the 128 m*M* K-induced one despite a smaller increase in the cytosolic Ca, as estimated by Fura 2. In fact, the K-induced contraction is smaller than the maximum contraction evoked by Ca in skinned muscle tissues (ITOH et al. 1986a). Furthermore, in skinned muscle tissues prepared by α-toxin or β-escin, NAd enhanced the contraction with no change in the Ca concentration, due to an increase in the Ca sensitivity (NISHIMURA et al. 1988; KITAZAWA et al. 1989). In addition, activation of protein kinase C synthesized together with IP_3 may also contribute to the enhancement of the contraction through activation of MLCK (ITOH et al. 1988; RASMUSSEN et al. 1990; KARAKI 1989). This phenomenon is thought to be due to an increase in the Ca sensitivity of contractile proteins, i.e. there is no change in the Ca concentration, but an increase in the phosphorylation and shortening velocity of the Ca-activated (dependent) contractile elements. However, such enhancement of contraction does not occur with the Ca-independent contraction (FUJIWARA et al. 1988, 1989).

VII. Ca Influx Induced by a Reversed Na-Ca Exchange Diffusion Mechanism

In many cells, including some smooth muscle cells, Ca can be transported both inwardly (Ca entry) and outwardly (Ca exit: reversed Na-Ca exchange diffusion) via the Na-Ca exchanger. Net Ca movements mediated by Na-Ca exchange diffusion (which mainly occur as 3 Na:1 Ca) are governed by the difference between the membrane potential and the reversal potential of this diffusion process and also by the kinetic parameters that control the rate of exchange (BLAUSTEIN 1977a; BRADING et al. 1980).

By contrast with the wide acceptance of the view that Na-Ca exchange plays an actual role in Ca regulation in the cytosol of cardiac muscle, there is substantial controversy about its physiological role in many smooth muscles (REUTER et al. 1973; BLAUSTEIN 1977a,b, 1987, 1989; VAN BREEMEN et al. 1979; BRADING and LATEGAN 1985; CASTEELS et al. 1985; MULVANY 1985; SOMLYO et al. 1986). To investigate Na-Ca exchange diffusion, mainly mechanographic recordings, cellular Ca contents and flux measurements have been carried out, but biophysical experiments (such as the patch-clamp method) have not been fully applied to investigate this mechanism in smooth muscle. To estimate the change in cytosolic Ca, contractions have been measured. However, as has been described, the amplitude of contraction does not occur in parallel with the change in the cytosolic Ca. Agonists especially increased the Ca sensitivity of contractile proteins in α-toxin-treated skinned smooth muscle (KITAZAWA et al. 1989). In addition, the Na-Ca exchange diffusion is also regulated by the Na-H exchange mechanism (AALKJAER and CRAGOE 1988; KAHN et al. 1988a,b; KORMBACHER et al. 1988; WRAY 1988). A reduction in the electrochemical gradient for Na may be expected to lower $[pH]_i$, thus attenuating contraction and producing a relaxation of tissues (WRAY 1988), and also to counteract the effect of a rise in the cytosolic Ca. A fall in $[pH]_i$ may also inhibit both Ca entry (BAKER and MCNAUGHTON 1977) and Ca exit (DIPOLO and BEAUGE 1982) mediated by the Na-Ca exchange diffusion system (BLAUSTEIN 1989).

In many smooth muscle tissues, the existence of the Na-Ca exchange diffusion has been postulated. In vascular tissues, experiments have employed tracer efflux or Ca transient measurement studies to examine the properties of the Na-Ca exchange diffusion in sarcolemmal vesicles (DANIEL 1985; MOREL and GODFRAIND 1984; MATLIB et al. 1987; MATLIB 1988; BLAUSTEIN 1987; SMITH et al. 1987; KAHN et al. 1988a,b; SLAUGHTER et al. 1987; BOVA et al. 1989). KAHN et al. (1988a,b) and MATLIB (1988) both reported that the Ca flux is a Na gradient- and voltage-dependent process and the calculated exchange stoichiometry for the depolarization-induced Na-Ca exchange diffusion is 3 Na:1 Ca (ASHIDA and BLAUSTEIN 1987a,b; RASGADO-FLORES and BLAUSTEIN 1987). The existence of Na-Ca exchange diffusion mechanisms in resistance vascular tissues has been reported as occurring under both physiological and pathological conditions (BLAUSTEIN 1989) in tracheal smooth muscle (SLAUGHTER et al. 1987; CHIDECKEL et al. 1987), myometrium (MATSUZAKA et al. 1987; MASAHASHI and TOMITA 1983), intestinal smooth muscle (HUDDART and SAAD 1978; BRADING et al. 1980; HIRATA et al. 1981; PRITCHARD and ASHLEY 1986, 1987), urinary tract smooth muscle (AICKIN et al. 1984, 1987; CALA et al. 1986; AICKIN 1987) and vas deferens (KATSURAGI and OZAWA 1978; WAKUI and INOMATA 1984).

By comparison with that in cardiac muscle, squid axon or barnacle muscle, the Na-Ca exchange diffusion mechanism in visceral and vascular smooth muscle has not been convincingly explained. Presumably, the active Ca-pump mechanism is a main mechanism for Ca efflux and in part the Na-Ca exchange mechanism may contribute to the reduction or increase in the cytosolic Ca concentration in different ionic environments, especially in different concentrations of Na. Further detailed investigations would be required to elucidate the role of Ca influx in the generation of contraction through activation of the reversed Na-Ca exchange diffusion mechanism.

VIII. Ca Influx Induced by Stretch

Stretches of smooth muscle cells produce depolarization of the membrane and induce a contraction. HISADA et al. (1991) reported that in the toad stomach a hyperpolarization-activated cation channel is stretch sensitive. In fact, this channel was more sensitive to stretch than to hyperpolarization of the membrane. There has not yet been a complete clarification of the features of this channel, but it is thought to be related to maintenance of the resting membrane potential (BENHAM et al. 1987a; EDMAN et al. 1987) or of the spontaneous electrical activity of cells (BADER and BERTRAND 1984; DIFRANCESCO et al. 1986; PAPE and MCCORMICK 1989).

C. Sarcoplasmic Reticulum as a Cytosolic Calcium Regulator in Visceral Smooth Muscles

In visceral smooth muscle, the T-system distributed in cardiac and skeletal muscles is not present and the amount of SR is relatively low (a few to 8% of total sarcolemmal dimensions) compared with that in other muscle systems (GABELLA 1981). Some of the SR vesicles distributed just beneath the surface membrane possess a foot structure as observed in the triad region (or diad region) in skeletal muscle. This means that Ca regulation in these SR vesicles may be directly coupled with the sarcolemmal membrane. It has also been reported that the SR vesicles distributed just beneath the cell membrane have a more causal relation to Ca mobilization than those located in the central region of the cell (SOMLYO et al. 1986).

The SR in visceral smooth muscles could accumulate and release Ca, and thus act as a regulator of the contraction-relaxation process. Ca accumulation into the SR requires the active transport system through hydrolysis of ATP. This Ca-pump process requires the presence of a 100-kDa protein, a Mg ATPase, and, in addition, calmodulin and phospholambane accelerate the Ca pump action. Second messengers, such as cyclic AMP and cyclic GMP, also help to accelerate the Ca pump through phosphorylation of the 100-kDa protein, which has different properties than the Mg ATPase distributed on the sarcolemma (140-kDa protein) in visceral smooth muscle. Cyclic nucleotides, such as cyclic AMP and cyclic GMP, also modify the functions of the SR through cyclic AMP-dependent protein kinase A and G, respectively. These substances modify the activity of Mg-ATPase (Ca pump), via the 140-kDa protein of MgATPase distributed in the sarcolemma and the 100-kDa protein of MgATPase in the SR. In the case of the 140-kDa protein, the Ca pump at the sarcolemma is thought to be activated indirectly through $PI\text{-}P_2$ and this function is accelerated by cyclic nucleotides. In the case of the 100-kDa protein in the SR, cyclic GMP and cyclic AMP accelerated the Ca pump (EGGERMONT et al. 1988a,b; RAEYMAEKERS et al. 1988; RAEYMAEKERS and JONES 1986; TADA and KATZ 1982; VROLIX et al. 1988). These actions of cyclic nucleotides, therefore, act to reduce the cytosolic Ca in smooth muscle cells.

I. Release of Ca from the SR as Estimated from the Ca Transient and Contraction

In Ca-free solution containing EGTA, NAd or ACh produced a contraction only once, and repetitive application of these agents failed to produce more contractions. These responses observed in Ca-free solution were not observed on application of high concentrations of K. Therefore, the Ca storage site which is responsible for the release of Ca induced by many agonists (spasmogenic agents) is thought to be present in smooth muscle cells (KURIYAMA 1970, 1981).

In visceral smooth muscle tissues, the cytosolic Ca has been measured using various procedures: namely, the Ca electrode (YAMAGUCHI 1986), aequorine-fluorescence measurement (FAY et al. 1979; MORGAN and MORGAN 1982, 1984a,b), and quin 2, Fura 2 and Indo I (DEFEO and MORGAN 1985; WILLIAMS et al. 1985; SHOGAKIUCHI et al. 1986; KOBAYASHI et al. 1986; PARKER et al. 1987; SUMIMOTO and KURIYAMA 1986; HIMPENS and SOMLYO 1988; BENHAM 1989). The basal amount of Ca in the cytosol in visceral smooth muscle cells is estimated to be 100–150 nM. The ability to release Ca from the SR has also been estimated from the contraction evoked in skinned muscle tissues using saponin, α-toxin, β-escin and others (ENDO et al. 1977; SAIDA and NONOMURA 1978; ITOH et al. 1982a,b; IINO 1981; KOBAYASHI et al. 1989). In skinned muscle preparations, to enable the release of Ca to be measured, cytosolic Ca was partly chelated with EGTA buffer, and then the minimum concentration of Ca required to produce contraction was estimated to be more than 100 nM and the maximum amplitude of contraction was evoked by 5–10 μM.

Saponin (20–50 μg/ml) makes small holes in the lipid layer of the sarcolemma without affecting the SR membrane and makes possible the permeation of Ca for its entry into the SR (ENDO et al. 1977; SAIDA and NONOMURA 1978; ITOH et al. 1982a,b). However, agonist receptors distributed on the sarcolemma lose their abilities to release Ca from the SR. Some intracellular constituents, such as calmodulin and others, may also be leaked out from the cells, as estimated from the occurrence of a rundown phenomenon in Ca-induced contraction which was prevented by the addition of calmodulin (ITOH et al. 1986). Unlike saponin, α-toxin and β-escin make small, Ca-permeable holes in the sarcolemma and do not damage the receptor activity in visceral smooth muscle (NISHIMURA et al. 1988; KITAZAWA et al. 1989; KOBAYASHI et al. 1989). This new procedure thus enabled us to investigate agonist-induced contractions in relation to Ca mobilization.

The IP_3-induced Ca release was blocked by heparin, as estimated from the amount of Ca release from the SR and the contraction (KOBAYASHI et al. 1988, 1989; KITAZAWA et al. 1989; SOMLYO et al. 1989). The minimum concentration of IP_3 required to release Ca has not yet been quantitatively investigated, but physiological concentrations of agonists (0.1–1 μM) synthesized micromolar orders of IP_3, and these figures were confirmed using caged IP_3. In skinned muscle tissues, 1 μM IP_3 produced an adequate amplitude of contraction in low-concentration EGTA-buffered solution. It is plausible to postulate that Ca released by IP_3 may be able to trigger the Ca-induced Ca release (CICR) mechanism, in the same manner as influx of Ca at the sarcolemma does. Therefore, the main mechanism underlying the release of Ca from the SR would be an IP_3-induced CICR mechanism. In addition, GTP-induced modification of Ca release and an increase in the Ca sensitivity of contractile proteins may occur. Here again, we should stress the fact that in some visceral smooth muscle the voltage-dependent Ca channel is not distributed on the sarcolemma.

During recent investigations concerning the P-M coupling process, SOMLYO and SOMLYO (1990) concluded their elegant experiments using caged substances, such as Ca, IP_3, phenylephrine, GTPγS or ATP. Using these substances, they measured the latency to evoked contraction after application of various caged substances which modulate the underlying steps. The delay between the photolysis of caged IP_3 and Ca release is 30 ms or less, while the latency of the contraction is 0.3–0.5 s, similar to the lag between the rise of cytosolic Ca and force development. The latency of the contraction following photolysis of caged ATP in permeabilized muscle in rigor (ATP-free contraction), in the presence of Ca and calmodulin, is similar, about 0.2–0.5 s at 22°C. Therefore, the major portion of the delay between agonist-receptor interaction and the resulting contraction is due to the activation of phospholipase C and IP_3 production. The delay (0.2–0.5 s) can be ascribed to the prephosphorylation reaction between Ca, calmodulin and MLCk, and/or to mechanical processes, or to the chemical kinetics of the two-step reaction (SOMLYO and SOMLYO 1990). They also reported that GTPγS produced an enhancement of the contraction required over 10 s, as measured using caged GTPγS, and these responses were similar to those observed by FUJIWARA et al. (1989).

II. Role of the SR in Relation to the Release of Ca

1. Caffeine-Induced and IP_3-Induced Ca Release

Some drugs can release Ca from the SR. For example, caffeine in intact and saponin-treated skinned muscle tissues can release Ca from the SR and this action of caffeine is inhibited by procaine (ITOH et al. 1981b). Caffeine is thought to release Ca through an activation of the CICR mechanism (ITOH et al. 1981b, 1982, 1983; SAIDA 1982; KOBAYASHI et al. 1986; IINO 1989). The CICR mechanism was initially found in skeletal muscle cells (FORD and PODOLSKY 1970; ENDO et al. 1970), and its existence later confirmed in smooth muscle cells (ITOH et al. 1981b, 1982b; SAIDA 1982; IINO 1989). ATP and adenosine activate this process in smooth muscle as also observed in skeletal muscle cells (ENDO 1977, 1985; IINO 1989). Furthermore, this Ca release mechanism in visceral smooth muscle cells is inhibited by ryanodine (SUTKO et al. 1985; MEISSNER 1986; ROUSSEAU et al. 1987; ITOH et al. 1986; HWANG and VAN BREEMEN 1987; SAKAI et al. 1988).

Inositol 1,4,5-triphosphate also releases Ca from the SR in visceral smooth muscle cells (SUEMATSU et al. 1984; YAMAMOTO and VAN BREEMEN 1985; Somlyo et al. 1985; HASHIMOTO et al. 1986; ABDEL-LATIF 1986, 1989). IP_3 is synthesized from hydrolysis of PI-P_2 by activation of phospholipase C. IP_3 is dephosphorlyated into inositol in a stepwise fashion and resynthesized into PI-P2 through several biochemical ATP-requiring steps. The synthesized IP_3 is readily dephosphorylated through one step to inositol, 1,4-bisphosphate (IP_2) or phosphorylated to inositol 1,3,4,5-tetrakisphosphate (IP_4) by IP_3-3 kinase. In visceral muscle, this process requires Ca and calmodulin (YAMAGUCHI et al. 1987). IP_4 is dephosphorylated into inositol 1,3,4-trisphosphate and is biologically less potent than IP_3 in its Ca-releasing action (BERRIDGE 1984, 1988; BERRIDGE and IRVINE 1984, 1989; IRVINE et al. 1986; IRVINE 1989; ABDEL-LATIF 1986, 1989). IP_3-induced Ca release is inhibited by heparin in vascular smooth muscle tissues (KOBAYASHI et al. 1988). It is now accepted that the release of Ca from the SR consists of at least two types: IP_3-induced Ca release and CICR (caffeine)-induced ryanodine-sensitive Ca release. The former is potently inhibited by heparin and the latter by procaine.

As a new tool for the investigation of Ca release from the SR, caged substances have been successfully introduced in this field. Somlyo and his group (WALKER et al. 1987; SOMLYO et al. 1989; SOMLYO and SOMLYO 1990) applied caged materials (IP_3, Ca, phenylephrine or GTPγS) to the investigation of the mechanical properties of smooth muscle. They reported that the force transient after photolysis of caged IP_3 is much faster than that induced by caged phenylephrine. Presumably, the time required for the essential metabolic process involved in the synthesis of IP_3 is responsible for such a delay.

IINO et al. (1988) and IINO (1989) postulated after experiments with ryanodine, caffeine and IP_3 (or agonists) that there are two Ca-releasing sources in intact and skinned smooth muscle tissues. One source is sensitive to ryanodine, caffeine and IP_3 and the other is only sensitive to IP_3, as estimated from differences in mechanical responses and Ca transient measurements. The contribution to the increase in the cytosolic Ca from each Ca source in smooth muscle tissues varied with the tissue. In various cell types and tissues, including vascular tissues, IP_3 released Ca from the SR and caused transient or repetitive rises of cytosolic Ca (BERRIDGE et al. 1988; COBBOLD et al. 1989). BERRIDGE and GALIONE (1988) suggested that, since only 30%–50% of the Ca stored in the ER may be mobilized by IP_3 in liver cells, there must be both IP_3-sensitive and -insensitive Ca pools. The two pools may interact with each other, in both space and time, to generate a complex pattern seen as Ca oscillations (BERRIDGE and GALIONE 1988; BERRIDGE and IRVINE 1989). Thus, IP_3 would provide a continuous supply of Ca which then primes, and ultimately triggers, the IP_3-sensitive pool to release Ca into the cytosol. Once all the stored Ca has been released, Ca is pumped out from the cell, and the intracellular Ca level returns close to the control level, thus generating a characteristic Ca transient. In the continued presence of low concentrations of agonist, this transient begins to repeat itself, thus setting up a Ca oscillation (cytoplasmic oscillator model). On the other hand, a receptor-controlled model has also been presented, largely on the basis of experiments on the SR of primary cultured rat aorta or the endoplasmic reticulum (ER) of hepatocytes, using aequorine in the presence of angiotensin II or ATP. The model proposed that IP_3 generation occurs transiently, and is curtailed at different rates by different inactivations of individual receptors or GTP-binding proteins (COBBOLD et al. 1989). WAKUI (1989) opposed the receptor-controlled model on the basis of measurements of the Ca-dependent Cl current activated by IP_3-induced Ca release from the ER. In these experiments, the oscillatory Ca release induced by agonists occurred even when the concentrations of IP_3 or IP_3S, a non-hydrolysable IP_3 derivative, were constant.

IRVINE (1989) proposed a different model for the action of IP_3 with joint participation of IP_4 on Ca release from the SR or ER, i.e. IP_3 has two pools, one type (A) has the IP_3 receptor, and the other (B) does not. Ca transfer from B to A would occur via a link which is triggered by a lowered Ca level in pool A. The function of IP_4 would be to desensitize the negative feedback, so that the activation of Ca transfer occurs at a high level of Ca in pool A. Furthermore, if the Ca level in pool B is lower than in pool A, Ca reuptake rather than release would be predicted.

2. Release of Ca from the SR Deduced from the Ca-Dependent K Current

As already briefly described previously, the K channel has been classified into several subtypes from unitary current recordings using the patch-clamp procedure, and some of them are Ca-dependent K channels. Using the whole-cell voltage-clamp procedure, the K channel can be classified into transient (I_t), oscillatory outward (I_{oo} or spontaneous transient outward current; STOC) and sustained (I_s) outward currents generated by application of depolarizing pulses (above -40 mV) from a holding potential of -60 to -80 mV. The I_t is generated successively after the generation of the transient Ca inward current and its generation is a prerequisite for the generation of the Ca inward current, but not vice versa. The I_t and I_{oo} require the presence of cytosolic Ca (Ca-dependent K current), but this is not the case for the I_s (Ca-independent K current). The amplitude of the I_t depends on the cytosolic Ca and is thought to be generated by activation of the CICR because depletion of Ca from the SR or destruction of the channel by A23187 blocks the I_t during generation of the transient Ca inward-current. Therefore, this current seems to have a close relation to the generation of the afterhyperpolarization of the action potential. On the other hand, the I_{oo} occurs on release of Ca from the SR (OHYA et al. 1986; BENHAM and BOLTON 1986; SAKAI et al. 1988; KOMORI and BOLTON 1990).

When the effects of ryanodine on the I_{oo} are observed, this agent blocks the generation of the I_{oo}. The ryanodine transient accelerates the generation of the unitary K current and then ceases, as observed using the cell-attached patch-clamp procedure. However, when the effects of ryanodine are observed on the unitary current recorded using the cell-free patch-clamp procedure (inside-out or outside-out patch), this agent has no effect on the channel activity at all (SAKAI et al. 1988). In Ca-free solution containing 0.2 m*M* EGTA, pre-application of ryanodine markedly inhibited the appearance of the I_{oo} following application of either caffeine or IP_3 in the guinea pig portal vein. This means that the actions of ryanodine on the ryanodine-sensitive and IP_3-sensitive Ca pools are not distinct, and presumably there may be only one source of Ca in the rabbit portal vein (SAKAI et al. 1988).

Inositol 1,4,5-triphosphate accelerates the generation of the I_{oo} and lowers the threshold required for its generation: thus, when the membrane was depolarized from the holding potential of -60 mV, the I_{oo} could scarcely be observed. However, following application of IP_3, the frequency of appearance of the I_{oo} was markedly enhanced. IP_4 showed much the same effects as IP_3, but with a much weaker potency (OHYA et al. 1988). Heparin is an inhibitor of Ca release from the SR (SOMLYO et al. 1988; KOBAYASHI et al. 1988, 1989) and this agent blocks the generation of the I_{oo}, although, when weak concentrations of heparin were used, the caffeine-induced I_{oo} was not modified. However, with increased concentrations of heparin (over 10 mU/mg) the caffeine-, IP_3- and agonist-induced I_{oo} were all blocked (XIONG et al. 1991). Recently, KOMORI and BOLTON (1991) in elegant observations on the role of IP_3 on the SR showed that caged IP_3, after light exposure, promptly accelerated the generation of STOCs (I_{oo}) due to release of Ca from the SR.

These results indicate that the I_{oo} is an indicator of Ca release from the SR, as observed for the mechanical responses in intact and skinned muscle tissues.

3. Receptor Proteins of the Ca-Release Channel

As a result of progress in molecular biological procedures, the primary structure and functional expression of ryanodine-sensitive and IP_3-sensitive proteins prepared from skeletal muscle and rat cerebellum have been reported from two laboratories, independently (TAKESHIMA et al. 1989; FURUICHI et al. 1989). TAKESHIMA et al. (1989) proposed that the sequence of 5037 amino acids comprising the ryanodine receptor prepared from rabbit skeletal muscle SR can be deduced using techniques for cloning and sequencing the complementary DNA. They noted that this receptor is located at the foot region and that the Ca release channel activity resides in the C-terminal region of the receptor molecule, contributed to by each of the four monomeric units which surround a central pore to form the Ca-release channel. Thus, the C-terminal portion of the ryanodine receptor, which includes both the channel-forming region and the modulator-binding sites, is responsible for Ca-release channel activity. The remaining portion constitutes the foot structure spanning the junctional gap between the SR and the T-system. TAKESHIMA et al. (1989) therefore postulated that the large cytoplasmic region of the ryanodine receptor may directly interact with the cytoplasmic site of the DHP receptor (Ca channel and voltage sensor; TANABE et al. 1987). Lai et al. (1988) also purified and reconstituted the Ca-release channel from skeletal muscle, and this portion (30S complex) was constituted in a planar lipid layer. This channel was opened by application of ryanodine. The structure observed under the electron microscope showed a four-leaf clover structure, as described for the foot in the transverse-SR junction. The two groups (Takeshima's and LAI's groups) thus agreed on the structural arrangements of the receptor protein.

From the primary structure and functional expression of the IP_3-binding protein prepared from Purkinje neurons, FURUICHI et al. (1989) reported that this receptor protein is the same as P_{400} (250-kDa protein) based on the features of glycosylation, combination with concanavalin A, cyclic AMP-dependent phosphorylation and tissue concentration. These features of P400 were also the same as those reported by SUPPATTAPONE et al. (1988a,b) for the protein excised from the rat cerebellum (260 kDa). FURUICHI et al.

(1989) proposed that positively charged amino acid residues in the large cytoplasmic N-terminal region or in the cytoplasmic loop region may be responsible for the bindings of negatively charged IP_3, and that this region of peptide has a role in increasing Ca from the ER (or SR). They discussed similarities and differences between the ryanodine- and IP_3-binding proteins, i.e. marked similarities in amino acid sequences occurred both between the putative transmembrane domains h and i and the successive C-terminal region in the P_{400} and between the transmembrane segments M3 and M4 and successive C-terminal region in the ryanodine receptor. This homology did not extend to the binding site for the modulator (Ca, nucleotide and calmodulin) as proposed by TAKESHIMA et al. (1989). Both receptors share a similar tertiary structure, and whereas the region encompassing segments M2 and M3 of the ryanodine receptor was claimed to have a sequence homologous with the M2 and M3 region of the nicotinic ACh receptor, the corresponding region of P400 did not. In addition, the ryanodine receptor has been proposed to have four membrane-spanning segments, with its C-terminal region on the cytoplasmic side of the SR (TAKESHIMA et al. 1989). On the other hand, the P400 has at least three (and at most seven) transmembrane domains and its C-terminal was therefore in the ER lumen or outside the cell (FURUICHI et al. 1989). Thus, distinctly different Ca-releasing sites and their mechanisms have been elucidated using molecular biological and biophysical procedures.

Unfortunately, in visceral smooth muscle cells, the primary amino acid sequences of the ryanodine- and IP_3-sensitive Ca channel proteins distributed on the SR have not yet been directly determined. However, in aortic smooth muscle cells, EHRLICH and WATRAS (1988) have observed direct evidence for the release of Ca from the IP_3-sensitive Ca channel. In the SR fraction incorporated in planar lipid bilayers, IP_3 opened the channel (slope conductance, 10 pS; corresponding to a T-type voltage-dependent Ca channel in conductance). Using the same procedure, they also measured the effects of IP_3 on the SR extracted from cardiac and skeletal muscle and found the unitary current conductance of the IP_3-sensitive channels in the SR in those tissues to be 105 pS and 125 pS, respectively. These observations suggest that the properties of the IP_3-sensitive Ca channel in smooth muscle differ from that in cardiac and skeletal muscles. More detailed investigations are required in smooth muscle cells.

D. Actions of Calcium on Contractile Proteins

Activation of the E-C coupling and P-M coupling processes increases the amount of Ca in the cytosol and induces a contraction. Formation of the 4 Ca-calmodulin-myosin light-chain kinase (MLCK)-myosin light chain (MLC) complex (phosphorylation through serine 19 of the 20-kDa protein of MLC; PERSON et al. 1984) causes hydrolysis of ATP via myosin·Mg-ATPase activation, thus causing contraction due to the activation of cross-bridge cycling through actin-myosin interaction. When the concentration of free Ca in the cytosol is reduced, myosin phosphatase dephosphorylated the phosphorylated myosin and thus relaxation occurs. Recently, many proteins which regulate the contractile proteins have been identified (SOBIESEZK 1977; ADELSTEIN and EISENBERG 1980; HARTSHORNE and GORECKA 1980; KAMM and STULL 1985, 1989; WALSH 1985; IKEBE and HARTSHORNE 1986; RÜEGG 1986; HARTSHORNE 1987; TAKAHASHI et al. 1988; ITO and HARTSHORNE 1990; WALSH 1990; IKEBE and HARTSHONE 1985).

I. Ca-Binding Proteins

There are many Ca-binding proteins in various tissues and their Ca-binding ability varies from 1 to 4; for example, calmodulin and calcineurin (4 Ca), troponin C (in skeletal and cardiac muscle this protein binds 4 Ca and 3 Ca, respectively), S-100a and β-chains (1 Ca) and palvalbumin and oncomodulin (2 Ca). In general, these proteins have been named "EF hand proteins" on the basis of their structures as determined using X-ray structure analysis (KRETSINGER and BARRY 1975). Recent investigations, however, have found Ca-binding proteins which have no such EF hand structure, for example, lipocortin I and II, calpactin, endonexin I (lipocortin III and IV) and II (lipocortin V), calelectrin (lipocortin VI), calcimezin and others. Recently, these proteins were named the annexin group. Some Ca-binding proteins related to smooth muscle contraction will now be briefly reviewed.

a) Calmodulin

KAKIUCHI and YAMAZAKI (1970) and CHEUNG (1970) first recognized that calmodulin was an activator of phosphodiesterase. Further investigations have indicated that this protein is a Ca-binding protein which has many properties in common with troponin C (Ca-binding protein of skeletal muscle; KAKIUCHI and YAMAZAKI 1970). Calmodulin is commonly distributed in cells in the animal and plant kingdoms and its molecular weight has been calculated to be that of a 17-kDa protein. Calmodulin (classified into calmodulin I, II and III) is the product of at least bona fide genes, and its ternary structure was determined by BABU et al. (1985, 1988). The length of the protein is about 65 Å and it has a dumbbell structure (composed of a total of four Ca-binding bells; two Ca-binding sites in the N-terminal region and the other two at the C-terminal region with the Ca-binding sites connected by seven fold rotated α-helical structures).

Calmodulin regulates the second messenger system through: (a) activation of protein kinases (MLCK, phosphorylase kinase, protein kinase II and others) and protein phosphatases (calcineurin and others) and (b) regulation of second messenger levels: it enhances the activities of guanylate cyclase, cyclic phosphodiesterase and IP_3 kinase, which converts IP_3 to IP_4.

Calmodulin binds four Ca molecules with different affinities, i.e. two Ca-binding sites located in the C domain strongly bind to Ca ($20\,\mu M$; each binding site has positive cooperativity) whereas the other two Ca-binding sites in the N domain bind weakly ($3-4\,\mu M$). Once the two Ca sites in the C domain are bound to Ca, the second two Ca sites are readily bound to the binding sites in the N domain. However, the K_m value measured in various concentrations of Ca needed to be much higher to provoke a physiological response ($5\,\mu M$ order), as estimated from the Ca-induced contraction in skinned muscle tissues (K_m value of $0.5\,\mu M$). By contrast, when the target enzyme and calmodulin were present, the Ca sensitivity of calmodulin was markedly increased and the relationship shifted to the right by one order of magnitude lower in the concentration of Ca (YAZAWA et al. 1987).

b) Annexin Family of Proteins

Lipocortin, synthesized after the application of glucocorticoids, is thought to inhibit phospholipase A_2 activity and to inhibit the synthesis of arachidonic acid from the phospholipid of the sarcolemmal membrane, thus inhibiting the synthesis of prostaglandins, thromboxane and leucotrienes (TSURUFUJI et al. 1979; HIRATA et al. 1984). At present, the inhibitory action of lipocortin is thought to be not directly on phospholipase A_2, but by binding to the phospholipid required for the release of arachidonic acid (DAVIDSON et al. 1987; AARSMAN et al. 1987; AHN et al. 1988).

Endonexin I (32–33 kDa; chromobinsin, p32.5, calelectrin, protein II, 35-kDa calcimedine) has the ability to bind Ca and phospholipid, as does lipocortin. Recently, this substance was reported to have the ability to inhibit phospholipase A_2, as observed for lipocortin. However, this substance has no ability to bind F-actin (GEISOW et al. 1986; CREUZ et al. 1987; SMITH and DEDMAN 1986). Endonexin II (FUNAKOSHI et al. 1987a,b; IWASAKI et al. 1987) is an anticoagulant protein and possesses properties in common with the lipocortin: Ca, phospholipid and F-actin binding properties. Calcimedine (MOORE and DEDMAN 1982) was extracted from chicken gizzard, and from the molecular weight of this protein it was named 67-kDa lipocortin VI or calcimedine (MOORE 1986). More detailed physiological studies of these proteins are needed before we can understand their physiological role.

II. Myosin Phosphorylation Model for Smooth Muscle Contraction

The degree of myosin phosphorylation needed to produce a contraction may occur with a parallel increase in free Ca in the cytosol. Phosphorylation of serine-19 in MLC leads to actin-activated myosin Mg-ATPase activity (and cyclic interaction of actin and myosin). A second phosphorylation site at threonine-18 has been described for light-chain kinase (IKEBE and HARTSHORNE 1986). However, phosphorylation of the MLC can also be induced by agents other than Ca. There are five potential phosphorylation sites within the first 19 N-terminal residues in 20-kDa proteins of the MLC and these sites can be phosphorylated by cyclic nucleotide kinases (cyclic AMP- and cyclic GMP-activated protein kinases; A and G kinase, respectively) and also by protein kinase C (C kinase) activated by diacylglycerol (DG) (NISHIKAWA et al. 1984; UMEKAWA et al. 1985a,b; IKEBE and HARTSHORNE 1986; RÜEGG 1986; BIALOJAN et al. 1988; HAI and MURPHY 1988a, b, 1989; KAMM and STULL 1989). The actual site of action of GTP is not yet known, but GTP enhanced the interaction of actin and myosin, and is thought to increase the sensitivity of contractile proteins to Ca (FUJIWARA et al. 1989).

Protein kinase C is activated in the presence of phospholipid and rather high concentrations of Ca (several hundred millimolar) (NISHIZUKA 1984). After the discovery of the inositol phospholipid metabolic path, it was found that activation of this enzyme could be induced by only low concentrations of Ca with DG. From these findings, the physiological role of this kinase became apparent (NISHIZUKA 1984, 1986, 1988). C kinase is composed of a 45-kDa catalytic domain, which is capable of binding ATP, and a 35-kDa regulating domain which binds to DG and Ca. C kinase has been reported to have several isozymes by a number of investigators. Each isozyme can phosphorylate many proteins and regulates the physiological roles of various cells. This C kinase is also activated by 12- *o*-tetradecanoyl-phorbol-13-β acetate (TPA) or oleoyl-2-acetyl glycerol (OAG). One of the specific features of this kinase is its translocation process in cells, i.e. after TPA stimulation, the activity of cytosolic C kinase is reduced and the activity of that distributed in fragmented membrane fractions transiently increased and then reduced due to the appearance of the rundown phenomenon.

In the case of smooth muscle, C kinase is also activated by DG, phosphatydyl serine and low concentrations of Ca. However, it is not yet clear what kind of isozyme contributes. It is thought that serine-1, serine-2 and threonine-9 in MLC can be phosphorylated and that these sites are distinct from the two sites which are phsophorylated by MLCK (threonine-18 and serine-19, IKEBE et al. 1987). Phosphorylation of MLC induced by C kinase decreases the subsequent phosphorylation induced by MLCK, and results in a decrease in the myosin Mg ATPase activity (NISHIKAWA et al. 1984). On the other hand, an activator of C kinase, phorbol esters and DG analogues enhanced the contraction with no change in the amount of Ca, as measured using quin 2 (ITOH et al. 1988). It was postulated by IKEBE et al. (1986) that when two phosphorylation sites (threonine-18 and serine-19) are phosphorylated these phosphorylations may cause further enforced actin-activated myosin ATPase activity. This idea was supported by COLBURN et al. (1988), who showed that increased concentrations of carbachol phosphorylated two phosphorylation sites (P1 and P2 corresponding to the above two sites) in tracheal smooth muscle and also in thoracic aorta. $PGF_{2\alpha}$, but not high K or histamine, produced two site phosphorylations, and this agent produced a larger contraction than did high K or histamine application (SETO et al. 1990). Activation of C kinase directly or indirectly modulates ion channels in the sarcolemma. However, the results in smooth muscle are not consistent, e.g. activation of C kinase either accelerates or inhibits the voltage-dependent Ca channel (see review of SHEARMAN et al. 1989).

III. Actin-Regulating Mechanisms in Smooth Muscle

In cardiac and skeletal muscle, Ca triggers contraction through activation of the actin filament by troponin C. As it does not possess troponin C, smooth muscle contraction is thought to be triggered by activation of MLCK through initiation of Ca-calmodulin complex formation. However, several factors can regulate the actions of actin filaments, thus also modifying the contractile process.

1. Caldesmon

In smooth muscle contraction, caldesmon, a 120- to 140-kDa actin-binding protein (following recent cDNA cloning procedures, the molecular weight is reported to be 87 000 or 88 000) is thought to be involved in the regulation of contractile mechanics (SOBUE et al. 1981, 1982; MARSTON and SMITH 1984; NGAI and WALSH 1985; WALSH 1985, 1987, 1990; MARSTON et al. 1988). The length of this protein has been established to be 75 nm (FÜRST et al. 1986; LYNCH et al. 1987). Thus, the length of one caldesmon corresponds to two tropomyosin lengths and 14 global actin molecules, and it would run parallel with the α-helix structure of the actin filament. In the absence of Ca, caldesmon would bind to actin and inhibit actin-myosin interaction, while, when the amount of Ca in the cytosol was increased, the caldesmon-calmodulin complex would be dissociated and produce a contraction (Ca-camodulin-MLCK-MLC complex, WALSH 1990). Therefore, this function corresponds with that of troponin I (inhibitory subunit of actin-myosin interaction) in skeletal muscle (TAKAHASHI et al. 1988, 1989). Recently, a new hypothesis has been presented: i.e. in the presence of ATP, caldesmon forms an actin-myosin bridge (LASH et al. 1986; IKEBE and REARDON 1988; HEMRIC and CHALOVICH 1988; SOUTHERLAND and WALSH 1989). The C-terminal of caldesmon would bind to an actin filament and an N-terminal would bind to the S_2 area (root region of myosin head). The binding of caldesmon and the S_2 area would cause inhibition of the ATPase area of the myosin head linked with actin; hence actin-activated myosin ATPase activity would be suppressed. Recently, caldesmon was thought to be related to the appearance of a catch phenomenon (RASMUSSEN et al. 1987, 1990). However, it is not yet clear whether the actin-caldesmon complex formation can produce tension development or not. Physiological evidence to support the above hypothesis has not yet been obtained.

2. Calponin

An additional actin-linked regulating protein for the contractile proteins, calponin, has been proposed (TAKAHASHI et al. 1986, 1987, 1988). This globular protein would have a function similar to that of troponin I in skeletal muscle, but from the tropomyosin interaction this protein seems to be a troponin T-like substance. Calponin is located 16–17 nm in distance from the C-terminal of tropomyosin and located at 38-nm intervals on the actin filament, i.e. one calponin regulates seven actin molecules through tropomyosin as also observed in the relationship between troponin T and actin (TAKAHASHI et al. 1988). Since calponin inhibited actin-ATPase activity triggered by phosphorylated myosin in the presence or absence of Ca, one of the important roles of calponin would be its inhibitory action on bridge formation between actin and myosin (WINDER and WALSH 1990). In addition, NISHIDA et al. (1990) have reported that the ATPase inhibitory action of calponin is thought to be due not to inhibition of cross-bridge formation but to inhibition of ADP or Pi release, which regulates cross-bridge cycling rate. For this action, troponin would not be required. This would mean that calponin in smooth muscle acts like troponin I and troponin T in skeletal muscle. The actions of calponin are thought to be related to the generation of the "latch" mechanism (WALSH 1990).

Thus, two proteins, caldesmon and calponin, regulate actin-myosin interaction at different stages, namely, cross-bridge formation (corresponds to the steric blocking model for the action of calponin in skeletal muscle; KRESS et al. 1986) and rotation (like the action of calponin in skeletal muscle; kinetic model, CHALOVICH and EISENBERG 1982), respectively. In smooth muscle, these two proteins may act as regulating factors in both processes. However, to understand the crucial roles of these proteins in relation to physiological function (muscle contraction), more detailed investigations are required.

In smooth muscle cells, in place of the myosin phosphorylation model, an actin-regulating model has been introduced following the discovery of leiotonin, a Ca-binding protein which corresponds to troponin C (MIKAWA et al. 1977; EBASHI 1990; EBASHI et al. 1987). Ca binds to leiotonin and regulates actin-myosin interaction through activation of an actin site. After more detailed investigations of leiotonin (leiotonin A; molecular weight 155K prepared from bovine stomach and aorta), a remodelled hypothesis (actin-regulating model through leiotonin action) for smooth muscle contraction has been presented. In this scheme, phosphorylation of MLCK may not directly activate the contractile proteins, but MLCK accelerates the Ca-dependent actin-myosin interaction. The amino acid sequences of leiotonin A have many similarities with the sequences of MLCK, and this protein may be a kind of MLCK (EBASHI et al. 1987). However, supporting evidence for the physiological role of leiotonin A in actin-myosin interaction is still awaited.

IV. Roles of Calcium as Estimated from Mechanical Responses

Although Ca plays an essential role in the production of smooth muscle contraction, it is now clear that the sustained (tonic) contraction following the phasic contraction evoked by agonists or stimulants does not completely correlate with the amount of cytosolic Ca (estimated from the Ca transient using Ca-sensitive dyes). DILLON et al. (1981) observed in the pig carotid artery that, after application of high K stimulation, phosphorylation of MLC, tension and shortening velocity (indicating velocity of actin-myosin interaction) are transiently enhanced in a proportional way, and then the tension is maintained after a marked reduction in the phosphorylation of MLC and the shortening velocity. This phenomenon has been called a "latch" and it is characteristic that, under low Ca or with a slight reduction in Ca, a low cycling rate of cross-bridge interaction between actin-myosin and a low phosphorylation of MLC occurs without a marked reduction in the amplitude of contraction (MURPHY et al. 1983; CHATTERJEE and MURPHY 1983; CHATTERJEE and TEJADA 1986; HAI and MURPHY 1988a,b, 1989). This high Ca-sensitive contraction (sustained contraction) occurred with low concentrations of Ca in the bath and could not be solely explained by Ca-calmodulin MLCK-MLC complex formation in smooth muscle contraction. The sustained contraction may occur in the presence of phosphorylated and dephosphorylated myosin. Recently, HAI and MURPHY (1989) noted that the "latch" phenomenon is a state characterized by moderate phosphorylation levels when high force is associated with reduced contraction rates and ATP consumption. Thus, the underlying mechanism of the "latch" phenomenon is postulated to involve phosphorylation and dephosphorylation of myosin during actin-myosin interaction (HAI and MURPHY 1988a,b, 1989), cooperation between phosphorylated and non-phosphorylated cross-bridges (HIMPENS et al. 1988; HIMPENS and SOMLYO 1988) or interaction with protein kinase C (RASMUSSEN et al. 1987, 1990; PARK and RASMUSSEN 1986; TAKUWA et al. 1988). On the other hand, the K-induced contraction (phasic and tonic components) may roughly be explained by the amount of cytosolic Ca in visceral smooth muscles. Another explanation is that phosphorylation of myosin is not required for the sustained contraction, but other factors, as yet unknown, may exist.

Thus, Ca may play an essential role in producing contraction through the 4 Ca-calmodulin-MLCK-MLC complex, but the contraction evoked by agonists or stimulants may not completely be explained by the amount of Ca in the cytosol, especially the enhancement of the phasic contraction induced by increased Ca sensitivity and the maintenance of the tonic component of contraction.

V. Pharmacology of Ca-Dependent Protein Phosphorylation

In visceral smooth muscle, cyclic nucleotide protein kinases such as A kinase and G kinase, and also C kinase, phosphorylate various cellular proteins and such actions induce a regulation of contractile proteins through the phosphorylation of MLCK or MLC. Recently, various agents which modify these Ca-dependent protein phosphorylations have been reported by Hidaka's group.

HIDAKA et al. (1980), TANAKA and HIDAKA (1980) and HIDAKA and TANAKA (1983) synthesized a calmodulin inhibitor, W7, [*N*-(6-aminohexyl)-5-chloro-1-naphthalene sulphonamide]. Subsequently, more specific inhibitors of individual protein kinases were synthesized such as the MLCK inhibitor ML-9 [1-(5-chloronaphthalenesulphonyl)-1*H*-hexahydro-1,4-diazepine] (K_i value of 3.8 m*M* and 4.1 m*M* after trypsin treatment). ML-9 is supposed to directly inhibit the catalytic action of MLCK since it inhibited both intact and trypsin-treated MLCK (SAITOH et al. 1987). The C kinase inhibitor, H-7 1-(5-isoquinolinesulphonyl)-2-methylpiperazine (HIDAKA et al. 1984), and an activator of C kinase, SC-9 (*N*-(6-phenylhexyl)-5-chloro-1-naphthalene-sulphonamide), have also been described (ITOH et al. 1986; HIDAKA and TANAKA 1985). H-9 can also be used as a affinity ligand for the purification of C kinase (INAGAKI et al. 1985). From H-8 (*N*-[2-(methylamino)ethyl]-5-isoquinoline sulphonamide), a W-7 derivative and an inhibitor of A kinase, a more selective A kinase inhibitor, H-89, [*N*-[*a*-(bromocinnamyl-amino)ethyl]-5-isoquinolinesulphonamide), was introduced (CHIJIWA et al. 1990). Recently, a calmodulin kinase II specific inhibitor, KN-62,(1-[*N*,*O*-bis(1,5-isoquinolinesulphonyl)-*N*-methyl-L-tyrosil]-4-phenylpiperazine) was reported (TOKUMITSU et al. 1990).

In addition, trifluoperazine and chlorpromazine are also well known to inhibit MLCK, but recently staurosporine derivatives (TAMAOKI et al. 1986) have also been introduced as C kinase inhibitors. Phorbol esters have been commonly used as accelerators of C kinase (NISHIZUKA 1984; KIKKAWA et al. 1989; SHEARMAN et al. 1989). From these investigations, we can say that ML-9 is selective as an MLCK inhibitor, H-7 is a selective inhibitor of C kinase, H-8 is a non-selective inhibitor for A and G kinase and H-89 is a selective inhibitor for A kinase (K_i value for A kinase 0.048 m*M*, 0.48 m*M* for G kinase and 31.7 m*M* for C kinase; CHIJIWA et al. 1990). The synthesis of such inhibitors or accelerators of individual protein kinases would merit further investigation with respect to the mechanism of second messenger mediated protein phosphorylation in this field.

E. Conclusion

The E-C and P-C coupling processes in visceral smooth muscle are essential as signal transducers between excitation of the sarcolemma and the contractile machinery via the actions of Ca. Some visceral tissues exhibit spontaneous discharges (E-C coupling may play an essential role in triggering contraction), but other tissues do not (here, P-M coupling may play an essential role). Therefore, the contribution of the E-C and P-M coupling mechanisms in triggering contraction differ by tissue, and either one or both mechanisms may contribute in individual tissues to the production of contraction. Furthermore, the distribution and densities of the receptor and ion channels in visceral smooth muscle may differ by region and species. In the SR in visceral smooth muscle, ryanodine-sensitive and IP_3-sensitive Ca-releasing sites are present. This implies that CICR and IP_3-induced Ca-releasing sites may play roles in the release of Ca from the SR. The former may be mainly related to activation of the voltage-dependent Ca channel and partly result from activation of the receptor-operated cation channel (depolarization, with some exceptions such as Ca-dependent Cl current or M current). In the latter, second messenger-induced Ca release may play a role. The contributions of the two factors differ by tissue. So-called increase or decrease in the Ca sensitivity of contractile proteins may occur through unknown steps during agonist-induced events in smooth muscle tissues.

At our present state of knowledge, it is difficult to review in detail the specific features of individual visceral smooth muscles cells. However, we can emphasize the differences in the E-C and P-M coupling mechanisms of visceral smooth muscle compared with those of cardiac and skeletal muscles and this may explain the regional differences in physiological responses to environmental changes which help to maintain homeostasis. Finally, we can say that the large black box preventing understanding of the E-C and P-M coupling mechanisms has been opened, but that many small unsolved problems are left for further investigation.

Acknowledgements. We thank Dr. R.J. Timms for the language editing. This work was supported by grant-in-Aid for Scientific Research from the Ministry of Science, Education and Culture, Japan.

References

Aalkjaer C, Cragoe EJ Jr (1988) Intracellular pH regulation in resting and contracting segments of rat mesenteric resistance vessels. J Physiol (Lond) 402: 391–410

Aarsman AJ, Mynbeek G, van den Bosch G, Rothhut B, Prieur B, Comera C, Jordan L, Russo-Marie F (1987) Lipocortin inhibition of extracellular and intracellular phospholipase A_2 is substrate concentration dependent. FEBS Lett 219:176–180

Abdel-Latif AA (1986) Calcium-mobilizing receptors, polyphospho-inositides, and the generation of second messengers. Pharmacol Rev 38:227–272

Abdel-Latif AA (1989) Calcium-mobilizing receptors, polyphospho-inositides, generation of second messengers and contraction in the mammalian iris smooth muscle: historical perspectives and current status. Life Sci 45:757–786

Adelstein RS, Eisenberg E (1980) Regulation and kinetics of the actin- myosin-ATP interaction. Annu Rev Biochem 49:921–956

Ahn NG, Teller DC, Bienkowski MJ, McMullen BA, Lipkin EW, de Haën C (1988) Sedimentation equilibrium analysis of five lipocortin-related phospholipase A_2 inhibitors from human placenta. J Biol Chem 263:18657–18663

Aickin CC (1987) Investigation of factors affecting the intracellular sodium activity in the smooth muscle of guinea-pig ureter. J Physiol (Lond) 385:483–505

Aickin CC, Brading AF, Burdyga TV (1984) Evidence of sodium-calcium exchange in the guinea-pid ureter. J Physiol (Lond) 347:411–430
Aickin CC, Brading AF, Walmsley D (1987) An investigation of sodium-calcium exchange in the smooth muscle of guinea-pig ureter. J Physiol (Lond) 391: 325–346
Allen SL, Boyle JP, Cortijo J, Foster RW, Morgan GP, Small RC (1986a) Electrical and mechanical effects of BRL 34915 in guinea-pig isolated trachealis. Br J Pharmacol 89:395–405
Allen SL, Foster RW, Morgan GP, Small RC (1986b) The relaxant action of nicorandil in guinea-pig isolated trachealis. Br J Pharmacol 87:117–127
Amédée T, Mironneau C, Mironneau J (1986) Isolation and contractile responses of single pregnant rat myometrial cells in short-term primary culture and the effects of pharmacological and electrical stimuli. Br J Pharmacol 88:873–880
Ashcroft FM (1988) Adenosine 5′-triphosphate-sensitive potassium channels. Annu Rev Neurosci 11:97–118
Ashida T, Blaustein MP (1987a) Regulation of cell calcium and contractility in mammalian arterial smooth muscle: the role of sodium-calcium exchange. J Physiol (Lond) 392:671–635
Ashida T, Blaustein MP (1987b) Control of contractility and the role of Na/Ca exchange in arterial smooth muscle. J Cardiovasc Pharmacol 10 [Suppl 10]: S65–S67
Auguet M, Delaflotte S, Chabrier P-E, Pirotzky E, Clostre F, Braquet P (1988) Endothelin and Ca^{++} agonist Bay K 8644: different vasoconstrictive properties. Biochem Biophys Res Commun 156:186–192
Babu YS, Sack JS, Greenhough TJ, Bugg CE, Means AR, Cook WJ (1985) Three-dimensional structure of calmodulin. Nature 315:37–40
Babu YS, Bugg CE, Cook WJ (1988) Structure of calmodulin refined at 2.2 Å resolution. J Mol Biol 204:191–204
Bader CR, Bertrand D (1984) Effect of changes in intra- and extra-cellular sodium on the inward (anomalous) rectification in salamander photoreceptors. J Physiol (Lond) 347:611–631
Baker PF, McNaughton PA (1977) Selective inhibition of the Ca-dependent Na efflux from intact squid axons by a fall in intracellular pH. J Physiol (Lond) 269:78P–79P
Beam KG, Knudson CM, Powell JA (1986) A lethal mutation in mice eliminates the slow calcium current in skeletal muscle cells. Nature 320:168–170
Bechem M, Schramm M (1987) Calcium-agonists. J Mol Cell Cardiol 19 [Suppl 2]:63–75
Beech DJ, Bolton TB (1989) Properties of the cromakalim-induced potassium conductance in smooth muscle cells isolated from the rabbit portal vein. Br J Pharmacol 98:851–864
Benham CD (1989) ATP-activated channels gate calcium entry in single smooth muscle cells dissociated from rabbit ear artery. J Physiol (Lond) 419:689–701
Benham CD, Bolton TB (1986) Spontaneous transient outward currents in single visceral and vascular smooth muscle cells of the rabbit. J Physiol (Lond) 381: 385–406
Benham CD, Tsien RW (1987) A novel receptor-operated Ca^{2+}-permeable channel activated by ATP in smooth muscle. Nature 328:275–278
Benham CD, Tsien RW (1988) Noradrenaline modulation of calcium channels in single smooth muscle cells from rabbit ear artery. J Physiol (Lond) 404:767–784
Benham CD, Bolton TB, Lang RJ, Takewaki T (1985) The mechanism of action of Ba^{2+} and TEA on single Ca^{2+}-activated K^+ channels in arterial and intestinal smooth muscle cell membrane. Pflugers Arch 403:120–127
Benham CD, Bolton TB, Lang RJ, Takewaki T (1986) Calcium-activated potassium channels in single smooth muscle cells of rabbit jejunum and guinea-pig mesenteric artery. J Physiol (Lond) 371:45–67

Benham CD, Bolton TB, Denbigh JS, Lang RJ (1987a) Inward rectification in freshly isolated single smooth muscle cells of the rabbit jejunum. J Physiol (Lond) 383:461–476
Benham CD, Hess P, Tsien RW (1987b) Two types of calcium channels in single smooth muscle cells from rabbit ear artery studied with whole-cell and single-channel recordings. Circ Res 61 [Suppl I]:10–16
Benndorf K, Boldt W, Nillius B (1985) Sodium current in single myocardial mouse cells. Pflugers Arch 404:190–196
Benoit E, Corbier A, Dubois JM (1985) Evidence for two transient sodium currents in the frog node of Ranvier. J Physiol (Lond) 361:339–360
Berridge MJ (1984) Inositol trisphosphate and diacylglycerol as second messengers. Biochem J 220:345–360
Berridge MJ (1988) Inositol trisphosphate-induced membrane potential oscillations in Xenopus oocytes. J Physiol (Lond) 403:589–599
Berridge MJ, Galione A (1988) Cytosolic calcium oscillators. FASEB J 2:3074–3082
Berridge MJ, Irvine RF (1984) Inositol trisphosphate, a novel second messenger in cellular signal transduction. Nature 312:315–321
Berridge MJ, Irvine RF (1989) Inositol phosphates and cell signalling. Nature 341:197–205
Berridge MJ, Cobbold PH, Cuthbertson KSR (1988) Spatial and temporal aspects of cell signalling. Philos Trans R Soc Lond [Biol] 320:325–343
Bevan JA, Brayden JE (1987) Nonadrenergic neural vasodilator mechanisms. Circ Res 60:309–326
Bevan JA, Bevan RD, Duckles SP (1980) Adrenergic regulation of vascular smooth muscle. In: Bohr DF et al. (eds) The cardiovascular system. Vascular smooth muscle. American Physiological Society, Bethesda, pp 515–566. (Handbook of physiology, section 2, vol 2)
Bialojan C, Rüegg JC, Takai A (1988) Effects of okadaic acid on isometric tension and myosin phosphorylation of chemically skinned guinea-pig taenia coli. J Physiol (Lond) 398:81–95
Biel M, Ruth P, Bosse E, Hullin R, Stühmer W, Flockerzi V, Hoffmann F (1990) Primary structure and functional expression of high voltage activated calcium channel from rabbit lung. FEBS Lett 269:409–412
Birnbaumer L (1990) G proteins in signal transduction. Annu Rev Pharmacol Toxicol 30:675–705
Blaustein MP (1977a) Sodium ions, calcium ions, blood pressure regulation and hypertension: a reassessment and a hypothesis. Am J Physiol 232:C165–C173
Blaustein MP (1977b) The role of Na-Ca exchange in the regulation of tone in vascular smooth muscle. In: Casteels R, Godfraind T, Rüegg JC (eds) Excitation-contraction coupling in smooth muscle. Elsevier North Holland, Amsterdam, pp 101–108
Blaustein MP (1987) Calcium and synaptic function. In: Baker PF (ed) (Calcium in drug actions. Springer, Berlin Heidelberg New York, pp 275–304 (Handbook of experimental pharmacology, vol 83)
Blaustein MP (1989) Sodium-calcium exchange in cardiac, smooth, and skeletal muscles: key to control of contractility. In: Hoffman JF, Giebish G, Schults SG (eds) Current topics in membranes and transport, vol 34. Academic, New York, pp 289–330
Bolton TB (1979) Mechanisms of action of transmitters and other substances on smooth muscle. Physiol Rev 59:606–718
Bolton TB, Lim SP (1989) Properties of calcium stores and transient outward currents in single smooth muscle cells of rabbit intestine. J Physiol (Lond) 409:385–401
Bosse E, Regulla S, Biel M, Ruth P, Meyer HE, Flockerzi V, Hofman F (1990) The cDNA and deduced amino acid sequence of the g subunit of the L-type calcium channel from rabbit skeletal muscle. FEBS Lett 267:153–156

Bournaud R, Mallart A (1987) An electrophysiological study of skeletal muscle fibers in the "muscular dysgenesis" mutation of the mouse. Pflugers Arch 409:468–476
Bova S, Goldman WF, Blaustein MP (1989) Na gradient reduction potentiates vasoconstriction-induced [Ca^{2+}] transients in vascular smooth muscle cells: a fura-2/digital imaging study. Biophys J 55(2/2):469a
Brading AF, Lategan TW (1985) Na-Ca exchange in vascular smooth muscle. J Hypertens 3:109–116
Brading AF, Burnett M, Sneddon P (1980) The effect of sodium removal on the contractile responses of the guinea-pig taenia coli to carbachol. J Physiol (Lond) 306:411–429
Brayden JE, Large WA (1986) Electrophysiological analysis of neurogenic vasodilation in isolated lingual artery of the rabbit. Br J Pharmacol 89:163–171
Brown DA, Adams OPR (1980) Muscarinic suppression of a novel voltage-sensitive K current in a vertebrate neuron. Nature 283:673–676
Brown AM, Kunze DL, Yatani A (1986) Dual effects of dihydropyridines on whole cell and unitary calcium currents in single ventricular cells of guinea-pig. J Physiol (Lond) 379:495–514
Bucher B, Bettermann R, Illes P (1987) Plasma concentration and vascular effect of b-endorphine in spontaneously hypertensive and Wistar Kyoto rats. Naunyn Schmiedebergs Arch Pharmacol 335:428–432
Bülbring E, Kuriyama H (1963a) Effect of changes in ionic environment on the action of acetylcholine and adrenaline on the smooth muscle cells of the guinea-pig taenia coli. J Physiol (Lond) 166:59–74
Bülbring E, Kuriyama H (1963b) The effect of adrenaline on the smooth muscle of guinea-pig taenia coli in relation to the degree of stretch. J Physiol (Lond) 169:198–212
Bult H, Boeckxstaens GE, Pelckmans PA, Jordaens FH, van Meercke YM, Herman AG (1990) Nitric oxide as inhibitory non-adrenergic non-cholinergic neurotransmitter. Nature 345:346–347
Burnstock G (1972) Purinergic nerves. Pharmacol Rev 24:509–581
Burnstock G (1980) Cholinergic and purinergic regulation of blood vessels. In: Bohr DF et al. (eds). The cardiovascular system. American Physiological Society, Bethesda, pp 567–612 (Handbook of physiology, sction 2, vol 2)
Burnstock G (1981a) Neurotransmitters and trophic factors in the autonomic nervous system. J Physiol (Lond) 313:1–35
Burnstock G (1981b) Purinergic receptor. Chapman and Hall, London
Burnstock G (1981c) Development of smooth muscle and its innervation. In: Bülbring E, Brading AF, Jones AW, Tomita T (eds) Smooth muscle: an assessment of current knowledge. Arnold, London, pp 431–458
Burnstock G (1982) The co-transmitter hypothesis, with special reference to the storage and release of ATP with noradrenaline and acetylcholine. In: Cucllo AC (ed) Co-transmission. Macmillan, London, pp 151–163
Burnstock G, Costa M (1975) Adrenergic neurons. Chapman and Hall, London
Burnstock G, Holman ME (1961) The transmission of excitation from autonomic nerve to smooth muscle. J Physiol (Lond) 155:115–133
Burnstock G, Kennedy C (1986a) A dual function for adenosine 5′-triphosphate in the regulation of vascular tone. Circ Res 58:319–350
Burnstock G, Kennedy C (1986b) Purinergic receptors in the cardiovascular system. Prog Pharmacol 6:111–1322
Caffrey JM, Josephson IR, Brown AM (1986) Calcium channels of amphibian and mammalian aorta smooth muscle cells. Biophys J 49:1237–1242
Cala PM, Mandel LJ, Murphy E (1986) Volume regulation by Amphiuma red blood cells: cytosolic free Ca and alkali metal-H exchange. Am J Physiol 250:C424–C429

Casteels R, Raeymaekers L, Droogmans G, Wuytack F (1985) Na^{2+}-K^+ ATPase, Na-Ca exchange, and excitation-contraction coupling in smooth muscle. J Cardiovasc Pharmacol 7:S103–S110

Catterall WA, Seagar MJ, Takahashi M (1988) Molecular properties of dihydropyridine-sensitive calcium channels in skeletal muscle. J Biol Chem 263:3535–3538

Chalovich JM, Eisenberg E (1982) Inhibition of actomyosin ATPase activity by troponin-tropomyosin without blocking the binding of myosin to actin. J Biol Chem 257:2432–2437

Chatterjee M, Murphy RA (1983) Calcium-dependent stress maintenance without myosin phosphorylation in skinned smooth muscle. Science 221:464–466

Chatterjee M, Tejada M (1986) Phorbol ester-induced contraction in chemically skinned vascular smooth muscle. Am J Physiol 251:C356–C361

Chen G, Suzuki H (1989) Some electrical properties of the endothelium-dependent hyperpolarization recorded from rat arterial smooth muscle cells. J Physiol (Lond) 410:91–106

Cheung DW (1982) Two components in cellular response of rat tail arteries to nerve stimulation. J Physiol (Lond) 328:461–468

Cheung DW, Fujioka M (1987) Inhibition of the junction potential in the guinea-pig saphenous artery by $ANAPP_3$. Br J Pharmacol 89:3–5

Cheung WY (1970) Cyclic 3,5′-nucleotide phosphodiesterase. Demonstration of an activator. Biochem Biophys Res Commun 38:533–538

Chideckel EW, Frost JL, Mike P, Fedan JS (1987) The effect of ouabain on tension in isolated respiratory tract smooth muscle of humans and other species. Br J Pharmacol 92:609–614

Chijiwa T, Mishima A, Hagiwara M, Sano M, Hayashi K, Inoue T, Naito K, Toshioka T, Hidaka H (1990) Inhibition of forskolin-induced neurite outgrowth and protein phosphorylation by a newly synthesized selective inhibitor of cyclic AMP-dependent protein kinase, N-[2-(p-Bromocinnamylamino)ethyl]-5-isoquinolinesulfonamide (H89), of PC12D pheochromocytoma cells. J Biol Chem 265:5267–5272

Clapp LH, Vivaudou MB, Wlash JV, SInger JJ (1987) Acetylcholine increases voltage-activated Ca^{2+} current in freshly dissociated smooth muscle cells. Proc Natl Acad Sci USA 84:2092–2096

Clapham JC, Wilson C (1987) Anti-spasmogenic and spasmolytic effects of BRL 34915: a comparison with nifedipine and nicorandil. J Auton Pharmacol 7:233–242

Cobbold P, Daly M, Dixon J, Woods N (1989) Repetitive calcium transient in hormone-stimulated cells. Biochem Soc Trans 17:69–12

Colburn JC, Michnoff CH, Hsu L-C, Slaughter CA, Kamm KE, Stull JT (1988) Sites phosphorylated in myosin light chain in contracting smooth muscle. J Biol Chem 263:19166–19173

Coldwell MC, Howlett DR (1987) Specificity of action of the novel antihypertensive agent, BRL 34915, as a potassium channel activator. Biochem Pharmacol 36: 3663–3669

Coldwell MC, Howlett DR (1988) Potassium efflux enhancement by cromakalim (BRL 34915) in rabbit mesenteric artery: an indirect effect independent of calcium? Biochem Pharmacol 37:4105–4110

Cook NS, Quast U, Manley P (1989) K^+ channel opening does not alone explain the vasodilator activity of pinacidil and its enantiomers. Br J Pharmacol 96: 181

Creed KE (1979) Functional diversity of smooth muscle. Br Med Bull 35:243–247

Creuz CE, Zaks WJ, Hamman HC, Crane S, Martin WH, Gould KL, Oddie KM, Parson SJ (1987) Identification of chromaffin granule-binding proteins. J Biol Chem 262:1860–1868

Cubbeddu L, Barnes E, Weiner N (1975) Release of norepinephrine and dopamine β-hydroxylase by nerve stimulation. IV. An evaluation of a role for cyclic adenosine monophosphate. J Pharmacol Exp Ther 193:105–127
Daniel EE (1985) The use of subcellular membrane fractions in analysis of control of smooth muscle function. Experientia 41:905–913
Davidson FF, Dennis EA, Powell M, Glenney JR Jr (1987) Inhibition of phospholipase A_2 by "lipocortins" and calpactins. An effect of binding to substrate phospholipids. J Biol Chem 262:1698–1705
DeFeo TT, Morgan KG (1985) Calcium-force relationships as detected with aequorin in two different vascular smooth muscles of the ferret. J Physiol (Lond) 369: 269–282
DeMay JG, Vanhoutte PM (1983) Anoxia and endothelium-dependent reactivity of the canine femoral artery. J Physiol (Lond) 335:65–74
DiFrancesco D, Ferroni A, Mazzanti M, Tromba C (1986) Properties of the hyperpolarization-activated current (i_f) in cells isolated from the rabbit sinoatrial node. J Physiol (Lond) 377:61–88
Dillon PF, Aksoy MO, Driska SP, Murphy RA (1981) Myosin phosphorylation and the cross bridge cycle in arterial smooth muscle. Science 211:495–497
DiPolo R, Beaugé L (1982) The effect of pH on Ca^{2+} extrusion mechanisms in dialyzed squid axons. Biochim Biophys Acta 688:237–245
Droogmans G, Declerck I, Casteels R (1987) Effects of adrenergic agonists on Ca^{2+}-channel currents in single vascular smooth muscle cells. Pflugers Arch 409:7–12
Ebashi S (1990) Development of Ca^{2+} concept in smooth muscle. In: Sperelakis N, Wood JD (eds) Frontiers in smooth muscle research. Liss, New York, pp 159–165
Ebashi S, Endo M (1968) Calcium ion and muscle contraction. Prog Biophys Mol Biol 18:123–183
Ebashi S, Mikawa T, Kuwayama H, Suzuki M, Ikemoto H, Ishizaki Y, Koga R (1987) Ca^{2+} regulation in smooth muscle; dissociation of myosin light chain kinase activity from activation of actin-myosin interaction. In: Siegman M, Somlyo AP, Stephens NL (eds) Regulation and contraction of smooth muscle. Liss, New York, pp 109–117
Edman A, Gestrelius S, Grampp W (1987) Current activation by membrane hyperpolarization in the slowly adapting lobster stretch receptor neuron. J Physiol (Lond) 384:671–690
Eggermont JA, Vrolix M, Raeymaekers L, Wuytack F, Casteels R (1988a) Ca^{2+}-transport ATPase of vascular smooth muscle. Circ Res 62:266–278
Eggermont JA, Vrolix M, Wuytack F, Raeymaekers L, Casteels R (1988b) The (Ca^{2+}-Mg^{2+})-ATPase of the plasma membrane and of the endoplasmic reticulum in smooth muscle cells and their regulation. J Cardiovasc Pharmacol 12 [Suppl 5]:S51–S55
Ehrlich BE, Watras J (1988) Inositol 1,4,5-trisphosphate activates a channel from smooth muscle sarcoplasmic reticulum. Nature 336:583–586
Ellis JL, Farmer SG (1989) Effects of peptidases on non-adrenergic, non-cholinergic inhibitory responses of tracheal smooth muscle; a comparison with effects on VIP- and PHI-induced relaxation. Br J Pharmacol 96:521–526
Ellis SB, Williams ME, Ways NR, Brenner R, Sharp AH, Leung AT, Campbell, KP, Mckenna E, Koch WJ, Hui A, Dchwartz A, Harpold MM (1988) Sequence and expression of mRNAs encoding the α_1 and α_2 subunits of a DHP-sensitive calcium channel. Science 241:1661–1664
Endo M (1977) Calcium release from the sarcoplasmic reticulum. Physiol Rev 57: 71–108
Endo M (1985) Calcium release from sarcoplasmic reticulum. Curr Top Membr Transp 25:181–230
Endo M, Tanaka M, Ogawa Y (1970) Calcium induced release of calcium from the sarcoplasmic reticulum of skinned muscle fibers. Nature 228:34–36

Endo M, Kitazawa T, Yagi S, Iino M, Kakuta Y (1977) Some properties of chemically skinned smooth muscle fibers. In: Casteels R, Godfraind T, Rüegg JC (eds) Excitation-contraction coupling in smooth muscle. Elsevier, Amsterdam, pp 199–209
Epelbaum J, Tapia-Arancibia L, Besson J, Rotsztejn WH, Korden C (1979) Vasoactive intestinal peptide inhibits release of somatostatin from hypothalamus in vitro. Eur J Pharmacol 58:493–495
Escande D, Thuringer D, Leguern S, Cavero I (1988) The potassium channel opener cromakalim (BRL 34915) activates ATP-dependent K^+ channels in isolated cardiac myocytes. Biochem Biophys Res Commun 154:620–625
Fay FS, Shlevin HH, Granger WC, Taylor SR (1979) Aequorin luminescence during activation of single isolated smooth muscle cells. Nature 280:506–508
Feletou M, Vanhoutte PM (1988) Endothelium-dependent hyperpolarization of canine coronary smooth muscle. Br J Pharmacol 93:515–524
Fleckenstein A (1985) Calcium antagonism in heart and vascular smooth muscle. Med Res Rev 5:395–425
Ford LE, Podolsky RJ (1970) Regenerative calcium release within muscle cells. Science 167:58–59
Fujii K, Ishimatsu T, Kuriyama H (1986) Mechanism of vasodilation induced by α-human atrial natriuretic polypeptide in rabbit and guinea-pig renal arteries. J Physiol (Lond) 377:315–332
Fujiwara T, Itoh T, Kubota Y, Kuriyama H (1988) Actions of a phorbol ester on factors regulating contraction in rabbit mesenteric artery. Circ Res 63:893–902
Fujiwara T, Itoh T, Kubota Y, Kuriyama H (1989) Effects of guanosine nucleotide on skinned smooth muscle tissue of the rabbit mesenteric artery. J Physiol (Lond) 408:535–547
Fukuda K, Kubo T, Maeda A, Akiba I, Bujo H, Nakai J, Mishima M, Higashida H, Neher E, Marty A, Numa S (1989) Selective effector coupling of muscarinic acetylcholine receptor subtypes. Trends Pharmacol Sci [Suppl] 4–10
Fukumitsu T, Hayashi H, Tokuno H, Tomita T (1990) Increase in calcium channel current by β-adrenoceptor agonists in single smooth muscle cells isolated from porcine coronary artery. Br J Pharmacol 100:593–599
Funakoshi T, Heimark RL, Hendrickson LE, McMullen BA, Fujikawa K (1987a) Human placental anticoagulant protein: isolation and characterization. Biochemistry 26:5572–5578
Funakoshi T, Hendrickson LE, McMullen BA, Fujikawa K (1987b) Primary structure of human placental anticoagulant protein. Biochemistry 26:8087–8092
Furchgott RF (1988) Studies on relaxation of rabbit aorta by sodium nitrate: the basis for the proposal that the acid-activatable factor from bovine retractor penis is inorganic nitrate and the endothelium-derived relaxing oxide. In: Vanhoutte PM (ed) Mechanisms of vasodilation. Raven, New York, pp 401–414
Furchgott RF, Zawadzki JV (1980) The obligatory role of endothelial cells in the relaxation of arterial smooth muscle by acetylcholine. Nature 288:373–376
Furchgott RF, Zawadzki JV, Cherry PD (1981) Role of endothelium in vasodilator response to acetylcholine. In: Vanhoutte PM, Leusen I (eds) Vasodilation. Raven, New York, pp 49–66
Furchgott RF, Khan MT, Jothianandan D, Khan AS (1988) Evidence that endothelium derived factor of rabbit aorta is nitric oxide. In: Bevan JA, Majewski H, Maxwell RA, Story DF (eds) Vascular neuroeffector mechanisms. IRL Press, Oxford, pp 77–84
Furness JB, Costa M (1989) Identification of transmitters of functionally defined enteric neurons. In: Woods JD, Schult SG (eds) The gastrointestinal system. American Physiological Society, Bethesda, pp 387–404 (Handbook of physiology, section 6, vol 1)
Fürst FO, Cross RA, DeMey J, Small JV (1986) Caldesmon is an elongated, flexible molecule localized in the actomyosin domains of smooth muscle. EMBO J 5:251–257

Furuichi T, Yoshikawa S, Miyawaki A, Wada K, Maeda N, Mikoshiba K (1989) Primary structure and functional expression of the inositol 1,4,5-trisphosphate-binding protein P400. Nature 342:32–38
Furukawa K, Itoh T, Kajiwara M, Kitamura K, Suzuki H, Ito Y, Kuriyama H (1981) Vasodilating actions of 2-nicotinamidoethyl nitrate on porcine and guinea-pig coronary arteries. J Pharmacol Exp Ther 218:248–259
Gabella G (1981) Structure of smooth muscles. In: Bübring E, Brading AF, Jones AW, Tomita T (eds) Smooth muscle: an assessment of current knowledge. Arnold, London, pp 1–46
Geisow MJ, Fritsche U, Hexham JM, Dash B, Johnson T (1986) A consensus amino-acid sequence repeat in Torpedo and mammalian Ca^{2+}-dependent membrane-binding proteins. Nature 320:636–638
Gibson A, Mirzazadeh S, Hoffs AJ, Moore PK (1990) L-N^G-monomethyl arginine and L-N^G-nitro arginine inhibit non-adrenergic, non-cholinergic relaxation of the mouse anococcygeus muscle. Br J Pharmacol 99:602–606
Gillespie JS (1964) Cholinergic junction potentials in intestinal smooth muscle. In: Bülbring E (ed) Pharmacology of smooth muscle. Pergamon, Oxford, pp 81–86
Gillespie JS, Liu X, Martin WÅ (1989) The effects of L-arginine and N^G-monomethyl L-arginine on the response of the rat anococcygeus muscle to NANC nerve stimulation. Br J Pharmacol 98:1080–1082
Gilman AG (1987) G proteins: transducers of receptor-generated signals. Annu Rev Biochem 56:615–649
Gluecksohn-Waelsch S (1963) Lethal genes and analysis of differentiation. In higher organisms lethal genes serve as tools for studies of cell differentiation and cell genetics. Science 142:1269–1276
Godfraind T, Miller R, Wibo M (1986) Calcium antagonism and calcium entry blockade. Pharmacol Rev 38:321–416
Göthert M, Hentrich F (1984) Role of cAMP for regulation of impulse-evoked noradrenaline release from rabbit pulmonary artery and its possible relationship to presynaptic ACTH receptor. Naunyn Schmiedebergs Arch Pharmacol 328: 127–134
Goto K, Kasuya Y, Matsuki N, Takuwa Y, Kurihara H, Ishikawa T, Kimura S, Yannagisawa M, Masaki T (1989) Endothelin activates the dihydropyridine-sensitive, voltage-dependent Ca^{2+} channel in vascular smooth muscle. Proc Natl Acad Sci USA 86:3915–3918
Greenberg SS, Diecke FFJ, Cantor E, Peevy K, Tanaka TP (1990), Inhibition of sympathetic neurotransmitter release by modulators of cyclic GMP in canine vascular smooth muscle. Eur J Pharmacol 187:409–423
Gurney AM, Nerbonne JM, Lester HA (1985) Photoinduced removal of nifedipine reveals mechanisms of calcium antagonist action on single heart cells. J Gen Physiol 86:353–379
Hagiwara S, Byerly L (1983) The calcium channel. Trend Neurosci 6:189–193
Hai C-M, Murphy RA (1988a) Cross-bridge phosphorylation and regulation of latch state in smooth muscle. Am J Physiol 254:C99–C106
Hai C-M, Murphy RA (1988b) Regulation of shortening velocity by crossbridge phosphorylation in smooth muscle. Am J Physiol 255:C86–C94
Hai C-M, Murphy RA (1989) Ca^{2+}, crossbridge phosphorylation, and contraction. Annu Rev Physiol 51:285–298
Hakoda H, Ito Y (1990) Modulation of cholinergic neurotransmission by the peptide VIP, VIP antiserum and VIP antagonists in dog and cat trachea. J Physiol (Lond) 428:133–154
Hamilton TC, Weir SW, Weston AH (1986) Comparison of the effects of BRL 34915 and verapamil on electrical and mechanical activity in rat portal vein. Br J Pharmacol 88:103–111
Hamilton TC, Weston AH (1989) Cromakalim, nicorandil and pinacidil: novel drugs which open potassium channels in smooth muscles. Gen Pharmacol 20:1–9

Hara Y, Kitamura K, Kuriyama H (1980) Actions of 4-aminopyridine on vascular smooth muscle tissues of the guinea-pig. Br J Pharmacol 68:99–106
Hartshorne DJ (1987) Biochemistry of the contractile process in smooth muscle. In: Johnson LJ (ed) Physiology of the gastrointestinal tract. Raven, New York, pp 423–482
Hartshorne DJ, Gorecka A (1980) Biochemistry of the contractile proteins of smooth muscle. In: Bohr DF et al. (eds) The cardiovascular system. Vascular smooth muscle. American Physiological Society, Bethesda, pp 93–120 (Handbook of physiology, section 2, vol 2)
Hashimoto T, Hirata M, Itoh T, Kanmura Y, Kuriyama H (1986) Inositol 1,4,5-trisphosphate activates pharmacomechanical coupling in smooth muscle of the rabbit mesenteric artery. J Physiol (Lond) 370:605–618
Hedqvist P, Fredholm BB, Qlundh S (1978) Antagonistic effects of theophylline and adenosine on adrenergic neuroeffector transmission in the rabbit kidney. Circ Res 43:592–598
Hemric ME, Chalovich J (1988) Effect of caldesmon on the ATPase activity and the binding of smooth and skeletal myosin subfragments to actin. J Biol Chem 263:1878–1885
Hering S, Beech DJ, Bolton TB (1987) Voltage dependence of the actions of nifedipine and Bay K 8644 on barium currents recorded from single smooth muscle ceffs from rabbit ear artery. Biomed Biochem Acta 467:S657–S661
Hering S, Kleppesh T, Timin EN, Boclewei R (1989) Characterization of the calcium channel state transitions induced by the enantiomers of the 1,4-dihydropyridine Sandoz 202 791 in neonatal rat heart cell. A nonmodulated receptor model. Pflugers Arch 414:690–700
Hermesmyer RK (1988) Pinacidil actions on ion channels in vascular smooth muscle. J Cardiovasc Pharmacol 12 [Suppl 2]:517–522
Hess P, Lansman JB, Tsien RW (1984) Different modes of Ca channel gating behaviour favoured by dihydropyridine Ca antagonists and antagonists. Nature 311:538–544
Hidaka H, Tanaka T (1983) Naphthalenesulfonamide as calmodulin antagonists. In: Means AR, O'Malley BW (eds) Methods in enzymology, vol 102. Academic, New York, pp 185–194
Hidaka H, Tanaka T (1985) Modulation of Ca^{2+}-dependent regulatory systems by calmodulin antagonists and other agents. In: Hidaka H, Hartshorne DJ (eds) Calmodulin antagonists and cellular physiology. Academic, New York, pp 13–22
Hidaka H, Yamaki T, Naka M, Tanaka T, Hayashi H, Kobayashi R (1980) Calcium-regulated modulator protein interacting agents inhibit smooth muscle calcium-stimulated protein kinase and ATPase. Mol Pharmacol 17:66–72
Hidaka H, Inagaki M, Kawamoto S, Sasaki Y (1984) Isoquinolinesulfonamides, novel and potent inhibitors of cyclic nucleotide dependent protein kinase and protein kinase C. Biochemistry 23:5036–5041
Hille B (1977) Local anesthetics: hydrophilic and hydrophobic pathways for the drug-receptor reaction. J Gen Physiol 69:497–515
Himpens B, Casteels R (1990) Different effects of depolarization and muscarinic stimulation on the Ca^{2+}/force relationship during the contraction, relaxation cycle in the guinea-pig ileum. Pflugers Arch 416:28–35
Himpens B, Somlyo AP (1988) Free-calcium and force transient during depolarization and pharmacomechanical coupling in guinea-pig smooth muscle. J Physiol (Lond) 395:507–530
Himpens B, Matthijs G, Somlyo AV, Butler T, Somlyo AP (1988) Cytosplasmic free calcium myosin light chain phosphorylation and force in phasic and tonic smooth muscle. J Gen Physiol 92:713–729
Himpens B, Matthijs G, Somlyo AP (1989) Desensitization to cytoplasmic Ca^{2+} and Ca^{2+} sensitivities of guinea-pig ileum and rabbit pulmonary artery smooth muscle. J Physiol (Lond) 413:489–503

Hirata F, Matsuda K, Notsu Y, Hattori T, del Carmine R (1984) Phosphorylation at a tyrosine residue of lipomodulin in mitogen-stimulated murine thymocytes. Proc Natl Acad Sci USA 81:4717–4721
Hirata M, Itoh T, Kuriyama H (1981) Effects of external cations on calcium efflux from single cells of the guinea-pig taenia coli and porcine coronary artery. J Physiol (Lond) 310:321–336
Hirata Y, Yoshimi H, Takata S, Watanabe TX, Kumagai S, Nakajima K, Sakakibara S (1988) Cellular mechanism of action by a novel vasoconstrictor endothelin in cultured rat vascular smooth muscle cells. Biochem Biophys Res Commun 154:868–875
Hirst GDS, Edwards FR (1989) Sympathetic neuroeffector transmission in arteries and arterioles. Physiol Rev 69:546–604
Hirst GDS, Neild TO (1980) Evidence for two population of excitatory receptors for noradrenaline on arteriolar smooth muscle. Nature 283:767–768
Hirst GDS, Neild TO (1981) Localization of specialized noradrenaline receptors at neuromuscular junctions on arterioles of the guinea-pig. J Physiol (Lond) 313:343–350
Hirst GDS, Neild TO, Silverberg GD (1982) Noradrenaline receptors on the rat basilar artery. J Physiol (Lond) 328:351–360
Hisada T, Ordway RW, Kirber MT, Singer JJ, Walsh JV jr (1991) Hyperpolarization-activated cationic channels in smooth muscle cells are stretch sensitive. Pflugers Arch 417:493–499
Holman ME (1958) Membrane potentials recorded with high-resistance micro-electrodes and the effects of changes in ionic environment on the electrical and mechanical activity of the smooth muscle of the taenia coli of guinea-pig. J Physiol (Lond) 141:464–488
Holman ME (1970) Junction potentials in smooth muscle. In: Bülbring E, Brading AF, Jones AW, Tomita T (eds) Smooth muscle. Arnold, London, pp 244–288
Holman ME (1981) The intrinsic innervation and peristaltic reflex of the small intestine. In: Bülbring E, Brading AF, Jones AW, Tomita T (eds) Smooth muscle. Arnold, London, pp 311–338
Hondegheim LM, Katzung BG (1977) A unifying molecular model for the interaction of antiarrhythmic drugs with cardiac sodium channels: application to guanidine and lidocaine. Proc West Pharmacol Soc 20:253–256
Hotta Y, Tsukui R (1968) Effects on the guinea-pig taenia coli of the substitution of strontium or barium ions for calcium ions. Nature 217:867–869
Huddart H, Saad KHM (1978) The effect of sodium and magnesium and their interaction with quinine and lanthanum on spontaneous activity and related calcium movements of rat ileal smooth muscle. J Comp Physiol 126:233–240
Hume JR, Leblane N (1989) Macroscopic K^+ currents in single smooth muscle cells of the rabbit portal vein. J Physiol (Lond) 413:49–73
Hwang KS, van Breemen C (1987) Ryanodine modulation of ^{45}Ca efflux and tension in rabbit aortic smooth muscle. Pflugers Arch 408:343–350
Ignarro LJ (1990) Biological actions and properties of endothelium-derived nitric oxide formed and released from artery and vein. Circ Res 65:1–21
Ignarro LJ, Buga GM, Wood KS, Byrns RE, Chaudhuri G (1987a) Endothelium-derived relaxing factor produced and released from artery and vein is nitric oxide. Proc Natl Acad Sci USA 84:9265–9269
Ignarro LJ, Byrnes RE, Buga GM, Wood KS (1987b) Endothelium-derived relaxing factor from pulmonary artery and vein possesses pharmacologic and chemical properties identical to those of nitric oxide radical. Circ Res 61:866–879
Iino M (1981) Tension responses of chemically skinned fibre bundles of the guinea-pig taenia caeci under varied ionic environments. J Physiol (Lond) 320:449–467
Iino M (1989) Calcium-induced calcium release mechanism in the guinea-pig taenia caeci. J Gen Physiol 94:363–383

Iino M, Kobayashi T, Endo M (1988) Use of ryanodine for functional removal of the calcium store in smooth muscle cells of the guinea-pig. Biochem Biophys Res Commun 152:417–422
Ikebe M, Hartshorne DJ (1985) Phosphorylation of smooth muscle myosin at two distinct sites by myosin light chain kinase. J Biol Chem 260:10027–10031
Ikebe M, Hartshorne DJ (1986) Proteolysis and actin-binding properties of 10S and 6S smooth muscle myosin: identification of a site protected from proteolysis in the 10S conformation and by the binding of actin. Biochemistry 25:6177–6185
Ikebe M, Reardon S (1988) Binding of caldesmon to smooth muscle myosin. J Biol Chem 263:3055–3058
Ikebe M, Hartshorne DJ, Elzinga M (1986) Identification, phosphorylation, and dephosphorylation of a second site for myosin light chain kinase on the 20000-dalton light chain of smooth muscle myosin. J Biol Chem 261:36–39
Ikebe M, Hartshorne DJ, Elzinga M (1987) Phosphorylation of the 20,000-dalton light chain of smooth muscle myosin by the calcium-activated, phospholipid dependent protein kinase. J Biol Chem 262:9569–9573
Illes P (1989) Modulation of transmitter and hormone release by multiple neuronal opioid receptors. Rev Physiol Biochem Pharmacol 112:141–233
Illes P, Bettermann R (1986) Classification of presynaptic opioid receptors in the rabbit ear artery by competitive antagonists. Eur J Pharmacol 122:153–156
Illes P, Ramme D, Starke K (1986) Presynaptic opioid δ-receptors in the rabbit mesenteric artery. J Physiol (Lond) 379:217–228
Imai S, Kitagawa T (1981) A comparison of the differential effects of nitroglycerin, nifedipine and papaverine on contractures induced in vascular and intestinal smooth muscle by potassium and lanthanum. Jpn J Pharmacol 31:193–199
Inagaki M, Watanabe M, Hidaka H (1985) N-(2-Aminoethyl)-5-isoquinoline-sulfonamide, a newly synthesized protein kinase inhibitor, functions as a ligand in affinity chromatography. J Biol Chem 260:2922–2925
Inomata H, Kao CY (1976) Ionic currents in an intestinal smooth muscle. In: Bülbring E, Shuba MF (eds) Physiology of smooth muscle. Raven, New York, pp 49–52
Inoue R, Isenberg G (1990a) Effect of membrane potential on ACh-induced inward current in guinea-pig ileum. J Physiol (Lond) 424:457–71
Inoue R, Isenberg G (1990b) Intracellular Ca ions modulate ACh-induced inward current in guinea-pig ileum. J Physiol (Lond) 424:73–92
Inoue R, Kitamura K, Kuriyama H (1987) ACh activates single sodium channels in smooth muscle cells. Pflugers Arch 410:69–74
Inoue T, Kanmura Y, Fujisawa K, Itoh T, Kuriyama H (1984) Effects of 2-nicotinamidoethyl nitrate (nicorandil; SG-75) and its derivatives on smooth muscle cells of the canine mesenteric artery. J Pharmacol Exp Ther 229: 793–802
Inoue Y, Xiong ZL, Kitamura K, Kuriyama H (1989) Modulation produced by nifedipine of the unitary Ba current of dispersed smooth muscle cells of the rabbit ileum. Pflugers Arch 414:534–542
Inoue Y, Oike M, Nakao K, Kitamura K, Kuriyama H (1990) Endothelin augments unitary Ca channel currents on the smoth muscle cell membrane of guinea-pig portal vein. J Physiol (Lond) 423:171–191
Irvine RF (1989) How do inositol 1,4,5-trisphosphate and inositol 1,3,4,5-tetrakisphosphate regulate intracellular Ca^{2+}? Biochem Soc Transact 17:6–9
Irvine RF, Letcher AJ, Heslop JP, Berridge MJ (1986) The inositol tris/tetrakisphosphate pathway-demonstration of Ins(1,4,5)P_3 3-kinase activity in animal tissues. Nature 320:631–634
Ishikawa S (1985) Actions of AAYP and α,β-methylene ATP in neuromuscular transmission and smooth muscle membrane of the rabbit and guinea-pig mesenteric arteries. Br J Pharmacol 86:777–787

Ishikawa S, Sperelakis N (1987) A novel class (H_3) of histamine receptors on perivascular nerve terminals. Nature 327:158–160
Ito K, Takamura S, Sato K, Sutko JL (1986) Ryanodine transients evoked by electrical stimulation of smooth muscle from guinea-pig ileum recorded by the use of Fura-2. J Physiol (Lond) 407:117–134
Ito M, Hartshorne DJ (1990) Phosphorylation of myosin as a regulatory mechanism in smooth muscle. In: Sperelakis N, Wood JD (eds) Frontiers in smooth muscle research. Liss, New York, pp 57–72
Ito S, Kajikuri J, Itoh T, Kuriyama H (1991) Effects of lemakalim on changes in Ca^{2+} concentration and mechanical activity induced by noradrenaline in the rabbit mesenteric artery. Br J Pharmacol 104:227–233
Ito Y, Tajima K (1981) Actions of indomethacin and prostaglandins on neuro-effector transmission in the dog trachea. J Physiol (Lond) 319:379–392
Ito Y, Takeda K (1982) Non-adrenergic inhibitory nerves and putative transmitters in the smooth muscle of cat trachea. J Physiol (Lond) 330:497–511
Ito Y, Kuriyama H, Sakamoto Y (1970) Effects of tetraethylammonium chloride on the membrane activity of guinea-pig stomach smooth muscle. J Physiol (Lond) 211:445–460
Ito Y, Suzuki H, Kuriyama H (1978) Effects of sodium nitroprusside on smooth muscle cells of rabbit pulmonary and portal vein. J Pharmacol Exp Ther 207: 1022–1031
Itoh M, Tanaka T, Inagaki M, Nakanishi K, Hidaka H (1986) N-(6-Phenylhexyl)-5-chloro-1-naphthalenesulfonamide, a novel activator of protein kinase C. Biochemistry 25:4179–4184
Itoh T, Furukawa K, Kajiwara M, Kitamura K, Suzuki H, Ito Y, Kuriyama H (1981a) Effects of 2-nicotinamidoethyl nitrate on smooth muscle cells and on adrenergic transmission in the guinea-pig and porcine mesenteric arteries. J Pharmacol Exp Ther 218:260–270
Itoh T, Kuriyama H, Suzuki H (1981b) Excitation-contraction coupling in smooth muscle cells of the guinea-pig mesenteric artery. J Physiol (Lond) 321:515–535
Itoh T, Izumi H, Kuriyama H (1982a) Mechanisms of relaxation induced by activation of β-adrenoceptors in smooth muscle cells of the guinea-pig mesenteric artery. J Physiol (Lond) 326:475–493
Itoh T, Kajiwara M, Kitamura K, Kuriyama H (1982b) Roles of stored calcium on the mechanical response evoked in smooth muscle cells of the porcine coronary artery. J Physiol (Lond) 322:107–125
Itoh T, Kuriyama H, Suzuki H (1983) Differences and similarities in the noradrenaline- and caffeine-induced mechanical responses in the rabbit mesenteric artery. J Physiol (Lond) 337:609–629
Itoh T, Kanmura Y, Kuriyama H (1986a) Inorganic phosphate regulates the contraction-relaxation cycle in skinned muscles of the rabbit mesenteric artery. J Physiol (Lond) 376:231–252
Itoh T, Kanmura Y, Kuriyama H, Sumimoto K (1986b) A phorbol ester has dual actions on the mechanical response in the rabbit mesenteric and porcine coronary arteries. J Physiol (Lond) 375:515–534
Itoh T, Kubota Y, Kuriyama H (1988) Effects of a phorbol ester on acetylcholine-induced Ca^{2+} mobilization and contraction in the porcine coronary artery. J Physiol (Lond) 397:401–419
Itoh T, Seki N, Kajikuri J, Kuriyama H (1990) Effects of pinacidil on electrical and mechanical activities in the rabbit mesenteric artery. Eur J Pharmacol 183:1739.
Itoh T, Suzuki S, Kuriyama H (1991) Effects of pinacidil on contractile proteins in high K^+-treated intact, and in β-escin-treated skinned smooth muscle of the rabbit mesenteric artery. Br J Pharmacol 103:1697–1702
Iwasaki A, Suda M, Nakao H, Nagaya T, Saino Y, Arai K, Mizoguchi T, Sato F, Yoshizaki H, Hirata M, Miyahara T, Shidara Y, Murata M, Maki M (1987) Structure and expression of cDNA for an inhibitor of blood coagulation isolated from human placenta: a new lipocortin-like protein. J Biochem 102:1261–1273

Jay SD, Ellis SB, McCue AF, Williams ME, Vedvick TS, Harpold MM, Campbell KP (1990) Primary structure of the γ subunit of the DHP-sensitive calcium channel from skeletal muscle. Science 248:490–292
Kahn AM, Allen JC, Cragoe EJ, Jr, Zimmer R, Shelat H (1988a) Sodium-lithium exchange in sarcolemmal vesicles from canine superior mesenteric artery. Circ Res 632:478–485
Kahn AM, Allen JC, Shelat H (1988b) Na-Ca exchange in sarcolemmal vesicles from bovine superior mesenteric artery. Am J Physiol 254:C441–C449
Kai H, Kanaide H, Nakamura M (1989) Endothelin-sensitive intracellular Ca^{2+} store overlaps with caffeine-sensitive one in rat aortic smooth muscle cells in primary culture. Biochem Biophys Res Commun 158:235–243
Kajioka S, Oike M, Kitamura K (1990) Nicorandil opens a Ca-dependent potassium channel in smooth muscle cells of the rat portal vein. J Pharmacol Exp Ther 254:905–913
Kajiwara M, Droogmans G, Casteels R (1984) Effects of 2-nicotinamidoethyl nitrate (Nicorandil) on excitation-contraction coupling in the smooth muscle cells of rabbit ear artery. J Pharmacol Exp Ther 230:462–468
Kakiuchi S, Yamazaki R (1970) Stimulation of the activity of cyclic 3′,5′-nucleotide phosphodiesterase by calcium ion. Proc Jpn Acad 46:387–392
Kameyama M, Hesheler J, Hofmann F, Trautwein W (1986) Modulation of Ca current during the phosphorylation cycle in the guinea pig heart. Pflugers Arch 407:123–128
Kamm KE, Stull JT (1985) The function of myosin and myosin light chain kinase phosphorylation in smooth muscle. Annu Rev Pharmacol Toxicol 25:593–620
Kamm KE, Stull JT (1989) Regulation of smooth muscle contractile elements by second messengers. Annu Rev Physiol 51:299–313
Karaki H (1989) Ca^{2+} localization and sensitivity in vascular smooth muscle. Trends Pharmacol Sci 10:320–325
Karashima T, Itoh T, Kuriyama H (1982) Effects of 2-nicotinamiedoethyl nitrate on smooth muscle cells of the guinea-pig mesenteric and portal veins. J Pharmacol Exp Ther 221:472–480
Kass KS (1987) Voltage-dependent modulation of cardiac calcium channel current by optical isomers of Bay K 8644: implications for channel gating. Circ Res 61:1–5
Kass KS, Krafte DS (1987) Negative surface charge densitiy near heart calcium channels. Relevance to block by dihydropyridines. J Gen Physiol 89:629–644
Katsuragi T, Ozawa H (1978) Ouabain-induced potentiation of Ca^{2+}-contraction in the depolarized vas deferens of guinea-pig. Arch Int Pharmacodyn Ther 231: 243–248
Kawasaki H, Takasaki K, Saito A, Goto K (1988) Calcitonin gene-related peptide acts as a novel vasodilator neurotransmitter in mesenteric vessels of the rat. Nature 335:164–167
Kawasaki H, Nuki C, Takasaki K (1990) Nonadrenergic, noncholinergic vasodilator innervation in the resistance blood vessel and its role in control of the vascular tone. Jpn J Phrmacol 52 [Suppl I]:56
Kawashima Y, Ochi R (1987) Two types of calcium channels in isolated vascular smooth muscles. J Physiol Soc Japan 49:369p
Kikkawa V, Kishimoto A, Nishizuka Y (1989) The protein kinase C family: heterogeneity and its implications. Annu Rev Biochem 58:31–44.
Kitamura K, Itoh T, Ueno H, Kanmura Y, Inoue R, Kuriyama H (1986) The stabilization of vascular smooth muscle by procaine. Drugs Exp Clin Res XII:773–784
Kitazawa T, Kobayashi S, Horiuchi K, Somlyo AV, Somlyo AP (1989) Receptor coupled, permeabilized smooth muscle: role of the phosphatidyl inositol cascade, G-proteins and modulation of the contractile response to Ca^{2+}. J Biol Chem 264:5339–5342
Klöckner U, Isenberg G (1985) Calcium activated potassium currents as an indicator for intracellular (i.c.) Ca-transients. Pflugers Arch 405:R61

Kobayashi M, Kondo S, Yasumoto T, Ohizumi Y (1986) Cardiotoxic effects of matitotoxin, a principal toxin of seafood poisoning, on guinea-pig and rat cardiac muscle. J Pharmacol Exp Ther 238:1077–1083
Kobayashi S, Somlyo AV, Somlyo AP (1988) Heparin inhibits the inositol 1,4,5-trisphosphate-dependent, but not the independent, calcium release induced by guanine nucleotide in vascular smooth muscle. Biochem Biophys Res Commun 153:625–631
Kobayashi T, Kitazawa T, Somlyo AV, Kitazawa AP (1989) Cytosolic heparin inhibits muscarinic and α-adrenergic Ca^{2+}-release in smooth muscle. J Biol Chem 264:1799–18004
Komori K, Suzuki H (1987a) Electrical responses of smooth muscle cells during cholinergic vasodilation in the rabbit sphenous artery. Circ Res 61:586–593
Komori K, Suzuki H (1987b) Heterogenous distribution of muscarinic receptors in the rabbit saphenous artery. Br J Pharmacol 92:657–664
Komori S, Bolton T (1990) Role of G-proteins in muscarinic receptor inward and outward currents in rabbit jejunal smooth muscle. J Physiol (Lond) 427:395–419
Komori S, Bolton T (1991) Calcium release induced by inositol 1,4,5-trisphosphate in single rabbit intestinal smooth muscle cells. J Physiol (Lond) 433:495–517
Kormbacher C, Helbig H, Stall F, Wiederholt M (1988) Evidence for Na/H exchange and Cl/HCO_3 exchange in A10 vascular smooth muscle cells. Pflugers Arch 412:29–36
Kress M, Huxley HE, Farugi AR (1986) Structural changes during activation of frog muscle: studies by time-resolved x-ray diffraction. J Mol Biol 188:325–342
Kretsinger RH, Barry CD (1975) The predicted structure of the calcium-binding component of troponin. Biochim Biophys Acta 405:40–52
Kreye VAW, Gerstheimer F, Weston AH (1987) Effects of the antihypertensive, BRL 34915, on membrane potential and ^{86}Rb efflux in rabbit tonic vascular smooth muscle. Pflugers Arch 408:R79
Kume H, Takai A, Tokuno H, Tomita T (1989) Regulation of Ca^{2+}-dependent K^+-channel activity in tracheal myocytes by phosphorylation. Nature 341:152–154
Kurachi Y, Ito H, Sugimoto T, Shimizu T, Miki I, Ui M (1989) Arachidonic acid metabolites as intracellular modulators of the G protein cardiac K^+ channel. Nature 337:555–557
Kuriyama H (1963a) The influence of potassium, sodium and chloride on the membrane potential of the smooth muscle of taenia coli. J Physiol (Lond) 166: 15–28
Kuriyama H (1963b) Electrophysiological observations on the motor innervation of the smooth muscle cells in the guinea-pig vas deferens. J Physiol (Lond) 169: 213–228
Kuriyama H (1970) Effects of ions and drugs on the electrical activity of smooth muscle. In: Bülbring E, Brading AF, Jones AW, Tomita T (eds) Smooth muscle. Arnold, London, pp 366–395
Kuriyama H (1981) Excitation-contraction coupling in various visceral smooth muscles. In: Bülbring E, Brading AF, Jones AW, Tomita T (eds) Smooth muscle, an assessment of current knowledge. Arnold, London
Kuriyama H, Makita Y (1982) Modulation of neuromuscular transmission by endogenous and exogenous prostaglandins in the guinea-pig mesenteric artery. J Physiol (Lond) 327:431–48
Kuriyama H, Makita Y (1984) The presynaptic regulation of noradrenaline release differs in mesenteric arteries of the rabbit and guinea-pig. J Physiol (Lond) 351:379–396
Kuriyama H, Ito Y, Suzuki H, Kitamura K, Itoh T (1982) Factors modifying contraction-relaxation cycle in vascular smooth muscles. Am J Physiol 243: H641–H662
Kusano K, Barros F, Katz G, Garcia M, Kaczorowski G, Reuben JP (1987) Modulation of K channel activity in aortic smooth muscle by BRL 34915 and a scorpion toxin. Biophys J 51:55a

Lai FA, Erickson HP, Rousseau E, Liu QY, Meissner G (1988) Purification and reconstitution of the calcium release channel from skeletal muscle. Nature 331:315–319
Lammel E, Deitmer P, Noack T (1991) Suppression of steady membrane currents by acetylcholine in single smooth muscle cells of the guinea-pig gastric fundus. J Physiol (Lond) 432:259–282
Langer SZ (1974) Presynaptic regulation of catecholamine release. Biochem Pharmacol 23:1793–1800
Langer SZ, Lehmann J (1988) Presynaptic receptors on catecholamine neurones. In: Trendeleuburg U, Weiner N (eds) Catecholamines I. Springer, Berlin Heidelberg New York, pp 419–507 (Handbook of experimental pharmacology, vol 85/1)
Lash JA, Sellers JR, Hathaway DR (1986) The effects of caldesmon on smooth muscle heavy actomeromyosin ATPase activity and binding of heavy meromyosin to actin. J Biol Chem 261:16155–16160
Latinen A, Partanen M, Hervonen A, Pelto-Huikko M, Latinen LA (1985) VIP like immunoreactive nerves in human respiratory tract. Histochemistry 82:313–319
Levitan IB (1988) Modulation of ion channels in neurons and other cells. Annu Rev Neurosci 11:119–136
Loirand G, Pacaud P, Baron A, Mironneau C, Mironneau J (1991) Large conductance calcium-activated non-selective cation channel in smooth muscle cells isolated from rat portal vein. J Physiol (Lond) 437:461–475
Lundberg JM, Hedlund B, Bartfai T (1982) Vasoactive intestinal polypeptide enhances muscarinic ligand binding in cat submandibular salivary gland. Nature 295:147–149
Lynch WP, Riseman VM, Bretscher A (1987) Smooth muscle caldesmon is an extended flexible monomeric protein in solution that can readily undergo reversible intra- and intermolecular sulfhydryl cross-linking. J Biol Chem 262: 7429–7437
Makita Y (1983) Effects of prostaglandin I_2 and carbo cyclic thromboxane A_2 on smooth muscle cells and neuromuscular transmission in the guinea-pig mesenteric artery. Br J Pharmacol 78:517–527
Marsden PA, Danthuluri NR, Brenner BM, Ballermann BJ, Brock TA (1989) Endothelin action on vascular smooth muscle involves inositol trisphosphate and calcium mobilization. Biochem Biophys Res Commun 158:86–93
Marston SB, Smith CWJ (1984) Purification and properties of Ca^{2+}-regulated thin filaments and F-actin from sheep aorta smooth muscle. J Muscle Res Cell Motil 5:559–575
Marston SB, Lehman W, Moody C, Pritchard K, Smith CWJ (1988) Caldesmon and Ca^{2+} regulation in smooth muscles. In: Gerday C, Gilles R, Bolis L (eds) Calcium binding proteins. Springer, Berlin Heidelberg New York, pp 69–81
Masahashi T, Tomita T (1983) The contracture produced by sodium removal in the non-pregnant rat myometrium. J Physiol (Lond) 334:351–363
Masaki T (1989) The discovery, the present state, and the future prospects of endothelin. J Cardiovasc Pharmacol 13 [Suppl 5]:S1–S4
Matlib MA (1988) Functional characteristics of a Na^+-Ca^{2+} exchange system in sarcolemmal membrane vesicles of vascular smooth muscle. Biophys J 53:346a
Matlib MA, Schwartz A, Yamori Y (1987) Solubilization and reconstitution of the sarcolemmal Na^+-Ca^{2+} exchange system of vascular smooth muscle. Biochim Biophys Acta 904:145–148
Matsuda JJ, Volk KA, Shibata EF (1990) Calcium currents in isolated rabbit coronary arterial smooth muscle myocytes. J Physiol (Lond) 427:657–680
Matsuzaka K, Masahashi T, Kihira M, Tomita (1987) Contracture caused by sodium removal in the pregnant rat myometrium. Jpn J Physiol 37:19–31
Matsuzaki Y, Hamasaki Y, Said SI (1980) Vasoactive intestinal peptide: a possible transmitter of nonadrenergic relaxation of guinea-pig trachea. Science 210: 1252–1253

Meissner G (1986) Ryanodine activation and inhibition of the Ca^{2+} release channel of sarcoplasmic reticulum. J Biol Chem 261:6300–6306
Miashiro T, Yamamoto H, Kanaide H, Nakamura M (1988) Does endothelin mobilize calcium from intracellular store sites in rat aortic vascular smooth muscle cells in primary culture. Biochem Biophys Res Commun 156:312–317
Mignery GA, Südhof TC, Takei K, De Camilli P (1989) Putative receptor for inositol 1,4,5-trisphosphate similar to ryanodine receptor. Nature 342:192–195
Mikami A, Imoto K, Tanabe T, Niidome T, Mori Y, Takeshima H, Narumiya S, Numa S (1989) Primary structure and functional expression of the cardiac dihydropyridine-sensitive calcium channel. Nature 340:230–233
Mikawa T, Nonomura Y, Ebashi S (1977) Does phosphorylation of myosin light chain have direct relation to regulation in smooth muscle? J Biochem 82: 1789–1791
Mishima S, Miyahara H, Suzuki H (1984) Transmitter release modulated by α-adrenoceptor antagonists in the rabbit mesenteric artery. Br J Pharmacol 83:537–547
Miyahara H, Suzuki H (1987) Pre- and post-junctional effects of adenosine triphosphate on noradrenergic transmission in the rabbit ear artery. J Physiol (Lond) 389:423–440
Moncada S, Vane JR (1979) Pharmacology and endogenous roles of prostaglandin endoperoxides, thromboxane A_2, and prostacyclin. Pharmacol Rev 30:293–331
Moncada S, Gryglewski R, Bunting S, Vane JR (1977) An enzyme isolated from arteries transformed prostaglandin endoperoxides to an unstable substance that inhibits platelet aggregation. Nature 263:663–665
Moore PB (1986) 67KDa calcimedin a new Ca^{2+} binding protein. Biochem J 238: 49–54
Moore PB, Dedman JR (1982) Calcium-dependent protein binding to phenothiazine columns. J Biol Chem 257:9663–9667
Morel N, Godfraind T (1984) Sodium/calcium concentration and pH on tension development of ATPase activity of the arterial actomyosin contractile system. Blood Vessels 11:277–286
Morgan JP, Morgan KG (1982) Vascular smooth muscle: the first recorded Ca^{2+} transients. Pflugers Arch 395:75–77
Morgan JP, Morgan KG (1984a) Stimulus-specific patterns of intracellular calcium levels in smooth muscle of ferret portal vein. J Physiol (Lond) 351:155–167
Morgan JP, Morgan KG (1984b) Alternations of cytoplasmic ionized calcium levels in smooth muscle by vasodilators in the ferret. J Physiol (Lond) 357:539–551
Mulvany MJ (1985) Changes in sodium pump activity and vascular contraction. J Hypertens 3:429–436
Murphy RA, Aksoy MO, Dillon PF, Gerthoffer WT, Kamm KE (1983) The role of myosin light chain phosphorylation in regulation of the cross-bridge cycle. Fed Proc 42:51–56
Nakao K, Okabe K, Kitamura K, Kuriyama H, Weston AH (1988) Characteristics of cromakalim-induced relaxation in the smooth muscle cells of guinea-pig mesenteric artery and vein. Br J Pharmacol 95:785–804
Nakashima M, Akata T, Kuriyama H (1992) Effects on the rabbit coronary artery of LP-805, a new type of release of endothelium-derived relaxing factor and a K^+ channel opener. Circ Res 71:859–869
Neild TO (1983) The relation between the structure and innervation of small arteries and arterioles and the smooth muscle membrane potential changes expected at different levels of sympathetic nerve activity. Proc R Soc Lond [Biol] 220: 237–249
Neild TO, Kotaecha N (1986) Effects of α,β-methylene ATP membrane potential, neuromuscular transmission and smooth muscle contraction in the rat tail artery. Gen Pharmacol 17:461–464

Nelson MT, Standen NB, Brayden JE, Worley III JF (1988) Noradrenaline contracts arteries by activating voltage-dependent calcium channels. Nature 336:382–385
Nelson MT, Huang Y, Brayden JE, Hescheler J, Standen NB (1990) Arterial dilatation in response to calcitonin gene-related peptide involves activation of K^+ channels. Nature 344:770–773
Ngai PK, Walsh MP (1985) Properties of caldesmon isolated from chicken gizzard. Biochem J 230:695–707
Nishida W, Abe M, Takahashi K, Hiwada K (1990) Do thin filaments of smooth muscle contain calponin? A new method for the preparation. FEBS Lett 268: 165–168
Nishikawa M, Sellers JR, Adelstein RS, Hidaka H (1984) Protein kinase C modulates in vitro phosphorylation of the smooth muscle heavy meromyosin by myosin light chain kinase. J Biol Chem 259:8808–8814
Nishimura N, Kolber M, van Breemen C (1988) Norepinephrine and GTPγS increase myofilament Ca^{2+} sensitivity in α-toxin permeabilized arterial smooth muscle. Biochem Biophys Res Commun 157:677–683
Nishiye E, Nakao K, Itoh T, Kuriyama H (1989) Factors inducing endothelium-dependent relaxation in the guinea-pig basilar artery as estimated from the action of haemoglobin. Br J Pharmacol 96:645–655
Nishizuka Y (1984) The role of protein kinase C in the cell surface signal transduction and tumor promotion. Nature 308:693–698
Nishizuka Y (1986) Studies and perspectives of protein kinase C. Science 233: 305–312
Nishizuka Y (1988) The molecular heterogeneity of protein kinase C and its implications for cellular regulation. Nature 334:661–665
Noma A (1983) ATP-regulated K^+ channels in cardiac muscle. Nature 305:147–148
Ohya Y, Terada K, Kitamura K, Kuriyama H (1986) Membrane currents recorded from a fragment of rabbit intestinal smooth muscle cells. Am J Physiol 251: C335–C346
Ohya Y, Kitamura K, Kuriyama H (1987a) Modulation of ionic currents in smooth muscle cells of the intestine by intracellularly perfused ATP and cyclic AMP. Pflugers Arch 408:465–473
Ohya Y, Kitamura K, Kuiryama H (1987b) Cellular Ca regulates inward currents in rabbit intestinal smooth muscle cell. Am J Physiol 252:C401–C410
Ohya Y, Terada K, Kitamura K, Kuriyama H (1987c) D600 blocks the Ca^{2+} channel from the outer surface of smooth muscle cell membrane of the rabbit intestinal and portal vein. Pflugers Arch 408:80–82
Ohya Y, Terada K, Yamaguchi K, Inoue R, Okabe K, Kitamura K, Hirata M, Kuriyama H (1988) Effects of inositol phosphates on the membrane activity of smooth muscle cells of the rabbit portal vein. Pflugers Arch 412:382–389
Oike M, Inoue Y, Kitamura K, Kuriyama H (1990) Dual actions of FRC8653, a novel dihydropyridine derivative, on the Ba^{2+} current recorded from the rabbit basilar artery. Circ Res 67:993–1006
Oike M, Kitamura K, Kuriyama (1992) Histamine H_3-receptor activation augments voltage-dependent Ca^{2+} current via GTP hydrolysis in rabbit saphenous artery. J Physiol (Lond) 448:133–152
Okabe K, Kitamura K, Kuriyama H (1987) Features of 4-aminopyridine sensitive outward current observed in single smooth muscle cells from the rabbit pulmonary artery. Pflugers Arch 409:561–568
Okabe K, Kitamura K, Kuriyama H (1988) The existence of a highly tetrodotoxin sensitive Na channel in freshly dispersed smooth muscle cells of the rabbit main pulmonary artery. Pflugers Arch 411:423–428
Okabe K, Kajioka S, Nakao K, Kitamura K, Kuriyama H, Weston AH (1990) Actions of cromakalim on ionic currents recorded from single smooth muscle cells of the rat portal vein. J Pharmacol Exp Ther 252:832–839

Ousterhout JM, Sperelakis N (1987) Cyclic nucleotides depress action potentials in cultured aortic smooth muscle cells. Eur J Pharmacol 144:7–14
Palmer RMJ, Ferrige AG, Moncada S (1987) Nitric oxide release accounts for the biological activity of endothelium-derived relaxing factor. Nature 327:524–526
Palmer RMJ, Ashton DS, Moncada S (1989) Vascular endothelial cells synthesize nitric oxide from L-arginine. Nature 33:664–666
Pape HC, McCormick DA (1989) Noradrenaline and serotonin selectively modulate thalamic burst firing by enhancing a hyperpolarization-activated cation current. Nature 340:715–718
Park S, Rasmussen H (1986) Carbachol-induced protein phosphorylation changes in bovine tracheal smooth muscle. J Biol Chem 261:15734–15739
Parker I, Ito Y, Kuriyama H, Miledi R (1987) β-Adrenergic agonists and cyclic AMP decrease intracellular resting free-calcium concentration in ileum smooth muscles. Proc R Soc Lond [Biol] 230:207–214
Person RB, Jakes R, Kendrick Johnes J, Kemp BE (1984) Phosphorylation site sequence of smooth muscle myosin light chain (Mr = 20000). FEBS Lett 168:108–112
Powell JA, Fambrough DM (1973) Electrical properties of normal and dysgenic mouse skeletal muscle in culture. J Cell Physiol 82:21–38
Pritchard K, Ashley CC (1986) Na^+/Ca^{2+} exchange in isolated smooth muscle cells demonstrated by the fluorescent calcium indicator fura-2. FEBS Lett 195:23–27
Pritchard K, Ashley CC (1987) Evidence for Na^+/Ca^+ exchange in isolated smooth muscle cells: a fura-2 study. Pflugers Arch 410:401–407
Purney TM, Hirning LD, Leeman SE, Miller RJ (1986) Multiple calcium channels mediate neurotransmitter release from peripheral neurons. Proc Natl Acad Sci USA 83:6656–6659
Quast U (1987) Effect of K^+ efflux stimulating vasodilator BRL 34915 on $^{86}Rb^+$ efflux and spontaneous activity in guinea-pig portal vein. Br J Pharmacol 91: 569–578
Quast U, Cook NS (1989) Moving together: K^+ channel openers and ATP-sensitive K^+ channels. Trends Pharmacol Sci 10:431–435
Raeymaekers L, Jones LR (1986) Evidence for the presence of phospho lamban in the endoplasmic reticulum of smooth muscle. Biochim Biophys Acta 882: 258–265
Raeymaekers L, Hoffmann F, Casteels R (1988) Cyclic GMP-dependent protein kinase phosphorylates phospholamban in isolated sarcoplasmic reticulum from cardiac and smooth muscle. Biochem J 252:269–273
Ramagopal MW, Leighton HJ (1989) Effects of N^G-monomethyl-L-arginine on field stimulation-induced decreases in cytosolic Ca^{2+} levels and relaxation in rat anococcygeus muscle. Eur J Pharmacol 174:297–299
Rasgado-Flores H, Blaustein MP (1987) Na/Ca exchange in barnacle muscle cells has a stoichiometry of $3Na^+:1Ca^{2+}$. Am J Physiol 252:C499–C504
Rasmussen H, Takuwa Y, Park S (1987) Protein kinase C in the regulation of smooth muscle contraction. FASEB J 1:177–185
Rasmussen H, Haller H, Takuwa Y, Kelly G, Park S (1990) Messenger Ca^{2+}, protein kinase C, and smooth muscle contration. In: Sperelakis N, Wood JD (eds) Frontiers in smooth muscle research. Liss, New York, pp 89–106
Resnik TJ, Scott-Burden T, Buhler FR (1988) Endothelin stimulates phospholipase C in cultured vascular smooth muscle cells. Biochem Biophys Res Commun 157:1360–1368
Reuter H, Blaustein MP, Haeusler G (1973) Na-Ca exchange and tension development in arterial smooth muscle. Philos Trans R Soc Lond [Biol] 265:87–94
Rieger F, Bournaud R, Shimahara T, Garcia L, Pincon-Raymond M, Romey G, Lazdunski M (1987) Restoration of dysgenic muscle contraction and calcium channel function by co-culture with normal spinal cord neurons. Nature 330: 563–566

Rousseau E, Smith JS, Meissner G (1987) Ryanodine conductance and gating behaviour of single Ca^{2+} release channel. Am J Physiol 253:C364–C368
Rüegg JC (1986) Calcium in muscle activation – a comparative approach. Springer, Berlin Heidelberg New York
Russel SN, Aaronson PI (1990) Carbachol inhibits the voltage-gated calcium current in smooth muscle cells isolated from the longitudinal muscle of the rabbit jejunum. J Physiol (Lond) 426:23
Ruth P, Röhrkasten AA, Biel M, Bosse E, Regulla S, Meyer HE, Flockerzi V, Hofman F (1989) Primary structure of β subunit of the DHP-sensitive calcium channel from skeletal muscle. Science 245:1115–1118
Sadoshima J, Akaike N, Tomoike H, Nakamura M (1988) Ca-activated K channel in cultured smooth muscle cells of rat aortic media. Am J Physiol 255:H410–H418
Saida K (1982) Intracellular Ca release in skinned smooth muscle. J Gen Physiol 80:191–202
Saida K, Nonomura Y (1978) Characteristics of Ca^{2+}- and Mg^{2+}-induced tension development in chemically skinned smooth muscle fibers. J Gen Physiol 72:1–14
Saitoh M, Ishikawa T, Matsushima S, Naka M, Hidaka H (1987) Selective inhibition of catalytic activity of smooth muscle myosin light chain kinase. J Biol Chem 262:7796–7801
Sakai T, Terada K, Kitamura K, Kuriyama H (1988) Ryanodine inhibits the Ca-dependent KL current after depletion of Ca stored in smooth muscle cells of the rabbit ileal longitudinal muscle. Br J Pharmacol 95:1089–1100
Sanders KM, Publicover NC (1989) Electrophysiology of the gastric musculature. In: Woods JD, Schultz SG (eds) The gastrointestinal system. American Physiological Society, Bethesda, pp 187–216 (Handbook of physiology, section 6, vol 1)
Sanders KM, Vogalis F (1989) Organization of electrical activity in the canine pyloric canal. J Physiol (Lond) 416:49–66
Sandow A (1965) Excitation-contraction coupling in skeletal muscle. Pharmacol Rev 17:265–320
Sanguinetti MC, Kass RS (1984) Voltage-dependent block of calcium channel current in the calf cardiac Purkinje fiber by dihydropyridine calcium channel antagonists. Circ Res 55:336–348
Sanguinetti MC, Krafle DS, Kass RS (1986) Voltage-dependent modulation of Ca channel current in heart cells by Bay K 8644. J Gen Physiol 88:389–392
Schneider MF, Chandler WK (1973) Voltage dependent charge movement in skeletal muscle: a possible step in excitation contraction coupling. Nature 242:244–246
Seelhorst A, Starke K (1986) Prejunctional opioid receptors in the pulmonary artery of the rabbit. Arch Int Pharmacodyn 281:298–860
Sekigawa K, Tamaoki J, Graff PD, Nadel JA (1988) Modulation of cholinergic neurotransmission by vasoactive intestinal peptide in ferret trachea. J Appl Physiol 64:2433–2437
Seto M, Sasaki Y, Sasaki Y (1990) Stimulus-specific patterns of myosin light chain phosphorylation in smooth muscle of rabbit thoracic aorta. Pflugers Arch 415: 484–489
Shearman NS, Sekiguchi K, Nishizuka Y (1989) Modulation of ion channel activity: a key function of the protein kinase C family. Pharmacol Rev 41:211–237
Shears SB (1989) Metabolism of the inositol phosphates produced upon receptor activation. Biochem J 260:313–324
Shirahase H, Usui H, Kurahashi K, Fujiwara M, Fukui K (1987) Possible role of endothelial thromboxane A_2 in the resting tone and contractile responses to acetylcholine and arachidonic acid in canine cerebral arteries. J Cardiovasc Pharmacol 10:517–522
Shogakiuchi Y, Kanaide H, Kobayashi S, Nishimura J, Nakamura M (1986) Intracellular free calcium transients induced by norepinephrine in rat aortic smooth muscle cells in primary culture. Biochem Biophys Res Commum 135:9–15

Silberberg SD, Poder TC, Lacerda AE (1989) Endothelin increases single-channel calcium currents in coronary arterial smooth muscle cells. FEBS Lett 247:68–72
Sims SM, Singer JJ, Walsh JV Jr (1985) Cholinergic agonists suppress a potassium current in freshly dissociated smooth muscle cells of the toad. J Physiol (Lond) 367:503–529
Sims SM, Walsh JV, Singer JJ (1986) Substance P and acetylcholine both suppress the same K^+ current in dissociated smooth muscle cells. Am J Physiol 251: C580–C587
Sims SM, Singer JJ, Walsh JV Jr (1988) Antagonistic adrenergic-muscarinic regulation of M current in smooth muscle cells. Science 239:190–193
Slaughter RS, Welton AF, Morgan DW (1987) Sodium-calcium exchange in sarcolemmal vesicles from tracheal smooth muscle. Biochim Biophys Acta 904: 92–104
Smith JB, Cracoe EJ Jr, Smith L (1987) Na^+/Ca^{2+} antiport in cultured arterial smooth muscle cells. Inhibition by magnesium and other divalent cations. J Biol Chem 262:11988–11994
Smith VL, Dedman JR (1986) An immunological comparison of several novel calcium-binding proteins. J Biol Chem 261:15815–15818
Sneddon P, Burnstock G (1984) Inhibition of excitatory junction potentials in guinea-pig vas deferens by α,β-methylene ATP: further evidence for ATP and noradrenaline as cotransmitters. Eur J Pharmacol 100:85–90
Sneddon P, Westfall DP (1984) Pharmacological evidence that adenosine triphosphate and noradrenaline are cotransmitters in guinea-pig vas deferens. J Physiol (Lond) 347:561–572
Sobieszek A (1977) Vertebrate smooth muscle myosin. Enzymatic and structural properties. In: Stephens NL (ed) The biochemistry of smooth muscle. University Park, Baltimore, pp 413–443
Sobue K, Muramoto Y, Fujita M, Kakiuchi S (1981) Purification of a calmodulin-binding protein from chicken gizzard that interacts with F-actin. Proc Natl Acad Sci USA 78:5652–5655
Sobue K, Morimoto K, Kanda M, Fukunaga E, Miyamoto E, Kakiuchi S (1982) Interaction of 13500 Mr calmodulin binding protein (myosin kinase) and F-actin: another Ca^{2+}- and calmodulin dependent flip flop switch. Biochem Infern 5:503–510
Somlyo AP, Somlyo AV (1990) Flash photolysis studies of excitation-contraction coupling, regulation, and contraction in smooth mucle. Annu Rev Physiol 52:857–874
Somlyo AP, Broderick R, Somlyo AV (1986) Calcium and sodium in vascular smooth muscle. Ann NY Acad Sci 488:228–239
Somlyo AP, Walker JW, Goldman YE, Trenthan DR, Kobayashi S, Kitazawa T, Somlyo AV (1988) Inositol trisphosphate, calcium and muscle contraction. Philos Trans R Soc Lond [Biol] 320:399–414
Somlyo AP, Kitazawa T, Himpens G, Matthijs B, Horiuti K, Kobayashi S, Goldman YE, Somlyo AV (1989) Modulation of Ca^{2+} sensitivity and of the time course of contraction in smooth muscle: a major role of protein phosphatases. In: Merlevede W, DiSalvo J (eds) Advances in protein phosphatases, vol 5. University Press, Leuven, pp 181–195
Somlyo AV, Somlyo AP (1968) Electromechanical and pharmacomechanical coupling in vascular smooth muscle. J Pharmacol Exp Ther 159:129–159
Somlyo AV, Bond M, Somlyo AP, Scarpa A (1985) Inositol trisphosphate-induced calcium release and contraction in vascular smooth muscle. Proc Natl Acad Sci USA 82:5231–5235
Southerland C, Walsh MP (1989) Phosphorylation of caldesmon prevents its interaction with smooth muscle myosin. J Biol Chem 264:578–583
Southerton JS, Weston AH, Bray KM, Newgreen DT, Taylor SG (1988) The potassium channel opening action of pinacidil; studies using biochemical, ion

flux and microelectrode techniques. Naunyn Schmiedebergs Arch Pharmacol 338:310–318

Speden RN (1970) Excitation of vascular smooth muscle. In: Bülbring E, Brading AF, Jones AW, Tomita T (eds) Smooth muscle. Arnold, London, pp 558–588

Sperelakis N, Ohya Y (1990) Cyclic nucleotides regulation of Ca^{2+} slow channels and neurotransmitter release in vascular muscle. In: Sperelakis N, Wood JD (eds) Frontiers in smooth muscle research. Liss, New York, pp 277–298

Standen NB, Quayle JM, Davies NW, Brayden JE, Huang Y, Nelson MT (1989) Hyperpolarizing vasodilators activate ATP-sensitive K^+-channels in arterial smooth muscle. Science 245:177–180

Starke K (1987) Presynaptic α-autoreceptors. Rev Physiol Biochem Pharmacol 107:73–146

Starke K, Göthert M, Kilbinger H (1989) Modulation of neurotransmitter release by presynaptic autoreceptors. Physiol Rev 69:864–989

Stjärne L (1989) Basic mechanisms and local modulation of nerve impulse-induced secretion of neurotransmitters from individual sympathetic nerve varicosities. Rev Physiol Biochem Pharmacol 112:1–137

Stjärne L, Brundin J (1976) β_2-Adrenoceptors facilitating noradrenaline secretion from human vasoconstrictor nerves. Acta Physiol Scand 97:88–93

Stjärne L, Bao J-X, Gordon FG, Mermet C, Msghino M, Stjärne E, Åstrand P (1990) Presynaptic receptors and modulation of noradrenaline and ATP secretion from sympathetic nerve varicosities. Ann NY Acad Sci 604:250–265

Stjernquist M, Owman C (1984) Vasoactive intestinal polypeptide (VIP) inhibits neurally evoked smooth muscle activity of rat uterine cervix in vitro. Regul Pept 8:161–167

Sturek M, Hermesmeyer K (1986) Calcium and sodium channels in spontaneously contracting vascular muscle cells. Science 233:475–478

Su C, Bevan JA, Ursillo RC (1964) Electrical quiescence of pulmonary artery smooth muscle during sympathomimetic stimulation. Circ Res 15:20–27

Suematsu E, Hirata M, Hashimoto T, Kuriyama H (1984) Inositol 1,4,5-trisphosphate releases Ca^{2+} from intracellular store sites in skinned single cells of porcine coronary artery. Biochem Biophys Res Commum 120:481–485

Sugiura M, Inagami T, Hare GMT, Johns JA (1989) Endothelin action: inhibition by a protein kinase C inhibitor and involvement of phosphoinorsitols. Biochem Biophys Res Commun 157:170–176

Sumimoto K, Kuriyama H (1986) Mobilization of free Ca^{2+} measured during contraction-relaxation cycles in smooth muscle cells of the porcine coronary artery using Quin 2. Pflugers Arch 406:173–180

Sumimoto K, Hirata M, Kuriyama H (1988) Characterization of [3H]nifedipine binding to intact vascular smooth muscle cells. Am J Physiol 254:C45–C52

Suppattapone S, Danoff SK, Theibert A, Joseph SK, Steiner J, Snyder SH (1988a) Cyclic AMP-dependent phosphorylation of a brain inositol trisphosphate receptor decreases its release of calcium. Proc Natl Acad Sci USA 85:8747–8750

Suppattapone S, Worley PF, Baraban JM, Snyder SH (1988b) Solubilization, purification, and characterization of an inositol trisphosphate receptor. J Biol Chem 263:1530–1534

Sutko JL, Ito K, Kenyon JL (1985) Ryanodine: a modifier of sarcoplasmic reticulum calcium release in striated muscle. Fed Proc 44:2984–2988

Suzuki H (1981) Effects of endogenous and exogenous noradrenaline on the smooth muscle of guinea-pig mesenteric vein. J Physiol (Lond) 321:495–512

Suzuki H (1985) Electrical responses of smooth muscle cells of the rabbit ear artery to adenosine triphosphate. J Physiol (Lond) 359:401–415

Suzuki H (1986) Increase in membrane resistance during noradrenaline-induced depolarization in arterial smooth muscle. Jpn J Physiol 36:433–440

Suzuki H (1989) Electrical activities of vascular smooth muscles in response to acetylcholine. Asia Pac J Pharmacol 4:141–150

Szabo B, Wichmann T, Starke K (1987) Presynaptic opioid receptors in the portal vein of the rabbit. Eur J Pharmacol 139:103–110
Tada M, Katz AM (1982) Phosphorylation of the sarcoplasmic reticulum and sarcolemma. Annu Rev Physiol 44:401–423
Takahashi K, Hiwada K, Kokubu T (1986) Isolation and characterization of a 34000-dalton calmodulin- and F-actin-binding protein from chicken gizzard smooth muscle. Biochem Biophys Res Commun 141:20–26
Takahashi K, Hiwada K, Kokubu T (1987) Occurrence of anti-gizzard P34K antibody cross-reactive components in bovine smooth muscles and non-smooth muscle tissues. Life Sci 41:291–296
Takahashi K, Hiwada K, Kokubu T (1988) Vascular smooth muscle calponin. A novel troponin-T-like protein. Hypertension 11:620–626
Takahashi K, Abe M, Hiwada K, Kokubu T (1989) Molecular organization of caldesmon, calponin and tropomyosin in smooth muscle thin filaments. Jpn Circ J 53:906
Takai A, Bialojan C,Troschka M, Rüegg JC (1987) Smooth muscle myosin phosphatase inhibition and force enhancement by black sponge toxin. FEBS Lett 217:81–84
Takai A, Troschka M, Mieskes G, Somlyo AV (1989) Protein phosphatase composition in the smooth muscle of guinea-pig ileum studied with okadaic acid and inhibitor 2. Biochem J 262:617–623
Takeshima H, Nishimura S, Matsumoto T, Ishida H, Kangawa K, Minamino N, Matsuo H, Ueda M, Hanaoka M, Hirose T, Numa S (1989) Primary structure and expression from complementary DNA of skeletal muscle ryanodine receptor. Nature 339:439–445
Takuwa Y, Kelly G, Takuwa N, Rasmussen H (1988) Protein phosphorylation changes in bovine carotid artery smooth muscle during contraction and relaxation. Mol Cell Endocrinol 60:71–86
Tamaoki T, Nomoto H, Takahashi I, Kato Y, Morimoto M, Tomita F (1986) Staurosporine, a potent inhibitor of phospholipid/Ca^{++} dependent protein kinase. Biochem Biophys Res Commun 135:397–402
Tanabe T, Takehsima H, Mikami A, Flockerzi V, Takahashi H, Kanagawa K, Kojima M, Matsuo H, Hirose T, Numa S (1987) Primary structure of the receptor for calcium channel blockers from skeletal muscle. Nature 328:313–318
Tanabe T, Beam KG, Powell JA, Numa S (1988) Restoration of excitation-contraction coupling and slow calcium current in dysgenic muscle by dihydropyridine receptor complementary DNA. Nature 336:134–139
Tanabe T, Beam KG, Adams BA, Niidome T, Numa S (1990a) Regions of the skeletal muscle dihydropyridine receptor critical for excitation-contraction coupling. Nature 346:567–569
Tanabe T, Mikami A, Numa S, Beam KG (1990b) Cardiac-type excitation-contraction coupling in dysgenic skeletal muscle injected with cardiac dihydropyridine receptor c DNA. Nature 344:451–453
Tanaka T, Hidaka H (1980) Hydrophobic regions function in calmodulin-enzyme(s) interactions. J Biol Chem 255:11078–11080
Tare M, Parkington HC, Coleman HA, Neild TO, Dusting GJ (1990) Hyperpolarization and relaxation of arterial smooth muscle caused by nitric oxide derived from endothelium. Nature 346:69–71
Terada K, Kitamura K, Kuriyama H (1987a) Blocking actions of Ca^{2+} antagonists on the Ca^{2+} channels in the smooth muscle cell membrane of rabbit small intestine. Pflugers Arch 408:552–557
Terada K, Nakao K, Okabe K, Kitamura K, Kuriyama H (1987b) Action of the 1,4-dihydropyridine derivative, KW-3049, on the smooth muscle cell membrane of the rabbit mesentric artery. Br J Pharmacol 92:615–625
Terada K, Ohya Y, Kitamura K, Kuriyama H (1987c) Actions of flunarizine, a Ca^{++} antagonist, on ionic currents in fragmented smooth muscle cells of the rabbit small intestine. J Pharmacol Exp Ther 240:978–983

Thuneberg L (1989) Interstitial cells of Cajal. In: Woods JD, Schultz SG (eds) The gatrointestinal system. American Physiological Society, Bethesda, pp 349–386 (Handbook of physiology, section 6, vol 1)
Tokumitsu H, Chijiwa T, Hagiwara M, Mizutani A, Terasawa M, Hidaka H (1990) KN-62, 1-[N,O-Bis(5-isoquinolinesulfonyl)-N-methyl-L-tyrosyl]-4-phenylpiperazine, a specific inhibitor of Ca^{2+}/calmodulin-dependent protein kinase II. J Biol Chem 265:4315–4320
Trautwein W, Hescheler J (1990) Regulation of cardiac L-type calcium current by phosphorylation and G proteins. Annu Rev Physiol 52:257–274
Tsien RW, Hess P, McCleskey EW, Rosenberg RL (1987) Calcium channel: mechanisms of selectivity, permeation and block. Annu Rev Biophys Chem 16:265–290
Tsurufuji S, Sugio K, Takemasa F (1979) The role of glucocorticoid receptor and gene expression in the anti-inflammatory action of dexamethasone. Nature 280:408–410
Umekawa H, Naka M, Inagaki M, Hidaka H (1985a) Interaction of W-7, a calmodulin antagonist, with another Ca^{2+} binding protein. In: Hidaka H, Hartshorne DJ (eds) Calcium antagonists and cellular physiology. Academic, London, pp 511–524
Umekawa H, Naka M, Inagaki M, Onishi H, Wakabayashi T, Hidaka H (1985b). Conformational studies of myosin phosphorylation by protein kinase C. J Biol Chem 260:9833–9837
van Breemen C, Aaronson P, Loutzenheiser R (1979) Sodium-calcium interactions in mammalian smooth muscle. Pharmacol Rev 30:167–208
Vane JR (1990) Endothelins come home to roost. Nature 348:673
Vanhoutte PM, Rubanyi GM, Miller JM, Houston DS (1986) Modulation of vascular smooth muscle contraction by the endothelium. Annu Rev Physiol 48:307–320
Villarroel A, Marrion NV, Lopez H, Adams PR (1989) Bradykinin inhibitis a potassium M-like current in rat pheochromocytoma PC12 cells. FEBS Lett 255:42–46
Vivaudou M, Clapp LH, Wlash JV Jr, Singer JJ (1988) Regulation of one type of Ca^{2+} current in smooth muscle cells by diacylglycerol and acetylcholine. FASEB J 2:2497–2504
von Kügelgen I, Illes P, Wolf D, Starke K (1985) Presynaptic inhibitory opioid δ- and κ-receptors in a branch of the rabbit ileocolic artery. Eur J Pharmacol 118:97–105
Vrolix M, Raeymaekers L, Wuytack F, Hoffmann F, Casteels R (1988) Cyclic GMP-dependent protein kinase stimulates the plasmalemmal Ca^{2+} pump of smooth muscle via phosphorylation of phosphatidylinositol. Biochem J 255:855–863
Wakui M (1989) The effects of acetylcholine, inositol trisphosphate and Ca^{2+} on Cl^- currents in single mouse pancreatic acinar cells. J Physiol (Lond) 410:80
Wakui M, Inomata H (1984) Requirements of external Ca and Na for the electrical and mechanical responses to noradrenaline in the smooth muscle of guinea-pig vas deferens. Jpn J Physiol 34:199–203
Walker JW, Somlyo AV, Goldman YE, Somlyo AP, Trentham DR (1987) Kinetics of smooth and skeletal muscle activation by laser pulse photolysis of caged inositol 1,4,5-trisphosphate. Nature 327:249–252
Walsh MP (1985) Calcium regulation of smooth muscle contraction. In: Marmé D (ed) Calcium and physiology. Springer, Berlin Heidelberg New York
Walsh MP (1987) Caldesmon, a major actin- and calmodulin-binding protein of smooth muscle. In: Siegman MJ, Somlyo AP, Stephens NL (eds) Progress in clinical and biological research, vol 245: regulation and contraction of smooth muscle. Liss, New York, pp 119–141
Walsh MP (1990) Smooth muscle caldesmon. In: Sperelakis N, Wood JD (eds) Frontiers in smooth muscle research. Liss, New York, pp 127–140

Webb DJ (1991) Endothelin receptors cloned, endothelin converting enzyme characterized and pathophysiological roles for endothelin proposed. Trend Pharmacol Sci 12:43–46

Weir SE, Weston AH (1986a) Effects of apamin on response to BRL 34915, nicorandil and other relaxants in the guinea-pig taenia caeci. Br J Pharmacol 88:113–120

Weir SW, Weston AH (1986b) The effects of BRL 34915 and nicorandil on electrical and mechanical activity and on ^{86}Rb efflux in rat blood vessels. Br J Pharmacol 88:121–128

Williams DA, Fogarty KE, Tsien RY, Fay FS (1985) Calcium gradients in single smooth muscle cells revealed by the digital imaging microscope using Fura-2. Nature 318:558–561

Winder SJ, Walsh MP (1990) Smooth muscle calponin. J Biol Chem 265:10148–10155

Wooten GF, Thoa NB, Kopin IJ, Axelrod J (1973) Enhanced release of dopamine β-hydroxylase and norepinephrine from sympathetic nerves by dibutyryl cyclic adenosine 3′5′-monophosphate and theophylline. Mol Pharmacol 9:178–183

Worley III JF, Deitmer JW, Nelson MT (1986) Single nisoldipine-sensitive calcium channels in smooth muscle cells isolated from rabbit mesenteric artery. Proc Natl Acad Sci USA 83:5746–5750

Wray S (1988) Smooth muscle intracellular pH: measurement, regulation and function. Am J Physiol 254:C213–C225

Xiong ZL, Kitamura K, Kuriyama H (1991) ATP activates cation currents and modulates the calcium current though GTP-binding protein in rabbit portal vein. J Physiol (Lond) 440:143–165

Yamaguchi H (1986) Recordings of intracellular Ca^{2+} from smooth muscle cells by submicron tip, double barrelled Ca^{2+} selective microelectrode. Cell Calcium 7:203–219

Yamaguchi K, Hirata M, Kuriyama H (1987) Calmodulin activates inositol 1,4,5-trisphosphate 3-kinase activity in pig aortic smooth muscle. Biochem J 244: 787–791

Yamamoto H, van Breenmen C (1985) Inositol-1,4,5-trisphosphate releases calcium from skinned cultured smooth muscle cells. Biochem Biophys Res Commun 130:270–274

Yamanaka K, Furukawa K, Kitamura K (1985) The different mechanisms of action of nicorandil and adenosine triphosphate on potassium channels of circular smooth muscle of the guinea-pig small intestine. Naunyn Schemiedebergs Arch Pharmacol 331:96–103

Yanagisawa M, Kurihara H, Kimura S, Tomobe Y, Kobayashi M, Mitsui Y, Yazaki Y, Goto K, Masaki T (1988) A novel potent vasoconstrictor peptide produced by vascular endothelial cells. Nature 332:411–415

Yatani A, Seidel CL, Allen JC, Brown AM (1987) Whole-cell and single-channel calcium currents of isolated smooth muscle cells from saphenous vein. Circ Res 60:523–533

Yazawa M, Ikuma M, Hikichi K, Ying L, Yagi K (1987) Communication between two globular domains of calmodulin in the presence of mastoparan or caldesmon fragment. J Biol Chem 262:10951–10954

Yoshino M, Someya T, Nishino A, Yabu H (1988) Whole-cell and unitary Ca channel currents in mammalian intestinal smooth muscle cells: evidence for existence of two types of Ca channels. Pflugers Arch 411:229–231

Yoshitomi T, Ito Y (1985) Adrenergic excitatory and cholinergic inhibitory innervations in the human iris dilator. Exp Eye Res 40:453–459

Section II
Endogenous Substances and Smooth Muscle

CHAPTER 4

Eicosanoids and Smooth Muscle Function

K. SCHRÖR and H. SCHRÖDER

A. Introduction

In 1934, Goldblatt and von Euler independently described the effects of seminal fluid on smooth muscle tone. Since the prostata, was held to be the source, VON EULER (1935) introduced the term "prostaglandin" to classify this new group of hitherto unknown acidic lipids. The description of these effects of prostaglandin(s) on smooth muscle tone marked the beginning of more than 50 years to date of intensive research on prostaglandins, thromboxanes, leukotrienes, and other members of this family of lipid mediators, now summarized under the more general term "eicosanoids." All eicosanoids are enzymatic oxygenation products from polyunsaturated fatty acids with a 20-carbon backbone and a characteristic steric configuration. Figure 1 gives an overview of the major pathways of eicosanoid generation.

Further advances in eicosanoid research were made with the description of the action of eicosanoids on the tone of selected smooth muscle preparations. Several eicosanoids, for example, the prostaglandin (PG) endoperoxides, thromboxane A_2, and cysteinyl leukotrienes, were originally described as "rabbit aorta contracting substance" (RCS) and "slow-reacting substance of anaphylaxis" (SRS-A) and were known for many years before their chemical structure was elucidated. In this respect, it must also be considered that most eicosanoids are local mediators that are only formed in minute amounts and in many cases are short-lived compounds, having a half-life of only a few seconds to minutes. Bioassay experiments in vivo have demonstrated the large pulmonary clearance of PGE- and PGF-type compounds, thus preventing circulatory prostaglandins, formed in inflammatory regions, from reaching the arterial circulation (VANE 1969). The different potencies of prostaglandins on smooth muscle in vitro have also proved useful in the classification of the prostaglandin receptors (KENNEDY et al. 1982).

Considerable progress has been made recently in our understanding of the cellular mechanisms of eicosanoid formation and action. This has been facilitated by our improved knowledge of cellular signal transduction pathways and the availability of selective analogs, which has made possible a more detailed pharmacological characterization of eicosanoid-related biological responses. More recently, molecular biology has provided a fas-

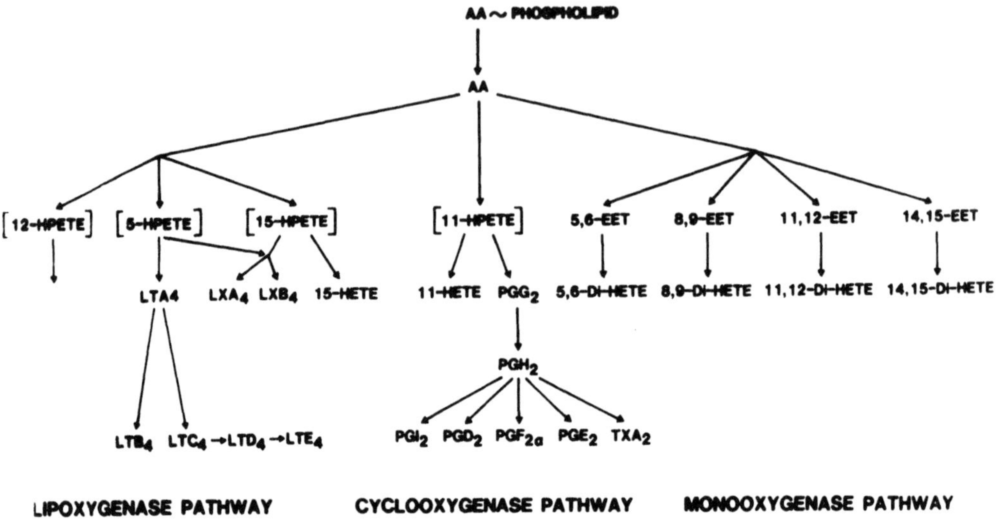

Fig. 1. Principal pathways of arachidonic acid metabolism via lipoxygenases, cyclooxygenase, and cytochrome P450-dependent pathways (epoxygenases) and their modification by drugs. *AA*, arachidonic acid; *EET*, epoxyeicosatraenoic acid; *HPETE*, hydroperoxyeicosatetraenoic acid; *HETE*, hydroxyeicosatetraenoic acid; *LT*, leukotriene; *LX*, lipoxin; *PG*, prostaglandin; *TX*, thromboxane. *Numbers* denote the position of the oxygen function(s) at the C-20 carbon backbone. Alternative pathways for some of these compounds (LX, 11-HETE) have been described. Only principal pathways are shown; interferences between several "main routes" have been omitted for the sake of clarity

cinating new and most helpful tool in the analysis of the genetic and biochemical aspects of eicosanoid formation and action. Using this technique, thromboxane A_2 (HIRATA et al. 1991) and the EP_3 receptor (SUGIMOTO et al. 1992) were structurally identified. Most recently, a PGI_2 receptor was identified by photoaffinity labeling (ITO et al. 1992). Cloning and sequencing of further receptors is underway. We may expect a considerable increase in our basic knowledge of eicosanoid formation and action in smooth muscle in the near future.

Several reviews on the subject are available (HEDQVIST 1977; SAMUELSSON 1983; GERRARD 1985; SMITH 1986; DRAZEN and AUSTEN 1987; JEREMY et al. 1988; HALUSHKA et al. 1990; SMITH et al. 1991; YAMAMOTO 1992) and are recommended to the interested reader.

B. Cellular Aspects of Formation and Action of Eicosanoids in Smooth Muscle Cells

I. Control of Local Eicosanoid Levels

It is generally accepted that local levels of eicosanoids are determined by local biosynthesis. Release of preformed, stored eicosanoids, with the possible exception of some monooxygenase products, does not occur. Two principally different mechanisms control biosynthesis: availability of substrate and mass and activity of synthesis-limiting enzymes. Under physiological conditions, availability of free, i.e., nonesterified, arachidonic acid (AA) as the natural substrate is controlled by the activity of phospholipases C and A_2. Under pathophysiological conditions, including inflammatory and allergic disorders, de novo synthesis of enzyme protein in inflammatory cells, such as macrophages, becomes more important.

Eicosanoids can either be generated within the smooth muscle cell or by cells in the vicinity, in particular vascular endothelium and epithelium of nonvascular smooth muscle. Phospholipases that determine eicosanoid formation in all of these cells are under hormonal control by chemical agonists (SMITH 1986). Peptides (kinins, angiotensin II), biogenic amines (serotonin, histamine), platelet-derived growth factor, platelet-activating factor (PAF), cytokines, and circulating hormones (oxytocin, norepinephrine) represent different classes of these mediators. Consequently, eicosanoid generation and smooth muscle function are the subject of multiple and complex interactions. It should also be noted that AA (and related fatty acids) have biological actions of their own, i.e., independent of oxygenation. These include stimulation of guanylyl cyclase (GLASS et al. 1977; GERZER et al. 1986) and the gating of ion channels (see Sect. B.III.5, this chapter).

1. Phospholipases and Arachidonic Acid Liberation

The natural source of precursor fatty acid, i.e., AA, are the phosphoglycerides of the plasma membrane, which contain up to 40% AA at the sn-2-position of the phospholipid backbone. In vascular smooth muscle, AA release appears to be a secondary response after primary stimulation of G-protein-coupled phospholipase(s) (PL) C. This results in inositol 1,4,5-triphosphate (IP_3) formation, increase in cytosolic Ca^{2+}, and PLA_2 activation with subsequent AA liberation from membrane phospholipids. There are only few reports suggesting a direct activation of PLA_2 as primary response. ABDEL-LATIF et al. (1991) showed that in rabbit iris sphincter smooth muscle low concentrations of endothelin-1 stimulated AA release without accumulation of PLC-related IP_3. This suggested a PLA_2-induced AA release from phosphoinositides. More recently, a cytosolic PLA_2 has been identified that binds to natural membrane vesicles in a Ca^{2+}-dependent fashion and selectively releases AA. Activation of this cytosolic PLA_2 by phosphorylation of a serin residue was prevented by inhibition of PLC and stimulated by phorbol esters, suggesting that, in addition to Ca^{2+}, phosphorylation of PLA_2 also plays an important role in agonist-induced stimulation of AA release (Lin et al. 1992).

2. Control of Enzyme Mass

Under physiological conditions, PLC and PLA_2 activity are regulated by hormonal (inter)actions. However, the increase in enzyme mass by de novo synthesis is a general feature of allergic/inflammatory diseases. Different classes of cytokine stimulate eicosanoid production via an increase in enzyme mass of PLA_2. Tumor necrosis factor (TNF) and interleukin (IL)-1β-induced PLA_2 (type II) cause PGH-synthase-dependent formation of PGE_2 (Kurihara et al. 1991)-induced PGI_2 (Marceau et al. 1990) in vascular smooth muscle.

IL-1β stimulates PGE_2 and cysteinyl leukotriene formation in the nonvascular smooth muscle of rat stomach (Mugridge et al. 1991). However, cytokines may act at additional steps of eicosanoid generation and these may differ even within smooth muscle of the same tissue (Yang et al. 1991). Thus, the issue is important but complex and clearly requires further investigation. This also includes the cytokine-induced synthesis of receptor protein (see Sect. B.III.1, this chapter).

Recent evidence suggests that in addition to the "classical" PGH-synthase, which is a constitutive enzyme (type I), another PGH-synthase exists. This type II PGH-synthase is inducible, has a 60% homology to the type I enzyme, and generates 15-hydro(pero)xyeicosatetraenoic acid (15-H(P)ETE) as a major product (Kujubu et al. 1991; Smith et al. 1991; Xie et al. 1991). The type II enzyme can be expressed in inflammatory cells in the presence of increased cytokine levels and is suppressed by endogenous glucocorticoids (Masferrer et al. 1992). The significance of this pathway for smooth muscle function is still unknown; however, a certain parallelism with the inducible and constitutive forms of the NO synthase is obvious.

Expression of the 15-lipoxygenase gene was demonstrated in pregnant human myometria and both mRNA and the catalytically active enzyme were lower at term pregnancy and during labor (Lei and Rao 1992). Thus, de novo synthesis of PLA_2 (type II), PGH-synthase (type II), and probably other enzyme proteins appears to be a more general pathway for the control of local eicosanoid generation at a site where it is needed.

3. Circulating Hormones and Chemical Messengers

According to their general function as local mediators of intercellular signal transduction, eicosanoids interact with almost every circulating hormone and chemical messenger that has been studied so far. Adrenergic and cholinergic neurotransmitter release and smooth muscle function are controlled by tissue-derived PGE (Hedqvist 1977) and $PGF_{2\alpha}$ (Takayanagi et al. 1990). Intact endothelium or epithelium may modify agonist-induced eicosanoid generation but is not essential for this response. Stimulation of vascular (see SCHRÖR 1992) and nonvascular (FARMER et al. 1991) smooth muscle cells by bradykinin results in enhanced PGI_2 formation and was blocked by a selective bradykinin B_2 antagonist. TAKEUCHI et al. (1991) have measured the level of cytosolic free calcium and the production rate of PGI_2 simultaneously in perfused monolayers of cultured vascular smooth muscle cells. Bradykinin and angiotensin II stimulated cytosolic Ca^{2+} and 6-keto-$PGF_{1\alpha}$ production. Acetylsalicylic acid abolished BK-induced PGI_2 formation but did not affect the BK-induced increase in cytosolic calcium. These results indicate that the peptide-induced calcium signal serves different purposes and the stimulation of cytosolic calcium can be separated from AA liberation (see Sect. B.III.3, this chapter). Interestingly, stimulation of PGE_2 in porcine aortic smooth muscle cells by angiotensin II was blocked by the selective AT_2 antagonist losartan, but not the stimulated PGI_2 formation (JAISWAL et al. 1991). However, losartan is also a competitive antagonist of thromboxane receptors (LIU et al. 1992). A detailed discussion of this issue is beyond the scope of this review and can be found elsewhere (HEDQVIST 1977; SCHRÖR 1992, 1993).

II. Additional Sources of Eicosanoids for Smooth Muscle

1. Endothelial/Epithelial Cells

Endothelium-derived relaxing factor (EDRF) and vasodilating prostaglandins are compounds which act synergistically with relaxing prostaglandins on smooth muscle tone. However, while EDRF, i.e., NO or NO-derived product, is considered as the dominant factor in control of vascular tone (MONCADA 1992), the situation in organ circulations is more complex. The endothelins (ETs) represent an important family of endothelium-derived vasoconstrictors. The contribution of endogenous eicosanoids to the vasomotor actions of ET-1 appears to depend on the preparation studied. For example, there was no reduced ET-1-induced vessel contraction after endothelial removal or indomethacin treatment in dog cerebral arteries (GARCIA et al. 1991) but evidence for PLA_2-dependent thromboxane $(TX)A_2$ formation by ET-1 and PAF in vascular smooth muscle cells (TAKAYASU-OKISHIO et al. 1990). ET-1-induced contractions were antagonized by a thromboxane receptor antagonist, suggesting that thromboxane release was involved (FILEP et al. 1991).

In isolated rabbit tracheal smooth muscle, ET-1 caused contractions. Interestingly, after precontraction by acetylcholine, low concentrations of ET-1 caused epithelium-dependent relaxation. These relaxations were accompanied by enhanced generation of PGI_2 and PGE_2 and were prevented by indomethacin treatment (GRUNSTEIN et al. 1991a,b). Similar results, i.e., an attenuation of phorbolester-induced contraction of guinea pig trachea by epithelium-derived bronchorelaxant prostaglandins, were obtained by SOUHRADA and SOUHRADA (1991). These data suggest that ET-1-induced contractions of vascular and nonvascular smooth muscle are attenuated by endothelium/epithelium-derived relaxing prostaglandins. In addition to prostaglandins (YU et al. 1992), an "epithelium-derived relaxing factor" (EpDRF) may also attenuate contractions of bronchial smooth muscle to several classes of bronchoconstrictive mediators. This includes muscarinergic stimuli (TESSIER et al. 1991) and cysteinyl leukotrienes (FERNANDES and GOLDIE 1991). The pathophysiological significance of these endothelium/epithelium-derived relaxing mediators for the underlying smooth muscle becomes apparent in epithelial/endothelial dysfunction, for example, during arterial vasospasm (Sect. D.II.1) and bronchial asthma (Sect. D.II.2).

2. Platelets

Platelets, sticking to the vessel wall or circulating in the blood, are an important source of vasoactive eicosanoids. UCHIDA and MURAO (1974) and FOLTS et al. (1976) originally described that platelet-derived vasoconstrictor cyclooxygenase products, probably PGH_2/TXA_2, can constrict coronary vessels. Coincubation experiments of aortic smooth muscle cells with thrombin-stimulated aspirin-treated platelets showed significant PGI_2 production, probably due to platelet-derived AA release. Moreover, aspirin-treated smooth muscle cells in cosuspension with platelets inhibited platelet aggregation in association with enhanced PGI_2 generation (HECHTMAN et al. 1991). These data confirm the original observation by MARCUS et al. (1980) that transcellular metabolism occurs between platelets and vascular endothelium and additionally shows that this interaction might result in generation of eicosanoids at vasoactive concentrations.

3. Leukocytes

Another significant source of vascular eicosanoids in circulating blood are white cells, in particular monocytes, macrophages, and polymorphonuclear leukocytes. These cells may bind to the endothelium via the expression of leukocyte-adhesive molecules and release their metabolites abluminally to the vascular smooth muscle cell. White cells are particularly prone to transcellular metabolism, eventually resulting in the generation of (vasoactive) products that cannot be formed by one single cell type alone, such as lipoxins and several Di-HETEs. For example, stimulation of isolated vessel preparations by chemotactic peptides resulted in the formation of vasoconstrictor products. This was prevented after treatment with indomethacin or a thromboxane receptor antagonist. This action was assumed to be due to macrophage-like cells present in the vessel wall (MARCEAU et al. 1990). Macrophages are also major targets for cytokine-induced PGH-synthase induction (see Sect. B.I.2).

III. Signal Transduction Pathways in Smooth Muscle Cells

Receptors for prostaglandins, TXA_2, and leukotrienes are located at the plasma membrane. Current evidence suggests that the multiple effects of these eicosanoids on smooth muscle are mediated by different receptor subtypes that in turn are coupled to intracellular signal transduction via different G proteins. Several recent reviews on this issue exist (JASCHONEK and MULLER 1988; HALUSHKA et al. 1990). Figure 2 gives a simplified overview of the major signal transduction pathways for prostaglandins and thromboxane A_2 in vascular smooth muscle.

1. Eicosanoid Receptors

a) Receptor Types

Prostaglandin receptors are divided into five main subclasses: thromboxane A_2 (TP), prostacyclin (EP), PGD_2 (BP), $PGF_{2\alpha}$ (FP), and three PGE-receptor subtypes for contractile (EP_1), relaxing (EP_2), and contractile effects of the PGE-mimetic sulprostone on the chicken ileum (EP_3) (KENNEDY et al. 1982; COLEMAN et al. 1987; BUNCE et al. 1991). Recent evidence suggests that this classification may not be the final one. For example, a "primary prostaglandin receptor" has been suggested from both radioligand

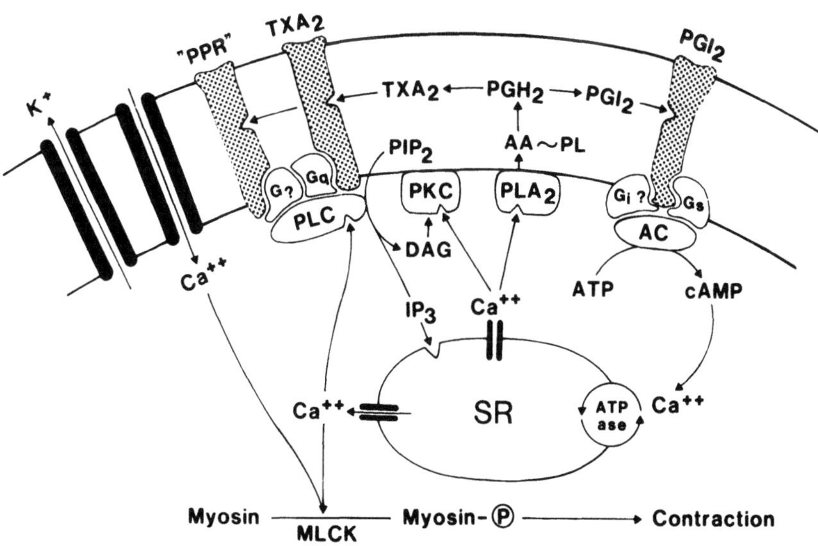

Fig. 2. Major signaling pathways of prostaglandins and thromboxane A_2 in vascular smooth muscle cell. *AA*, arachidonic acid; *AC*, adenylyl cyclase; *MLCK*, myosin light-chain kinase; *PIP_2*, phosphatidylinositol bisphosphate; *PL*, phospholipase; "*PPR*", primary prostaglandin receptor. For further explanations see text

binding studies and in vitro experiments in cell cultures and isolated vessel preparations (HANASAKI et al. 1988; HANASAKI and ARITA 1989; DORN et al. 1992). Pharmacological data suggest that the EP_1 receptor is similar to the thromboxane receptor. Tissues usually have more than one receptor and it is this lack of selectivity of PG-related agonists and antagonists that limits their clinical use (KEEN et al. 1989).

So far there have been no general proposals for a (sub)classification of receptors for lipoxygenase products. Current data suggest the existence of a distinct LTB_4 receptor, and separate LTC_4 and LTD_4 receptors (FLEISCH et al. 1982; SMITH et al. 1989). LTC_4-binding sites are widely distributed throughout the body, including tissues with no known biological function for this eicosanoid (HALUSHKA et al. 1990). Recent reports suggest that subtypes of the LTD_4 receptor exist. LTE_4 reactions appear to be mediated via a subset of LTD_4 receptors. The biological actions of LTE_4 are similar to those of LTC_4 and LTD_4, although their molar potency is 10- to 30-fold less.

Most data on vascular smooth muscle eicosanoid receptors are available for thromboxane receptor(s). One reason for this is the availability of specific receptor antagonists, such as SQ 29,548 (OGLETREE et al. 1985) and I-SAP (MORINELLI et al. 1992) with k_D values of $<1\,nM$. There is no doubt that vascular smooth muscle contains significant amounts of functionally active TXA_2 receptors that mediate increase in tone and are also occupied by PGH_2. The current discussion is mainly focussed on the PGH_2/TXA_2 receptor subtype(s) in the vessel wall compared with platelets and the cellular signal transduction pathways.

Pharmacological studies in a series of 13-Aza-Pinane-TXA_2 analogs yielded a different sequence of IC_{50} values for platelets and saphenous vein preparations in dog and men (MAIS et al. 1985; HALUSHKA et al. 1990). These data and findings in other tissues (see HALUSHKA et al. 1990; MASUDA et al. 1991a,b) suggested that the thromboxane receptor(s) in blood vessels are different from that in platelets. VERHEGGEN and SCHRÖR (1986) have compared the inhibition of TXA_2-induced 5-hydroxytryptamine (5-HT) secretion of human platelets with the inhibition of contractile effects of TXA_2 on bovine coronary arteries, using the PGEP/TXA_2-receptor antagonist sulotroban. There was a clear dissociation in inhibition of these responses by the thromboxane antagonist, also suggesting different properties of TXA_2 receptors in platelets and the vessel wall. Evidence for different binding sites in platelets and the vasculature came also from synthetic thromboxane mimetics, such as carbocyclic thromboxane A_2 (CTA_2). This compound was an agonist of vascular smooth muscle but an antagonist in human platelets (LEFER et al. 1980). FURCI et al. (1991) used the thromboxane receptor antagonist vapiprost (LUMLEY et al. 1989) to discriminate two forms of the TXA_2 receptor. They showed that homologous desensitization of phospholipase C and Ca^{2+} transients occurred in rat aortic smooth muscle cells after repeated exposure to a thromboxane mimetic (U-46619). In contrast to platelets, no heterologous desensitization to PAF or thrombin was observed. Moreover, there was only one reversible binding site for vapiprost in vascular smooth muscle which was identical to that for SQ 29,548. These data suggest that vascular smooth muscle solely contains the reversible PGH_2/TXA_2-binding site which is also present in platelets but lacks the irreversible binding site that causes platelet aggregation.

This reversible, high-affinity TXA_2 receptor in vascular smooth muscle has been further characterized in binding studies. Using the thromboxane A_2 agonist [^{125}I]BOP (MORINELLI et al. 1989), a single class of high-affinity binding site was identified in smooth muscle cells from rabbit aorta and human saphenous vein. The k_D was 0.4 nM in both cases and the B_{max} 5322 and 2017 binding sites/cell, respectively. There was a displacement of the ligand by stable TXA_2/PGH_2 analogs but not by other prostaglandins (SESSA et al. 1990; DORN 1991). Recent studies by MORINELLI et al. (1992) identified high-affinity binding sites in guinea pig coronary artery and aortic smooth muscle cells. The k_D was 0.13 and 0.15 nM, respectively, and was in the same range as the biological response (stimulation of mitogenesis). Interspecies differences between vascular thromboxane receptors rather than different thromboxane receptor subtypes may explain the different behavior of smooth muscles to thromboxane mimetics (OGLETREE and ALLEN 1992).

"Primary" prostaglandins (PGE, $PGF_{2\alpha}$, PGD_2) may contract vascular smooth muscle at concentrations that are about 2 orders of magnitude higher than those for TXA_2. This leads to the question of whether these prostaglandins cross-react with the vascular thromboxane receptor or act via different, low-affinity binding sites. Hanasaki and collegues were the first to address this question. Binding studies on cultured vascular smooth muscle cells demonstrated high-affinity specific binding sites for TXA_2 (HANASAKI et al. 1988). However, in addition to this high-affinity receptor for TXA_2 (k_D, 1 nM), a second, low-affinity binding site (k_D, 100–200 nM) was found for prostaglandins PGE_1 and $PGF_{2\alpha}$ (HANASAKI and ARITA 1989). Later work by these authors (HANASAKI et al. 1990a) demonstrated that these two receptors are also differentially regulated by PKC-dependent pathways (see below).

More detailed studies on the biological significance of different binding sites for prostaglandins in smooth muscle were carried out by DORN et al. (1992), who showed that rat aortic contraction and myosin light-chain phosphorylation induced by $PGF_{2\alpha}$ and PGE_2 at high, micromolar concentrations were blocked by the specific thromboxane receptor antagonists SQ 29,548, suggesting that these effects were mediated by cross-reaction with the vascular TXA_2 receptor. In contrast, $PGF_{2\alpha}$- and PGE_2-induced PI hydrolysis and calcium signaling at low, nanomolar concentrations were not antagonized by SQ 29,548, suggesting that these responses were not TX receptor mediated. These data were consistent with a high-affinity interaction of these prostaglandins at the primary prostaglandin receptor. This raised the question of whether the calcium signal induced by nanomolar concentrations of $PGF_{2\alpha}$ may have other physiological effects. DORN et al.

(1992) showed that $PGF_{2\alpha}$ stimulated de novo protein synthesis in vascular smooth muscle cells (leucin incorporation) which was sensitive to cycloheximide but insensitive to SQ 29,548. There was no change in DNA synthesis. From these data it was suggested that the physiological function of $PGF_{2\alpha}$ is stimulation of cell proliferation and not vasocontraction. In this context, it is interesting to remember the failure of a selective thromboxane receptor antagonist (vapiprost) to prevent late restenosis, i.e., growth processes, after percutaneous transluminal coronary angioplasty (PTCA) in man (SERRUYS et al. 1991).

A high-affinity (k_D, 6.2 nM) PGH_2/TXA_2 receptor was recently identified in cultured guinea pig tracheal smooth muscle cells (MIKI et al. 1992). JANSSEN and DANIEL (1991) have investigated the effects of a thromboxane mimetic (U-46619) on nonvascular smooth muscle of canine bronchial airway. U-46619 elicited tonic contractions, membrane depolarization, and oscillations in membrane potential which were antagonized by thromboxane receptor antagonists. The mechanical response was insensitive to nitrendipine, or exposure to Ca^{2+}-free media. U-46619 also potentiated electrical field stimulation-elicited contractions and excitatory junction potentials. This suggests both prejunctional and postjunctional thromboxane receptors in this preparation. The former potentiate cholinergic neurotransmission. The latter stimulate smooth muscle through a mechanism that utilizes intracellular Ca^{2+}.

There is currently little knowledge about the PGD_2 receptor(s) in vascular and nonvascular smooth muscle cells. The relaxing effects of the compound and its antiplatelet effects are generally considered to be due to DP-receptor-mediated elevation in cAMP levels (NARUMIYA and TODA 1985) whereas contractile effects of the compound and its PGF metabolites (see below) are assumed to be due to stimulation of vascular (SHIKADA et al. 1991) and nonvascular (MCKENNIFF et al. 1991; NOREL et al. 1991) thromboxane receptors. The evidence is mainly based on antagonism of these responses by thromboxane receptor antagonists. Alternatively, PGD_2 might also stimulate the vascular receptor for primary prostaglandins. Comparisons between a number of structurally related compounds suggested that platelet and vascular DP receptors are similar (LEFF and GILES 1992). Interestingly, PGD_2 relaxed bovine coronary arteries by a receptor-mediated, endothelium-dependent, and cGMP-mediated mechanism, probably involving nitric oxide formation. This response could be separated from direct contractile effects of PGD_2 after endothelial removal that were blocked by a thromboxane receptor antagonist (BRAUN and SCHRÖR 1992).

Three types of PGE receptors (EP_{1-3}) are currently known and all of them have been characterized in terms of smooth muscle contraction (EP_1, EP_3) and relaxation (EP_2), respectively. High-affinity binding sites for PGE_2/PGE_1 (k_D, 0.27 nM) have been described in bovine myometrial plasma membranes (LERNER et al. 1990). Similar data were obtained for

PGE_2 (k_D, 1.5 nM) in slices of human myometrium (LOPEZ BERNAL et al. 1991). There is little information on PGE receptors in vascular smooth muscle, which is at least partially due to the absence of selective antagonists. Contractile effects of PGE can also be mediated by vascular thromboxane receptors (ARNER et al. 1991) or the primary prostaglandin receptor because they can be blocked by TP-receptor antagonists.

Although an FP receptor has been claimed for vascular tissue (KEITH and SALAMA 1987), it is frequently found that contractile actions of $PGF_{2\alpha}$ on vascular and nonvascular smooth muscle can be blocked by thromboxane receptor antagonists (MCKENNIFF et al. 1991; NOREL et al. 1991), suggesting a cross-reactivity with the TXA_2 receptor (DORN et al. 1992) or the involvement of the "primary" vascular receptor for prostaglandins (HANASAKI and ARITA 1989). Moreover, differences may exist between different vascular beds of the same species. NAKAJIMA and UEDA (1990) have compared several isolated endothelium-denuded vessel preparations of the cat, finding that contractions of coronary arteries by $PGF_{2\alpha}$ were much less affected by a specific TX-receptor antagonist (S-145) than those in renal and mesenteric arteries, but were antagonized in a competitive manner by diphloretin phosphate, a nonspecific inhibitor of contractile prostaglandins. This suggests that in coronary vessels of the cat $PGF_{2\alpha}$ may cause contractions that are not mediated by the TXA_2 receptor but by the "primary" receptor for prostaglandins. Evidence for two different receptors, mediating contractile and relaxing effects of $PGF_{2\alpha}$, was obtained in preparations from human hand veins (ARNER et al. 1991).

A PGI_2 receptor was recently characterized in membranes of mastocytoma cells by photoaffinity labeling (ITO et al. 1992). According to displacement studies, a similar type of PGI_2 receptor may also exist in smooth muscle cells. Ligand-binding studies with stable analogs suggested that different PGI_2 receptor subtypes exist on a neuronal cell line (NCB-20) and platelets (WILKENS and MACDERMOT 1987). In vascular tissue, binding studies with chemically stable PGI_2 analogs have identified two binding sites in membrane preparations of the bovine coronary artery. The high-affinity receptor had a k_D of 21 nM at a capacity of 40 fmol/mg protein. This (sub)set of PGI_2 receptors was thought to be responsible for the biological effects of PGI_2 in this tissue (TOWN et al. 1982), because these required comparable concentrations (SCHRÖR et al. 1981). A high-affinity binding site with similar properties was also found in membrane preparations of the pig aorta (RÜCKER and SCHRÖR 1983).

At concentrations $\geqslant 1\,\mu M$, PGI_2 becomes a vasoconstrictor in large arteries and veins (SCHRÖR et al. 1981; NOLL et al. 1991). These effects can be blocked by thromboxane-receptor antagonists (NOLL et al. 1991), again suggesting some cross-reactivity of PGI_2 with the thromboxane receptor in vascular smooth muscle as discussed above. Interestingly, TXA_2-receptor-mediated contractile actions in porcine coronary arteries were antagonized by PGI_2 at low concentrations ($<1\,\mu M$) but not those induced by potassium chloride, i.e., receptor-independent stimulation of $Ca^{2+}{}_i$ via Ca^{2+}-channel-mediated influx of Ca^{2+}.

b) Regulation of Receptors

Prolonged exposure of vascular smooth muscle to an excitatory agonist may result in reduction of receptor density. The response is slow in onset and reaches peak values after 24 h in cultured rat aortic smooth muscle cells (Furci et al. 1991). As an alternative to this agonist-induced homologous down-regulation of receptors, there is also evidence for upregulation of receptors by de novo synthesis of receptor protein. For example, expression of the low-affinity binding site for contractile prostaglandins in vascular smooth muscle was stimulated eightfold by active phorbol esters. This effect was blocked by the PKC inhibitor H-7, cycloheximide, and actinomycin D. No such changes were seen with the high-affinity TXA_2 receptor. These data strongly suggest a selective, PKC-mediated stimulation of receptor protein synthesis for the low-affinity binding site of thromboxane in vascular tissue (Hanasaki et al. 1990a).

Dorn et al. (1992) tested whether serum or defined growth factors could regulate TXA_2 receptors in cultured rabbit aorta smooth muscle cells. Fetal bovine serum (10%) stimulated cell proliferation and DNA synthesis in subconfluent cell cultures. Binding studies showed a 41% decrease in TXA_2 receptors in cells treated with 10% serum compared with serum-deprived (0.1%) controls. This receptor downregulation by serum was gradually reversible upon serum withdrawal. Furthermore, low concentrations of platelet-derived growth factor and basic fibroblast growth factor also decreased TXA_2 receptor density without stimulating cell proliferation of DNA synthesis. This suggests a regulatory action of growth factors on a vascular smooth muscle vasoconstrictor receptor which is independent of effects on cell proliferation or DNA synthesis.

There is also one report demonstrating an increase in thromboxane receptor number in cultured rat aortic smooth muscle cells after testosterone treatment (Masuda et al. 1991b). These receptors were coupled to cellular signal transduction pathways as shown by the increased IP_3 formation and contractile responses of the aorta. The opposite, i.e., TXA_2-receptor down-regulation by 25% after treatment with glucocorticoids, has also been shown in vascular tissue (Sessa et al. 1990) and may also have functional significance (Sessa and Nasjletti 1990). Disappearance of the human myometrial adenylyl cyclase activation by prostaglandins has been reported at the end of pregnancy (Breuiller et al. 1991).

2. G Proteins

Heterotrimeric GTP-binding proteins (G proteins) couple receptor activation to signaling pathways in the cell membrane, including phospholipases C and

A, adenylyl cyclase, and ionic channels (see Fig. 2). Several distinct classes of G proteins are involved in coupling of eicosanoid receptors at the plasma membrane to cellular signal transduction pathways. G_q probably mediates the actions of TXA_2/PGH_2 receptor stimulation. Relaxing actions of PGE_1, PGE_2, and PGI_2 are thought to be mediated by G_s-coupled adenylyl cyclase activation and involve cAMP as a second messenger in some but not all types of vascular smooth muscle (VEGESNA and DIAMOND 1986) and selected intestinal smooth muscle preparations (NAKAHATA and SUZUKI 1981). In human erythroleukemia cells, prostacyclin stimulates $[Ca^{2+}]_i$ via pertussis-toxin-insensitive G protein (SCHWANER et al. 1992). However, there is evidence for involvement of a pertussis-toxin-sensitive G protein, probably G_i, that mediates inhibition of isoprenaline-induced increase in cAMP by prostaglandins in the pregnant rat myometrium (GOUREAU et al. 1990) and flow-induced PGI_2 production (BERTHIAUME and FRANGOS 1992).

Short-term exposure of vascular smooth muscle to stimulatory agonists, such as thromboxane mimetics, results in a desensitization, i.e., reduced phosphatidylinositol breakdown (PI response) and calcium signaling, probably reflecting receptor/G protein uncoupling (FURCI et al. 1991). This reaction is similar to that in platelets (see JASCHONEK and MULLER 1988) and may also occur in endothelial cells (SCHRÖDER and SCHRÖR 1993) and human erythroleukemia cells (SCHWANER et al. 1992). There is currently much research being undertaken in this field and fresh insights may be expected from molecular biology.

3. Phosphoinositol Metabolism, Protein Kinase C, and Cytosolic Ca^{2+}

Receptor activation in probably every cell type including smooth muscle of blood vessels, urinary bladder, penis, trachea, and myometrium results in G-protein-coupled stimulation of phospholipase C (PLC) as a primary response. This is followed by hydrolysis of membrane-associated phosphatidyl inositol by the phosphodiesteratic action of the enzyme and results in the formation of two additional classes of second messenger, diacyl glycerol (DAG) and inositol 1,4,5-trisphosphate (IP_3). The generated IP_3 acts as a receptor agonist on intracellular Ca^{2+} storage sites causing a transient initial increase in Ca_i^{2+}, which in concert with DAG synergistically stimulates protein kinase C (PKC), resulting in sustained responses, i.e., tonic contractions of smooth muscle. DAG is also a source for AA after cleavage by diacyl glycerol lipase.

There are two pathways of increasing cytosolic Ca^{2+}: (IP_3-induced) release from intracellular storage sites and stimulated Ca^{2+} entry from the extracellular space. Both mechanisms appear to be involved in contractile actions of eicosanoids. Ca_i^{2+} could then act at different levels: (a) phospholipase activation and eicosanoid generation, (b) stimulation of second messenger systems within the smooth muscle cell, i.e., PKC and other kinases, and (c) stimulation of myosin phosphorylation. In addition, some eicosanoids may act as calcium sensitizers.

a) Phospholipase Activation and Eicosanoid Generation

Liberation of AA via PLA_2 is Ca^{2+}-dependent. Association with G proteins may lower the Ca^{2+} requirements for PLA_2 in a way similar to PLC (SMITH et al. 1986). This suggests that the level of free cytosolic Ca^{2+} determines cellular eicosanoid production. TAKEUCHI et al. (1991) have shown that an increase in Ca_i^{2+} is associated with bradykinin (BK) as well as angiotensin II-induced PGI_2 synthesis in rat vascular smooth muscle cells. Thus, an interaction between PLA_2 (prostaglandin synthesis) and phospholipase C (IP_3-induced Ca^{2+} mobilization) is suggested.

STEWART et al. (1984) originally demonstrated that contractile agonists (norepinephrine) stimulate vascular prostaglandin synthesis by a pathway that was sensitive to Ca^{2+}-channel blockers. This suggested that (a) receptor-linked Ca^{2+} channels were involved and that (b) significant amounts of the Ca_i^{2+} elevation following agonist-induced receptor stimulation were from the extracellular space because the responses were inhibited by removal of extracellular Ca^{2+} and treatment with Ca^{2+} channel blockers. Similar results were obtained in nonvascular tissue of the rabbit bladder (GOTOH et al. 1986).

JOHNSON et al. (1988) have shown that the calcium channel agonist Bay K 8644 stereoselectively competed for binding with a thromboxane mimetic in washed human platelets. Stimulation of vascular PGI_2 synthesis by the thromboxane mimetic U-46619 was prevented in the absence of external Ca^{2+} and dose-dependently antagonized by verapamil (JEREMY et al. 1985). These data suggest a relationship between the TXA_2/PGH_2 receptor signaling pathways and dihydropyridine/verapamil-sensitive calcium channels. However, enhanced cytosolic Ca^{2+} per se appears not to be sufficient, since K^+-induced stimulation in contrast to that evoked by chemical agonists was not accompanied by enhanced prostaglandin formation (STEWART et al. 1984; JEREMY et al. 1988). Nor was there any stimulation of vascular PGI_2 formation by a calcium agonist, Bay K 8644 (RITTER et al. 1987).

b) Stimulation of Cellular Messenger Systems

Contractile eicosanoids, such as TXA_2 or $PGF_{2\alpha}$, cause G-protein-coupled stimulation of PLC and enhance cytosolic Ca^{2+} concentrations (FUKUO et al. 1986; DORN et al. 1987; TAYLOR et al. 1989). IP_3 is probably the most relevant stimulator of Ca^{2+} release from intracellular sources, in particular the sarcoplasmic reticulum (SUGA and ROTH 1987). $PGF_{2\alpha}$-induced increase in vessel tone and rises in Ca_i^{2+} are accompanied by a rapid increase in IP_3 and Ca_i^{2+} which precedes the increase in contractile tone in rat aorta (MORIMOTO et al. 1990), suggesting that stimulation of phosphoinositol breakdown and subsequent Ca^{2+} increase from both extracellular and (to a small extent) intracellular sources is the major cause of the contractile effects of this agonist. Experiments in permeabilized rat aortic smooth muscle cells by the same authors additionally showed that part of the

$PGF_{2\alpha}$-induced calcium signal was still maintained, suggesting an additional receptor-independent component. A similar conclusion was reached by FRANTZIDES et al. (1992), who demonstrated a direct, verapamil-sensitive stimulation of intestinal smooth muscle contraction by $PGF_{2\alpha}$. Prostaglandin stimulation in nonvascular smooth muscle of guinea pig myometrium also resulted in an increase in IP_3 that was possibly receptor mediated (SCHIEMANN et al. 1991).

Contractions of rabbit basilar arteries (TOWART and PERZBORN 1981) and cat coronaries (SMITH et al. 1981) by the thromboxane mimetic carbocyclic thromboxane A_2 were dose-dependently antagonized by calcium channel blockers whereas contractions of the saphenous vein remained largely unchanged (TOWART and PERZBORN 1981). However, rat aortic phosphoinositol hydrolysis was nifedipine-insensitive (LEGAN et al. 1985), whereas that in the rabbit iris was blocked by calcium antagonists and calcium removal and restored by small amounts ($20\,\mu M$) of Ca^{2+} in microsomal preparations of this tissue (AKHTAR and ABDEL-LATIF 1980). Thus, a general rule concerning the contribution of different calcium sources for cytosolic calcium levels does not appear to exist. LEGAN et al. (1985) have shown that a heterogeneous set of $\text{adrenergic}_{\alpha 1}$ receptors exists in the rat aorta: One class mobilizes Ca_c^{2+}, is antagonized by Ca^{2+}-antagonists, and may mediate the tonic phase of contraction. A second class of recognition sites mobilizes Ca_i^{2+} via enhanced PI turnover and IP_3 accumulation independent of Ca^{2+}-channel blockers and may mediate the acute phasic component.

Another important messenger system for eicosanoids in smooth muscle is PKC. Several isoenzymes, both cytosolic and membrane associated, have been identified and suggested to be involved in smooth muscle activation and AA release after exposure to chemical agonists. LANG and VALLOTTON (1989) have shown that two different Ca^{2+}-sensitive isoforms of PKC exist in cell membranes and cytosol of rat aortic smooth muscle cells. These cells are known to respond to vasoactive peptides by producing PGI_2. Recent studies in rat renal mesangial cells have additionally suggested that PKC_α mediates feedback inhibition of phosphoinositole hydrolysis, whereas PKC_ε is a candidate for regulating PG synthesis in mesangial cells (HUWILER et al. 1991).

Phorbol esters mimick the activating effects of DAG on PKC in vitro. Phorboldibutyrate and other active phorbolesters cause marked contractions of isolated rabbit vascular smooth muscle (RASMUSSEN et al. 1984), rat and rabbit aorta (DANTHULURI and DETH 1984), and cat pial arteries (HORSBURGH et al. 1990). These contractions were slow in onset (about 1 h) and irreversible after washout. However, reduction of cytosolic Ca^{2+} by cAMP elevation with forskolin (RASMUSSEN et al. 1984) and removal of external Ca^{2+} reversed these contractions while calcium ionophore potentiated these responses. This implied that PKC activation elicits calcium entry from external sources associated with excitation-contraction coupling which may be responsible for the tonic phase of agonist-induced contractions (RASMUSSEN et al. 1984; DANTHULURI and DETH 1984).

Protein kinase C activation in vascular smooth muscle is associated with enhanced AA liberation via PLA_2 stimulation, which can be potentiated by calcium ionophore (CHAKRABORTI et al. 1991). Receptor-activated calcium channels, linked to prostaglandin synthesis in vascular tissue, can also be activated by PKC. It is possible that an increase in receptor density at the protein synthesis and transcriptional level contributes to these stimulatory effects of PKC (HANASAKI et al. 1990a) (see Sect. B.III.1).

c) Stimulation of Myosin Phosphorylation

An increase in cytosolic calcium allows binding to calmodulin, the ubiquitous and multifunctional Ca^{2+}-binding protein. This results in conformational changes of Ca^{2+}/calmodulin and expression of interaction sites with target proteins, most notably myosin light-chain kinase and phospholipase(s) (STEWART et al. 1984). This is followed by myosin phosphorylation and smooth muscle contraction (see WALSH 1991 for review). Thromboxane mimetics may also stimulate smooth muscle contraction by acting as Ca^{2+} sensitizers. This was shown in permeabilized rabbit pulmonary artery smooth muscle cells, where intracellularly stored Ca^{2+} was eliminated by Ca^{2+} ionophore. The reaction appeared to be G-protein-coupled because it could also be elicited by GTPγS (HIMPENS et al. 1990). A similar conclusion was reached by OZAKI et al. (1990), studying the $PGF_{2\alpha}$-induced changes in Ca_i^{2+} in vascular smooth muscle of the rat aorta. $PGF_{2\alpha}$ stimulated mono- and diphosphorylation of myosin light chain in a smooth muscle cell line. The diphosphorylation was significantly inhibited by staurosporin, an inhibitor of PKC as well as by downregulation of PKC after long-term treatment with phorbolester, suggesting PKC involvement in myosin phosphorylation by $PGF_{2\alpha}$ (SASAKI et al. 1990). Similar results were obtained by the same group in rabbit aorta (SETO et al. 1991).

4. Cyclic Nucleotide Related Pathways

In addition to excitatory pathways, G-protein-coupled and PLC-dependent mechanisms appear to be involved in inhibitory actions of eicosanoids on smooth muscle function. G_S-coupled adenylyl cyclase stimulation and subsequent formation of cAMP appears to be the dominant cellular second messenger pathway. It is generally accepted that inhibition of smooth muscle contraction by cAMP via decreases in myosin light chain phosphorylation may be related to decreases in free cytosolic calcium concentrations (TAYLOR et al. 1989).

In contrast to the established role of PLC in eicosanoid-induced changes in smooth muscle function and eicosanoid formation the involvement of cyclic nucleotides and its role in regulation in smooth muscle activity is far less understood and in general appears to be more complex. For example, PGI_2 was shown to stimulate vascular cAMP in endothelial-denuded preparations of the rabbit aorta and bovine coronary artery. PGI_2 contracted the rabbit aorta but relaxed the coronary vessel (VEGESNA and DIAMOND 1986). Stimulation of cAMP by forskolin also caused relaxations of cat pial arteries (HORSBURGH et al. 1990). SCHRÖR and RÖSEN (1979) found a reduced cAMP level after low concentrations ($\geqslant 30\,nM$) in nonprecontracted bovine coronary arteries under conditions in which isoprenaline caused marked increases in cAMP, while DEMBINSKA-KIEC et al. (1980) reported a significant increase in cAMP in bovine artery smooth muscle cells. Vasopressin and angiotensin II stimulated cAMP accumulation in rat mesenteric artery smooth muscle cells. The effects on cAMP were transient and weak (about twofold) (HASSID 1986). Additionally, cAMP-dependent protein kinases may increase the Ca^{2+} sensitivity of the Ca^{2+}-activated K^+ channels in some vascular smooth muscles (SADOSHIMA et al. 1988) (see below). Thus, it remains to be determined whether changes in tissue cAMP are the cause or effect of prostacyclin-mediated reductions in smooth muscle tone.

In nonvascular tissue, forskolin was found to act synergistically with isoproterenol and relaxing prostaglandins (PGE_2, PGI_2) in terms of cAMP elevation and relaxation (MOKHTARI et al. 1985). This suggests the involvement of a common G protein, probably G_S. However, studies in the pregnant rat myometrium showed an inhibition of isoprenaline-induced cAMP stimulation by PGE_2 and $PGF_{2\alpha}$. This effect was pertussis toxin-sensitive and insensitive to Ca^{2+} depletion, suggesting the involvement of G_i (GOUREAU et al. 1990).

Moreover, $PGF_{2\alpha}$, PGE_2, and PGI_2 stimulated cAMP levels in nonvascular smooth muscle from the rat myometrium (VESIN et al. 1979) but uniformly caused contraction. PGE_2 inhibited receptor-mediated adenylyl cyclase stimulation in bovine myometrial plasma membranes. This was taken as evidence that this receptor may be involved in regulation of myometrial contractility (LERNER et al. 1990). However, the opposite, i.e., stimulation of cAMP formation, was reported in slices of human myometrium (LOPEZ BERNAL et al. 1991).

All of these data suggest that the relationship between changes in cAMP and smooth muscle function are tissue-dependent and may not be tightly correlated.

5. Ionic Channels

In addition to PLC-related cell stimulation via IP_3, chemical agonists may also modify smooth muscle tone by a direct action on ionic channels. As discussed above, G-protein-related L-type calcium channels may become activated after agonist binding to its membrane recognition site. Alternatively, the receptor may be directly connected with an ionic channel. In fact, there is evidence that even the same chemical agonist, e.g., acetylcholine or GABA, may act on G-protein-dependent and non-G-protein-dependent signal transduction pathways.

G-protein-independent increase of Ca_i^{2+} by enhanced Ca^{2+} influx from the extracellular space through potassium-activated Ca^{2+} channels could provide another mechanism to explain the contractile effects of eicosanoids on smooth muscle tone (SCORNIK and TORO 1992). These authors have investigated the contractile action of the thromboxane mimetic U-46619 on tone and K^+-Ca^{2+} channel activity in pig coronary artery. They found a significant inhibition of channel activity by the thromboxane mimetic, which was antagonized by the receptor blocker SQ 29,548. Indirect evidence suggested that G proteins were not involved. Earlier work by this group had already identified different types of K^+-Ca^{2+} channels in membrane preparations of coronary smooth muscle (TORO et al. 1991). Studies in isolated human myometrial cells have shown that $PGF_{2\alpha}$ stimulated the cellular calcium signal only in the presence of external Ca^{2+}, whereas in the absence of external Ca^{2+} the $PGF_{2\alpha}$ dose-response curve was shifted to the right by two log units. The contractile response was inhibited by verapamil but not by pertussis toxin or phorbol ester and was obtained at concentrations of the agonist that did not require enhanced IP_3 production. These findings were clearly different from calcium stimulation by oxytocin and suggested a primary action of $PGF_{2\alpha}$ on receptor-independent promotion of calcium entry via voltage-operated calcium channels (MOLNAR and HERTELENDY 1990).

There is also evidence that the vasorelaxing properties of PGI_2 may be due to its action on potassium channels, i.e., an activation resulting in membrane hyperpolarization and secondary inhibition of activation of voltage-operated Ca^{2+} channels (SIEGEL et al. 1987, 1989). These authors studied changes in membrane potential and tension of isolated dog carotid arteries and found a marked hyperpolarization and decrease in tension, both effects being dose dependent at nanomolar concentrations of the PGI_2 mimetic iloprost. The K^+ permeability was increased by 340% as shown in tracer flux experiments. From these data it was concluded that iloprost may act as K^+-channel opener.

Similar results were also obtained in nonvascular smooth muscle. SHIKADA et al. (1991) investigated the relaxant effect of potassium channel openers on spontaneous and spasmogen-induced tone in the isolated guinea pig trachea. Cromakalim reversed the contraction induced by the TXA_2 mimetic U-46619, suggesting that potassium channels were involved. YAMADA et al. (1984) and SATOH et al. (1991) reported that contractions of isolated dog coronary arteries evoked by the thromboxane mimetic U-46619 and TXA_2 were prevented by KF4939 and cromakalim. TORO et al. (1991) have demonstrated the existence of calcium-activated potassium in coronary arteries. Later work by this group (SCORNIK and TORO 1992) provided evidence that inhibition of this channel occurs after treatment with the thromboxane mimetic U-46619. These data collectively suggest the presence of functionally relevant potassium channels for vasoconstrictor prostaglandins.

A specific type of potassium channel was recently detected in gastric smooth muscle cells of the toad. This channel was directly activated by AA and a number of nonsubstrate fatty acids. The activation did not require Ca^{2+}, nucleotides, or any other water-soluble cytosolic cofactor, suggesting a direct action of the fatty acid at a protein or lipid site in the membrane. However, the possible contribution of conversion of AA to active metabolites was not eliminated (ORDWAY et al. 1992). Similar channels were also found in smooth muscle of porcine tracheobronchial lymphatic vessels (FERGUSON and TZENG 1991). Contractile responses following exposure to histamine alone and in the presence of AA demonstrated a 40%–55% inhibition of histamine-induced contraction by AA. These inhibitory actions of AA (and other fatty acids) were only partially, by about 50%, reduced by indomethacin and nordihydroguaiaretic acid but eliminated by exposure to a high potassium concentration.

Additionally, large-conductance Ca^{2+}-activated K^+ channels have been recently characterized in rabbit pulmonary artery smooth muscle cells. These channels are activated by mechanical stretch, AA, and other fatty acids. Activation does not require cytosolic constituents. This suggests that fatty acids and membrane stretch either directly affect the channel itself or a membrane component closely associated with it (KIRBER et al. 1992).

Opening of these large-conductance potassium channels (270 pS at 130 mM K^+) will produce outward rectifying currents, resulting in rapid hyperpolarization of the smooth muscle cell. This will antagonize contractile activity because of reducing Ca^{2+} influx through voltage-operated Ca^{2+} channels. This effect will be more pronounced if the muscle is in an activated state. There is evidence for this type of action with AA (DE MEY and VANHOUTTE 1982).

SIEGEL et al. (1989) were the first to demonstrate that the stable PGI_2 mimetic iloprost at vasoactive concentrations ($1\,nM - 1\,\mu M$) causes marked hyperpolarization of vascular smooth muscle cells in the canine carotid artery. This was appreciated with a marked increase in K^+ efflux, suggesting that prostacyclin will exert its vasodilating actions via opening of K^+ channels.

C. Eicosanoids and Smooth Muscle Function

Biosynthesis of eicosanoids occurs in two stages: (a) conversion of arachidonate to the prostaglandin endoperoxide PGH_2 by PGH synthase, lipoxygenases, or cytochrome P450-dependent monooxygenases and (b) formation of end products, for example, isomerization of PGH_2 to the active end products PGD_2, PGE_2, PGI_2, TXA_2, or others.

I. Cyclooxygenase Products

1. Prostaglandin Endoperoxides

The prostaglandin endoperoxides PGG_2 and PGH_2 are unstable intermediates in prostaglandin synthesis. Their main phypsiological function is that of precursors for the subsequent isomerase steps. Moreover, PGH_2 may accumulate under certain conditions, for example, after inactivation of major catabolizing enzymes in vitro, such as the prostacyclin synthase, resulting in nonenzymatic isomerization to PGD_2, PGE_2, TXA_2, and $PGF_{2\alpha}$ (SMITH 1986). PGG_2 and PGH_2 contract vascular smooth muscle (ELLIS et al. 1976; SVENSSON and HAMBERG 1976) after binding to the TXA_2/PGH_2 receptor (TESFAMARIAM and COHEN 1992). These direct effects of PGH_2 are superimposed on the activities of various metabolites, and are, therefore, only detected if significant metabolization of PGH_2 does not occur. This can be demonstrated by injection of high doses (5–50 μg) of AA into a rabbit aorta (PAGANO et al. 1991). PGH_2 may attenuate endothelium-dependent relaxations in vitro by a mechanism involving (endothelium-derived) generation of superoxide anion (TESFAMARIAM and COHEN 1992).

2. Prostaglandin D_2

Prostaglandin D_2 is a significant AA metabolite in the gastrointestinal tract of rodents and some small blood vessels, including bovine coronary microvessels (GERRITSEN and PRINTZ 1981). A major source for PGD_2 in men is the mast cell, which releases PGD_2 together with histamine and the cysteinyl leukotrienes (LTC_4, LTD_4). The initial product of PGD_2 catabolism, 9α-11β-$PGF_{2\alpha}$, has also been found in human blood vessels (ROBERTS et al. 1987). PGF-type compounds represent about two-thirds of all PGD_2 metabolites in primates. Thus, several of the actions of PGD_2 on smooth muscle, including broncho- and vasoconstriction, may be caused by formation of PGF metabolites (SHIKADA et al. 1991). In bovine coronary arteries, PGD_2 was found to cause both direct contractions and endothelium-dependent relaxations that were mediated by NO-induced cGMP stimulation (BRAUN and SCHRÖR 1992). Relaxing effects were also obtained with PGD_2 in rabbit transverse stomach strips (NARUMIYA and TODA 1985).

Because of the involvement of different signal transduction pathways, the biological actions of PGD_2 on smooth muscle function are difficult to predict (NARUMIYA and TODA 1985; TODA 1982; GILES and LEFF 1988). Major targets of PGD_2 are smooth muscles of the vasculature and bronchi. Coronary vessels are contracted in some species, whereas the tone of many other vessels is not changed. PGD_2 is a potent bronchoconstrictor in men (HARDY et al. 1984). PGD_2 is involved in inflammatory and allergic erythema (cold urticaria) by dilatation of arterioles. Intradermal administration of PGD_2 results in increased skin blood flow and erythema (GILES and LEFF 1988). The primary metabolite of PGD_2, 9α,11β-PGD_2, is also biologically active and causes vasoconstriction of coronary arteries as well as inhibition of platelet function at a molar activity similar to that of PGD_2 (GILES and LEFF 1988).

3. Prostaglandin E_2

Prostaglandin E_2 is a major cyclooxygenase product in the microcirculation (GERRITSEN and CHELI 1983). Generation of PGE_2 (and other PGH isomerase products) in freshly isolated smooth muscle (aorta) may result from non-enzymatic breakdown of accumulating PG endoperoxides since homogenates of aorta lack PGH-PGE isomerase activity (SALMON et al. 1978; AGER et al. 1982). However, subcultured arterial smooth muscle cells can generate PGE_2 from added AA or PG endoperoxides (AGER et al. 1982). Relaxation of arterial vessels, i.e., hypotension, is the most prominent effect of PGE_2 on vascular smooth muscle in vivo. In vitro, the type of vasomotor action is different, less predictable, and frequently biphasic, consisting of a relaxation at low and contraction at high ($\geqslant 1\,\mu M$) concentrations. However, the opposite behavior has also been observed in cat coronary arteries (NAKAJIMA and UEDA 1990).

Venous tissue is usually relaxed by PGE_2 at low and contracted at higher concentrations (ARNER et al. 1991). In human hand veins, these contractions are antagonized by thromboxane receptor antagonists, suggesting the involvement of the thromboxane receptor (ARNER et al. 1991). Similar results, i.e., a pronounced contraction subsequent to PGE_1, which was blocked by the thromboxane receptor antagonist SQ 29,548, was seen in isolated human saphenous vein (Braun and Schrör, unpublished). Alternatively, the primary prostaglandin receptor could be involved (see above).

Prostaglandin E_2 also markedly influences smooth muscle tone in non-vascular tissue. BENNETT et al. (1968) have shown that E-type prostaglandins contract the longitudinal musculature of the small intestine in several species including men while the circular musculature was relaxed. PGE also contracts the trachea of different species in vitro but antagonizes spasm induced by histamine at nanomolar concentrations (MAIN 1964).

4. Prostaglandin $F_{2\alpha}$

Biosynthesis of $PGF_{2\alpha}$ from the prostaglandin endoperoxide PGH_2 occurs via two different pathways: direct reduction of PGH_2 and isomerization of PGE_2 (or PGD_2) with subsequent reduction of the 9-keto group by the prostaglandin 9-keto-reductase. Like PGE_2, $PGF_{2\alpha}$ can also be formed non-enzymatically by spontaneous hydrolysis of prostaglandin endoperoxides. This mechanism is particularly relevant in vivo during accumulation of prostaglandin endoperoxides at low enzymatic conversion rates of PG endoperoxides, e.g., in the central nervous system or in the presence of thromboxane synthase inhibitors. $PGF_{2\alpha}$ stimulates vascular smooth muscle although less potent than many other eicosanoids. These actions are probably mediated by the contractile prostaglandin receptor. $PGF_{2\alpha}$ might become an important vasoactive substance because of the high chemical stability in areas with low metabolic degradation, i.e., cerebral circulation and vasospasm after hemorrhage. $PGF_{2\alpha}$ is a potent contractile agent for nonvascular smooth muscle in human bronchi (MCKENNIFF et al. 1991; NOREL et al. 1991) and the uterus.

5. Prostaglandin I_2

Prostaglandin I_2 is the major eicosanoid synthesized in smooth muscle and endothelial cells of large arteries and veins (AGER et al. 1982; DEWITT et al. 1983; JEREMY and DANDONA 1989). The PGI_2 production of arteries is significantly higher than that of veins. Interestingly, if vein graft-arterialization was performed, the low PGI_2-forming capacity of the vein bypass at 12 weeks remained the same, amounting to about 50% of that from the arteries (FANN et al. 1990). Stimulation of vascular smooth muscle by any type of contractile agonist results in marked PGI_2 production which is independent of the presence of endothelium (JEREMY and DANDONA 1989). PGI_2 is generated at different subcellular sites within smooth muscle cells, including plasma membrane, endoplasmic reticulum, and nuclear membranes. According to immunocytochemical studies, the enzymes are probably located at the cytoplasmic surface of the respective membrane (SMITH et al. 1983).

PGI_2 relaxes most vascular and nonvascular smooth muscle preparations at low (nanomolar) concentrations, but exhibits the constrictor activity at high (micromolar) concentrations. The relaxing response is probably due to activation of K^+ channels and may be partially antagonized at higher concentrations of the compound by increased Na^{2+} permeability (SIEGEL et al. 1989) or an action on the primary prostaglandin receptor. The contractile effect of PGI_2 dominates in vein and lymphatic vessels, including bovine coronary veins (SCHRÖR et al. 1981). PGI_2 is also a weak contractor in isolated pig coronary arteries (DUSTING et al. 1977). PGI_2 contracts rabbit detrusor muscle at a potency comparable to that of $PGF_{2\alpha}$ and PGE_2 (GOTOH et al. 1986).

6. Thromboxane A_2

Thromboxane A_2 (TXA_2) is the predominant AA metabolite in human platelets. It is also synthesized in smaller amounts in some blood vessels, most notably (embryonic) vessels from umbilical cord and lung. Using immunohistochemical methods, thromboxane synthase was localized within the smooth muscle layer of the uterus, possibly within dendritic or monocytotic cells (NÜSING et al. 1990). TXA_2 is a constrictor of vascular (arteries, veins, and lymphatic vessels) and nonvascular smooth muscles (ELLIS et al. 1976; SVENSSON and HAMBERG 1976; SMITH et al. 1987). Chemically stable thromboxane mimetics may be more potent because of their resistance to degradation. For example, the EC_{50} for contractions of canine saphenous veins by I-BOP was as low as 38 p*M* (MORINELLI et al. 1989). In the presence of platelets, direct effects of TXA_2 may be superimposed on the release of platelet-derived vasoconstrictors, such as 5-HT (VERHEGGEN and SCHRÖR 1986). TEMPLETON et al. (1991) showed that the increase in oxygen partial pressure caused contraction of the human umbilical arteries that was antagonized by thromboxane receptor antagonists.

II. Leukotriene Pathways

Leukotrienes (LTs) are formed via the 5- and 15-lipoxygenase pathways of AA. The intermediate epoxide LTA_4 is the precursor of LTB_4, synthesized from AA as a major product in polymorphonuclear neutrophils. In mast cells, monocyte/macrophages, and the lung, LTA_4 is converted to LTC_4 with further stepwise degradation to LTD_4 and LTE_4. The biological actions of LTA_4 on smooth muscle are determined primarily by its metabolites, i.e., the cysteinyl leukotrienes. Leukotrienes are important mediators of inflammatory and immune reactions as well as potent spasmogens in vascular and nonvascular smooth muscle. This has stimulated considerable pharmacological research into selective inhibitors of leukotriene formation or blockers of leukotriene action with putative use as antiinflammatory, antiallergic, or antiasthmatic agents (SNYDER and FLEISCH 1989).

1. Leukotriene B_4

Leukotriene B_4 has no significant effect on smooth muscle tone at concentrations that markedly influence neutrophil chemotaxis. Nevertheless, LTB_4 may cause indirect changes of contraction because of precursor exchange after accumulation of polymorphonuclear leukocytes because of its chemotactic properties. In addition, FEINMARK and CANNON (1986) have shown that transcellular metabolism between endothelium and adherent polymorphonuclear neutrophil (PMN) results in profound LTC_4 formation by the combined action of PMN and endothelium: LTA_4, generated by the PMN, may be converted into LTC_4 by endothelial cell γ-glutamyl transferase. LTB_4 also stimulates smooth muscle cell proliferation in vitro by a mechanism that might involve stimulation of endogenous PG production (PALMBERG et al. 1991).

2. Cysteinyl Leukotrienes

Leukotriene C_4 (LTC_4) is the conjugation product of LTA_4 and glutathione and a major AA metabolite in macrophages and human mast cells as well as in antigen-sensitized lung tissue. LTD_4 is generated from LTC_4 after partial hydrolysis of the peptide chain. Whether LTC_4 or LTD_4 dominate depends upon the presence of a peptidase, converting LTC_4 into LTD_4.

The cysteinyl leukotriene LTC_4 (and its metabolites LTD_4 and LTE_4) contracts human bronchi at a potency that is several orders of magnitude higher than that of histamine (WEISS et al. 1982). Cysteinyl leukotrienes are the biologically active principle of "slow-reacting substance of anaphylaxis" (SRS-A). The compounds also stimulate mucus secretion, and both actions are probably important for allergic-inflammatory pulmonary disorders (DRAZEN and AUSTEN 1987). After antigen challenge, a biphasic contraction is obtained in different types of smooth muscle, including airways (see BURKA 1988) and intestine (BARNETTE and GROUS 1992), where the late contractile response appears to be cysteinyl leukotriene-mediated. LTC_4 and LTD_4 are also considered as mediators in hypoxic pulmonary vasoconstriction (VOELKEL et al. 1987) and hypoxia-induced contractions of canine isolated basilar artery (GU et al. 1991). LTC_4 appears to be involved in circulatory and pulmonary disorders during septic shock and in anaphylactic reactions with the possible exception of intestinal anaphylaxis (SCOTT and MARIC 1991).

III. Other Lipoxygenase Products

Hydro(pero)xyeicosatetraenoic acids (H(P)ETEs) are products of arachidonic metabolism via the several lipoxygenase pathways. The unstable hydroperoxides are reduced either spontaneously or by the action of peroxidases to the corresponding hydroxyeicosatetraenoic acid (HETE) (SPECTOR et al. 1988).

1. 5-Hydro(pero)xyeicosatetraenoic Acid

5-Hydro(pero)xyeicosatetraenoic acid (5-H(P)ETE) can be converted either to 5-HETE or into the epoxide leukotriene $A_4(LTA_4)$. There is no evidence for significant 5-HETE production by smooth muscle cells. 5-HETE is a contractile agent for human myometrium in vitro, the potency being tenfold less than $PGF_{2\alpha}$ (BENNETT et al. 1987).

2. 12-Hydro(pero)xyeicosatetraenoic Acid

12-Hydro(pero)xyeicosatetraenoic acid (12-H(P)ETE) formation is a major metabolic pathway of AA in many tissues, including vascular smooth muscle. 12-HETE is the substrate for formation of further vasoactive products, i.e., the lipoxins. 12-HETE stimulates the phenotypic modulation and motility of aortic smooth muscle cells in culture. No such effect is seen with 15-HETE and 13-HODE, which appear as potent prodifferentiating molecules (RAMBOER et al. 1992). 12-HETE, like other fatty acid peroxides, is a vasoconstrictor in many blood vessels (TRACHTE et al. 1979). Although less active on a molar basis than 5-HETE, 12-HETE has to be considered as a major modulator of the inflammatory process because of its multiple sources in mammalian tissues, in particular the microvasculature.

3. 15-Hydro(pero)xyeicosatetraenoic Acid

15-Hydro(pero)xyeicosatetraenoic acid (15-H(P)ETE) is the enzymatic product of arachidonic acid conversion via the 15-lipoxygenase and/or the inducible type II PGH synthase (see above). 15-HETE is the major enzymatic product of AA metabolism in human airway epithelial cells (SALARI and SCHELLENBERG 1991) and is the major HETE in the vasculature. There is evidence for an inverse relationship between smooth muscle derived 15-HETE and PGI_2 generation during enhanced lipid peroxidation in vitro (EK and HUMBLE 1991). There is also evidence for an inverse relationship between 15-lipoxygenase and 5-lipoxygenases, which may be important for induction of labor (LEI and RAO 1992). 15-HETE is probably a key mediator of airway mucous secretion in allergic respiratory diseases. It enhances endothelial cell migration and proliferation, suggesting its involvement in angiogenesis. 15-H(P)ETE is a potent inhibitor of endothelial prostacyclin formation and of other lipoxygenases, eventually resulting in reduced 12-H(P)ETE and leukotriene formation. In some vascular beds, most notably the cerebral and coronary circulation, 15-HETE is a vasoconstrictor (TRACHTE et al. 1979).

Cultured human airway epithelial cells, stimulated with AA or PAF for periods of 30 min to 24 h, generated 15-HETE as the dominating product but no leukotrienes, thromboxane, or prostaglandins aside from minimal amounts of PGE_2. Maximal production of 15-HETE was 235 and 153 ng/mg protein subsequent to stimulation by AA or PAF, respectively. 15-HETE at concentrations of $\geqslant 0.1\,\mu M$ contracted isolated human bronchus on a time course similar to that of endogenous 15-HETE generation, beginning 30–60 min after 15-HETE challenge and reaching a maximum at approximately 2 h. These results demonstrate that PAF may induce delayed airway smooth muscle contraction by the generation of 15-HETE from epithelial cells. The kinetics of 15-HETE generation and its contractile activity are compatible with it being a mediator of the late asthmatic reaction (SALARI and SCHELLENBERG 1991).

In general, the biological actions of 5-, 12-, and 15-H(P)ETE are much less well understood than those of leukotrienes or cyclooxygenase products. One reason for this is the absence of selective antagonists and the uncertainty about structure and function of cellular binding sites. Recent evidence suggests that all of these fatty acid peroxides can be taken up by smooth muscle cells (human umbilical artery) and may inhibit cell proliferation under certain experimental conditions (BRINKMAN et al. 1991).

4. Lipoxins

Lipoxins (LXs) ("lipoxygenation interaction products") are trihydroxy fatty acids formed by a unique interaction of 15- and 5-lipoxygenases. Alternatively, the compounds may be generated from 5,15-di-H(P)ETE via a 12-lipoxygenase. LXA_4 and LXB_4 have a broad spectrum of biological activities on vascular and nonvascular smooth muscle which in some tissues may involve additional stimulation of other eicosanoids (DAHLÉN 1989). The compounds contract airway smooth muscle cells, suggesting an involvement in pathologic immune reactions, including bronchospasm.

5. Hepoxilins

Hepoxilins are formed from 12-H(P)ETE as the precursor of 8-hydroxy-11,12-epoxyeicosa-5,9,14-traenoic acid (hepoxilin A_3) and 10-hydroxy-11,12-epoxyeicosa-5,8,14-traenoic acid (hepoxilin B_3), respectively. This pathway, originally described in the pancreas, has also recently been demonstrated in the rat aorta (LANEUVILLE et al. 1991). Its biological significance is still unknown.

IV. Monooxygenase Pathways

A third pathway of arachidonic acid oxygenation involves cytochrome P450-dependent monooxygenases, which have been most extensively studied in the kidney (MCGIFF 1991). However, this pathway of AA is also present in the vasculature, in particular in the intima, but also in the media (ABRAHAM et al. 1985). Initially, epoxides are formed which are then converted by epoxide hydrolases to the corresponding diols, i.e., dihydroxyeicosatraenoic acids. This pathway has also been demonstrated in microsomes of vascular smooth muscle cells (HASUNUMA et al. 1991).

The biological significance and function of monooxygenase-derived AA metabolites is not yet completely understood. In cholesterol-fed rabbits, expression of the epoxygenase pathway was found (PFISTER et al. 1991). The four EET regioisomers were found to relax the precontracted rabbit aorta with greater effectivity in vessels prepared from hypercholesterolemic animals where expression of this pathway was found (PFISTER et al. 1991). A particularly interesting aspect is that monooxygenase products may act as substrate for subsequent conversion by cyclooxygenases into vasoactive products (see MCGIFF 1991).

These metabolites exert a variety of vasomotor effects which are dependent on the type of vessel studied (PROCTOR et al. 1989). Interestingly, some compounds, such as 20-HETE and 5,6-EET, are only expressed after conversion by cylooxygenase to prostaglandin analogs (CARROLL et al. 1990). ESCALANTE et al. (1990) have shown that 19- and 20-hydroxylated HETEs may relax vascular smooth muscle via stimulation of Na^+/K^+-ATPase, which appears to be mediated by cyclooxygenase product formation, stimulated by these compounds. A cytochrome P450 link also appears to be important for the vascular responses in changes to oxygen tension, for example in closure of the ductus arteriosus after birth (COCEANI et al. 1984) or in hypoxic pulmonary vasoconstriction (SYLVESTER and MCGOWAN 1978). The pathophysiology of this system is just beginning to be explored (MCGIFF 1991).

D. Eicosanoids and Smooth Muscle Dysfunction

I. Smooth Muscle Cell Proliferation

Proliferation of vascular smooth muscle cells is an important event in wound healing. Smooth muscle cell proliferation and changes of the phenotype are also important in vessel injury in atherosclerosis as is growth of media smooth muscle cells in hypertension (CAMPBELL and CAMPBELL 1986); both events are probably related to endothelial dysfunction. U-46619 stimulated DNA synthesis in primary culture of smooth muscle cells from rat aorta. This was accompanied by activation of PLC and generation of IP_3, resulting in increased cytosolic free Ca^{2+} levels. The effects of the thromboxane mimetic were antagonized by thromboxane receptor antagonists, suggesting that they were receptor-mediated (HANASAKI et al. 1990b). However, no mitogenic effect was seen in these cells by Dorn et al. (1992), using a different experimental protocol. MORINELLI et al. (1992) have studied the thromboxane-induced mitogenesis in guinea pig vascular smooth muscle cells using the stable thromboxane mimetic I-BOP. They also found a receptor-mediated potent mitogenic activity for this compound, which involved stimulation of intracellular protein kinases.

Vascular smooth muscle cells from spontaneously hypertensive rats exhibit rapid cell growth as shown by an increased [^{3}H]thymidine uptake. The capacity for formation of vasodepressor prostaglandins was significantly reduced. Supplementation of PGI_2 by a stable PGI_2 analog reduced the enhanced cell growth to values in nonhypertensive animals, suggesting that impairments in biosynthesis of vasodilating prostaglandins are involved in the hypertrophy of vascular smooth muscle in hypertension (UEBARA et al. 1991). Stimulation of cAMP in smooth muscle may mediate these antiproliferative effects of PGI_2 (WILLIS et al. 1986; SINZINGER et al. 1987; SHIROTANI et al. 1991).

II. Smooth Muscle Tone in Pathologic Situations

1. Arterial Vasospasm

The role of eicosanoids in arterial vasospasm is often discussed in terms of a disturbed "balance" between vasoconstrictor and vasodilator prostaglandins, in particular an insufficient amount of PGI_2. It should be noted, however, that PGI_2 in all vascular tissues studied so far, including coronary vessels (SCHRÖR et al. 1981) and human cerebral arteries (SCHRÖR and VERHEGGEN 1986), exerts vasocontractile actions at high ($\geqslant 1\,\mu M$) concentrations. Moreover, contractions evoked by potassium chloride depolarization have been shown to be not antagonized by PGI_2 (NOLL et al. 1991).

The significance of eicosanoids and endothelium-derived factors in hypoxia-induced coronary vasospasm is species-dependent (TODA et al.

1992). In monkey and human coronary artery strips contracted with prostaglandin $PGF_{2\alpha}$ or K^+, exchange of N_2 for O_2 produced an endothelium-independent contraction. Indomethacin, and ONO 3708, a PG receptor antagonist, markedly inhibited the hypoxia-induced contraction, whereas superoxide dismutase and ozagrel were ineffective, suggesting that TXA_2- or oxygen-centered radicals were not involved. Diltiazem depressed the contraction. Hypoxia increased the release of PGE_2 but not of PGI_2 and TXB_2, suggesting that the hypoxia-induced contraction of monkey and human epicardial coronary arteries is associated with release of vasoconstrictor cyclooxygenase products from subendothelial sources which are not identical to TXA_2 (TODA et al. 1992). $PGF_{2\alpha}$ or PG endoperoxides appear to be likely candidates.

2. Bronchospasm

Interactions between eicosanoids, i.e., cysteinyl leukotrienes and histamine, are most prominent in anaphylaxis, including bronchospasm (DOUGLAS and BRINK 1987). Endogenous eicosanoids may markedly alter the tone of airway smooth muscle (BURKA 1988). Inhalation of LTD_4 (or LTC_4) causes bronchospasm in healthy volunteers. In asthmatics, only one-tenth of the dose in healthy volunteers is required for the same bronchoconstrictor response. This suggests that the bronchi of asthmatics are about tenfold more sensitive against these agents. However, this finding of specific bronchial hyperreactivity against leukotrienes in asthmatics has been challenged in more recent investigations. The contribution of other spasmogenic mediators, such as histamine and platelet-activating factor in the overall spasmogenic effect of LTD_4, is unknown as is the major site of action, i.e., large airways or small bronchi and bronchioli. According to DRAZEN and AUSTEN (1987), the major pathology of asthmatic attacks consists of a more general sensitization of the hyperreactive bronchial system against endogenous bronchoconstrictors. Any direct effect of leukotrienes on these responses is probably antagonized by rapid tachyphylaxis. However, human airway epithelial cells are a rich source for the bronchoconstrictor eicosanoid 15-HETE after platelet activating factor (PAF) challenge (SALARI and SCHELLENBERG 1991). There is also a clearly increased generation of broncho constrictor leukotrienes in the airways of patients suffering from bronchial asthma (SEEGER and GRIMMINGER 1990). Alternatively, a nonselective airway hyperreactivity to spasmogens in asthmatics may also result from impaired barrier function of central airway epithelium i.e., reduced generation and release of smooth muscle relaxant factors (GOLDIE et al. 1988).

Potent (pKB, 9.3 in human bronchi), selective and orally active LTD_4 antagonists, such as Ro 24-5913, [(*E*)-4-[3-[2-(4-cyclobutyl-2-thiazolyl) ethenyl] phenylamino]-2,2-diethyl-4-oxobutanoic acid], are now available that are effective antagonists of LTD_4-induced contractions in human bronchi (O'DONNELL et al. 1991). Clinical studies with these compounds are presently underway and should provide fresh insights into the role of these substances in bronchospasm.

E. Summary and Conclusions

Vascular and nonvascular smooth muscle are an important source and target for eicosanoids. Prostaglandins, thromboxanes, and leukotrienes affect smooth muscle tone and cell proliferation by receptor-mediated interference with intracellular signal transduction pathways that are similar to those for other hormones and chemical agonists. Consequently, there are numerous interactions between eicosanoids, circulating hormones (kinins, catecholamines), and other local chemical mediators. In almost every smooth muscle studied so far, there is a modification of contractile states under physiological conditions by endothelium/epithelium-derived relaxing (EDRF, EpDRF) and contractile (endothelins) factors.

High-affinity receptors for prostaglandins ($k_D \leqslant 1\,nM$) are located at the plasma membrane. Five main classes and several subtypes of receptors have been pharmacologically separated. A more detailed classification, also including leukotriene receptors, is expected from molecular biology in the near future. Receptors for excitatory prostaglandins and thromboxane A_2 are coupled via G proteins (Gq?) to phospholipase C. Stimulation causes a rapid rise in IP_3 with subsequent elevation of cytosolic calcium from both intra- (SR) and extracellular sources. Thromboxane A_2 also stimulates a calcium-activated K^+ channel that causes prolonged depolarization and vasoconstriction. In addition, there appears to exist a "primary prostaglandin receptor" to which all primary prostaglandins bind at higher ($\geqslant 1\,\mu M$) concentrations.

Receptors for inhibitory prostaglandins (PGI_2) are bound to both G_s and probably G_i proteins. Stimulation also results in generation of a calcium signal and reduction of cytosolic calcium concentrations by stimulation of its uptake into cellular storage sites via stimulation of adenylyl cyclase and increase in cAMP. There is also evidence for activation of KATP (index) channel at the cell membrane by prostacyclins, resulting in a marked increase in K^+ permeability, hyperpolarization, and relaxation. The opposite, i.e., inhibition of K^+-channels, was reported for thromboxane mimetics. There appear to be considerable differences between the several types of smooth muscle regarding these and other cellular signal transduction pathways. Thus, even the direction of responses in particular with PGI_2 is difficult to predict.

Typical diseases with a pathology in arachidonic acid metabolism are arterial vasospasm and bronchospasm. In both cases there is a markedly increased contractile tone and a hyperreactivity against contractile eicosanoids and other stimulatory factors. Numerous compounds have been synthesized that modulate these contractile responses. Preliminary data suggest that these agents will also be helpful for clinical treatment.

References

Abdel-Latif AA, Zhang Y, Yousufzai SY (1991) Endothelin-1 stimulates the release of arachidonic acid and prostaglandins in rabbit iris sphincter smooth muscle: activation of phospholipase A_2. Curr Eye Res 10:259–265

Abraham BG, Pinto A, Mullane KM, Levere RD, Spokas RG (1985) Presence of cytochrome P-450-dependent monooxygenase in intimal cells of the hog aorta. Hypertension 7:899–904

Ager A, Gordon JL, Moncada S, Pearson JD, Salmon J, Trevethick MA (1982) Effect of isolation and culture on prostaglandin synthesis by porcine aortic endothelial and smooth muscle cells. J Cell Physiol 110:9–16

Akhtar RA, Abdel-Latif AA (1980) Requirement for calcium ions in acetylcholine-stimulated phosphodiesteratic cleavage of phosphatidyl-myoinositol 4,5-biphosphate in rabbit iris smooth muscle. Biochem J 192:783–791

Arner M, Hoegestaett ED, Uski TK (1991) Characterization of contraction-mediating prostanoid receptors in human hand veins: effects of the thromboxane receptor antagonists BM 13,505 and AH 23848. Acta Physiol Scand 141:79–86

Barnette MS, Grous M (1992) Characterization of the antigen-induced contraction of colonic smooth muscle from sensitized guinea pigs. Am J Physiol 262: G144–G149

Bennett A, Eley G, Scholes GB (1968) Effects of prostaglandins E_1 and E_2 on human, guinea-pig and rat isolated small intestine. Br J Pharmacol 34:630–638

Bennett PR, Murdoch G, Elder MD, Myatt L (1987) The effects of lipoxygenase metabolites of arachidonic acid on human myometrial contractility. Prostaglandins 33:837–844

Berthiaume F, Frangos JA (1992) Flow-induced prostacyclin production is mediated by a pertussis toxin-sensitive G-protein. FEBS Lett 308:277–279

Braun M, Schrör K (1992) Prostaglandin D_2 relaxes bovine coronary arteries by endothelium-dependent nitric oxide-mediated cGMP formation. Circ Res 71:1305–1313

Breuiller M, Doualla-Bell F, Litime MH, Leroy MJ, Ferre F (1991) Disappearance of human myometrial adenylate cyclase activation by prostaglandins at the end of pregnancy. Comparison with beta-adrenergic response. Adv Prostaglandin Thromboxane Leukotriene Res 21B:811–814

Brinkman HJ, van Buul-Wortelboer MF, van Mourik JA (1991) Selective conversion and esterification of monohydroxyeicosatetraenoic acids by human vascular smooth muscle cells: relevance to smooth muscle cell proliferation. Exp Cell Res 192:87–92

Bunce KT, Clayton NM, Coleman RA, Collington EW, Finch H, Humphray JM, Humphrey PP, Reeves JJ, Sheldrick RL, Stables R (1991) GR63799X – a novel prostanoid with selectivity for EP_3 receptors. Adv Prostaglandin Thromboxane Leukotriene Res 21A:379–382

Burka JF (1988) Role of eicosanoids in airway smooth muscle tone. Prog Clin Biol Res 263:35–46

Campbell JH, Campbell GR (1986) Endothelial cell influences on vascular smooth muscle phenotype. Annu Rev Physiol 48:295–306

Carroll MA, Garcia MP, Falck JR, McGiff JC (1990) 5,6-Epoxyeicosatrienoic acid, a novel arachidonate metabolite: mechanism of vasoactivity in the rat. Circ Res 67:1082–1088

Chakraborti S, Michael JR, Patra SK (1991) Protein kinase C dependent and independent activation of phospholipase A_2 under calcium ionophore (A23187) exposure in rabbit pulmonary arterial smooth muscle cells. FEBS Lett 285: 104–107

Coceani F, Hamilton NC, Labuc J, Olley PM (1984) Cytochrome P-450-linked monooxygenase: involvement in the lamb ductus arteriosus. Am J Physiol 246:H640–H643

Coleman RA, Kennedy I, Sheldrick PLG (1987) Evidence for the existence of three subtypes of PGE_2 sensitive (EP) receptors in smooth muscle. Br J Pharmacol 91:323P

Dahlén SE (1989) Pharmacological activities of lipoxins and related compounds. Adv Prostaglandin Thromboxane Leukotriene Res 19:122–127

Danthuluri NR, Deth RC (1984) Phorbol ester induced contraction of arterial smooth muscle and inhibition on α-adrenergic response. Biochem Biophys Res Commun 125:1103–1109

Dembinska-Kiec A, Rücker W, Schönhöfer PS (1980) Effects of PGI_2 and PGI_2 analogues on cAMP levels in cultured smooth muscle cells derived from bovine arteries. Naunyn Schmiedebergs Arch Pharmacol 311:67–70

De Mey JG, Vanhoutte PM (1982) Heterogenous behaviour of the canine arterial and venous wall. Importance of the endothelium. Circ Res 51:439–447

DeWitt DL, Day JS, Sonnenburg WK, Smith WL (1983) Concentrations of prostaglandin endoperoxide synthase and prostaglandin I_2 synthase in the endothelium and smooth muscle of bovine aorta. J Clin Invest 72:1882–1888

Dorn GW II (1991) Tissue- and species-specific differences in ligand binding to thromboxane A_2 receptors. Am J Physiol 261:R145–R153

Dorn GW II, Becker MW (1992) Growth factors downregulate vascular smooth muscle thromboxane receptors independent of cell growth. Am J Physiol 262:C927–C933

Dorn GW II, Sens D, Chaikhouini D, Halushka PV (1987) Cultured human vascular smooth muscle cells with functional thromboxane A_2 receptors: measurement of U 46619-induced ^{45}Ca efflux. Circ Res 60:952–956

Dorn GW II, Becker MW, Davis MG (1992) Dissociation of the contractile and hypertrophic effects of vasoconstrictor prostanoids in vascular smooth muscle. J Biol Chem 267:24897–24905

Drazen JM, Austen KF (1987) Leukotrienes and airway responses. Am Rev Resp Dis 136:985–998

Douglas JS, Brink C (1987) Airway smooth muscle and disease workshop: histamine and prostanoids. Am Rev Respir Dis 136:S21–S24

Dusting GJ, Moncada S, Vane JR (1977) Prostacyclin (PGI_2) is a weak contractor of coronary arteries of the pig. Eur J Pharmacol 45:301–304

Ek B, Humble L (1991) Correlation between oxidation of low density lipoproteins and prostacyclin synthesis in cultured smooth muscle cells. Biochem Pharmacol 41:695–699

Ellis EF, Oelz O, Roberts II LJ, Payne NA, Sweetman BJ, Nies AS, Oates JA (1976) Coronary arterial smooth muscle contraction by a substance released from platelets: evidence that it is thromboxane A_2. Science 193:1135–1137

Escalante B, Sessa WC, Falck JR, Yadagiri P, Schwartzman ML (1990) Cytochrome P-450-dependent arachidonic acid metabolites, 19- and 20-hydroxyeicosatetraenoic acids, enhance sodium-potassium ATPase activity in vascular smooth muscle. J Cardiovasc Pharmacol 16:438–443

Fann JI, Sokoloff MH, Sarris GE, Yun KL, Kosek JC, Miller DC (1990) The reversibility of canine vein-graft arterialization. Circulation 82 Suppl 5: IV9–IV18

Farmer SG, Ensor JE, Burch RM (1991) Evidence that cultured airway smooth muscle cells contain bradykinin B_2 and B_3 receptors. Am J Respir Cell Mol Biol 4:273–277

Feinmark SJ, Cannon JP (1986) Endothelial cell leukotriene C_4 synthesis results from intercellular transfer of leukotriene A_4 synthesized by polymorphonuclear leukocytes. J Biol Chem 261:16466–16472

Ferguson MK, Tzeng E (1991) Attenuation of histamine-induced lymphatic smooth muscle contractility by arachidonic acid. J Surg Res 51:500–505

Fernandes LB, Goldie RG (1991) Antigen-induced release of airway epithelium-derived inhibitory factor. Am Rev Respir Dis 143:567–571

Filep JG, Battistini B, Sirois P (1991) Pharmacological modulation of endothelin-induced contraction of guinea-pig isolated airways and thromboxane release. Br J Pharmacol 103:1633–1640
Fleisch JH, Rinkema LE, Baker SR (1982) Evidence for multiple leukotriene D_4 receptors in smooth muscle. Life Sci 31:577–581
Folts JD, Crowell ED, Rowe GG (1976) Platelet aggregation in partially obstructed vessels and its elimination with aspirin. Circulation 54:365–370
Frantzides CT, Frantzides EAL, Wittmann D, Greenwood B, Edmiston CE (1992) Prostaglandins and modulation of small bowel myoelectric activity. Am J Physiol 262:G488–G497
Fukuo K, Morimoto S, Koh E, Yukawa S, Tsuchiya H, Imanaka S, Yamamoto H, Onishi T, Kumahara Y (1986) Effects of prostaglandins on the cytosolic free calcium concentration in vascular smooth muscle cells. Biochem Biophys Res Commun 136:247–252
Furci L, Fitzgerald DJ, FitzGerald GA (1991) Heterogeneity of prostaglandin H_2/thromboxane A_2 receptors: distinct subtypes mediate vascular smooth muscle contraction and platelet aggregation. J Pharmacol Exp Ther 258:74–81
Garcia JL, Monge L, Gomez B, Dieguez G (1991) Response of canine cerebral arteries to endothelin-1. J Pharm Pharmacol 43:281–284
Gerrard JM (ed) (1985) Prostaglandins and leukotrienes. Blood and vascular cell function. Dekker, New York
Gerritsen ME, Cheli CD (1983) Arachidonic acid metabolism and prostaglandin endoperoxide metabolism in isolated rabbit coronary microvessels and isolated and cultivated coronary microvessel endothelial cells. J Clin Invest 72: 1658–1671
Gerritsen ME, Printz MP (1981) Sites of prostaglandin synthesis in the bovine heart and isolated bovine coronary microvessels. Circ Res 49:1152–1163
Gerzer R, Brash AR, Hardman JG (1986) Activation of soluble guanylate cyclase by arachidonic acid and 15-lipoxygenase products. Biochim Biophys Acta 886: 383–389
Giles H, Leff P (1988) The biology and pharmacology of PGD_2. Prostaglandins 35:277–300
Glass DB, Frey W II, Carr DW, Goldberg ND (1977) Stimulation of platelet guanylate cyclase by fatty acids. J Biol Chem 252:1279–1285
Goldie RG, Fernandes LB, Rigby PJ, Paterson JW (1988) Epithelial dysfunction and airway hyperreactivity in asthma. Prog Clin Biol Res 263:317–329
Gotoh M, Hassouna M, Elhilali MM (1986) The mode of action of prostaglandin E_2, $F_{2\alpha}$ and prostacyclin on vesicourethral smooth muscle. J Urol 135:431–437
Goureau O, Tanfin Z, Harbon S (1990) Prostaglandins and muscarinic agonists induce cyclic AMP attenuation by two distinct mechanisms in the pregnant-rat myometrium. Interaction between cyclic AMP and Ca^{2+} signals. Biochem J 271:667–673
Grunstein MM, Chuang ST, Schramm CM, Pawlowski NA (1991a) Role of endothelin 1 in regulating rabbit airway contractility. Am J Physiol 260:L75–L82
Grunstein MM, Rosenberg SM, Schramm CM, Pawlowski NA (1991b) Mechanisms of action of endothelin 1 in maturing rabbit airway smooth muscle. Am J Physiol 260:L434–L443
Gu M, Elliott DA, Ong BY, Bose D (1991) Possible role of leukotrienes in hypoxic contraction of canine isolated basilar artery. Br J Pharmacol 103:1629–1632
Halushka PV, Morinelli TA, Mais DE (1990) Radioligand binding assays for thromboxane A_2/prostaglandin H_2 receptors. Methods Enzymol 187:397–405
Hanasaki K, Arita H (1989) A common binding site for primary prostanoids in vascular smooth muscle: a definite discrimination of the binding for thromboxane A_2/prostaglandin H_2 receptor agonist from its antagonist. Biochim Biophys Acts 1013:28–35

Hanasaki K, Nakano T, Ksai H, Arita H, Ohtani K, Doteuchi M (1988) Specific receptors from thromboxane A_2 in cultured vascular smooth muscle cells of rat aorta. Biochem Biophys Res Commun 150:1170–1175
Hanasaki K, Kishi M, Arita H (1990a) Phorbol ester-induced expression of the common, low-affinity binding site for primary prostanoids in vascular smooth muscle cells. J Biol Chem 265:4871–4875
Hanasaki K, Nakano T, Arita H (1990b) Receptor-mediated mitogenic effect of thromboxane A_2 in vascular smooth muscle cells. Biochem Pharmacol 40: 2535–2542
Hardy CC, Robinson C, Tattersfield AE, Holgate ST (1984) The bronchoconstrictor effect of inhaled prostaglandin D_2 in normal and asthmatic men. N Engl J Med 311:209–213
Hassid A (1986) Increase of cyclic AMP concentrations in cultured vascular smooth muscle cells by vasoactive peptide hormones. Role of endogenous prostaglandins. J Pharmacol Exp Ther 239:334–339
Hasunuma K, Terano T, Tamura Y, Yoshida S (1991) Formation of epoxyeicosatrienoic acids from arachidonic acid by cultured rat aortic smooth muscle cell microsomes. Prostaglandins Leukot Essent Fatty Acids 42:171–175
Hechtman DH, Kroll MH, Gimbrone MA Jr, Schafer AI (1991) Platelet interaction with vascular smooth muscle in synthesis of prostacyclin. Am J Physiol 260: H1544–H1551
Hedqvist P (1977) Basic mechanisms of prostaglandin action on autonomic neurotransmission. Annu Rev Pharmacol Toxicol 17:259–279
Himpens B, Kitazawa T, Somlyo AP (1990) Agonist-dependent modulation of Ca^{2+} sensitivity in rabbit pulmonary artery smooth muscle. Pflugers Arch 417:21–28
Hirata M, Hayashi Y, Ushikubi F, Yokota Y, Kageyama R, Nakanishi S, Narumiya S (1991) Cloning and expression of cDNA for a human thromboxane A_2 receptor. Nature 349:617–619
Horsburgh K, Jansen I, Edvinsson L, McCulloch J (1990) Second messenger systems: functional role in cerebrovascular smooth muscle regulation. Eur J Pharmacol 191:205–211
Huwiler A, Fabbro D, Pfeilschifter J (1991) Possible regulatory functions by protein kinase C-$_{\alpha}$ and -$_{\varepsilon}$ isoenzyme in rat renal mesangial cells. Biochem J 279:441–445
Ito S, Hashimoto H, Negishi M, Suzuki M, Koyano H, Noyori R, Ichikawa A (1992) Identification of the prostacyclin receptor by use of [15-^{3}H1]19-(3-azidophenyl)20-norisocarbacyclin, an irreversible specific photoaffinity probe. J Biol Chem 267:20326–20330
Jaiswal N, Diz DI, Tallant EA, Khosla MC, Ferrario CM (1991) The nonpeptide angiotensin II antagonist DuP 753 is a potent stimulus for prostacyclin synthesis. Am J Hypertens 4:228–233
Janssen LJ, Daniel EE (1991) Pre- and postjunctional effects of a thromboxane mimetic in canine bronchi. Am J Physiol 261:L271–L276
Jaschonek K, Muller CP (1988) Platelet and vessel associated prostacyclin and thromboxane A_2/prostaglandin endoperoxide receptors. Eur J Clin Invest 18: 1–8
Jeremy JY, Dandona P (1989) Effect of endothelium removal on inhibitory modulation of rat aortic prostacyclin synthesis. Br J Pharmacol 96:243–250
Jeremy JY, Mikhailidis DP, Dandona P (1985) The thromboxane A_2 analogue U46619 stimulates vascular prostacyclin synthesis. Eur J Pharmacol 107:259–262
Jeremy JY, Mikhailidis DP, Dandona P (1988) Excitatory receptor-prostanoid synthesis coupling in smooth muscle: mediation by calcium, protein kinase C and G proteins. Prostaglandins Leukot Essent Fatty Acids 34:215–227
Johnson GJ, Dunlop PC, Leis LA, From AHL (1988) Dihydropyridine agonist Bay K 8644 inhibits platelet activation by competitive antagonism of thromboxane A_2-prostaglandin H_2 receptor. Circ Res 62:494–505

Keen M, Kelly E, MacDermot J (1989) Prostaglandin receptors in the cardiovascular system: potential selectivity from receptor subtypes or modified responsiveness. Eicosanoids 2:193–197
Keith RA, Salama AI (1987) Individual variations of prostanoid agonist responses in rabbit aorta. Evidence for the independent regulation of prostanoid receptor subtypes. Br J Pharmacol 92:133–148
Kennedy I, Coleman RA, Humphrey PPA, Levy GP, Lumley P (1982) Studies on the characterisation of prostanoid receptors: a proposed classification. Prostaglandins 24:667–689
Kirber MT, Ordway RW, Clapp LH, Walsh JV Jr, Singer JJ (1992) Both membrane stretch and fatty acids directly activate large conductance $CA^{(2+)}$-activated K^+ channels in vascular smooth muscle cells. FEBS Lett 297:24–28
Kujubu DA, Fletcher BS, Varnum BC, Lim RW, Herschman HR (1991) TIS10, a phorbol ester tumor promoter-inducible mRNA from Swiss 3T3 cells, encodes a novel prostaglandin synthase/cyclooxygenase homologue. J Biol Chem 266: 12866–12872
Kurihara H, Nakano T, Takasu N, Arita H (1991) Intracellular localization of group II phospholipase A_2 in rat vascular smooth muscle cells and its possible relationship to eicosanoid formation. Biochim Biophys Acta 1082:285–292
Laneuville O, Corey EJ, Couture R, Pace-Asciak CR (1991) Hepoxilin A_3 (HxA_3) is formed by the rat aorta and is metabolized into HxA_3-C, a glutathione conjugate. Biochim Biophys Acta 1084:60–68
Lang U, Vallotton MB (1989) Effects of angiotensin II and of phorbol ester on protein kinase C activation and on prostaglandin production in cultured rat aortic smooth muscle cells. Biochem J 259:477–483
Lefer AM, Smith EF III, Araki H, Smith JB, Aharony D, Claremon D, Magolda RL, Nicolaou KC (1980) Dissociation of vasoconstrictor and platelet aggregatory activities of thromboxane by carbocyclic thromboxane A_2, a stable analog of thromboxane A_2. Proc Natl Acad Sci USA 77:1706–1710
Leff P, Giles H (1992) Classification of platelet and vascular prostaglandin D_2 (DP) receptors: estimation of affinities and relative efficacies for a series of novel bicylic ligands. With an appendix on goodness-of-fit analyses. Br J Pharmacol 106:996–1003
Legan E, Chernow B, Parrillo J, Roth BL (1985) Activation of phosphatidylinositol turnover in rat aorta by $alpha_1$-adrenergic receptor stimulation. Eur J Pharmacol 110:389–390
Lei ZM, Rao CV (1992) The expression of 15-lipoxygenase gene and the presence of functional enzyme in cytoplasm and nuclei of pregnancy human myometria. Endocrinology 130:861–870
Lerner RW, Lopaschuk GD, Olley PM (1990) High-affinity prostaglandin E receptors attenuate adenylyl cyclase activity in isolated bovine myometrial membrane. Can J Physiol Pharmacol 68:1574–1580
Lin L-L, Lin AY, Knopf JL (1992) Cytosolic phospholipase A_2 is coupled to hormonally regulated release of arachidonic acid. Proc Natl Acad Sci USA 89:6147–6151
Liu ECK, Hedberg A, Goldenberg HJ, Harris DN, Webb ML (1992) DUP 753, the selective angiotensin II blocker, is a competitive antagonist to human platelet thromboxane A_2/prostaglandin H_2 (TP) receptor. Prostaglandins 44: 89–99
Lopez Bernal A, Buckley S, Rees CM, Marshall JM (1991) Meclofenamate inhibits prostaglandin E binding and adenylyl cyclase activation in human myometrium. J Endocrinol 129:439–445
Lumley P, White BP, Humphrey PPA (1989) GR 32191, a highly potent and specific thromboxane A_2 receptor blocking drug on platelets and vascular and airway smooth muscle. Br J Pharmacol 97:783–794
Main IHM (1964) The inhibitory actions of prostaglandins on respiratory smooth muscle. Br J Pharmacol 22:511–519

Mais DE, Sauss DL Jr, Chaikhouni A, Kochel PJ, Knapp DR, Hamanaka N, Halushka PV (1985) Pharmacologic characterization of human and canine thromboxane A_2/prostaglandin H_2 receptors in platelets and blood vessels. evidence for different receptors. J Pharmacol Exp Ther 233:418–424

Marceau F, deBlois D, Laplante C, Petitclerc E, Pelletier G, Grose JH, Hugli TE (1990) Contractile effect of the chemotactic factors f-Met-Leu-Phe and C5a on the human isolated umbilical artery. Role of cyclooxygenase products and tissue macrophages. Circ Res 67:1059–1070

Marcus AJ, Weksler BB, Jaffe EA, Broekman MJ (1980) Synthesis of prostacyclin from platelet-derived endoperoxides by cultured human endothelial cells. J Clin Invest 66:979–986

Masferrer JL, Seibert K, Zweifel B, Needleman P (1992) Endogenous glucocorticoids regulate an inducible cyclooxygenase enzyme. Proc Natl Acad Sci USA 89: 3917–3921

Masuda A, Mais DE, Oatis JE, Halushka PV (1991a) Platelet and vascular thromboxane A_2/prostaglandin H_2 receptors. Evidence for different subclasses in the rat. Biochem Pharmacol 42:537–544

Masuda A, Mathur R, Halushka PV (1991b) Testosterone increases thromboxane A_2 receptors in cultured rat aortic smooth muscle cells. Circ Res 69:638–643

McGiff JC (1991) Cytochrome P-450 metabolism of arachidonic acid. Annu Rev Pharmacol Toxicol 31:339–369

McKenniff MG, Norman P, Cuthbert NJ, Gardiner PJ (1991) Bay u3405, a potent and selective thromboxane A_2 receptor antagonist on airway smooth muscle in vitro. Br J Pharmacol 104:585–590

Miki I, Nonaka H, Ishii A (1992) Characterization of thromboxane A_2/prostaglandin H_2 receptors and histamine H_1 receptors in cultured guinea-pig tracheal smooth muscle cells. Biochim Biophys Acta 1137:107–115

Mokhtari A, Do Khac L, Tanfin Z, Harbon S (1985) Forskolin modulates cyclic AMP generation in the rat myometrium. Interactions with isoproterenol and prostaglandins E_2 and I_2. J Cyclic Nucleotide Protein Phosphor Res 10:213–227

Molnar M, Hertelendy F (1990) Regulation of intracellular free calcium in human myometrial cells by prostaglandin $F_{2\alpha}$: comparison with oxytocin. J Clin Endocrinol Metab 71:1243–1250

Moncada S (1992) The L-arginine:nitric oxide pathway. (The 1991 Ulf von Euler Lecture). Acta Physiol Scand 145:201–227

Morimoto S, Koh E, Kim E, Morita R, Fukuo K, Ogihara T (1990) Effects of prostaglandin $F_{2\alpha}$ on the mobilization of cytosolic free calcium in vascular smooth muscle cells and on the tension of aortic strips from rats. Am J Hypertens 3:241S–244S

Morinelli TA, Oatis JE Jr, Okwu AK, Mais DE, Mayeux PR, Matsuda A, Knapp DR, Halushka PV (1989) Characterization of an ^{125}I-labeled thromboxane A_2/prostaglandin H_2 receptor agonist. J Pharmacol Exp Ther 251:557–462

Morinelli TA, Meier KE, Zhang L-M, Newman WE (1992) Thromboxane A_2/prostaglandin H_2 (TXA_2/PGH_2) stimulated mitogenesis of guinea pig vascular smooth muscle cells (SMC) is associated with activation of MAP-kinase and S6 kinase (Abstr). 8th International Conference on Prostaglandins and Related Compounds, Montreal

Mugridge KG, Perretti M, Becherucci C, Parente L (1991) Persistent effects of interleukin-1 on smooth muscle preparations from adrenalectomized rats: implications for increased phospholipase-A_2 activity via stimulation of 5-lipoxygenase. J Pharmacol Exp Ther 256:29–37

Nakahata T, Suzuki T (1981) Effects of prostaglandin E_1, I_2 and isoproterenol on the tissue cyclic AMP content in longitudinal muscle of rabbit intestine. Prostaglandins 22:159–165

Nakajima M, Ueda M (1990) Regional differences in the prostanoid receptors mediating prostaglandin $F_{2\alpha}$-induced contractions of cat isolated arteries. Eur J Pharmacol 191:359–368

Narumiya S, Toda N (1985) Different responsiveness of prostaglandin D_2-sensitive systems to prostaglandin D_2 and its analogues. Br J Pharmacol 85:367–375

Noll G, Buehler FR, Yang Z, Lüscher TF (1991) Different potency of endothelium-derived relaxing factors against thromboxane, endothelin, and potassium chloride in intramyocardial porcine coronary arteries. J Cardiovasc Pharmacol 18: 120–126

Norel X, Labat C, Gardiner PJ, Brink C (1991) Inhibitory effects of BAY u3405 on prostanoid-induced contractions in human isolated bronchial and pulmonary arterial muscle preparations. Br J Pharmacol 104:591–595

Nüsing R, Lesch R, Ullrich V (1990) Immunohistochemical localization of thromboxane synthase in human tissues. Eicosanoids 3:53–58

O'Donnell M, Crowley HJ, Yaremko B, O'Neill N, Welton AF (1991) Pharmacologic actions of Ro 24-5913, a novel antagonist of leukotriene D_4. J Pharmacol Exp Ther 259:751–758

Ogletree ML, Allen GT (1992) Interspecies differences in thromboxane receptors: studies with thromboxane receptor antagonists in rat and guinea pig smooth muscles. J Pharmacol Exp Ther 260:789–793

Ogletree ML, Harris DN, Greenberg R, Haslanger MF, Nakane M (1985) Pharmacological actions of SQ 29,548, a novel selective thromboxane antagonist. J Pharmacol Exp Ther 234:435–441

Ordway RW, Walsh JV, Singer HH (1992) Arachidonic acid and other fatty acids directly activate potassium channels in smooth muscle cells. Science 244: 1176–1179

Ozaki H, Ohyama T, Sato K, Karaki H (1990) Ca^{2+}-dependent and independent mechanisms of sustained contraction in vascular smooth muscle of rat aorta. Jpn J Pharmacol 52:509–512

Pagano PJ, Lin L, Sessa WC, Nasjletti A (1991) Arachidonic acid elicits endothelium-dependent release from the rabbit aorta of a constrictor prostanoid resembling prostaglandin endoperoxides. Circ Res 69:396–405

Palmberg L, Lindgren JA, Thyberg J, Claesson HE (1991) On the mechanism of induction of DNA synthesis in cultured arterial smooth muscle cells by leukotrienes. Possible role of prostaglandin endoperoxide synthase products and platelet-derived growth factor. J Cell Sci 98:141–149

Pfister SL, Falck JR, Campbell WB (1991) Enhanced synthesis of epoxyeicosatrienoic acids by cholesterol-fed rabbit aorta. Am J Physiol 261:H843–H852

Proctor KG, Capdevila JH, Falck JR, Fitzpatrick FA, Mullane KM, McGriff JC (1989) Cardiovascular and renal actions of cytochrome P-450 metabolites of arachidonic acid. Blood Vessels 26:53–64

Ramboer I, Blin P, Lacape G, Daret D, Lamaziere JM, Larrue J (1992) Effects of monohydroxylated fatty acids on arterial smooth muscle cell properties. Kidney Int Suppl 37:S67–S72

Rasmussen H, Forder J, Kojima I, Scriabine A (1984) TPA induced contraction of isolated rabbit vascular smooth muscle. Biochem Biophys Res Commun 122: 776–784

Ritter JM, Frazer CE, Taylor GW (1987) pH-dependent stimulation by calcium of prostacyclin synthesis in rat aortic rings: effects of drugs and inorganic ions. Br J Pharmacol 91:439–446

Roberts LJ II, Seibert K, Liston TE, Tantengoo MV, Robertson RM (1987) PGD_2 is transformed by human coronary arteries to $9\alpha,11\beta$-PGF_1, which contracts human coronary artery rings. Adv Prostaglandin Thromboxane Leukotriene Res 17:427–429

Rücker W, Schrör K (1983) Evidence for high affinity prostacyclin binding sites in vascular tissue: radioligand studies with a chemically stable analog. Biochem Pharmacol 32:2405–2410

Sadoshima J-I, Akaike N, Kanaide H, Nakamura M (1988) Cyclic AMP modulates Ca-activated K channel in cultured smooth muscle cells of rat aortas. Am J Physiol 255:H754–H759

Salari H, Schellenberg RR (1991) Stimulation of human airway epithelial cells by platelet activating factor (PAF) and arachidonic acid produces 15-hydroxyeicosatetraenoic acid (15-HETE) capable of contracting bronchial smooth muscle. Pulm Pharmacol 4:1–7

Salmon J, Smith DR, Flower RJ, Moncada S, Vane JR (1978) Further studies on the enzymatic conversion of prostaglandin endoperoxide into prostacyclin by porcine aorta microsomes. Biochim Biophys Acta 523:250–262

Samuelsson B (1983) Leukotrienes: mediators of immediate hypersensitivity reactions and inflammation. Science 220:568–575

Sasaki Y, Seto M, Komatsu K (1990) Diphosphorylation of myosin light chain in smooth muscle cells in culture. Possible involvement of protein kinase C. FEBS Lett 276:161–164

Satoh K, Yamada H, Taira N (1991) Differential antagonism by glibenclamide of the relaxant effects of cromakalim, pinacidil and nicorandil on canine large coronary arteries. Naunyn Schmiedebergs Arch Pharmacol 343:76–82

Schiemann WP, Doggwiler KO, Buxton IL (1991) Action of adenosine in estrogen-primed nonpregnant guinea pig myometrium: characterization of the smooth muscle receptor and coupling to phosphoinositide metabolism. J Pharmacol Exp Ther 258:429–437

Schröder H, Schrör K (1993) Prostaglandin-dependent cyclic AMP formation in endothelial cells. Naunyn Schmiededbergs Arch Pharmacol 347:101–104

Schrör K (1992) Role of prostaglandins in the cardiovascular effects of bradykinin and angiotensin-converting enzyme inhibitors. J Cardiovasc Pharmacol 20 Suppl 9:S68–S73

Schrör K (1993) Prostaglandin-mediated actions of the renin-angiotensin systems. Arzneimittelforsch 43:236–241

Schrör K, Rösen P (1979) Prostacyclin (PGI_2) decreases the cyclic AMP levels in coronary arteries. Naunyn Schmiedebergs Arch Pharmacol 306:101–103

Schrör K, Verheggen R (1986) Prostacyclins are only weak antagonists of coronary vasospasm indiced by authentic thromboxane A_2 and serotonin. J Cardiovasc Pharmacol 8:607–613

Schrör K, Darius H, Matzky R, Ohlendorf R (1981) The antiplatelet and cardiovascular actions of a new carbacyclin derivative (ZK36374) – equipotent to PGI_2 in vitro. Naunyn Schmiedebergs Arch Pharmacol 316:252–255

Schwaner I, Seifert R, Schultz G (1992) The prostacyclin analogues, cicaprost and iloprost, increase cytosolic CA^{2+}-concentration in human erythroleukemia cell line, HEL, via pertussis toxin-insensitive G-proteins. Eicosanoids 5 (Suppl): 10–12

Scornik FS, Toro L (1992) U46619, a thromboxane A_2 agonist, inhibits K_{Ca} channel activity from pig coronary artery. Am J Physiol 262:C708–C713

Scott RB, Maric M (1991) A limited role for leukotrienes and platelet-activating factor in food protein induced jejunal smooth muscle contraction in sensitized rats. Can J Physiol Pharmacol 69:1841–1846

Seeger W, Grimminger F (1990) Die Rolle von Arachidonsäuremetaboliten in der Pathogenese des Asthma bronchiale. Verh Dtsch Ges Inn Med 96:685–693

Serruys PW, Rutsch W, Heyndrickx GR, Danchin N, Gijs Mast E, Mijns W, Rensing BJ, Vos J, Stibbe J (1991) Prevention of restenosis after percutaneous transluminal coronary angioplasty with thromboxane A_2 receptor blockade. A randomized, double-bind placebo-controlled trial. Circulation 84:1568–1580

Sessa WC, Nasjletti A (1990) Dexamethasone selectively attenuates prostanoid-induced vasoconstrictor responses in vitro. Circ Res 66:383–388

Sessa WC, Halushka PV, Okwu A, Nasjletti A (1990) Characterization of the vascular thromboxane A_2/prostaglandin endoperoxide receptor in rabbit aorta. Regulation by dexamethasone. Circ Res 67:1562–1569

Seto M, Sasaki Y, Hidaka H, Sasaki Y (1991) Effects of HA1077, a protein kinase inhibitor, on myosin phosphorylation and tension in smooth muscle. Eur J Pharmacol 195:267–272

Shikada K, Yamamoto A, Tanaka S (1991) NIP-121 and cromakalim, potassium channel openers, preferentially suppress prostanoid-induced contraction of the guinea-pig isolated trachea. Eur J Pharmacol 209:69–73
Shirotani M, Yui Y, Hattori R, Kawai C (1991) U-61,431F, a stable prostacyclin analogue, inhibits the proliferation of bovine vascular smooth muscle cells with little antiproliferative effect on endothelial cells. Prostaglandins 41:97–110
Siegel G, Stock G, Schnalke F, Litza B (1987) Electrical and mechanical effects of prostacyclin in the canine carotid artery. In: Gryglewski RJ, Stock G (eds) Prostacyclin and its stable analogue iloprost. Springer, Berlin Heidelberg New York, pp 143–149
Siegel G, Carl A, Adler A, Stock G (1989) Effect of the prostacyclin analogue iloprost on K^+ permeability in the smooth muscle cells of the canine carotid artery. Eicosanoids 2:213–222
Sinzinger H, Zidek T, Fitscha P, O'Grady J, Wagner O, Kaliman J (1987) Prostaglandin I_2 reduces activation of human arterial smooth muscle cells in-vivo. Prostaglandins 33:915–918
Smith CD, Cox CC, Snyderman R (1986) Receptor coupled activation of phosphoinositides specific phospholipase C by an N-protein. Science 232:97–100
Smith EF III, Lefer AM, Nicolaou KC (1981) Mechanism of coronary vasoconstriction induced by carbocyclic thromboxane A_2. Am J Physiol 240:H493–H497
Smith EF III, Slivjak MJ, Eckardt RD, Newton JF (1989) Antagonism of leukotriene C_4, leukotriene D_4 and leukotriene E_4 vasoconstrictor responses in the conscious rat with the peptidoleukotriene receptor antagonist SK&F 104353: evidence for leukotriene D_4 receptor heterogeneity. J Pharmacol Exp Ther 249:805–811
Smith JB, Yanagisawa A, Ziplin R, Lefer AM (1987) Constriction of cat coronary arteries by synthetic thromboxane A_2 and its antagonism. Prostaglandins 33: 777–782
Smith WL (1986) Prostaglandin biosynthesis and its compartmentation in vascular smooth muscle and endothelial cells. Annu Rev Physiol 48:251–262
Smith WL, DeWitt DL, Allan ML (1983) Bimodal distribution of the prostaglandin I_2 antigen in smooth muscle cells. J Biol Chem 258:5922–2926
Smith WL, Marnett LJ, DeWitt DL (1991) Prostaglandin and thromboxane biosynthesis. Pharmacol Ther 49:153–179
Snyder DW, Fleisch JH (1989) Leukotriene receptor antagonists as potential therapeutic agents. Annu Rev Pharmacol Toxicol 29:123–143
Souhrada M, Souhrada JF (1991) Respiratory epithelium-dependent inhibition of protein kinase C of airway smooth muscle cells. J Appl Physiol 70:2137–2144
Spector AA, Gordon JA, Moore SA (1988) Hydroxyeicosatetraenoic acids (HETEs). Prog Lipid Res 27:271–323
Stewart D, Poutney E, Fitchett D (1984) Norepinephrine-stimulated vascular prostacyclin synthesis. Receptor-dependent calcium channels control prostaglandin synthesis. Can J Physiol Pharmacol 62:1341–1347
Suga EA, Roth BL (1987) Prostaglandins activate phosphoinositide metabolism in rat aorta. Eur J Pharmacol 136:325–332
Sugimoto Y, Namba T, Honda A, Hayashi Y, Negishi M, Ichikawa A, Narumiya S (1992) Cloning and expression of a cDNA for mouse prostaglandin E receptor EP_3 subtype. J Biol Chem 267:6463–6466
Svensson J, Hamberg M (1976) Thromboxane A_2 and prostaglandin H_2: potent stimulators of the swine coronary artery. Prostaglandins 12:943–950
Sylvester JT, McGowan C (1978) The effects of agents that bind to cytochrome P-450 on hypoxic pulmonar vasoconstriction. Circ Res 43:429–437
Takayanagi I, Kawano K, Koike K (1990) Evidence for release of prostaglandin $F_{2\alpha}$ in contractile response of guinea pig trachea to norepinephrine. Jpn J Pharmacol 54:330–332
Takayasu-Okishio M, Terashita Z, Kondo K (1990) Endothelin-1 and platelet activating factor stimulate thromboxane A_2 biosynthesis in rat vascular smooth muscle cells. Biochem Pharmacol 40:2713–27122

Takeuchi K, Abe K, Maeyama K, Sato M, Yasujima M, Watanabe T, Yoshinaga K (1991) Simultaneous measurements of cytosolic free calcium level and prostaglandin synthesis reveal a correlation between them in perfused monolayer of cultured rat vascular smooth muscle cells: effects of bradykinin and angiotensin II. Tohoku J Exp Med 165:183–192
Taylor DA, Bowman BF, Stull JT (1989) Cytoplasmic Ca^{2+} is a primary determinant for myosin phosphorylation in smooth muscle cells. J Biol Chem 264:6207–6213
Templeton AG, McGrath JC, Whittle MJ (1991) The role of endogenous thromboxane in contractions to U46619, oxygen, 5-HT and 5-CT in the human isolated umbilical artery. Br J Pharmacol 103:1079–1084
Tesfamariam B, Cohen RA (1992) Role of superoxide anion and endothelium in vasoconstrictor action of prostaglandin endoperoxide. Am J Physiol 262: H1915–H1919
Tessier GJ, Lackner PA, O'Grady SM, Kannan MS (1991) Modulation of equine tracheal smooth muscle contractility by epithelial-derived and cyclooxygenase metabolites. Respir Physiol 84:105–114
Toda N (1982) Different responses of a variety of isolated dog arteries to prostaglandin D_2. Prostaglandins 23:99–112
Toda N, Matsumoto T, Yoshida K (1992) Comparison of hypoxia-induced contraction in human, monkey, and dog coronary arteries. Am J Physiol 262:H678–H683
Toro L, Vaca L, Stefani E (1991) Ca^{2+} activated K^+ channels from coronary smooth muscle reconstituted in lipid bilayers. Am J Physiol 260:H1779–H1789
Towart R, Perzborn E (1981) Nimodipine inhibits carbocyclic thromboxane induced contraction of cerebral arteries. Eur J Pharmacol 69:213–215
Town M-H, Schillinger E, Speckenbach A, Prior G (1982) Identification and characterisation of a prostacyclin-like receptor in bovine coronary arteries using a specific and stable prostacyclin analogue, ciloprost, as radioactive ligand. Prostaglandins 24:61–72
Trachte GJ, Lefer AM, Aharony D, Smith JB (1979) Potent constriction of cat coronary arteries by hydroperoxides of arachidonic acid and its blockade by anti-inflammatory agents. Prostaglandins 18:909–914
Uchida Y, Murao S (1974) Cyclic changes in peripheral blood pressure of partially constricted coronary artery. Jpn Coll Angiol 14:383
Uehara Y, Numabe A, Kawabata Y, Nagata T, Hirawa N, Ishimitsu T, Matsuoka H, Ikeda T, Sugimoto T (1991) Rapid smooth muscle cell growth and endogenous prostaglandin system in spontaneously hypertensive rats. Am J Hypertens 4:806–814
Vane JR (1969) The release and fate of vasoactive hormones in the circulation. (Second Gaddum Memorial Lecture). Br J Pharmacol 35:209–242
Vegesna RVK, Diamond J (1986) Elevation of cyclic AMP by prostacyclin is accompanied by relaxation of bovine coroanry arteries and contraction of rabbit aortic rings. Eur J Pharmacol 128:25–31
Verheggen R, Schrör K (1986) The modification of platelet-induced vasoconstriction by a thromboxane receptor antagonist. J Cardiovasc Pharmacol 8:483–490
Vesin MF, Khac LO, Harbon S (1979) Prostacyclin as endogenous modulator of adenosine 3′5′-monophosphate levels in rat endometrium and myometrium. Mol Pharmacol 15:823–840
Voelkel NF, Chang SW, McDonnell TJ, Westcott JY, Haynes J (1987) Role of membrane lipids in the control of normal vascular tone. Am Rev Respir Dis 136:214–217
von Euler US (1935) Über die spezifische blutdrucksenkende Substanz des menschlichen Prostata- und Samenblasensekrets. Klin Wochenschr 14:1182–1185
Walsh MP (1991) Calcium-dependent mechanisms of regulation of smooth muscle contraction. Biochem Cell Biol 69:771–780
Weiss JW, Drazen JM, Coles N, McFadden ER Jr, Weller PF, Corey EJ, Lewis RA, Austen KF (1982) Bronchoconstrictor effects of leukotriene C in humans. Science 216:196–198

Wilkens AJ, MacDermot J (1987) The putative prostacyclin receptor antagonist (FCE-22176) is a full agonist on human platelets and NMCB-20 cells. Eur J Pharmacol 127:117–119
Willis AL, Smith DL, Vigo C (1986) Suppression of principal atherosclerotic mechanisms by prostacyclin and other eicosanoids. Prog Lipid Res 25:645–666
Xie W, Chipman JG, Robertson DL, Erikson RL, Simmons DL (1991) Expression of a mitogen-responsive gene encoding prostaglandin synthase is regulated by mRNA splicing. Proc Natl Acad Sci USA 88:2692–2696
Yamada K, Kubo K, Shuto KI, Nakamizo N (1984) Inhibition of thromboxane A_2 induced vasoconstriction by KF4939, a new anti-platelet agent, in rabbit mesenteric and dog coronary arteries. Jpn J Pharmacol 36:283–290
Yamamoto S (1992) Mammalian lipoxygenases: molecular structures and functions. Biochim Biophys Acta 1128:117–131
Yang SG, Saifeddine M, Chuang M, Severson DL, Hollenberg MD (1991) Diacylglycerol lipase and the contractile action of epidermal growth factor-urogastrone: evidence for distinct signal pathways in a single strip of gastric smooth muscle. Eur J Pharmacol 207:225–230
Yu XY, Hubbard W, Spannhake EW (1992) Inhibition of canine tracheal smooth muscle by mediators from cultured bronchial epithelial cells. Am J Physiol 262:L229–L234

CHAPTER 5

Angiotensin, the Kinins, and Smooth Muscle

D. REGOLI and N.-E. RHALEB

A. Introduction

Angiotensin and the kinins are peptidic hormones contained in large precursors, the angiotensinogens and kininogens, originating from the liver. Precursors are α-globulin substrates for the proteolytic enzymes renin and kallikreins, which are produced by specialized structures (renin by the juxtaglomerular cell) or circulate as inactive precursors readily activated by various types of stimuli (the kallikreins). Molecular biology studies have been instrumental in elucidating the structures and in localizing the substrates and the enzymes of the two systems (CAMPBELL 1985; MULLER-ESTERL et al. 1986). Release of angiotensin I and of bradykinin occurs in the blood stream, while kallidin (Lys-BK) is produced in the extracellular fluid, by the action of tissue kallikreins on low molecular weight kininogens. All these peptides undergo metabolic changes in the lung (VANE 1969), angiotensin I (AT_I) being activated to angiotensin II (AT_{II}) and the kinins being transformed into inactive fragments by kininase II (VANE 1969), which is the same enzyme that converts AT_I (Fig. 1) (ERDÖS 1975).

The biologically active products of the two systems, angiotensin II and the two kinins, bradykinin and kallidin, act on two different receptor types, named AT_1 and AT_2 (BUMPUS 1991), B_1 and B_2 (REGOLI and BARABÉ 1980).

B. Renin-Angiotensin System

The octapeptide angiotensin II has been found to increase blood pressure (PAGE and BUMPUS 1961), stimulate the release of aldosterone from the adrenal glands (LARAGH et al. 1960; GENEST et al. 1960), the contraction of vascular, intestinal, uterine, and other smooth muscles (PRADO et al. 1954; REGOLI et al. 1974), the release of catecholamines from intact animals (FEDELBERG and LEWIS 1964) and isolated perfused adrenals (PEACH et al. 1971), and potentiate the effect of sympathetic nerve stimulation (BENELLI et al. 1964; BLUMBERG et al. 1975). Moreover, AT_{II} is a potent dipsogenic (FITZSIMONS 1972), which also facilitates the release of vasopressin (MALVIN 1971), inhibits renin release (VAN DONGEN et al. 1974), and in some experimental conditions induces an increase of the glomerular filtration rate by constricting the efferent glomerular artery (REGOLI 1972). Depending on the

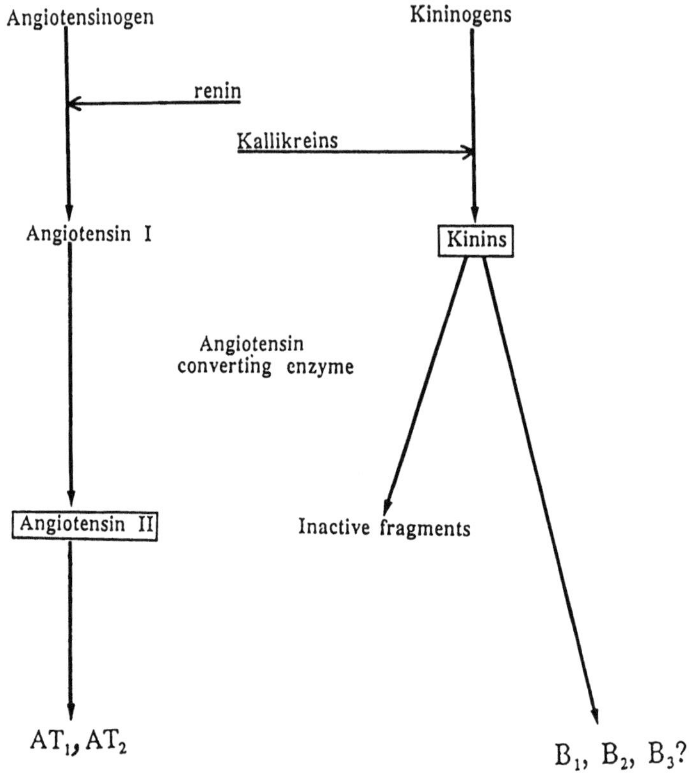

Fig. 1. The renin-angiotensin system and the kallikrein-kinin system

experimental conditions, angiotensin has been shown to promote either the reabsorption or the excretion of sodium (BARRACLOUGH 1965; BONJOUR et al. 1967). By its central and peripheral effects that concur to increase peripheral vascular resistance and expand the extracellular fluid volume, angiotensin plays a major role in the regulation of blood pressure (McCUBBIN 1974; BUNAG 1974). Early studies in the dog (SCHAFFENBURG et al. 1960) and in rats (GROSS et al. 1963) suggested that renin may play a very important role in the pathogenesis of experimental hypertension. The importance of the RAS in human physiopathology has since been demonstrated by the success of the converting enzyme inhibitors in the treatment of hypertension and congestive heart failure (GAVRAS and GAVRAS 1991) and more recently by the efficacy of a non-peptide angiotensin receptor antagonist in pharmacology and experimental pathology (TIMMERMANS et al. 1991).

I. Effects of Angiotensin on the Cardiovascular System and on Smooth Muscles

Some important cardiovascular effects of AT_{II} depend on activation of smooth muscles in various sections of the cardiovacular system: indeed,

Table 1. Activities of angiotensins in various preparations

Preparation	Order of potency of angiotensins			References
	AT_{II}	AT_{III}	AT_{I}	
Rat blood pressure	100	38	55	REGOLI et al. (1974)
Rat vas deferens	100	18	22	MAGNAN and REGOLI (1979)
Rat stomach strip	100	56	22	REGOLI et al. (1974a)
Rat colon	100	8	5	KHOSLA et al. (1974a)
Rat uterus	100	18	–	KHOSLA et al. (1974a)
Rabbit aorta strip	100	5.6	5.1	REGOLI et al. (1974a)
Rabbit adrenal cortical cells	100	130	2	PEACH and ACKERLY (1976)
Rabbit adrenal medulla	100	15	100	PEACH and ACKERLY (1976)
Rabbit kidney				
Perfusion pressure	100	21	5.0	BLUMBERG et al. (1977b)
Prostaglandin release	100	50[a]	10[a]	BLUMBERG et al. (1977b)
Rabbit mesentery	100	150[a]	20[a]	BLUMBERG et al. (1977b)

[a] These values have been calculated from the dose-response curves published by BLUMBERG et al. (1977) and are approximate. The table has been reproduced from an article by REGOLI (1979).

angiotensin is a potent stimulant of arteriolar smooth muscle (REGOLI et al. 1974), whose tone is determinant for the peripheral vascular resistance: angiotensin contracts the veins (KHOSLA et al. 1974a; RHALEB et al. 1991a) and may facilitate venous return: angiotensin has a positive chronotropic and inotropic effect (BONNARDEAUX and REGOLI 1974) on the heart and produces coronary vasoconstriction and cardiac hypertrophy (SCHELLING et al. 1991). These effects are mediated by receptors which have been studied with pharmacological or biochemical assays, using the same criteria as for other endogenous agents (for instance, the catecholamines). Pharmacological data have been obtained in various tissues or perfused organs (Table 1), particularly the rabbit aorta, which is considered the most reliable pharmacological preparation for angiotensin-related peptides. The relative activities (expressed in percent of that of AT_{II}), observed with the three most important natural products of the renin-angiotensin system, namely AT_{I}, AT_{II} and AT_{III} (desAsp^{1}-AT_{II}) in various organs, are presented and compared in Table 1.

AT_{II} is more active than the other two peptides, except on the rabbit adrenal cortical cells, where AT_{III} is more potent than AT_{II}, and on the rabbit adrenal medulla, where AT_{I} is as active as AT_{II} (PEACH and ACKERLY 1976). These differences were taken by PEACH (1977) as indicative of three different receptor types, a hypothesis that has not been supported by experimental findings, since the same functional receptor is found in vascular smooth muscles and in the adrenal cortex (REGOLI 1979; CAPPONI and CATT 1979) (see below). Angiotensin (in concentrations ranging from 10^{-7} to

$10^{-9}M$) has been found to stimulate vascular (generally arterial) and other smooth muscles of all animal species in which it has been tested: these include the *dog* carotid, renal and coronary arteries (WALTER and BASSENGE 1969; BOHR 1974), *cat* aorta, carotid artery and vena cava (NANTEL et al. 1991) as well as the cat spleen (BENELLI et al. 1964), the *guinea pig* aorta (PALAIC and LE MORVAN 1971), ileum (REGOLI and VANE 1964) and vas deferens (BENELLI et al. 1964), the *rat* aorta and portal vein, as well as the stomach (REGOLI et al. 1974), duodenum, ileum (REGOLI and VANE 1964), the colon (REGOLI and VANE 1964), urinary bladder (RHALEB et al. 1991a), and vas deferens (MAGNAN and REGOLI 1979), the *rabbit* aorta, pulmonary artery (REGOLI et al. 1974), jugular and cava veins (KHAIRALLAH et al. 1966; NANTEL et al. 1991; RHALEB et al. 1991), various human arteries and the human ileum, colon and urinary bladder and also arteries from cows, pigs and sheep (RHALEB et al. 1991; LIBEAU et al. 1965; FISHLOCK and GUNN 1970; KHAIRALLAH et al. 1966).

Functional studies have indicated that angiotensin receptors may be present and act on various cell components of the smooth muscle wall, particularly the smooth muscle fibre, the autonomic nerves and the endothelium: in fact, AT_{II} exerts direct effects on the smooth muscle fibre (and this is the most common effect) (REGOLI et al. 1974), but it also promotes noradrenaline release from the sympathetic nerves (MAGNAN and REGOLI 1979) and the release of endothelium-dependent relaxing factor (EDRF) from the endothelium (DE NUCCI et al. 1988). A recent report connects angiotensin with endothelin (EMORI et al. 1989). Release of noradrenaline has been measured in isolated vessels stimulated electrically (LIBEAU et al. 1965; ZIMMERMAN 1972), as well as in other organs (JUMBLATT and HACKMILLER 1990), and has been shown to contribute to smooth muscle contraction in vessels (PEACH 1977) and in peripheral organs such as the rat vas deferens (MAGNAN and REGOLI 1979; TRACHTE 1988). Release of relaxant prostaglandins has been observed in a few isolated vessels (the rabbit mesenteric artery; BLUMBERG et al. 1977b; WEBB 1982; TODA and MIYAZAKI 1981), and release of EDRF in some vessels, such as the dog renal artery (TODA 1984), where it might play some functional role in the regulation of local blood flow by reducing the peripheral resistance or attenuating the vasoconstrictor effect of AT_{II}. Direct effect on the smooth muscle fibre has been used for extensive structure-activity studies and the receptor characterization, as described below.

II. Receptor Characterization: Pharmacological Assays

In the early 1970s, RIOUX et al. (1973) applied the concepts and the equations of the drug-receptor theories, elaborated by ARIENS (1966), SCHILD (1968) and other pharmacologists (SCHILD 1973; SPETH and KIM 1990) to angiotensin and related peptides. It was found that "receptors for angiotensin can be classified with the same criteria accepted and used for biologically active

Table 2. Biological activities of angiotensin-related peptides in four pharmacological preparations

	RBP RP	pD_2	RA RA	α^E	pD_2	RSS RA	α^E	RVD RA
$[Asp^1,Ile^5]AT_{II}$	100	8.86	100	1.0	7.75	100	1.0	100
$[Asp^1,Val^5]AT_{II}$	100	8.85	100	1.0	7.75	100	1.0	100
$AT_{II}(2-8)$	38	7.62	5.6	1.0	7.50	56	1.0	18
$AT_{II}(3-8)$	0.25	4.82	<0.1	0.9	5.20	0.3	1.0	–
$AT_{II}(4-8)$	0.04	<4.0	<0.1	0.5	–	–	–	Inact.
$AT_{II}(5-8)$	0	–	–	0	–	–	0	–
$AT_{II}(1-7)$	0	–	–	0	–	–	0.05	Inact.
$AT_{II}(1-4)$	0	–	–	0	–	–	0	–
$[Sar^1]AT_{II}$	110			1.0			1.0	–
$[pNH_2Phe^6]AT_{II}$[a]	0	–	–	0	–	–	–	0
$[Ala^3]AT_{II}$	80	7.73	7.4	1.0	7.33	38	1.0	98
$[Ala^5]AT_{II}$	5.0	6.42	0.4	1.0	5.50	0.6	1.0	4.6
$[Ala^8]AT_{II}$	0	–	–	0 Ant.	–	–	0 Ant.	0 Ant.

RBP, rat blood pressure; RSS, rat stomach strip; RA, rabbit aorta; RVD, rat vas deferens; pD_2, log of molar concentration of agonist that produces 50% of the maximal response; RA, relative affinity; RP, relative potency expressed as a percentage of those of AT_{II}; α^E, intrinsic activity expressed as a fraction of that of AT_{II}; Inact., inactive; Ant., antagonist. Data are from REGOLI et al. (1974), MAGNAN and REGOLI (1979), RIOUX et al. (1973).
[a] Inactive in anaesthetized dogs bilaterally nephrectomized (Rhaleb, personal communication).

amines. . . ." (REGOLI 1979) by comparing the effects and calculating relative affinities of agonists analogues or by measuring the apparent affinities of antagonists. Such comparisons were made on isolated smooth muscle preparations (taken from the cardiovascular, the gastrointestinal, the genitourinary systems of various animals) and on the rat blood pressure (test *in vivo*), both with the naturally occurring peptides and with analogues selected because of their resistance to enzymatic degradation or because of their marked differences in myotropic activity (REGOLI 1979). One such comparison is shown in Table 2 and is used to determine the order of potency of agonists as a first criterion for receptor characterization, according to Schild (SCHILD 1973).

Among the angiotensins, $[Asp^1,Ile^5]AT_{II}$ is the most potent and is taken as standard since it is the naturally occurring sequence: the British Medical Research Council (MRC) (Millhill, London) standard $[Asp^1,Val^5]AT_{II}$ is equally active. The active sequence of angiotensin is to be found at the C-terminal part of the octapeptide, since AT(2–8) and AT(3–8) maintain some activity while AT(1–7) is inactive: however, a minimum of 5 residues

Table 3. Activities of angiotensin antagonists on isolated organs

Compound	Rat stomach strip			Rabbit aorta		
	α^E	pD_2	pA_2	α^E	pD_2	pA_2
AT_{II}	1.0	7.75	–	1.0	8.86	–
$[Tyr^8]AT_{II}$	1.0	7.16	–	1.0	7.94	–
$[Nle^8]AT_{II}$	0.5	7.55	8.44	0.55	8.60	8.66
$[Leu^8]AT_{II}$	0.1	–	8.60	0	–	8.78
$[Val^8]AT_{II}$	0.1	–	7.94	0	–	8.34
$[Ala^8]AT_{II}$	0	–	7.22	0	–	7.70
$[Sar^1,Ala^8]AT_{II}$	0.001	–	8.40	0	–	8.60
$[Sar^1,Leu^8]AT_{II}$	0.002	–	9.20	0	–	9.0
DUP 753	0	–	7.96[a]	0	–	8.27
PD 123117	0.22[b]	<4.0	–	0	Inact.	–

α^E, pD_2, Same as in Table 2; pA_2, The -log of the concentration of antagonist that reduces the effect of a double to that of a single dose of agonist (SCHILD 1947).
[a] From RHALEB et al. (1991a).
[b] This agonistic effect is not inhibited with high doses of DUP 753 ($10^{-5}M$).
All other results are taken from REGOLI et al. (1974).

(AT(4–8)) is required for activity. $[Sar^1]AT_{II}$ shows very good activity, suggesting that Asp^1 is not essential, while $[pNH_2Phe^6]AT_{II}$, the selective AT_2 receptor ligand, identified by SPETH and KIM (1990), is inactive. The replacement of some natural residues with L-Ala is associated with different changes of affinities or intrinsic activity, as illustrated by the three compounds at the bottom of Table 3: replacement of Val^3 by L-Ala is compatible with good activity *in vivo* and *in vitro*, while that of Ile^5 reduces affinity and that of Phe^8 changes the pharmacological spectrum and leads to antagonism. The changes of agonistic activities are very similar in the various preparations (with some minor differences of relative affinities) and the order of potency is the same: this suggests the existence of a single functional site, the AT_1 receptor which is insensitive to $[pNH_2Phe^6]AT_{II}$ (SPETH and KIM 1990).

The data obtained in these pharmacological studies led to the hypothesis (REGOLI et al. 1974; KHOSLA et al. 1974a) that a distinction can be made between the groups (side chains) involved in the occupation (binding) of the receptor and the group(s) which triggers the receptor molecule to initiate a chain of events leading to the biological effect (active site(s)). Antagonists should have been obtained by modifying the active site (RIOUX et al. 1973; REGOLI et al. 1971; TÜRKER et al. 1971). The data in Table 3 demonstrate that the active site is Phe^8, whose replacement modifies the pharmacological spectrum from that of an agonist (AT_{II}) to that of an antagonist (e.g. $[Leu^8]AT_{II}$), passing through a partial agonist $[Nle^8]AT_{II}$. Antagonists were instrumental for receptor characterization. In fact, studies with agonists in various preparations had raised the possibility that angiotensin acts on multiple receptors (PEACH 1977; DOUGLAS 1987; GOODFRIEND and PEACH 1977), since the naturally occurring peptides AT_I, AT_{II} an AT_{III} showed different activity ratios in smooth muscles (REGOLI et al. 1974), adrenal medulla (PEACH and ACKERLY 1976) and adrenal cortex (PEACH and ACKERLY 1976). Such a possibility was excluded, by considering that different activity ratios of naturally occurring agonists may be primarily due to metabolism (REGOLI 1979), while the similarity of the apparent affinities of antagonists indicated that the same functional receptor is present in the three organs.

Recent data obtained in our laboratory with two non-peptide ligands of AT_{II} receptors, DUP 753 and PD 123177, are also included in Table 3, for comparison. While the first compound is a potent inhibitor of the angiotensin effects in the two preparations, PD 123177 is completely inactive, suggesting the existence of a single receptor entity.

III. Receptor Characterization by the Binding

Multiple angiotensin receptors have been proposed in the last 20 years, based on results obtained with binding assays (TIMMERMANS et al. 1991; CRISCIONE et al. 1990; PEACH and DOSTAL 1990; GASPARO et al. 1990). However, careful studies in which binding was measured in parallel with biological activity (ALESSANDRO and CATT 1979) showed that the functional site mediating the release of aldosterone from the dog adrenal glands is the same as the site of the dog uterus and is very similar to the receptor of vascular (rabbit aorta) and other smooth muscle of various species (RHALEB et al. 1991a). The recent discovery of specific angiotensin receptor ligands, DuP 753 and PD 123177 (chemical structures in Fig. 2), has contributed to the clarification of this issue. In fact, two different sites AT_1 and AT_2, of which the first (AT_1) is responsible for all biological activities of AT_{II} (TIMMERMANS et al. 1991), while the second (AT_2) is a binding site, apparently without functional role (TIMMERMANS et al. 1991), have been demonstrated, using as ligand the non-selective compound $[^{125}I]AT_{II}$. AT_1 and AT_2 binding sites have been found together in rat adrenal cortex, medulla, uterus, brain, rabbit adrenal cortex, kidney, human adrenal cortex and renal artery (PEACH and DOSTAL 1990; DUDLEY et al. 1990; WHITEBREAD et al. 1989) and in brain homogenates of various species (BENNETT and SNYDER 1980; WEYHENMEYER and HWANG 1985). The two sites are therefore present in many organs, but cannot be differentiated with peptide agonists (AT_{II}) or with antagonists such as saralasin (Fig. 2), since these compounds are not selective. The two sites are well differentiated by the selective ligands, DuP 753, the AT_1 antagonist and PD 123177, and AT_2 ligand which is inappropriately called an antagonist (WONG et al. 1990). Moreover, AT_1 is inhibited by DDT (dithiothreitol) while AT_2 is not (WHITEBREAD et al. 1989). DuP 753 is the result of a systematic pharmacological study, performed by the Dupont

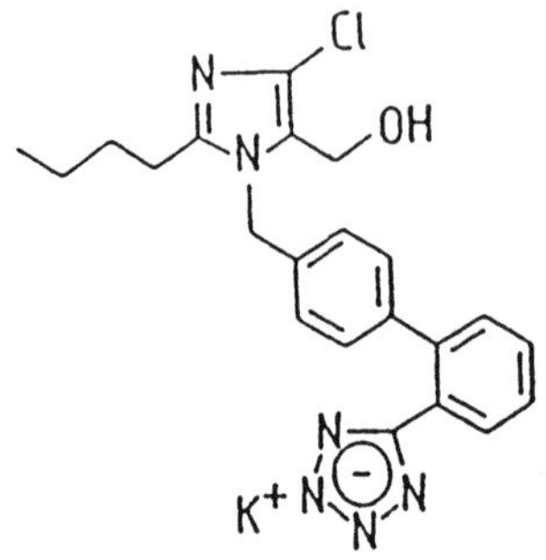

DuP 753

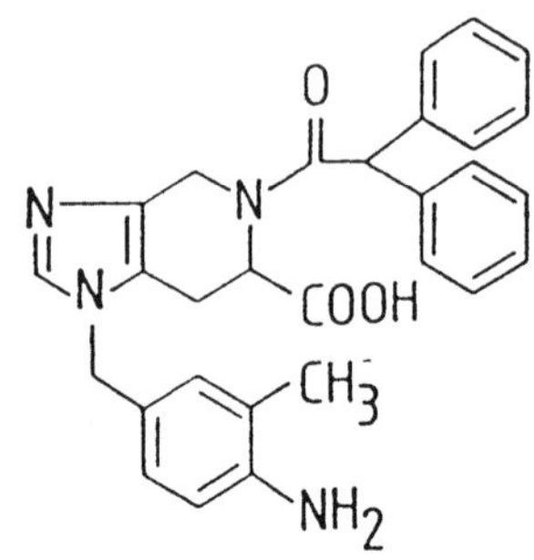

PD123177

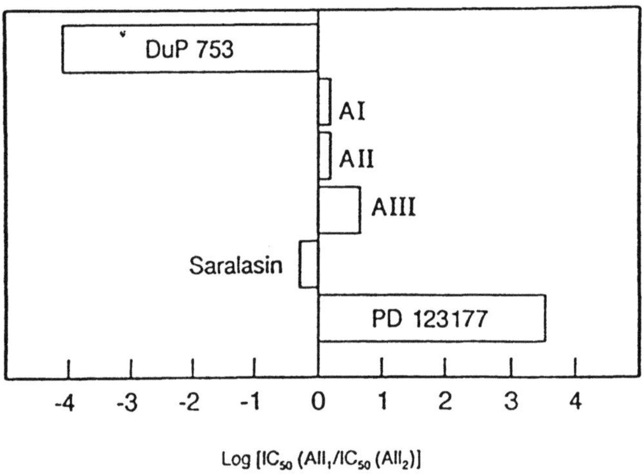

Selectivity of peptide and nonpeptide ligands in binding to (DuP753-sensitive) AII_1 and (PD123177-sensitive) AII_2 receptors/binding sites in rat adrenal cortical microsomes. The ratio of IC_{50} (AII_1)/IC_{50} (AII_2), obtained by radioligand binding using ^{125}I-labeled angiotensin II, was calculated as a measure of selectivity.

Activities of nonpeptide angiotensin II receptor antagonists

	IC_{50} (M)†	pA_2‡	ED_{30} (mg kg^{-1} i.v.)§
S8307	4.0×10^{-5}	5.49	30
S8308	1.3×10^{-5}	5.74	30
EXP6155	1.6×10^{-6}	6.54	10
EXP6803	1.4×10^{-7}	7.20	11
EXP7711	3.0×10^{-7}	6.90	3.7
EXP9654	3.3×10^{-7}	7.32	1.5
EXP9020	5.5×10^{-7}	7.65	2.0
EXP9270	8.0×10^{-8}	7.93	1.0
DuP753	1.9×10^{-8}	8.48	0.78

† Inhibition of specific binding of [^{3}H]angiotensin II (2 nM) to rat isolated adrenal cortical microsomes. ‡ Antagonism of angiotensin II-induced constriction of rabbit isolated aorta. § Antihypertensive potency (i.v.) in conscious renal hypertensive rats; ED_{30}, dose to decrease mean arterial pressure by 30 mm Hg.

Fig. 2. Chemical structures of DuP 753 and PD 127177 and AT receptor/binding site characterization. (From TIMMERMANS et al. 1991)

investigators and summarized in Fig. 2, by showing the almost perfect correlation existing between binding (IC_{50} on rat adrenals), biological activity in vitro (pA_2 on rabbit aorta) and antihypertensive effect in vivo (ED_{30} in renal hypertensive rats). These data suggest that the same functional site is present in the adrenal cortex, the smooth muscle of large arteries and in peripheral resistance vessels.

IV. Cloning and Expression of the Angiotensin Receptor

The functionally active site of the renin-angiotensin system has been recently identified by two groups of investigators (Jackson et al. 1988; Carson et al. 1987). Sasaki et al. (1991) have cloned and expressed a complementary DNA encoding the AT_1 receptor from bovine adrenals and Murphy et al. (1991) have isolated a cDNA encoding for the AT_1 receptor of the rat aorta. The two receptor molecules consist of 359 residues and have very similar sequences and molecular weights (41 093 the adrenal and 40 889 the vascular sequence). Competition binding profiles measured with a variety of peptide and non-peptide antagonists against the ligand $[^{125}I]AT_{II}$ in the adrenals (Whitebread et al. 1989) and against I^{125}-labelled $[Sar^1,Ile^8]AT_{II}$ in the vascular cells (Criscone et al. 1990) are practically the same. DuP 753 shows very high affinities on the two systems, while PD 123177 is inactive (Wong et al. 1990). The angiotensin receptor belongs to the rhodopsin group and shows a transmembrane topology similar to other G protein coupled receptors. The third intracellular loop is rich in positively charged residues (6 Lys, 1 Arg) (Jackson et al. 1988; Carson et al. 1987), which may subserve the interaction with G proteins (Sasaki et al. 1991).

V. Mechanism of Action of Angiotensin: Second Messenger(s)

Many investigator reports suggest that AT_{II} receptors are coupled to at least two transducing enzymes. Phosphoinositide-phospholipase C (Guillon et al. 1988) or adenylcyclase (Cerione et al. 1985), by GTP-binding proteins (Barrett et al. 1989).

The first pathway is present in smooth muscles and is involved in the contractile response of smooth muscles to angiotensin (Smith et al. 1984), as well as in the chronotropic and inotropic effect of AT_{II} on the heart (Bonnardeaux and Regoli 1974) or the contractile effect on the uterus and even in the stimulation of aldosterone secretion from the adrenals (Barrett et al. 1989). The interaction of AT_{II} with its receptor leads to activation of a G protein (Barrett et al. 1989) and of a phosphoinositide-specific phospholipase C (PLC) which catalyses the hydrolysis of the polyphosphoinositide, phosphatidylinositol-4,5-bisphosphate (PIP_2), to produce inositol-1,4,5-triphosphate (IP_3) and diacylglycerol (DAG).

IP_3 elicits the release of Ca^{2+} from a non-mitochondrial site, presumably the endoplasmic reticulum (83, 84, 85), and DAG activates a protein kinase C (Barrett et al. 1989).

In general, vascular and other smooth muscles respond to AT_{II} with a rapid initial phasic contraction that is followed by a prolonged, persistent tonic plateau. Different mechanisms appear to be involved in the two phases: activation of PLC and production of IP_3 lead to: (a) release of Ca^{2+} from intracellular stores, (b) increase of Ca^{2+} in the cytoplasm, (c) activation of caldmodulin and (d) muscle contraction. During the tonic phase, cytosolic Ca^{2+} concentration changes very little, but there is an increase of influx and efflux of Ca^{2+} through the membrane. In general, Ca^{2+} efflux balances influx, but there is an increased cycling of Ca^{2+} through the membrane that activates membrane-associated protein kinase C (PKC) (BARRETT et al. 1989). Phosphorylation of membrane proteins may be the basic mechanism involved in the sustained response of smooth muscle to AT_{II} as a result of AT_{II}-induced generation of DAG. It is known that protein kinase C is already active at low concentrations of Ca^{2+} such as those that are maintained during the sustained response. This PKC, activated by low concentrations of Ca^{2+}, promotes the phosphorylation of a different subset of cellular proteins that serve to sustain the response to AT_{II}. DAG may in fact function physiologically as a second messenger within the membrane to increase the affinity of PKC for calcium and for membrane phospholipids (BARRETT et al. 1989). It was well demonstrated that, in prelabelled cultured vascular smooth muscle cells, AT_{II} induced a biphasic production of DAG with the first peak (within 5 min) originating from PIP_2 hydrolysis and the second one (within 20 min) from phosphatidylinositol (PI) (GRIENDLING et al. 1986).

The mechanism by which AT_{II} inhibits adenylate cyclase activity in the smooth muscle (ANAND-SRIVASTAVA 1983) seems to be identical to that observed in adrenal glomerulosa (BARRETT et al. 1989). In fact, the complex AT_{II}-receptor activates an inhibitory guanine nucleotide regulatory protein (Gi) (CERIONE et al. 1985; ANAND-SRIVASTAVA 1983) which in turn reduces both basal and the adenylate cyclase activity (BARRETT et al. 1989). However, this inhibition probably has no physiological significance, as regards to known AT_{II} actions, since it has been linked to: (a) the attenuation of the AT_{II}-mediated increase in the aldosterone secretory rate which is observed only at very high doses of AT_{II} ($1\,\mu M$) (BARRETT et al. 1989) and (b) the maximal inhibition of adenylate cyclase activity by almost 20% obtained with $10\,\mu M$ (ANAND-SRIVASTAVA 1983) of AT_{II}. Adenylcyclase inhibition by AT_{II} has also been demonstrated in the liver (CERIONE et al. 1985). Both second messengers activation are prevented by DuP 753 and therefore depend on the activation of the AT_1 receptor.

C. Kallikrein-Kinin System

Bradykinin (H.Arg-Pro-Pro-Gly-Phe-Ser-Pro-Phe-Arg.OH) and kallidin (Lys-bradykinin) are released in body fluids from large precursors (kininogens) following the activation of kallikreins, a group of serine-proteases that

Table 4. In vivo effect of bradykinin in mammals

Species	Assay	Reference
Rat	Blood pressure	REGOLI and BARABÉ (1980), COSTELLO and HARGREAVES (1989), STERANKA et al. (1988)
	Hyperalgesia Hyperthermia Paw oedema	HARGREAVES et al. (1988)
Guinea pig	Bronchoconstriction	
	Heart:	
	Positive inotropism	COLLIER et al. (1960), RIOUX et al. (1987)
	Positive chronotropism	GEPPETTI et al. (1988)
Sheep	Bronchoconstriction	SOLER et al. (1989)
Man	Bronchoconstriction cough	SIMMONSSON et al. (1973), WHALLEY et al. (1987)
	Blister base (pain)	LIM et al. (1967), FULLER et al. (1987)
	Visceral pain Rhinorrhea Rhinitis	PROUD et al. (1988)

are present in most tissues (Fig. 1). Kinins evoke a variety of biological responses, including peripheral vasodilatation and hypotension, plasma extravasation and oedema, pain and hyperalgesia, bronchoconstriction and cough (Table 4). When applied to the skin of animals and men, bradykinin induces local inflammatory reactions, characterized by hyperemia, pain and oedema. In physiopathology, the kallikrein-kinin system has been implicated in inflammatory diseases such as rheumatoid arthritis, psoriasis, rhinitis, burns and chemical injuries. The system is also activated during the endotoxic shock and may contribute to worsening of the shock syndrome. In allergic and inflammatory diseases, such as asthma, the kinins could play an important pathological role through their local pro-inflammatory effects and their potent stimulatory actions on bronchial smooth muscle and sensory fibres (REGOLI et al. 1991a). Attempts have therefore been made to control the activities of this system by inhibiting the kallikreins in order to prevent the release of kinins or by blocking the effects of kinins with antagonists. Two types of kinin receptors (B_1 and B_2) have been identified, B_2 being considered the most important in physiological conditions, and B_1 in some physiopathological states (REGOLI and BARABÉ 1980; MARCEAU and REGOLI 1991; REGOLI et al. 1989).

In vitro, kinins have been shown to be potent stimulants or inhibitors of a variety of smooth muscles from various organs of men and laboratory animals, in particular *human* intestine, bronchus (SIMONSSON et al. 1973) and urinary bladder (RHALEB et al. 1992), *rabbit* aorta, jugular and cava veins (REGOLI et al. 1990; NANTEL et al. 1991), perfused heart (NEEDLEMAN et al. 1975), kidney and ear (MALIK and NASJLETTI 1980; JUAN and LEMBECK 1974), *guinea pig* ileum (DAY and VANE 1963), pulmonary artery (HOCK et al. 1991), trachea (RHALEB et al. 1988), perfused heart (RIOUX et al. 1987), *rat*, mesenteric vein (NORTHOVER 1967), colon, duodenum (WALASZEK 1970), urinary bladder and vas deferens (WALASZEK 1970; TOUSIGNANT et al. 1987), isolated perfused kidney, hind quarter and isolated heart (ROBLERO et al. 1976; ROWLEY 1964; RÖSEN et al. 1983). Among the pharmacological tests most commonly used for evaluating kinins activities worthy of mention are: the plasma extravasation in rats (RÖSEN et al. 1983), guinea pigs (WILLIAMS and MORLEY 1973) and rabbits (WILLIAMS and PECK 1977), rat paw oedema and hind limb pressure test (HADDY et al. 1970; COSTELLO and HARGREAVES 1989), rat and rabbit blood pressure, and rat and guinea pig bronchoconstriction (REGOLI and BARABÉ 1980; FARMER 1991).

As mentioned before, the myotropic effects, as well as the in vivo activities of kinins, are mediated by at least two receptor types, B_1 and B_2, of which B_2 is considered to be ubiquitous and to subserve the contractile, secretory and other types of biological actions of the kinins.

Pharmacological characterization of kinin receptors has been carried out using sensitive and selective (with either the B_1 or the B_2 receptor) preparations, such as the rabbit jugular vein (RJV), the guinea pig ileum (GPI), the hamster urinary bladder (HUB), the rat vas deferens (RVD) and the rabbit aorta (RA) (Table 5). Data obtained in five classic preparations with a series of kinin agonists are presented in Table 5 and indicate that the order of potency of agonists on the B_1 receptor is as follows: desArg^{9}BK = Lys-BK > D-Arg-BK > BK > [Hyp^3]BK, while that on the B_2 receptor (the rabbit jugular vein) is: [Hyp^3]BK > Lys-BK > BK >>> desArg^{9}BK. Similar orders of potency of agonists are observed in the other B_2 receptor systems, the GPI, the HUB and the RVD (Table 5). The two sites differ with respect to agonists; the B_1 receptor is preferentially activated by the metabolites des-Arg^9-BK and des-Arg^{10}-kallidin, while B_2 are sensitive to the naturally occurring kinins Lys-BK and BK. Specific and selective antagonists have also been identified for both B_1 (REGOLI et al. 1977) and B_2 (REGOLI et al. 1989; VAVREK and STEWART 1985; HOCK et al. 1991) receptors and have been tested in some of the preparations used for determining the agonists order of potency. The results are shown in Table 6.

The first chemical change that modified the pharmacological spectrum and led to antagonism (on the GPI) was the replacement of Pro^7 by D-Phe (VAVREK and STEWART 1985). However [D-Phe^7]BK was found to act as partial agonist in some preparations and as antagonist in others (REGOLI et al. 1989; RHALEB et al. 1991b). Antagonist affinity was improved, but only on some preparations by replacing Phe^5 and Phe^8 with Thi: the compound still showed partial agonistic activity in other tissues, particularly the RVD (RHALEB et al. 1991b; RIFO et al. 1987). A major advance in antagonist affinity was accomplished by adding a D-Arg residue at the N-terminal and replacing Pro^3 with Hyp^3, the amino acid that is present in human kinins

Table 5. Pharmacological activities of kinins and their analogues on five isolated smooth muscle preparations

Peptide	Preparation														
	pD_2	RJV RA	α^E	pD_2	GPI RA	α^E	pD_2	HUB RA	α^E	pD_2	RVD RA	α^E	pD_2	RA RA	α^E
BK	8.48	100	1.0	7.90	100	1.0	7.70	100	1.0	6.72	100	1.0	6.22	9	1.0
KD	8.63	141	1.0	7.88	95	1.0	7.18	30	1.0	6.62	79	0.9	7.27	96	1.0
[Hyp^3]BK	8.88	254	1.1	7.27	23	1.1	7.81	129	0.9	6.62	79	0.9	6.17	8	1.1
Ile-Ser-BK	8.18	50	0.7	7.58	47	1.0	7.60	79	1.0	5.85	135	0.8	5.97	5	0.7
[Hyp^3,Tyr(Me)^8]BK	8.56	123	1.1	7.82	84	1.1	8.0	199	1.0	6.80	120	0.8	Inactive		
D-Arg-BK	8.70	166	1.1	7.41	33	0.9	7.85	141	1.2	6.38	46	0.9	6.56	19	1.0
D-Arg[Hyp^3]BK	8.55	117	0.9	7.13	17	0.9	7.94	174	1.0	6.82	126	0.9	6.30	12	0.4
[Aib^7]BK	6.28	0.6	0.6	7.90	100	1.3	–	–	–	7.0	191	1.1	<4.5	–	<0.1
[Phe$^8\psi$(CH$_2$-NH)Arg^9]BK	8.29	65	0.8	7.45	35	1.0	7.85	142	0.9	7.38	457	1.2	Inactive		
DesArg^{9}BK	Inactive			Inactive			Inactive			Inactive			7.29	100	1.0

pD_2, negative logarithm of the concentration of agonist required to evoke 50% of the maximum response; RA, relative affinities as a percentage of that of desArg^{9}BK on rabbit aorta and of BK in the others; α^E, intrinsic activity (desArg^{9}BK = 1.0 on rabbit aorta and BK = 1.0 on the others); RJV, rabbit jugular vein; GPI, guinea pig ileum; HUB, hamster urinary bladder; RVD, epidydimal rat vas deferens; RA, rabbit aorta.

Table 6. From kinin agonists to antagonists

Peptide	Rabbit jugular vein			Guinea pig ileum			Rabbit aorta		
	α^E	pD_2	pA_2	α^E	pD_2	pA_2	α^E	pD_2	pA_2
BK	1.0	8.48	–	1.0	7.90	–	1.0	6.22	–
[D-Phe^7]BK	0.4	6.53	P-Ag	0.2	–	5.85	0	–	5.94
[Thi$^{5.8}$,D-Phe^7]BK	0.3	–	6.71	0.3	–	5.90	0	–	6.23
D-Arg[Hyp^3]BK	0.9	8.55	–	0.9	7.13	–	0.4	6.30	–
D-Arg[Hyp^3,D-Phe^7]BK	0.7	–	8.01	0.1	–	5.41	0	–	6.41
[Hyp^3,D-Phe^7,Leu^8]BK	0.5	–	8.40	0.2	–	6.56	–	–	5.10
D-Arg[Hyp^3,Leu^8]BK	–	–	7.45	0.2	–	<5.0	0	–	<5.0
D-Arg[Hyp^3,D-Phe^7,Leu^8]BK	0.1	–	8.86	0	0	6.77	–	–	5.77
D-Arg[Hyp^3,Thi$^{5.8}$,D-Phe^7]BK	0.4	–	7.86	0	0	6.34	–	–	6.16
D-Arg[Hyp^3,Thi^5,D-Tic^7,Oic^8]BK (HOE-140)	–	0	9.19	0	–	8.94	0	Inact.	Inact.
[Leu^8]desArg^{9}BK	0.05	Inact.	0	Inact.	–	0	–	7.29	–

α^E, pD_2 as in Table 5; pA_2, -log of the concentration of antagonist reducing the effect of a double dose of the agonist (BK for the RJV and GPI; desArg^{9}BK for the RA) to that of a single dose (SCHILD 1947).

(SASAGURI et al. 1989). Further increase of affinity and a significant reduction of the partial agonistic activity in several tissues, excluding the guinea pig trachea, was obtained by replacing Phe^8 with Leu in the sequence D-Arg[Hyp^3,D-Phe^7]BK. D-Arg[Hyp^3,D-Phe^7,Leu^8]BK is an antagonist of high affinity, quite selective for B_2 since its pA_2 on the B_1 system of the rabbit aorta is at least 3 log units lower than on the B_2 receptor of the rabbit jugular vein. The most potent and selective antagonist for the B_2 receptor is however HOE 140, which contains two unnatural amino acids in its sequence and is resistant to metabolic degradation (HOCK et al. 1991; WIRTH et al. 1991). In the rabbit aorta, the anti B_1 [Leu^8]des Arg^9BK is inactive on the B_2 receptor of the rabbit jugular vein.

The two antagonists, [Leu^8] desArg^9BK and D-Arg[Hyp^3,D-Phe^7, Leu^8]BK, as well as other similar compounds (RHALEB et al. 1991b), are unable to antagonize the kinins myotropic effect on the guinea pig trachea (RHALEB et al. 1991b; FARMER et al. 1989) and this has been taken as an indication (FARMER et al. 1989) that kinins may act through another receptor type, "B_3", on this tissue. Recently, however, it has been shown (RHALEB et al. 1992; FARMER et al. 1991) that Hoe 140 or its analogue (NPC 16731) are able to inhibit the effect of BK on the guinea pig trachea without exerting any direct myotropic effect, contrary to the other antagonists. A systematic study has therefore been performed by Farmer et al. (FARMER 1991; FARMER et al. 1989, 1991) on this tissue, by measuring binding of [^3H]BK, biological activities in vitro and the Ca^{2+} efflux from the smooth muscle cells in culture (FARMER et al. 1991). The results of this extensive investigation are summarized as follows:

1. Forty per cent and 27% of not displaceable binding of [^3H]BK to guinea pig lung and sheep lung.
2. D-Arg[Hyp^3,$Thi^{5,8}$,D-Phe^7]BK is agonist instead of antagonist in guinea pig trachea in vitro.
3. Absence of block by D-Arg[Hyp^3,D-Phe^7]BK of Ca^{2+} efflux from guinea pig smooth muscle in culture.
4. pA_2 value of Hoe 140 significantly lower in guinea pig trachea (pA_2, 7.4) than in other B_2 receptor systems (e.g. rabbit jugular vein, pA_2, 9.2).

The existence of a third receptor for kinins has been initially proposed on the basis of events observed in binding assays, whereby D-Arg[Hyp^3,D-Phe^7]BK, a fairly strong antagonist of the B_2 receptor, is able to displace only 60% and 70% of [^3H]BK binding to guinea pig (FARMER et al. 1989) and sheep (BURCH et al. 1990) lung membranes respectively. The remaining (approximately 40% and 30%) sites occupied by [^3H]BK are considered to be B_3 receptors, since they are not displaced by the B_2 receptor antagonist (BURCH et al. 1990). However, the in vivo bronchoconstrictor effects of BK in the guinea pig and the sheep are completely blocked by the same antagonist. It thus appears that the "B_3" site is a binding, non-functional site. This topic has been discussed in a recent review (REGOLI et al. 1993).

A second item of evidence in favour of a "B_3" receptor has come from the finding that a number of antagonists ([D-Phe^7]BK, [Thi^{5,8},D-Phe^7]BK, D-Arg[Hyp^3,Thi^{5,8},D-Phe^7]BK, D-Arg[Hyp^3,D-Phe^7]BK, D-Arg[Hyp^3,D-Phe^7, Leu^8]BK) are fairly potent stimulants of the trachea and do not block the effect of BK. It has been argued that, while these compounds are antagonists in several B_2-containing tissues, they act as agonists on the trachea, because this tissue has a different receptor. A careful analysis of the literature however reveals that B_2 antagonists have partial agonistic effects in various organs (REGOLI et al. 1989, 1993) and such an effect is particularly strong on the trachea (RHALEB et al. 1991b). Therefore, they cannot be used as antagonists on this tissue. Recent data obtained with HOE 140 have contributed to the clarification of this issue by showing that HOE 140 or its analogue is inactive as an agonist on the trachea and efficiently antagonize the myotropic effects of kinins in this tissue (RHALEB et al. 1991b, 1992). Both B_2 and B_3 receptors are therefore blocked by HOE 140, a finding that challenges the B_3 receptor hypothesis.

Other evidence for a B_3 receptor site has been suggested from data obtained on guinea pig isolated tracheal smooth muscle cells in culture. D-Arg[Hyp^3,D-Phe^7]BK is competing with [^{3}H]BK for the tracheal site and blocks the stimulant effect of BK on prostaglandin release: it does not, however, affect the Ca^{2+} efflux, which is not necessarily associated with biological activity.

Recently, the expression of poly(A)$^+$m RNA for bradykinin receptors from 3T3 cells was obtained in oocytes from *Xenopus laevis* (MAHAN and BURCH 1990). This expression was examined by bradykinin but not desArg^{9}BK-mediated stimulation of Ca^{2+} efflux from transfected oocytes. This effect was efficiently inhibited with the bradykinin receptor antagonist (D-Arg[Hyp^3,D-Phe^7]) but not by the B_1 receptor antagonist ([Leu^8]desArg^{9}BK). Therefore, it was suggested that the bradykinin receptor appearing on the oocytes was of the B_2 type (MAHAN and BURCH 1990).

I. Mechanism of Action of Bradykinin: Second Messenger(s)

Binding of a kinin agonist to B_1, B_2 receptors can lead to activation of several second messenger systems. The B_2 receptor is the most commonly used entity to study the bradykinin transduction mechanisms and a large body of data is accumulating in the literature on the mechanism of action of bradykinin and related peptides (MARGOLIUS 1989). The system(s) activation depend(s) upon which effector enzymes are present in a particular tissue. In a large number of tissues, bradykinin activates the release of arachidonic acid and, depending on the production site, this activation leads to a variety of lipooxygenase and/or cycloxygenase products (REGOLI and BARABÉ 1980). These include prostaglandins (REGOLI and BARABÉ 1980), leukotrienes (WILLIAMS and PECK 1977), hydroxyeicosatetraenoic acids (HETEs) and platelet-activating factor (PAF) (WHATLEY et al. 1988). Generally, the release of arachidonic acid is mediated by phospholipase A_2 (BURCH and AXELROD 1987). This enzyme releases arachidonate directly from a phospholipid or indirectly from the diacylglycerol (a phosphatidylcholine-specific phospholipase C product) which can be cleaved by monoacylglycerol lipase (MAJERUS et al. 1986) or by a phosphatidic acid-specific phospholipase A_2 (BILLAH et al. 1981). Therefore, the inhibition of this pathway may interfere with the biological response elicited by kinins, for instance in the guinea pig trachea (REGOLI et al. 1990; FARMER et al. 1989; RHALEB et al. 1991c), the dog renal artery (RHALEB et al. 1989), the hamster urinary bladder (RHALEB et al. 1991c), or the rabbit urinary bladder (NAKAHATA et al. 1987).

Another metabolic pathway involved in tissue responses to kinins is the adenyl-cyclase pathway. It has been elegantly demonstrated that B_2 receptor activation in guinea pig lungs may lead to cellular accumulation of cyclic AMP (STONER et al. 1973; BRONTON et al. 1976) and this effect appears to be due to stimulation of the biosynthesis of prostaglandins (PGE_1 and PGE_2) or PGI_2, which subsequently bind to their receptors and activate adenylate cyclase (BRONTON et al. 1976). The effect of bradykinin is obviously blocked by indomethacin and aspirin.

In other studies performed over the last 10 years, bradykinin has been shown to activate the production or release of a variety of endogenous mediators, e.g., acetylcholine (DIENER et al. 1988) in rat intestine, noradrenaline and purines (TOUSIGNANT et al. 1987; RIFO et al. 1987) in rat vas deferens, prostacyclin (D'ORLÉANS-JUSTE et al. 1989; SUNG et al. 1988), especially in endothelial cells in culture or in renal vascular cells (RHALEB et al. 1989), substance P and neurokinin A or calcitonin-gene related peptide from sensory nerves, for instance in the rabbit iris (UEDA et al. 1984; GRIESBACHER and LEMBECK 1987) or in the guinea pig heart (GEPPETTI et al. 1988), and interleukin-1 (TIFFANY and BURCH 1989) in vascular tissues.

These various secretory effects of BK appear to be dependent on the activation of PI-PLC and the release of inositol phosphates and diacylglycerol. Inositol-1,4,5-triphosphate (IP_3) is involved in the release of calcium from the endoplasmic reticulum (BERRIDGE 1987). The increased intracellular calcium concentration serves to modulate the activity of several cellular enzymes such as those implicated in the stimulus-secretion or the stimulus-contraction coupling (see below), as well as in the production of endothelial-derived-relaxing factor (EDRF) from endothelial cells (VOYNO-YASENETSKAYA et al. 1989) and the opening of ion channels. Inositol phosphates also appear to activate plasma membrane calcium channels to allow influx of extracellular calcium (SCHILLING et al. 1988). Diacylglycerol in combination with Ca^{2+} activates a protein kinase C and modulates the function of ion channels.

Bradykinin may activate G_i protein to inhibit adenylate cyclase (MURAYAMA and UI 1985) or conversely induce an increase of intracellular cAMP in several tissues (BRONTON et al. 1976), including the rat duodenum (LIEBMANN et al. 1987), rat thymocytes (WHITFIELD et al. 1970) and guinea pig lung slices (STONER et al. 1973). Bradykinin activates the intracellular accumulation of cGMP (Stoner et al. 1973; Snider and Richelson 1984) through B_2 receptors.

The transduction pathway activated by B_1 receptors is only poorly understood. However, desArg^9-BK can induce prostaglandin synthesis in some tissues, e.g., dog renal artery (Rhaleb et al. 1989; Marceau and Tremblay 1986; Cahill et al. 1988), also prostacyclin and EDRF release from bovine endothelial cells (D'Orléans-Juste et al. 1989; Sung et al. 1988), and prostaglandin-dependent bone resorption (Lerner et al. 1987). It is conceivable that accumulation of cAMP by prostanoids is the basic mechanism of the prostaglandin-dependent effects of B_1 receptors. Smooth muscle contraction by B_1, on the other hand, may be obtained through the activation of the PIP and DAG systems, as for B_2, but this remains to be demonstrated.

Acknowledgements. The secretarial work of H. Morin has been greatly appreciated as well as the technical help of M. Boussougou for the biological assays and of R. Laprise for the peptide synthesis. We also acknowledge the contributions of two graduate students, S. Télémaque and D. Jukic, to biological and chemical data, particularly on kinins. This work has been performed with the financial help of the Medical Research Council of Canada (MRCC) and the Heart and Stroke Foundation of Canada (HSFC). D.R. is a career investigator of the MRCC and N.-E.R. is a student of the HSFC.

References

Alessandro MC, Catt KJ (1979) Angiotensin II receptors in adrenal cortex and uterus. J Biol Chem 254:5120–5127

Anand-Srivastava MB (1983) Angiotensin II receptors negatively coupled to adenylate cyclase in rat aorta. Biochem Biophys Res Commun 117:420–428

Ariens EJ (1966) Receptor theory and structure action relationships. Adv Drug Res 3:235–285

Barraclough MA (1965) Dose-dependent opposite effects of angiotensin on renal sodium excretion. Lancet 2:987–988

Barrett PQ, Bollag WB, Isales CM, McCarthy RT, Rasmussen H (1989) Role of calcium in angiotensin II-mediated aldosterone secretion. Endocr Rev 10: 496–518

Benelli G, Della Bella D, Gandini A (1964) Angiotensin and peripheral sympathetic nerve activity. Br J Pharmacol 27:211–219

Bennett JP, Snyder SH (1980) Receptor binding interactions of the angiotensin II antagonist, ^{125}I[Sarcosine^1,Leucine^8]-angiotensin II, with mammalian brain and peripheral tissues. Eur J Pharmacol 67:11–25

Berridge MJ (1987) Inositol trisphosphate and diacylglycerol: two interacting second messengers. Annu Rev Biochem 56:159–193

Billah MM, Lapetina EG, Cuatrecasas P (1981) Phospholipase A_2 activity specific for phosphatidic acid: a possible mechanism for the production of arachidonic acid in platelets. J Biol Chem 256:5399–5403

Blumberg AL, Ackerly JA, Peach MJ (1975) Differentiation of neurogenic and myocardial angiotensin II receptors in isolated rabbit atria. Circ Res 36: 719–726

Blumberg AL, Denny SE, Marshall GR, Needleman P (1977a) Blood vessels hormone interactions: angiotensin, bradykinin and prostaglandins. Am J Physiol 232:H305–H310

Blumberg AL, Nishikawa K, Denny SE, Marshall GR, Needleman P (1977b) Angiotensin (A_I, A_{II}, A_{III}) receptor characterization: correlation of prostaglandin release with peptide degradation. Circ Res 41:154–158
Bohr D (1974) Angiotensin on vascular smooth muscle. In: Page IH, Bumpus FM (eds) Angiotensin. Springer, Berlin Heidelberg New York, pp 424–440 (Handbook of experimental pharmacology, vol 37)
Bonjour JP, Peters G, Chomety F, Regoli D (1967) Renal effects of Val^5-angiotensin II amide, vasopressin and diuretics in the rat, as influenced by water diversis and ethanol anesthesia. Eur J Pharmacol 2:88–105
Bonnardeaux J, Regoli D (1974) Action of angiotensin in the heart. Can J Physiol Pharmacol 52:50–60
Bronton LL, Wiklund RA, Van Arsdale PM, Gilman AG (1976) Binding of [^{3}H] prostaglandin E_1 to putative receptors linked to adenylate cyclase of culture cell clones. J Biol Chem 251:3037–3044
Bumpus FM (1991) Nomenclature for angiotensin receptors. A report of the nomenclature committee of the council for high blood pressure research. Hypertension 17:720–721
Bunag RD (1974) Circulatory effects of angiotensin. In: Page IH, Bumpus FM (eds) Angiotensin. Springer, Berlin Heidelberg New York, pp 441–454 (Handbook of experimental pharmacology, vol 37)
Burch RM, Axelrod J (1987) Dissociation of bradykinin-induced prostaglandin formation from phosphatidyl inositol turnover in Swiss 3T3 fibroblasts: evidence for G protein regulation of phospholipase A_2. Proc Natl Acad Sci USA 84: 6374–6378
Burch RM, Farmer SG, Steranka LR (1990) Bradykinin receptor antagonists. Med Res Rev 10:237–269
Burgess GM, Godfrey PP, McKinney JS, Berridge MJ, Irvine RF, Putney JW Jr (1984) The second messenger linking receptor activation to internal Ca^{2+} release in liver. Nature 309:63–66
Cahill M, Fishman JB, Polgar P (1988) Effect of des-$Arginine^9$-bradykinin and other bradykinin fragments on the synthesis of prostacyclin and the binding of bradykinin by vascular cells in culture. Agent Actions 24:224–231
Campbell DJ (1985) The site of angiotensin production. J Hypertens 3:199–207
Capponi AM, Catt KJ (1979) Angiotensin II receptors in adrenal cortex and uterus. J Biol Chem 254:5120–5127
Carson MC, Harper CML, Baukal AJ, Aguilera G, Catt KJ (1987) Physiochemical characterization of photoaffinity-labled angiotensin II receptors. Mol Endocrinol 1:147–154
Cerione RA, Staniszewski C, Caron MG, Lefkowitz RJ, Codina J, Birnbaumer L (1985) A role for N_i in the hormonal stimulation of adenylate cyclase. Nature 318:293–295
Collier HOJ, Holgate JA, Schachter M, Shorley PJ (1960) The bronchoconstrictor actin of bradykinin in the guinea pig. Br J Pharmacol 15:290–297
Costello AH, Hargreaves KM (1989) Suppression of carrageenan-induced hyperalgesia, hypertermia and edema by a bradykinin antagonist. Eur J Pharmacol 171:259–263
Criscione L, Thomann H, Whitebread S, Gasparo M, Bühlmayer P, Herold P, Ostermayer F, Kamber B (1990) Binding characteristics and vascular effects of various angiotensin II antagonists. J Cardiovasc Pharmacol 16(Suppl 4):S56–S59
D'Orléans-Juste P, de Nucci G, Vane JR (1989) Kinins act on B_1 and B_2 receptors to release conjointly endothelium-derived relaxing factor and prostacyclin from bovine aortic endothelial cells. Br J Pharmacol 96:920–926
Day M, Vane JR (1963) An analysis of the direct and indirect actions of drugs on the isolated guinea pig ileum. Br J Pharmacol 20:150–170
De Nucci G, Gryglewski RJ, Warner TD, Vane JR (1988) Receptor-mediated release of endothelium-derived relaxing factor and prostacyclin from bovine aortic endothelial cells is coupled. Proc Natl Acad Sci USA 85:2334–2338

Diener M, Bridges RJ, Knobloch SF, Rummel W (1988) Indirect effects of bradykinin on ion transport in rat colon descendens: mediated by prostaglandins and enteric neurons. Naunyn Schmiedebergs Arch Pharmacol 337:69–73

DiMarzo V, Tippins JR, Morris HR (1988) Bradykinin- and chemotactic peptide fMLP-stimulated leukotriene biosynthesis in rat lungs and its inhibition by vasoactive intestinal peptide. Biochem Int 17:235–242

Douglas JG (1987) Angiotensin receptor subtypes of the kidney cortex. Am J Physiol 253:F1–F7

Dudley DT, Panek RL, Major TC, Lu GH, Bruns RG, Klinkefus BA, Hodges JC, Weishaar RE (1990) Subclasses of angiotensin II binding sites and their functional significance. Mol Pharmacol 38:370–377

Emori T, Hirata Y, Ohta K, Shichiri M, Marumo F (1989) Secretory mechanism of immunoreactive endothelin in cultured bovine endothelial cells. Biochem Biophys Res Commun 160:93–100

Erdös EG (1975) Angiotensin I converting enzyme. Circ Res 36:247–255

Farmer SG (1991) Airway pharmacology of bradykinin. In: Burch RM (ed) Bradykinin antagonists, basic and clinical research. Dekker, New York, pp 213–236

Farmer SG, Burch RM, Meeker SN, Wilkins DE (1989) Evidence for a pulmonary B_3 bradykinin receptor. Mol Pharmacol 36:1–8

Farmer SG, Burch RM, Kyle DJ, Martin JA, Meeker SN, Togo J (1991) D-Arg[Hyp^3-Thi^5-D-Tic^7-Tic^8]-bradykinin, a potent antagonist of smooth muscle BK_2 receptors and BK_3 receptors. Br J Pharmacol 102:785–787

Feldberg W, Lewis GP (1964) The action of peptides on the adrenal medulla. Release of adrenaline by bradykinin and angiotensin. J Physiol (Lond) 171: 98–108

Fishlock DJ, Gunn A (1970) The action of angiotensin on the human colon in vitro. Br J Pharmacol 39:34–41

Fitzsimons JT (1972) Thirst. Physiol Rev 52:468–561

Fuller RW, Dixon CMS, Cuss FMC, Barnes PJ (1987) Bradykinin-induced bronchoconstriction in humans: mode of action. Am Rev Respir Dis 135:176–180

Gasparo M, Whitebread S, Mele M, Motani AS, Whitcombe PJ, Ramjoué HP, Kamber B (1990) Biochemical characterization of two angiotensin II receptor subtypes in rat. J Cardiovasc Pharmacol 16(Suppl 4):531–535

Gavras H, Gavras I (1991) Cardioprotective potential of angiotensin converting enzyme inhibitors. J Hypertens 9:385–392

Genest J, Koiw E, Nowaczynski W, Sander T (1960) Study of urinary adrenocortical hormones in human arterial hypertension. 1st International Congress of Endocrinology, Copenhagen, p 173

Geppetti P, Maggi CA, Perretti F, Frilli S, Manzini S (1988) Simultaneous release by bradykinin of substance P and calcitonin gene-related peptide immunoreactivities from capsaicin-sensitive structures in guinea pig hearts. Br J Pharmacol 94: 288–290

Goodfriend TL, Peach MJ (1977) Specific functions of angiotensins I, II and III. In: Genest J, Koiw E, Kuchel O (eds) Hypertension: physiopathology and treatment. McGraw-Hill, New York, pp 168–173

Griendling KK, Rittenhouse SE, Brock TA, Ekstein LS, Gimbrone MA Jr, Alexander RW (1986) Sustained diacylglycerol formation form inositol phospholipids in angiotensin II-stimulated vascular smooth muscle cells. J Biol Chem 261:2901–2907

Griesbacher T, Lembeck F (1987) Effect of bradykinin antagonists on bradykinin-induced plasma extravasation, venoconstriction, prostaglandin E_2 release, nociceptor stimulation and contraction of the iris sphincter muscle in the rabbit. Br J Pharmacol 92:333–340

Gross F, Regoli D, Schaechtelin G (1963) Renal content and blood concentration of renin. In: Williams PC (ed) Hormones and the kidney. Academic, London, pp 293–300

Guillon G, Gallo-Payet N, Balestre M-H, Lambard C (1988) Choleratoxin and corticotropin modulation of inositol phosphate accumulation induced by vasopressin and angiotensin II in rat glomerulosa cells. Biochem J 253:765–775
Haddy FJ, Emerson TE, Scott JB, Daugherty RM (1970) The effect of the kinins on the cardiovascular system. In: Erdös EG (ed) Bradykinin, Kallidin and Kallikrein. Springer, Berlin Heidelberg New York, pp 362–384 (Handbook of experimental pharmacology, vol 25)
Hargreaves K, Troullos E, Dionne R, Schmidt E, Schafer S, Joris J (1988) Bradykinin is increased during acute and chronic inflammation: therapeutic implications. Clin Pharmacol Ther 44:613–618
Hock FJ, Wirth K, Albus M, Linz W, Gerhards HJ, Wiemer G, Henke ST, Breipohl G, König W, Knolle J, Shölkens BA (1991) HOE 140 a new potent and long-acting bradykinin antagonist: in vitro studies. Br J Pharmacol 102:769–773
Jackson TR, Blair LAC, Marshall J, Goedert M, Hanley MR (1988) The mass oncogene encodes an angiotensin receptor. Nature 335:437–440
Joseph SK, Thomas AP, Williams RJ, Irvine RF, Williamson JR (1984) Myo-inositol 1,4,5-trisphosphate. J Biol Chem 259:3077–3081
Juan H, Lembeck F (1974) Action of peptides and other algesic agents on paravascular pain receptors of the isolated perfused rabbit ear. Naunyn Schmiedebergs Arch Pharmacol 283:151–164
Jumblatt JE, Hackmiller RC (1990) Potentiation of norepinephine secretion by angiotensin II in the isolate rabbit iris-ciliary body. Curr Eye Res 9:169–176
Khairallah PA, Page IH, Bumpus FM, Türker RK (1966) Angiotensin tachyphylaxis and its reversal. Circ Res 19:247–254
Khosla MC, Smeby RR, Bumpus FM (1974a) Structure-activity relationship in angiotensin II analogs. In: Page IH, Bumpus FM (eds) Angiotensin. Springer, Berlin Heidelberg New York, pp 126–161 (Handbook of experimental pharmacology, vol 37)
Khosla MC, Hall MM, Smeby RR, Bumpus FM (1974b) Synthesis of analogs of [8-isolecine]angiotensin II with variations in position 1. J Med Chem 17:431–434
Laragh JH, Angers M, Kelly WG, Lieberman S (1960) Hypotensive agents and pressure substances. The effect of epinephrine, norepinephrine, angiotensin II and others on the secretory rate of aldosterone in man. JAMA 174:234–240
Lerner UH, Jones IL, Gustafsson GT (1987) Bradykinin, a new potential mediator of inflammation-induced bone resorption. Arthritis Rheum 30:530–540
Libeau H, Distler A, Wolff HP (1965) The noradrenaline releasing effect angiotensin II in isolated blood vessels. Acta Endocrinol (Copenh) Suppl 100:138
Liebmann C, Riessmann S, Robberech P, Arold H (1987) Bradykinin action in the rat duodenum: receptor binding and influence on the cyclic AMP system. Biomed Biochim Acta 46:469–478
Lim RKS, Miller DG, Guzman F, Rodgers DW, Rogers RW, Wang SK, Chao PY, Shih TY (1967) Pain and analgesia evaluated by the intraperitoneal bradykinin-evoked pain method in man. Clin Pharmacol Ther 8:521–542
Magnan J, Regoli D (1979) Characterization of receptors for angiotensin in the rat vas deferens. Can J Physiol Pharmacol 57:417–423
Mahan LC, Burch RM (1990) Functional expression of B_2 bradykinin receptors from Balb/c cell mRNA in xenopus oocytes. Mol Pharmacol 37:785–789
Majerus PW, Connolly TM, Deckmyn H, Ross TS, Ross TE, Ishii H, Bansal VS, Wilson DB (1986) The metabolism of phosphoinositide-derived messenger molecules. Science 234:1519–1526
Malik KU, Nasjletti A (1980) Effect of bradykinin on the vascular neuroeffector junction. In: Bevan JA, Godfraind T, Maxwell RA, Vanhoutte, PM (eds) Vascular neuroeffector mechanisms. Raven, New York, pp 76–82
Malvin RL (1971) Possible role of the renin angiotensin system in the regulation of antidiuretic hormone secretion. Proc Am Soc Exp Biol 30:1383–1386

Marceau F, Regoli D (1991) Kinin receptors of the B_1 type and their antagonists. In: Burch RM (ed) Bradykinin antagonists, basic and clinical research. Dekker, New York, pp 33–50
Marceau F, Tremblay B (1986) Mitogenic effect of bradykinin and of desArg^9 bradykinin on cultured fibroblasts. Life Sci 39:2351–2358
Margolius HS (1989) Tissue kallikreins and kinins: regulation and roles in hypertensive and diabetic diseases. Annu Rev Pharmacol Toxicol 29:343–364
McCubbin JW (1974) Peripheral effects of angiotensin on the antonomic nervous system. In: Page IH, Bumpus FM (eds) Angiotensin. Springer, Berlin Heidelberg New York, pp 417–423 (Handbook of experimental pharmacology, vol 37)
Muller-Esterl W, Iwanaga S, Nakanishi S (1986) Kininogens revisited, Trends Biochem Sci 11:336–339
Murayama T, Ui M (1985) Receptor mediated inhibition of adenylate cyclase and stimulation of arachidonic acid release in 3T3 fibroblasts. Selective susceptibility to islet-activating protein perfusis toxin. J Biol Chem 260:7226–7233
Murphy TJ, Alexander RW, Grinedling KK, Runge MS, Bernstein KE (1991) Isolation of a cDNA encoding the vascular type-1 angiotensin II receptor. Nature 351:233–236
Nakahata N, Ono T, Nakanishi H (1987) Contribution of prostaglandin E_2 to bradykinin-induced contraction in rabbit urinary detrusor. Jpn J Pharmacol 43:351–357
Nantel F, Rouissi N, Rhaleb NE, Jukic D, Regoli D (1991) Mechanism of action of vasoactive peptides on venous smooth muscle. J Cardiovasc Pharmacol 18: 398–405
Needleman P, Marshall GR, Sobel BE (1975) Hormone interactions in the isolated rabbit heart: synthesis and coronary vasomotor effects of prostaglandins, angiotensin and bradykinin. Circ Res 37:802–805
Northover BJ (1967) The antagonism between anti-inflammatory drugs and substances that constric veins. J Pathol Bacteriol 94:206–212
Page IH, Bumpus FM (1961) Angiotensin. Physiol Rev 41:331–390
Palaic D, Le Morvan P (1971) Angiotensin tachyphylaxis in guinea pig aortic strips. J Pharmacol Exp Ther 179:522–531
Palmer MA, Piper PJ, Vane JR (1973) Release of rabbit aorta contracting substance (RCS) and prostaglandins induced by chemical or mechanical stimulation of guinea pig lungs. Br J Pharmacol 49:226–242
Peach MJ (1977) Renin-angiotensin system: biochemistry and mechanisms of action. Physiol Rev 57:313–370
Peach MJ, Ackerly JA (1976) Angiotensin antagonists and the adrenal cortex and medulla. Proc Fed Am Soc Exp Biol 35:2502–2507
Peach MJ, Dostal DE (1990) The angiotensin II receptor and the actions of angiotensin II. J Cardiovasc Pharmacol 16(Suppl 4):525–530
Peach MJ, Bumpus FM, Khairallah PA (1971) Release of adrenal catecholamines by angiotensin I. J Pharmacol Exp Ther 176:366–376
Pobiner BF, Hewlett EL, James CG (1985) Role of N_i in coupling angiotensin receptors to inhibition of adylate cyclase in hepatocyts. J Biol Chem 260:16200–16209
Prado JL, Valle JR, Picarelli ZP (1954) Observations concerning the unitage system and sensitivity of some biological preparations to hypertensin. Acta Physiol Lat Am 4:104–120
Proud D, Reynolds CJ, LaCapra S, Kagey-Sobotka A, Lichtenstein LM, Naclerio RM (1988) Nasal provocation with bradykinin induces symptoms of rhinits and a sore throat. Am Rev Respir Dis 137:613–616
Regoli D (1972) Studies on the intrarenal action of the renin-angiotensin system. In: Genest J, Koiw E (eds) Hypertension. Springer, Berlin Heidelberg New York, pp 72–81
Regoli D (1979) Receptors for angiotensin: a critical analysis. Can J Physiol Pharmacol 57:129–139

Regoli D, Barabé J (1980) Pharmacology of bradykinin and related kinins. Pharmacol Rev 32:1–46
Regoli D, Vane JR (1964) A sensitive method for the assay of angiotensin. Br J Pharmacol 23:351–359
Regoli D, Park WK, Rioux F, Chan CS (1971) Antagonists of angiotensin. Substitution of an aliphatic chain to phenyl ring in position 8. Rev Can Biol 30:319–329
Regoli D, Park WK, Rioux F (1974) Pharmacology of angiotensin. Pharmacology 26:69–123
Regoli D, Barabé J, Park WK (1977) Receptors for bradykinin in rabbit aortae. Can J Physiol Pharmacol 55:855–867
Regoli D, Drapeau G, Rhaleb NE, Jukic D, Dion S (1989) Kinin receptors and their antagonists. In: Fritz H, Schmidt I, Dietze G (eds) The kallikrein-kinin system in health and disease. Limbach, Braunschweig, pp 205–214
Regoli D, Jukic D, Gobeil F, Rhaleb NE (1993) Receptors for bradykinin and related kinins: a critical analysis. Can J Physiol Pharmacol 71:556–567
Regoli D, Rhaleb N-E, Rouissi N, Tousignant C, Jukic D, Drapeau G (1991a) Activation of sensory nerves by kinins: pharmacologic tools for studying kinin receptors. Adv Exp Med Biol 298:63–74
Regoli D, Rhaleb N-E, Dion S, Drapeau G (1991b) New selective bradykinin receptor antagonists and bradykinin B_2 receptor characterization. TIPS 11: 156–161
Rhaleb N-E, Dion S, D'Orléans-Juste P, Drapeau G, Regoli D, Browne R (1988) Bradykinin antagonism: differentiation between peptide antagonists and anti-inflammatory agents. Eur J Pharmacol 151:275–279
Rhaleb N-E, Dion S, Barabé J, Rouissi N, Drapeau G, Regoli D (1989) Receptors for kinins in dog isolated arterial vessels. Eur J Pharmacol 162:419–427
Rhaleb N-E, Rouissi N, Nantel F, Regoli D (1991a) DUP 753 is a specific antagonist for the angiotensin receptor. Hypertension 17:480–484
Rhaleb N-E, Télémaque S, Rouissi N, Dion S, Jukic D, Drapeau G, Regoli D (1991b) Structure-activity studies of bradykinin and related peptides. B_2 receptor antagonists. Hypertension 17:107–115
Rhaleb N-E, Rouissi N, Drapeau G, Jukic D, Regoli D (1991c) Characterization of bradykinin receptors in peripheral organs. Can J Physiol Pharmacol 69:938–943
Rhaleb N-E, Rouissi N, Jukic D, Regoli D, Henke S, Breipohl G, Knolle J (1992) Pharmacological characterization of a new highly-potent B_2 receptor antagonist (HOE 140: D-Arg-[Hyp^3,Thi^5,D-Tic^7,Oic^8]-BK). Eur J Pharmacol 210:115–120
Rifo J, Pourrat M, Vavrek RJ, Stewart JM, Huidobrotoro JP (1987) Bradykinin receptor antagonists used to characterize the heterogeneity of bradykinin-induced responses in rat vas deferens. Eur J Pharmacol 142:305–312
Rioux F, Park WK, Regoli D (1973) Application of drug receptor theories to angiotensin. Can J Physiol Pharmacol 51:665–672
Rioux F, Bachelard H, St-Pierre S, Barabé J (1987) Epicardial appliaction of bradykinin elicits pressor effects and tachycardia in guinea pigs, possible mechanisms. Peptides 8:863–869
Roblero J, Croxatto H, Garcia R, Corthorn J, De Vito E (1976) Kallikrein-like activity in perfusates and urines of isolated rat kidneys. Am J Physiol 231: 1383–1389
Rösen P, Eckel J, Reinaver H (1983) Influence of bradykinin on glucose uptake and metabolism studied in isolated cardiac myocytes and isolated perfused rat hearts. Hoppe Seylers Z Physiol Chem 364:431–438
Rowley DA (1964) Venous constriction as the cause of increased permeability produced by 5-hydroxytryptamine, histamine, bradykinin and 48/80 in the rat. Br J Pathol 45:56–62
Sasaguri M, Ikeda M, Ideishi M, Arakawa K (1989) Isolation of [hydroxyproline^3] lysylbradykinin formed by kallikrein from human plasma protein. Adv Exp Med Biol 247A:539–544

Sasaki K, Yamano Y, Bardhan S, Iwai N, Murray JJ, Hasegawa M, Mattuda Y, Inagami T (1991) Cloning and expression of a complementary DNA encoding a bovine adrenal angiotensin II type-1 receptor. Nature 351:230–232
Schaffenburg CA, Haas E, Goldblatt H (1960) Concentration of renin in kidneys and angiotensinogen in serum of various species. Am J Physiol 199:788–792
Schelling P, Fisher H, Ganten D (1991) Angiotensin and cell growth: a link to cardiovascular hypertrophy. J Hypertens 9:3–15
Schild HO (1947) $_pA_2$1 a new scale for the measurement of drug antagonism. Br J Pharmacol 2:189–206
Schild HO (1968) A pharmacological approach to drug receptors. In: Tedeschi RE, Tedeschi DH (eds) The importance of fundamental principles in drug evaluation. Raven, New York, pp 257–276
Schild HO (1973) Receptor classification with special reference to beta adrenergic receptors. In: Rang HP (ed) Drug receptors. University Park Press, Baltimore, pp 29–36
Schilling WP, Ritchie AK, Navarro LT, Eskin SG (1988) Bradykinin-stimulated calcium influx in cultured bovine aortic endothelial cells. Am J Physiol 255: H219–H227
Simonsson BG, Skoogh BE, Bergh NP, Andersson R, Svedmyr N (1973) In vivo and in vitro effect of bradykinin on brochnial motor tone in normal subjects and patients with airways obstruction. Respiration 30:378–388
Smith JB, Smith L, Brown ER, Barnes D, Sabir MA, Davis JS, Farese RV (1984) Angiotensin II rapidly increases phosphatidate-phosphoinositide synthesis and phosphoinositide hydrolysis and mobilizes intracellular calcium in cultured arterial muscle cells. Proc Natl Acad Sci USA 81:7812–7816
Snider RM, Richelson E (1984) Bradykinin receptor-mediated cyclic GMP formation in a nerve cell population (murine neuroblastoma clone N1E-115). J Neurochem 43:1749–1754
Soler M, Sielczak M, Abraham WM (1989) A bradykinin antagonist blocks antigen-induced airway hyperresponsiveness and inflammation in sheep. Pulm Pharmacol 3:9–15
Speth RC, Kim KH (1990) Discrimination of two angiotensin II receptor subtypes with a selective agonist analogue of angiotensin II, p-aminophenylalamine^6 angiotensin II. Biochem Biophys Res Commun 169:997–1006
Steranka LR, Burch RM (1991) Bradykinin antagonists in pain and inflammation. In: Burch RM (ed) Bradykinin antagonists. Dekker, New York, pp 191–212
Steranka LR, Manning DC, Dehaas CJ, Ferkany JW, Borosky SA, Connor JR, Vavrek RJ, Stewart JM, Snyder H (1988) Bradykinin as a pain mediator: receptors are localized to sensory neurons and antagonists have analgesic actions. Proc Natl Acad Sci USA 85:3245–3249
Stoner J, Manganiello VC, Vaughan M (1973) Effects of bradykinin and indomethacin on cyclic GMP and cyclic AMP in lung slices. Proc Natl Acad Sci USA 70:3830–3833
Streb H, Irvine RF, Berridge JJ, Schultz F (1983) Release of Ca^{2+} from a non mitochondrial intracellular store in pancreatic acinar cells by inositol-1,4,5-triphosphate. Nature 306:67–69
Sung CP, Arleth AJ, Kazuhisa S, Berkowitz BA (1988) Characterization and function of bradykinin receptors in vascular endothelial cells. J Pharmacol Exp Ther 247:8–13
Tiffany CW, Burch RM (1989) Bradykinin stimulates tumor necrosis factor and interlukin-1 release from macrophages. FEBS Lett 247:189–192
Timmermans PBMWM, Wong PC, Chiu AT, Herblin WF (1991) Non peptide angiotensin-II receptor antagonists. TIPS 12:55–62
Toda N (1984) Endothelium-dependent relaxation induced by angiotensin II and histamine in isolated arteries of dog. Br J Pharmacol 81:301–307
Toda N, Miyazaki M (1981) Angiotensin-induced relaxation in isolated dog renal and cerebral arteries. Am J Physiol 240:H247–H254

Tousignant C, Dion S, Drapeau G, Regoli D (1987) Characterization of pre- and post-junctional receptors for neurokinins and kinins in the rat vas deferens. Neuropeptides 9:333–343

Trachte GJ (1988) Angiotensin effects on vas deferens adrenergic and purinergic neurotransmission. Eur J Pharmacol 146:261–269

Türker RK, Yamamato M, Khairallah PA, Bumpus FM (1971) Competitive antagonism of 8-Ala-angiotensin II to angiotensin I and II on isolated rabbit aorta and rat ascending colon. Eur J Pharmacol 15:285–291

Ueda N, Muramatsu I, Fujiwara M (1984) Capsaicin and bradykinin-induced substance P-ergic responses in the iris sphincter muscle of the rabbit. J Pharmacol Exp Ther 230:469–473

Van Dongen R, Peart WS, Boyd GW (1974) Effect of angiotensin II and its non-pressor derivatives on renin secretion. Am J Physiol 226:277–282

Vane JR (1969) The release and fate of vasoactive hormones in the circulation. Br J Pharmacol 35:409–242

Vavrek RJ, Stewart JM (1985) Competitive antagonists of bradykinin. Peptides 6:161–164

Voyno-Yasenetskaya TA, Tkacuk VA, Cheknyova EG, Panchenko MP, Grigorian GY, Vavrek RJ, Stewart JM, Ryan US (1989) Guanine nucleotide-dependent pertussis-insensitive regulation of phosphoinositide turnover by bradykinin in bovine pulmonary artery endothelial cells. FASEB J 3:44–51

Walaszek EJ (1970) The effect of bradykinin and kallidin on smooth muscle. In: Erdös EG (ed) Bradykinin, Kallidin and Kallikrein. Springer, Berlin Heidelberg New York, pp 421–433 (Handbook of experimental pharmacology, vol 25)

Walter P, Bassenge E (1969) Effect of angiotensin on vascular smooth muscle. Pflugers Arch Gen Physiol 307:70–79

Webb RC (1982) Angiotensin II-induced relaxation of vascular smooth muscle. Blood Vessels 19:165–176

Weyhenmeyer JA, Hwang CJ (1985) Characterization of angiotensin II binding sites on neuroblastoma x glisma hybrid cells. Brain Res Bull 14:409–444

Whalley ET, Clegg S, Stewart JM, Vavrek RJ (1987) The effect of kinin agonists and antagonists on the pain response of the human blister base. Naunyn Schmeidebergs Arch Pharmacol 336:652–655

Whatley RE, Zimmerman GA, McIntyre TM, Presscott SM (1988) Endothelium from diverse vascular sources synthesizes platelet-activating factor. Arteriosclerosis 8:321–331

Whitebread S, Mele M, Kamber B, Gasparo M (1989) Preliminary biochemical characterization of two angiotensin II receptor subtypes. Biochem Biophys Res Commun 163:284–291

Whitfield JF, MacManus JP, Gillan DJ (1970) Cyclic AMP mediation of bradykinin-induced stimulation of mitotic activity and DNA synthesis in thymocytes. Proc Soc Exp Biol Med 133:1270–1274

Williams TJ, Morley J (1973) Prostaglandins as potentiators of increased vascular permeability in inflammation. Nature 246:215–217

Williams TJ, Peck MJ (1977) Role of prostaglandin-mediated vasodilation in inflammation. Nature 270:530–532

Wirth K, Hock FJ, Albus M, Linz W, Alpermann HG, Anagnostopoulos H, Henke ST, Breipohl G, König W, Knolle J, Schölkens BA (1991) HOE 140 a new potent and long-acting bradykinin antagonist: in vivo studis. Br J Pharmacol 102:774–777

Wong PC, Hart SD, Taspel AM, Chiu AT, Ardecky RJ, Smith RD, Timmermans PBMWM (1990) Functional studies of non peptide angiotensin II receptor subtype-specific ligands: DuP 753 (A_{II}-1) and PD 123177 (A_{II}-2). J Pharmacol Exp Ther 255:584–592

Zimmerman B (1972) Action of angiotensin on vascular adrenergic nerve endings: facilitation of norepinephrine release. Fed Proc 31:1344–1350

CHAPTER 6

Effect of Histamine on Smooth Muscle

E. MASINI and P.F. MANNAIONI

A. Introduction

Since early in this century we have known that histamine has effects on the smooth muscle of various tissues in a wide variety of species. DALE and LAIDLAW (1910, 1911) showed that histamine stimulates the smooth muscle of many tissues such as intestine, bronchioles, arterioles, uterus, and spleen. Moreover, although the actions of histamine on smooth muscle were studied at an early date, identification of the receptor involved had to await the development of specific antagonists. The first effects of antihistamines were studied by BOVET and STAUB (1937); in the following years, numerous antihistamines of greater potency and specificity were synthesized and studied.

In general, these compounds inhibited the effects of histamine on smooth muscle with some notable exceptions. ASH and SCHILD (1966) suggested that mepyramine-sensitive receptors for histamine should be distinguished from histamine receptors which mediate gastric secretion and relaxation of the rat uterus, since these actions were not antagonized by mepyramine. This was one of the observations which has led to the hypothesis that histamine acts via two populations of receptors (ASH and SCHILD 1966). With the development of burimamide (BLACK et al. 1972), this hypothesis was confirmed and histamine receptors were subdivided into H_1, which are blocked by the classical antihistaminics such as mepyramine, and H_2, which are blocked by burimamide and subsequently developed H_2-receptor antagonists (BLACK et al. 1973; BRIMBLECOMBE et al. 1975). With the advent of specific H_1-agonists such as 2-pyridylethylamine and 2-thiazolylethylamine (DURANT et al. 1975) and specifc H_2-agonists such as dimaprit (PARSONS 1977) and impromidine (DURANT et al. 1978), pharmacologic tools became available to identify the nature of the receptors mediating the effects of histamine on smooth muscle.

A variety of preparations have been used to study the effect of histamine on smooth muscle, ranging from in vivo experiments on whole animals and studies on isolated whole organs to the use of muscle strips and isolated cells in culture. Conflicting results have frequently been obtained from these different preparations: in fact, factors such as the species used, the tone of the tissue, and methods of measurement are clearly important. To characterize the type of receptors involved in a typical response to histamine, different concentrations of agonists in the absence and presence of a range of concentrations of appropriate antagonists must be tested. This is particularly important since many H_1-receptor antagonists possess anti-cholinergic and local anesthetic properties; moreover, histamine may act on certain muscle tissues indirectly, involving acetylcholine or catecholamine release.

In this chapter some aspects of the action of histamine on smooth muscle will be selected in order to identify the possible physiologic and pathologic roles of histamine in this area.

B. The Gastrointestinal Tract

The role of histamine in the gastrointestinal tract has not yet been completely clarified. Its distribution is consistent with the idea that histamine is involved in secretion rather than in motility. However, there are many important aspects which cannot be disregarded such as the rate of histamine turnover, the number and affinity of receptors, and interaction with humoral and neural mediators.

Much of what was known about histamine before the discovery of H_2-receptors is contained in two exhaustive volumes edited by ROCHA and SILVA (1966, 1978), and the specific effect of the amine on gut motility was carefully reviewed by DANIEL in 1968.

The discovery of H_2-receptor antagonists, which have been so useful to the understanding of the physiology of gastric secretion, probably initially contributed in a negative way to knowledge of the physiologic role of histamine in gastrointestinal motility. In fact, some effects of the first H_2-blockers were immediately attributed to an antagonistic effect at the receptorial level without any consideration of the possible intrinsic unspecific action which was found with some H_2-antagonists, such as burimamide, metiamide, cimetidine, and ranitidine (BERTACCINI and DOBRILLA 1980).

I. Activity of the Esophagus

The lower esophageal sphincter (LES) provides a good example of the problem of species and preparation differences in the analysis of the action of histamine. Most published papers are not about the action of histamine but about that of cimetidine, due to clinical interest in this drug in gastrointestinal disorders such as reflux esophagitis. Conflicting results have been obtained in anesthetized animals. In the pig-tailed macaque, *Macaco nemestrina*, histamine relaxes the LES, acting on both H_1- and H_2-receptors, and this effect is blocked by a combination of H_1- and H_2-antagonists (DE CARLE and GLOVER 1975). Different results were obtained in a LES preparation of North American opossum on which exhaustive studies were performed with H_1- and H_2-receptor agonists and antagonists and also with tetrodotoxin, a neural inhibitor, suggesting that the stimulation of H_1-receptors on sphincter muscle causes contraction, whereas activation of H_1-receptors on intramural inhibitory neurons causes inhibition of the sphincter (RATTAN and GOYAL 1978; GOYAL and RATTAN 1978). The data reported suggested that in these species the LES contains mostly excitatory H_1- and also inhibitory H_2-histaminergic receptors; the same was observed by WALDMAN et al. (1977) and by SCHOETZ et al. (1978) in the guinea pig and baboon gallbladder, respectively.

Similar results were obtained with isolated LES in the guinea pig (TAKAYANAGI and KASUYA 1977) and rat (BERTACCINI et al. 1981), in which histamine was able to contract the LES (threshold dose $3 \times 10^{-6} M$ and $1.6 \times 10^{-5} M$, respectively). These effects were blocked in both species by chlorpheniramine, an H_1-antagonist. In the conscious baboon, histamine causes an increase in LES pressure. This effect was blocked by chlorpheniramine, without any effect on basal tone (BROWN et al. 1978). Cimetidine has no effect, suggesting that only H_1-receptors are involved in the stimulant action of histamine (BROWN et al. 1978).

Contrasting results were reported on the effects of histamine on human LES in vitro. MISIEWICZ et al. (1969) reported that histamine provokes either contractions or biphasic response; BURLEIGH (1979) reported only a relaxant effect of histamine in similar experimental conditions. CORUZZI and BERTACCINI (1982) demonstrated that 2-aminoethylthiazole, a selective H_1-agonist, evokes a contraction, an effect which was inhibited by an H_1-antagonists. In this experimental model, both H_2-agonists and -antagonists had erratic effects, causing contraction or relaxation and/or biphasic effects only at very high concentrations, suggesting an action related to the specific molecules and mediated by H_2-receptors.

Histamine, at a dose of 2–40 μg/kg per h i.v., was shown to increase the pressure of human LES in vivo (KRAVITZ et al. 1978), an effect which was completely reversed by cimetidine without modifying the basal pressure of LES; diphenhydramine was completely inactive. On the other hand, other authors (BAILEY et al. 1976; ROESCH et al. 1976) found that cimetidine causes a slight increase in LES basal pressure and a potentiation of the effect of pentagastrin. Ranitidine, an H_2-antagonist with a nitrofuran ring exerts a significant increase in pressure in human LES, an effect which is completely reversed by small doses of subcutaneous atropine, indicating stimulation of the cholinergic system (BERTACCINI et al. 1981).

In conclusion, endogenous histamine seems to have no important effect on human LES, while exogenous histamine seems to stimulate H_2-receptors in an opposite direction than in experimental animals. It is probable that the effect of histamine on the LES represents the effect of stimulation of different receptors.

II. Action on the Stomach

Histamine has been shown to contract the stomach in several animal species. The most comprehensive study on the effects of histamine on the isolated whole stomach was carried out by PATON and VANE (1963) using isolated preparations of guinea pig, kitten, rat, and mouse. Histamine caused a contraction of the guinea pig and kitten stomach which was inhibited by hyoscine and hexamethonium. High concentrations of histamine still caused contraction even in the presence of hexamethonium and hyoscine, so the authors suggested that histamine acted on ganglionic cells of cholinergic neurons at low doses and had a direct effect on the smooth muscle at high doses, although the effect of an H_1-antagonist was not studied. In contrast, BENNET and WHITNEY (1966), using strips from the corpus and antrum of human stomach, observed that histamine caused contraction or relaxation or had no effect. However, whatever response was obtained, it was consistently antagonized by mepyramine, but not by hyoscine or hexamethonium, suggesting species differences between rat and man.

In the guinea pig fundus and antrum, histamine caused an increase in the baseline without an alteration of amplitudes, an effect which was blocked by mepyramine but not by cimetidine, suggesting that the motor response to histamine is mediated by H_1-receptors (GERNER et al. 1979). In other studies, FARA and BERKOWITZ (1978) found that both H_1- and H_2-receptor selective agonists contracted strips from the circular muscle of dog antrum; however, high doses were necessary for the maximum response. Peculiar results were reported by BERTACCINI et al. (1977) in the rat pylorus, in which histamine causes a dose-related contraction of the gastroduodenal junction, an effect which was mimicked by H_1- and H_2-agonists but not abolished by H_1- or H_2-antagonists; the authors (BERTACCINI and CORUZZI 1981) hypothesized that this effect was mediated by an unknown histamine receptor or different subclasses of the classical histamine receptors.

Experiments performed on gastric fundus from kittens and young ferrets show that histamine produces secretory effects as well as modulation of motor activity, with an initial increase in the magnitude of spontaneous contraction followed by sustained increase in tone (YATES et al. 1978).

Histamine has been shown to have direct excitatory and inhibitory effects in bovine ruminal and reticular smooth muscle (OHGA and TANEIKE 1978). The dose–response curve to histamine and the pA_2 values for the antagonists suggested that H_1-receptors mediated the stimulatory and H_2-receptors the inhibitory responses to histamine. Moreover, H_2-receptors seem to be involved in the regulation of gastric emptying, as suggested by the fact that dimaprit, a selective H_2-receptor agonist, increased gastric emptying in the monkey, whereas cimetidine decreased it (DUBOIS et al. 1978). In contrast, SCARPIGNATO et al. (1981) showed that in the rat, H_1-receptors are responsible for the delay in gastric emptying; in fact, the effect was mimicked by 2-aminoethylthiazole and inhibited by chlorpheniramine in a dose-dependent way, suggesting that H_1-receptors are responsible for the delay in gastric emptying in the rat.

III. Action on the Intestine

Predictably, it was Dale and coworkers who first showed that histamine caused intestinal smooth muscle to contract (DALE and LAIDLAW 1910). Histamine determined contraction of the bowel in most species, though with significant differences between species. The sensitivity and reproducibility of the response of the terminal guinea pig ileum to histamine and the complete lack of tachyphylaxis has led to its use as a routine test preparation for studying the effects of histamine and its antagonists (SCHILD 1949; ARUNLAKSHANA and SCHILD 1959). The whole ileum and the more sophisticated longitudinal muscle preparation (PATON and VIZI 1969) have been used satisfactory with overlapping results. In the guinea pig longitudinal muscle, histamine has both an indirect and more important direct action (PARROT and THOUVENOT 1966): the first action, observed after low doses of histamine and represented by fast, unsustained contraction mediated by stimulation of the nerve cells or fibers in the intestinal wall, was depressed by hexamethonium and atropine; the second, represented by a sustained, more constant atropine-resistant response, was shown to be mediated by stimulation of specific receptors. However, even this apparently simple response has been the subject of controversy.

Some of the problems of the site of action of histamine on the guinea pig ileum may have been due to the fact that the tissue contains both longitudinal and circular muscle. The elegant studies of HARRY and coworkers (BROWNLEE and HARRY 1963) have shown that histamine acts directly on the longitudinal, but not on the circular muscle. The circular muscle, relatively insensitive to histamine, becomes sensitive after preincubation with anticholinesterase agents, indicating that histamine acts on the postsynaptic intramural nerve plexuses. In the intact ileum, the greater sensitivity of the longitudinal muscle to the action of histamine makes it unlikely that any indirect effect exerted by histamine can contribute to its response in normal conditions. The direct effect of histamine on guinea pig ileum, based on competitive antagonism by mepyramine (ASH and SCHILD 1966), received further confirmation from the specific binding of labeled mepyramine (HILL et al. 1977). The potency of histamine in competing for specific [^{3}H]mepyramine binding was less than the biological potency of histamine: this discrepancy might be explained by the existence of spare receptors for histamine (CHANG et al. 1979).

The affinity of histamine for the binding of labeled mepyramine was less in rat than in guinea pig ileum, in accordance with the lower biological activity of the amine. The age of the tissues can also influence their responsiveness to histamine. For example, only neonatal rabbit ileum was contracted dose dependently by histamine, whose action was inhibited by mepyramine (pA_2, 8.9). In the adult animal, a large amount of histamine (3×10^{-3} M) produced a notable effect that was never greater than 25% of the maximal contraction of acetylcholine (BOTTING 1975). A similar situation was found for H_2-receptors (EKELAND and OBRINK 1976). In in vivo experiments in the dog, histamine caused contractions of duodenum, which were prevented by atropine and hexamethonium; the motor activity was slightly increased both in amplitude and in tone with a rather short-lasting effect (BERTACCINI 1982). However, according to other authors (KONTUREK and SIEBERS 1980), histamine alters small-bowel motility in the dog, via stimulation of H_1-receptors. According to the early papers on human bowel strips in vitro, histamine relaxes both longitudinal and circular smooth muscle in the ileum; it has a biphasic effect on the circular colonic muscle, while the effect in *Taenia coli* is more erratic.

There is evidence both for and against the occurrence of H_2-receptors in the intestinal smooth muscle. In one of their early papers, AMBACHE and ABOO ZAR (1970) found that in guinea pig longitudinal ileal muscle preparations, histamine, in the presence of mepyramine, produced a dose-dependent inhibition of tetanic spasm induced by field stimulation. This inhibitory effect was blocked by burimamide, an H_2-antagonist, but was not mimicked by 4-(5)-methylhistamine, an H_2-agonist (AMBACHE et al. 1973), suggesting that the receptor involved resembles an H_2-receptor with a different sensitivity to H_2-agonists, leading to the hypothesis of the existence of a heterogeneous population of H_2-receptors.

Further evidence for the existence of an H_2-receptor subtype in guinea pig ileum was suggested by FJALLAND (1979), who showed that histamine was shown to suppress electrically induced twitches of ileal smooth muscle, although the order of potency of H_2-antagonists in inhibiting this action was not correlated with the known activity of the compounds at the H_2-receptor of guinea pig right atrium.

Bertaccini and coworkers (BERTACCINI and ZAPPIA 1981; BERTACCINI and CORUZZI 1981) described the presence of an H_2-receptor subtype on guinea pig duodenum which mediates relaxation of precontracted muscle. Clonidine and tolazoline, which are known to stimulate the H_2-receptors, caused a relaxant effect quite similar to that elicited by dimaprit and impromidine. All the H_2-receptor agonists tested induced relaxation of the guinea pig duodenum contracted not only by histamine, but also by acetylcholine or high $[K^+]$ (BERTACCINI and ZAPPIA 1981); however, this effect was not inhibited by the H_2-antagonists or by tetrodotoxin, propranolol, or phentolamine.

In a recent paper LEURS et al. (1991) reported that guinea pig intestinal muscle strips, when precontracted with KCl, were fairly resistant to relaxation by agents acting on histamine receptors. This might indicate that the observed relaxation is due to activity at the potassium channels (WEIR and WESTON 1986), since similar findings have been published for another potassium channel opener, cromokalin (BUCHHEIT and BERTHOLET 1988).

The classification of histamine receptors has recently been extended (ARRANG et al. 1983), and the presence of an H_3-receptor which is involved in the regulation of histamine release and synthesis both in brain and peripheral tissues (TIMMERMAN 1990) is currently accepted. Since some H_2-antagonists can also act as H_3-antagonists (SCHWARTZ et al. 1986), the findings of FJALLAND (1979) were reevaluated by JRZECIAKOWSKI, who concluded that the observed inhibition of electrically induced twitches was due to an H_3-receptor-mediated reduction of acetylcholine release (JRZECIAKOWSKI 1987). However, the myenteric plexus contains a fairly heterogeneous neuronal population, which might release various substances (STERNINI 1988). Besides acetylcholine, noradrenaline, and 5-hydroxytryptamine, several neuropeptides and excitatory amino acids are probably involved in the regulation and modulation of intestinal motility (STERNINI 1988; NEMETH and GULLIKSON 1989; SHANNON and SAWYER 1989). LEURS et al. (1991) have recently reported that the twitch responses of intestinal preparations were inhibited by *R*-α-methylhistamine, a highly selective H_3-receptor agonist, in the presence of atropine, with pD_2 values ranging from 8.10 ± 0.06 for the ileum to 8.27 ± 0.03 for the colon. These effects were competitively antagonized by thioperamide, an H_3-receptor antagonist, with pA_2 values not significantly different among preparations tested, indicating a homogeneous population

Table 1. Histamine receptors in intestinal smooth muscle (adapted from BERTACCINI 1982)

Species	Site	Receptors and effect	Reference[b]
Guinea pig	Duodenum	H_1 contraction H_2 relaxation	BAREICHA and ROCHA E SILVA (1976)
Guinea pig	Duodenum	H_2[a] relaxation	BERTACCINI and ZAPPIA (1981) BERTACCINI and CORUZZI (1981)
Dog	Duodenum	H_1 contraction	EKELAND and OBRINK (1976)
Guinea pig	Ileum	H_2[a] relaxation	AMBACHE et al. (1973)
Guinea pig	Ileum	H_2[a] relaxation	FJALLAND (1979)
Guinea pig	Ileum	H_1 contraction H_2 relaxation	BAREICHA and ROCHA E SILVA (1975)
Guinea pig	Ileum	H_1 contraction H_2 relaxation	BERTACCINI et al. (1979)
Chicken	Ileum	H_1 contraction	CHAND and DE ROTH (1978)
Rat	Ileum	H_1 contraction	PARROT and THOUVENOT (1966)
Guinea pig	*Taenia coli*	H_1 contraction	PARROT and THOUVENOT (1966)
Rabbit	Ileum	H_1 contraction	BOTTING (1975)
Dog	Small intestine in vivo	H_1 contraction	KONTUREK and SIEBERS (1979)
Guinea pig	Small and large intestine	H_1 contraction H_2 relaxation H_3 relaxation	LEURS et al. (1991)

[a] Possible subtype of classical receptor.
[b] Only representative papers are cited.

of H_3-receptors in the intestinal tract. These receptors are involved in the modulation of the release of acetylcholine (JRZECIAKOWSKI 1987) and probably of other noncholinergic contractile substances, indicating that the H_3-receptor has a quite general regulatory mechanism in the intestine. The receptors involved in the effects of histamine on the intestine are reported in Table 1.

The problem of molecular mechanism of histamine action on motility beyond amine–receptor interaction has not yet been completely resolved. In the intestinal smooth muscle, the modulation of action potential discharge is one of the most important mechanisms of varying tension, since action potentials are propagated in these cells and can generate tension in cells remote from those in which they originated (BOLTON 1979; BOLTON et al. 1981). OHMURA (1976) reported that histamine's effect on guinea pig ileum consists of phasic contractions with subsequent tonic contractions, initiated by the influx of Ca^{2+} and/or its release from the storage sites.

In the longitudinal smooth muscle of guinea pig intestine, low concentrations of histamine activate H_1-receptors, leading to an increase in the frequency of action potential discharge accompanied by an elevation in the intracellular concentration of free calcium ions (BOLTON 1979), suggesting histamine H_1-receptor as a purported "calcium-mobilizing receptor." At higher concentrations a appreciable depolarization occurs and action potential discharge ceases (BOLTON et al. 1981). Consistent with these findings is the fact that the contractile response to H_1-receptor stimulation in intestinal smooth muscle is largely sensitive to inhibition by dihydropyridine antagonists of voltage-dependent calcium channels (MOREL et al. 1987).

The response to histamine appeared to be more dependent than the response to acetylcholine on the presence of external calcium, whereas the release and transport of this ion from binding stores did not change significantly. Indirect studies on [^{32}P]-labeled phosphate ([^{32}Pi]) incorporation into phosphatidylinositol have indicated that histamine increases inositol phospholipid hydrolysis (JAFFERJI and MICHELL 1976a,b) as the transducting mechanism for controlling calcium permeability of plasma membrane (MICHELL 1975, 1979; BERRIDGE 1981, 1984; MICHELL et al. 1981) and for the mobilization of intracellular calcium stores (BERRIDGE 1983; BURGESS et al. 1984). Although calcium is very important for the development of H_1-receptor responses, this ion is not involved in the loss of H_1-receptor responsiveness after prolonged exposure to histamine.

In the process of H_1-receptor desensitization, a role for protein kinase C has been suggested and substantiated by experimental results (TIMMERMAN et al. 1988). In the longitudinal smooth muscle of guinea pig jejunum, histamine directly contracts the smooth muscle via interaction with H_1-receptors. Moreover, in nerve-stimulated preparations, histamine can indirectly depress neurogenic contraction via H_3-receptors (LEURS et al. 1991), but the mechanism for this modulation is not yet clear. In addition, COWLEN et al. (1990) also reported discrepant findings on receptor desensitization and protein kinase C activation for H_1-receptor-mediated inositol phospholipid hydrolysis.

IV. Action on the Gallbladder

Studies performed with smooth muscle preparations of the gallbladder obtained from guinea pig (Waldman et al. 1977) and baboon (Schoetz et al. 1978) have shown that both histamine H_1- and H_2-receptors are present. Under basal conditions, histamine produces a dose-related contraction of strips of gallbladder muscle, an effect which was blocked by diphenhydramine and potentiated by cimetidine. In vivo studies in the anesthetized guinea pig confirm the dual action of histamine on gallbladder: a spasmogenic effect mediated by H_1-receptors and a relaxant effect mediated by H_2-receptors (Impicciatore 1978). This observation is in agreement with the studies carried out by Waldman and coworkers (1977), showing that blockade of H_2- but, not H_1-receptors augmented the response of gallbladder muscle to cholecystokinin, a gut hormone which may play a physiologic role in the control of gallbladder emptying.

V. Conclusion

Although a large number of papers on the action of histamine on gastrointestinal smooth muscle are available, the problem of the possible physiologic role of histamine in gastrointestinal motility has not yet been clearly resolved (Bertaccini 1982). It is probable that histamine does not play a primary role in peristalsis, as it does in gastric secretion; moreover, most of the early results concerning histamine activity on gastrointestinal muscle should be reconsidered according to the presence of different subtypes of histamine receptors.

In the gastrointestinal tract, H_1-receptors are predominant and mediate contraction; H_2-receptors usually mediate relaxation, but in some case also contraction; and H_3-receptors probably subserve an inhibitory role, modulating neurotransmission at the presynaptic level. The role of many endogenous substances in gut motility and the possible hormonal, neuronal, or vascular influence on histamine activity are not completely defined. However, the possibility that this amine may contribute to the regulation of gastrointestinal motility is widely accepted.

C. The Genitourinary Tract

I. Action on the Ureter, Bladder, and Penile Tissue

Pioneering work by Chen and coworkers (1957) demonstrated that histamine provokes contraction of isolated urethral segment in the dog. The effect was blocked by H_1-antagonists, but the high concentrations used also inhibited spontaneous urethral contractions. More recent studies have shown that histamine is able to contract detrusor muscle strips from rabbit bladder through activation of H_1-receptors (FREDERICKS 1975). The pA_2 value was 9.3, clearly similar to that obtained in guinea pig ileum. KHANNA and coworkers (1977) studied the contractile response to histamine of guinea pig, dog, and human bladder smooth muscle tissue. The contractile response to histamine was more evident in the body of the bladder and very slight in the proximal urethra, and the sensitivity of the urethrovesical smooth muscle of the guinea pig to histamine was greater than that of the dog and human.

In the rabbit bladder, part of the contraction may be mediated by the release of acetylcholine, because the histamine-induced contraction was effectively inhibited by atropine or propantheline (FREDERICKS 1975). KONDO and coworkers (1985) verified the contractile effect of the amine in the guinea pig and characterized, by means of radioligand binding, the H_1-receptor involved in the response. In strips of sheep intravesical junction, histamine produced a concentration-dependent contraction through activation of H_1-receptors (BENEDITO et al. 1991). Part of the effect may have been mediated by a cholinergic mechanism; in fact, hyoscine had an inhibitory effect on the contractions induced by the amine. Interestingly, in the guinea pig bladder, histamine was found to inhibit noncholinergic contraction through stimulation of H_2-receptors (TANIYAMA et al. 1984). However, POLI and coworkers (1988) did not confirm these results; they suggest that prejunctional histamine receptors are also of the H_1-subtype and that there is heterogeneity in the H_1-receptor population between the pre- and postjunctional receptors (POLI et al. 1988).

Variable responses to histamine have also been demonstrated in human cavernous tissue: contraction, relaxation, or contraction followed by relaxation (ADAIKAN and KARIM 1977). Stimulation of H_1-receptors by 2-methylhistamine mainly caused contraction, and stimulation of H_2-receptors by 4-methylhistamine produced relaxation. These authors reported that H_1-antagonists enhanced relaxant effects, while burimamide, an H_2-antagonist, increased the contractile effect of histamine. Further support for the existence of contraction-mediating H_1-receptors and relaxation-mediating H_2-receptors was found in vivo in anesthetized baboons (ADAIKAN et al. 1991).

KIRKEBY and coworkers (1989), investigating the effects of histamine on human cavernous tissue and circumflex veins, found that histamine did not produce any effect at basal tension, but the amine relaxed both preparations when precontracted with noradrenaline. This effect, which was not altered by H_1- or H_2-antagonists seems to be due to a histamine-induced release of nitric oxide (NO) from the endothelium. In fact, the relaxation induced by histamine of noradrenaline-precontracted preparations is partly blocked by N^{ω}-nitro-L-arginine (L-NNA, ANDERSSON and HEDLUND, unpublished results, reported in ANDERSSON 1993). NO produces relaxation of the corpus cavernosum through stimulation of a soluble guanylate cyclase, leading to an increase in the tissue levels of cyclic guanosine monophosphate (cGMP), as firstly demonstrated by IGNARRO and coworkers (1990) in rabbit corpus cavernosum. The selective inhibition of cGMP phosphodiesterase enhanced the relaxant effect evoked by electrical stimulation (IGNARRO et al. 1990; BUSH et al. 1992a,b).

Histamine given to intracorpus cavernosum tissue has been shown to induce erection (NAHOUM et al. 1988; AIKADAM et al. 1991). The mechanism of the relaxant effect is unknown but, as indicated before, may be due to stimulation of relaxation-mediating H_2-receptors as well as to the release of NO. Moreover, since mast cells have been demonstrated in human corpus cavernosum tissue, it cannot be ruled out that histamine plays a role in penile erection.

II. Action on the Uterus

The effect of histamine on uterine smooth muscle, particularly in vitro, is markedly species dependent. In guinea pigs, histamine causes a contraction (DALE and LAIDLAW 1910), an effect blocked by H_1-antagonists (DEWS and GRAHAM 1946). In contrast, histamine exerts an inhibitory effect on rat uterus. Histamine has been shown to inhibit the spontaneously contracting uterus (MITZNEGG et al. 1975), an effect mediated by the increase in cyclic adenosine monophosphate (cAMP); moreover, the amine also inhibits the contractile response evoked by serotonin (AMIN et al. 1954) and by electrical stimulation (BLACK et al. 1972). The inhibitory action of histamine on rat uterus was mediated by H_2-receptors (BLACK et al. 1972). Pretreatment of the tissue with reserpine shifted the histamine dose–response curve to the right, suggesting that in the rat uterus the effect of histamine was mediated by the release of endogenous catecholamines (TOZZI 1973; MCNEILL and VERMA 1975).

As in the rat uterus, histamine has been shown to inhibit electrically evoked contractions of the mouse vas deferens (MARSHALL 1978), an effect which is mediated by H_2-receptors. In contrast, the twitch responses of both the epididymal and prostatic portions of the vas deferens of the guinea pig and rabbit were potentiated by histamine by means of the activation of an H_1-receptors (BHALLA and MARSHALL 1980).

The possible physiologic and/or pathologic role of histamine in the smooth muscle tissues of genitourinary tract is still under discussion.

D. The Cardiovascular System

I. Overview and Historical Background

Histamine is present in high concentrations in cardiac tissues in most animal species (MANNAIONI 1972; LEVI et al. 1982) and elicits a wide range of responses in several areas of the cardiovascular system, some of which have had a key role in the historical development of histamine receptors and in the identification of histamine receptor antagonists.

In most animal species, including humans, histamine was found to increase heart rate and contractility, to decrease atrioventricular nodal conduction velocity, and increase automaticity and coronary flow. Moreover, histamine is highly arrhythmogenic (GIOTTI and ZILLETTI 1976; LEVI et al. 1981).

The intravenous administration of histamine causes a fall in peripheral vascular resistance, mediated by both H_1- and H_2-receptors, which determines a decrease in systemic blood pressure (LEVI et al. 1982, 1991). The fall in blood pressure activates a sympathoadrenal reflex associated with a marked rise in plasma epinephrine levels, particularly in the cat and guinea pig (BLACK et al. 1975; MACQUIN-MAVIER et al. 1988). The effects of histamine on specific regional vascular beds are quite variable. Depending on the dose, route of administration, animal species, anatomical location, caliber, and tone of the vessels, histamine can elicit either vasoconstriction, vasodilation, or both.

A variety of exhaustive reviews have appeared in recent years on the action of histamine on the mammalian heart and vasculature (ALTURA and HALEVY 1978; WOLFF and LEVI 1986; LEVI et al. 1982, 1991), rendering the task of reviewing the physiopathologic effects of histamine on vascular smooth muscle apparently repetitive. However, the recent discovery that endothelial cells synthesize vasoconstrictor and vasodilator substances (FURCHGOTT and ZAWADZKI 1980; GRYGLEWSKI et al. 1986; YANAGISAWA et al. 1988) has contributed substantially to a better understanding of the actions of histamine on vascular smooth muscle. Therefore, we have chosen to direct our attention to the most recent developments on this subject.

In general, the effect of histamine on a given regional vasculature can best be understood as the result of multiple actions on the smooth muscle and underlying endothelium. H_1- and H_2-receptors on vascular smooth muscle mediate constriction and relaxation, respectively, while endothelial H_1-receptors indirectly promote vasorelaxation via the release of endothelium-derived relaxing factor (EDRF), which was discovered by FURCHGOTT and ZAWASKI (1980) and identified as NO (PALMER et al. 1987), and/or of prostacyclin.

When a vessel is denuded of its endothelial lining, it will respond to histamine with a contractile or dilatatory response, depending on which of two histamine receptor subtypes predominates in the smooth muscle. However, when the endothelium is intact, the release of the above-mentioned relaxing factors may offset the vasoconstriction mediated by H_1-receptors, resulting in vasodilatation. Vasodilatation may also result from inhibition of the sympathetic tone by activation of histamine H_3-receptors on perivascular nerve terminals (ISHIKAWA and SPERELAKIS 1987). Thus, the effects of histamine on the vasculature are the results of the various components and

Table 2. Effect of histamine on vascular smooth muscle (adapted from LEVI et al. 1991)

Species	Site		Endothelium	Receptors and effect	Reference
Rabbit	Aorta	IR	+	H_2 relaxation	ABACIOGLU et al. (1987)
Rat	Thoracic aorta		+	H_1/H_2 relaxation	RAPOPORT and MURAD (1983)
Mouse	Thoracic aorta	IR	+	H_1 contraction	VAN DE VOORDE and LEUSEN (1984)
Cat	Thoracic aorta	IR	+	H_1 contraction	VAN DE VOORDE and LEUSEN (1984)
Rabbit	Coronary artery	IH	+	H_1 contraction	SAARI (1986)
Dog	Epicardial coronary arteries	IN	+	H_1 contraction H_2 relaxation	MILLER and BOVE (1988)
Dog	Distal coronary arteries	IN	+	H_1/H_2 relaxation	MILLER and BOVE (1988)
Dog	Epicardial coronary arteries	IR	+/−	H_2 relaxation	NAKANE and CHIBA (1987)
Dog	Coronary arteries	HS	+/−	H_2 relaxation	TODA (1986)
Pig	Coronary arteries	HS	+	? contraction	YAMAMOTO et al. (1987)
Monkey	Coronary arteries	IS	+/−	H_2 relaxation	TODA (1986)
Human	Coronary arteries	IN	+	H_1 contraction	VIGORITO et al. (1986)
Human	Coronary arteries	HS	+	H_1/H_2 relaxation	TODA (1983, 1987)
Human	Coronary arteries	IR	+	H_2 contraction	GODFRAIND and MILLER (1983)
Human	Proximal epicardial coronary arteries	IR	+	H_2 relaxation (decreased effect)	GINSBURG et al. (1984)

Table 2. *Continued*

Species	Site		Endothelium	Receptors and effect	Reference
Rabbit	Middle cerebral arteries	IR	+	H_1 contraction	SERCOMBE et al. (1986)
			+	H_2 relaxation	KIM et al. (1988)
			+	H_3 relaxation	KIM and OUDART (1988)
Dog	Middle cerebral arteries	HS	+	H_1 contraction	TODA et al. (1985)
Human	Cerebral arteries	IS	+	H_1 contraction	TAKAGI et al. (1989)
			+	H_2 relaxation	
Guinea pig	Pulmonary arteries	HS	+	H_1/H_2 relaxation	ABACIOGLU et al. (1987)
Guinea pig	Pulmonary arteries	HS	+	H_1 contraction	SUZUKI and KOU (1983)
Guinea pig	Pulmonary arteries	IR	+	? contraction	SAKUMA and LEVI (1988)
Human	Pulmonary arteries:				
	small	HS	+/–	H_1 contraction	SCHELLENBERG et al. (1986)
	large	HS	+	H_1 relaxation	
Human	Pulmonary arteries	HS	+	? contraction	BOE (1983)
Human	Pulmonary arteries and veins	IR	+	H_1 contraction	MIKKELSEN et al. (1984)
Rabbit	Renal arteries	IR	+/–	H_1 contraction	ROBINSON and MAXSON (1982)

HS, helical strips; IR, isolated rings; IS, isolated segments; IN, infusion.

responses observed in different animal species and reflect the predominance of one component over another. Table 2 summarizes the effect of histamine on vascular smooth muscle.

II. Aorta

Histamine relaxes rat aorta in vitro with an endothelium-dependent mechanism and is associated with an increase in cGMP within the smooth muscle cells (RAPOPORT and MURAD 1983), involving NO formation (MONCADA et al. 1991).

Helical strips or rings prepared from thoracic aorta of rabbit, guinea pig, cat, and mouse are contracted by histamine; this effect is H_1 mediated and enhanced by endothelium removal. Anoxia, renal and spontaneous hypertension, diabetes, and aging (SIM and CHUA 1985; TANZ et al. 1989) attenuate histamine-induced, endothelium-dependent relaxation through the reduced ability of endothelium to produce EDRF and/or to activate the guanylcyclase activity (LEE et al. 1987; HONGO et al. 1988; VANHOUTTE and EBER 1991).

III. Coronary Vessels

Remarkable species differences have been reported concerning the responsiveness of coronary vessels to histamine in vitro. Coronary arteries from monkey (TODA 1986) and dog (TODA 1986; NAKANE and CHIBA 1987; MILLER and BOVE 1988) respond to the amine with relaxation, while human (TODA 1983, 1987; GODFRAIND and MILLER 1983; KALSNER and RICHARDS 1984), pig (YAMAMOTO et al. 1987; DAVIES et al. 1987), cattle (OBI et al. 1991), rabbit (SAARI 1986), and guinea pig (LEVI 1972; BROADLEY 1975; FLYNN et al. 1979) respond with contraction.

In humans and pigs, the coronary constriction evoked by histamine is decreased by H_1-receptor blockade and potentiated by H_2-antagonists or by endothelium removal (TODA 1987; YAMAMOTO et al. 1987; DAVIES et al. 1987). This evidence suggests that histamine-induced vasoconstriction is mainly dependent on the H_1-receptors in the smooth muscle cells, while the activation of H_2-receptors on the smooth muscle and of H_1-receptors on the endothelium dampens this effect (Fig. 1).

The response of monkey and dog coronary vasculature to histamine has been shown to consist of relaxation, converted to vasoconstriction in the presence of H_2-receptor-blocking drugs. Neither H_1-blockers nor endothelium removal modifies the relaxation of the dog coronary artery in response to histamine (TODA 1986; NAKANE and CHIBA 1987), while both treatments abolish the vasodilating effect of histamine on monkey coronary arteries (TODA 1986). Thus, the relaxation of the isolated coronary artery of the dog is exclusively mediated by smooth muscle H_2-receptors, while in monkey, the vasodilating action of histamine is mediated by endothelial

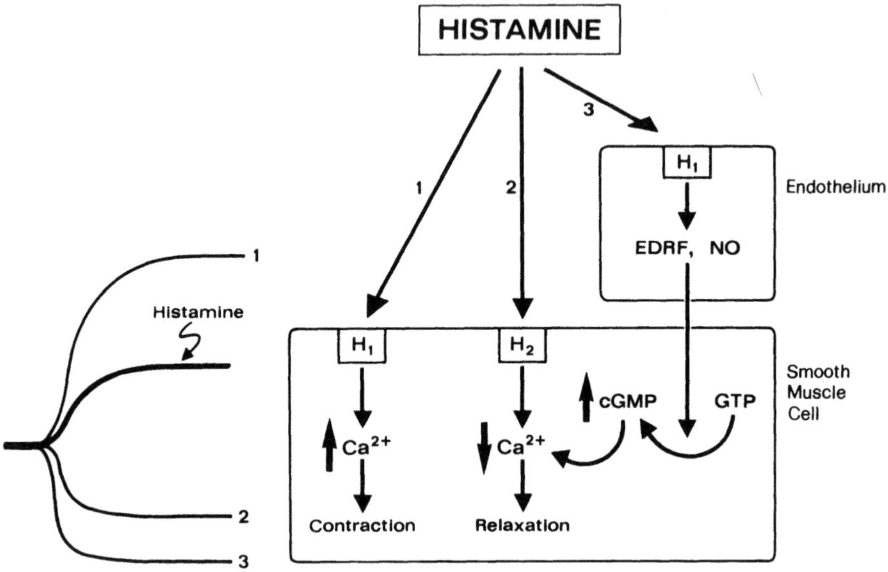

Fig. 1. Actions of histamine in human coronary arteries. *Squares* in endothelium and smooth muscle represent histaminergic H_1- and H_2-receptors. Possible responses mediated by different sites of action (*1–3* on the *right* side) are reported on the *left* side. *EDRF*, Endothelium-derived relaxing factor; *NO*, nitric oxide; *GTP*, guanosine triphosphate; *cGMP*, cyclic guanosine monophosphate. Adapted from TODA (1987)

H_1-receptors which promote the release of NO. This unstable molecule, generated during arginine metabolism (MONCADA et al. 1988; PALMER et al. 1988), induces vasodilation by increasing cGMP levels in vascular smooth muscle (IGNARRO 1989; FURCHGOTT and VANHOUTE 1989). The contribution of NO to the coronary vasomotor action of histamine varies not only with the animal species, but also with the vascular caliber. Histamine-induced relaxation is inversely correlated with the thickness of intima (KEITOKU et al. 1988) and may explain why the amine constricts the proximal coronary artery more than the distal in humans (GINSBURG et al. 1984). The H_1-mediated spasm of large epicardial vessels can be induced by histamine administration either in normal subjects or in patients with variant angina pectoris (GINSBURG et al. 1981). Moreover, in a porcine model, in which atherosclerotic plaques were produced in the epicardial coronary artery, histamine was the only vasoactive agent which induced coronary spasm in these areas (SHIMOKAWA et al. 1983). This spasm has been attributed to impaired endothelium-dependent relaxation and to an H_1-mediated enhanced Ca^{2+} influx into vascular smooth muscle (TOMOIKE et al. 1989).

In human coronary artery with moderate atherosclerotic lesions, the histamine dose–response curve was shifted to the left, indicating that atherosclerosis induced histamine supersensitivity, probably consequent to an increase in the H_1-receptor density in the diseased coronary vessels

(BRISTOW et al. 1981). In contrast, human distal coronary vessels are dilated by histamine, with an increase in coronary blood flow (VIGORITO et al. 1987) mediated by H_1-receptors and associated with the release of catecholamine (VIGORITO et al. 1986). Also in the anesthetized dog, the intracoronary infusion of histamine induced an increase in coronary flow (MILLER and BOVE 1988).

The clinical significance of the effect of histamine on the modulation of coronary tone is unknown. Histamine is released by antigen–antibody reaction and by endogenous substances and therapeutic agents including basic polypeptides, alkaloids, and others (MASINI et al. 1985; BARRETT and PEARCE 1991). Mast cells, the major source of endogenous histamine, are increased around zones of recent thrombosis, in the atheromatous arteries and also in edematous areas of the vascular wall (FORMAN et al. 1985; DOVRAK 1986). Endogenous histamine, like exogenously applied histamine, may be one of the key substances responsible for localized coronary vasospasm (GINSBURG et al. 1981).

IV. Cerebral Vessels

Histamine dilates cerebral arteries in primates and humans when administered intravenously (LEVI et al. 1982). The influence of histamine on cerebral arterial and venous smooth muscles has been evaluated by studies in vitro with isolated vessels and in situ by the measurement of cerebral blood flow.

In isolated human cerebral arteries, the response to histamine is the result of predominant H_1-mediated constriction and weak H_2-mediated relaxation (TAKAGI et al. 1989), which appears to be endothelium dependent. In the dog, the systemic administration of histamine decreases arterial blood pressure and increases blood flow (LEVI et al. 1982, 1991). However, histamine elicits an H_1-mediated contraction of isolated canine cerebral and basilar artery strips (TODA et al. 1982). This effect is potentiated by aspirin, suggesting that cyclooxygenase products attenuate H_1-mediated vasoconstriction.

In the middle cerebral artery of the rabbit, histamine evokes vasoconstriction, which is potentiated by removal of the endothelium or H_2-blockade, suggesting a vasorelaxant component in the action of histamine (SERCOMBE et al. 1986). These effects may involve the release of prostacyclin, since indomethacin potentiates histamine-induced constriction (KIM et al. 1988) and/or EDRF production. Moreover, histamine H_3-receptors, which mediate the autoinhibition of histamine release in the human brain (ARRANG et al. 1991), may also play a role in this relaxation (KIM and OUDART 1988).

V. Pulmonary Vessels

Histamine has a dual effect on pulmonary vessels in vitro: constriction mediated via H_1-receptors and vasodilatation via H_2-receptors (BOE et al.

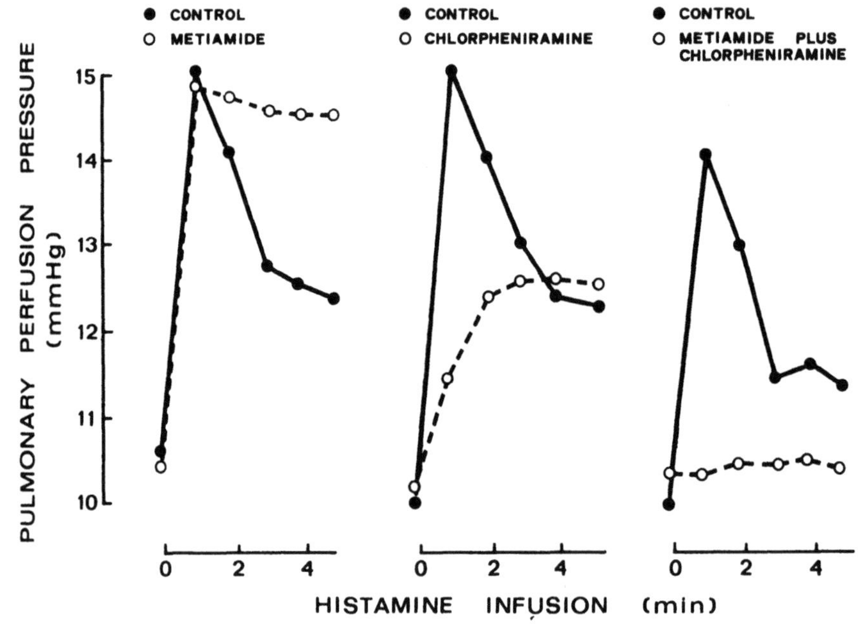

Fig. 2. Histamine receptor subtypes in dog pulmonary artery; response to histamine infusion on pulmonary perfusion pressure. Adapted from EYRE and CHAND (1982)

1980). The pulmonary vasodilator response to histamine is more apparent in the presence of H_1-antagonists and it is blocked by H_2-antagonists (TURKER et al. 1975), confirming that the response to injected histamine reflects the balance between vasoconstriction mediated by H_1-receptors and vasodilation mediated by H_2-receptors (Fig. 2). Human bronchial vessels in vitro are released by low and contracted by high concentrations of histamine (LIU et al. 1990). Pretreatment with histamine H_1-receptor antagonists inhibits both relaxation and constriction (MIKKELSEN et al. 1984). Removal of the endothelium or treatment with indomethacin abolishes histamine-induced relaxation, suggesting that prostacylin and/or EDRF mediate the vasodilating effect of histamine (SCHELLEMBERG et al. 1986; SAKUMA et al. 1988), while the constrictor effect is due to direct action of H_1-receptors on the vascular smooth muscle.

Histamine increases airway blood flow in vivo, but there is some doubt about whether this effect is mediated by H_1- or H_2-receptors, since even in the same species different effects of H_1- and H_2-blockers are reported (LONG et al. 1985; WEBBER et al. 1988). Although H_2-vasodilating receptors have clearly been demonstrated in pulmonary arteries and veins, the role of H_2-receptors in bronchial circulation is still under investigation.

There is evidence that histamine evokes both contraction and relaxation in isolated rings of guinea pig pulmonary artery. Contraction has been attributed to H_1-receptors present on vascular smooth muscle cells (SAKUMA and LEVI 1988) and on perivascular adrenergic nerves (SUZUKI and KOU 1983). In contrast, H_2-receptors elicit relaxation (ABACIOGLU et al. 1987) by inducing smooth muscle hyperpolarization (SUZUKI and KOU 1983). In the sheep and dog, histamine induces an increase in bronchial blood flow, mediated by H_2-receptors (LONG et al. 1985; LAITINEN et al. 1987); however, in human bronchial vessels H_2-antagonists have no effect on the vasodilator response to histamine (LIU et al. 1990).

VI. Portal, Mesenteric and Renal Vessels

Portal and mesenteric vessels respond to histamine both with contraction and/or relaxation. In vitro preparations from mesenteric rabbit and guinea pig arteries respond to histamine with an H_1-mediated contraction, while relaxation of preconstricted arteries is an H_2 mediated event (ROBINSON and MAXSON 1982). In the dog, relaxation is mediated by H_1- and H_2-receptors and partly mediated by the release of endothelial relaxing factors (prostacyclin, EDRF). The endothelium-dependent relaxation of canine mesenteric veins mediated by histamine is not affected by indomethacin but by methylene blue, suggesting that EDRF, rather than prostacyclin, contributes to this relaxation.

It has been recently proposed that histamine acts as a chemical mediator of renal autoregulation in the dog. Histamine induces relaxation in the preconstricted dog renal artery (TODA 1984); this process is H_2 mediated and also involves the release of prostacyclin. The renal artery of the rat does not respond to histamine in vitro; however, when preconstricted, a relaxant response mediated by H_1- and H_2-receptors is observed (KRSTIČ et al. 1989).

VII. Vascular Physiopathology

Although there is evidence that histamine plays a role in vascular physiology and pathology, the function of vascular histamine is presently unknown. It is significant that the heart contains large quantities of histamine, and mast cells, mostly located in perivascular tissues, are the main repository of cardiac histamine (RYDZYNSKI et al. 1988; WOLFF and LEVI 1986). Degranulation of cardiac mast cells by an immunoglobulin E (IgE)-mediated mechanism has been demonstrated (ASSEM et al. 1986). Cardiac anaphylaxis is characterized by severe tachycardia, tachy- and bradyarrhythmias, and coronary vasoconstriction. These responses are H_2 mediated (LEVI et al. 1982), potentiated by leukotriene D_4 (BURKE et al. 1982) and by complement-derived C3a and C5a anaphylatoxins (DEL BALZO et al. 1989). The release of NO stimulated by histamine during the initial phase of cardiac anaphylaxis in the isolated guinea pig heart has a protective role (MASINI et al. 1994). Histamine is also implicated in coronary vasoconstriction (TODA 1987), in atherogenesis, and as a possible mediator of vasospasm in atherosclerotic coronary vessels (FORMAN et al. 1985; ATKINSON et al. 1987).

Summing up, a role for histamine in vascular pathophysiology appears to be firmly established, although evidence in support of a possible homeostatic role is still being gathered.

E. Pulmonary Tract

I. Introductory Notes

A variety of stimuli, including neurotransmitters, arachidonic acid metabolites, hormones, and histamine, are known to modulate the contractile state of the respiratory tract myocytes, and histamine has long been regarded as an important mediator of the bronchoconstrictor response in pulmonary hypersensitivity. It has been known since the pioneering work of DALE and LAIDLAW (1910) that exogenous administration of histamine to guinea pigs evokes signs and symptoms of respiratory distress very much like those appearing after challenge with the specific antigen. This historic finding has been confirmed in many animal species during the past 80 years. BARTOSCH and collegues (1932) first demonstrated that histamine is released during anaphylaxis from the isolated guinea pig and human lung, and bronchial tissues from asthmatic patients were first shown to release histamine when challenged with specific antigen in 1951 (SCHILD et al. 1951). In recent years, it has become clear that the effect of histamine on smooth airway muscle varies considerably according to the region of the airways, the animal specie examined, and the experimental conditions employed.

The main histologic site of histamine storage is the mast cells (RILEY and WEST 1953), which are principally located in perivascular lung tissue. Mast cells are also present within the bronchial and nasal submucosa in the alveolar septal connective tissue and in the pleura (CUTZ and ORANGE 1977). Histamine released locally in the respiratory tract has several effects; these are mediated through histamine receptors mainly of the H_1- and H_2-subtypes. Recently, H_3-receptors have also been detected in human lung in binding studies (ARRANG et al. 1987), but only a few studies have investigated their functional role in airways.

II. Effects of Histamine on Airway Smooth Muscle

The best-recognized action of histamine in the respiratory tract is bronchoconstriction, which is mediated by H_1-receptors. Mepyramine and other related H_1-receptor antagonists block histamine-induced contraction of the trachea, bronchi, and bronchioles in most species. A large body of evidence is available on this point and has been thoroughly reviewed by EYRE and CHAND (1982). Histamine contracts large and small human airways in vitro (FINNEY et al. 1985) and is more potent in contracting lung strips, suggesting that the peripheral airways are more sensitive to histamine (DRAZEN and SCHNEIDER 1978), although this increasing response may be partially due to contraction of vascular smooth muscle.

However, the entire airways of the rat, rabbit, and cat fail to contract to histamine in vitro (FLEISCH and CALKINS 1976; CHAND and EYRE 1977, 1978). In these species, histamine may cause relaxation of the tracheobronchial muscle contracted by carbachol, an effect sensitive to H_2-antagonists. H_2-receptors which mediate bronchodilation have been demonstrated in several species, including the cat, rat, rabbit, sheep, and horse (CHAND and EYRE 1975). In some species, such as the rabbit, the H_2-receptor-mediated response predominates, since histamine itself causes bronchodilation. However, there is some debate as to whether H_2-receptors are present in human airways.

Human peripheral lung strips show a relaxant response to histamine via H_2-receptors (VINCENC et al. 1983); this response probably also reflects the relaxant effect on pulmonary vessels which are present in lung strip preparations.

H_2-selective blockers do not cause any bronchoconstriction in vivo, nor do they increase the bronchoconstrictor response to inhaled histamine (THOMPSON and KERR 1980), although it has been reported that in asthmatic patients H_2-receptor function may be impaired (GONZALEZ and AHMED 1986). Histamine H_2-receptors in the lung have been recently reviewed by FOREMAN (1991).

In bovine and canine tracheal smooth muscle, H_1-receptor stimulation elicits a marked transient increase in intracellular free Ca^{2+} followed by a lower plateau for the following 5 min (MATSUMOTO et al. 1986; PANETTIERI et al. 1989). The initial, transient calcium increase is due to a release of calcium from intracellular stores (MATSUMOTO et al. 1986; TAKUWA et al. 1987), while the second sustained phase of contraction involves a calcium influx insensible to the hydropyridine calcium entry blockers (TAKUWA et al. 1987). In these tissues, histamine can also elicit inositol phospholipid hydrolysis (HALL and HILL 1988; HALL and CHILVERS 1989), and a transient increase in inositol-1,4,5-triphosphate in bovine tracheal smooth muscle has been demonstrated (CHILVERS et al. 1989).

Histamine causes tachyphylaxis in smooth airway muscle of some species such as guinea pigs, and this seems to be mediated by both the release of prostaglandins such as PGE_2, since it is reduced by indomethacin (HAYE-LEGRAND et al. 1986), and by H_2-receptors, since cimetidine pretreatment prevents this effect (JACKSON et al. 1988). In several species, including humans, the contraction of airway smooth muscle induced by histamine is increased after mechanical removal of the epithelium (CUSS and BARNES 1987; KNIGHT et al. 1990). This is likely to be due to release of a relaxant factor from epithelial cells (CUSS and BARNES 1987).

In keeping with the observations relevant to the pulmonary vascular system, the bronchorelaxant effect of histamine is completely resistant to the action of all the H_2-receptor antagonists presently available, thus suggesting the presence of H_2-isoreceptors or H_3-receptors mediating the atypical effects. It has been shown that activation of H_3-receptors inhibits vagally mediated contraction of the guinea pig trachea (Ichinose and Barnes 1989), suggesting that presynaptic H_3-receptors can modulate the release of contracting substances from the vagus.

In fact, besides the H_1/H_2 moiety located on cell membranes of pulmonary vascular and smooth airway muscle and of mast cells, the activation of which is responsible for direct effects, a further histaminergic receptor, which may mediate the cholinergic reflex actions of histamine, is especially relevant to the pathophysiology of asthma (Fig. 3). In light of these observations, H_3-agonists could be of some therapeutic benefit in asthma and other diseases where neurogenic inflammation has been implicated (Barnes et al. 1990). In addition, it is worth remembering that there are also cholinergic mechanisms that may facilitate the release of mediators themselves. In fact, in isolated rat serosal mast cells, acetylcholine stimulates the secretion of histamine (Fantozzi et al. 1978). The process is blocked by atropine, adrenaline, and H_2-receptor agonists and is enhanced by the presence of IgE on mast cell membranes (Blandina et al. 1980; Masini et al. 1982, 1985). H_2-mediated bronchodilation is mediated by an increase in intracellular levels of cAMP (Hall and Hill 1988; Hall et al. 1989). Elevation of tissue cAMP content probably leads to tissue relaxation through a range of different mechanisms including phosphorylation of myosin light-chain kinase (Silver and Stull 1982), membrane hyperpolarization (Fujiwara et al. 1988), effects upon Ca^{2+}-gated K^+ channels (Kume et al. 1989), and sequestration of intracellular calcium (Mueller and Van Breeman 1979).

It is therefore conceivable that acetylcholine released at the parasympathetic nerve endings would in turn evoke secretion of histamine from mast cells, leading to bronchoconstriction. The functional relevances of inhibiting H_3-receptors on airway nerves may be considered as a protective feedback inhibitory mechanism (Barnes and Ichinose 1989). Moreover, there is a close relationship between airway mast cells and nerves (Bienenstock et al. 1988). Therefore, it is possible that histamine released from mast cells by parasympathetic stimulation may act on H_3-receptors on cholinergic nerve

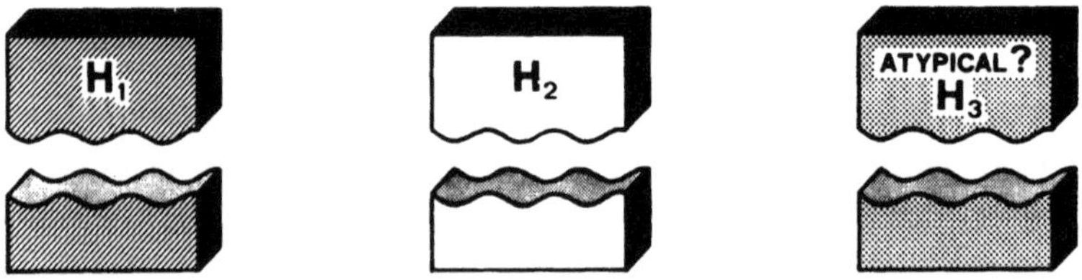

Fig. 3. Airway histamine receptor subtypes. *Left*, contraction of smooth muscle in trachea, primary bronchus, and lung strips; *middle*, relaxation of precontracted tracheobronchial smooth muscle; *right*, bronchodilation evoked by H_2-agonists insensitive to H_2-antagonists

terminals and parasympathetic ganglia to inhibit transmission and thus prevent activation of bronchoconstrictor reflexes. The receptorial effects of histamine on smooth airway muscle are schematically shown in Fig. 3.

In conclusion, histamine has several effects on respiratory smooth muscle. Histamine-induced bronchoconstriction is mediated by H_1-receptors. In most species, H_2-receptors which cause relaxation have been clearly demonstrated by in vitro experiments.

Some atypical receptors in the lung have been reported, and it is unclear whether these are really H_2-receptors which appear to be atypical because of the experimental conditions or whether they are H_3-receptors. H_2- and H_3-receptors also have certain effects which are more subtle and tend to counteract the actions of histamine mediated through H_1-receptors.

F. Concluding Remarks

Histamine exerts both direct and indirect effects on smooth muscle. Three classes of histamine receptors have now been clearly identified.

H_1-receptors are predominant and mediate contraction; H_2-receptors mediate relaxation; and H_3-receptors modulate cholinergic neurotransmission, the release of neuropeptides from sensory nerves and histamine from mast cells.

References

Abacioglu N, Ercan ZS, Kanzik I, Zengil H, Demirÿrek T, Türker RK (1987) Endothelium-dependent relaxing effect of histamine on the isolated guinea-pig main pulmonary artery strips. Agents Actions 22:30–35

Adaikan PG, Lau LC, NG, SC, Ratnam SS (1991) Physiopharmacology of human penile erection – autonomic/nitrergic neurotransmission and receptors of the human corpus cavernosum. Asian Pac J Pharmacol 6:213–227

Adaikan PG, Karim SMM (1977) Effect of histamine on the human penis muscle in vitro. Eur J Pharmacol 45:261–265

Altura BM, Halevy S (1978) Cardiovascular actions of histamine. In: Rocha e Silva M (ed) Histamine and anti-histaminics. Springer, Berlin Heidelberg New York, pp 1–39 (Handbook of experimental pharmacology, vol XVIII/2)

Ambache N, Aboo Zar M (1970) An inhibitory action of histamine on the guinea-pig ileum. Br J Pharmacol 38:229–240

Ambache N, Killick SW, Aboo Zar M (1973) Antagonism by burimamide of inhibition induced by histamine in plexus containing longitudinal muscle preparations from guinea-pig ileum. Br J Pharmacol 48:362P–363P

Amin AH, Crawford TBR, Gaddum JH (1954) Distribution of substance P and 5-hydroxytriptamine in the central nervous system of the dog. J Physiol 126: 596–618

Andersson KE (1993) Pharmacology of lower urinary tract smooth muscles and penile erectile tissues. Pharmacol Rev 45:253–308

Arrang JM, Garbarg M, Schwartz JC (1983) Autoinhibition of brain histamine release by a novel class (H_3) of histamine receptors. Nature 327:117–123

Arrang JM, Garbarg M, Lancelot JC, Lecomte JM, Pollar H, Bobba M, Schunak W, Schwartz JC (1987) Highly potent and selective ligands for histamine H_3-receptors. Nature 327:117–123

Arrang JM, Garbarg M, Schwartz JC, Lipp R, Stark H, Schunack W, Lecomte JM (1991) The histamine H_3-receptor: pharmacology, roles and clinical implications studied with agonists. In: Timmerman H, van der Goot H (eds) New prospective in histamine research. Birkhauser, Basel, pp 55–67 (Agents actions supplement, vol 33)
Arunlakshana O, Schild HO (1959) Some quantitative uses of drug antagonists. Br J Pharmacol Chemother 14:48–58
Ash ASF, Schild HO (1966) Receptors mediated some actions of histamine. Br J Pharmacol 27:427–439
Assem ESK, Machad FRD's, Ghanem MS (1986) Cardiac mast cells, partial purification of guinea pig atrial mast cells and release from them of histamine and leukotriene C_4 by immune and non-immune stimuli. Agents Actions 18:167–171
Atkinson JB, Roberts LJ, Aulsebrook KA, Harlan CW, Virmani R (1987) Role of histamine in atherogenesis. Atherosclerosis 7:530–539
Bailey RJ, Sullivan SN, MacDougall BRD, Williams R (1976) Effect of cimetidine on lower esophageal sphincter. Br Med J 2:678–683
Bareicha I, Rocha e Silva M (1976) H_1- and H_2-receptors for histamine in the guinea pig intestine. Biochem Pharmacol 24:1215–1219
Barnes PJ, Ichinose M (1989) H_3 receptors in airway. Trends Pharmacol Sci 10: 264–267
Barnes PJ, Belvisi MG, Rogers DF (1990) Modulation of neurogenic inflammation: novel approaches to inflammatory diseases. Trends Pharmacol Sci 11:185–189
Barrett KE, Pearce FL (1991) Heterogeneity of mast cells. In: Uvnäs B (ed) Histaine and histamine antagonists. Springer, Berlin Heidelberg New York, pp 93–117 (Handbook of experimental pharmacology, vol 97)
Bartosch R, Feldberg W, Nagel E (1932) Das Freiwerden eines histaminähulichen Stoffes bei der Anaphylaxie der Meerschweinchens. Pfluegers Arch Gesamte Physiol 230:129–163
Benedito S, Prieto D, Rivera L, Costa G, García-Sacristín A (1991) Mechanisms implicated on the histamine response of the sheep ureterovescical junction. J Urol 146:184–187
Bennett A, Whitney B (1966) A pharmacological investigation of human isolated stomach. Br J Pharmacol Chemother 26:286–298
Berridge MJ (1981) Phosphatidylinositol hydrolysis: a multifunctional transducing mechanism. Mol Cell Endocrinol 24:115–172
Berridge MJ (1983) Rapid accumulation of inositol triphosphate reveals that agonists hydrolyse polyphosphoinositides instead of phosphatidylinositol. Biochem J 212: 849–858
Berridge MJ (1984) Inositol triphosphate and diacylglycerol as second messangers. Biochem J 220:345–360
Bertaccini G (1982) Amines: histamine. In: Bertaccini G (ed) Mediators and drugs in gastrointestinal motility. Springer, Berlin Heidelberg New York, pp 201–218 (Handbook of experimental pharmacology, vol LIX/2)
Bertaccini G, Coruzzi G (1981) Evidence for and against heterogeneity in the histamine H_2 receptor population. Pharmacology 23:1–13
Bertaccini G, Dobrilla G (1980) Histamine H_2-receptor antagonists: old and new generation. Ital J Gastroenterol 12:297–302
Bertaccini G, Zappia L (1981) Histamine receptors in the guinea pig duodenum. J Pharm Pharmacol 33:590–593
Bertaccini G, Coruzzi G, Molina E, Chiavarini M (1977) Action of histamine and related compounds on the pyloric sphincter of the rat. Rend Gastroenterol 9:163–168
Bertaccini G, Molina E, Zappia L, Zseli J (1979) Histamine receptors in the guinea pig ileum. Naunyn Schmiederbergs Arch Pharmacol 309:65–68
Bertaccini G, Coruzzi G, Scarpignato C (1981) Exogenous and endogenous compounds which affect the contractility of the lower esophageal sphincter (LES).

In: Stipa S, Belsey RHR, Moraldi A (eds) Medical and surgical problems of the esophagus. Academic, London, pp 22–29
Bhalla I, Marshall I (1980) Histamine H_1- and H_2-receptor-mediated effects in vasa deferentia from the rat, guinea-pig and rabbit. Br J Pharmacol 60:304P–305P
Bienenstock J, Perdue M, Blennerhassett M, Stead R, Kakuta N, Sestini P, Vancheri C, Marshall J (1988) Inflammatory cells and epithelium: mast cell/nerve interactions in lung "in vitro" and "in vivo." Am Rev Respir Dis 138:31–34
Black JW, Duncan WAM, Durant CJ, Ganellin CR, Parsons EM (1972) Definition and antagonism of histamine H_2-receptors. Nature 236:385–390
Black JW, Duncan WAM, Emmett JC, Ganellin CR, Hesselbo T, Parsons ME, Wyllie JH (1973) Metiamide – an orally active histamine H_2-receptor antagonist. Agents Actions 3:133–137
Black JW, Owen DAA, Parsons ME (1975) An analysis of the depressor responses to histamine in the cat and dog: involvement of both H_1 and H_2-receptors. Br J Pharmacol 54:319–324
Blandina P, Fantozzi R, Mannaioni PF, Masini E (1980) Characteristics of histamine release evoked by acetylcholine in isolated rat mast cells. J Physiol (Lond) 301:281–293
Boe J (1983) Hypoxic and embolic pulmonary vasoconstriction. Gen Pharmacol 14:149–151
Boe J, Boe MA, Simonsson BG (1980) A dual action of histamine in isolated human pulmonary arteries. Respiration 40:117–122
Bolton TB (1979) Mechanism of action of transmitters and other substances on smooth muscle. Physiol Rev 59:606–618
Bolton TB, Clark JP, Kitmaura K, Lang RJ (1981) Evidence that histamine and carbachol may open the same ion channels in longitudinal smooth muscle of guinea pig ileum. J Physiol (Lond) 320:363–379
Botting JH (1975) Sensitivity of neonatal rabbit ileum to histamine. Br J Pharmacol 53:428–429
Bovet D, Staub AM (1937) Action protective des ethers phenoliques an cours de l'intoxication histaminique. C R Soc Biol (Paris) 124:547–549
Brimblecombe RW, Duncan WAM, Durant GJ (1975) The pharmacology of cimetidine, a new histamine H_2-receptor antagonist. Br J Pharmacol 53:435–436
Bristow MR, Ginsburg R, Minobe WA, Sageman WS, Harrison DC (1981) Receptor density differences may account for the focal alterations in vascular reactivity. Circulation 64:IV-288
Broadley KJ (1975) The role of H_1- and H_2-receptors in the coronary vascular response to histamine of isolated perfused hearts of guinea-pig and rabbits. Br J Pharmacol 54:511–521
Brown FC, Dubois A, Castell DO (1978) Histaminergic pharmacology of primate lower esophageal sphincter. Am J Physiol 235:E42–E46
Brownlee G, Harry J (1963) Some pharmacological properties of the circular and longitudinal muscle strips from the guinea-pig isolated ileum. Br J Pharmacol 21:544–554
Buchheit K, Bertholet A (1988) Inhibition of small intestine motility by cromokalin (BRL 34915). Eur J Pharmacol 154:335–337
Burke JA, Levi R, Guo ZG, Corey EJ (1982) Leukotrienes C_4, D_4 and E_4: effects on human and guinea pig cardiac preparations "in vitro." J Pharmacol Exp Ther 221:235–241
Burleigh DE (1979) The effects of drugs and electrical field stimulation on the human lower esophageal sphincter. Arch Int Pharmacodyn Ther 240:169–176
Burgess GM, Godfrey PP, McKinney JS, Berridge MJ, Irvine RF, Putney JW (1984) The second messenger linking receptor activation to internal calcium release in liver. Nature 309:63–66
Bush PA, Aronson WJ, Buga GM, Rajfer J, Ignarro LJ (1992a) Nitric oxide is a potent relaxant of human and rabbit corpus cavernosum. J Urol 147:1650–1655

Bush PA, Aronson WJ, Rajfer J, Buga GM, Ignarro LJ (1992b) Comparison of nonadrenergic, noncholinergic and nitric oxide – mediated relaxation of corpus cavernosum. Int J Impotence Res 4:85–93
Chand N, De Roth L (1978) Occurrence of H_2-inhibitory histamine receptors in chicken ileum. Eur J Pharmacol 52:143–145
Chand N, Eyre P (1975) Classification and biological distribution of histamine receptor sub-types. Agents Actions 5:277–295
Chand N, Eyre P (1977) Atypical (relaxant) response to histamine in cat bronchus. Agents Actions 7:183–190
Chand N, Eyre P (1978) Action of histamine on airway smooth muscle of ferret and rat. Res Commun Chem Pathol Pharmacol 21:55–65
Chang RSL, Tran VT, Snyder SH (1979) Caracteristic of histamine H_1 receptors in peripheral tissues labeled with [^{3}H]-mepyramine. J Pharmacol Exp Ther 209: 437–442
Chen PS Jr, Emmel VM, Benjamin JA, Di Stefano V (1957) Studies on the isolated dog ureter. The pharmacological action of histamine, levartezenol and antihistaminics. Arch Int Pharmacodyn Ther 110:131–141
Chilvers ER, Challiss RAJ, Barnes PJ, Nahorski SR (1989) Mass changes of inositol (1,4,5)triphosphate in trachealis muscle following agonist stimulation. Eur J Pharmacol 164:587–590
Coruzzi G, Bertaccini G (1982) Histamine receptors in the lower esophageal sphincter (LES). Agents Actions 12:1–5
Cowlen MS, Barnes MR, Toews ML (1990) Regulation of histamine H_1-receptor-mediated phosphoinositide hydrolysis by histamine and phorbol esters in DDT_1 MF-2 cells. Eur J Pharmacol 188:105–112
Cuss FM, Barnes PJ (1987) Epithelial mediators. Am Rev Respir Dis 136:S32–S35
Cutz E, Orange RP (1977) Mast cells and apocrine (APUD) cells of the lung. In: Lichtenstein LM, Austen KF (eds) Asthma, physiology, immunopharmacology and treatment, vol 2. Academic, New York, p 52
Dale HH, Laidlaw PP (1910) The physiological action of imidazolylethylamine. J Physiol 41:318–344
Dale HH, Laidlaw PP (1911) Further observations on the action of beta-imidazolylethyamine. J Physiol 43:182–195
Daniel E (1968) Pharmacology of the gastrointestinal tract. In: Code CF (ed) Handbook of physiology, vol IV. American Physiological Society, Washington DC, pp 2267–2324
Davies JM, Epstein SE, Pierce JR, Ramwell PW, Sprecher D (1987) Low density lipoprotein modulation of porcine coronary artery conctractile response to histamine. Atherosclerosis 64:21–25
De Carle DJ, Glover WE (1975) Independence of gastrin and histamine receptors in the lower oesophageal sphincter of the monkey and opossum. J Physiol (Lond) 245:78P–79P
Del Balzo U, Polley MJ, Levi R (1989) Cardiac anaphylaxis, complement activation as an amplification system. Circ Res 65:847–857
Dews PB, Graham JDP (1946) The antihistamine substance 2786RP. Br J Pharmacol Chemother 1:278–286
Drazen JM, Schneider MW (1978) Comparative responses of tracheal spirals and parenchymal strips to histamine and carbachol "in vitro." J Clin Invest 61: 1441–1448
Dubois A, Nompleggi D, Myers L, Castell DO (1978) Histamine H_2-receptor stimulation increases gastric emptying. Gastroenterology 74:A1028
Durant GJ, Ganellin CR, Parsons ME (1975) Chemical differentiation of histamine H_1- and H_2-receptor agonists. J Med Chem 18:830–833
Durant GJ, Duncan WAM, Ganellin CR, Parsons ME, Blakemore RC, Rasmussen AC (1978) Impromidine is a very potent and specific agonist for H_2 receptors. Nature 276:403–404

Dvorak AM (1986) Mast cell degranulation in human hearts. N Engl J Med 315: 969–970
Ekeland M, Obrink KKJ (1976) Histamine sensitivity of isolated gastric mucosa from growing rats. Acta Physiol Scand 96:34–39
Eyre P, Chand N (1982) Histamine receptor mechanisms of the lung. In: Ganellin GR, Parsons ME (eds) Pharmacology of histamine receptors. Wright, Bristol, pp 299–322
Fantozzi R, Masini E, Blandina P, Mannaioni PF, Bani Sacchi T (1978) Release of histamine from rat mast cells by acetylcholine. Nature 273:473–474
Fara JW, Berkowitz JM (1978) Effects of histamine and gastrointestinal hormones on dog antral smooth muscle in vitro. Scand J Gastroenterol [Suppl] 49(13):60
Finney MJB, Karlsson JA, Persson CGA (1985) Effects of bronchoconstrictors and bronchodilators on a novel human small airway preparation. Br J Pharmacol 85:29–36
Fjalland B (1979) Evidence for the existence of another type of histamine H_2-receptor in guinea-pig ileum. J Pharm Pharmacol 31:50–51
Fleisch JH, Calkins PJ (1976) comparison of drug-induced responses of rabbit trachea and bronchus. J Appl Physiol 41:62–66
Flynn SB, Gristwood RW, Owen DAA (1979) Differentiation of the roles of histamine H_1- and H_2-receptors in the mediation of the effects of histamine in the isolated working heart of the guinea-pig. Br J Pharmacol 65:127–137
Foreman JC (1991) Histamine H_2 receptors and lung functions. In: Uvnäs B (ed) Histamine and histamine antagonists. Springer, Berlin Heidelberg New York, pp 285–304 (Handbook of experimental pharmacology, vol XCVII)
Forman MB, Oates JA, Robertson D, Robertson RM, Roberts LJ, Virmani R (1985) Increased adventitial mast cells in a patients with coronary spasm. N Engl J Med 313:1138–1141
Fredericks CM (1975) Characterization of the rabbit detrusor response to histamine through pharmacologic antagonism. Pharmacology 13:5–11
Fujiwara T, Suminoto K, Itoh T, Suzuki K, Kuriyama H (1988) Relaxing actions of procaterol, a β_2 adrenoceptor stimulant, on smooth muscle cells of the dog trachea. Br J Pharmacol 93:199–209
Furchgott RF, Vanhoutte PM (1989) Endothelium-derived relaxing and contracting factors. FASEB J 3:2007–2018
Furchgott RF, Zawadzki JV (1980) The obligatory role of endothelial cells in the relaxation of arterial smooth muscle by acetylcholine, Nature 288:373–376
Gerner T, Haffner JFW, Norstein J (1979) The effect of mepyramine and cimetidine on the motor responses to histamine, cholecystokinin and gastrin in the fundus and antrum of isolated guinea pig stomach. Scand J Gastroenterol 14:65–72
Ginsburg R, Bristow MR, Kantrowitz N, Baim DS, Harrison DC (1981) Histamine provocation of clinical coronary artery spasm: implications concerning pathogenesis of variant angina pectoris. Am Heart J 102:819–822
Ginsburg R, Bristow MR, Davis K (1984) Receptor mechanisms in the human epicardial coronary artery heterogeneous pharmacological response to histamine and carbachol. Circ Res 55:416–421
Giotti A, Zilletti L (1976) Antiarrhythmic actions of anti-histamines. In: Szekeres L (ed) Pharmacological therapy, vol 2. Pergamon, New York , pp 863–900
Godfraind T, Miller RC (1983) Effects of histamine and the histamine antagonists mepyramine and cimetidine on human coronary arteries in vitro. Br J Pharmacol 79:979–984
Gonzalez H, Ahmed T (1986) Suppression of gastric H_2-receptor mediated function in patients with bronchial asthma and ragweed allergy. Chest 89:491–496
Goyal RK, Rattan S (1978) Neurohumoral, hormonal and drug receptors for the lower esophageal sphincter. Gastroenterology 74:598–619
Gryglewski RJ, Moncada S, Palmer RMJ (1986) Bioassay of prostacyclin and endothelium-derived relaxing factor (EDRF) from porcine aortic endothelial cells. Br J Pharmacol 87:685–694

Hall IP, Chilvers ER (1989) Inositol phosphate and airways smooth muscles. Pulmon Pharmacol 2:113–120
Hall IP, Hill SJ (1988) β_2 adrenoceptor stimulation inhibits histamine-stimulated inositol phospholipid hydrolysis in bovine tracheal smooth muscle. Br J Pharmacol 95:1204–1212
Hall IP, Donaldson J, Hill SJ (1989) Inhibition of histamine-stimulated inositol phospholipid hydrolysis by agents which increase cyclic AMP levels in bovine tracheal smooth muscle. Br J Pharmacol 97:603–616
Haye-Legrand I, Cerrina J, Raffestin B, Labat C, Boullet C, Bayol A, Benveniste J, Brink C (1986) Histamine contraction of isolated human airway muscle preparations: role of prostaglandins. J Pharmacol Exp Ther 239:546–541
Hill SJ, Young JM, Marrian DH (1977) Specific binding of [^{3}H]-mepyramine to histamine H_1 receptors in intestinal smooth muscle. Nature 252:54–55
Hongo K, Nakagomi T, Kassel NF, Sasak T, Lehman M, Vollmer DG, Tsukuhara T, Ogawa H, Torner J (1988) Effects of aging and hypertension on endothelium-dependent vascular relaxation in rat carotid artery. Stroke 19:892–897
Ichinose M, Barnes PJ (1989) Inhibitory histamine H_3-receptors on cholinergic nerves in human airways. Eur J Pharmacol 163:383–386
Ignarro LJ (1989) Endothelium-derived nitric oxide: actions and properties. FASEB J 3:31–36
Ignarro LJ, Bush PA, Buga GM, Wood KS, Fukoto JM, Rajfer J (1990) Nitric oxide and cyclic GMP formation upon eletrical field stimulation cause relaxation of corpus cavernosum smooth muscle. Biochem Biophys Res Commun 170: 1843–1850
Impicciatore M (1978) Occurrence of H_1- and H_2-histamine receptors in the guinea-pig gall bladder "in situ." Br J Pharmacol 64:219–222
Ishikawa S, Sperelakis N (1987) A novel class (H_3) of histamine receptors on perivascular nerve terminals. Nature 327:158–160
Jackson PJ, Manning PJ, O'Byrne PM (1988) A new role for histamine H_2-receptors in asthmatic airways. Am Rev Respir Dis 138:784–788
Jafferji SS, Michell RH (1976a) Stimulation of phosphatidylinositol turnover by histamine, 5-hydroxytriptamine and adrenaline in the longitudinal smooth muscle of guinea-pig ileum. Biochem Pharmacol 25:1429–1430
Jafferji SS, Michell RH (1976b) Effects of calcium-antagonistic drugs on the stimulation by carbamylcholine and histamine of phosphatidylinositol turnover in longitudinal smooth muscle of guinea-pig ileum. Biochem J 160: 163–169
Jrzeciakowki JP (1987) Inhibition of guinea-pig ileum contraction mediated by a class of histamine receptor resembling the H_3 subtype. J Pharmacol Exp Ther 243: 874–880
Kalsner S, Richards R (1984) Coronary arteries of cardiac patients are hyperreactive and contain stores of amines: a mechanism for coronary spasm. Science 223: 1435–1437
Keitoku M, Okayama H, Satoh Y, Maruyama Y, Jakishima T (1988) Does diffuse intimal thickening in human coronary artery act as a diffusion barrier to endothelium-derived relaxing factor? J Exp Med 154:413–414
Khanna OP, De Gregorio GJ, Sample RC (1977) Histamine receptors in urethrovescical smooth muscle. Urology 10:375–381
Kim LE, Oudart N (1988) A high potent and selective H_3 agonist relaxes rabbit middle cerebral artery in vitro. Eur J Pharmacol 150:393–396
Kim LE, Sercombe R, Oudart N (1988) Relaxation of rabbit middle cerebral arteries in vitro by H_1 histaminergic agonists is inhibited by indomethacin and tranylcypromine. Fundam Clin Pharmacol 2:463–475
Kirkeby HJ, Forman A, Sorensen S, Andersson KE (1989) Effects of noradrenaline, 5-hydroxytriptamine and histamine on human penile cavernosus tissue and circumflex venis. Int J Impotence Res 1:181–188

Knight DA, Adcock JA, Phillips MJ, Thompson PJ (1990) The effect of epithelium removal on human bronchial smooth muscle responsiveness to acetylcholine and histamine. Pulmon Pharmacol 3:198–202
Kondo M, Taniyama K, Tanaka C (1985) Histamine H_1-receptors in the giunea pig urinary bladder. Eur J Pharmacol 114:89–92
Konturek SJ, Siebers R (1979) Role of histamine H_1 and H_2 receptors in the myoelectric activity of the small bowel. Gastroeneterology 76:A1174
Konturek SJ, Siebers R (1980) Role of histaminic H_1- and H_2-receptors in myoeletric activity of the small bowel in the dog. Am J Physiol 238:G50–G56
Kravitz JJ, Snape WJ Jr, Cohen S (1978) Effect of histamine and histamine antagonists on human lower esophageal sphinter function. Gastroenterology 74:435–440
Krstic MK, Stepanovic RM, Krstic SK, Katusic ZS (1989) Endothelium-dependent relaxation of the rat renal artery caused by activation of histamine H_1-receptors. Pharmcaology 38:113–120
Kume H, Takai A, Tokuno H, Tomita T (1989) Regulation of Ca^{2+} dependent K^+ channel activity in tracheal myocytes by phosphorylation. Nature 341:152–154
Laitinen LA, Laitinen A, Widdicombe JG (1987) Effects of inflammatory and other mediators on airway vascular beds. Am Rev Respir Dis 135:S67–S70
Lee TJF, Shirasaki Y, Nicholson GA (1987) Altered endothelial modulation of vascular tone in aging and hypertension. Blood Vessels 24:132–136
Leurs R, Brozius MM, Smit MJ, Bast A, Timmerman H (1991) Effects of histamine H_1-, H_2- and H_3-receptor selective drugs on the mechanical activity of guinea-pig small and large intestine. Br J Pharmacol 102:179–185
Levi R (1972) Effects of exogenous and immunologically released histamine on the isolated heart: a quantitative comparison. J Pharmacol Exp Ther 182:227–238
Levi R, Malm JR, Bowman FO, Rosen MR (1981) The arrhythmogenic actions of histamine on human atrial fibers. Circ Res 49:545–550
Levi R, Owen DAA, Trzeciakowski J (1982) Actions of histamine on the heart and vasculature. In: Ganellin CR, Parsons ME (eds) Pharmacology of histamine receptors. Wright, London, pp 236–297
Levi R, Rubin LE, Gross SS (1991) Histamine in cardiovascular function and dysfunction: recent developments. In: Uvnäs B (ed) Histamine and histamine antagonists. Springer, Berlin Heidelberg New York (Handbook of experimental pharmacology, vol 97)
Liu SF, Yacoub M, Barnes PJ (1990) Effect of histamine on human bronchial arteries "in vitro." Naunyn Schmiedebergs Arch Pharmacol 342:90–93
Long WM, Sprung CL, Elfawal H (1985) Effects of histamine on bronchial artery blood flow and bronchomotor tone. J Appl Physiol 59:254–261
Macquin-Mavier I, Clerici C, Franco-Montoya ML, Harf A (1988) Mechanism of histamine-induced epinephrine release in guinea pig. J Pharmacol Exp Ther 247:706–709
Mannaioni PF (1972) Physiology and pharmacology of cardiac histamine. Arch Int Pharmacodyn 196:64–78
Marshall I (1978) An inhibitory histamine H_2-receptor in the mouse vas deferens. Br J Pharmacol 62:447P
Masini E, Blandina P, Mannaioni PF (1982) Mast cell receptors controlling histamine release: influence on the mode of actions of drugs used in the treatment of adverse drug reactions. Klin Wochenschr 60:1031–1038
Masini E, Fantozzi R, Blandina P, Brunelleschi S, Mannaioni PF (1985) The riddle of cholinergic histamine release. Prog Med Chem 22:267–291
Masini E, Pistelli A, Gambassi F, Di Bello MG, Mannaioni PF (1994) The role of nitric oxide in anaphylactic reaction of isolated guinea pig hearts and mast cells. In: Nisticò G, Masek K, Higgs AE, Moncada S (eds) Nitric oxide: brain and immune system. Portland, London, pp 277–287
Matsumoto T, Kanaide H, Nishimura J, Shogakiuchi Y, Kobayashi S, Nakumura M (1986) Histamine activates H_1-receptors to induce cytosolic free calcium tran-

sient in cultured vascular smooth muscle cells from rat aorta. Biochem Biophys Res Commun 135:172–177
McNeill JH, Verma SC (1975) Histamine receptors in rat uterus. Res Commun Chem Pathol Pharmacol 11:639–644
Michell RH (1975) Inositol phospholipids and cell surface receptor function. Biochem Biophys Acta 415:81–147
Michell RH (1979) Inositol phospholipids in membrane function. Trends Biochem Sci 4:128–131
Michell RH, Kirk CJ, Jones LM, Downes C, Creba JB (1981) The stimulation of inositol lipid metabolism that accompanies calcium mobilization in stimutated cells: defined characteristics and answered questions. Phil Trans R Soc B 296: 123–137
Mikkelsen E, Sakr AM, Jespersen LT (1984) Studies on the effect of histamine in isolated human pulmonary arteries and veins. Acta Pharmacol Toxicol 54:86–93
Miller WL, Bove AA (1988) Differential H_1- and H_2-receptor-mediated histamine responses of canine epicardial conductance and distal resistence of coronary vessels. Circ Res 62:226–232
Misiewicz JJ, Waller SL, Anthony PP, Gummer JWP (1969) Achalasia of the cardia: pharmacology and histopathology of isolated cardiac sphinter muscle from patients with and without achalasia. Q J Med 38:17–30
Mitznegg P, Schubert E, Fuchs J (1975) Relations between the effects of histamine, pheniramine and metiamide on spontaneous motility and formation of cyclic AMP in the isolated rat uterus. Naunyn Schmiedebergs Arch Pharmacol 287: 321–327
Moncada S, Radomski MW, Palmer RMJ (1988) Endothelium-derived relaxing factor: identification as nitric oxide and role in the control of vascular tone and platelet function. Biochem Pharmacol 37:2495–2501
Moncada S, Palmer RMJ, Higgs EA (1991) Nitric oxide: physiology, pathophysiology and pharmacology. Pharmacol Rev 43:109–142
Morel N, Hardy JP, Godfraind T (1987) Histamine-operated calcium channel in intestinal smooth muscle of the guinea pig. Eur J Phramacol 135:69–75
Mueller E, Van Breeman C (1979) Role of intracellular Ca sequestration in beta adrenergeic relaxation of a smooth muscle. Nature 281:682–683
Nahoum CRD, Hadler WA, Santana CAR (1988) In: Proceedings of the international society for impotence research. Boston, p 43
Nakane T, Chiba S (1987) Characteristics of histamine receptors in the isolated and perfused canine coronary arteries. Arch Int Pharmacodyn Ther 290:92–103
Nemeth PR, Gullikson GW (1989) Gastrointestinal motility stimulating drugs and 5-HT receptors on myenteric neurons. Eur J Pharmacol 166:387–391
Obi T, Matsumoto M, Nishio A (1991) Vasomotor effect of histamine on pig and cattle coronary artery in vitro. Jpn J Pharmacol 55:311–320
Ohga A, Taneike T (1978) H_1- and H_2-receptors in the smooth muscle of the ruminant stomach. Br J Pharmacol 62:333–337
Ohmura I (1976) Action mechanisms of the contracting drugs, K, acetylcholine, histamine and Ba and of the antispasmodics, isoproterenol, and papaverine in the isolated guinea pig ileum, particularly in relation to Ca. Folia Pharmacol Jpn 72:201–210
Palmer RMJ, Ferrige AG, Moncada S (1987) Nitric oxide release accounts for the biological activity of endothelium-derived relaxing factor. Nature 327:524–526
Palmer RMJ, Rees DD, Ashton DS, Moncada S (1988) L-Arginine is the physiological precursor of the formation of nitric oxide in endothelium-dependent relaxation. Biochem Biophys Res Commun 153:1251–1256
Panettieri RA, Murray RK, Depalo LR, Yadvish TA, Kotlikoff MI (1989) A human airway smooth muscle cell line that retains physiological responsiveness. Am J Physiol 256:C329–C335

Parrot JL, Thouvenot J (1966) Action de l'histamine sur les muscles lisses. In: Rocha e Silva M (ed) Histamine. Its chemistry, metabolism and physiological and pharmacological actions. Springer, Berlin Heidelberg New York, pp 202–204 (Handbook of experimental pharmacology, vol XVIII/1)
Parsons ME (1977) The antogonism of histamine H_2-receptors "in vitro" and "in vivo" with particular reference to the actions of cimetidine. In: Burland WL, Alison Simkins M (eds) Cimetidine. Exerpta Medica, Amsterdam, pp 13–20
Paton WDM, Vane JR (1963) An analysis of the response of the isolated stomach to electrical stimulation and to drugs. J Physiol 165:10–46
Paton WDM, Vizi ES (1969) The inhibitory action of noradrenaline and adrenaline on acetylcholine output by guinea-pig ileum longitudinal muscle strip. Br J Pharmacol 35:10–29
Poli E, Coruzzi G, Bertaccini G (1988) Pre- and post-junctional effects of histamine on the guinea-pig urinary bladder: evidence for heterogenity in the H_1 population? Agents Actions 23:241 243
Rapoport RM, Murad F (1983) Agonist-induced endothelium-dependent relaxation in rat thoracic aorta may be mediated through cGMP. Circ Res 52:352–357
Rattan S, Goyal RK (1978) Effects of histamine on the lower esophageal sphinter "in vivo": evidence for action of three different sites. J Pharmacol Exp Ther 204:334–342
Riley JF, West GB (1953) The presence of histamine in tissue mast cells. J Physiol 120:528–531
Robinson CP, Maxson S (1982) Differences in histamine H_1 and H_2 receptor responses in several rabbit arteries. Res Commun Chem Pathol Pharmacol 36: 355–366
Rocha e Silva M (ed) (1966) Histamine. Its chemistry, metabolism physiological and pharmacological agents. Springer, Berlin Heidelberg New York (Handbook of experimental pharmacology, vol XVIII/1)
Rocha e Silva M (ed) (1978) Histamine and anti-histaminics. Springer, Berlin Heidelberg New York (Handbook of experimental pharmacology, vol XVIII/2)
Roesch W, Lux G, Schittenhelm W, Demlin L (1976) Stimulation of lower esophageal sphincter (LES) pressure by cimetidine. A double blind study. Acta Hepatogastroenterol (Stuttg) 23:423–425
Rydzynski K, Kolago B, Zaslonka J, Kuroczynski W (1988) Distribution of mast cells in human heart auricles and correlation with tissue histamine. Agents Actions 23:273–275
Saari JT (1986) Characterization of the coronary vascular response to histamine in rabbit hearts using cimetidine. Pharmacology 32:80–89
Sakuma I, Levi R (1988) Vasomotor effects of leukotrienes C_4 and D_4 on cavian pulmonary artery and aorta: characterization and mechanism. Ann NY Acad Sci 524:91–102
Sakuma I, Stuehr DJ, Gross SS, Nathan C, Levi R (1988) Identification of arginine as a precursor of endothelium-derived relaxing factor. Proc Natl Acad Sci USA 85:8664–8667
Scarpignato C, Coruzzi G, Bertaccini G (1981) Effect of histamine and related compounds on gastric emptying in the rat. Pharmacology 23:185–191
Schellenberg RR, Duff MJ, Foster A, Paddon HB (1986) Histamine releases PGI_2 from human coronary artery. Prostaglandins 32:201–209
Schild HO (1949) pAx and competitive drug antagonism. Br J Pharmacol Chemother 4:277–280
Schild HO, Hawkins DR, Mongar JL, Herxheimer H (1951) Reactions of isolated human asthmatic lung and bronchial tissue to a specific antigen. Histamine release and muscolar contraction. Lancet 261:376–382
Schoetz DJ, Wise WE, La Morte WW, Bickett DH, Williams LF (1978) Histamine receptors in the primate gallbladder. Gastroenterology 74:A1090

Schwartz JC, Arrang JM, Garbarg M (1986) Three classes of histamine receptors in brain. Trends Phramacol Sci 7:24–28
Sercombe R, Verrecchia C, Philipson V, Oudart N, Dimitriadov V, Bouchaud C, Seylaz J (1986) Histamine-induced constriction and dilatation of rabbit middle cerebral arteries in vitro: role of endothelium. Blood Vessels 23:137–153
Shannon HE, Sawyer BD (1989) Glutamate receptors of the N-methyl-D-aspartate subtype in the myenteric plexus of the guinea-pig ileum. J Pharmacol Exp Ther 251:518–523
Shimokawa H, Tomoike H, Nabeyama S, Yamamoto H, Araki H, Nakamura M, Ishii Y, Tanaka K (1983) Coronary artery spasm induced in atherosclerotic miniature swine. Science 221:560–562
Silver PJ, Stull JT (1982) Regulation of myosin light chain kinase and phosphorylase phosphorylation in tracheal smooth muscle. J Biol Chem 257:6145–6150
Sim MK, Chua ME (1985) Altered responsiveness of the aorta of hypertensive rats to histamine and acetylcholine. Jpn J Pharmacol 39:551–553
Sternini S (1988) Structural and chemical organisation of the myenteric plexus. Ann Rev Physiol 50:81–93
Suzuki H, Kou K (1983) Direct and indirect effects of histamine on the smooth muscle cells of the guinea-pig main pulmonary artery. Pflugers Arch 399:46–53
Takagi T, Tan EC, Nagai H, Usui H, Satake N, Shibata S (1989) Characteristics of histamine response in isolated human cerebral arteries. FASEB J 3:A845
Takayanagi I, Kasuya Y (1977) Effects of some drugs on the circular muscle of the isolated lower oesophagus. J Pharm Pharmacol 29:559–560
Takuwa Y, Takuwa N, Rasmussen H (1987) Measurement of cytoplasmic free Ca concentration in bovine tracheal smooth muscle using aequorin. Am J Physiol 253:C817–C827
Taniyama K, Kusunoki M, Tanaka C (1984) Inhibitory histamine H_2-receptors on the guinea pig urinary bladder. Eur J Pharmacol 105:113–119
Tanz RD, Change KSK, Weller TS (1989) Histamine relaxation of aortic rings from diabetic rats. Agents Actions 28:1–8
Thompson NC, Kerr JW (1980) The effect of inhaled H_1- and H_2-receptor antagonists in normal and asthmatic subjects. Thorax 35:428–434
Timmerman H (1990) Histamine H_3 ligands: just pharmacological tools or potential therapeutic agents? J Med Chem 33:4–11
Timmerman H, Haaksma EEJ, Leurs R (1988) Selective legands for the three type of histamine receptors. In: Melchiorre C, Giannella M (eds) Recent advances in receptor chemistry. Elsevier, Amsterdam
Toda N (1983) Isolated human coronary arteries in response to vasoconstrictor substances. Am J Physiol 245:H937–H941
Toda N (1984) Endothelium-dependent relaxation induced by angiotensin II and histamine in isolated arteries of dog. Br J Pharmacol 81:529–535
Toda N (1986) Mechanisms of histamine-induced relaxation in isolated monkey and dog coronary arteries. J Pharmacol Exp Ther 239:529–535
Toda N (1987) Mechanism of histamine actions in human coronary arteries. Circ Res 61:280–286
Toda N, Konishi M, Miyazaki M (1982) Involvement of endogenous prostaglandin I in the vascular action of histamine in dogs. J Pharmacol Exp Ther 223:257–262
Toda N, Okamura T, Miyazaki M (1985) Heterogeneity in the response to vasoconstrictors of isolated dog proximal and distal middle cerebral arteries. Eur J Pharmacol 106:291–299
Tomoike H, Egashira K, Yamamoto Y, Nakamura M (1989) Enhanced responsiveness of smooth muscle, impaired endothelium-dependent relaxation and the genesis of coronary spasm. Am J Cardiol 63:33E–39E
Tozzi S (1973) The mechanism of action of histamine on the isolated rat uterus. J Pharmacol Exp Ther 187:511–517

Turker A, Reeves JT, Grover RF (1975) Cardiovascular actions of histamine H_1- and H_2-receptor stimulation in the dog. Fed Proc 34:438–442
Van de Voorde J, Leusen I (1984) Effect of histamine on aorta preparations of different species. Arch Int Pharmacodyn Ther 268:95–105
Vanhoutte PM, Eber B (1991) Endothelium-derived relaxing and contracting factors. Wien Klin Wochenschr 14:405–411
Vigorito C, Poto S, Picotti CB, Triggiani M, Marone G (1986) Effect of activation of the H_1 receptor on coronary hemodynamics in man. Circulation 73:1175–1182
Vigorito C, Giordano A, De Caprio L, Vitale DF, Manrea N, Silvestri P, Tuccillo B, Ferrara N, Marone G, Rengo F (1987) Effects of histamine on coronary hemodynamics in humans: role of H_1 and H_2 receptors. J Am Cell Cardiol 10: 1207–1213
Vincenc K, Black J, Yan K, Armour CL, Donnelly PD, Woolcock AJ (1983) Comparison of in vivo and in vitro responses to histamine in human airways. Am Rev Respir Dis 128:875–879
Waldman DB, Zfass AM, Makhlouf JM (1977) Stimulating (H_1) and inhibitory (H_2) histamine receptors in gall bladder muscle. Gastroenterology 72:932–936
Webber SE, Salonen RO, Widdicombe JG (1988) H_1 and H_2-receptor characterization in the tracheal circulation of sheep. Br J Pharmacol 95:551–561
Weir SW, Weston AH (1986) The effect of BRL 34915 and nicorandil on electrical and mechanical activity and on 86Rb efflux in rat blood vessels. Br J Pharmacol 88:121–128
Wolff AA, Levi R (1986) Histamine and cardiac arrhythmias. Circ Res 58:1–16
Yamamoto Y, Tomoike H, Egashira K, Nakamura M (1987) Attenuation of endothelium-related relaxation and enanced responsiveness of vascular smooth muscle to histamine in spastic coronary arterial segments from miniature pigs. Circ Res 61:772–778
Yanagisawa M, Kurihara H, Kimura S, Tomobe Y, Kobayashi M, Mitsui Y, Yazaki Y, Goto K, Masaki T (1988) A novel potent vasoconstrictor peptide produced by vascular endothelial cells. Nature 332:411–415
Yates JC, Schofield B, Roth SH (1978) Acid secretion and motility of isolated mammalian gastric mucosa and attached muscolaris externa. Am J Physiol 234:E319–E326

CHAPTER 7

Angiohypotensin

J. KNOLL

A. Introduction

The term "essential" hypertension was coined by Traube in 1856 to label a condition characterized as persistently elevated blood pressure without a discernible underlying causative factor. Almost 150 years have passed by and we are still no nearer to understanding the cause of this condition. What we do know is that peripheral vascular resistance increases, this being the only consistent change in this disease (PEART 1980). Since the sympathetic nervous system controls smooth muscle tone, one would expect an increased concentration of noradrenaline in the plasma of patients suffering from "essential" hypertension. Indeed, LOUIS et al. (1973) demonstrated that the mean value of the concentration of noradrenaline in the plasma of normotensives is significantly lower than in patients with "essential" hypertension.

In the search for an explanation as to why increased peripheral vascular resistance is the only detectable change in "essential" hypertension, the working hypothesis that a hitherto unknown highly selective regulation may control the release of noradrenaline in the resistance vessel, the disturbance of which leads to hypertension, has led to the discovery of angiohypotensin (AH), a chemically still unidentified blood-borne substance, which is highly potent and selective in inhibiting the release of noradrenaline in vascular smooth muscle (KNOLL 1977, 1978, 1979, 1980, 1987, 1988).

In "essential" ("benign") hypertension, which afflicts hundreds of millions all over the world, the majority of patients are medicated with antihypertensive drugs.

The goal of reducing increased peripheral vascular resistance can be achieved by:

1. Decreasing the sympathetic tone
2. Changing the sodium balance
3. Reducing the angiotensin II concentration in plasma
4. Direct action on vascular smooth muscle

Ad 1. The sympathetic tone can be decreased by:

a) Centrally acting α_2-adrenergic receptor agonists, which inhibit the release of noradrenaline

b) α_1-Adrenergic receptor antagonists, which inhibit the noradrenaline-induced vasoconstriction
c) β-Adrenergic receptor antagonists, which reduce cardiac output and inhibit renin secretion

Ad 2. Sodium balance can be changed by diuretics.
Ad 3. The concentration of angiotensin II can be decreased by angiotensin-converting enzyme (ACE) inhibitors.
Ad 4. Vascular smooth muscle tone can be reduced directly by calcium-channel blockers or by vasodilators, which may relax arteriolar smooth muscle only or both arteriolar and venous smooth muscle. Highly potent and specific endothelium derived substances are now known which enhance (endothelin) or diminish (nitric oxide, endothelium-derived relaxing factor) smooth muscle tone in the resistance vessels.

Though the available medicines for the pharmacotherapy of hypertension are efficient, there is still much room for improvement. The antihypertensive drugs interfere to differing degrees with everyday living in terms of side effects and measure of satisfaction. More than 10% of patients stop taking medication because of serious side effects. It is evident that for the successful treatment of "benign" hypertension the true cause of the disease must be found.

Angiohypotensin, a physiological substance which inhibits noradrenaline release from the sympathetic nerve terminals of the resistance vessels with unique selectivity, represents a previously unknown mechanism for reducing increased peripheral vascular resistance. This sustains the hope that further developments in AH research will finally lead to a safe and efficient new medication for hypertension.

B. Angiohypotensin in Human and Mammalian Sera

The working hypothesis that an unknown selective regulator of noradrenaline release exists in vascular smooth muscle was substantiated by a study on enkephalins in 1976 (KNOLL 1976). Enkephalins (methionine-enkephalin methylester and leucine-enkephalin were used) proved to be potent inhibitors of the vasoconstrictor responses to nerve stimulation in the perfused ear artery of rabbit. This was the first evidence that enkephalin-sensitive receptors are present at the noradrenergic nerve terminals in vascular smooth muscle. It was also the first paper which demonstrated that a subtype of opioid receptors exists which is neither stimulated nor inhibited by morphine, the classic opiate agonist.

The unexpected observation of this previously unknown, although non-specific, peptide receptor in resistance artery initiated the search in human serum for the blood-borne substance which was hoped would *selectively* inhibit noradrenaline release in vascular smooth muscle. Experimental studies aiming to find the predicted substance, named angiohypotensin (AH), started in 1977. The existence of AH in human serum was first demonstrated in 1977 and was verified in 1978–1979 (KNOLL 1977, 1978, 1979).

Angiohypotensin activity was assayed on a highly stable, reliable and sensitive resistance vessel preparation, the central (3–4 cm) portion of the rabbit ear artery, prepared, cannulated, mounted in an organ bath (4 ml) and perfused with Krebs solution at a constant rate at 2.7 ml/min. Vasoconstriction was elicited by field stimulation (supramaximal voltage, about 60 V) and the resulting increase in pressure was measured and recorded. Trains of 6–12 pulses, 1 ms in duration, at a frequency of 3 Hz, were delivered every minute. AH activity was expressed in biological units. The smallest amount of a preparation which, when given to the organ bath, stopped the vasoconstriction of the perfused rabbit ear artery to nerve stimulation, was taken to contain one unit of AH.

The existence of AH in human serum was first proved in a butanol extract of serum which was filtered through an Amicon UM-10 membrane and fractionated on a Sephadex G-15 column. The predicted substance capable of blocking neuromuscular transmission in the rabbit ear artery was clearly detectable in a couple of fractions (KNOLL 1978).

A more sophisticated method was then developed using fresh blood and omitting the butanol extraction. A series of studies (42 experiments) clearly demonstrated the presence of two forms of AH (named AHI and II) in human serum (KNOLL 1979). AH activity was also detected in mammalian sera (KNOLL 1979). The AH concentrations in human and mammalian sera were found to be very low.

The characteristic profile of the effect of AH on noradrenergic transmission was established with moderately purified material prepared from human serum. AH preparations containing 100–300 units/mg were used in these studies (KNOLL 1978, 1979).

The main findings were as follows:

1. Human serum AH inhibited in a dose-dependent manner the field stimulation induced vasoconstriction in different vascular smooth muscle preparations (perfused central ear artery of the rabbit, saphenous artery strip of the rabbit, etc.) and no tolerance to this effect developed.
2. The amount of AH which completely inhibited the vasoconstriction of the vascular smooth muscle preparation to field stimulation left the effect of exogenous noradrenaline, as well as the effect of the indirectly acting amines such as tyramine, unchanged.
3. Angiohypotensin preparations (25–100 units/kg) induced a dose-dependent systemic hypotension in the anaesthetized rat, rabbit and cat and no tolerance to this effect developed.

4. High amounts of AH left both the electrically induced contractions of the papillary muscle and the amount of noradrenaline released to field stimulation unchanged.
5. High amounts of AH left the α-adrenergic transmission in both the guinea pig vas deferens and the rat vas deferens unchanged.
6. Angiohypotensin left the field stimulation induced contraction of the Auerbach plexus-longitudinal muscle preparation unchanged, proving the inefficiency of this subtance on cholinergic transmission and indicating that AH preparations were free from biologically active impurities of histamine and histamine-like material.
7. The known extremely potent endogenous hypotensive peptides are without exception highly active smooth muscle stimulants on the isolated uterus of oestradiol-treated rat. AH was found to be devoid of any effect in this test.
8. Except for vascular smooth muscle, only a few preparations with noradrenergic transmission, the cat splenic strip, the cat nictitating membrane and the rat rectococcygeus muscle showed some sensitivity to AH.

In all the tests moderately purified bovine and horse serum AH preparations acted like those prepared from human serum.

Table 1 summarizes the effects of AH in different noradrenergically innervated isolated organs. The results clearly show the remarkable responsiveness of those preparations towards AH, which are evidently the appropriate tests for checking its effects on resistance vessels.

Table 1. Effect of moderately purified human serum angiohypotensin preparations on field stimulation induced contractions in different noradrenergically innervated smooth muscle preparations

Preparation	Source	Effect of angiohypotensin
Perfused central ear artery	Rabbit	DDI, HP
Helical strip of saphenous artery	Rabbit	DDI, HP
Strip of portal vein	Rabbit	DDI, HP
Strip of main pulmonal artery	Rabbit	VDI, VDD
Strip of aorta	Rat	0
Rectococcygeus muscle	Rat	VDI, VDD
Papillary muscle	Cat	0
Splenic strip	Cat	VDI, VDD
Nictitating membrane	Cat	VDI, VDD
Uterus	Rat	0
Vas deferens	Rat	0
	Rabbit	0
	Guinea pig	0
	Mouse	0

DDI, dose-dependent inhibition; HP, high potency; VDI, variably detectable inhibition; VDD, vague dose dependency; 0, without noticeable effect.

Angiohypotensin II preparations (see KNOLL 1979), containing 100–300 units per milligram activity, were used in the experiments.

C. Inhibition by Angiohypotensin of the Release of Noradrenaline to Field Stimulation in Vascular Smooth Muscle

As to the mechanism of action of AH it was proved that the substance inhibits the release of [^{3}H]noradrenaline to nerve stimulation in vascular smooth muscle. AH was found to be more potent in this respect than either clonidine or prostaglandins (KNOLL 1978, 1979).

The release of noradrenaline from the noradrenergic nerve terminals is highly sophisticated, being locally inhibited by a variety of endogenous substances acting by different mechanisms. The best-known and probably most extensively studied local modulation of transmission is the inhibition of noradrenaline release by the transmitter itself via the α_2-adrenergic receptors. The continuously acting autoinhibition of the transmitter release is a non-specific mechanism operating in an essentially similar manner everywhere in the central nervous system and in the periphery at the noradrenergic nerve terminals.

This mechanism can be studied precisely on the perfused rabbit ear artery, which is one of the best experimental models for following the effect of endogenous substances and drugs on resistance vessels. In this test the increase in pressure to the stimulation with 3 Hz and 6–12 pulses is completely inhibited by 25–40 ng/ml clonidine, which by stimulating the α_2-adrenergic receptors of the noradrenergic nerve terminals inhibits the release of noradrenaline. In this dose range only a slight increase in the tone of the smooth muscle is a sign of mild α_1-adrenergic receptor stimulation caused by this very small dose of clonidine. With intensive stimulation of the nerves (e.g., 20 Hz, 20 pulses) clonidine is unable to inhibit the release of noradrenaline. As α_2-adrenergic receptors are ubiquitous at the noradrenergic nerve terminals, the presynaptic noradrenaline release inhibitory effect of small doses of clonidine is from a functional point of view a non-specific effect.

Figure 1 shows that the effect of clonidine in the perfused rabbit ear artery is inhibited by yohimbine, which in small doses blocks the α_2-adrenergic receptors. The inhibitory effect of AH, however, remains unchanged, demonstrating its independency from α_2-adrenergic receptors.

The second, widely but not evenly distributed, local inhibitor of the release of noradrenaline is a transjunctional control mechanism, operating via the release of prostaglandins. This transjunctional inhibitory modulation is widely enough distributed in the noradrenergic system for it to be looked upon as a non-selective mechanism. One of the most sensitive noradrenergic organs to PGE_1 and PGE_2 is the guinea pig vas deferens. In this preparation 1–2 ng/ml PGE_1 or PGE_2 is sufficient to inhibit the field stimulation induced contractions. AH is completely ineffective in this test (KNOLL 1978).

The third, perplexingly complicated inhibitory modulation of noradrenergic transmission is the different interneuronal connections involved,

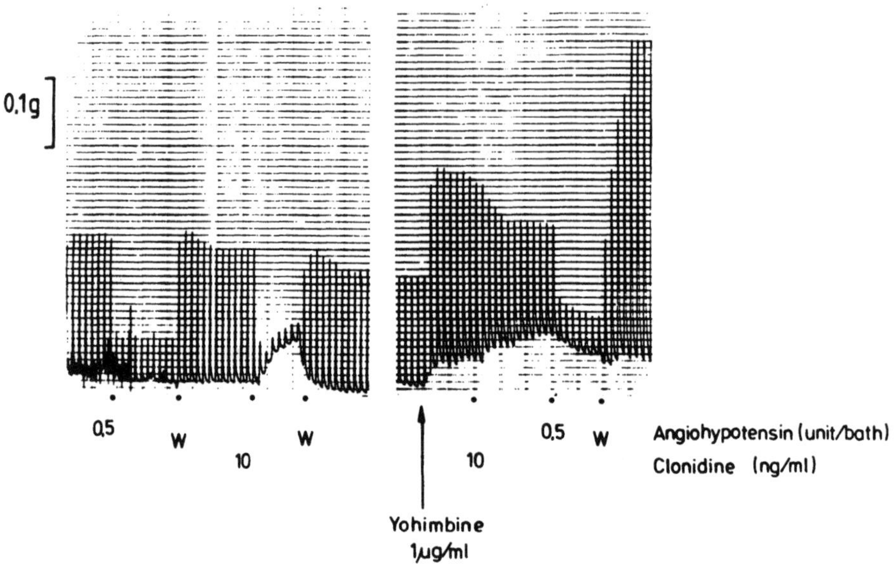

Fig. 1. Proof that the effect of angiohypotensin on the isolated, perfused, central ear artery of the rabbit is unrelated to α_2-adrenergic receptors. A highly purified bovine liver angiohypotensin preparation (300 ng = 1 unit) was used. Stimulation, 3 Hz, 12 pulses

of which dopamine, serotonin, acetylcholine, GABA and even histamine are worth mentioning; though none is either specific or very efficient in modulating noradrenergic transmission.

Finally, local hormonal inhibitors of noradrenaline release, adenosine or the enkephalins, are also known. Adenosine is of negligible potency in inhibiting the field-stimulated pressure increase in the isolated rabbit ear artery. Enkephalins are highly potent inhibitors for the release of noradrenaline in this test (KNOLL 1976). Their effect is inhibited by naloxone, which leaves the effect of AH unchanged, showing that its effect is unrelated to enkephalin-sensitive receptors.

Though the overwhelming majority of the local physiological mechanisms which modulate noradrenergic transmission are inhibitory, some facilitatory modulators (e.g., prostaglandin $F_{2\alpha}$, angiotensin II, in some places GABA, cyclic AMP, even adrenaline) are also known, but none play a major role in influencing noradrenergic transmission.

In summary, AH represents a unique endogenous substance, the only one which inhibits the release of noradrenaline in resistance vessels with high selectivity. Thus, its characteristic profile would fit in with the predicted significant role of this substance in the regulation of peripheral vascular resistance.

D. Angiohypotensin in Pig and Bovine Liver

Because of the very low amounts of AH in the sera of different species, more appropriate sources for the production of AH-containing preparations were needed. Pig and bovine liver were finally selected for this purpose. An improved gel chromatographic method using Sephadex G-15, Bio-Gel P-2 and carboxymethylcellulose (CMC) columns for the elaboration of purified angiohypotensin preparations was developed (KNOLL 1988).

Using this procedure, 5000–25 000 units AH/kg liver were prepared, the preparations containing 1 unit AH in 250–400 ng of material. The purified pig and bovine liver samples were found to possess essentially the same spectrum of pharmacological activity as those prepared from human and mammalian sera. The highly purified bovine liver AH preparations showed even higher selectivity to the resistance vessels than the human serum AH preparations of moderate purity.

Figure 2 shows the dose-dependent effect of a highly purified bovine liver AH preparation on the central ear artery of the rabbit. This still inhomogeneous preparation completely blocked the field stimulation induced increase in pressure at a concentration of 60 ng/ml. The preparations were found to be equally active on the saphenous artery of the rabbit (Fig. 3).

The highly purified pig or bovine liver AH preparations proved to be ineffective on the rabbit main pulmonal artery strip (Fig. 4). They were either ineffective on the rectococcygeus muscle of the rat (Fig. 5A), or a slight, non-dose-dependent inhibitory effect was detectable (Fig. 5B). The preparations acted similarly on the nictitating membrane of the cat.

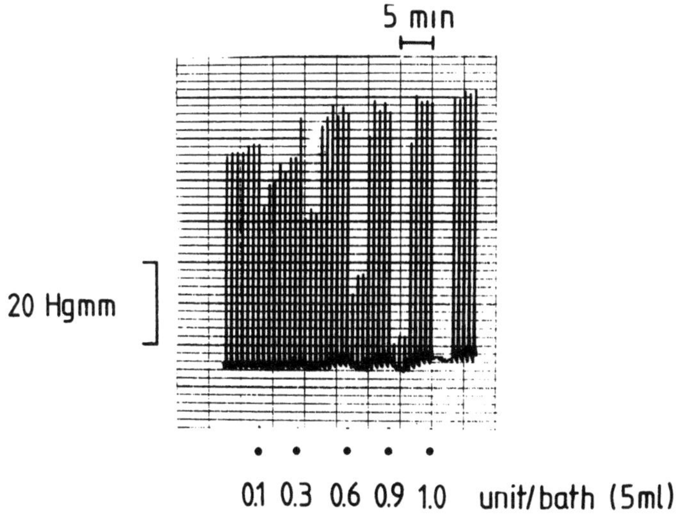

Fig. 2. Dose-dependent inhibition of field stimulation induced vasoconstriction by angiohypotensin on the isolated, perfused, central ear artery of the rabbit. The AH preparation characterized in Fig. 1 was used. Stimulation, 3 Hz, 12 pulses

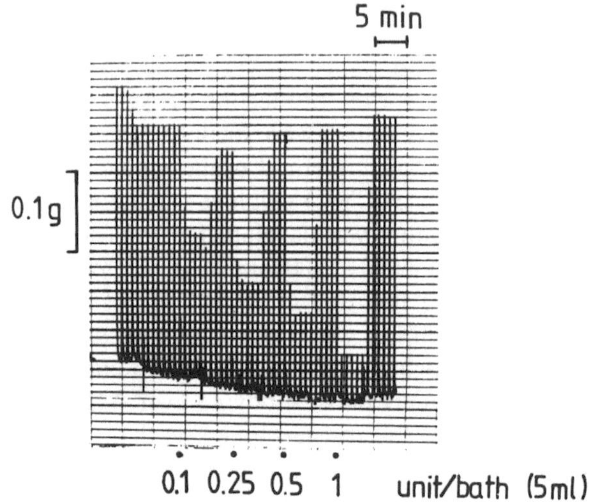

Fig. 3. Dose-dependent inhibition of field stimulation induced vasoconstriction by angiohypotensin on the helical strip of the saphenous artery of the rabbit. The AH preparation characterized in Fig. 1 was used. Stimulation, 5 Hz, 10 pulses

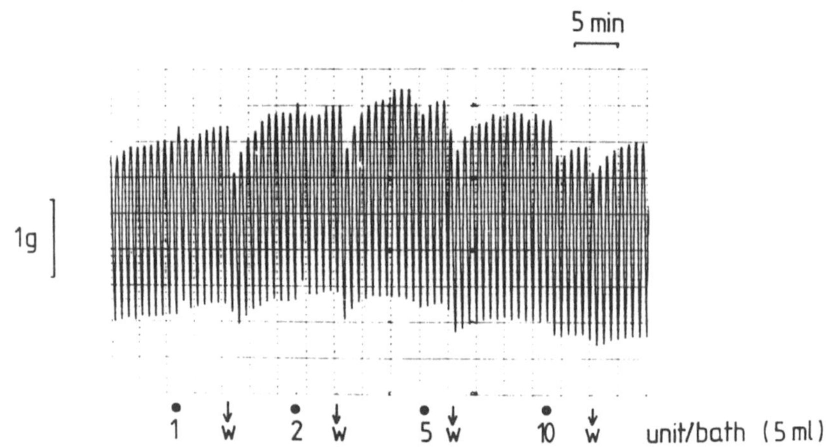

Fig. 4. The ineffectiveness of angiohypotensin on a strip of the main pulmonal artery of the rabbit. The AH preparation characterized in Fig. 1 was used. Stimulation, 3 Hz, 21 pulses

The unique specificity of AH came strikingly apparent on the portal vein strip. In this preparation the vascular smooth muscle contractions, elicited by long-lasting field stimulation, were dose dependently inhibited by highly purified AH. Figure 6 shows that neither clonidine nor PGE_1 or methionine enkephalin methylester changed the field stimulation induced contraction of the rabbit portal vein strip, whereas a very small amount of AH (0.2

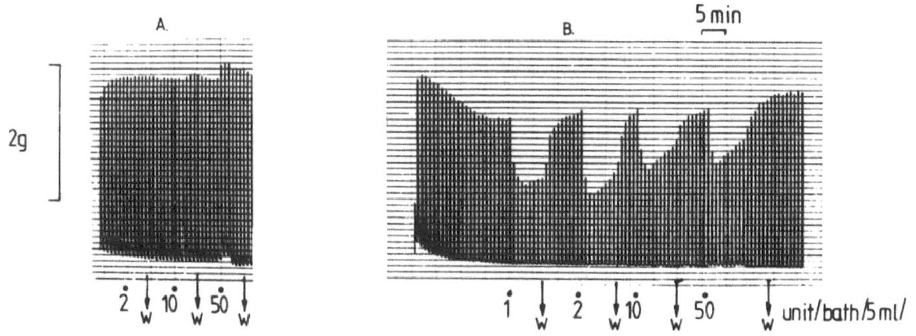

Fig. 5A,B. The ineffectiveness (**A**) or a slight non-dose-dependent effect (**B**) of angiohypotensin on the isolated rectococcygeus muscle preparation of the rat. The AH preparation characterized in Fig. 1 was used. Stimulation, 3 Hz, 12 pulses

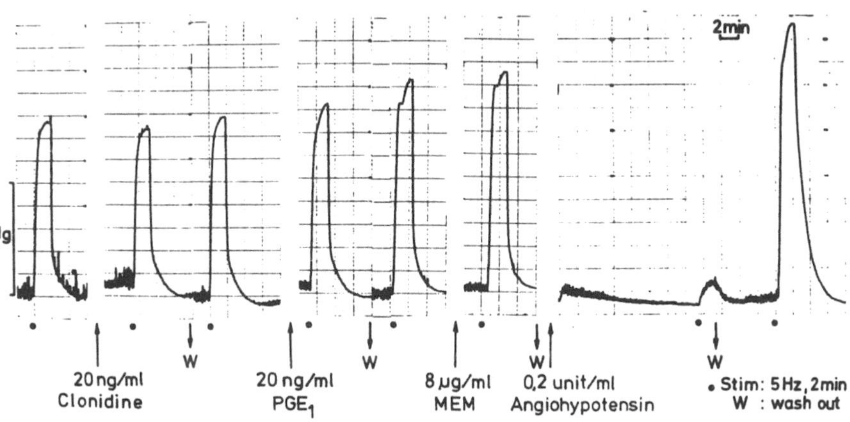

Fig. 6. The high potency of angiohypotensin in inhibiting field stimulation induced vasoconstriction in the portal vein strip of the rabbit and the inefficiency of clonidine, PGE_1 and methionine enkephalin methylesther (*MEM*) in the test. The AH preparation characterized in Fig. 1 was used

units/ml) almost completely inhibited the response of the vascular smooth muscle to nerve stimulation and this effect was readily abolished by a single washing.

The highly purified pig and liver AH preparations did not influence the field stimulation induced contractions of the rat aorta strip and proved to be completely ineffective on the vas deferens preparations of the rat, mouse, rabbit and guinea pig.

The highly purified pig and liver AH preparations inhibited the release of [^{3}H]noradrenaline to field stimulation on the central ear artery of the rabbit and in this context the preparations proved to be more potent than those of clonidine (Knoll 1988).

E. Hypotensive Effect of Moderately Purified Angiohypotensin Preparations in Anaesthetized Animals

Moderately purified AH preparations from human, bovine or horse sera were found to exert a dose-dependent systemic vasodilatation in the anaesthetized rat, rabbit and cat (KNOLL 1978, 1979). The vasodilatory dose of AH (100 units/kg) did not influence the pressor effect of adrenaline in the rat (KNOLL 1978), being in agreement with the finding that the mechanism of action of AH is the inhibition of the release of noradrenaline from the nerve terminals in the resistance vessels. Tolerance to the hypotensive effect of AH developed neither on the isolated vascular smooth muscle preparations nor in the anaesthetized animals (KNOLL 1978, 1979).

F. Blood Pressure Reducing Effect of Highly Purified Angiohypotensin Preparations in Anaesthetized Spontaneous Hypertensive Rat (SHR strain)

With highly purified pig liver preparations the hypotensive effect was studied on spontaneous hypertensive rats (SHR strain). The effect of AH was measured on anaesthetized animals. An infusion of as low as 0.5–5 units/kg per minute of the highly purified pig liver AH preparations into the femoral veins reduced the blood pressure of the rats within a few seconds. The effect remained unchanged during the infusion and disappeared within a few seconds following the discontinuation of the infusion. This effect of AH could be elicited repeatedly in the same animal without significant changes in the onset, offset and magnitude of the effect, clearly showing that no tolerance to the influence of AH on peripheral resistance develops (KNOLL 1987, 1988).

The infusion of AH decreased blood pressure in the SHR strain dose dependently to a reasonable maximal level beyond which even drastic increases of the dose, up to a level as high as 200 units/kg per minute, remained ineffective. Thus, a state of hypotension with grave consequences could not be elicited with AH. This phenomenon is in good agreement with the assumption that AH plays a major regulatory role in tuning peripheral resistance.

That only a reasonable maximal level of blood pressure decrease can be reached with administration of AH seems to show that, even in the case of complete saturation of the AH receptors in vascular smooth muscle with exogenous AH, compensatory mechanisms, such as the renin-angiotensin system, remain sufficently active to maintain the peripheral resistance needed for the adequate perfusion of the vital organs. The physiological role of endothelin, the endothelium-derived, hitherto known most potent local stimulator of vascular smooth muscle, is still obscure in this context.

G. Blood Pressure Reducing Effect of Highly Purified Angiohypotensin Preparations in Freely Moving Awake, Spontaneously Hypertensive Rats and in Their Normotensive Control Peers

The effect of AH on the blood pressure of awake rats was studied in male spontaneously hypertensive rats (SHR strain) and their normotensive control peers (Wistar-Kyoto, WKY) supplied with indwelling vascular catheters, implanted chronically, to allow subsequent recording of blood pressure and heart rate (performed by L. Kerecsen; for methodological details see KERECSEN and BUNAG 1989).

For these experiments AH preparations further purified by the HPLC technique were used. Two preparations (B1 and B2) were studied, which were found to be nearly equally active by bioassay; each of the vacuum-dried preparations contained about 10000 units in 1 mg.

Figure 7 shows the effect of i.v. bolus injection of AH (B1) on the pulsatile blood pressure of a freely moving awake normotensive rat. The figure clearly demonstrates that AH does have a blood pressure reducing effect, which is not due to any direct effect on the heart (no arrhythmia, no pulse pressure change) and therefore it can be attributed to the peripheral, vasodilating effect of the preparation. Repeated i.v. bolus injections result in similar decreases in the mean blood pressure as demonstrated in Table 2. The change in blood pressure shows a definitive dose-response relationship.

Figure 8 shows the dose-response curve to AH (B1) in freely moving awake normotensive rats. The i.v. bolus injections produced a dose-dependent decrease in blood pressure; however, the variability of the effect of the lower doses was substantial.

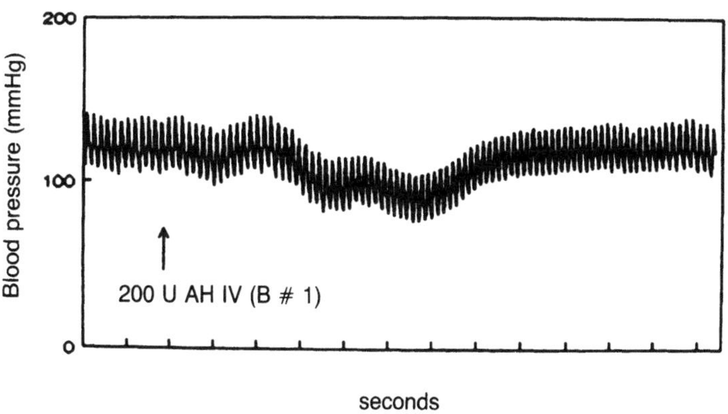

Fig. 7. Effect of intravenous bolus injection of angiohypotensin on the pulsatile blood pressure of a freely moving awake rat. The *arrow* shows the intravenous bolus injection of 200 units angiohypotensin. A batch of highly purified material (B1, see text) was used

Table 2. Effect of repeated intravenous bolus injections of angiohypotensin in freely moving awake rat. Each dose was injected three times, and at the end of the series the 100-U dose was again injected three times

Injected i.v. bolus (units)	Mean blood pressure (mmHg)		Decrease of BP (mmHg)
	Before	At peak effect	
50	118	96	22
	110	95	15
	109	86	23
		Average	20 ± 4
100	105	85	20
	106	86	20
	104	81	23
		Average	21 ± 2
200	116	91	25
	115	86	29
	110	88	22
		Average	25 ± 4
400	109	76	33
	118	85	33
	119	85	34
		Average	33 ± 1
100 (Repeated)	107	86	21
	106	87	19
	110	87	23
		Average	21 ± 2

A batch of highly purified material (B1, see text) was used.

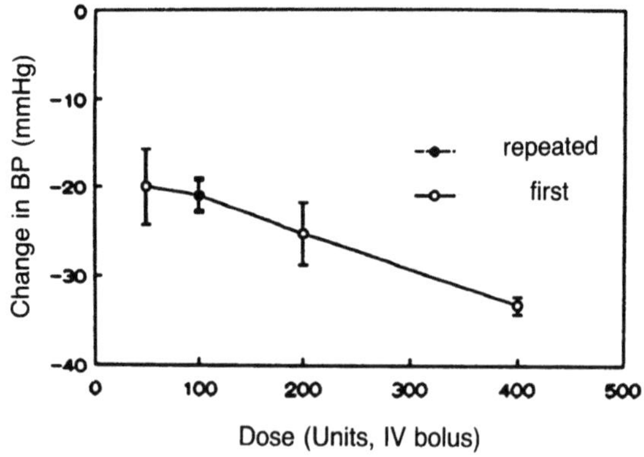

Fig. 8. Dose response curve to intravenous bolus injections of angiohypotensin in freely moving awake rat. Batch B1 (see Fig. 7) was used

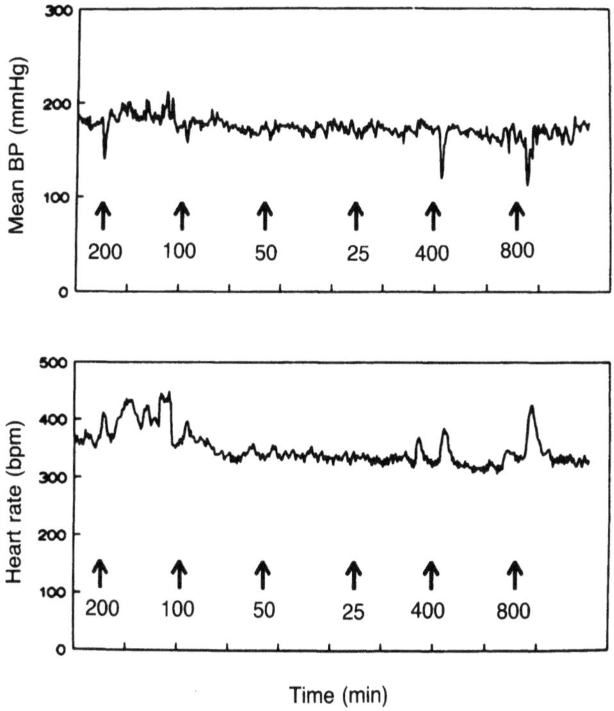

Fig. 9. Dose-related effect of 8-s intravenous infusions of angiohypotensin in freely moving awake rats. Batch B1 (see Fig. 7) was used

The observation was made that AH adsorps for unknown reasons to the polyethylene tube. In order to eliminate the effect of adsorption of AH to the polyethylene tubing instead of the i.v. bolus injections, a relatively short i.v. infusion (8 s duration) was applied and the infusion speed was changed according to the dose requirements. This method eliminated the effect of adsorption and decreased the stress of the overall dosing procedure (no more moving activities around the rat, injecting, rinsing with saline, etc.).

Figure 9 shows a typical blood pressure and heart rate recording using the infusion technique. The actual time between injections is longer (about 10 min), the figure only showing 20 s before and 80 s after the beginning of each infusion. This recording demonstrates the main characteristics of the other effects of AH. Rapid onset, short duration and decrease in blood pressure are followed by a moderate reflex tachycardia.

Figure 10 shows dose response curves for both AH preparations, B1 and B2, in freely moving, awake, spontaneously hypertensive rats and in their normotensive controls. The figure shows a linear dose-response curve on a semilogarithmic scale. B2 is slightly more effective than B1, and both are slightly more effective in SHR than in WKY rats. The slight difference in the potency between the two batches was not detected in the bioassay. The

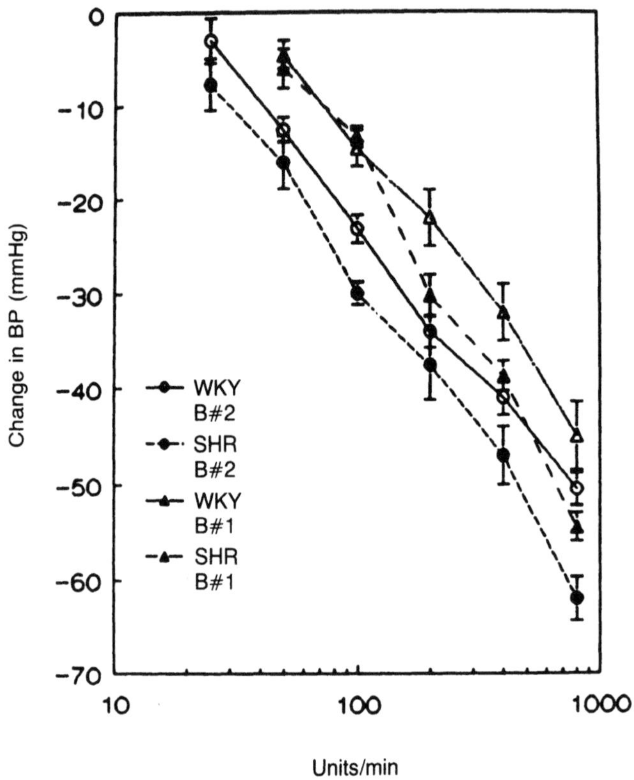

Fig. 10. Dose response curves of two batches of angiohypotensin preparation (B1 and B2) in freely moving awake spontaneously hypertensive (*SHR*) and normotensive (*WKY*) rats, using the short (8-s) intravenous infusion technique

fact that in both series of experiments AH was found to be more potent in decreasing blood pressure in the SHR strain than in the normotensive one might be a promising sign for the future.

In a series of experiments male Wistar rats with indwelling catheters in the abdominal aorta, through the femoral artery and the femoral vein, were prepared. One week after surgery their mean blood pressure and heart rate were recorded. A series of 8-s infusions with different infusion speeds (200, 400, 800, 100, 50 and 25 μl/min) of AH (1000 units/ml; B1) was infused. The same infusions in the same rat were repeated 20 min after 50 mg/kg sodium pentobarbital injection i.p. The maximum decreases in blood pressure were extracted from the files containing the experimental data. Figure 11 shows the results. Sodium pentobarbital anaesthesia greatly diminished the compensatory tachycardia following the drop in blood pressure. In these experiments, although there was a significant ($p < 0.003$) difference between the awake and the anaesthetized groups, it seems clear that the short duration of action of AH cannot be attributed to the counterregulatory effects of the baroreflexes.

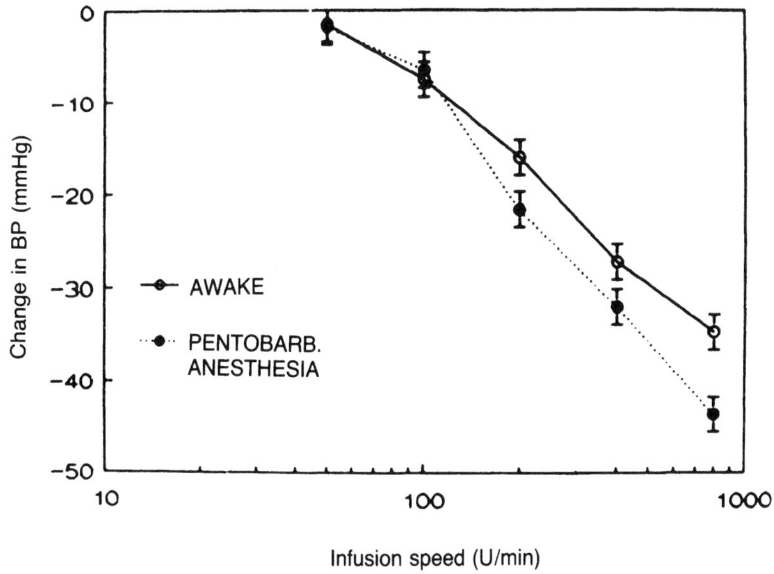

Fig. 11. Changes in blood pressure following 8-s intravenous angiohypotensin (B1) infusions in freely moving awake and then pentobarbital anaesthetized male Wistar rats. The error bars indicate the standardized error calculated from the repeated measured analysis of variance. Over whole data set, the effect of anaesthesia is significant ($F = 10.69$, $p < 0.003$) although not markedly so

The rapid offset of the AH effect in vivo favours the assumption that efficient mechanism(s) take care of the elimination of AH activity, and that the physiological level of this highly active endogenous substance may be carefully controlled.

H. Working Hypothesis of Angiohypotensin as a Major Regulator of Peripheral Vascular Resistance

The *selectivity* and high *potency* of AH in inhibiting the field stimulation induced release of noradrenaline in the resistance vessels suggests that even a small long-lasting change in AH regulation has a significant impact on blood pressure.

Angiohypotensin may act like a break and the slightest decrease in this function (e.g., a deficit in AH: hypoangiohypotensinaemia) will necessarily increase the amount of continuously released noradrenaline from the nerve terminals in the resistance vessels and peripheral resistance will increase ("essential" hypertension). Thus, proper AH substitution therapy may counter hypertension.

On the other hand, a change in the opposite direction (e.g., a slight increment in AH; hyperangiohypotensinaemia) will decrease peripheral vascular resistance ("essential" hypotension), and we may some time learn to correct this state by lowering the AH level properly. Nevertheless, identification of the chemical nature of AH, its verification via synthesis, a basic understanding of its origin and metabolism, specification of the AH receptor and its interaction with endogenous and synthetic ligands seem to be the minimum requirements for further progress in AH research.

I. Summary and Conclusion

The existence of a chemically still unidentified substance, named angiohypotensin (AH), in human serum and in mammalian sera and liver was demonstrated which inhibits the release of noradrenaline in the resistance vessels with high selectivity. AH decreases blood pressure in a dose-dependent manner in awake normotensive and hypertensive rats. As increased peripheral vascular resistance of unknown origin is the only consistent change in "essential" hypertension and AH is the first endogenous substance found to inhibit noradrenaline release in the resistance vessels with high potency and selectivity, the working hypothesis is proposed that a disturbance in AH regulation might be the underlying causative factor in the disease.

References

Kerecsen L, Bunag RD (1989) Selective pressor enhancement by monoamine oxidase inhibitors in conscious rats. J Pharmacol Exp Ther 251:645–649

Knoll J (1976) Neuronal peptide (enkephalin) receptors in the ear artery of the rabbit. Eur J Pharmacol 39:403–407

Knoll J (1977) Angiohypotensin: a highly specific vasodilator substance in human serum. Proc IUPS 13:394

Knoll J (1978) Selective inhibition of noradrenergic transmission in vascular smooth muscle. Pol J Pharmacol Pharm 3:269–280

Knoll J (1979) Angiohypotensin: a selective endogenous inhibitor of neuromuscular transmission in vascular smooth muscle. In: Usdin E, Kopin IJ, Barchas J (eds) Catecholamines: basic and clinical frontiers, vol 2. Pergamon, New York, pp 1152–1154

Knoll J (1980) Highly selective peptide-chalones in human serum. A concept on the control of physiological functions by blood-borne selective inhibitors of neurochemical transmission. In: Knoll J, Vizi ES (eds) Modulation of neurochemical transmission. Akadémiai Kiadó, Budapest; Pergamon, New York, pp 97–125

Knoll J (1987) Angiohypotensin: an interpretation why peripheral resistance in essential hypertension is increased. In: Papp JG (ed) Cardiovascular pharmacology '87. Akadémiai Kiadó, Budapest, pp 245–252

Knoll J (1988) Angiohypotensin: selective inhibitor of the release of norephinephrine from vascular smooth muscle. In: Dahlström A, Belmaker RH, Sandler M (eds) Basic aspects and peripheral mechanisms. Liss, New York, pp 13–16 (Progress in catecholamine research, part A)

Louis WJ, Doyle AE, Anavelear S (1973) Plasma norephinephrine levels in essential hypertension. N Engl J Med 288:599–601

Peart WS (1980) Concepts in hypertension. (The Croonian Lecture). J R Coll Physicians Lond 14:141–152

CHAPTER 8

Neuropeptides (Neurokinins, Bombesin, Neurotensin, Cholecystokinins, Opioids) and Smooth Muscle

D. REGOLI, N. ROUISSI and P. D'ORLÉANS-JUSTE

A. Introduction

Five groups of linear biopolymers, composed of amino acid residues and called neuropeptides, are considered in this chapter. The primary structure of a prototype of each group is shown in Table 1. The sequences were initially identified in animals and subsequently all of them were also found in man.

Knowledge about neuropeptides has advanced rapidly since the introduction of solid-phase synthesis by Merrifield in 1963 and high-performance liquid chromatography (HPLC) by GREEN et al. in 1966. With these techniques, series of analogues of each peptide were prepared for structure-activity studies and for receptor identification (e.g., for neurokinins (REGOLI et al. 1987), bombesin (ROUISSI et al. 1991a; JENSEN and COY 1991), neurotensin (ST-PIERRE et al. 1984) and opioids (MARTIN 1984). Chemical and immunological methods (ERSPAMER and MEHCH TORRI 1980; MUTT 1983) and radioimmunoassays with high sensitivity and specificity were successfully used to detect and quantify the naturally occurring peptides in tissue extracts and in biological fluids (MUTT 1983), as well as to assess neuropeptide presence and distribution in brain regions (SCHUTZBERG et al. 1982), in ganglia, sensory or motor nerve fibres and in endocrine (paracrine) cells that are present in the gut, the pancreas, and the hypothalamus. In some of these organs, neuropeptides are colocalized with classical neurotransmitters (SCHUTZBERG et al. 1982), to subserve a variety of physiological functions, especially smooth muscle stimulation or inhibition. In the 1980s, the utilisation of recombinant cDNA techniques led to identification of the genes, the messenger RNAs and the precursor proteins of several neuropeptides (e.g. the neurokinins; NAKANISHI 1986) and to the knowledge that their biosynthesis follows the same general pathways demonstrated for other secretory proteins (RÖKAEUS 1991). In this way, the mature peptides are obtained and are stored in granules or vesicles, ready for release (HÖKFELT 1986; FURNESS and COSTA 1987; EKBLAD et al. 1991).

I. Receptors for Neuropeptides

When released by nervous or other types of stimuli, the mature peptides accumulate into synaptic clefts or near the target cells, where they find

Table 1. Primary structures of the neuropeptides

Group	Prototype	Primary structure of prototype	Year of discovery	Reference
Neurokinin	Substance P	Arg-Pro-Lys-Pro-Gln-Gln-Phe-Phe-Gly-Leu-Met[a]	1970	CHANG and LEEMAN (1970)
Bombesin	$hGRP_{18-27}$	Gly-Asn-His-Trp-Ala-Val-Gly-His-Leu-Met[a]	1984	ORLOFF et al. (1984)
Neurotensin	Neurotensin	pGlu-Leu-Tyr-Glu-Asn-Lys-Pro-Arg-Arg-Pro-Tyr-Ile, Leu	1973	CARRAWAY and LEEMAN (1973)
Cholecystokinin	CCK-8	Asp-Tyr(S)-Met-Gly-Trp-Met-Asp-Phe[a]	1984	MILLER et al. (1984)
Opioid	Met-enkephalin	Tyr-Gly-Gly-Phe-Met	1975	HUGHES et al. (1975)

hGRP, human gastrin-releasing peptide; CCK-8, C-terminal octapeptide of cholecystokinin.
[a] C-terminal amide.

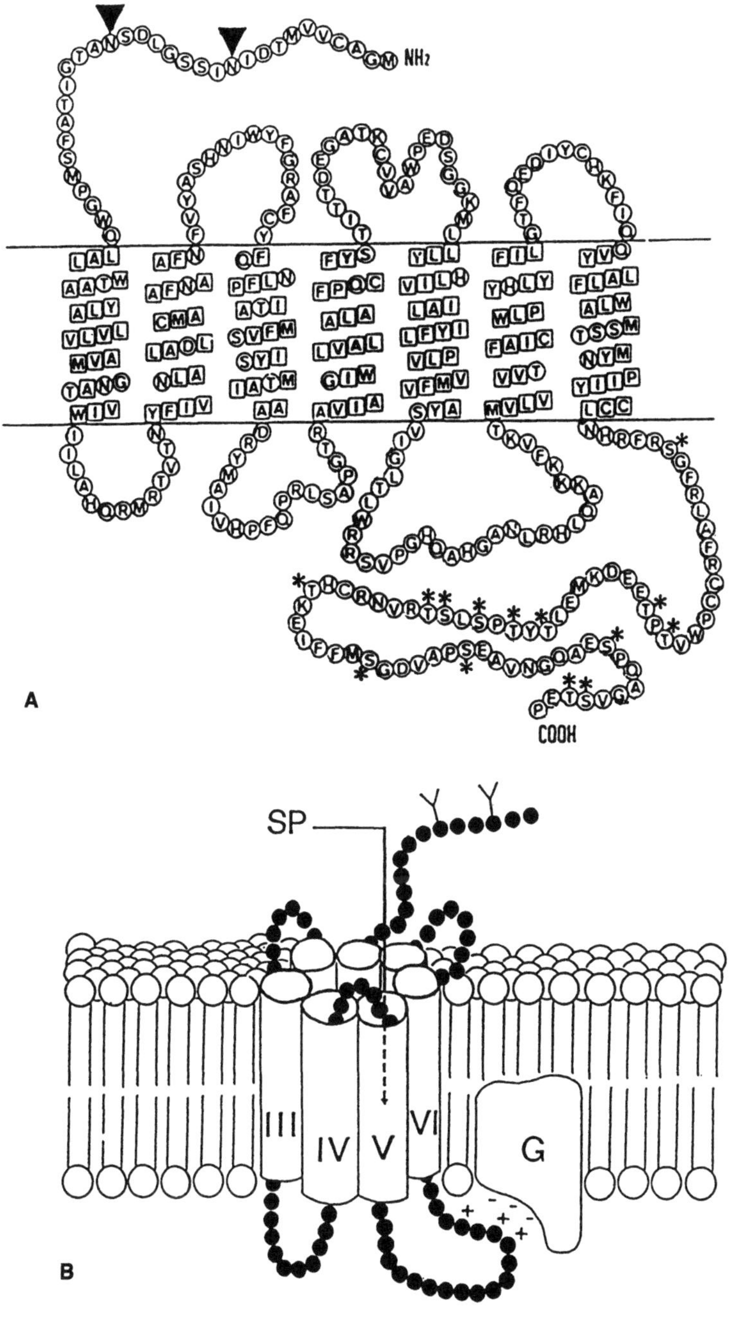

Fig. 1. A Model of transmembranous receptor for neurokinin A (*A*) (from MASU et al. 1987). The *triangles* represent the putative glycosylation sites and the *asterisks* the putative phosphorylation sites. The *squares* represent the hydrophobic amino acids in the seven transmenbranous domains. **B** Schematic representation of the substance P receptor and its hypothetical contact with the G protein. (Modified from MOUSLI et al. 1990)

specific receptors, whose occupation and activation lead to the biological effects. Neuropeptides have receptors in various cells, in neurons, endocrine or exocrine glands and particularly in smooth muscles. Receptor identification has been systematically pursued, using same approaches as for the naturally occurring agents. Functional sites have been demonstrated in vivo and characterized in isolated organs [examples may be quoted for neurokinins (REGOLI et al. 1987) or bombesin (ROUISSI et al. 1991; JENSEN and COY 1991)] by measuring contractile or secretory phenomena. Binding sites in target organs have been demonstrated with labelled peptides: number of sites and affinity constants have been estimated: labelled peptides have been used successfully in historadiographic studies to demonstrate receptor distribution and localization, even in complex anatomical structures such as the brain (HELKE et al. 1990; WAMSLEY 1983). During the last decade, DNAs encoding for neuropeptide receptors and/or mRNAs have been identified and, from the nucleotide sequences of the cDNA, receptor primary structures have been established (NAKANISHI 1986).

One example is given in Fig. 1, by showing the amino acid sequence of the NK-2 receptor, identified by MASU et al. (1987) in bovine brain. The receptor is composed of a series of seven hydrophobic intramembrane domains, connected by external and internal loops, the most important of which appears to be the third internal one, which contains several basic residues (Arg, Lys) and eventually makes contact with the acid residues of the G protein (LEFKOWITZ et al. 1989; REGOLI and NANTEL 1990). By interacting with its active site, deeply into the membrane, the neuropeptide initiates the cascade of events that lead to activation of enzymes (adenylcyclase, guanylcyclase, phospholipases, etc.) and to the biological activity.

The numerous data obtained on the localization of neuropeptides and their receptors support the theory that receptors may be particularly abundant in the proximity of the nerve endings from which neuropeptides are released, in order to allow maximum efficiency, as it is to be expected by the combination of very high concentrations of endogenous agents (agonist) and of functional sites (receptors) occurring in synapses and junctions (e.g., the skeletal muscle neuromuscular junction). The relationship between neuroterminals and smooth muscles differs from organ to organ and at least four different patterns have been identified by BURNSTOCK (1970). Smooth muscles contain the genes of neuropeptide receptors and express these receptors in their plasma membranes. Focus will therefore be placed on receptors, since they are the smooth muscle components of the system. Receptors for neurokinins, bombesin, neurotensin and cholecystokinin have been identified: the most important findings in the field of neuropeptide receptor genetics are summarized in Table 2.

Table 2. Expression of receptors for neuropeptides

Receptor	Preferred ligand	Receptor category	No. of amino acids/mol. wt.	Species of origin	Reference
NK-1	Substance P	G protein coupled	407	Rat	Yokota et al. (1989)
NK-2	Neurokinin A	G protein coupled	384	Hog	Masu et al. (1987)
	Neurokinin A	G protein coupled	391	Rat	Sasai and Nakanishi (1989)
NK-3	Neurokinin B	G protein coupled	452	Rat	Shigemoto et al. (1990)
BB-1	Neuromedin B	–	390	Rat	Wado et al. (1991)
BB-2	Bombesin	Ca^{2+} mobilization[a]	75–95 kDa	AR42J cells	Williams et al. (1988)
NT	Neurotensin	G protein coupled	424	Rat	Tanaka et al. (1990)
CCK	CCK_8	Ca^{2+} mobilization[a]	75–95 kDa	AR42J cells	Williams et al. (1988)

[a] CCK and bombesin receptors expressed following rat liver poly $(A)^+$ RNA incorporation in AR42.

II. Smooth Muscle as Target for Neuropeptides

Like those of other cells, the plasma membrane of smooth muscle fibres is a mosaic of receptors for endogenous agents (the biologically active amines, the peptides, the prostanoids, etc.). Smooth muscle tone is therefore a steady state resulting from excitatory and inhibitory stimuli and the ionic exchanges across the membrane. Such a steady state can be influenced by neuropeptides in different ways: in fact, neuropeptide receptors are found not only on the smooth muscle membrane, but also on nerve endings, endothelia, epithelia, other cell types (mastocytes, macrophages), where they modulate the synthesis and/or the release of other agents such as histamine, prostanoids, noradrenaline and acetylcholine and other factors that are able to influence smooth muscle tone. Thus, in addition to their direct effects, mediated by receptors on the smooth muscle fibre, neuropeptides exert indirect effects through other smooth muscle stimulants or relaxants. Specificity studies, whereby indirect effects are revealed by the use of specific antagonists or inhibitors of other endogenous agents, have been extremely useful to support the conclusion that receptors for neuropeptides (for instance, neurokinins) are indeed present on smooth muscle fibres (COUTURE and REGOLI 1982; NANTEL et al. 1990). The existence of direct and indirect effects makes the situation so complex that the physiological role of both peptide hormones which are secreted into the blood stream (e.g., angiotensin, vasopressin, the kinins) and of neuropeptides that appear to act strictly through local release mechanisms has not been established. A definite assessment of such roles, especially for locally acting peptides, appears to be possible only with selective and specific receptor antagonists (see examples for kinins: REGOLI et al. 1990b); and neurokinins: REGOLI et al. 1988). Potent and selective non-peptide antagonists are available for opioids (MARTIN 1984), cholecystokinin (FRIEDLINGER 1989) and substance P (SNIDER et al. 1991; ROUISSI et al. 1991b). Peptide antagonists for one of the bombesin receptors have also been identified (HEIMBROOK et al. 1989) and a weak antagonist for neurotensin was reported some years ago (ST-PIERRE et al. 1984).

III. Neuropeptide Receptors in Smooth Muscle

Smooth muscles were extensively used in early studies on neuropeptide isolation and identification. Neuropeptides were assigned to different classes, according to a pattern of biological activity evaluated qualitatively on pieces, or strips of peripheral organs, as illustrated in Table 3, which is taken from BERTACCINI (1976).

In the early 1970s, attempts were made to apply to peptides the concepts of the drug-receptor theories (RIOUX et al. 1973; BARAPÉ et al. 1975) and in the following years isolated vessels became the preparations of choice (REGOLI 1982) for studying receptors for some neuropeptides (neurokinins:

Table 3. Relative potency of eledoisin, physalaemin, bradykinin, bombesin and caerulein on isolated smooth muscle preparations[a]. (From BERTACCINI 1976)

	Tachykinins		Bradykinin	Bombesin	Caerulein
	Eledoisin	Physalaemin			
Rabbit colon	++	+++	(+)	0	+++
Guinea pig ileum	++	+++	++	+	++
Guinea pig large intestine	+++	+++	+++	+++	(+)
Kitten small intestine	+	+	+++	+++	+++
Rat duodenum	++	+++	−−−	−	(+)
Rat uterus	(+)	(+)	+++	+++	(+)
Rat urinary bladder	+++	+++	++	+++	(+)
Guinea pig urinary bladder	+++	+++	++	+++	(+)
Hamster urinary bladder	+++	(+)	0	0	0
Rabbit urinary bladder	+	(+)	+++	0	0
Dog urinary bladder	++	(+)	+++	0	0
Cat urinary bladder	(+)	(+)	+++	++	0
Mouse urinary bladder	+++	+	+++	0	0
Human uterus[b]	+++	+++	+	0	+
Human gastrointestinal tract[c]	(+)	(+)	(+)	+++	+

[a] +++, stimulant action of the most potent polypeptides; ++, +, (+), 0, 20%–60%, 1–5% and 1% respectively, of the action of the most potent polypeptide(s); −−−, −, inhibitory action.
[b] Non-pregnant uterus.
[c] Mean values obtained from different strips removed from stomach, small intestine (excluding the appendix) and large intestine.

REGOLI and NANTEL 1991; D'ORLÉANS-JUSTE et al. 1986; MASTRANGELO et al. 1987; for neurotensin: QUIRION et al. 1980a) as well as for angiotensin (RIOUX et al. 1973) and the kinins (BARABÉ et al. 1975). Non-vascular smooth muscle preparations are still extensively used for bombesin (FALCONIERI-ERSPAMER et al. 1988) and cholecystokinin (CHANG and LOTTI 1988), whereas preparations of non-vascular and vascular origins are used for opioids (TAKEMORI and PROTOGHESE 1985; OKA et al. 1980; SCHANG et al. 1986). For years, contractions or relaxations of smooth muscles in response to peptides were the parameters used to assess the biological activity of neuropeptides and evaluate neuropeptide-receptor function; later, binding assays and the measurement of second messenger activation were utilized for characterizing neuropeptide receptors, albeit in other tissues (brain, liver, intestine) than smooth muscles. Very recently, the introduction of ion imagery has been applied successfully to demonstrate the movements of

Ca^{2+} during changes of smooth muscle activity induced by peptides (MORGAN and MORGAN 1982). Because of the existence of receptors for neuropeptides in a variety of cells, in each smooth muscle preparation, changes of smooth muscle tone may result from the activation of different receptor types (multiple receptor systems) or of a receptor in different cells, including the smooth muscle fibre and secretory cells, from which the neuropeptide might promote release of other myotropic agents (multiple effector systems). Such preparations are unsuitable for precise analysis of neuropeptide-receptor interaction and particularly for evaluation of antagonist affinities (CHANG and LOTTI, 1988; REGOLI et al. 1988). Systematic studies have therefore been undertaken to identify and characterize pharmacologically preparations which contain a single receptor type for a given neuropeptide and where the receptor is located in the smooth muscle cell

Table 4. Preparations of choice for pharmacological evaluation of neuropeptide-receptor function

Neuropeptide	Smooth muscle preparation	Receptor type	Reference
Neurokinins	Rabbit jugular vein	NK-1	NANTEL et al. (1990)
	Rabbit pulmonary artery	NK-2	D'ORLÉANS-JUSTE et al. (1986)
	Rat portal vein	NK-3	MASTRANGELO et al. (1987)
Neurotensins	Rat portal vein	NT-1	QUIRION et al. (1980a)
	Rat stomach strip	NT-2	KATAOKA et al. (1978)
Bombesins	Rat urinary bladder	BB-1	ROUISSI et al. (1991a)
	Guinea pig urinary bladder	BB-2	ROUISSI et al. (1991a)
Cholecystokinins	Guinea pig ileum[a]	CCK-A	CHANG and LOTTI (1988)
	Guinea pig ileum[b]	CCK-B	CHANG and LOTTI (1988)
Opioids	Guinea pig ileum	μ	TAKEMORI and PORTOGHESE (1985), ALLESCHER and AHMAD (1991), BAUER and SZURSWESKI (1989)
	Rabbit vas deferens	κ	OKA et al. (1980), ALLESCHER and AHMAD (1991), LORD et al. (1977)
	Mouse vas deferens	δ	TAKEMORI and PORTOGHESE (1985), ALLESCHER and AHMAD (1991), AHMAD et al. (1991)
	Rat vas deferens	ε	WUSTER et al. (1980), WOORDRUFF and HUGHES (1991), FOX and DANIEL (1986)

[a] GPI, stimulated by CCK-8.
[b] GPI, stimulated by pentagastrin.

membrane and mediates a direct effect. A list of such preparations in given in Table 4.

In the following sections, dedicated to each group of neuropeptides, these and other smooth muscle preparations will be analysed and evaluated in regard to their usefulness for physiological and pharmacological purposes.

IV. Neuropeptide Metabolism

Biological activity of neuropeptides is modulated by proteolytic enzymes such as aminopeptidases, post-proline proteases, endopeptidases (trypsin, chymotrypsin, endopeptidase 24-11, etc.) and carboxypeptidases of various categories (Rouissi et al. 1990a). The naturally occurring sequences presented in Table 1 are generally transformed into inactive products by the enzymatic hydrolysis of peptide bonds: in some cases, however, these sequences serve as precursors of biologically active metabolites: examples of such products are SP(6–11), NKA(4–10), NT(8–13) or BB(9–14). Prevention of peptide degradation by peptidase inhibitors (e.g., captopril and congeners, thiorphan and analogues, phosphoramidon) enhances the activity of neurokinins, the gastrin-releasing peptides, cholecystokinins (see review by Checler 1991) or opioids (see review by Roques et al. 1980; Borson 1991).

Proteolytic enzymes interfering with neurokinin and other peptide activities in smooth muscles have been studied by Rouissi et al. (1990a) and Nau et al. (1986), using a variety of peptidase inhibitors in several preparations. Experiments have also been performed with bombesin, neurotensin and cholecystokinin-related peptides (Bunnet et al. 1983, 1985; Cohen et al. 1983). Although indirect and strictly dependent on the specificity and selectivity of the peptidase inhibitors, such an approach has made possible the conclusion that (a) only inhibitors of the angiotensin-converting enzyme (ACE) are able to potentiate the effects of neurokinins (particularly substance P) and of other peptides (e.g., bradykinin) in arterial and venous smooth muscles (Nau et al. 1986); (b) other peptidase inhibitors (e.g., phosphoramidon) have been shown to potentiate the biological activity of neurokinins in smooth muscle preparations derived from the urogenital (rat or guinea pig urinary bladder: Rouissi et al. 1990; or the respiratory systems of various animals: Atkins and Hoster 1989). These data suggest that ACE is the most important functional protolytic enzyme in in vitro isolated vessels, while other proteases are contributing to peptide degradation in peripheral organs. Metabolism of neurokinins appears to be influenced by aminopeptidases and is prevented by bestatin in membrane vescicles (Checler et al. 1988) of the myenteric plexus longitudinal smooth muscle of guinea pig small intestine. Neurokinins and neurotensin appear to be substrates of endopeptidase 24–11 in the dog ileum (Checler et al. 1988). Endopeptidase 24–11 has also been shown to be involved in the inactivation of GRP (18–27) (Gafford et al. 1983), while hydrolysis of CCK-8 and

CCK-7 by enkephalinase-like enzymes isolated from human kidney were blocked by thiorphan (COHEN et al. 1983).

As to the opioids, the effect of enkephalins has been shown to be potentiated by bestatin (OPMEER et al. 1982) and thiorphan (AOKI et al. 1986) on the guinea pig ileum and the mouse vas deferens (ROQUES and BEAUMONT 1990). Recent findings suggest that endopeptidase 24–11 may also take part in opioid degradations since inhibitors of this enzyme have been shown to act as analgesics as potent as thiorphan (ROQUES and FOURNIE-ZALUSKI 1986; ROQUES et al. 1980). Peptide metabolism in various organs has been extensively studied and the reader is referred to the following review articles for further details (ROQUES and FOURNIE-ZALUSKI 1986; ROQUES et al. 1980; ATKINS and HOSTER 1989).

B. Substance P and Related Neurokinins

Three mammalian neurokinins whose primary structures are shown in Table 5 have been identified. Substance P (SP) was first described in 1931 by Von Euler and Gaddum and its peptide nature was recognized in 1936 by Von Euler, who found that SP is inactivated by trypsin. SP was purified by STUDER et al. (1973) and identified as an undecapeptide (tachykinin-like) in 1970, by CHANG and LEEMAN, using conventional purification procedures. The undecapeptide SP was also deducted from a cDNA clone of bovine brain, coding for the precursor, preprotachykinin A (PPT-A) (NAWA et al. 1983). Furthermore, NAWA et al. (1983) demonstrated that, from a single *PPT* gene, two PPT-AmRNAs are produced as a consequence of alternate RNAs splicing, namely αPPT-A mRNA, which encodes only for SP, and β-PPT-A mRNA, which encodes for SP and for the decapeptide neurokinin A (NKA). The sequence of NKA was identified also at the C-terminal of neuropeptide K (NPK), a larger (36 residues) peptide extracted from porcine brain (TAKEMOTO et al. 1985). Thus, different proteolytic processes lead to different end products (e.g., NKA, NPK) in various cells possessing the β-PPT-A mRNA. A different gene (Table 5) coding for PPT-B, the precursor of neurokinin B (NKB), was found in bovine brain (KOTANI et al. 1986) and shown to have a different distribution than the other precursor. It is therefore assumed that the end products of neurokinin-containing nerves are: SP alone, SP, NKA and/or NPK and NKB alone. The three neurokinins show a similar but not identical spectrum of biological activity, in vivo and in vitro (Tables 6, 7). Neurokinins have in common the C-terminal pentapeptide Phe-X-Gly-Leu-Met-NH_2, which appears to be essential for biological activity. The presence of an aromatic residue in X (Phe, Tyr) favours occupation of NK-1 and that of an aliphatic (Val, Ile) the occupation of NK-2 and NK-3 receptor sites.

Table 5. Primary structures of neurokinins

Peptide	Primary structure	Precursor	Gene	Reference
Substance P	Arg-Pro-Lys-Pro-Gln-Gln-Phe-Phe-Gly-Leu-Met[a]	α-Preprotachykinin β-Preprotachykinin	*PPT* gene I[b]	Yokota et al. (1989)
Neurokinin A	His-Lys-Thr-Asp-Ser-Phe-Val-Gly-Leu-Met[a]	β-Preprotachykinin	*PPT* gene I	Masu et al. (1987), Sasai and Nakanishi (1989)
Neurokinin B	Asp-Met-His-Asp-Phe-Phe-Val-Gly-Leu-Met[a]	Preprotachykinin B	*PPT* gene II	Kotani et al. (1986)
Neuropeptide K	Asp-Ala-Asp-Ser-Ser-Ile-Glu-Lys-Gln-Val-Ala Leu-Leu-Lys-Ala-Leu-Tyr-Gly-His-Gly-Gln-Ile Ser-His-Lys-Arg-His-Lys-Thr-Asp-Ser-Phe-Val-Gly-Leu-Met[a]	β-Preprotachykinin	*PPT* gene I	Sasai and Nakanishi (1989)

[a] C-terminal amide.
[b] The number of the human chromosome containing the gene.

Table 6. Biological activity of neurokinins in vivo

Target organ (species)	Effect	Potency of agonists	Receptor type	Reference
Heart and vessels (Rat)	Hypotension	SP > NKA > NKB	NK-1	HUA et al. (1984)
	Tachycardia	NKA > NKB > SP	NK-2	COUTURE et al. (1989)
	Bradycardia	NKB > NKA > SP	NK-3	COUTURE et al. (1989)
Airways (Guinea pig)	Bronchoconstriction	NKA > NKB > SP	NK-2	HUA et al. (1984)
Intestine (Dog)	Increase of motility	NKA > NKB > SP	NK-2	ALLESCHER et al. (1989)
	Reduction of motility	SP		SPINDEL et al. (1990)
Spinal cord (Rat)	Analgesia[a]	NKB > SP > NKA	NK-3	LANEUVILLE et al. (1988)
	Hyperalgesia	SP > NKA > NKB	NK-1	LANEUVILLE et al. (1988)
Peripheral tissues (Rat)	Plasma extravasation	SP	NK-1	COUTURE and CUELLO (1984)
		NKB	NK-3	
Exocrine glands (Rat)	Salivation	SP > NKA > NKB	NK-1	MAGGI et al. (1987a)
Urinary bladder (Rat)	Micturition	NKB > NKA > SP	NK-3	MAGGI et al. (1987b)

SP, substance P; NKA, neurokinin A; NKB, neurokinin B.
[a] Neurokinins were administered intravenously or intrathecally.

Table 7. Biological activity of neurokinins in vitro

System	Tissue	Effect	Potency of agonists	Receptor type	Reference
Cardiovascular	Human epicardial coronary artery	R	SP	NK-1	CROSSMAN et al. (1989)
(a) With endothelium	Pig coronary artery	R	SP > NKA > NKB	NK-1	GULATI et al. (1987)
(b) With endothelium	Dog carotid artery	R	SP > NKA > NKB	NK-1	D'ORLÉANS-JUSTE et al. (1985)
	Renal artery	R	SP > NKA > NKB	NK-1	RHALEB (personal communication)
	Rabbit aorta	R	SP > NKA > NKB	NK-1	REGOLI et al. (1984), BRIZZOLARA and BUENSTOCK (1991)
	Hepatic artery	R	SP		
	Pulmonary artery	R	SP > NKA > NKB	NK-1	D'ORLÉANS-JUSTE et al. (1986)
		C[a]	*NKA > NKB > SP*	*NK-2*	D'ORLÉANS-JUSTE et al. (1986)
	Mesenteric artery	R	SP		DEBLOIS and MARCEAU (1987)
	Jugular vein	*C*	*SP > NKA > NKB*	*NK-1*	NANTEL et al. (1990)
	Mesenteric vein	C	NKA > NKB > SP	NK-1/NK-2	DION et al. (1987)
	Vena cava	C	SP > NKA > NKB	NK-1	NANTEL et al. (1991)
	Rat *mesenteric vein*	*C*	*NKB > NKA > SP*	*NK-3*	MASTRANGELO et al. (1987)
Urogenital	Rat vas deferens	C	NKA > NKB > SP	NK-2	TOUSIGNANT et al. (1987)
	Urinary bladder	C	NKA = SP > NKB	NK-2	DION et al. (1987)
	Guinea pig vas deferens	C	SP > NKA > NKB	NK-1	REGOLI et al. (1990a)
	Urinary bladder	C	SP = NKB > NKA	NK-1	DION et al. (1987)
	Human urinary bladder	C	NKA > NKB > SP	NK-2	DION et al. (1988)
	Uterus	C	SP		
	Hamster urinary bladder	C	NKA = NKB > SP	NK-2/NK-3	REGOLI et al. (1989b), COUTURE et al. (1981)

Table 7. *Continued*

System	Tissue	Effect	Potency of agonists	Receptor type	Reference
Gastrointestinal	Human colon	C	SP		
	Rat stomach	C	NKA > SP > NKB	NK-2	PATACCHINI et al. (1991), IRELAND et al. (1991)
	Colon	C	SP		PATACCHINI et al. (1991)
	Duodenum	C	NKA > NKB > SP	NK-2	IRELAND et al. (1990)
	Jujenum	C	NKA > NKB > SP	NK-2	
	Guinea pig gall bladder	C	NKA > NKB > SP	NK-2	PATACCHINI et al. (1991)
	Ileum	C[b]	SP > NKB = NKA	NK-1	REGOLI et al. (1988)
	Ileum	C	NKB > NKA > SP	NK-3	LAUFER et al. (1985)
Respiratory	Human bronchus	C	NKA > SP > NKB	NK-2	NALINE et al. (1989)
	Guinea pig trachea	C	NKA > SP > NKB	NK-2	DION et al. (1987)
	Hamster trachea	C	NKA > NKB > SP	NK-2	MAGGI et al. (1991)
Other smooth muscle	Human spincter of Oddi	C	SP		GUO et al. (1986)
	Rabbit iris sphincter	C	SP > NKA		VEDA et al. (1986)

[a] Without endothelium.
[b] Guinea pig ileum treated with atropine, diphenhydramine and indomethacine (REGOLI et al. 1987).

I. Biological Activity of Neurokinins In Vivo

When given intravenously or intrathecally in laboratory animals, neurokinins induce a variety of biological effects (described in Table 6), some of which (for instance the marked, yet transient, decrease of blood pressure; the constriction of the airways) appear to be due to modulation of smooth muscle tone, while others (tachycardia, bradycardia, hyperalgesia, micturition, modulation of intestinal motility and secretion) involve other cell types, presumably neurons. Peripheral vasodilatation (hypotension) is attributed to the activation of NK-1 receptors (MAGGI et al. 1987b) in the endothelium and the subsequent release of nitric oxide, the endothelium-derived relaxing factor (EDRF) (FURCHGOTT 1981), which is a potent relaxant of the underlying vascular smooth muscle. Changes of heart rate induced by NKA and NKB have been referred to modulation of sympathetic (tachycardia) and parasympathetic (bradycardia) nerve activities by NK-2 and NK-3 receptors (COUTURE et al. 1989). NK-2 receptors appear to be present in the small airways (NALINE et al. 1989; RODGERS et al. 1989) and their activation leads to bronchoconstriction, while NK-1 receptors may be involved in the plasma extravasation, the increased mucous secretion, the oedema and the local inflammatory reactions associated with asthma (RODGERS et al. 1989). Modulation of smooth muscle tone (generally by NK-2 receptors in man) and activation of sensory or motor nerves [through NK-3 receptors (COUTURE et al. 1989)] appear to be the most important mechanisms by which neurokinins can influence gastrointestinal motility, micturition, other urogenital functions, and uterine motility and also contribute to neurogenic inflammation (see Table 6 for references). Analgesia has been attributed to activation of NK-3 receptors in the spinal cord and the subsequent release of opioids [since it is blocked by naloxone (REGOLI et al. 1988; LANEUVILLE et al. 1988)] and hyperalgesia to that of NK-1 receptors in the spinal cord neurons that transmit painful stimuli (REGOLI et al. 1988; LANEUVILLE et al. 1988). Salivary secretion is increased by the stimulation of NK-1 receptors in the acinal cell and the subsequent protein hypersecretion (ROLLANDY et al. 1993).

Receptors involved in the regulation of NKs effects in vivo have been characterized with the naturally occurring peptides (generally the three neurokinins) and in some recent studies also with selective agonists (COUTURE et al. 1989), while antagonists are under investigation. The use of specific and selective antagonists should contribute substantially to confirming the implication of the three neurokinin receptors in the various physiological phenomena analysed in Table 6.

II. Neurokinin-Receptor Activity in Isolated Smooth Muscles In Vitro

Early studies by Erspamer and coworkers (reviewed by BERTACCINI 1976) had shown that neurokinins exert myotropic effects in a variety of organs (see Table 4). In the last 10 years, several other tissues, especially isolated vessels, have been found to be sensitive to neurokinins and some of them have been used successfully for pharmacological characterization of NKs receptors (see below). A fairly complete list of preparations belonging to different systems and organs responding to neurokinins with contraction (C) or relaxation (R) is analysed in Table 7. The responses of the various tissues appear to be mediated (in general) by a single receptor type, but a few tissues (e.g., the rabbit aorta and pulmonary artery) have two receptors with opposite effects (inhibitory NK-1, stimulatory NK-2) and other tissues (rabbit mesenteric vein, hamster urinary bladder, guinea pig ileum) have at least two receptors, both contractile. NK-1 receptors are found in arterial vessels localized in endothelia and in the smooth muscle of some veins (NANTEL et al. 1990, 1991; D'ORLÉANS-JUSTE et al. 1985), while NK-2 receptors are predominant, very often alone, in the smooth muscles of the gastrointestinal, respiratory, genitourinary systems, especially in man. NK-1 and NK-2 receptors have been demonstrated in the rabbit eye (VEDA et al. 1986). NK-3 receptors are only present in some veins (the rat portal vein) and, together with other receptor types, in other organs (e.g., the hamster urinary bladder) (REGOLI et al. 1989b). Some of the preparations analysed in Table 7 have been extensively used in neurokinin pharmacology for receptor characterization, as well as for studying the neurokinin mechanism of action on smooth muscles and the second messengers involved in the stimulatory or inhibitory effects of substance P and congeners. The most frequently used preparation is the guinea pig ileum, which contains at least two (NK-1, NK-3) and possibly three contractile receptors: isolated vessels have been preferred to the ileum for receptor characterization with agonists and (most recently) with antagonists. The basic data obtained in the rabbit jugular vein (NK-1), the rabbit pulmonary artery (NK-2) and the rat portal vein (NK-3) are shown in Table 8. Apparent affinities have been evaluated in terms of pD_2 for the agonists and in terms of pA_2 for antagonists, while the maximal responses to SP (in the RJV), NKA (in the RPA) or NKB (in the RPV) are taken as a reference (equal to unity) to express the intrinsic activities of the other agonists and the potential residual effects of antagonists, according to REGOLI et al. (1988). The two criteria for receptor characterization, namely the order of potency of agonists and the apparent affinity of antagonists, recommended by SCHILD (1947), have been applied (Table 8) in the present investigation: moreover, data obtained with agonists selective for each one of the three neurokinin receptors have been added to further support the existence of three receptor types.

NK-1 receptors show the following order of potency of neurokinins and selective agonists:

SP > NKA > NKB;
[Sar^9, $Met(O_2)^{11}$]SP > [$\beta$$Ala^8$]NKA(4–10) > [$MePhe^7$]NKB

and are blocked selectively by the antagonist CP 96345 (SNIDER et al. 1991; ROUISSI et al. 1991b).

Table 8. Activity of neuroninins, selective agonists and antagonists on isolated organs: Receptor classification

Agonists	Preparations								
	RJV			RPA			RPV		
	pD_2	RA	α^E	pD_2	RA	α^E	pD_2	RA	α^E
Substance P	8.83	100	1.0	6.13	0.8	0.8	5.82	1.4	0.5
Neurokinin A	7.65	6.7	0.9	8.22	100	1.0	6.45	5.8	1.2
Neurokinin B	7.84	10	0.7	7.45	16	1.0	7.68	100	1.0
[Sar^9, Met$(O_2)^{11}$]SP	8.86	107	1.1	Inac.	–	–	Inac.	–	–
[β-Ala^8]NKA(4–10)	6.23	0.3	0.8	8.6	239	1.1	6.13	2.8	1.0
[MePhe^7]NKB	6.23	0.3	0.9	5.24	0.1	0.5	8.30	416	1.1
Antagonists				pA_2			pA_2		pA_2
Spantide				6.78			5.78		6.18
R-544				7.00			5.40		5.80
CP. 96,345				8.90			<4.32		5.62
MEN 10,207				nd			7.9		4.9
R-396				Inac.			5.6		Inac.
[MePhe^7,β-Ala^8]-NKA(4–10)				Inac.[9]			Inac.[a]		7.26

RJV, rabbit jugular vein; RPA, rabbit pulmonary artery; RPV, rat portal vein; pD_2, apparent affinity; RA, relative affinity; α^E, intrinsic activity; pA_2, log of the molar concentration of antagonist required to reduce the effect of a double concentration of agonist to the effect of a single concentration.
Spantide, [D-Arg^1,D-Trp^{7,9},Leu^{11}]-substance P.
R-544, Ac-Thr-D-Trp(For)-Phe-*N*-MeB$_z$.
CP 96.345, (±)-*cis*-3-(2-methoxybenzylamino)-2-benzhydrylquinuclidine.
MEN 10207, Asp-Tyr-D-Trp-Val-D-Trp-D-Trp-Arg-NH_2.
R-396, Ac-Leu-Asp-Gln-Trp-Phe-Gly-NH_2.
[a] Inactive.

NK-2 receptors show high sensitivity to NKA and to [βAla^8]NKA(4–10) and the orders of potency of the agonists are as follows:

NKA > NKB > SP;
[βAla^8]NKA(4–10) > [MePhe^7]NKB > [Sar^9, Met$(O_2)^{11}$]SP

NK-2 receptors are antagonized by compounds MEN-10207 and by R-396. Recently, Maggi et al. (1990, 1991), Patacchini et al. (1991), using various antagonists, described an NK-2 receptor subtype (NK-2b) in the hamster trachea, which is blocked by compound R-396, while compound MEN-10207 blocks the NK-2a receptor of the rabbit pulmonary artery and acts as an agonist on the NK-2b (Table 8).

NK-3 receptors are more sensitive to NKB than to the other neurokinins and respond selectively to [MePhe^7]NKB in the following order:

NKB > NKA > SP
[MePhe^{7}]NKB > [βAla]NKA(4–10) > [Sar^{9}, Met(O_2)^{11}]SP

The effect of both NKB and [Me-Phe^{7}]NKB on the RPV (NK-3) is antagonized by compound [Trp^{7}, βAla^{8}]NKA(4–10), which exerts fairly potent stimulatory effects on the other two preparations (DRAPEAU et al. 1990).

III. Mechanisms of Action of Neurokinins: Mediators and Second Messengers

As illustrated in Tables 6 and 7, substance P and related neurokinins induce marked myotropic effects in smooth muscle of vascular and non-vascular origin via the activation of at least three receptor types, namely NK-1, NK-2 and NK-3.

The mediators and second messengers involved in the biological effects of neurokinins are illustrated in Table 9. Substance P and related neurokinins induce arterial vasodilatation via an indirect mechanism that involves the release of the endothelium-derived relaxing factor (EDRF, D'ORLÉANS-JUSTE et al. 1985). Indeed, removal of the endothelium from isolated arterial preparations abolishes the vasodilatory responses of neurokinins, suggesting that receptors for these peptides are present in the endothelium (D'ORLÉANS-JUSTE et al. 1985). Activation of these receptors generates EDRF via a calcium-calmodulin mechanism (BUSSE and MULSCH 1990). EDRF activates the soluble guanylate cyclase in the smooth muscle and leads to increase of intracellular cyclic GMP (IGNARO 1989), which is responsible for membrane hyperpolarization and smooth muscle relaxation.

In a large variety of smooth muscles, both arterial (e.g., the rabbit pulmonary artery) and venous (rabbit jugular vein, rat portal vein), as well as intestinal smooth muscle (guinea pig ileum, rat ileum), neurokinins exert contractile effects directly by activation of the smooth muscle fibres or indirectly by stimulating autonomic nerve terminals (e.g., in the guinea pig ileum). Some of these direct and indirect effects (e.g., on the guinea pig ileum) have been attributed to a similar transducing mechanism, the phospholipase C-mediated increase of phosphoinositides (PI_3) (GUARD and WATSON 1991). However, the experimental evidence to prove this mechanism is still insufficient because of the complexity of isolated organs, such as the intestine, which consist of different cell types. Furthermore, the experimental conditions utilized in some of the studies (GUARD and WATSON 1991) are considered inadequate because measurement of PI_3 was made in the presence of extracellular calcium, thus not excluding the possibility that such an effect could be in part due to stimulated calcium influx (POWELL 1981). The mechanisms of action by which substance P and related neurokinins contracts arterial (through NK-2 receptors) or venous (through NK-

Table 9. Mediators and second messengers of neurokinin receptors

Smooth muscle	Preparation	Receptor	Target tissue		Effector system			
			Cell type	Mediator	Cell type	Second Messenger	Effect	Reference
Vascular	Dog carotid artery	NK-1	Endothelium	EDRF [77,100]	Smooth muscle	cGMP	R	D'ORLÉANS-JUSTE et al. (1985)
	Rabbit pulmonary artery	NK-2	Smooth muscle		Smooth muscle	Ca^{2+}	C	NANTEL (personal communication)
	Rabbit vena cava	NK-1	Smooth muscle		Smooth muscle	Ca^{2+}	C	NANTEL et al. (1991)
	Rat portal vein	NK-3	Smooth muscle		Smooth muscle	Ca^{2+}	C	SUTTER (1990)
Gastrointestinal	Guinea pig ileum	NK-1	Smooth muscle		Smooth muscle	PI_3	C	GUARD and WATSON (1991)
	Guinea pig ileum	NK-3	Parasym-pathetic	Acetylcholine [94]	Smooth muscle	PI_3	C	GUARD and WATSON (1991)
	Rat ileum	NK-2	Smooth muscle?		Smooth muscle	PI_3	C	WATSON (1984)

EDRF, endothelium-derived relaxing factor; R, relaxation; C, contraction.

1 or NK-3 receptors) smooth muscle have not been investigated and the suggestion that they depend on calcium (see Table 9) is purely hypothetical.

Worthy of note is the fact that the three neurokinin receptors, which stimulate the gastrointestinal system of mammals, activate the same transducing pathway (phospholipase C and calcium mobilization) (GUARD and WATSON 1991; WATSON 1984). Selectivity occurs therefore at the receptor level.

C. Gastrin-Releasing Peptides and Neuromedins B

Identification of bombesin-related peptides occurred in the early 1970s, first with ranatensin (NAKAJIMA et al. 1970) and shortly after with bombesin and alytesin (ERSPAMER et al. 1970), three peptides originating from amphibian skins. The mammalian peptides were discovered much later: first the porcine gastrin-releasing peptide (GRP) (McDONALD et al. 1979), which was extracted from the non-antral stomach in 1979, then the avian GRP in 1980 and the human GRP in 1984 (ORLOFF et al. 1984). The three GRPs consist of 27 amino acids. Porcine and human GRP show a great homology, differing only by four residues at the N-terminal and by the residue in position 12. Biological activities are completely maintained by the C-terminal decapeptide GRP 18–27 (Table 10), a naturally occurring peptide that has been isolated from human (ORLOFF et al. 1984) and porcine (MINAMINO et al. 1984b) tissues. This is the same in man, hog, and dog; but there is a difference of one residue in chicken, rat and dogfish (McDONALD 1991). Molecular biology studies demonstrated the presence of GRP in a single gene (in human chromosome 18), and of multiple types of Pre-pro GRP mRNA (148 residues) (SPINDEL et al. 1986), with sequence differences occurring albeit only in the C-terminal extension peptide.

In 1983, Minamino et al. identified neuromedin B (NMB), the mammalian counterpart of ranatensin in porcine spinal cord, and suggested

Table 10. The bombesin family

Members	Human peptides	Minimum sequences for biological activity	Human gene chromosome
Bombesin	GRP_{1-27}	GRP_{18-27} GNHWAVGHLM	18
Neuromedin B	NMB_{1-32}	NMB_{23-32} GNLWATGHFM	15
Phyllolitorin	-	QLWAVGHFM	
Primary structures of biologically active peptides			
GRP_{18-27}	Gly-Asn-His-Trp-Ala-Val-Gly-His-Leu-Met-NH_2		
NMB_{23-32}	Gly-Asn-Leu-Trp-Ala-Thr-Gly-His-Phe-Met-NH_2		
Phyllolitorin	Glu-Leu-Trp-Ala-Val-Gly-His-Phe-Met-NH_2		
Bombesin	Pyr-Gln-Arg-Leu-Gly-Asn-Gln-Trp-Ala-Val-Gly-His-Leu-Met-NH_2		

that NMB was the most important bombesin-related peptide in the brain. Human NMB was discovered 5 years later (KRANE et al. 1988) in a human hypothalamic cDNA library. Similar to GRPs, the biologically active sequence is contained in the C-terminal decapeptide, which shows perfect homology in various mammals (men, hog) and has been isolated from tissue extracts (McDONALD 1991). The gene encoding for neuromedin B is present in chromosome 15 and codes for a large precursor of NMB (RÖKAEUS 1991). The primary structures of a representative of each bombesin class, variety GRP_{18-27}, NMB_{23-32} and phyllolitorin as well as the structure of bombesin are shown in Table 10.

Bombesin-like peptides are found predominantly in central and peripheral neurons: they appear to act as paracrine hormones, being released from nerve endings or varicosities directly onto, or in the vicinity of, target cells: the target cells for GRP are presumably the G (gastrin) cells of the stomach antrum, the endocrine cells containing CCK in the duodenal mucosa, the mesenteric plexus and the gastrointestinal smooth muscles (TACHÉ 1988).

I. Biological Activity of Bombesin-Related Peptides In Vivo and In Vitro

Early studies showed that ranatensin, bombesin and litorin injected intravenously at high doses are hypertensive in dogs and rats: it was soon realized, however, that the cardiovascular effects of these peptides were "erratic and unpredictable" (BERTACCINI 1976). On the other hand, these peptides were found to be potent stimulants of the release of gastrointestinal hormones (gastrin, CCK, perhaps enterogastrone) (BERTACCINI et al. 1973, 1974; VARNER et al. 1981) and to enhance gastric and pancreatic secretion, or modulate gastrointestinal motility (CAPRILLI et al. 1975). Thus, a large part of the observed effects were considered to be "indirect", mediated by gastrin or CCK (LHOSTE et al. 1985). Today, the most important biological effects of GRP and congeners in vivo are on three major areas: the gastrointestinal tract, the brain, and cell growth and multiplication (see Table 11). Studies in humans and animals have shown that GRP has receptors on G cells and promotes gastrin secretion (BERTACCINI et al. 1974; VARNER et al. 1981); GRP is released from nerves or paracrine cells by electrical stimulation (SEVERI et al. 1988). Through stimulation of CCK release, as well as by direct activation of receptors in the target cells, GRP stimulates pancreatic exocrine or endocrine secretions and growth as well as gall bladder motility (FALCONIERI-ERSPAMER et al. 1988) and modulates gastrointestinal movements (CAPRILLI et al. 1975) (see Table 11). In the brain GRP intervenes in the regulation of body temperature by acting on nervous (sympathetic) or hormonal (thyroid, other glands) pathways (BROWN et al. 1977); it also inhibits gastric secretion, feeding and drinking (satiety) (GIBBS et al. 1979)

Table 11. Biological effects of bombesin-related peptides in vivo

Species	Biological effect	Functional site	Reference
Rat	Hypothermia	Central	BROWN et al. (1977)
Dog, rat	Glycemia modulation	Central and peripheral	BROWN et al. (1979)
Rat	Food intake – Gastric motility – (Satiety)	Central and peripheral	GIBBS et al. (1979) NEGRI (1986)
Rat	Grooming +	Central	BROWN et al. (1977)
Human, dog, rat,	Gastric secretion +	Peripheral	BERTACCINI et al. (1973), BERTACCINI et al. (1974), VARNER et al. (1981)
Human, dog, chicken	Pancreatic secretion +	Peripheral	ERSPAMER et al. (1972)
Human, dog, guinea pig	Gall bladder +	Peripheral	SEVERI et al. (1988)
Human, dog	Intestinal motility –	Peripheral	CAPRILLI et al. (1975)
Human, dog, rat	Trophic effect + G cell and pancreas	Peripheral	LEHY et al. (1983), LHOSTE et al. (1985)

+, stimulation; –, inhibition.

and stimulates grooming-scratching behaviour in the rat (NEGRI 1986) through central effects expressed peripherally by the reduced activity of the vagus nerve. The simultaneous inhibition of the vagal and activation of the sympatho-adrenal pathways lead to changes in the pancreas, endocrine system, liver and adrenal functions and to modulation of the glycemia (BROWN et al. 1979).

GRP-stimulated cell growth is expressed first by hypertrophy and then by hyperplasia of G cells (LEHY et al. 1983), pancreatic exocrine cells (LHOSTE et al. 1985), bronchial epithelial cells and others. Some tumour cells, especially from the lung, the thyroid and malignant neuroendocrine structures express GRP mRNAs and are the source of bombesin-related peptides (POLAK et al. 1988).

In vitro, bombesin, GRP and congeners have been shown to contract a variety of isolated smooth muscle (Table 12), to promote the secretion of gastrin and pancreatic enzymes (LHOSTE et al. 1985) and to act on isolated cell lines in culture (e.g., Swiss 3T3 fibroblasts) (POLAK et al. 1988). The effects of these peptides on exocrine secretion and cell growth have been the subject of extensive reviews (ERSPAMER et al. 1974; LEHY et al. 1983). As to smooth muscles, the most important target and assays organs are analysed in Table 12. GRP and congeners are potent stimulants of the intestine, the stomach (GIRARD et al. 1984), the duodenum, and the colon of several species (FALCONIERI-ERSPAMER et al. 1988), including man (REGOLI et al.

Table 12. Effect of bombesin-related peptides in various gastrointestinal and urogenital smooth muscle preparations

Species	Organ	Active agonist	Receptor	Effect	Reference
Rat	Urinary bladder	A = C > B	BB_1	Contraction	Rouissi et al. (1991), Mizrahi et al. (1985), Regoli (1987), Falconieri-Erspamer et al. (1988) Regoli et al. (1989b)
	Colon	A, B, C	?	Contraction	Falconieri-Erspamer et al. (1988)
	Stomach	A > B ≫ C	BB_2	Contraction	Rouissi et al. (1991a), Mizrahi et al. (1985), Regoli et al. (1989b), Girard et al. (1984)
	Duodenum	A	?	Relaxation	Erspamer et al. (1972), Cantor et al. (1987)
	Uterus	A, B, C	?	Contraction	Erspamer et al. (1972)
Guinea pig	Urinary bladder	A > B ≫ C	BB_2	Contraction	Rouissi et al. (1991a), Mizrahi et al. (1985), Regoli (1987), Falconieri-Erspamer et al. (1988), Regoli et al. (1989b)
	Ileum	A	?	Contraction	Falconieri-Erspamer et al. (1988)
	Gall bladder	A > B ≫ C	BB_2	Contraction	Rouissi et al. (1991a), Erspamer et al. (1972), Falconieri-Erspamer et al. (1988), Regoli et al. (1989b)
	Ureter	A	?	Contraction	Erspamer et al. (1972)
Kitten	Intestine	A	?	Contraction	Falconieri-Erspamer et al. (1988)
Chick	Intestine	A = C > B	BB_1	Contraction	Erspamer et al. (1972)
Rabbit	Intestine	A	?	Contraction	Erspamer et al. (1972)

A, peptide from bombesin family; B, peptide from ranatensin; C, peptide from phyllolitorin family.

Table 13. Pharmacological characterization of bombesin receptors in isolated smooth muscles. (From ROUISSI et al. 1991a)

Agonists	Preparations					
	RUB			GPUB		
	pD_2	RA	α^E	pD_2	RA	α^E
Bombesin	8.33	100	1.0	8.83	100	1.0
Litorin	8.20	74	0.8	8.36	34	1.3
Neuromedin C	7.80	30	0.7	7.91	12	0.6
[Phe^6]Neuromedin C	7.24	8	0.6	6.24	0.3	0.5
Neuromedin B	8.29	91	1.1	6.71	0.8	0.5
GRP (human)	7.13	6	0.8	8.20	23	0.6
Bombesin (6–14)	8.52	155	1.0	8.92	123	1.0
Ac. bombesin (6–14)	8.48	141	1.2	8.83	100	1.3
Antagonist	pA_2			pA_2	pA_{10}	pA_2-pA_{10}
AcGRP(20–26)OCH_3	Inactive			7.95	6.96	0.99
AcGRP(20–26)OC_2H_5	Inactive			7.95	6.94	1.01

RUB, rat urinary bladder; GPUB, guinea pig urinary bladder; pD_2, RA, α^E, pA_2, same as in Table 8; GRP, gastrin-releasing peptide.
Ac, acetyl.

1989a), and exert myotropic effects also on the urogenital tract (urinary bladder, ureter, uterus), (ERSPAMER et al. 1972; MIZRAHI et al. 1985; FALCONIERI-ERSPAMER et al. 1988); they also stimulate gall bladder in vitro (FALCONIERI-ERSPAMER et al. 1988). Quantitative pharmacological studies have been performed in some of the preparations analysed in Table 13 in an attempt to characterize and clarify the receptors subserving the myotropic effects in peripheral organs (FALCONIERI-ERSPAMER et al. 1988) and compare them with those mediating exocrine secretions (BERTACCINI et al. 1973, 1974; VARNER et al. 1981). As for other peptides (see section on neurokinins), naturally occurring agonists, some fragments and analogues as well as antagonists have been used for receptor characterization in various tissues (ROUISSI et al. 1991a; REGOLI et al. 1989a; GIRARD et al. 1984). All organs studied in Table 12 respond to GRP and bombesin with concentration-dependent contractions and several tissues show a higher sensitivity to GRP than to NMB (ROUISSI et al. 1991a; GIRARD et al. 1984). There are, however, a few exceptions, for instance the rat urinary bladder and the chick intestine, which are more sensitive to NMB than to GRP (ROUISSI et al. 1991a; FALCONIERI-ERSPAMER et al. 1988).

Some of the preparations analysed in Table 12 have been used in a systematic structure-activity study (ROUISSI et al. 1991a) intended to characterize and classify the two receptor sites on which GRP, NMB and congeners appear to act in mammals. Data taken from this study (ROUISSI et al. 1991a) with agonists and antagonists are analysed in Table 13. Two different functional sites BB_1 and BB_2 are well demonstrated by the different orders of potency of agonists and by the use of an antagonist. The essential data can be summarized as follows:

Receptor type	Order of potency of agonist	pA_2 of AcGRP(20–26)OCH_3
BB_1	NMB = BB >> GRP	0
BB_2	BB > GRP >> NMB	7.95

The first receptor system (the rat urinary bladder, BB_1) is sensitive to NMB more than to GRP and the AcGRP(20–26)OCH_3 antagonist does not affect the action of the bombesin-like peptides; conversely, in the guinea pig urinary bladder (BB_2 receptor) and in various other organs (ROUISSI et al. 1991a), GRP is much more potent than NMB and the effects of all bombesin-related peptides are blocked by the antagonist. The data in Table 13 indicate that bombesin-related peptides act on two different sites, of which the second (BB_2) is particularly abundant in peripheral tissues, while the other (BB_1) may be important in the brain, where NMB is the most abundant naturally occurring peptide (MINAMINO et al. 1983). Data obtained in two preparations, the rat urinary bladder (RUB) and the guinea pig urinary bladder (GPUB), are shown in Table 13.

II. Second Messengers of Bombesin Receptors

Specific binding sites for labelled bombesin, $[I^{125}Tyr^4]$BB, have been found in G cells, pancreatic acini, brain and perennial cell lines [Swiss 3T3 fibroblasts or small cell lung cancer (SCLC)]. The action mechanism and the second messenger involved in the biological effect of GRP and bombesin have been studied in human and guinea pig pancreas (acini) (JENSEN et al. 1984). It has been shown that these peptides induce membrane depolarization, breakdown of phosphatidylinositol, mobilization of cytoplasmic Ca^{2+}, activation of a Ca^{2+}/calmodulin-dependent kinase and activation of protein kinase C. In the same tissues, a marked increase of cGMP has been measured, but the physiological significance of this finding remains obscure (JENSEN et al. 1988).

The action mechanism of GRP has been investigated by ROZENGURT (1988) in Swiss 3T3 fibroblasts, where the peptide stimulates DNA synthesis and cell division. Stimulation of receptors by GRP induces a rapid influx of Na^+ that leads to cytoplasmic alkalinization and to Ca^{2+} mobilization from intracellular stores; this event may be mediated by the increase of phosphoinositides that rapidly increases protein kinase C activity and stimulation of DNA synthesis (ROZENGURT 1988).

A similar mechanism of action has been identified (SANSVILLE et al. 1988) in human small cell lung cancer (SCLC) cells, where bombesin stimulates the influx of extracellular Ca^{2+} and the release of Ca^{2+} from intracellular stores: this leads to *myc*-family gene expression and to cell proliferation.

The mechanism by which bombesin stimulates smooth muscle contraction has not yet been clarified; it is, however, conceivable that activation of Ca^{2+} flux through the membrane and/or of Ca^{2+} release from intracellular stores are the basic mechanism of action by which this family of peptides induce smooth muscle contractions. Furthermore, since some of the effects of bombesin on smooth muscles are mediated by gastrin, CCK or other endogenous agents, whose release is promoted by bombesin, these indirect effects and their mechanisms will be discussed later (see Sect. E).

D. Neurotensin

The tridecapeptide neurotensin (pGlu-Leu-Tyr-Glu-Asn-Lys-Pro-Arg-Pro-Tyr-Ile-Leu) was first isolated from bovine hypothalamus (CARRAWAY and LEEMAN 1973) during the isolation of substance P. The same peptide was extracted from bovine intestine (KITABGI et al. 1976) and deduced from cDNA sequences of other mammals, the dog (DOBNER et al. 1987), and the rat (KISLAUSKIS et al. 1988); it was also found in human intestine (HAMMER et al. 1980). The C-terminal part of neurotensin shows a great homology with the amphibian xenopsin (ARAKI et al. 1973), an octapeptide that has NT-like activities. Early investigation indicated that the entire tridecapeptide sequence of NT is not required for biological activity, since the most important of them are also exerted by the C-terminal hexapeptide Arg-Arg-Pro-Tyr-Ile-Leu (NT_{8-13}) (CARRAWAY and LEEMAN 1973). NT_{8-13} is also found in chicken neurotensin, while another biologically active hexapeptide, neuromedin N (NMD-N) (Lys-Ile-Pro-Tyr-Ile-Leu) (MINAMINO et al. 1984a) extracted from porcine brain differs from NT_{8-13} by the presence of Lys-Ile instead of Arg-Arg at the N-terminal; NMD-N shows weaker biological activity than neurotensin.

Neurotensin is present in structures of the central and peripheral nervous systems and in a population of mucosal cells, the N cells. The most consistent biological effects are hypotension, stimulation of gastrointestinal motility, inhibition of gastric acid and stimulation of other exocrine and endocrine secretions (ARMSTRONG et al. 1986; ALLESCHER and AHMAD 1991). Like other neuropeptides, neurotensin exerts a variety of central effects (NEMEROFF 1980; QUIRION et al. 1980) and appears to act on at least two receptor types which have been partially characterized using agonists and a rather weak antagonist, [D-Trp^{11}]NT (QUIRION et al. 1980a).

Eighty percent of neurotensin is present in the intestine and the rest is in the brain; there is little in other organs. When released, especially by nutrients or hormonal stimuli, NT is rapidly inactivated to NT 1–11 and NT 1–8 (ALLESCHER and AHMAD 1991), two biologically inactive fragments;

however, NT 8–13 appears to be highly preserved and it maintains full biological activity. The presence of such large quantities of NT in the intestine suggests that NT may be a paracrine hormone involved in the regulation of intestinal motility and secretions (ALLESCHER and AHMAD 1991). NT is present in various parts of the brain, especially in the hypothalamus and the dopaminergic regions (dopamine-containing neurons); it has therefore been suggested that NT takes part in the regulation of dopaminergic functions (NEMEROFF 1980).

I. Biological Activity of Neurotensin In Vivo

Neurotensin is hypotensive in the rat, the rabbit, the dog, the pig, and the goat and induces a biphasic pressor response (initial hypotension followed by increase of blood pressure) in the guinea pig and the cat (Table 14). The hypotensive effect of NT in the rats is reduced by [D-Trp^{11}]NT and is partially prevented by 48/80: it has therefore been suggested that NT-induced hypotension is in part due to histamine release (CARRAWAY et al. 1982). Indeed, NT is a potent releaser of histamine from rat and human mast cells in vitro (SELBEKK et al. 1980) and this effect is blocked by the antagonist [D-Trp^{11}]NT (QUIRION et al. 1980b), suggesting that it might be receptor mediated. This contrasts with a non-specific effect observed in the same test with neurokinins, the kinins and other peptides (MOUSLI et al. 1990). The release of histamine is also involved in another in vivo effect of neurotensin, the increase of vascular permeability (plasma extravasation) in the rat (QUIRION et al. 1980b).

When administered into the mesenteric circulation in the dog, neurotensin affects the intestinal vessels, the smooth muscle and the secretory organs. As pointed out by ALLESCHER and AHMAD (1991): "Ingestion of fat decreases lower esophageal sphincter pressure, delays gastric emptying and produces a strong increase in colonic motility in humans. Similar responses were obtained after i.v. infusion of [Gln^4] neurotensin (ROSELL 1977) at doses resulting in plasma concentrations similar to those observed after fat ingestion. Therefore, it has been speculated that neurotensin might act as a hormone to coordinate postprandial gastrointestinal activity."

When injected into the brain, neurotensin produces hypothermia and this effect has been attributed to inhibition of dopaminergic fibres (NEMEROFF et al. 1980). NT acts also as an analgetic in the mouse and this effect appears to be independent of opioids, since it is not modified by naloxone (NEMEROFF et al. 1979). Applied on the surface of viscera, neurotensin activates capsaicin-sensitive fibres and reflex pathways leading to tachycardia (RIOUX et al. 1989). Central effects of neurotensin include hypothermia, probably by inhibition of dopaminergic fibres (PRANGE and NEMEROFF 1982; NEMEROFF et al. 1980), muscle relaxation, decrease of locomotor activity and potentiation of barbiturate sedation (NEMEROFF et al. 1980). These various effects have been attributed to changes of brain mono-

Table 14. Activity of neurotensin in vivo

System	Pharmacological effect	Receptor type	Species	Reference
Cardiovascular	Hypotension	NT_1[a]	Rabbit, dog	St-Pierre et al. (1984), Nemeroff (1980)
	Biphasic change of BP	NT_1[a]	Cat, sheep, guinea pig	St-Pierre et al. (1984), Nemeroff (1980)
	Tachycardia	NT_2	Guinea pig	Prange and Nemeroff (1982)
	Plasma extravasation	NT_1[a]	Rat, guinea pig	Quirion et al. (1979), Mousli et al. (1990)
Gastrointestinal	Gastric secretion and motility	–	Dog, man, rat	Armstrong et al. (1986), Allescher and Ahmad (1991)
	Stimulation of colonic activity	NT_1[a]	Dog, man	Armstrong et al. (1986), Allescher and Ahmad (1991)
Central nervous system	Hypothermia	NT_2	Rat	St-Pierre et al. (1984), Allescher and Ahmad (1991), Nemeroff et al. (1980)
	Analgesia		Mouse	St-Pierre et al. (1984), Allescher and Ahmad (1991), Nemeroff et al. (1979)
	Muscle relaxation	NT_2	Rat	St-Pierre et al. (1984), Allescher and Ahmad (1991), Quirion et al. (1980a)
	Reduced motility	NT_1[a]	Rat	St-Pierre et al. (1984), Allescher and Ahmad (1991), Quirion et al. (1980)
	Increase of brain monoamine turnover		Rat	Araki et al. (1973)

[a] Blocked by [D-Trp^{11}]NT.

amine turnover, especially in the dopaminergic system (QUIRION et al. 1980a). Only the decrease of motility has been shown to be blocked by [D-Trp^{11}]NT and can be attributed to activation of NT_1 receptors. Of the other central nervous functions, [D-Trp^{11}]NT acts as an agonist and the receptor involved may be of the NT_2 type. The complex central effects of neurotensin have been reviewed by NEMEROFF et al. (1979, 1980); who suggested that neurotensin may have a possible "neuroleptic-like" role. In fact, the effect of neurotensin appears to be similar to those of synthetic neuroleptic drugs, such as chlorpromazine and haloperidol. In the clinic, subnormal levels of neurotensin have been measured in the cerebrospinal fluid of schizophrenic patients (WILDERLÖV et al. 1982) and these changes have been reversed by treatment with neuroleptics.

In vivo and in vitro, neurotensin has been shown to produce excitation and inhibition of smooth muscle in various organs, particularly the gut. A large number of pharmacological preparations have been used since 1971 to evaluate the activity of neurotensin-like peptides in vitro. The majority of these preparations are isolated smooth muscles taken either from the cardiovascular (rat portal vein, rat isolated heart or hind quarter) or the gastrointestinal systems (segments of ileum, colon, duodenum, stomach of various species).

II. Pharmacological Effects of Neurotensin In Vitro: Receptor Characterization

Neurotensin has been shown to stimulate or inhibit smooth muscle activity in a variety of isolated perfused organs and of isolated tissues (Table 15); it is also a potent stimulant of histamine release from human and rat mast cells (SELBEKK et al. 1980; CARRAWAY et al. 1982; MORGAN-BOYD et al. 1987). Infused into the main artery in the rat isolated heart or hind quarter, NT increases the perfusion pressure probably by contracting the arterial smooth muscle and this effect can be attributed to activation of NT_1 receptors, because it is blocked by [D-Trp^{11}]NT. On the contrary, the positive inotropic and chronotropic effects observed after application of NT on the epicardium in anaesthetized guinea pigs (RIOUX et al. 1989) or on the guinea pig isolated beating atria appear to be mediated by the NT_2 receptors, because they are not inhibited by the antagonist (RIOUX personal communication). These effects are capsaicin sensitive and have been attributed to stimulation by NT of sensory fibres (RIOUX et al. 1989). The effect on the histamine release has been already discussed.

In classic pharmacological assays on isolated organs, NT has been shown to stimulate or inhibit various parts of the guinea pig or the rat intestine. NT is a potent stimulant of the guinea pig ileum and tenia coli as well as of the rat stomach, probably through an NT_2 receptor type since [D-Trp^{11}]NT is a weak stimulant and not an antagonist in these organs. NT is a potent relaxant of the guinea pig colon, the rat duodenum and small intestine, but

Table 15. Effects of neurotensin on in vitro perfused organs and isolated tissues

Preparation	Effect	Block by [D-Trp[11]]	Receptor	Reference
Rat isolated heart	Coronary vasoconstriction	Yes	NT-1	St-Pierre et al. (1984), Quirion et al. (1979)
Rat hindquarter	Decrease of perfusion pressure	Yes	NT-1	St-Pierre et al. (1984)
Human mast cells	Degranulation (histamine)	Yes	NT-1	St-Pierre et al. (1984), Selbekk et al. (1980)
Rat mast cells	Degranulation (histamine)	Yes	NT-1	St-Pierre et al. (1984), Carraway et al. (1982), Morgan-Boyd et al. (1987)
Guinea pig atria	+ inotropic; + chronotropic	No	NT-2	St-Pierre et al. (1984)
Guinea pig ileum	Contraction (acetylcholine)	No[a]	NT-2	Edkins (1905)
Guinea pig colon	Relaxation	–	–	Kitabgi and Vincent (1981)
Guinea pig tenia coli	Contraction	No[a]	NT-2	Kitabgi and Freychet (1978)
Rat stomach fundus	Contraction	No	NT-2	St-Pierre et al. (1984)
Rat duodenum	Relaxation	–	–	
Rat ileum	Relaxation	–	–	
Rat portal vein	Contraction (prostaglandins)	Yes	NT-1	St-Pierre et al. (1984), Rioux et al. (1980)

[a] Rioux (personal communication).

Table 16. Pharmacological parameters used for characterization of neurotensin receptors with agonists and antagonists on isolated tissues

Peptide	Preparation										
	Rat isolated heart				Rat portal vein				Rat stomach fungus		
	$-\log(EC_{45})$	RA	α^E	pA_2	pD_2	RA	α^E	pA_2	pD_2	RA	α^E
Neurotensin (NT)	8.56	100	1.0		7.79	100	1.0		7.96	100	1.0
NT (8–13)	8.33	58	1.0		7.77	95	1.0		7.80	69	1.0
NT (9–13)	6.66	1.26	1.0		6.17	2.39	1.0		7.0	11	1.0
[Ala^9]-NT	5.82	0.2	0.9						6.12	1.5	
[Gly^{10}]-NT		<0.01	0.2						4.42	0.03	
[Ala^{11}]-NT	4.64	0.01	0.9			<0.01	0.4		4.07	0.01	1.0
[Trp^{11}]-NT	8.33	58	1.0		7.85	108	1.0		8.0	109	
[D-Trp^{11}]-NT		<0.01	–	6.8	5.02	0.2	–	6.5	5.79	0.64	1.0
[-Tyr(Me)^{11}]-NT		<0.01	–	6.6	5.27	0.3	–	6.2	5.62	0.5	

pD_2, RA, α^E and pA_2, same as in Table 13.
Data taken from Wolfe and McGuigan (1984); Morgan-Boyd et al. (1987); Christinck et al. (1989).

the type of receptor involved in these inhibitory effects remains to be determined, since none of the known NT antagonists was tested in these tissues. NT contracts the rat isolated portal vein by activating an NT_1 receptor which is inhibited by [D-Trp^{11}]NT. Some of the preparations briefly described in Table 15 have been used to further characterize the two hypothetical receptor sites for NT by the use of agonists and of [D-Trp^{11}]NT. The results are summarized in Table 16. Activities of agonists are expressed in terms of EC_{50} or of pD_2 (the -log of the concentration of agonist producing 50% of the maximum effect) and the maximum effect (α^E) as a fraction of that of NT, the reference compound. NT is the most potent peptide in the three preparations and NT (8–13) maintains more than 50% (in the rat portal vein 95%) of activity while NT (9–13) is much weaker. The biologically active sequence of NT is found at the C-terminal since C-terminal truncated NT fragments are inactive (ROSSIE and MILLER 1982). Changes in positions 9, 10 or 11 have marked effects on activity, as demonstrated by the drastic reduction of activity observed by replacing Arg^9 by Ala, Pro^{10} by Gly or Tyr^{11} by Ala. It appears that the positive charge of Arg^9, the conformational function of Pro^{10} and the aromaticity of Tyr^{11} are essential to receptor occupation and this is true for both NT_1 and NT_2; agonists do not discriminate between NT_1 and NT_2; the only compound that shows important differences between NT_1 and NT_2 is [D-Trp^{11}]NT, which acts as antagonist on the vascular preparations (rat isolated heart and portal vein) while it is a weak (although full) agonist on the intestinal tissues (rat stomach and other tissues analysed in Table 15).

III. Mechanism of Action of Neurotensin and Second Messengers

As mentioned above, the major role of neurotensin appears to be the coordination of the postprandial gastrointestinal activity which involves local vasodilatation, inhibition of gastrointestinal smooth muscles, and stimulation of gastrointestinal and pancreatic secretions (ROSSIE and MILLER 1982). Some of these effects are due, at least in part, to direct or indirect activation of smooth muscles in the intestinal arterial vessels and the gut wall. The most important phenomenon appears to be the smooth muscle inhibition which has been demonstrated in vivo on field-stimulated circular canine smooth muscle (GOEDERT et al. 1984), and in vitro on the phasic contractions of gastric, duodenal and intestinal preparations in the dog. The direct inhibitory effects of NT on the muscle appear to be due to an initial hyperpolarization that follows the opening of K^+ channels as it is blocked by apamin (CHRISTINCK et al. 1989). In some parts of the intestinal tract of the guinea pig (e.g., longitudinal muscle of the ileum), neurotensin exerts TTX-insensitive relaxation (EDKINS 1905). In other tissues, e.g., the rat portal vein, the effect is indirect and appears to be mediated entirely by prostanoids (RIOUX et al. 1980). However, the mechanism of action of the direct myotropic effect and that subserving the prostanoid release may be similar and may involve calcium transport through the plasma membrane or release of calcium from intracellular stores (POWELL 1981). From the above, it is evident that the two major mechanisms of action of neurotensin involve changes of ion transport through the plasma membranes and Ca^{2+} transport. As to the peripheral vasodilatation produced by a reduction of arterial smooth muscle tone, this may be brought about by the release of EDRF or of PGI_2, since it is prevented e.g., in the dog isolated carotid artery (D'ORLÉANS-JUSTE et al. 1985)] by the removal of the endothelium. Stimulation of EDRF or PGI_2 release by NT is assumed to be a Ca^{2+}-dependent phenomenon similar to those of other peptides [e.g. bradykinin (CHRISTINCK et al. 1989)].

E. Cholecystokinin

Early studies on factors promoting gastroduodenal motility and exocrine secretions suggested the existence of gastrin (IVY and OLDBERG 1928), cholecystokinin (HARPER and RAPER 1943) and pancreozymin (JORPES and MUTT 1968). Later, it was found that cholecystokinin and pancreozymin are the same hormone (ANASTASI et al. 1967), which was identified as a polypeptide of 33 residues. One year previously, Anastasi et al. (see ENG et al. 1983) had isolated a decapeptide (caerulein) whose pharmacological activity was the same as that of cholecystokinin. It was soon realized that the large peptide sequence (33 residues) was not required for biological activity, since shorter sequences down to the C-terminal tetrapeptide (for gastrin) still exerted strong biological activity. This led to the identification of the C-terminal octapeptide of CCK (CCK-8) in pig (EYSSELEIN et al. 1984), dog (MILLER et al. 1984; DOCKRAY et al. 1989) and human (WOLFE and MCGUIGAN 1984) brain. The primary structure of CCK-8 is compared with that of caerulein and that of the C-terminal gastrin in Table 17.

The three peptides have the same C-terminal pentapeptide sequence, but gastrin has the Tyr-sulfate residue in position 3 and this residue appears to be irrelevant for gastric secretion (MCDONALD 1991; TAKAHASHI et al.

Table 17. Primary structure of C-terminal gastrin and CCK octapeptides and of caerulein

	1 2 3 4 5 6 7 8
	SO_3H
$Gastrin_{10-17}$	-Gln-Ala-Tyr-Gly-Trp-Met-Asp-Phe[a]
	SO_3H
CCK_{26-33} (CCK-8)	-Asp-Tyr-Met-Gly-Trp-Met-Asp-Phe[a]
	SO_3H
Caerulein	(Pyr-Gln)-Asp-Tyr-Thr-Gly-Trp-Met-Asp-Phe[a]

[a] C-terminal amide.

1985), while it is essential (in position 7) for gall bladder motility and pancreatic secretion, exerted by CCK and caerulein. CCK-8 and caerulein-8 differ only by the residue in position 3 (DOCKRAY et al. 1989).

The application of molecular biology techniques has enabled the identification of the CCK precursor (115 amino acid residues in human; RENFELD 1986), which undergoes Tyr-sulfation, cleavage at pairs of basic residues and the amidation reaction (MUTT 1988; CHANG and LOTTI 1986).

I. Biological Activity of Cholecystokinin In Vivo

Cholecystokinin and CCK-8 have been found in the gut and the brain (RENFELD 1986; MUTT 1988; CHANG and LOTTI 1986), where they exert a variety of biological activity (Table 18). In the gut, these peptides are found in endocrine cells and in the myenteric plexus, from which they are secreted into the blood after a meal to inhibit stomach and duodenum motility, to inhibit Oddi's sphincter at the same time as they stimulate contractions of the gall bladder and of the distal tract of the intestine. Gastrin stimulates the secretions of the stomach and CCK promotes the production of bile and the secretion of the exocrine pancreas (PENDLETON et al. 1987; MORLEY 1980). Through its gastrointestinal effects and possibly also by activating hypothalamic receptors (RENFELD et al. 1985), CCK induces satiety and takes part in the control of feeding behaviour (RENFELD et al. 1985). The majority of gastrin and CCK-like immunoreactivity in the brain is contributed by CCK-8, either the sulfated (hypocampus, nucleus accumbens, cerebral cortex) or the non-sulfated form (other brain regions) (HÖKFELT et al. 1988; CRAWLEY 1991) which, in some neurons, coexist with dopamine, 5-HT or neuropeptides (LOTTI et al. 1987; PENDLETON et al. 1987). The presence of CCK in various brain nuclei and in dopaminergic neurons led to the hypothesis that CCK may act as modulator of the dopaminergic system and either potentiate or inhibit this system (DE MONTIGNY 1989; MCROBERTS 1986) – CCK has therefore been involved in the pathogenesis of schizophrenia, which is thought to be linked to hyperactivity of the midbrain dopamine system or in drug addition, since cocaine, amphetamine and morphine sensitize dopaminergic pathways. Most recently, it has been suggested that CCK may be involved in anxiety, after the observation (DOURISH et al. 1988) that the CCK tetrapeptide (30–33) induces anxiety in healthy volunteers.

CCK has also been involved in pain control (YU et al. 1990) after the observation that some CCK antagonists (e.g. L-364-718, proglumide) potentiate the analgesic effects of opioids (WATKINS et al. 1985; INNIS and SNYDER 1980).

Peripheral and central CCK activity is due to the activation of two different receptor types, mainly CCK-A in the gastrointestinal area and both CCK-A and CCK-B in the brain. The two receptor types have been characterized with biochemical and pharmacological assays (WALSH 1987;

Table 18. Biological activity of cholecystokinin in vivo

System	Pharmacological effect	Receptor	Species	Reference
Gastrointestinal	Inhibition of gastric motility	A	Rat, dog, man	Innis and Snyder (1980), Walsh (1987)
	Inhibition of duodenal motility	A	Rat, dog, man	Innis and Snyder (1980), Walsh (1987)
	Gall bladder contraction	A	Cat, dog, guinea pig	Innis and Snyder (1980), Blundell (1991)
	Inhibition of Oddi's sphincter	?	Cat, dog, guinea pig	Innis and Snyder (1980), Blundell (1991)
	Contraction of distal intestine	A	Rabbit	Innis and Snyder (1980), Blundell (1991)
	Stimulation of pancreatic secretion	A	Cat, dog	Innis and Snyder (1980), Altar and Boyar (1989)
	Stimulation of gastric secretion	B	Mouse, dog	Innis and Snyder (1980), Marshall et al. (1991)
Central nervous	Satiety, feeding behaviour	?	Various species	Andersson et al. (1972)
	Dopamine potentiation	A	Rat	McRoberts (1986), Lord et al. (1977)
	Dopamine inhibition	B	Rat	McRoberts (1986), Lord et al. (1977)
	Nociception (opioid antagonism)	A?	Rat	Mualleni et al. (1989), Rogers et al. (1988), Martin et al. (1976)
	Anxiety	B	Rat	Lord et al. (1977)

ID_{50} of L-364.718 i.v. 25.4 mg/kg p.o. versus 0.04 mg/kg p.o. for the CCK-A-mediated functions.

BLUNDELL 1991) and by the use of agonists and antagonists (LORD et al. 1977). The essential pharmacological features of the two receptor types can be summarized below, together with the features of the receptor for gastrin, according to ALLESCHER and AHMAD (1991).

	Agonist order of potency	Antagonist
CCK-A	CCK-8 >>> des (SO_3) CCK-8 > gastrin	L-364-718
CCK-B	CCK-8 > gastrin = des (SO_3) CCK-8	L-365-260
Gastrin R	CCK-8 = gastrin, des (SO_3) CCK-8	L-365-260

Agonist potency differs between CCK-A and CCK-B by the fact that CCK-8 is much more active on CCK-A than on CCK-B, compared to desulfated CCK-8 and gastrin. There is, however, no difference between CCK-B and the gastrin receptor. As to the antagonists, L-364-718 is a potent and selective blocker of CCK-A, inactive on CCK-B, while L-365-260 blocks both CCK-B and the gastrin receptor, with similar affinities (PANDOL et al. 1985). Therefore, CCK-B and gastrin receptors cannot be differentiated pharmacologically and should be considered the same entity (LORD et al. 1977), despite the claims by YU et al. (of CHANG and LOTTI 1991) that CCK binds to three different sites in guinea pig pancreatic acini. Binding studies with agonists ([^{125}I.BH]-CCK-8) usually lead to the demonstration of high- and low-affinity sites, which are, however, not found with labelled antagonists (PANDOL et al. 1985; LORD et al. 1977) and should therefore be considered allosteric states of the same receptor molecule rather than different receptor types.

II. Activity of Cholecystokinin In Vitro

Receptors for CCK, especially the CCK-A type, are found in a variety of smooth muscles, mainly in the gastrointestinal system, where they subserve contractile and possibly inhibitory effects. Early observations by ERSPAMER and his coworkers (see Table 3) indicated that caerulein is a potent stimulant of the rabbit colon and cat ileum, the guinea pig ileum, rat duodenum and human intestine, although much weaker than bombesin in the human preparation. A variety of other organs have been shown (Table 19) to be sensitive to CCK-related peptides. In general, gastric smooth muscle and gall bladder of various animals (cat, pig, dog, rat) and of man are very sensitive to cholecystokinin and CCK-8, the most commonly used agonist in in vitro studies and appear to contain CCK-A receptors, except perhaps for the human gastric muscle, which is sensitive to pentagastrin (Table 19). The effects of CCK in these preparations appear to be due primarily to activation of receptors in the smooth muscles (direct effect) while the contractile responses of the duodenum, the intestine and the pyloric muscle of the dog are mainly indirect and appear to be mediated by acetylcholine. Relaxing effects of CCK-A are observed in various sphincters,

Table 19. Effect of cholecystokinin in various gastrointestinal smooth muscle preparations

Species	Organ	Agonist	Biological effect	Receptor type	Reference
Cat	Gastric muscle	CCK-8	Contraction	CCK-A	Rattan and Gayal (1983), Scheurer et al. (1983)
	Gall bladder	CCK-8 > gastrin	Contraction	CCK-A	Behar and Biancani (1977)
	Lower oesophageal sphincter	CCK-8	Relaxation	–	Behar and Biancani (1980) Yau et al. (1973)
	Sphincter of Oddi	CCK = CCK-8	Relaxation	CCK-B	Rattan and Gayal (1983)
Guinea pig	Gall bladder	CCK-8 > gastrin	Contraction	CCK-A	Gridger and Makhlouf (1987)
	Ileum muscle	CCK-8	Contraction (Ach)	CCK-A	Vizi et al. (1972), Kimura et al. (1983) Yan et al. (1983), Kimura et al. (1989)
	Duodenal muscle	CCK-8	Relaxation (Ach)	–	Vizi et al. (1972), Gerner and Haffner (1977)
	Gastric smooth muscle	CCK-8	Contraction	CCK-A	Morgan and Szurszewski (1980)
	Colon	CCK-8	Contraction	CCK-A	Heimbrook et al. (1989)
Dog	Gastric smooth muscle	CCK	Contraction	–	Valenzuela (1976), Pozo et al. (1989)
	Gall bladder	CCK-8	Contraction	CCK-A	Rakovska et al. (1986)
	Pyloric muscle	CCK-8	Contraction (Ach)	–	Cameron et al. (1967)
Human	Gastric muscle	Pentagastrin	Contraction	CCK-B	Yau et al. (1974), Checler (1991)
	Gall bladder muscle	CCK	Contraction	CCK-A	Resin et al. (1973)
	Lower oesophageal	CCK-8	Contraction	–	Li and Chang (1976)
Rat	Pylorus	CCK-8	Contraction	–	Ohkawa and Watanabe (1977)
	Gastric muscle	CCK-8	Contraction	–	Isenberg and Csendes (1972)

(Ach), effect mediated by endogenous acetylcholine.

the lower oesophageal sphincter, Oddi's sphincter and the duodenum. The great majority of the myotropic effects of CCK in various sections of the gastrointestinal tract are mediated by CCK-A receptors, with the exception of the human stomach. The type of receptor involved in the relaxing effects of CCK-8, particularly on Oddi's sphincter, is CCK-B. The availability of potent and selective CCK antagonists (non-peptidic), such as L-364-718 (3*S*(−)-*N*-(2,3-dihydro-1-methyl-2-oxo-5-phenyl-1*H*-1, 4-benzodiazepine-3-yl)-1*H*-inodole-2-carboxamide) and L-365-260 [(3*R*-(+)-2,3-dihydro-1-methyl-2-oxo-5-phenyl-1*H*-1,4-benzodiazepine-3-yl)-*N*′-(3-methylphenyl) urea], which discriminates between CCK-A (L-364-718) and CCK-B (L-365-260) receptors by at least two log units of difference in affinities, has made it possible to establish that CCK-A and CCK-B receptors are present both in the brain and in different peripheral tissues (LOTTI et al. 1987; WALSH 1987; PANDOL et al. 1985). Further studies are needed to determine, by the use of antagonists, the type of receptors subserving the various pharmacological effects of cholecystokinin and gastrin.

III. Mechanism of Action of CCK and Second Messengers

Similar to neurokinins and bombesin-related peptides, the cholecystokinins belong to the category of Ca^{2+}-mobilizing mediators (MUALLENI et al. 1989; ROGERS et al. 1988). Activation of receptors by CCK has been extensively studied in dispersed pancreatic acinar cells by measuring intracellular levels of inositol 1,4,5-triphosphate and cGMP (PANDOL et al. 1985). It has been found that CCK increases cytoplasmic Ca^{2+} exclusively by mobilizing Ca^{2+} from the intracellular stores, since amylase secretion occurs in the absence of extracellular Ca^{2+}. Amylase release persisted after Ca^{2+} returned to basal levels and the CCK effect could be separated into an acute initial phase that is sustained by IP_3 and the late phase mediated by diacylglycerol-activated protein kinase C (MUALLENI et al. 1989; ROGERS et al. 1988). In parallel with the increase of cytoplasmic C^{2+}, CCK increases the intracellular levels of cGMP, but the role of this mucleotide remains unknown (BITAR et al. 1986).

A similar Ca^{2+}-dependent mechanism of signal transduction appears to be involved in the mobilization of intracellular Ca^{2+} that mediates the contractile response of intestinal smooth muscle (Tables 4, 19) to CCK (CROCHELT and PEIKIN 1986). The contractile response occurs in the absence of extracellular Ca^{2+} and in the presence of Ca^{2+} channel inhibitors. In some preparations, however (e.g., bovine and guinea pig gall bladder) (ANDERSSON et al. 1972), the contractile response is sensitive to inhibitors of Ca^{2+} channels. Activation of smooth muscle receptors by CCK results in changes of intracellular levels of cAMP, generally a decrease that may be due to activation of phosphodiesterases. This mechanism is expected to facilitate contraction and has been demonstrated in guinea pig gall bladder (LISTON et al. 1983) and in rabbit tissues. According to KOSTKA (1991), "it is likely that the CCK-induced modulation of cAMP level is a secondary transducing event which occurs subsequently to the generation of phospholipase C-derived second messengers".

F. Opioids

In the mid-1970s the first endogenous opioid peptides, enkephalins and endorphins were discovered (Hughes et al. 1975; Mudge et al. 1979). Since then, more than 20 different endogenous opioids have been isolated. All these peptides derive from three distinct percursor molecules and hence they have been classified into three distinct families: the proopiomelanocortin (POMC) family, the common percursor of β-endorphin among others, the proenkephalin A family, the common precursor of Met and Leu enkephalins, and the proenkephalin B family, the precursor of Leu-enkephalin, the dinorphins and neoendorphins.

The major source of POMC is the pituitary (Allescher and Ahmad 1991). Proenkephalin A is processed mostly in the adrenal medulla (Martin 1967) but it is also found in the brain (bovine caudate nucleus) (Watson et al. 1981) and the digestive tract (Allescher and ahmad 1991). Proenkephalins B are most abundant in the neural lobe of the rat pituitary as they are colocalized with vasopressin in neurons in the magnocellular nuclei of the hypothalamus (Martin et al. 1976). Endogenous opioids share the N-terminal sequence of enkephalins (R-Met-enkephalin or R-Leu-enkephalin). The primary structures of some of the opioid prototypes are illustrated below:

Opioid peptides	Primary structure
β-Endorphin (bovine)	Tyr-Gly-Gly-Phe-Met-Thr-Ser-Glu-Lys-Ser-Gln-Thr-Pro-Leu-Val. Thr-Leu-Phe-Lys-Asn-Ala-Ile-Ile-Lys-Asn-Ala-Lys-Lys-Gly-Gln
Met-enkephalin	Tyr-Gly-Gly-Phe-Met
Leu-enkephalin	Tyr-Gly-Gly-Phe-Leu
Dynorphin A	Tyr-Gly-Gly-Phe-Leu-Arg-Arg-Ile-Arg-Pro-Lys-Leu-Lys-Trp-Asp-Asn-Gln
Dynorphin B	Tyr-Gly-Gly-Phe-Leu-Arg-Arg-Gln-Phe-Lys-Val-Val-Tyr
α-Neo-endorphin (porcine)	Tyr-Gly-Gly-Phe-Leu-Arg-Lys-Tyr-Pro-Lys
β-Neo-endorphin	Tyr-Gly-Gly-Phe-Leu-Arg-Lys-Tyr-Pro

Table 20. Effects of opiates in vivo and on isolated smooth muscles

	Agonist	Receptor	Effect	Antagonist	Reference
In vivo					
Gastrointestinal tract		μ	Constipation	Naloxone	McGilliard and Takemori (1978)
Pain	Morphine	μ	Supraspinal analgesia	Naloxone	McClane and Martin (1967)
		κ	Spinal analgesia	Naloxone	Kadani et al. (1982)
Respiratory centres		μ ·	Depressor	Naloxone	McQueen (1983)
Cardiovascular system	Morphine	μ	Decrease of peripheral resistance and heart rate	Naloxone	Clark (1981)
Temperature regulation			Hypo, hyperthermia	Naloxone	Jessel and Iversen (1977)
Pupillary muscles	Morphine	κ	Miosis (human, dog)	Naloxone	Milligan et al. (1987)
Cough centre	Dextromethorphan		Suppressant	Naloxone	Campbell et al. (1961)
Uterus	Morphine		Antioxytocic	Naloxone	Lemaire et al. (1978)
In vitro					
Rat vas deferens	Endorphins	ε	Inh. contraction	Naloxone	Woordruff and Hughes (1991)

Rabbit vas deferens	Endorphins	κ	Inh. contraction	TENA	ALLESCHER and AHMAD (1991), LORD et al. (1977)
Mouse vas deferens	Leu-enk	δ	Inh. contraction	Morphiceptin	ALLESCHER and AHMAD (1991), SCHANG et al. (1986)
Human colon	Morphine		Inh. chol. nerv. stim.	Naloxone	PATON (1957)
Guinea pig ileum	Met-enk	μ	Inh. chol. nerv. stim.	CTP	ALLESCHER and AHMAD (1991), BAUER and SZURSWESKI (1989)
Dog duodenum	Dynorphin	δ	Inh. nerve transm.	Naloxone	GRUNDY (1971)
Rabbit mesenteric vein	Morphine		Inh. spontaneous act.	Naloxone	YAMAMOTO et al. (1984)
Rat mesenteric vein	Met-enkephalin		Stim. spontaneous act.	Naloxone	HANKO and HARDBEO (1978)
Rabbit ear artery	Met, Leu-enk	δ,κ	Inh. nerve stim.	Naloxone MR 226	FUKUDA et al. (1985) ALLESCHER and AHMAD (1991), SCHANG et al. (1986)
Cat pial artery	Leu-enk		Vasodilatation	Naloxone	BORISON (1971)

TENA, $6\beta,6'\beta$(ethylene-bis[oxyethyleneimino])bis(17-[cyclopropylmethyl]-4,SQ-epoxymorphinan-3,14-diol); CTP, cyclic-D-Pen-Cys-Tyr-D-Trp-Lys-Thr-Pen-Thr-NH_2; MR 226, ([−]-2-(3-furylmethyl)-5,9-diethyl-2′-hydroxy-6,7-benzomorphan; Inh, inhibits chol. nerv. stim., cholinergic nerve stimulation; chol. nerv. transm., cholinergic nerve transmission; act., activity.

I. Biological Activity of Opioids In Vivo

As illustrated in the first part of Table 20, the administration of opioids in vivo in man and animals induces a variety of effects. Opioids delay gastrointestinal transit and hence act as antidiarrheal drugs. Opioids have been demonstrated to inhibit the release of various neurotransmitters including substance P, serotonin, VIP and opioids themselves, and reduce neuronal functions, for instance in the electrically contracted guinea pig ileum (YAKSH et al. 1980; CHENG and PRUSOFF 1973). Endogenous opioids, such as endorphins (MARTIN 1984), and opiates, such as morphine (GEN and VALDMAN 1967), have been extensively studied for their analgesic properties, both at the spinal and supraspinal levels. The most documented property of opioids is their capacity to relieve clinical pain and increase the threshold of experimentally induced pain (GILLAN et al. 1981). Analgesia induced by opioids is due to the activation of receptors in several sites within the central nervous system (CHOWDHURY et al. 1975) and involves inhibition of the release of various neurotransmitters. A classical example of such a modulation is the opioid-induced inhibition of substance P release in areas involved in the modulation of pain (SALEM and AVIASTO 1970; SMITH et al. 1968) in the spinal cord.

Opioids depress respiratory centres in animals by reducing the response of these centres to CO_2 and selectively depress neuronal modulation of respiration (MCQUEEN 1983). Morphine has also been reported to induce a significant decrease of peripheral vascular resistance and heart rate by increasing vagal and decreasing sympathetic tone in various species (FRAZER et al. 1954).

Morphine and endogenous opioids also exert complex effects on temperature regulation (hypo- and hyperthermia) and act as cough suppressants (dextrometorphan). The "pinpoint" (pupils) reported in all morphine users illustrates well the myotic effect of opioids in humans and dogs among other species (LEE and WANG 1975; MILLIGAN et al. 1987). Intravenous administration of morphine has also been reported to induce antioxytoxic effects in pregnant women (LEMAIRE et al. 1978).

The in vivo effects of opioids are extremely complex as these peptides may act on different receptor sites, which are now well characterized.

II. Effects of Opioids In Vitro

The effects of various endogenous opioids and of morphine on in vitro isolated preparations are analysed in Table 20 (bottom section). The use of in vitro preparations, as well as of binding assays, led to the identification and characterization of four receptor types, namely μ, δ, κ and ε.

Among the most useful preparations for receptor characterization and classification is the vas deferens of various species (rat, rabbit, mouse). Pharmacological assays in these tissues, as well as the binding assays in brain

Table 21. Pharmacological characterization of opioid receptors

Receptor type	μ	δ	κ	ε
Typical isolated preparations	Guinea pig heum	Mouse vas deferens	Rabbit vas deferens	Rat vas deferens
Order of potency of agonists	Morphine β-Endorphin Met-enkephalin	Leu-enk = Met-enk β-Endorphin Dynorphin	Dynorphin Leu-enkephalin β-Endorphin	β-Endorphin Leu-enkephalin Dynorphin
Selective agonist	Syndiphalin	DPDPE	EKC	–
Antagonist (dissociation constant k_I)[a]	Naloxone (1.4)	Diprenorphin (0.9)	MR 2266	Bremazocine
Selective antagonist	CTP	ICI 174,864	TENA	–

Met-enk, methionine^5-enkephalin; Leu-enk, leucine^5-enkephalin; DPDPE, *d*-penicillamine^{2,5}-enkephalin; EKC, ethylketocycloazocine; ICI 174,864, (*N*,*N*-dially-Tyr-Aib-Phe-Leu-OH); CTP, TENA and MR 2266, refer to legend of Table 18.
[a] Taken from CHENG and PRUSOFF (1973).

homogenates, have contributed to the assessment of the pharmacological properties of opioids and further development of selective agonists and antagonists for the various receptor types. In general, endogenous opioids exert inhibitory effects in isolated smooth muscle preparations. For instance, endorphins, enkephalins and morphine are potent inhibitors of the response of the rat, rabbit and mouse vas deferens to electrical stimulation. Opioids act via the activation of different receptor sites, namely ε (rat vas deferens), κ (rabbit vas deferens), δ (mouse vas deferens); selective antagonists have been recently developed to further characterize the latter two tissues (Table 21). In general, the pharmacological effects of opioids in vitro are mainly indirect (inhibition of intrinsic and autonomic nervous systems activities) since opioids modulate the release of neurotransmitters, such as noradrenaline (vasa deferentia) and acetylcholine (human colon, guinea pig ileum and dog duodenum). Opioids have also marked vasodilatory or inhibitory effects in isolated vascular preparations with the exception of the rat portal vein (HANKO and HARDEBEO 1978), where Met-enkephalin has been found to be a potent stimulant of the spontaneous activity. Here again, the effects of opioids are indirect, as they interfere with vascular sympathetic innervation in these isolated vessels, with the exception of the feline pial artery, where Leu-enkephalin has been reported to directly induce vasodilatation by relaxing the smooth muscle fibres (BORISON 1971). This exception indicates that postsynaptic opioid receptors in the smooth muscle of some vessels may be found.

III. Characterization of Opioid Receptors by Endogenous Agonists, Antagonists and Selective Analogues

A pharmacological characterization of opioid receptors is presented in Table 21. Recent studies suggest that the biological activity of opioids depends on the activation of three receptor types, μ, δ and κ, and that ε receptor type could be a mixture of μ and δ (FOX and DANIEL 1986). Basic pharmacological studies performed in three preparations that contain a single opioid receptor type, namely the guinea pig ileum (μ), the mouse vas deferens (δ) and the rabbit vas deferens (κ), have led to the establishment of different orders of potency for opioid agonists in the three sites (see Table 21) and the identification of selective agonists for each of the three sites. For example, syndiphalin has been shown to be selective for μ-receptors (MOSBERG et al. 1983), [D-Pen^{2,5}]-enkephalin is the selective agonist for δ-receptors (GILBERT and MARTIN 1976), whereas ethylketazocine-like compounds have been shown to be selective for κ-receptors in in vitro assays (PELTON et al. 1985).

Work with antagonists has confirmed the existence of three receptor sites, the first of which (μ) is sensitive to naloxone, the second to diprenorphin (AHMAD et al. 1991), and the third to a recently identified κ-selective compound, MR2266 (ALLESCHER and AHMAD 1991). Even more selective antagonists have been reported for μ (CTP) (COTTON et al. 1984), δ

(ICI174864) (Portoghese and Takemori 1985) and κ (TENA) (Gintzier and Scalisi 1982) receptors (Table 21).

IV. Mechanism of Action of Opioids on Smooth Muscle and Second Messengers Involved

Although the mechanism of signal transduction involved in the response of neuronal tissues has been well documented for opioids, the characterization of the intracellular messengers involved in the response of the smooth muscle, following the activation of opioid receptors, is poorly documented. In the central and periphral nervous systems, the activation of opioid receptors is coupled to adenylate cyclase and to ionic channels (Powell 1981). According to Kostka (1991), the relative contribution of these two second-messenger systems to the signal transmission depends on the receptor subtype. Overall, the opioids induce a negative modulation of adenylate cyclase in neuronal tissues and in cell lines rich in δ-receptors and coupling mechanisms involve the intermediate G_i protein (Powell 1981). The activation of μ-receptors also induced an inhibition of adenylate cyclase yet with a much lower potency than when δ-receptors are activated (Laws et al. 1983; North and Williams 1985).

Opioids have been reported to induce a hypopolarization of plasma membrane (activation of μ- and δ-receptors), inducing an activation of potassium channels in the central neurons and the enteric ganglion (North et al. 1987; Cherubini and North 1985). Opioid receptor activation led to inhibition of firing of the myenteric plexus of the guinea pig ileum by attenuating the inward calcium currents (Kostka 1991).

References

Ahmad S, Allescher HD, Kwan CY (1991) Receptors for neuropeptides: ligand-binding studies. In: Daniel EE (ed) Neuropeptide function in gastrointestinal tract. CRC, Boca Raton, pp 209–229

Allescher HD, Ahmad S (1991) Postulated physiological and pathophysiological roles in motility. In: Daniel EE (ed) Neuropeptide function in the gastrointestinal tract. CRC, Boca Raton, pp 309–400

Allescher HD, Kostolanska F, Tougas G, Fox JET, Regoli D, Drapeau D, Daniel EE (1989) The action of neurokinin and substance P receptors in canine pylorus, antrum and duodenum. Peptides 10:671–679

Altar CA, Boyar WC (1989) Brain CCK-B receptors mediate the suppression of dopamine release by cholecystokin. Brain Res 483:321–326

Anastasi A, Erspamer V, Endean R (1967) Isolation and structure of caerulein, an active decapeptide from the skin of Hyla coerulea. Experientia 23:699–703

Andersson KE, Andersson R, Hedner R (1972) Cholecystokinetic effect and concentration of cyclic AMP in gall bladder in vitro. Acta Physiol Scand 85:511–516

Aoki K, Kajiwara M, Oka T (1986) The inactivation of [Met^5]-enkephalin by bestatin-sensitive aminopeptidase, captopril-sensitive peptidyl dipeptidase A and thiorphan-sensitive endopeptidase-24.11 in mouse vas deferens. Jpn J Pharmacol 40:297–302

Araki K, Tachibana S, Uchiyama M, Nakajima T, Yasuhara T (1973) Isolation and structure of a new active peptide "Xenopsin" on the smooth muscle especially on a strip of fundus from a rat stomach, from the skin of Xenopus laevis. Chem Pharm Bull (Tokyo) 21:2801–2804
Armstrong MJ, Parker MC, Ferris CF, Leeman SE (1986) Neurotensin stimulates (^{3}H) oleic acid translocation across rat small intestine. Am J Physiol 251: G823–G829
Atkins HL, Oster ZH (1989) Asymetric gall bladder contraction following cholecystokinin hepatobiliary imaging. Chem Nucl Med 14:82–86
Barabé J, Park WK, Regoli D (1975) Application of drug-receptor theories to the analysis of the myotropic effects of bradykinin. Can J Physiol Pharmacol 53: 345–353
Bauer AJ, Szursweski JH (1989) Dynorphin presynaptically inhibits neuro muscular transmission via delta opioid receptors in circular muscle of canine duodenum (Abstr). Gastroenterology 96:A33
Behar J, Biancani P (1977) Effect of cholecystokinin octapeptide on lower esophageal sphincter. Gastroenterology 73:57–61
Behar J, Biancani P (1980) Effect of cholecystokinin and the octapeptide cholecystokinin on the feline sphincter of Oddi and gall bladder; Mechanism of action. J Clin Invest 66:1231–1239
Bennett A, Misiewicz JJ, Waller SL (1967) Analysis of the motor effect of gastrin and pentagastrin on the human alimentary tract in vitro. Gut 8:470–474
Bertaccini G (1976) Active polypeptides of non-mammalian origin. Pharmacol Rev 28:127–177
Bertaccini G, Erspamer V, Impicciatore H (1973) The actions of bombesin on gastric secretion of the dog and rat. Br J Pharmacol 49:437–444
Bertaccini G, Erspamer V, Melchiorri P, Sopranzi N (1974) Gastrin release by bombesin in the dog. Br J Pharmacol 52:219–225
Bitar KN, Burgess GM, Putney JW Jr, Makhlouf GM (1986) Source of activator calcium in isolated guinea pig and human gastric smooth muscle cells. Am J Physiol 250:G280
Blundell J (1991) Pharmacological approaches to appetite suppression. Trends Pharmacol Sci 12:147–157
Borison HL (1971) Sites of action of narcotic analgesic drugs: the nervous system. In: Clouet DE (ed) Narcotic drugs: biochemical pharmacology. Plenum, New York
Borson DB (1991) Roles of neutral endopeptidases in airways. Am J Physiol 260: L212–L225
Brizzolara A, Burnstock G (1991) Endothelium dependent and endothelium independent vasodilatation of the hepatic artery of the rabbit. Br J Pharmacol 103:1206–1212
Brown MR, River J, Vale WW (1977) Bombesin: potent effect on thermoregulation in the rat. Science 196:988–990
Brown MR, Taché Y, Fisher D (1979) Central nervous system action of bombesin: mechanism to induce hyperglycemia. Endocrinology 105:660–665
Bunnet NW, Reeve JR, Walsh JH (1983) Catabolism of bombesin in the interstitial fluid of the rat stomach. Neuropeptides 4:55–64
Bunnet NW, Kobayashi R, Orloff MS, Reeve JR, Turner AJ, Walsh JH (1985) Catabolism of gastrin-releasing peptide and substance P by gastric membrane-bound peptidases. Peptides 6:277–283
Burnstock G (1970) Structure of smooth muscle and its innervation. In: Bulbring E, Brading A, Jones A, Tomita T (eds) Smooth muscle. Arnold, London, pp 1–69
Busse R, Mulsch A (1990) Calcium-dependent nitric oxide synthesis in endothelial cytosol is mediated by calmodulin. FEBS Lett 265:133–134
Cameron AJ, Phillips SF, Summerskill WHJ (1967) Effect of cholecystokinin on motility of human stomach and gall bladder muscle in vitro. Clin Res 15:416

Campbell C, Phillips DC, Frazier TM (1961) Analgesia during labor: a comparison of pentobarbital, meperidine and morphine. Obstet Gynecol 17:714–718
Cantor P, Holst JJ, Knuhtsen S, Rehfeld JF (1987) Effect of neuroactive agents in cholecystokinin release from the isolated perfused porcine duodenum. Acta Physiol Scand 130:627–632
Caprilli R, Melchiorri P, Improta G, Vernia P, Frieri G (1975) Effects of bombesin and bombesin-like peptides on gastrointestinal myoelectric activity in the dog. Gastroenterology 68:1228–1235
Carraway RE, Leeman SE (1973) The isolation of a new hypotensive peptide, neurotensin, from bovine hypothalami. J Biol Chem 248:6854–6861
Carraway RE, Leeman SE (1975) Structural requirements for the biological activity of neurotensin, a new vasoactive peptide. In: Walter R, Meienhofer J (eds) Peptide chemistry, structure and biology. Ann Arbor Sci, Ann Arbor, p 679
Carraway RE, Cochrane DE, Lansoman JB, Leeman SE, Paterson BM, Welch HJ (1982) Neurotensin stimulates exocytic histamine secretion from rat mast cells and elevates plasma histamine levels. J Physiol (Lond) 323:404–414
Chang MM, Leeman SE (1970) Isolation of a sialogogic peptide from bovine hypothalamic tissue and its characterization as substance P. J Biol Chem 245:4784–4790
Chang RS, Lotti VJ (1986) Biochemical and pharmacological characterization of a new extremely potent and selective non peptide cholecystokinin antagonist. Proc Natl Acad Sci USA 83:4923–4926
Chang RS, Lotti VJ (1988) A review of the biological and pharmacological characterization of a highly peripherally selective CCK antagonists L-364;718. In: Wang RY, Schoenfeld R (eds) Cholescytokinin Antagonists. Liss, New York, pp 13–28
Chang RS, Lotti VJ (1991) Ligands for cholecystokinin A and choecystokinin B/gastrin receptors. In: Conn PM (ed) Methods in neurosciences. Academic, Orlando, 5:479–493
Checler F (1991) Peptidases and neuropeptide-inactivating mechanisms in the circulation and gastrointestinal tract. In: Daniel EE (ed) Neuropeptide function in the gastrointestinal tract. CRC, Boca Raton, pp 273–307
Checler F, Kostolanska B, Fox JET (1988) In vivo inactivation of neurotensin in dog ileum: major involvement of endopeptidase 24-11. J Pharmacol Exp Ther 244:1040–1044
Cheng YC, Prusoff WH (1973) Relationship between the inhibition constant (Ki) and the concentration of inhibitor which causes 50 percent inhibition (I_{50}) of an enzymatic reaction. Biochem Pharmacol 22:3099–3108
Cherubini E, North RA (1985) Mu and kappa opioids inhibit transmitter release by different mechanisms. Proc Natl Acad Sci USA 82:1860–1863
Chowdhury JR, Berkowitz JM, Praissman M, Fara JW (1975) Interaction between octapeptide cholecystokinin, gastrin and secretin on cat gall bladder in vitro. Am J Physiol 229:1311–1315
Christinck F, Daniel EE, Fox JET (1989) Electrophysiological responses of canine ileal circular muscle to electrical stimulation and neurotensin (NT) (Abstr). Gastroenterology 96:A680
Clark WG (1981) Effects of opioid peptides on thermoregulation. Fed Proc 40: 2754–2759
Cohen ML, Geary LE, Wiley KS (1983) Enkephalin degradation in the guinea pig ileum: effect of aminopeptidase inhibitors, puromycin and bestatin. J Pharmacol Exp Ther 224:379–385
Constantine JW, Lebel WS, Woody HA (1990) Smooth muscle of rabbit isolated aorta contains the NK-2 tachykinin receptor. Naunyn Schmiedebergs Arch Pharmacol 342:722–724
Cotton R, Giles MG, Miller L, Shaw JS, Timms D (1984) ICI 174864: a highly-selective antagonist for the opioid δ-receptor. Eur J Pharmacol 97:331–332

Couture R, Cuello AC (1984) Plasma protein extravasation induced by mammalian tachykinins in the rat skin: influence of anaesthetic agents and an acetylcholine antagonist. Br J Pharmacol 91:265–273
Couture R, Regoli D (1982) Smooth muscle pharmacology of substance P. Pharmacology 24:1–25
Couture R, Mizrahi J, Regoli D, Devroede G (1981) Peptides and the human colon: an in vitro pharmacological study. Can J Physiol Pharmacol 59:957–964
Couture R, Laneuville O, Guimond C, Drapeau G, Regoli D (1989) Characterization of the peripheral action of neurokinins and neurokinin receptor selective agonists on the rat cardiovascular system. Naunyn Schmiedebergs Arch Pharmacol 340:547–557
Crawley JN (1991) Cholecystokinin-dopamine interactions. Trends Pharmacol Sci 12:232–236
Crochelt RF, Peikin SR (1986) Characterization of Ca channels mediating gall bladder contraction in the guinea pig (Abstr). Gastroenterology 90:1787
Crossman DC, Larkin S, Fuller RW, Davies GJ, Maseri A (1989) Substance P dilates epicardial coronary arteries and increases coronary blood flow in humans. Circulation 80:475–484
D'Orléans-Juste P, Dion S, Mizrahi J, Regoli D (1985) Effects of peptides and non-peptides on isolated arterial smooth muscles: role of endothelium. Eur J Pharmacol 114:9–21
D'Orléans-Juste P, Dion S, Drapeau G, Regoli D (1986) Different receptors are involved in the endothelium-mediated relaxation and the smooth muscle contraction of the rabbit pulmonary artery in response to substance P and related neurokinins. Eur J Pharmacol 125:37–44
De Montigny C (1989) Cholecystokinin tetrapeptide induces panick attacks in healthy volunteers: preliminary findings. Arch Gen Psychiatry 46:511–517
Deblois D, Marceau F (1987) The ability of des-Arg^9-bradykinin to relax rabbit isolated mesenteric arteries is acquired during in vitro incubation. Eur J Pharmacol 142:141–144
Dion S, D'Orléans-Juste P, Drapeau G, Rhaleb NE, Rouissi N, Tousignant C, Regoli D (1987) Characterization of neurokinin receptors in various isolated organs by the use of selective agonists. Life Sci 41:2269–2278
Dion S, Corcos J, Carmel M, Drapeau G, Regoli D (1988) Substance P and neurokinins as stimulants of the human isolated urinary bladder. Neuropeptides 11:83–87
Dobner PR, Barber DL, Villa-Komaroff L, McKiernan C (1987) Cloning and sequence analysis of cDNA from the canine neurotensin/neuromedin N precursor. Proc Natl Acad Sci USA 84:3516–3520
Dockray GJ, Dimaline R, Pauwels S, Varro A (1989) Gastrin and CCK-related peptides. In: Martinez J (ed) Peptide hormones and prohormones. Harwood, Chichester, pp 244–284
Dourish CT, Hawley D, Iversen SD (1988) Enhancement of morphine analgesia and prevention of morphine tolerance in the rat by cholecystokinin antagonist L 364 718. Eur J Pharmacol 147:469–472
Drapeau G, Rouissi N, Nantel F, Rhaleb NE, Tousignant C, Regoli D (1990) Antagonists for the neurokinin NK-3 receptor evaluated in selective receptor systems. Regul Pept 31:125–135
Edkins JS (1905) On the chemical mechanism of gastric secretion. Proc R Soc Lond [B] 76:376
Ekblad E, Hakanson R, Sundler F (1991) Microanatomy and chemical coding of peptide containing neurons in the digestive tract. In: Daniel EE (ed) Neuropeptide function in the gastrointestinal tract. CRC, Boca Raton, pp 131–179
Eng J, Shiina Y, Pan YC, Blacher R, Chang M, Stein S, Yalow R (1983) Pig brain contains cholecystokinin octapeptide and several cholecystokinin desoctapeptides. Proc Natl Acad Sci USA 80:6381–6385

Erspamer V, Melchiorri P (1980) Active polypeptides: from amphibian skin to gastrointestinal tract and brain of mammals. Trends Pharmacol Sci 1:391–395
Erspamer V, Falconieri Erspamer G, Inselvini M (1970) Some pharmacological actions of alytesin and bombesin. J Pharm Pharmacol 22:875–876
Erspamer V, Falconieri-Erspamer G, Inselvini M, Negri L (1972) Occurence of bombesin and alytesin in extracts of the skin of three European discoglossid frogs and pharmacological actions of bombesin on extravascular smooth muscle. Br J Pharmacol 45:333–348
Erspamer V, Improta G, Melchiorri P, Sopranzi N (1974) Evidence of cholecystokinin release by bombesin in dog. Br J Pharmacol 52:227–232
Eysselein VE, Reeve JR Jr, Shively JE, Miller C, Walsh JH (1984) Isolation of a large cholecystokinin precursor from canine brain. Proc Natl Acad Sci USA 81:6565–6568
Falconieri-Erspamer G, Serverini C, Erspamer V, Melchiorri P, Dellefave G, Nakajima J (1988) Parallel bioassay of 27 bombesin like peptides on 9 smooth muscle preparations. Structure-activity relationships and bombesin receptor subtypes. Regul Pept 21:1–11
Fox JET, Daniel EE (1986) Substance P a potent inhibitor of the canine small intestine in vivo. Am J Physiol 250:G21–G27
Frazer HF, Nash TL, Vanhorn GD, Isbell M (1954) Use of miotic effect in evaluating analgesic drugs in man. Arch Int Pharmacodyn Ther 98:443–451
Friedlinger RM (1989) Nonpeptide ligands for peptide receptors. Trends Pharmacol Sci 10:270–274
Fukuda H, Hosoki E, Ishida Y, Moritoki H (1985) Opioids receptor types on adrenergic nerve terminals of rabbit ear artery. Br J Pharmacol 86:539–545
Furchgott RF (1981) The requirement of endothelial cells in the relaxation of arteries by acetylcholine and some other vasodilators. Trends Pharmacol Sci 3:173–175
Furness JB, Costa M (1987) The enteric nervous system. Churchill Livingstone, Edinburgh
Gafford JT, Skidgel RA, Erdos EG, Hersh LBC (1983) Human kidney "enkephalinase", a neutral metalloendopeptidase that cleaves active peptides. Biochemistry 22:3265–3271
Gen MC, Valdman AV (1967) Experimental observations on the pharmacology of the pontine respiration center. Prog Brain Res 20:148–170
Gerner T, Haffner JFW (1977) The role of local cholinergic pathways in the motor responses to cholecystokinin and gastrin in isolated guinea pig fundus and antrum. Scand J Gastroenterol 12:751–757
Gibbs J, Fauser DJ, Rowe EA, Rolls BJ, Rolls ET, Maddison SP (1979) Bombesin suppresses feeding in rats. Nature 282:208–210
Gilbert PE, Martin WR (1976) The effects of morphine and nalorphine-like drugs in the non-dependent, morphine-dependent and cycazocine-dependent chronic spinal dog. J Pharmacol Exp Ther 198:66–82
Gillan MG, Kosterlitz HW, Magnan J (1981) Unexpected antagonism in the rat vas deferens by benzo-morphans which are agonists in other pharmacological tests. Br J Pharmacol 72:13–15
Gintzler AR, Scalisi JA (1982) Effects of opioids on non-cholinergic excitatory responses of the guinea pig isolated ileum: inhibition of the release of enteric substance P. Br J Pharmacol 75:199–205
Girard F, Bachelard H, St-Pierre S, Rioux F (1984) The contractile effect of bombesin, gastrin releasing peptide and various fragments in the rat stomach strip. Eur J Pharmacol 102:489–497
Goedert M, Hunter JC, Ninkovic M (1984) Evidence for neurotensin as a non-adrenergic, non-cholinergic neurotransmitter in guinea pig ileum. Nature 311:59–62
Green J, Nunley C, Anderson NG (1966) High-pressure column chromatography. I. Design of apparatus and separation of bases, nucleosides and nucleotides. Natl Cancer Inst Monogr 21:431–440

Gridger JR, Makhlouf GM (1987) Regional and cellular heterogeneity of CCK receptors mediating muscle contraction in the gut. Gastroenterology 92:175–180
Grundy HF (1971) Cardiovascular effects of morphine, pethidine, dismorphine and nalorphine on the cat and rabbit. Br J Pharmacol 42:159–178
Guard S, Watson SP (1991) Tachykinin receptor types: classification and membrane signaling mechanisms. Neurochem Int 18:149–165
Gulati N, Mathison R, Huggel H, Regoli D, Beny JL (1987) Effects of neurokinins on the isolated pig coronary artery. Eur J Pharmacol 137:149–154
Guo YS, Singh P, Lluis F, Gomez G, Thompson JC (1986) Contractile response of gallbladder and sphincter of Oddi to substance P and related peptides compared to CCK-8 (Abstr). Fed Proc 45:291
Hammer RA, Leeman SE, Carraway R, Williams RH (1980) Isolation of human intestinal neurotensin. J Biol Chem 255:2476–2480
Hanko JM, Hardebeo JE (1978) Enkephalin-induced dilatation of pial arteries probably mediated by opiate receptors. Eur J Pharmacol 51:295–297
Harper AA, Raper HS (1943) Pancreazymin, a stimulant of the secretion of pancreatic enzymes in extracts of the small intestine. J Physiol (Lond) 102:115–125
Heimbrook DC, Saari WS, Balishin NL, Friedman A, Moore KS, Riemen MW, Kiefer DM, Rotberg NS, Wallen JW, Oliff A (1989) Carboxyl-terminal modification of a gastrin derivative generates potent antagonists. J Biol Chem 264: 11258–11262
Helke CJ, Krause JE, Mantyh PW, Couture R, Bannon MJ (1990) Diversity in mammalian tachykinin peptidergic neurons: multiple peptides, receptors and regulatory mechanisms. FASEB J 4:1606–1615
Hökfelt T (1986) Coexistence of neuronal messengers: a new principle in chemical transmission. In: Fuxe K, Pernow B (eds) Progress in brain research, 68, Elsevier, Amsterdam
Hökfelt T, Herrera-Marschitz M, Seroogy K, Ju G, Staines WA, Holets V, Schalling M, Ungerstedt U, Post C, Rehfeld JF (1988) Immunohistochemical studies on cholecystokinin (CCK)-immunoreactive neurons in the rat using sequence specific antisera and with special reference to the candate nucleus and primary sensory neurons. J Chem Neuroanat 1:11–51
Hua X, Lundberg JM, Theodorsson-Norheim E, Brodin E (1984) Comparison of cardiovascular and bronchosonstrictor effects of substance P, substance P, substance K and other tachykinins. Naunyn Schmiedebergs Arch Pharmacol 328: 196–201
Hughes J, Smith TW, Kosterlitz HW, Fathergill LH, Morgan BA, Morris HR (1975) Identification of two related pentapeptides from the brain with potent opioid agonist activity. Nature 258:577–580
Ignarro LJ (1989) Endothelium-derived nitric oxide actions and properties. FASEB J 3:31–36
Innis RB, Snyder SH (1980) Distinct cholecystokinin receptors in brain and pancreas. Proc Natl Acad USA 77:6917–6921
Ireland SJ, Hagan RM, Bailey F, Jordan CC, Stephens-Smith ML (1990) Receptors mediating neurokinin-induced contractions of the guinea-pig trachea. Br J Pharmacol 99:63P
Ireland SJ, Bailey F, Cook A, Hagan RM, Jordan CC, Stephens-Smith ML (1991) Receptors mediating tachykinin-induced contractile responses in guinea pig trachea. Br J Pharmacol 103:1463–1469
Isenberg JI, Cseudes A (1972) Effect of octapeptide of cholecystokinin on canine pyloric pressure. Am J Physiol 222:428–431
Ivy AC, Oldberg E (1928) A hormone mechanism for gallbladder contraction and evacuation. Am J Physiol 65:599–613
Jensen RT, Coy DH (1991) Progress in the development of potent bombesin receptor antagonists. Trends Pharmacol Sci 12:13–19

Jensen RT, Jones SW, Folkers K, Gardner JD (1984) A synthetic peptide that is a bombesin-receptor antagonist. Nature 309:61–63
Jensen RT, Coy DH, Saeed ZA, Heinz-Erian P, Montey S, Gardner JD (1988) Interaction of bombesin and related peptides with receptors on pancreatic acinar cells. Ann NY Acad Sci 547:138–149
Jessell TM, Iversen LL (1977) Opiates analgesics inhibit substance P release from rat trigeminal nucleus. Nature 268:549–551
Jorpes JE, Mutt V (1968) Structure of porcine cholecystokinin-pancreozymin. I. Cleavage with thrombin and with trypsin. Eur J Biochem 6:156–162
Kakidani H, Furutani Y, Takahashi H, Noda M, Morimoto Y, Hirose T, Asai M, Inayama S, Nakanishi S, Numa S (1982) Cloning and sequence analysis of cDNA for porcine β-neo endorphine/dynorphin precursor. Nature 298:245–249
Kataoka K, Taniguchi A, Shimizu H, Sods D, Okumo S, Yajima H, Kitawaga K (1978) Biological activity of neurotensin and its C-terminal partial sequences. Brain Res Bull 3:555–557
Kimura I, Kimura M, Nakayama N, Kimura M (1983) Relaxation and Ca spike suppression in circular and longitudinal muscles of hog bile duct ampulla by CCK-C-terminal peptide. Arch Int Pharmacodyn 265:320–334
Kimura I, Kondoh T, Kimura M (1989) Effects of tetrodotoxin on relaxation of pig duodenal circular muscle induced by C-terminal fragments of cholecystokinin. Br J Pharmacol 96:739–745
Kislauskis E, Bullock B, McNeil S, Dobner PR (1988) The rat gene encoding neurotensin and neuromedin. J Biol Chem 263:4963–4968
Kitabgi P, Freychet P (1978) Effects of neurotensin on isolated intestinal smooth muscles. Eur J Pharmacol 50:349–357
Kitabgi P, Vincent JP (1981) Neurotensin is a potent inhibitor of guinea pig colon contractile activity. Eur J Pharmacol 74:311–318
Kitabgi P, Carraway R, Leeman SE (1976) Isolation of a tridecapeptide from bovine intestine tissue and its partial characterization as neurotensin. J Biol Chem 251:7053–7058
Kostka P (1991) Gastrointestinal neuropeptides and second messenger systems. In: Daniel ED (ed) Neuropeptide function in the gastrointestinal tract. CRC, Boca Raton, pp 249–271
Kotani H, Hoshimaru M, Nawa H, Nakanishi S (1986) Structure and gene organization of bovine neuromedin K precursor. Proc Natl Acad Sci USA 83:7074–7078
Krane IM, Naylor SL, Helin-Davis D, Chin WW, Spindel ER (1988) Molecular cloning of cDNAs encoding the human bombesin-like peptide neuromedin B. J Biol Chem 263:13317–13323
Laneuville O, Dorais J, Couture R (1988) Characterization of the effects produced by neurokinins and three agonists selective of neurokinin receptor subtypes in a spinal nocieptive reflex of the rat. Life Sci 42:1295–1305
Laufer R, Wormser V, Friedman ZY, Gilon C, Chorev M, Selinger Z (1985) Neurokinin B is a prefered agonist for a neuronal substance P receptor and its action is antagonized by enkephalins. Proc Natl Acad Sci USA 82:7444–7448
Laws PY, Hom DS, Loh HH (1983) Opiate regulation of adenosine $3^1:5^1$-cyclic monophosphate level in neuroblastoma x glioma NG 108–15 hybrid cells. Relationship between receptor occupancy and effect. Mol Pharmacol 23:26–35
Lee HK, Wang SC (1975) Mechanism of morphine-induced miosis in the dog. J Pharmacol Exp Ther 192:415–431
Lefkowitz RJ, Kobilka BK, Caron MG (1989) The new biology of drug receptors. Biochem Pharmacol 38:2941–2948
Lehy T, Accary JP, Labeille D, Dubrasquet M (1983) Chronic administration of bombesin stimulates antral gastric cell proliferation in rat. Gastroenterology 84:914–919

Lemaire S, Magnan J, Regoli D (1978) Rat vas deferens: a specific bioassay for endogenous opioid peptides. Br J Pharmacol 64:327–329
Lhoste E, Aprahamian M, Pousse A, Hoeltzel A, Stock-Damgé C (1985) Trophic effect of bombesin on the rat pancreas: is it mediated by the release of gastrin or cholecystokinin? Peptides 6:89–97
Li Ch, Chung D (1976) Isolation and structure of an untriakonta peptide with opiate activity from camel pituitary glands. Proc Natl Acad Sci USA 73:1145–1148
Liston DR, Vanderhaeghen JJ, Rossier J (1983) Presence in brain of synenkephalin, a proenkephalin-immunoreactive protein which does not contain enkephalin. Nature 302:62–65
Lord JAH, Waterfield AA, Hughes J, Kosterlitz HW (1977) Endogenous opioid peptide: multiple agonists and receptors. Nature 267:495–499
Lotti VJ, Pendleton RG, Gould RJ, Hanson HM, Chang RSL, Clineschmidt BV (1987) In vivo pharmacology of L-364.718, a new potent non peptide peripheral cholecystokinin antagonist. J Pharmacol Exp Ther 241:103–109
Maggi CA, Giuliani S, Santiciolli P, Abelli L, Regoli D, Meli A (1987a) Further studies on the mechanism of the tachykinin-induced activation of micturition reflex in rats: evidence for the involvement of the capsaicin-sensitive bladder mechanoreceptors. Eur J Pharmacol 136:189–205
Maggi CA, Giuliani S, Santicioli P, Regoli D, Meli A (1987b) Peripheral effects of neurokinins: functional evidence for the existence of multiple receptors. J Auton Pharmacol 7:11–32
Maggi CA, Patacchini R, Giuliani S, Rovero P, Dion S, Regoli D, Giachetti A, Meli A (1990) Competitive antagonists discriminate between NK_2 tachykinin receptor subtypes. Br J Pharmacol 100:583–592
Maggi CA, Patacchini R, Astolfi M, Rovero P, Giuliani S, Giachetti A (1991) NK-2 receptor agonists and antagonists. Ann NY Acad Sci 632:184–191
Marshall FH, Barnes S, Hughes J, Woodruff GN, Hunter JC (1991) Cholecystokinin modulates the release of dopamine from the anterior and posterior nucleus accumbens by two different mechanisms. J Neurochem 56:917–922
Martin WR (1967) Opioid antagonists. Pharmacol Rev 19:463–521
Martin WR (1984) Pharmacology of opioids. Pharmacol Rev 35:283–323
Martin WR, Eades CG, Thompson JA, Huppler RE, Gilbert PE (1976) The effects of morphine and neomorphine-like drugs in the non-dependent and morphine-dependent chronic spinal dog. J Pharmacol Exp Ther 197:517–532
Mastrangelo D, Mathison R, Huggel HJ, Dion S, D'Orléans-Juste P, Rhaleb NE, Drapeau G, Rovero P, Regoli D (1987) The rat isolated portal vein: a preparation sensitive to neurokinins, particularly neurokinin B. Eur J Pharmacol 134:321–336
Masu Y, Nakayama K, Tamaki H, Harada Y, Kuno M, Nakanishi S (1987) cDNA cloning of bovine substance K receptor through oocyte expression system. Nature 329:836–838
McClane TK, Martin WR (1967) Antagonism of the spinal cord effects or morphine and cyclazocine by naloxone and thebaine. Int J Neuropharmacol 6:325–327
McDonald TJ, Jörnvall H, Nilsson G, Vagne M, Ghatei M, Bloom SR, Mutt V (1979) Characterization of gastrin-releasing peptide from porcine non antral gastric tissue. Biochem Biophys Res Commun 90:227–233
McDonald TJ (1991) Gastroenteropancreatic regulatory peptide structures: an overview. In: Daniel EE (ed) Neuropeptide function in the gastrointestinal tract. CRC, Boca Raton, pp 19–86
McGilliard KL, Takemori AE (1978) Antagonism by naloxone of narcotic-induced respiratory depression and analgesia. J Pharmacol Exp Ther 207:494–503
McQueen DS (1983) Opioid peptide interactions with respiratory and circulatory systems. Br Med Bull 39:77–82
McRoberts JW (1986) Cholecystokinin and pain: a review. Anesth Prog 33:87–90
Merrifield RB (1963) Solid-phase peptide synthesis. I. The synthesis of a tetrapeptide. J Am Chem Soc 85:2149–2152

Miller LJ, Jardine I, Weissman E, Go VL, Speicher D (1984) Characterization of cholecystokinin from the human brain. J Neurochem 43:835–840
Milligan G, Streaty RA, Gierschik P, Spiegel AM, Klee WA (1987) Development of opiate receptors and GTP-binding regulatory proteins in neonatal rat brain. J Biol Chem 262:8626–8630
Minamino N, Kangawa K, Matsuo H (1983) Neuromedin B: a novel bombesin-like peptide identified in porcine spinal cord. Biochem Biophys Res Commun 114: 541–548
Minamino N, Kangawa K, Matsuo H (1984a) Neuromedin N: a novel neurotensin-like peptide identified in porcine spinal cord. Biochem Biophys Res Commun 122:542–549
Minamino N, Kangawa K, Matsuo H (1984b) Neuromedin C: a bombesin-like peptide identified in porcine spinal cord. Biochem Biophys Res Commun 119: 14–20
Mizrahi J, Dion S, D'Orléans-Juste P, Regoli D (1985) Activities and antagonism of bombesin on urinary smooth muscles. Eur J Pharmacol 111:339–345
Morgan JP, Morgan KG (1982) Vascular smooth muscle: the first recorded Ca^{2+} transients Pflugers Arch 395:75–77
Morgan KG, Szurszewski JH (1980) Mechanisms of phasic and tonic actions of pentagastrin on gastric smooth muscle. J Physiol (Lond) 301:229–242
Morgan-Boyd R, Stewart JM, Vavrek RJ, Hassis A (1987) Effects of bradykinin and angiotensin II intracellular Ca^{2+} dynamics in endothelial cells. Am J Physiol 253:C588–C598
Morley JE (1980) The neuroendocrine controls of appetite: the role of endogenous opiates, cholecystokinin, TRH, gamma-amino-butiric acid and diazepam receptors. Life Sci 27:355–368
Mosberg HI, Hurst R, Hruby VG, Gee K, Yamamura HI, Galligan JJ, Burks PF (1983) Bispenicillamine enkephalins possess highly-improved specificity towards δ opioid receptors. Proc Natl Acad Sci USA 80:5871–5874
Mousli M, Bueb JL, Bronner C, Rouot B, Landry Y (1990) G-protein activation: a receptorindependent mode of action for cationic amphilic neuropeptides and venom peptides. TIPS 11:358–362
Muallem S, Pandol SJ, Beeker TG (1989) Hormone-evoked calcium release from intracellular stores is a quantal process. J Biol Chem 264:205–212
Mudge AW, Leeman SE, Fischbach GD (1979) Enkephalin inhibits release of substance P from sensory neurons in culture and decreases action potential duration. Proc Natl Acad Sci USA 76:526–530
Murphy RB, Smith GP, Gibbs J (1987) Pharmacological examination of CCK-8-induced contractile activity in the rat isolated pylorus. Peptides 8:127–134
Mutt V (1983) New approaches to the identification and isolation of hormonal polypeptides. Trends Neurosci 6:357–360
Mutt V (1988) Secretin and cholecystokinin. In: Mutt V (ed) Gastrointestinal hormones, vol 2. Academic, New York, p 251
Nakajima T, Tanimura T, Pisano JJ (1970) Isolation and structure of a new vaso-active peptide (Abstr) Fed Proc 29:282
Nakanishi S (1986) Structure and regulation of the preprotachykinin gene. Trends Neurosci 9:41–44
Naline E, Devillier P, Drapeau G, Toty L, Bakdach H, Regoli D, Advenier C (1989) Characterization of neurokinins effects and receptor selectivity in human isolated bronchi. Am Rev Respir Dis 140:679–686
Nantel F, Rouissi N, Rhaleb NE, Dion S, Drapeau G, Regoli D (1990) The rabbit jugular vein is a contractile NK-1 receptor system. Eur J Pharmacol 179:457–462
Nantel F, Rouissi N, Rhaleb NE, Jukic D, Regoli D (1991) Pharmacological evaluation of the angiotensin, kimin and neurokinin receptors on the rabbit vena cava. J Cardiovasc Pharmacol 18:398–405

Nau R, Schafer G, Deacon CF, Cole T, Agaston DV, Conlon JM (1986) Proteolytic inactivation of substance P and neurokinin A in the longitudinal muscle layer of guinea pig small intestine. J Neurochem 47:856–864
Nawa H, Hirose T, Takashima H, Inayama S, Nakanishi S (1983) Nucleotide sequences of cloned cDNAs for two types of bovine brain substance P precursor. Nature 306:32–36
Negri L (1986) Satiety and scratching effects of bombesin-like peptides. Eur J Pharmacol 132:207–212
Nemeroff CB (1980) Neurotensin: perchance an endogenous neuroleptic? Biol Psychiatry 15:283–302
Nemeroff CB, Osbahr AJ, Manberg PJ, Ervin GN, Prange AJ (1979) Alteration in nociceptive and body temperature after intracisternally-administered neurotensin, β-endorphine, other endogenous peptides and morphine. Proc Natl Acad Sci USA 76:5368–5371
Nemeroff CB, Binette G, Manberg PJ, Osbahr AJ, Breese GR, Prange AJ (1980) Neurotensin induced hypothermia: evidence for an interaction with dopaminergic system and the hypothalamic-pituitary-thyroid axis. Brain Res 195:69–84
North RA, Williams JT (1985) On the potassium conductance increased by opioids in rat locus coeruleus neurones. J Physiol (Lond) 364:265–280
North RA, Williams JT, Surprenant A, Christie MJ (1987) μ and δ receptors belong to a family of receptors that are coupled to potassium channels. Proc Natl Acad Sci USA 84:5487–5491
Ohkawa H, Watanabe M (1977) Effect of gastrointestinal hormones on the electrical and mechanical activities on cat small intestine. Jpn J Physiol 27:271–279
Oka T, Negishi K, Suda M, Matsumiya T, Inazu T, Ueki M, Vekimasaaki V (1980) Rabbit vas deferens: a specific bioassay for opioid K-receptor agonists. Eur J Pharmacol 73:235–236
Opmeer FA, Peter J, Burbach H, Wiegant VM, Van Ree JM (1982) β-endorphin proteolysis by guinea pig ileum myenteric plexus membranes: increased γ-endorphin turnover after chronic exposure to morphine. Life Sci 31:323–328
Orloff MS, Reeve JR, Ben Avram C, Shively JE, Walsh JH (1984) Isolation and sequence analysis of human bombesin-like peptides. Peptides 5:865–870
Pandol SJ, Schoeffield MS, Sachs G, Muallem S (1985) Role of free cytosolic calcium in secretagogue-stimulated anylase release from dispersed acini from guinea pig pancreas. J Biol Chem 260:10081–10086
Patacchini R, Astolfi M, Quartara L, Rovero P, Giachetti A, Maggi CA (1991) Further evidence for the existence of NK-2 tachykinin receptor subtypes. Br J Parmacol 104:91–96
Paton WDM (1957) The action of morphine and related substances on contraction and on acetylcholine output of coaxially stimulated guinea pig ileum. Br J Pharmacol 12:119–127
Pelton JT, Gulya K, Hruby VG, Duckles SP, Yamamura HI (1985) Comformationally restricted analogues of somatostatin with high μ opioid receptor specificity. Proc Natl Acad Sci USA 82:236–239
Pendleton RG, Bendesky RJ, Schaffer L, Nolan TE, Gould RJ, Clineschmidt BV (1987) Roles of endogenous cholecystokinin in biliary, pancreatic and gastric function: studies with L-364-718 specific cholecystokinin receptor antagonist. J Pharmacol Exp Ther 241:110–116
Polak JM, Hamid, Springall DR, Cuttita F, Spindel E, Ghatei MA, Bloom SR (1988) Localization of bombesin-like peptides in tumors. Ann NY Acad Sci 547:322–335
Portoghese PS, Takemori AE (1985) TENA, a selective κ opioid receptor antagonist. Life Sci 36:801–805
Powell DW (1981) Muscle or mucosa: the site of action of antidiarrheal opiates? Gastroenterology 80:406–408

Pozo MJ, Salido GM, Madrid JA (1989) Cholecystokinin-induced gall bladder contraction is influenced by nicotinic acid and muscarinic receptors. Arch Int Physiol Biochim 97:403–408
Prange A, Nemeroff CB (1982) The manyfold actions of neurotensin: a first synthesis. Am NY Acad Sci 400:368–375
Quirion R, Rioux F, Regoli D, St-Pierre S (1979) Neurotensin-induced coronary vessel constriction in perfused rat hearts. Eur J Pharmacol 55:221–223
Quirion R, Regoli D, Rioux F, St-Pierre S (1980a) Structure activity studies with neurotensin: analysis of positions 9, 10 and 11. Br J Pharmacol 69:689–692
Quirion R, Rioux F, Regoli D, St-Pierre S (1980b) Compound 48/80 inhibits neurotensin-induced hypotension in rats. Life Sci 27:1889–1895
Rakovska A, Milenov K, Yanev S (1986) Mode of action of cholecystokinin octapeptide on smooth muscles of stomach, ileum and gall bladder. Methods and Findings in Experimental and Clinical Pharmacology 8:697–703
Rattan S, Goyal RK (1983) Pharmacological differences between neural (inhibitory) and muscle (excitatory) CCK receptors in the cat lower esophageal sphincter (Abstr). Gastroenterology 84:1281
Regoli D (1982) Vascular receptors for polypeptides. Trends Pharmacol Sci 3: 286–288
Regoli D (1987) Peptides, receptors, antagonists. TIPS 8: Centrefold January issue
Regoli D, Nantel F (1990) Receptor-independent action of bradykinin. Direct activation of G proteins. Trends Pharmacol Sci 11:400–401
Regoli D, Nantel F (1991) Pharmacology of neurokinin receptors. Biopolymers 31:777–783
Regoli D, Escher E, Drapeau G, D'Orléans-Juste P, Mizrahi J (1984) Receptors for substance P. III. Classification by competitive antagonists. Eur J Pharmacol 97:179–189
Regoli D, Drapeau G, Dion S, D'Orléans-Juste P (1987) Pharmacological receptors for substance P and neurokinins. Life Sci 40:109–117
Regoli D, Drapeau G, Dion S, Couture R (1988) New selective agonists for neurokinin receptors: pharmacological tools for receptor characterization. Trends Pharmacol Sci 9:290–295
Regoli D, Dion S, Rhaleb NE, Drapeau G, Rouissi N, D'Orléans-Juste P (1989a) Receptors for neurokinins, tachykinins and bombesin: a pharmacological study. Ann NY Acad Sci 547:158–173
Regoli D, Drapeau G, Dion S, D'Orléans-Juste P (1989b) Receptors for substance P and related neurokinins. A mini-review. Pharmacology 38:1–15
Regoli D, Rhaleb NE, Dion S, Tousignant C, Rouissi NE, Jukic D, Drapeau G (1990a) Neurokinin A. A pharmacological study. Pharmacol Res 22:1–14
Regoli D, Rhaleb NE, Dion S, Drapeau G (1990b) New selective bradykinin receptor antagonists and bradykinin B_2 receptor characterization. Trends Pharmacol Sci 11:156–161
Rehfeld JF (1986) Accumulation of nonamidated preprogastrin and preprocholecystokinin products in porcine pituitary corticotrophs. J Biol Chem 261: 5841–5847
Renfeld JF, Hansen HF, Marley PD, Stenjard-Pedersen K (1985) Molecular forms of cholecystokinin in the brain and the relationship to neuronal gastrin. Ann N Y Acad Sci 448:11–23
Resin H, Stern DH, Sturdevant R, Isenberg JL (1973) Effect of the C-terminal of cholecystokinin on lower esophageal sphincter pressure in man. Gastroenterology 64:946–949
Rioux F, Park WK, Regoli D (1973) Application of drug-receptor theories to angiotensin. Can J Physiol Pharmacol 51:665–672
Rioux F, Quirion R, Regoli D, Leblanc MA, St-Pierre S (1980) Possible interactions between neurotensin and prostaglandins in the isolated rat portal vein. Life Sci 27:259–267

Rioux F, Lemieux M, Kerouac R, Bernoussi A, Roy G (1989) Local application of neurotensin to abdominal organs triggers cardiovascular reflexes in guinea pigs: possible mechanisms. Peptides 10:647–655
Rodgers DF, Aursudkij B, Barnes PJ (1989) Effects of tachykinins on mucus secretion in human bronchi in vitro. Eur J Pharmacol 174: 283–286
Rogers J, Hughes RG, Mathews EK (1988) Cyclic GMP inhibits protein kinase C-mediated secretion in rat pancreatic acini. J Biol Chem 263:3713–3719
Rökaeus A (1991) Regulation of gastrointestinal neuropeptide gene expression and processing. In: Daniel EE (ed) Neuropeptide function in the gastrointestinal tract. CRC, Boca Raton, pp 87–129
Rollandy I, Guillemain I, Imhoff V, Drapeau G, Regoli D, Rossignol B (1991) Involvement of NK-1 receptors and importance of the N-terminal sequence of substance P in the stimulation of protein secretion in rat parotid glands. Eur J Pharmacol 209:95–100
Roques BP, Beaumont A (1990) Neutral endopeptidase-24.11 inhibitors: from analgesics to antihypertensives? Trends Pharmacol Sci 11:245–249
Roques BP, Fournie-Zaluski MC (1986) Enkephalin-degrading enzyme inhibitors: a physiological way to new analgesics and psychoactive agents. NIDA Res Monogr Ser 70:128–154
Roques BP, Fournié-Zaluski MC, Soroca E, Lecomte JH, Malfroy B, Llorens C, Schwartz JC (1980) The enkephalinase inhibitor thiorphan shows anti nociceptive activity in mice. Nature 288:286–288
Rosell S (1977) Actions of neurotensin and [Gln^4]-neurotensin on isolated tissues. Acta Pharmacol Toxicol (Copenh) 41:141–147
Rossie SS, Miller RJ (1982) Regulation of mast cell histamine release by neurotensine. Life Sci 31:509–516
Rouissi N, Nantel F, Drapeau G, Rhaleb N-E, Dion S, Regoli D (1990a) Inhibitors of peptidases: how they influence the biological activities of substance P, neurokinins, bradykinin and angiotensin in guinea pig, hamster and rat urinary bladders. Pharmacology 40:196–204
Rouissi N, Nantel F, Drapeau G, Rhaleb NE, Dion S, Regoli D (1990b) Inhibitors of peptidases: how they influence the biological activities of substance P, neurokinins, kinins and angiotensins in isolated vessels. Pharmacology 40:185–195
Rouissi N, Rhaleb NE, Nantel F, Dion S, Drapeau G, Regoli D (1991a) Characterization of bombesin receptors in peripheral contractile organs. Brit J Pharmacol 103:1141–1147
Rouissi N, Gitter BD, Waters DC, Howbert JJ, Nixon JA, Regoli D (1991b) Selectivity and specificity of new, non peptide, quinuclidine antagonists of substance P. Biochem Biophys Res Commun 176:894–901
Rozengurt E (1988) Bombesin-induction of cell prolipheration in 3T3 cells: specific receptors and early signaling events. Ann NY Acad Sci 547:277–292
Salem H, Aviasto DM (eds) (1970) Antitussive agents, vols 1–3. Pergamon, Oxford International encyclopedia of pharmacology and therapeutics, sect 27
Sausville EA, Moyer JD, Heikkila R, Neckers LM, Trepel JB (1988) A correlation of bombesin responsiveness with myc-family gene expression in small cell lung carcinoma cell lines. Ann NY Acad Sci 547:310–321
Sasai Y, Nakanishi S (1989) Molecular characterisation of rat substance K receptors and its mRNAs. Biochem Biophys Res Commun 165:695–702
Schang JC, Hemond M, Hebert M, Pilote M (1986) How does morphine work on colonic motility? An electro myographic study in the human left and sigmoid colon. Life Sci 38:671–676
Scheurer UL, Varga L, Drack E, Burk HR, Halter F (1983) Mechanisms of action of CCK-octapeptide on rat antrum pylorus and duodenum. Am J Physiol 244: G266–G272
Schild HO (1947) pAx, a new scale for the measurement of drug antagonism. Br J Pharmacol 2:189–206

Schutzberg M, Hökfelt T, Lundberg JM (1982) Co-existence of classical transmitters and peptides in the central and peripheral nervous system. Br Med Bull 38: 309–318
Selbekk BH, Flaten O, Hanssen LE (1980) The in vitro effect of neurotensin on human jejunum mast cells. Scand J Gastroenterol 15:457–460
Severi C, Grider JR, Makhlouf GH (1988) Identification of separate bombesin and substance P precursors on isolated muscle cells from canine gallbladder. J Pharmacol Exp Ther 245:195–198
Shigemoto R, Yokota Y, Tsuchida K, Nakanishi S (1990) Cloning and expression of rat neuromedin K receptor cDNA. J Biol Chem 265:623–628
Smith GM, Lowenstein E, Hubbard JH, Beecher HK (1968) Experimental pain produced by the submaximum effort tourniquet technique: further evidence of validity. J Pharmacol Exp Ther 163:468–474
Snider RM, Constantine JW, Lowe JA, III Longo KP, Lebel WS, Woody HA, Drozda SE, Desai MC, Vinick FJ, Spencer RW, Hess HJ (1991) A potent non peptide antagonist of the substance P (NK-1) receptor. Science 251:435–437
Spindel ET, Zilberberg MD, Habener JF, Chin WW (1986) Two prohormones for gastrin releasing peptide are encoded by two mRNAs differing by 19 nucleotides. Proc Natl Acad Sci USA 83:19–23
Spindel ET, Giladi E, Brehm P, Goodman RH, Segerson TP (1990) Cloning and functional characterization of a complementary DNA encoding the murine fibroblast bombesin/gastrin-releasing peptide receptor. Mol Endocrinol 4:1956–1963
St-Pierre S, Kerouac R, Quirion R, Jolicoeur FB, Rioux F (1984) Neurotensin. Pept Protein Rev 2:83–171
Studer RO, Trzeciak A, Lergier W (1973) Isolierung und Aminosäuresequenz von Substanz P aus Pferdedarm. Helv Chim Acta 56:860–866
Sutter MC (1990) The mesenteric-portal vein in research. Pharmacol Rev 42:287–325
Taché Y (1988) Summary and concluding remarks. Bombesin-like peptides in health and disease. Ann NY Acad Sci 547:429–437
Takahashi Y, Kato K, Hayashizaki Y, Wakabayashi T, Ohtsuka E, Matsuki S, Ikehara M, Matsubara K (1985) Molecular cloning of the human cholecystokinin gene by use of a synthetic probe containing deoxynosine. Proc Natl Acad Sci USA 82:1931–1935
Takemori AE, Portoghese PS (1985) Receptors for opioid peptides in the guinea pig ileum. J Pharmacol Exp Ther 235:389–392
Takemoto K, Lundberg JM, Jörnvall H, Mutt V (1985) Neuropeptide K: isolation, structure and biological activities of a novel brain tachykinin. Biochem Biophys Res Commun 128:947–953
Tanaka K, Masu M, Nakanishi S (1990) Structure and functional expression of the cloned rat neurotensin receptor. Neuron 4:847–854
Tousignant C, Dion S, Drapeau G, Regoli D (1987) Characterization of pre and postfunctional receptors for kinins and neurokinins in the rat vas deferens. Neuropeptides 9:333–343
Ueda N, Muramatsu I, Taniguchi T, Nakanishi S, Fujiwaea M (1986) Effects of neurokinin A, substance P and electrical stimulation on the rabbit iris sphincter muscle. J Pharmacol Exp Ther 239:494–499
Valenzuela JE (1976) Effect of intestinal hormones and peptides on intragastric pressure in dogs. Gastroenterology 71:766–769
Varner AA, Modlin IM, Walsh JH (1981) High potency of bombesin for stimulation of human gastrin release and gastric acid secretion. Regul Pept 1:289–296
Vizi ES, Bertaccini G, Impicciatore M, Knoll J (1972) Acetylcholine-releasing effect of gastrin and related polypeptides. Eur J Pharmacol 17:175–178
Von Euler US (1936) Preparation of substance P. Scand Arch Physiol 73:142–144
Von Euler US, Gaddum JH (1931) An unidentified depressor substance in certain tissue extracts. J Physiol (Lond) 72:74–87

Wado E, Way J, Shapira H, Kusano K, Lebacq-Verheyden AM, Coy D, Jensen R, Battery J (1991) cDNA cloning, characterization and brain region specific expression of a neuromedin-B preferring bombesin receptor. Neuron 6:421–430
Walsh JH (1987) Gastrointestinal hormones – cholecystokinin. In: Johnson LR (ed) Physiology of the gastrointestinal tract. Raven, New York, p 195
Wamsley JK (1983) Opioid receptors: autoradiography. Pharmacol Rev 35:69–83
Watkins LR, Kinscheck IB, Mayer DJ (1985) Potentiation of morphine analgesia by the cholecystokinin antagonist proglumide. Brain Res 327:169–180
Watson SJ, Akil H, Ghazarossian VE, Goldstein A (1981) Dynorphin immunocytochemicals localization in brain and peripheral nervous sytem: preliminary studies. Proc Natl Acad Sci USA 78:1260–1263
Watson SJ (1984) The action of substance P on contraction, inositol phospholipids and adenylate cyclase in rat small intestine. Biochem Pharmacol 33:3733–3737
Widerlöv E, Lindstrom LH, Besev G, Manberg PJ, Nemeroff CB, Breese GR, Kizer JS, Prange AJ (1982) Subnormal CSF levels of neurotensin in a subgroup of schizophrenic patients: normalization after neuroleptic treatment. Am J Psychiatry 139:1122–1126
Williams JA, McChesney DJ, Calayag MC, Lingappa VR, Logsdon CD (1988) Expression of receptors for cholecystokinin and other Ca^{2+} mobilizing hormones in xenopus oocytes. Proc Natl Acad Sci USA 85:4939–4943
Wolfe M, McGuigan JE (1984) Immunochemical characterization of gastrin and cholecystokininlike peptides released in dogs in response to a peptone meal. Gastroenterology 87:323–334
Woodruff GN, Hughes J (1991) Cholecystokinin antagonists. Annu Rev Pharmacol Toxicol 31:469–501
Wuster M, Schulz R, Herz A (1980) The direction of opioids agonists towards μ, δ and κ receptors in the vas deferens of the mouse and the rat. Life Sci 27:163–170
Yaksh TL, Jessel TM, Gamse R, Mudge AW, Leeman E (1980) Intrathecal morphine inhibits substance P release from mammalian spinal cord. Nature 286:155–157
Yamamoto Y, Holts K, Matanota T (1984) Effect of methionine-enkephalin on the spontaneous electrical and mechanical activity of the smooth muscle of the rat portal vein. Life Sci 34:993–999
Yamamura T, Takahashi T, Kusonoki M, Kantoh M, Ishikawa Y, Utsunomiya J (1986) Cholecystokinin octapeptide-evoked [^{3}H] acetylcholine release from guinea pig gall bladder. Neurosci Lett 65:167–170
Yau WM, Makhlouf GM, Edwards LE, Ferrar JT (1973) Mode of action of cholecystokinin and related peptides on gall bladder muscle. Gastroenterology 65:451–456
Yau WM, Makhlouf GM, Edwards LE, Farrar JT (1974) The action of cholecystokinin and related peptides on guinea pig small intestine. Can J Physiol Pharmacol 52:298–303
Yau WM, Lingle PF, Youther ML (1983) Interaction of enkephalin and caerulein on guinea pig small intestine. Am J Physiol 244:G65–G70
Yokota Y, Sasai Y, Tanaka K, Fujiwara T, Tsuchida K, Shigemato R, Kakiguka A, Ohkubo H, Nakanishi S (1989) Molecular characterization of a functional cDNA for rat ésubstance P receptor. J Biol Chem 264:17649–17652
Yu DH, Huang SC, Wank SA, Mantey S, Gardner JD, Jensen RT (1990) Pancreatic receptors for cholecystokinin: evidence for three receptor classes. Am J Physiol 258:G86–G95

CHAPTER 9

Serotonin and Smooth Muscle

A.L. KILLAM and M.L. COHEN

A. Introduction

Serotonin as an endogenous neurotransmitter has been implicated in the pathophysiology of diseases involving many smooth muscle systems including hypertension, migraine, Raynaud's phenomenon, vasospasm of the coronary and the cerebral vasculature, gut motility disorders, and dysfunction of genitourinary smooth muscle. Compared to our knowledge of the cholinergic and adrenergic systems, the precise involvement of serotonin in both the physiological and pathological functions of smooth muscle has remained elusive.

Although serotonin was isolated and identified as 5-hydroxy tryptamine in the late 1940s, delineation of the role of serotonin has been hampered until recently by: (a) the lack of highly selective drugs for serotonin receptor subtypes and (b) the multiple effects of serotonin on smooth muscle and other cells such as nerves also found in isolated tissues. Significant progress has been made in the last 10 years toward understanding serotonergic systems in smooth muscle. Such progress has been intimately intertwined with the development of new compounds and more extensive identification of serotonin receptor subtypes.

B. History

The focus of this chapter, i.e., smooth muscle responses to serotonin, has a strong historical basis. The first description of what was probably the action of serotonin appeared during the late nineteenth and early twentieth century with the demonstration of vascular contraction in isolated vessels exposed to clotting blood (BRODIE 1900/1901; see also WANG and PEROUTKA 1988). While the initial observations on the actions of "serotonin" were primarily a compilation of isolated vascular smooth muscle studies underway at the Cleveland clinic in the 1930s, simultaneously ERSPAMER and his colleagues found a substance in the gut which they termed "enteramine" (for a review of early history, see WANG and PEROUTKA 1988). The realization that the blood factor (thrombocytin or serotonin) studied at the Cleveland clinic and "enteramine" were the same entity (ERSPAMER and ASERO 1952) was delayed until the isolation and identification of the chemical composition of serotonin as 5-hydroxytryptamine by Rapport and his colleagues (RAPPORT et al. 1948; RAPPORT 1949). We will use serotonin or its chemical name 5-hydroxytryptamine (abbreviated 5-HT) interchangeably throughout this chapter. The synthesis and availability of synthetic 5-hydroxytryptamine sparked a burst of pharmacological studies using both in vivo and in vitro preparations. The early findings on multiple physiological effects of 5-hydroxytryptamine have been exhaustively reviewed (ERSPAMER 1966).

C. Serotonin Receptor Identification and Classification

As with other neurotransmitters, serotonin acts on smooth muscle via the activation of specific receptors. Early evidence for specific receptors for 5-HT on smooth muscle was based on the lack of cross protection to irreversible alkylation in blood vessels between 5-HT and other neurotransmitters such as norepinephrine (FURCHGOTT 1955). Shortly thereafter, GADDUM and PICARELLI (1957), using the isolated guinea pig ileum, provided evidence for two distinct 5-HT receptor subtypes. These "subtypes" of the 5-HT receptor were referred to as the 5-HTM (neuronal) and the 5-HTD receptor (smooth muscle) based on sensitivity to blockade of 5-HT-induced contraction by morphine and dibenzylamine, respectively.

Although evidence for functional 5-HT receptors existed, the current classification of 5-HT receptor subtypes is based primarily on radioligand binding profiles in CNS tissues. Using ligand binding techniques in CNS membranes, with [^{3}H]LSD, [^{3}H]5-HT, and [^{3}H]spiperone, PEROUTKA and SNYDER (1979) observed that [^{3}H]spiperone differentially labeled sites in the rat frontal cortex from those labeled by [^{3}H]5-HT, whereas [^{3}H]d-LSD labeled both the spiperone and 5-HT sites. PEROUTKA and SNYDER (1979) proposed the classification of central 5-HT receptors into 5-HT$_1$ and 5-HT$_2$ receptor subtypes, a designation accepted for both binding sites and functional receptors (BRADLEY et al. 1986).

Currently there are at least seven generally accepted and characterized 5-HT receptors based upon cloning techniques and pharmacological profiles of agonists and antagonists in both binding and functional studies (HUMPHREY et al. 1993; BRANCHEK 1993). 5-HT$_1$ receptors, characterized by high affinity for 5-HT (K_D in the 1–10 nM range), have at least five recognized subtypes (PEROUTKA 1988). Using a β-adrenergic receptor probe, the 5-HT$_{1A}$ receptor from human genomic DNA has been cloned and sequenced (KOBILKA et al. 1987; FARGIN et al. 1988). The 5-HT$_1$ class of receptors contains the putative "autoreceptor," the 5-HT$_{1B}$ receptor in rats and mice and the 5-HT$_{1D}$ receptor in other species (PEROUTKA 1988; ADHAM et al. 1992). A sequence for the human 5-HT$_{1D}$ receptor (HAMBLIN and METCALF 1991) has been elucidated. Multiple related structures have recently been cloned from the human genome and termed 5-HT$_{1D\alpha}$ and 5-HT$_{1D\beta}$ (WEINSHANK et al. 1992). More recently, two novel receptors that, like the other 5-HT receptors, are negatively coupled to adenylate cyclase have been cloned and labeled 5-

HT_{1E} and 5-HT_{1F} (see HUMPHREY et al. 1993 and BRANCHEK 1993 for review). The function of these receptors has not been elucidated.

The identification of 5-HT_{1C} receptors as receptors having high affinity for serotonin and found predominantly in the choroid plexus as well as other areas of the brain has led to the cloning and sequencing of this receptor (JULIUS et al. 1988; PAZOS et al. 1984; YU et al. 1991). Although serotonin has high affinity at the 5-HT_{1C} receptor, this receptor has an excellent sequence homology with the 5-HT_2 receptor and utilizes a similar signal transduction mechanism (increases in phosphoinositide turnover) as the 5-HT_2 receptor. Thus, some have proposed the reclassification of the 5-HT_{1C} receptor as a member of the 5-HT_2 receptor family (HARTIG 1989), renaming the 5-HT_{1C} receptor as the 5-HT_{2C} receptor (HUMPHREY et al. 1993; BRANCHEK 1993). Another recently cloned receptor that has been added to the 5-HT_2 class is the rat stomach fundus receptor (KURSAR et al. 1992), to be designated the 5-HT_{2B} receptor (HUMPHREY et al. 1993).

The classical 5-HT_2 receptor (5-HT_{2A} receptor by new nomenclature) has low affinity for serotonin (300–1000 n*M*) and high affinity for certain antagonists such as ketanserin. In general, the 5-HT_2 receptor is most closely related to, and probably identical to, the 5-HTD receptor of GADDUM and PICARELLI (1957; PEROUTKA 1988). Because of the availability of high-affinity and fairly selective antagonists at the 5-HT_2 receptor, considerable functional data are available linking the presence of the 5-HT_2 receptor to contractile responses in several smooth muscles. Most recently, 5-HT_2 receptors from the human and rat (KAO et al. 1989; PRITCHETT et al. 1988) have been cloned. Subtle, but possibly important differences in amino acid sequences of the 5-HT_2 receptor among species coupled to distinctive pharmacological profiles of functional responses attributed to activation of 5-HT_2 receptors have led to speculations on the existence of multiple vascular 5-HT_2 receptor subtypes (KILLAM et al. 1990), although validation of this concept has not yet occurred.

Unlike the 5-HT_1 and 5-HT_2 receptors, which were initially characterized with radioligand binding studies, the 5-HT_3 receptor was identified in the absence of a radioligand, based on functional effects in peripheral systems. Serotonin has intermediate affinity at the 5-HT_3 receptor and many potent and selective 5-HT_3 receptor antagonists have been identified. The 5-HT_3 receptor is a ligand-gated receptor found primarily on neurons (both sensory and autonomic; HAMON 1992). This receptor has been partially cloned (MARICQ et al. 1991). The role of 5-HT_3 receptors in the periphery has recently been reviewed (COHEN 1992). These receptors may play a role in vascular pain (GIORDANO and ROGERS 1989) and the control of neurotransmitter release in smooth muscle. The 5-HT_3 receptor is now thought to be identical with the 5-HTM receptor previously described (GADDUM and PICARELLI 1957).

A recently characterized member of the serotonin family of receptors is the 5-HT_4 receptor identified initially in mouse embryo neurocolliculi and associated with stimulation of adenylate cyclase (DUMUIS et al. 1988). Since that time, studies have suggested the presence of 5-HT_4 receptors in gastrointestinal (BAXTER et al. 1991; BUNCE et al. 1991; REEVES et al. 1991) and cardiac (KAUMANN et al. 1991) muscle. The development of highly potent and selective agonists and antagonists at the 5-HT_4 receptor will expedite our understanding of the role of this receptor in smooth muscle function.

Most recently, molecular biological approaches have led to the cloning of three additional receptor subtypes, the 5-HT_5 (ERLANDER et al. 1993), 5-HT_6 (RUAT et al. 1993), and 5-HT_7 (SHEN et al. 1993). Association of these cloned receptors to a function is now under active research.

The extensive history and evolutionary conservation of serotonin as it impacts smooth muscle continues to evolve with the identification and cloning of multiple serotonin receptors mediating different smooth muscle responses (HEN 1992). The development of highly selective agonists and antagonists at each of the characterized serotonergic receptors is paramount to our understanding of the role for each of these serotonin receptors in both normal function and the pathophysiology of smooth muscle. The remainder of this chapter will detail our current knowledge of the presence and role of serotonin receptors in peripheral smooth muscle.

D. Vascular Smooth Muscle

I. Source of Serotonin

The multiple effects of serotonin in vascular smooth muscle are well documented. Less well understood, however, is the source of serotonin that is available to affect blood vessel caliber. Blood platelets which possess high concentrations of serotonin, although lacking the synthetic enzymes capable of synthesizing serotonin, are thought to be the major source of serotonin impacting blood vessel tone. Under conditions of platelet activation, such as blood vessel injury, serotonin released from platelets is thought to play an important facilitative role in several vascular processes including thrombosis, vasoconstriction, and hemostasis. The diffusion of serotonin, once released by activated platelets from the lumen of vessels into the media and adventitia, may be taken up by adrenergic nerves innervating such vessels (COHEN et al. 1987). The ability of adrenergic nerves to accumulate serotonin which can subsequently be co-released with norepinephrine and possibly substance P (SZABO et al. 1991) has been documented. Evidence that serotonin taken up by adrenergic nerves is derived from platelets has been suggested from the observation that the content of serotonin in perivascular nerves is markedly increased after balloon injury to the endothelium, a process resulting in platelet activation and serotonin release (COHEN et al. 1987; VERBEUREN et al. 1988).

Another potential source for serotonin, in addition to platelets and adrenergic nerves that may have taken up serotonin from the circulation, would include serotonergic nerves capable of directly synthesizing serotonin. However, to date, the existence of serotonergic innervation of vascular smooth muscle has not been convincingly demonstrated in peripheral blood vessels. Serotonergic innervation has only been documented in certain cerebral blood vessels (CHEDOTAL and HAMEL 1990), an observation that remains controversial (SAITO and LEE 1986). Thus, for most blood vessels, the inability to demonstrate clearly the existence of serotonergic innervation suggests that serotonergic nerves are not likely to provide an important source of serotonin for interaction with vascular smooth muscle.

II. General Comments on Vascular Responses

The effects of serotonin on vascular smooth muscle are complex and may be mediated by the activation of receptors located on (a) vascular smooth muscle, (b) vascular endothelium, and (c) autonomic nerves innervating blood vessels (COHEN 1988). An overview of these sites for serotonin activation of vascular tone is shown in Fig. 1. Evidence exists to suggest that 5-HT_1-like and 5-HT_2 receptors may co-exist in blood vessels and that the relative proportions of each receptor may depend on species, strain, tissue handling, and experimental as well as pathological state of the vessel (SAXENA and VILLALON 1990; VANHOUTTE 1991).

Our understanding of the relative importance of each of these receptors to vascular responses is critically dependent on the availability of potent and highly selective agonists and antagonists for each of these receptors and their subtypes. Because of the limitations of the available "selective" agonists and antagonists for each of these sites, our understanding of the precise receptors that mediate each of the responses to serotonin is incomplete and continues to evolve.

III. Neuronal Receptors in Vascular Smooth Muscle

Serotonin (either released from nerves or from platelets) can regulate the further release of neurotransmitters (COHEN et al. 1987; MEEHAN and STORY 1989; SZABO et al. 1991) by interacting with presynaptic neuronal receptors located on adrenergic, cholinergic, or peptidergic nerves innervating smooth muscle (SAITO and LEE 1986). The inhibition of norepinephrine release by serotonin has been documented in several peripheral blood vessels including the canine coronary and carotid arteries, canine saphenous vein, rat vena cava, rabbit ear artery, and human saphenous vein (COHEN 1988; SAXENA and VILLALON 1990).

General agreement exists that such presynaptic serotonergic receptors are not 5-HT_2 receptors based on the inability of potent 5-HT_2 receptor

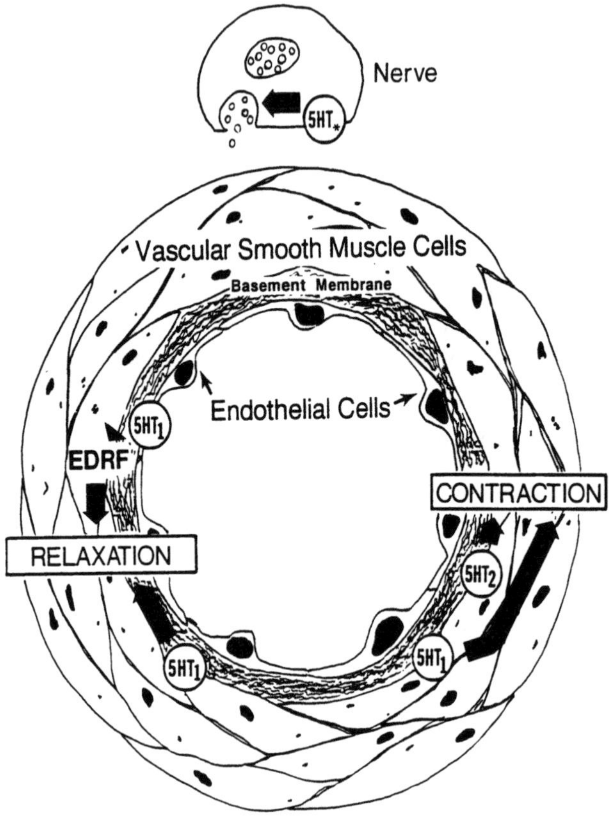

Fig. 1. Cross-sectional representation of a blood vessel containing all identified serotonin receptors. Individual 5-HT_1 receptors have been identified as "5-HT_1-like," "5-HT_{1A}," or "5-HT_{1D}" (please see text sections on contraction and relaxation of vascular smooth muscle). 5-HT_1 receptors have been proposed to produce both contraction and relaxation (endothelial dependent or independent). 5-HT_2 receptors on smooth muscle have not been subdivided and have been associated with contraction of vascular smooth muscle. Neuronal 5-HT receptors (5-HT*) proposed for autonomic nerve and regulation of firing or neurotransmitter release are "5-HT_1-like," "5-HT_{1A}," or "5-HT_3." *EDRF*, endothelial derived relaxing factor

antagonists LY53857 and ketanserin (SAXENA and VILLALON 1990) to block inhibition of neurotransmitter release by serotonin. The ability of non-selective 5-HT_1 receptor antagonists, such as metergoline and methiothepin (MEEHAN and KREULEN 1991), to block serotonin-induced inhibition of norepinephrine release in the guinea pig mesenteric artery has been used to characterize these neuronal receptors as 5-HT_1-like. Data in the dog carotid vasculature supports the idea that the neuronal 5-HT_1-like receptor mediating serotonin-induced inhibition of norepinephrine release from sympathetic neurons is the 5-HT_{1A} subtype (HONG and VILLALON 1984). In blood vessels from the rat, the 5-HT_{1B} receptor may also modulate neurotransmitter release from nerves (SZABO and HARDEBO 1990), a role possibly subserved by the 5-$HT_{1D\beta}$ receptor in nerves from other species and humans (MOLDERINGS et al. 1990).

In addition to 5-HT_{1A} receptors and possibly 5-HT_{1B} receptors (or 5-$HT_{1D\beta}$ receptors in species other than the rat) involved in the modulation of neurotransmitter release from nerves, 5-HT_3 receptors on the somata of sympathetic efferent nerve fibers can also modulate sympathetic activity in the rabbit ear artery (MEEHAN and STORY 1989) and guinea pig superior mesenteric ganglion (MEEHAN and KREULEN 1991) and artery. Additionally, the presence of 5-HT_3 receptors (presumably on nerves) has been postulated in the human forearm, based on blood flow studies demonstrating the ability of 5-HT_3 receptor antagonists to block a vasodilatory effect of serotonin (BLAUW et al. 1988). The possibility that 5-HT_3 receptors can exist on nerves responsible for the modulation of vascular tone in humans is an intriguing, but far from proven one and awaits additional confirmation.

IV. Serotonin Receptors Responsible for Vascular Smooth Muscle Contraction

The 5-HT_2 receptor (also termed the 5-HT_{2A} receptor) is well established to be the primary serotonin receptor involved in vascular contraction of most peripheral blood vessels (COHEN 1988; SAXENA and VILLALON 1990). Stimulation of vascular 5-HT_2 receptors has been associated with an increase in phosphoinositide turnover and the subsequent release of intracellular calcium (SANDERS-BUSH 1988; WANG et al. 1991), effects thought to mediate serotonin-induced vascular contraction.

Activation of 5-HT_2 receptors has been associated with increases in phosphoinositide turnover. In some vascular tissues activation of 5-HT_2 receptors and the subsequent contraction has been associated with influx of extracellular calcium (ZHANG and DYER 1991) based on the ability of classical calcium channel antagonists to block serotonin-induced contraction in some vessels.

Certain blood vessels appear to possess exclusively 5-HT_2 receptors as the contractile serotonergic receptor, such as the rabbit aorta (CLANCY and MAAYANI 1985), the rat aorta (COHEN et al. 1981; KILLAM et al. 1990), and the rat jugular vein (COHEN et al. 1986; COHEN et al. 1981). However, studies with multiple ligands have revealed inconsistencies in both the kinetic behavior and apparent affinities of several 5-HT_2 receptor antagonists at certain 5-HT_2 vascular receptors. Such observations have led to the speculation that multiple vascular 5-HT_{2A} receptors or subtypes of the 5-HT_{2A} receptor may exist (KILLAM et al. 1990; CUSHING and COHEN 1993). KAUMANN and FRENKEN (1985) have postulated the presence of an allosteric regulatory site for the 5-HT_{2A} receptor in several bovine and rat blood vessels. In addition, the possibility of subtypes of this 5-HT_{2A} receptor must be enterained. In support of this possibility is the observation that 5-HT_{2A} receptors cloned from the brain of different species have shown differences in amino acid sequences (HARTIG 1989). Whether these 5-HT_{2A} receptor sequences are different in vascular smooth muscle from each species and whether differences in structure can account for some of the pharmacological and physiological disparity remains to be determined. The development of selective agonists and antagonists that can discriminate between such sites will be important in furthering our understanding of the possibility for multiple 5-HT_{2A} receptor subtypes in vascular tissue.

In addition to the 5-HT_2 receptor mediating vascular smooth muscle contractility, some blood vessels also possess non-5-HT_2 receptors which can mediate contraction, i.e., serotonin-induced contraction that cannot be blocked by selective 5-HT_2 receptor antagonists (COHEN 1988; SAXENA and VILLALON 1990). Evidence that 5-HT_1 receptor activation can induce vascular contraction in certain blood vessels is derived from the high potency of serotonin and the nonselective 5-HT_1 receptor agonist 5-carboxamidotryptamine in contracting certain canine blood vessels. Because of the lack of highly selective tools to identify such receptors, considerable controversy exists regarding the precise 5-HT_1-like receptor involved. For example, contraction to serotonin in the canine basilar artery has been suggested to be mediated by the 5-HT_{1A} receptor (PEROUTKA et al. 1986; TAYLOR et al. 1986), by a receptor similar to that found in the rat stomach fundus (COHEN 1988), or by a 5-HT_2 receptor rather than a 5-HT_1-like receptor (MULLER-SCHWEINITZER and ENGEL 1983).

The possibility that a non-5-HT_2 receptor mediates vascular contraction has received considerable attention with the development of sumatriptan. This agent potently contracted the canine saphenous vein and canine cerebral arteries (HUMPHREY et al. 1989) and possesses relatively high affinity for the 5-HT_{1D} receptor in brain (SCHOEFFTER and HOYER 1989a; PEROUTKA and MCCARTHY 1989). Since sumatriptan is not highly selective for the 5-HT_{1D} receptor in brain, and because 5-benzyloxytryptamine, another high affinity 5-HT_{1D} receptor ligand, did not show similar agonist effects in the canine saphenous vein (COHEN et al. 1992), the receptor mediating sumatriptan-induced contraction in the canine saphenous vein, while similar, may not be identical with the 5-HT_{1D} receptor defined in brain. Sumatriptan-induced contraction of canine and human coronary arteries has also been documented (HUMPHREY et al. 1989; CHESTER et al. 1990).

The identification of multiple receptors that mediate serotonin-induced contraction in blood vessels has led to the observation that more than one serotonin receptor may mediate contraction in any given blood vessel. Most recently, the suggestion has been made that the relative proportion of 5-HT_2 and 5-HT_1-like receptors (possibly 5-HT_{1D}, based on sensitivity to sumatriptan) may vary in cardiovascular disease. For example, CHESTER et al. (1990) have demonstrated a relative reduction in 5-HT_2 receptor mediated contractility and a relative increase in sumatriptan or 5-HT_{1D}-like receptor-mediated contractility of coronary arteries from ischemic hearts. Such studies have led to a greater appreciation for the possibility that damaged blood vessels may respond differently to serotonin via the regulation of contractile serotonin receptors in vascular smooth muscle.

V. Vascular Relaxation

The relaxation of vascular smooth muscle by serotonin may occur as a result of its ability to release factors from the endothelium or by a direct effect on the smooth muscle itself. Several studies have documented the ability of serotonin to relax vascular tissue by a mechanism involving the endothelium.

Serotonin, by induction of endothelial-dependent relaxation of certain vascular tissues, can modulate 5-HT-induced contraction (CONNOR et al. 1989). In other tissues, notably the canine basilare artery (TSUJI and CHIBA 1987) and the feline middle cerebral artery (LOPEZ DE PABLO et al. 1991), although 5-HT-induced relaxation can be demonstrated, removal of the endothelium had no effect on the 5-HT contractile response. In pigs which have undergone balloon angioplasty (SHIMOKAWA et al. 1989) and in clinical studies in which human coronary arteries were damaged by balloon angioplasty or atherosclerosis, a normal vasodilatory response to serotonin (endothelium-dependent) can be converted to a vasoconstrictor response mediated by smooth muscle 5-HT receptors (GOLINO et al. 1991; MCFADDEN et al. 1991). Endothelial-dependent relaxation to 5-HT in several vessels including the porcine coronary artery and vena cava (MOLDERINGS et al. 1989; SUMNER 1991) and rat coronary vessels (MANKAD et al. 1991) has been definitively linked to release of endothelial-derived relaxant factor (EDRF) nitroso compounds. Serotonin-evoked release of EDRF has been associated with a pertussis toxin-sensitive G protein in the porcine coronary artery (FLAVAHAN et al. 1989).

The endothelial serotonin receptor associated with relaxation has been classified as 5-HT_1-like in the bovine coronary artery (COCKS and ANGUS 1983), rabbit jugular vein (LEFF et al. 1987), canine coronary artery (HOUSTON and VANHOUTTE 1988; TODA and OKAMURA 1990), and initially in the procine coronary artery (MOLDERINGS et al. 1989). SCHOEFFTER and HOYER (1989a, 1990) have suggested, based on rank order potencies of several compounds, including the 5-HT_{1D} selective agonist sumatriptan, that the porcine coronary artery endothelial 5-HT_1-like receptor strongly resembles the recently characterized 5-HT_{1D} receptor. However, not all endothelial responses to serotonin are mediated by the 5-HT_{1D} receptor. The endothelial receptor involved in 5-HT-induced relaxation of the porcine vena cava does not resemble any of the known 5-HT receptors (SUMNER 1991). Two recent studies (BODELSON et al. 1993; GLUSA and RICHTER 1993) have identified 5-HT-induced relaxation in rat jugular vein and porcine pulmonary artery that shows pharmacology homologous to either the 5-HT_{1C} (5-HT_{2C}) or the 5-HT_{2B} receptor.

Endothelial-independent relaxation of vascular smooth muscle has also been described in several species. FENUIK et al. (1983) showed a "direct" 5-

HT-induced relaxant effect on feline saphenous vein helical strips. SUMNER (1991; SUMNER et al. 1989) has demonstrated a $5\text{-HT}_{1\text{-like}}$ smooth muscle receptor in the porcine vena cava which is distinct from the endothelial receptor. Recent evidence in the canine coronary artery shows a relaxant smooth muscle receptor that appears to be distinct (CUSHING and COHEN 1992). Thus, 5-HT-induced relaxation of blood vessels has been most often associated with 5-HT_1-like receptors and further characterized as 5-HT_{1D}-like in the porcine coronary endothelium. However, the pharmacological profile for 5-HT-induced relaxation in some vascular tissues is not consistent with any of the well-characterized 5-HT receptors. 5-HT-induced relaxation of vascular smooth muscle can be both endothelium dependent and independent, a fact that must be taken into account when studying vascular smooth muscle contractile responses to serotonin.

E. Gastrointestinal Smooth Muscle

I. Source of Serotonin

Of the several smooth muscle preparations discussed in this chapter, gastrointestinal smooth muscle provides the richest source for serotonin. Aside from the central nervous system, neuronal tissue in the gastrointestinal tract is known to possess serotonergic nerves capable of synthesizing serotonin for subsequent release upon neuronal activation. Most serotonin utilized within the periphery is thought to be derived from the gastrointestinal enterochromaffin cells which are capable of synthesizing the amine in large quantities. Serotonin, localized in the circulating blood platelets, is derived predominantly from a gastrointestinal source (DE CHAFFOY DE COURCELLES and DECLERCK 1990). Thus, gastrointestinal tissue with its rich supply of innervation and secretory cells is the major site for the synthesis of serotonin that affects smooth muscles by its release into the circulation and subsequent uptake and release from platelets and peripheral neurons.

II. General Comments on Gastrointestinal Responses

As mentioned in the introduction, gastrointestinal smooth muscle provided the first evidence for multiple serotonin receptor subtypes (GADDUM and PICARELLI 1957). More recent evidence suggests that the existence of only two serotonin receptor subtypes in the gastrointestinal system is an underestimation of the complexity of serotonergic responses in the gastrointestinal tract. As with our understanding of responses and receptors to serotonin in vascular smooth muscle, our understanding of the receptors mediating the multiple effects of serotonin in gastrointestinal tissue is continuously evolving. Most recently the novel 5-HT_4 receptor has been identified in gastrointestinal smooth muscle (see below). In addition, the presence of yet another novel serotonin receptor responsible for the contractile effects of serotonin in rat stomach fundus has been established (BAEZ et al. 1990; KURSAR et al. 1992).

Because of the complexities of this organ system, we present this section with reference first to neuronal receptors for serotonin found in the gastrointestinal tract and then with the state of knowledge of serotonin receptors identified in each of the tissues associated with the gastrointestinal system. At this time it is reasonable to say that most known serotonin receptor classes and some, yet uncharacterized serotonin receptors reside within the gastrointestinal tract.

III. Neuronal Receptors in Gastrointestinal Smooth Muscle

Of the serotonin receptors found in the gastrointestinal tract, the 5-HT_3 receptor has been most extensively studied. Activation of 5-HT_3 receptors regulates release of neurotransmitters (most commonly acetylcholine, but substance P has also been implicated; BUCHHEIT et al. 1985). In the guinea pig ileum, activation of 5-HT_3 receptors on enteric nerves results in a contractile response, in the activation of chloride ion secretion (BAIRD and CUTHBERT 1987) and acetylcholine release (FOX et al. 1990). Such excitatory 5-HT_3 receptor-mediated responses have also been suggested in the small bowel (BUTLER et al. 1990; EGLEN et al. 1990).

The effectiveness of 5-HT_3 receptor antagonists to inhibit cancer chemotherapy-induced emesis has led some (BUCHHEIT et al. 1990) to suggest that the 5-HT_3 receptor responsible for the emetic response is located in the periphery. Others (ROBERTSON et al. 1990) have suggested that 5-HT_3 receptor antagonists act through central 5-HT_3 receptors since charged compounds were ineffective in blocking cisplatin-induced emesis in dogs. The site of 5-HT_3 receptors participating in emesis is likely to be species dependent (GIDDA et al. 1991).

In addition to 5-HT_3 receptors, neuronal 5-HT_{1A} receptors have also been identified in gastrointestinal tissue. Activation of neuronal 5-HT_{1A} receptors has been associated with an inhibition of the twitch response in the guinea pig ileum (FOZARD and KILBINGER 1985) by 5-HT_{1A} selective agonists, such as 8-OH-DPAT. The extent to which 5-HT_{1A} receptor activation can modulate such a response is limited since maximal effectiveness attributed to 5-HT_{1A} receptor activation ranges between 30% and 40% inhibition. This is in marked contrast to the virtual abolition of a twitch response with α_2 receptor agonists. In high concentrations (see Fig. 2), 8-OH DPAT can abolish the twitch response in the guinea pig ileum, an effect resulting from its ability to block cholinergic receptors [$-\log_{KB} = 5.61 \pm 0.07(4)$], emphasizing the need to know the selectivity of tools used to probe serotonergic receptors.

Another neuronal receptor has been described by GERSHON et al. (1990) in gastrointestinal tissue. This receptor has been designated the 5-HT_{1P}

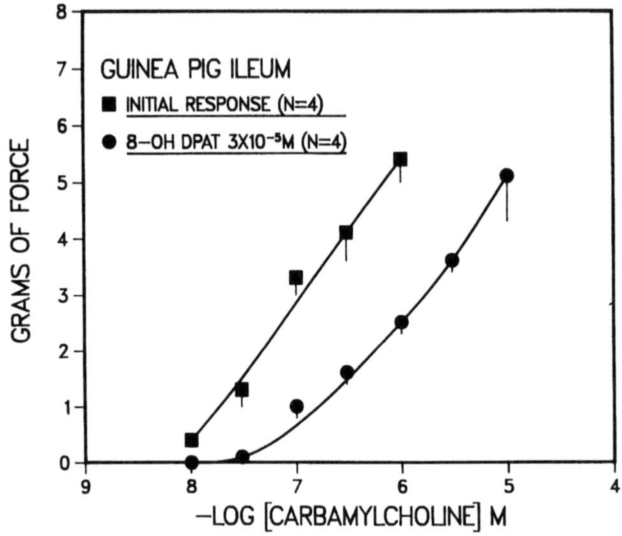

Fig. 2. Competitive inhibition of the contractile response to carbamylcholine by 8-OH DPAT ($3 \times 10^{-5} M$) in guinea pig ileum, documenting anticholinergic activity of high concentrations of the "5-HT_{1A} selective agonist" 8-OH DPAT. *Points* are mean values and *vertical bars* represent the standard error of the mean for four tissues

receptor and its activation can induce slow and long-lasting depolarization of neurons within the myenteric plexus. This receptor possesses a unique serotonergic pharmacology not associated with any of the receptors described by BRADLEY et al. (1986) or, more recently, HUMPHREY et al. (1993).

IV. Esophageal Smooth Muscle

Unlike the other tissues found within the gastrointestinal tract, the esophagus is composed of both smooth and striated muscle. The relative proportions of smooth versus striated muscle vary from species to species. In general, the smooth muscle is located in the muscularis mucosa layer closest to the epithelium surrounding the esophageal lumen. Striated muscle is found most prominently in the muscularis externa located more distally from the lumen.

Most recently, esophageal smooth muscle from the rat has been shown to possess a 5-HT_4 receptor which mediates relaxation in this tissue (BAXTER et al. 1991; REEVES et al. 1991). The 5-HT_4 receptor is characterized primarily by interaction with several benzamide derivatives and antagonism by high concentrations of the 5-HT_3 receptor antagonist ICS 205,930 (BAXTER et al. 1991).

Although the relaxant 5-HT_4 receptor has been established in the rat esophagus, rabbit and guinea pig esophagus do not relax to serotonin, but rather contract to this biogenic amine (Fig. 3), a contraction which is mediated via activation of 5-HT_2 receptors. Thus, contractile 5-HT_2 receptors

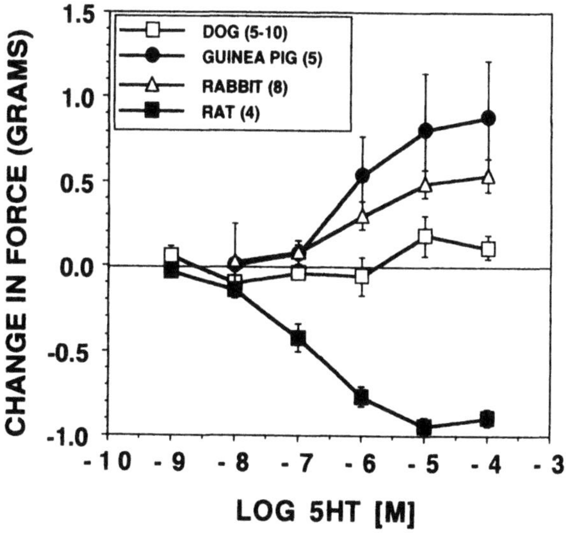

Fig. 3. Comparison of the effect of serotonin (5-HT) on tone in the esophagus from four species; dog, guinea pig, rabbit and rat. Serotonin contracted the esophagus from guinea pig and rabbit, relaxed carbamylcholine ($10^{-6}\,M$)-contracted esophagus from the rat and neither relaxed nor contracted canine esophagus. *Points* are mean values and *vertical bars* represent the standard error of the mean for the number of tissues indicated in parentheses

are present in the esophagus of certain species. In contrast to the esophagus from the rat which relaxes to serotonin, and the rabbit and guinea pig esophagus which contract to serotonin, esophageal smooth muscle from the dog neither contracted nor relaxed to serotonin, suggesting marked species differences in the serotonergic responsiveness of esophageal smooth muscle. Serotonin-induced contraction of the esophageal sphincter has also been studied and attributed to 5-HT_2 receptor activation (Lombardi et al. 1987).

V. Gastric Smooth Muscle

Serotonin activation of gastric smooth muscle, like other peripheral tissues, is complex. Serotonin can modify gastric acid secretion via receptors that are not well defined (Mate et al. 1985; Cho and Ogle 1986). In addition, serotonin can modify neuronal influences in gastric smooth muscle. With regard to its direct effects on smooth muscle, there is some evidence that serotonin can produce a contractile response via activation of 5-HT_2 receptors, in the stomach from some species (Baez et al. 1991).

Perhaps the most well studied gastric preparation is the rat stomach fundus strip which responds to exceptionally low concentrations of serotonin. Contractile responses in this tissue do not appear to be mediated by activation of 5-HT_2 receptors (Cohen and Fludzinski 1987; Van Nueten et al.

1982; CLINESCHMIDT et al. 1985). Similarities in pharmacology between the fundus receptor and 5-HT_{1C} receptors have been documented. However, in situ hybridization studies and further pharmacology have provided evidence that this receptor is not identical to the 5-HT_{1C} receptor but represents a novel, serotonergic receptor (BAEZ et al. 1990). Cloning of the rat fundus receptor has recently been published and confirms the sequence homology with the known 5-HT_2 and 5-HT_{1C} receptor clones (KURSAR et al. 1992). The most recent nomenclature for the rat stomach fundus receptor has been the 5-HT_{2B} receptor (HUMPHREY et al. 1993).

VI. Ileal Smooth Muscle

Much of the pharmacology studied in ileal smooth muscle may be attributed to effects of serotonin on neuronal components in this preparation. Clearly, activation of 5-HT_3 receptors can result in a profound contractile response in the guinea pig ileum (RICHARDSON et al. 1985). However, this effect of serotonin is known to be mediated by an indirect action via activation of 5-HT_3 receptors on presynaptic nerves resulting in acetylcholine release. A relatively weak contractile response elicited by activation of 5-HT_2 receptors on ileal smooth muscle has also been reported (ENGEL et al. 1984; BUCHHEIT et al. 1985).

In addition to the presence of 5-HT_3 receptors on nerves and the presence of 5-HT_2 receptors on smooth muscle, there is evidence that 5-HT_4 receptors may be present in the guinea pig ileum and responsible for an enhancement of the twitch response produced by serotonin (CRAIG and CLARKE 1990). This effect is not well understood with regard to the precise location of the 5-HT_4 receptors, i.e., whether 5-HT_4 receptors are located on nerves or directly on the smooth muscle.

VII. Colonic Smooth Muscle

As with other segments of the gastrointestinal tract, serotonin can produce both excitation or inhibition of smooth muscle responses in the colon. Thus, multiple receptors are present both on the nerves and smooth muscle in the colon. For example, serotonin-induced contraction in the guinea pig colon can be blocked by the 5-HT_2 receptor antagonists ketanserin and spiperone, consistent with the possibility of a 5-HT_2 receptor mediating contraction in this tissue (KOJIMA and SHIMO 1986). In the rabbit colon, contraction of the circular smooth muscle has been attributed to 5-HT_1 receptors (NG et al. 1991). Serotonin-induced contraction has also been documented in the rat colon and attributed to the ability of serotonin to interact with a 5-HT_4 receptor (BUNCE et al. 1991), which may also be present in guinea pig colon (ELSWOOD et al. 1991). Human colonic smooth muscle has also been studied with evidence to support the presence of several serotonergic receptors (BURLEIGH 1977).

F. Pulmonary Smooth Muscle

The lungs are a major source of monoamine oxidase for the degradation and clearance of serotonin from the blood (GILLIS and RAOTH 1976). In spite of the rapid clearance of serotonin by pulmonary tissue, serotonin is known to exert multiple effects in pulmonary tissue including effects on pulmonary vasculature, smooth muscle, and nerves.

Serotonin is known to constrict in vitro preparations of pulmonary arteries (BUCKNER et al. 1987; FRENKEN and KAUMANN 1984) by activation of 5-HT_2 receptors. Furthermore, using an isolated perfused guinea pig lung, serotonin was also shown to increase pulmonary arterial pressure, documenting a constrictor effect in smaller pulmonary vessels, again mediated by activation of 5-HT_2 receptors (SELIG et al. 1988). Interestingly, tachyphylaxis to the vascular smooth muscle effects of serotonin, but not to its effects on pulmonary smooth muscle, were apparent (SELIG et al. 1988). Thus, as with other peripheral blood vessels, the primary effect of serotonin on pulmonary vasculature appears to be a constrictor response mediated via activation of 5-HT_2 receptors.

Studies on the effect of serotonin in pulmonary smooth muscle have suggested the presence of bronchoconstrictor 5-HT_2 receptors in tracheal smooth muscle (COHEN et al. 1985; LEMOINE and KAUMANN 1986; WATTS and COHEN 1992) and in smaller airways (SELIG et al. 1988; BUCKNER et al. 1991). Although the pharmacological profile of the serotonergic receptor in the guinea pig trachea matches the 5-HT_2 receptor, the link to phosphoinositide turnover indicative of 5-HT_2 receptors has not been convincingly shown (COHEN and WITTENAUER 1987; BAUMGARTNER et al. 1990).

Other studies have documented the presence of 5-HT_3 receptors presumably on neuronal elements in pulmonary tissue that may participate in reflex effects of serotonin. Activation of 5-HT_3 receptors results in bronchoconstriction due to acetylcholine release (AAS 1983; ISLAM et al. 1974; KAY and ARMSTRONG 1991; SHELLER et al. 1982).

Thus, in several animal species, serotonin has been shown to exert a bronchoconstrictor effect that appears to be mediated via activation of 5-HT_2 receptors on smooth muscle and 5-HT_3 receptors on nerves.

G. Genitourinary Smooth Muscle

I. Vas Deferens

Serotonin receptors are also located in the vas deferens where the effects of serotonin appear to be both species and age dependent. As with other peripheral smooth muscle, serotonin may activate a presynaptic neuronal receptor as well as a postsynaptic smooth muscle receptor in the rat, mouse, and guinea pig vas deferens (YOSHIDA and KUGA 1986; MORITOKI et al.

1986a; SEONG et al. 1990). In the rat vas deferens, serotonin can inhibit responses to electrical stimulation, an effect presumably mediated by the 5-HT_{1B} receptor on presynaptic nerves (SMITH and BENNETT 1990). In the mouse vas deferens, a 5-HT_1-like receptor (with characteristics similar to the 5-HT_{1A} receptor) has been suggested (SEONG et al. 1990). In addition, in the mouse vas deferens, contraction to electrical stimulation has been shown to involve activation of a neuronal 5-HT_3 receptor (SEONG et al. 1990). In rats, the prejunctorial inhibitory effect of 5-HT diminishes with age, leading to speculation of increasing loss of 5-HT_{1B} receptors in the rat vas deferens in maturation and aging (BORTON and DOCHERTY 1990; MORITOKI et al. 1986b).

The 5-HT_2 receptor has also been reported in the rat and guinea pig vas deferens based on the ability of 5-HT_2 receptor antagonists, such as ketanserin, to increase serotonin-induced inhibition of electrical stimulus-evoked responses (MORITOKI et al. 1986a; KARASAWA et al. 1985). The proposed vas deferens 5-HT_2 receptor is suggested to be postsynaptic and excitatory (MORITOKI et al. 1986a).

Although the vas deferens clearly responds to serotonin via multiple receptors, characterization of the serotonergic responses in the vas deferens has been complicated by strong tachyphylaxis to serotonin and age-related changes in serotonin receptors in several species (LUCCHELLI et al. 1984; MORITOKI et al. 1986b; COHEN 1989).

II. Bladder

Serotonin also exerts effects in the bladder via activation of multiple receptors. In the bladder from most species, serotonin-induced contractions are biphasic consisting of an initial rapid contraction followed by a sustained, tonic response (LONGHURST et al. 1984; SAXENA et al. 1985; KLARSKOV and HORBY-PETERSEN 1986). The serotonergic receptors involved in the biphasic contractile response have not been studied in many species. In the cat, the initial rapid contraction was blocked by the 5-HT_3 receptor antagonist MDL 72222 and is likely to be mediated by activation of neuronal 5-HT_3 receptors effecting neurotransmitter release (SAXENA et al. 1985). Presynaptic 5-HT_3 receptors have also been identified on purinergic nerves in the rabbit bladder (CHEN 1990). Furthermore, the tonic phase of contraction to serotonin in the cat bladder was blocked by 5-HT_2 receptor antagonists and is likely to be mediated by activation of postsynaptic 5-HT_2 receptors on smooth muscle (SAXENA et al. 1985). 5-HT_2 receptor-activated contractile responses to serotonin have been documented in bladder from several species (GYERMEK 1962; KLARSKOV and HORBY-PETERSEN 1986). In addition to these receptor mechanisms, high concentrations of serotonin may contract mouse bladder via activation of a neuronal 5-HT_{1B} receptor, since pindolol, but not ketanserin or MDL 72222, blocked the response and 8-OH DPAT, the 5-HT_{1A} selective agonist, had no effect (HOLT et al. 1986).

In addition to serotonin's ability to induce contraction in the bladder, certain segments of the urinary tract may relax in response to serotonin (HILLS et al. 1984; KLARSKOV and HORBY-PETERSEN 1986). Such an effect has been documented in the human and porcine trigone, bladder neck, and urethra. The serotonin receptor mediating relaxation in these tissues has not been characterized. Such characterization may be complicated by wide species variability in urinary tissue responsiveness to serotonin (COHEN and DREY 1989).

III. Prostate

Significant concentrations of 5-HT have been documented in the adult prostate from several species including humans (DAVIS 1987; DI SANT'AGNESE et al. 1987; CROWE et al. 1991b). Serotonin appears to be present in both neurons and neuroendocrine cells and may increase with age and prostatic hyperplasia (DI SANT'AGNESE et al. 1987; CHAPPLE et al. 1991). Serotonin can contract the rat prostate (although with a lower efficacy and potency than carbamylcholine; COHEN and DREY 1989). However, as with other smooth muscle responses, considerable species variability exists in the contractile response to serotonin; for example, serotonin had no effect on guinea pig prostate (COHEN and DREY 1989). Because the prostatic capsule may contribute to a contractile response in the rat prostate, studies separating prostatic smooth muscle from the capsule showed that the contractile responses to serotonin resulted from its ability to contract prostatic smooth muscle rather than the capsule (STEIDLE et al. 1989). Receptors subserving the contractile response to serotonin in the rat prostate have not been fully characterized; however, preliminary evidence with the 5-HT_2 antagonist LY53852 suggests involvement of 5-HT_2 receptors (unpublished observations). The possibility that other serotonin receptors may be involved in the prostate remains to be established.

IV. Penile Smooth Muscle

Erectile responses in the mammalian penis are governed by (a) activity of the smooth muscles of the corpus cavernosum, (b) activity of the extensor, erector, and retractor muscles in some species, (c) penile blood flow, and (d) neural inputs from the CNS. The CNS component of penile smooth muscle function in response to serotonin has been associated with activation of the 5-HT_{1D} receptor in rats (BERENDSON and BROEKKAMP 1987) and activity at the 5-HT_{1C} receptor (BERENDSON et al. 1990). In addition, 5-HT_{1A} and/or 5-HT_2 receptors may also modulate serotonergically induced central effects on penile smooth muscle (BERENDSON et al. 1990). CNS effects on erectile function have been identified in primates where the central receptor was classified as 5-HT_1-like without complete characterization (SZELE et al. 1988). A role for 5-HT_2 receptors in the regulation of sexual activity has been suggested (FOREMAN et al. 1989) and reinforced by the observation that a potent 5-HT_2 receptor antagonist, LY237733 (amesergide), can enhance sexual performance in rats (FOREMAN et al. 1992). Whether such effects are centrally or peripherally mediated has not been resolved.

At present, there is no evidence for serotonergic innervation of either vascular or nonvascular penile smooth muscle in mammals including humans (CROWE et al. 1991a). However, because serotonin can have an effect on neurotransmitter release from adrenergic nerves, the effects of serotonin on penile smooth muscle may involve an effect on neural adrenergic components in this tissue as well as a direct effect on the smooth muscle. In in vivo studies with pithed rats, where i.v. infusion of 5-HT inhibits penile erection via a fall in corpus cavernosum pressure, serotonin effects can be mimicked by 5-carboxamidotryptamine, but 8-OH-DPAT, RU 24969, or MCPP have no effect (FINBERG and VARDI 1990). These authors suggest the presence of the 5-HT_{1D} receptor subtype (FINBERG and VARDE 1990). While the fall in corpus cavernosum pressure is a peripheral effect of 5-HT, it is unclear whether the inhibitory effect of 5-HT on erection is on cavernosa vessels or peripheral nerve endings.

In species with penile sheaths, such as the cow, relaxation of the penile retractor muscle is an integral part of the penile erectile response (KLINGE and SJOSTRAND 1974). The paired retractor muscle in the cow is potently relaxed by acetylcholine and contracted by 5-HT and histamine (KLINGE et al. 1988). 5-HT contractions can be blocked by cyproheptadine and methysergide, but the 5-HT receptor subtype has not been further characterized (KLINGE et al. 1988).

Thus, while there is evidence for the involvement of serotonin in penile smooth muscle regulation, both in the central nervous system and locally, further studies will be necessary to identify specifically the location, role and receptor subtypes involved in these responses to serotonin.

V. Uterine Smooth Muscle

Uterine smooth muscle contracts in response to serotonin via activation of 5-HT_2 receptors (MILLAR et al. 1982; ICHIDA et al. 1983; COHEN et al. 1986). Contractile responses can be competitively antagonized by selective 5-HT_2 receptor antagonists LY53857 and ketanserin (COHEN et al. 1986). Activation of 5-HT_2 receptors in rat uterus has been linked to increases in phosphoinositide turnover and influx of extracellular calcium (COHEN et al. 1986; COHEN and WITTENAUER 1987). In addition, serotonin-induced contraction in rat uterus may be dependent on estrous cycle stage (possibly through an estrogen-dependent mechanism; HAYASHI et al. 1984; ZMIRA et al. 1991). In fact, increased sensitivity to serotonin has been proposed to be a "late estrogenic response" (CAMPOS-LARA et al. 1990).

In addition to having a direct effect on uterine smooth muscle, serotonin is a powerful and potent constrictor of umbilical and placental arteries

(Zhang and Dyer 1991), and in high concentrations (m*M*) serotonin can produce fetal death when delivered intraperitoneally to pregnant rats (Furahasi et al. 1991). Similar to uterine smooth muscle, uterine vascular effects of serotonin may be hormonally regulated (Zhang and Dyer 1991). Serotonin may also play a role in regulating uterine responses to prostaglandins (Bhattacharyya et al. 1990).

H. Summary

In the last decade, the identification of highly potent and selective agonists and antagonists coupled to advances in radioligand binding studies has led to the discovery of multiple classes of serotonin receptor subtypes. All identified serotonin-receptor subtypes have been found in smooth muscle, providing an explanation for the diversity and complexity of serotonin's effects in smooth muscle. More recently, advances in molecular biology have permitted the cloning and structural elucidation of many of the serotonin receptors. Molecular biological techniques have also led to the identification of novel serotonin receptors not yet associated with functional responses either in the brain or smooth muscle. These new developments will permit identification of even more selective and potent receptor agonists and antagonists to enhance further our understanding of the role of serotonin in the physiology as well as the pathophysiology of smooth muscle. Based on the functional complexity of responsiveness to serotonin (i.e., effects on multiple cell types in smooth muscle and ability to modulate both contractile and relaxant receptors), the likely possibility exists that smooth muscle will provide a source for the further identification and subsequent characterization of novel serotonergic receptors.

References

Aas P (1983) Serotonin induced release of acetylcholine from neurons in the bronchial smooth muscle of the rat. Acta Physiol Scand 117:477–480

Adham N, Romanienko P, Hartig P, Weinshank RL, Branchek T (1992) The rat 5-hydroxytryptamine$_{1B}$ receptor is the species homologue of the human 5-hydroxytryptamine$_{1D}$ receptor. Mol Pharmacol 41:1–7

Baez M, Yu L, Cohen ML (1990) Pharmacological and molecular evidence that the contractile response to serotonin in rat stomach fundus is not mediated by activation of the 5-hydroxytryptamine$_{1C}$ receptor. Mol Pharmacol 38:31–37

Baez M, Yu L, Cohen ML (1991) Is contraction to serotonin mediated via 5-HT$_{1C}$ receptor activation in rat stomach fundus? In: Fozard JR, Saxena PR (eds) Serotonin molecular biology, receptors and functional effects. Birkhäuser, Basel, pp 144–152

Baird AW, Cuthbert AW (1987) Neuronal involvement in type 1 hypersensitivity reactions in gut epithelia. Br J Pharmacol 92:647–655

Baumgartner RA, Wills-Karp M, Kaufman MJ, Munakatos M, Hirshman C (1990) Serotonin induces constriction and relaxation of the guinea pig airway. J Pharmacol Exp Ther 255:165–173

Baxter GS, Craig DA, Clarke DE (1991) 5-Hydroxytryptamine$_4$ receptors mediate relaxation of the rat oesophageal tunica muscularis mucosa. Naunyn Schmiedebergs Arch Pharmacol 345:439–446
Berendsen HHG, Broekkamp CLE (1987) Drug-induced penile erections in rats: indications of serotonin$_{1B}$ receptor mediation. Eur J Pharmacol 135:279–287
Berendsen HHG, Jenck F, Broekkamp CLE (1990) Involvement of 5-HT$_{1C}$-receptors in drug-induced penile erections in rats. Psychopharmacology (Berlin) 101:57–61
Bhattacharyya D, Bhattacharjee S, Naq BN, Maity CR, Bandyopodhyay SK (1990) Biochemical evidence for a mediatory role of 5-hydroxytryptamine in the uterotonic action of prostaglandin $F_{2\alpha}$. Mater Med Pol 22:194–197
Blauw GJ, von Brummelen P, van Zwieten PA (1988) Serotonin induced vasodilation in the human forearm is antagonized by the selective 5-HT$_3$ receptor antagonist ICS 205-930. Life Sci 43:1441–1449
Bodelsson M, Torenbrandt K, Arneklo-Nobin (1993) Endothelial relaxing 5-hydroxytryptamine receptors in the rat vas deferens. Naunyn Schmiedebergs Arch Pharmacol 342:130–135
Borton M, Docherty JR (1990) The effects of ageing on prejunctional 5-hydroxytryptamine receptors in the rat vas deferens. Naunyn Schmiedebergs Arch Pharmacol 342:130–135
Bradley PB, Engel G, Fenuik W, Fozard JR, Humphrey PPA, Middlemiss DN, Mylecharane EJ, Richardson BP, Saxena PR (1986) Proposals for the classification and nomenclature of functional receptors for 5-hydroxytryptamine. Neuropharmacology 25:563–576
Branchek T (1993) More serotonin receptors? Current Biol 3:315–317
Brodie TG (1900–1901) The immediate action of an intravenous injection of blood serum. J Physiol 26:48–71
Buchheit K-H, Engel G, Mutschler E, Richardson B (1985) Study of the contractile effect of 5-hydroxytryptamine (5-HT) in the isolated longitudinal muscle strip from guinea-pig ileum. Naunyn Schmiedebergs Arch Pharmacol 329:36–41
Buchheit KH, Buscher HH, Gourese R (1990) The antiemetic profile of the 5-HT$_3$ receptor antagonist, ICS 205-930 and its quaternary derivative. The second IUPHAR satellite meeting on serotonin. Abstr #148.
Buckner CK, Dea D, Liberati N, Krell RD (1991) A pharmacologic examination of receptors mediating serotonin-induced bronchoconstriction in the anesthetized guinea pig. J Pharmacol Exp Ther 257:26–34
Buckner CK, Saban R, Hand JM, Laravuso RM, Will JA (1987) Pharmacological studies of contractile and relaxant responses to serotonin in extralobar pulmonary arteries isolated from the guinea pig. In: Dawson CA, Weir EK, Buckner CH (eds) The pulmonary circulation in health and disease. New York Academy Press, New York, pp 109–128
Bunce KT, Elswood CJ, Ball MT (1991) Investigation of the 5-hydroxytryptamine receptor mechanism mediating the short-circuit current response in rat colon. Br J Pharmacol 102:811–816
Burleigh DW (1977) Evidence for more than one type of 5-hydroxytryptamine receptor in the human colon. J Pharm Pharmacol 29:538–541
Butler A, Elswood CJ, Burridge J, Ireland SJ, Bunce KT, Kilpatrick GJ, Typers MB (1990) The pharmacological characterization of 5-HT$_3$ receptors in three isolated preparations derived from guinea-pig tissues. Br J Pharmacol 101:591–598
Campos-Lara G, Caracheo F, Valencia-Sanchez A, Ponce-Monter H (1990) The sensitivity of rat uterus to serotonin in vitro is a late estrogenic response. Arch Invest Med 21:71–75
Chapple CR, Crowe R, Gilpin SA, Gosling J, Burnstock G (1991) The innervation of the human prostate gland – the changes associated with benign enlargement. J Urol 146:1637–1644
Chedotal A, Hamel E (1990) Serotonin-synthesizing nerve fibers in rat and cat cerebral arteries and arterioles: immunohistochemistry of tryptophan-5-hydroxylase. Neurosci Lett 116(3):269–274

Chen H-I (1990) Evidence for the presynaptic action of 5-hydroxytryptamine and the involvement of purinergic innervation in the rabbit lower urinary tract. Br J Pharmacol 101:212–216

Chester AH, Martin GR, Bodelsson M, Arneklo-Nobin B, Tadjkarimi S, Tornebrandt K, Yacoub MH (1990) 5-hydroxytryptamine receptor profile in healthy and diseased human coronary arteries. Cardiovasc Res 24:932–937

Cho CH, Ogle CW (1986) The inhibitory action of 5-hydroxytryptamine on gastric secretory function in rats. Br J Pharmacol 87:371–377

Clancy BM, Maayani S (1985) 5-hydroxytryptamine receptor in isolated rabbit aorta: characterization with tryptamine analogues. J Pharmacol Exp Ther 233:761–769

Clineschmidt BV, Reiss DR, Pettibone DJ, Robinson JL (1985) Characterization of 5-hydroxytryptamine receptors in rat stomach fundus. J Pharmacol Exp Ther 235:696–708

Cocks TM, Angus JA (1983) Endothelium-dependent relaxation of coronary arteries by noradrenaline and serotonin. Nature 305:627–630

Cohen ML (1988) Serotonin receptors in vascular smooth muscle. In: Sanders-Bush E (ed) The serotonin receptors. Humana, Clifton, pp 295–318

Cohen ML (1989) 5-hydroxytryptamine and non-vascular smooth muscle contraction and relaxation. In: Fozard JR (ed) The peripheral actions of 5-hydroxytryptamine. Oxford University Press, New York, pp 201–217

Cohen ML (1992) 5-HT_3 receptors in the periphery. In: Hamon M (ed) Central and peripheral 5-HT_3 receptors. Academic, New York, pp 19–32

Cohen ML, Carpenter R, Schenck K, Wittenauer L, Mason N (1986) Effect of nitrendipine, diltiazem, trifluoperazine and pimozide on serotonin_2 (5-HT_2) receptor activation in the rat uterus and jugular vein. J Pharmacol Exp Ther 238:860–867

Cohen ML, Colbert WE (1986) Relationship between receptors mediating serotonin (5-HT) contractions in the canine basilar artery to 5-HT_1, 5-HT_2 and rat stomach fundus 5-HT receptors. J Pharmacol Exp Ther 237(3):713–718

Cohen ML, Drey K (1989) Contractile responses in bladder body, bladder neck and prostate from rat, guinea pig and cat. J Pharmacol Exp Ther 248(3):1063–1068

Cohen ML, Fludzinski LA (1987) Contractile serotonergic receptor in rat stomach fundus. J Pharmacol Exp Ther 243(1):264–269

Cohen ML, Fuller RW, Wiley KS (1981) Evidence for 5-HT_2 receptors mediating contraction in vascular smooth muscle. J Pharmacol Exp Ther 218(2):421–425

Cohen ML, Schenck KW, Colbert W, Wittenauer (1985) The role of 5-HT_2 receptors in serotonin-induced contractions of peripheral non-vascular smooth muscle. J Pharmacol Exp Ther 232:770–774

Cohen ML, Schenck K, Nelson D, Robertson DW (1992) Sumatriptan and 5-benzyloxytryptamine: contractility of two 5-HT_{1D} receptor ligands in canine saphenous veins. Eur J Pharmacol 211:43–46

Cohen ML, Wittenauer LA (1987) Serotonin receptor activation of phosphoinositide turnover in uterine, fundal, vascular, and tracheal smooth muscle. J Cardiovasc Pharmacol 10:176–181

Cohen RA, Zitnay KM, Weisbrod RM (1987) Accumulation of 5-hydroxytryptamine leads to dysfunction of adrenergic nerves in canine coronary artery following intimal damage in vivo. Circ Res 61:829–833

Connor HE, Feniuk W, Humphrey PPA (1989) 5-Hydroxytryptamine contracts human coronary arteries predominantly via 5-HT_2 receptor activation. Eur J Pharmacol 161:91–94

Craig DA, Clarke DE (1990) Pharmacological characterization of a neuronal receptor for 5-hydroxytryptamine in guinea pig ileum with properties similar to the 5-$\text{hydroxytryptamine}_4$ receptor. J Pharmacol Exp Ther 252(3):1378–1386

Crowe R, Burnstock G, Dickinson IK, Pryor JP (1991a) The human penis: an unusual penetration of NPY-immunoreactive nerves within the medial muscle coat of the deep dorsal vein. J Urol 145:1292–1296

Crowe R, Chapple CR, Burnstock G (1991b) The human prostate gland: a histochemical and immunohistochemical study of neuropeptides, serotonin, dopamine β-hydroxylase and acetylcholinesterase in autonomic nerves and ganglia. Br J Urol 68:53–61

Cushing DJ, Cohen ML (1992) Relaxant serotonin receptor in canine coronary smooth muscle. J Pharmacol Exp Ther 263:123–129

Davis NS (1987) Determination of serotonin and 5-hydroxyindoleacetic acid in guinea pig and human prostate using HPLC. Prostate 11:353–360

De Chaffoy de Courcelles D, deClerck F (1990) The human platelet 5-HT_2-receptor: an up-date. In: Saxena PR, Wallis DI, Wouters W, Bevan P (eds) Cardiovascular pharmacology of 5-hydroxytryptamine. Kluwer, Dordrecht, pp 445–457

Di Sant'Agnese PA, Davis NS, Chen M, De Mesy Jensen KL (1987) Age-related changes in the neuroendocrine (endocrine-paracrine) cell population and the serotonin content of the guinea pig prostate. Lab Invest 57(6):729–736

Dumuis A, Bohelal R, Sebben M, Cory R, Bockaert J (1988) A nonclassical 5-hydroxytryptamine receptor positively coupled with cyclase in the central nervous system. Mol Pharmacol 34:880–887

Edvinsson L, Birath E, Uddman R, Lee TJ-F, Duverger D, MacKenize ET, Scatton B (1984) Indoleaminergic mechanisms in brain vessels: localization, concentration, uptake and in vitro responses of 5-hydroxytryptamine. Acta Physiol Scand 121:291–299

Eglen RM, Swank SR, Walsh LKM, Whiting RL (1990) Characterization of 5-HT_3 and "atypical" 5-HT receptors mediating guinea-pig ileal contractions in vitro. Br J Pharmacol 101:513–520

Elswood CJ, Bunce KT, Humphrey PPA (1991) Identification of putative 5-HT_4 receptors in guinea-pig ascending colon. Eur J Pharmacol 196:149–155

Engel G, Hoyer D, Kalkman HO, Wick MB (1984) Identification of $5HT_2$-receptors on longitudinal muscle of the guinea pig ileum. J Recept Res 4:113–126

Erlander MG, Lovenberg TW, Baron BM, Delecea L, Danielson PE, Racke M, Slone AL, Siegel BW, Foye PE, Cannon K, Burns JE, Sutcliffe JG (1993) Two members of a distinct subfamily of 5-hydroxytryptamine receptors differentially expressed in rat brain. Proc Natl Acad Sci USA 90:3452–3456

Erspamer V (1966) Peripheral physiological and pharmacological actions of indolealkylamines. In: Erspamer V (ed) Handbook of experimental pharmacology, vol 14. Springer, Berlin Heidelberg New York, pp 245–359

Erspamer V, Asero B (1952) The identification of enteramine, the specific hormone of the enterochromaffin cell system, as 5-hydroxytryptamine. Nature 169:800–801

Fargin A, Raymind JR, Lohse MJ, Kobilka BK, Caron MG, Lefkowitz RJ (1988) The genomic clone G-21 which resembles a beta-adrenergic receptor sequence encodes the 5-HT_{1A} receptor. Nature 335:358–360

Fenuik W, Humphrey PPA, Watts AD (1983) 5-Hydroxytryptamine-induced relaxation of isolated mammalian smooth muscle. Eur J Pharmacol 96:71–78

Finberg JPM, Vardi Y (1990) Inhibitory effect of 5-hydroxytryptamine on penile erectile function in the rat. Br J Pharmacol 101(3):698–702

Flavahan NA, Shimokawa H, Vanhoutte PM (1989) Pertusis toxin inhibits endothelium-dependent relaxations to certain agonists in porcine arteries. J Physiol (Lond) 408:549–560

Foreman MM, Hall JL, Love RL (1989) The role of the $5HT_2$ receptor in the regulation of sexual performance of male rats. Life Sci 45:1263–1270

Foreman MM, Fuller RW, Nelson DL, Calligaro DO, Kurz KD, Misner JW, Garbrecht WL, Parli CJ (1992) Preclinical studies on LY237733, a potent and selective serotonergic antagonist. J Pharmacol Exp Ther 260(1):51–57

Fox AJ, Morton IKM (1990) An examination of the 5-HT_3 receptor mediating contraction and evoked [^{3}H]-acetylcholine release in the guinea-pig ileum. Br J Pharmacol 101:553–558

Fozard JR, Kilbinger H (1985) 8-OH DPAT inhibits transmitter release from guinea pig enteric cholinergic neurons by activating 5-HT_{1A} receptors. Br J Pharmacol 86:601P

Frenken M, Kaumann AJ (1984) Interaction of ketanserin and its metabolite ketanserinol with $5HT_2$ receptors in pulmonary and coronary arteries of calf. Naunyn Schmiedebergs Arch Pharmacol 326:334–339

Furchgott RF (1955) Dibenamine blockade in strips of rabbit aorta and its use in differentiating receptors. J Pharmacol Exp Ther 111:265–284

Furuhashi N, Tsujiei M, Kimura H, Yajima A (1991) Effects of ketanserin-a serotonin receptor antagonist-on placental blood flow, placental weight and fetal weight of spontaneously hypertensive rats and normal Wistar Kyoto rats. Gynecol Obstet Invest 32:65–67

Gaddum JH, Picarelli ZP (1957) Two kinds of tryptamine receptor. Br J Pharmacol 12:323–328

Gershon MD, Wade PR, Kirchgessner AL, Tamir H (1990) 5-HT receptor subtypes outside the central nervous system: roles in the physiology of the gut. Neuropsychopharmacology 3(5–6):385–395

Gidda JS, Evans DC, Krushinski JH, Robertson DW (1991) Differential effects of quaternized $5HT_3$ antagonists in cisplatin-induced emesis in dogs and ferrets. Serotonin 1991, Birmingham University, July 14–17

Gillis CN, Roth JA (1976) Pulmonary disposition of circulating vaso-active hormones. Biochem Pharmacol 25:2547–2553

Giordano J, Rogers L (1989) Peripherally administered serotonin 5-HT_3 receptor antagonists reduce inflammatory pain in rats. Eur J Pharmacol 170:83–86

Glusa E, Richter M (1993) Endothelium-dependent relaxation of porcine pulmonary arteries via 5-HT_{1C}-like receptors. Naunyn Schmiedebergs Arch Pharmacol 347: 471–477

Golino P, Piscione F, Willerson JT, Cappelli-Bigazzi M, Focaccio A, Villari B, Indolfi C, Russolillo E, Condorelli M, Chiariello M (1991) Divergent effects of serotonin on coronary-artery dimensions and blood flow in patients with coronary atherosclerosis and control patients. N Engl J Med 324(10):641–648

Gothert M, Molderings GJ, Fink K, Schlicker E (1991) Heterogeneity of presynaptic serotonin receptors on sympathetic neurons in blood vessels. Blood Vessels 28:11–18

Gyermek L (1962) Action of 5-hydroxytryptamine on the urinary bladder of the dog. Arch Int Pharmacodyn 137(1–2):137–144

Hamblin M, Metcalf M (1991) Primary structure and characterization of a human 5-HT_{1D} receptor. Mol Pharmacol 40:143–148

Hamon M (ed) (1992) Central and peripheral 5-HT_3 receptors. Academic, New York

Hartig PR (1989) Molecular biology of 5-HT receptors. Trends Pharm Sci 10:64–69

Hayashi T, Iwana A, Nagai K, Ichida S (1984) Increase of serotonin receptors (specific [^{3}H]ketanserin binding) in rat uterine membrane fractions induced by estradiol. Jpn J Pharmacol 36:180

Hen R (1992) Of mice and flies: commonalities among 5-HT receptors. Trends Pharm Sci 13:160–165

Hills J, Meldrum LA, Klarskov P, Burnstock G (1984) A novel non-adrenergic, non-cholinergic nerve-mediated relaxation of the pig bladder neck: an examination of possible neurotransmitter candidates. Eur J Pharmacol 99:287–293

Holt SE, Cooper M, Wyllie JH (1986) On the nature of the receptor mediating the action of 5-hydroxytryptamine in potentiating responses of the mouse urinary bladder strip to electrical stimulation. Naunyn Schmiedebergs Arch Pharmacol 334:333–340

Hong E, Villalon CM (1988) External carotid vasodilation induced by serotonin and indorenate. Proc West Pharmacol Soc 31:99–101

Houston DS, Vanhoutte PM (1988) Comparison of serotonergic receptor subtypes on the smooth muscle and endothelium of the canine coronary artery. J Pharmacol Exp Ther 244:1–10

Humphrey PP, Feniuk W, Perren MJ, Conner HE, Oxford AW (1989) The pharmacology of the novel 5-HT_1-like receptor agonist, GR 43175. Cephalalgia 9(9):23–33

Humphrey PP, Hartig P, Hoyer D (1993) A proposed new nomenclature for 5-HT receptors. Trends Pharm Sci 14:233–236

Ichida S, Hayashi T, Terao M (1983) Selective inhibition by ketanserin and spiroperidol of 5-HT-induced myometrial contraction. Eur J Pharmacol 96:155–158

Islam MS, Melville GN, Ulmer WT (1974) Role of atropine in antagonizing the effect of 5-hydroxytryptamine (5-HT) on bronchial and pulmonary vascular systems. Respiration 31:47–59

Julius D, MacDermott AB, Axel R, Jessell TM (1988) Molecular characterization of a functional cDNA encoding the serotonin 1c receptor. Science 241:558–564

Julius D, Huang KN, Livelli TJ, Axel R, Jessell TM (1990) The $5HT_2$ receptor defines a family of structurally distinct but functionally conserved serotonin receptors. Proc Natl Acad Sci USA 87:928–932

Kao H-T, Olsen MA, Hartig PR (1989) Isolation and characterization of a human 5-HT_2 receptor clone. Soc Neurosci Abstr 15:486

Karasawa A, Kubo K, Shuto K, Nakamizo N (1985) Interaction of 5-hydroxytryptamine and ketanserin in rat vas deferens subjected to low frequency field stimulation. Jpn J Pharmacol 37:285–291

Kaumann AJ, Frenken M (1985) A paradox: the 5-HT_2-receptor antagonist ketanserin restores the 5-HT induced contraction depressed by methysergide in large coronary arteries of calf. Allosteric regulation of 5-HT_2-receptors. Naunyn Schmiedebergs Arch Pharmacol 328:295–300

Kaumann AJ, Sanders L, Brown AM, Murray KJ, Brown MJ (1991) A 5-HT_4-like receptor in human right atrium. Naunyn Schmiedebergs Arch Pharmacol 344:150–159

Kay IS, Armstrong DJ (1991) MDL 72222 (a selective $5HT_3$ receptor antagonist) prevents stimulation of intrapulmonary c-fibres by pulmonary embolization in anaesthetized rabbits. Exp Physiol 76:213–218

Killam AL, Nikam SS, Lambert GM, Martin AR, Nelson DL (1990) Comparison of two different arterial tissues suggests possible 5-$hydroxytryptamine_2$ receptor heterogeneity. J Pharmacol Exp Ther 252:1083–1089

Klarskov P, Horby-Petersen J (1986) Influence of serotonin on lower urinary tract smooth muscle in vitro. Br J Urol 58:507–513

Klinge E, Sjostrand NO (1974) Contraction and relaxation of the retractor penis muscle and the penile artery of the bull. Acta Physiol Scand 93(420):1–88

Klinge E, Alaranta S, Sjostrand NO (1988) Pharmacological analysis of nicotinic relaxation of bovine retractor penis muscle. J Pharmacol Exp Ther 245(1): 280–286

Kobilka BK, Frielle T, Collins S, Yang-Fang T, Kobilka TS, Francke U, Lefkowitz RJ, Caron MG (1987) An intronless gene encoding a potential member of the family of receptors coupled to guanine nucleotide regulatory proteins. Nature 329:75–79

Kojima S, Shimo Y (1986) The sites of action of 5-Hydroxytryptamine in the longitudinal muscle of the guinea pig proximal colon. Asian Pac J Pharmacol 1:111–116

Kursar JD, Nelson DL, Wainscott DB, Cohen ML, Baez M (1992) Molecular cloning, functional expression and pharmacological characterization of a novel serotonin receptor (5-HT_{2F}) from rat stomach fundus. Mol Pharmacol 42: 549–557

Leff P, Martin GR, Morse JM (1987) Differential classification of vascular smooth muscle and endothelial cell 5-HT receptors by use of tryptamine analogues. Br J Pharmacol 91:321–331

Lemoine H, Kaumann AJ (1986) Allosteric properties of 5-HT_2 receptors in tracheal smooth muscle. Naunyn Schmiedebergs Arch Pharmacol 333:91–97

Lombardi DM, Barone FC, Ormsbee HS (1987) Excitatory serotonin receptors mediate contraction of isolated canine lower esophageal sphincter (LES). Gastroenterology 92:1509

Longhurst PA, Belis JA, O'Donnell JP, Galie JR, Westfall DP (1984) A study of the atropine-resistant component of the neurogenic response of the rabbit urinary bladder. Eur J Pharmacol 99:295–302

Lopez de Pablo AL, Marco EJ, Benito JM, Fraile ML, Sanz ML, Moreno MJ, Conde MV (1991) Response of isolated cat middle cerebral artery to platelets: effect of endothelial removal. Gen Pharmacol 22:353–358

Lucchelli A, Santagostino-Barbone MG, Modesto F, Grana E (1984) Direct and indirect actions of 5-hydroxytryptamine on the rat isolated vas deferens. Arch Int Pharmacodyn 269:236–251

Mankad PS, Chester AH, Yacoub MH (1991) 5-Hydroxytryptamine mediates endothelium dependent coronary vasodilatation in the isolated rat heart by the release of nitric oxide. Cardiovasc Res 25:244–248

Maricq AV, Peterson AS, Brake AJ, Myers RM, Julius D (1991) Primary structure and functional expression of the $5HT_3$ receptor, a serotonin-gated ion channel. Science 254:432–437

Mate L, Sakamoto T, Greeley GH, Thompson JC (1985) Regulation of gastric acid secretion by secretion and serotonin. Am J Surg 149:40–45

McFadden EP, Clarke JG, Davies GJ, Kaski JC, Haider AW, Maseri A (1991) Effect of intracoronary serotonin on coronary vessels in patients with stable angina and patients with variant angina. N Engl J Med 324(10):648–654

Meehan AG, Kreulen DL (1991) Electrophysiological studies on the interaction of 5-hydroxytryptamine with sympathetic transmission in the guinea pig inferior mesenteric artery and ganglion. J Pharmacol Exp Ther 256(1):82–87

Meehan AG, Story DF (1989) Interaction of the prejunctional inhibitory action of 5-hydroxytryptamine on noradrenergic transmission with neuronal amine uptake in rabbit isolated ear artery. J Pharmacol Exp Ther 248:342–347

Millar JA, Facoory BD, Laverty R (1982) Mechanism of action of ketanserin. Lancet 2:1154

Molderings GJ, Engel G, Roth E, Gothert M (1989) Characterization of an endothelial 5-hydroxytryptamine (5-HT) receptor mediating relaxation of the porcine coronary artery. Naunyn Schmiedebergs Arch Pharmacol 340:300–308

Molderings GJ, Werner K, Likunger J, Gothert M (1990) Inhibition of noradrenaline release from the sympathetic nerves of the human saphenous vein via presynaptic 5-HT receptors similar to the 5-HT_{1D} subtype. Naunyn Schmiedebergs Arch Pharmacol 342:371–377

Moritoki H, Fukuda H, Kanaya J, Ishida Y (1986a) Ketanserin potentiates the prejunctional inhibitory effect of 5-hydroxytryptamine on rat vas deferens. J Pharm Pharmacol 38:737–741

Moritoki H, Iwamoto T, Kanaya J, Ishida Y, Fukuda H (1986b) Age-related change in serotonin-mediated prejunctional inhibition of rat vas deferens. Eur J Pharmacol 132:39–46

Muller-Schweinitzer E, Engel G (1983) Evidence for mediation by 5-HT_2 receptors of 5-hydroxytryptamine-induced contraction of canine basilar artery. Naunyn Schmiedebergs Arch Pharmacol 324:287–292

Ng WW, Jing J, Hyman PE, Snape WJ Jr (1991) Effect of 5-hydroxytryptamine and its antagonists on colonic smooth muscle of the rabbit. Dig Dis Sci 36(2):168–173

Pazos A, Hoyer D, Palacios JM (1984) The binding of serotonergic ligands to the porcine choroid plexus, characterization of a new type of serotonin recognition site. Eur J Pharmacol 106:539–546

Peroutka SJ (1988) 5-Hydroxytryptamine receptor subtypes. Ann Rev Neurosci 11:45–60

Peroutka SJ, McCarthy BG (1989) Sumatryptan (GR 43175) interacts selectively with 5-HT_{1B} and 5-HT_{1D} binding sites. Eur J Pharmacol 163:133–136

Peroutka SJ, Snyder SH (1979) Multiple serotonin receptors: differential binding of [^{3}H]15-hydroxytryptamine, [^{3}H]lysergic acid diethylamide and [^{3}H]spiroperidol. Mol Pharmacol 16(3):687–699

Peroutka SJ, Huang S, Allen GS (1986) Canine basilar artery contractions mediated by 5-hydroxytryptamine$_{1A}$ receptors. J Pharmacol Exp Ther 237:901–906

Pritchett DB, Bach AWJ, Wozny M, Taleb O, DalToso R, Shih JC, Seeburg PH (1988) Structure and functional expression of cloned rat serotonin $5HT_2$ receptor. EMBO J 7:4135–4140

Rapport MM (1949) Serum vasoconstrictor (serotonin). V. The presence of creatinine in the complex. A proposed structure of the vasoconstrictor principle. J Biol Chem 180:961–969

Rapport MM, Green AA, Page IH (1948) Serum vasoconstrictor (serotonin). IV. Isolation and characterization. J Biol Chem 176:1243–1251

Reeves JJ, Bunce KT, Humphrey PP (1991) Investigation into the 5-hydroxytryptamine receptor mediating smooth muscle relaxation in the rat oesophagus. Br J Pharmacol 103:1067–1072

Richardson BP, Engel G, Donatsch P, Stadler PA (1985) Identification of serotonin M-receptor subtypes and their specific blockade by a new class of drugs. Nature 316:126–131

Robertson DW, Cohen ML, Krushinski JH, Wong DT, Parli J, Gidda JS (1990) LY191617, A 5-HT_3 receptor antagonist which does not cross the blood brain barrier. The second IUPHAR satellite meeting on serotonin, Abstr #149

Roth BL, Nakaki T, Chuang DM, Costa E (1984) Aortic recognition sites for serotonin (5HT) are coupled to phospholipase C and modulate phosphatidylinositol turnover. Neuropharmacology 23:1223–1225

Ruat M, Traiffort E, Arrang J, Tardivel-Lacombe J, Diaz J, Leurs R, Schwartz J (1993) A novel rat serotonin (5-HT_6) receptor: molecular cloning, localization and stimulation of cAMP accumulation. Biochem Biophys Res Commun 193:268–276

Saito A, Lee J-F (1986) Serotonin as an alternative transmitter in sympathetic nerves of large cerebral arteries of the rabbit. Circ Res 60:220–228

Sanders-Bush E (1988) 5-HT receptors coupled to phosphoinositide hydrolysis. In: Sanders-Bush E (ed) The serotonin receptors. Humana, Clifton, pp 181–198

Saxena PR, Villalon CM (1990) Cardiovascular effects of serotonin agonists and antagonists. J Cardiovasc Pharmacol 15(7):S17–S34

Saxena PR, Heiligers J, Mylecharane EJ, Tio R (1985) Excitatory 5-hydroxytryptamine receptors in the cat urinary bladder are of the M- and 5-HT_2-type. J Auton Pharmacol 5:101–107

Schoeffter P, Hoyer D (1989a) Is the sumatriptan (GR43175)-induced endothelium-dependent relaxation of pig coronary arteries mediated by 5-HT_{1D} receptors? Eur J Pharmacol 166:117–119

Schoeffter P, Hoyer D (1989b) 5-Hydroxytryptamine 5-HT_{1B} and 5-HT_{1D} receptors mediating inhibition of adenylate cyclase: pharmacological comparison with special reference to the effects of yohimbine, rauwolscine and some β-adrenoceptor antagonists. Naunyn Schmiedebergs Arch Pharmacol 340:285–292

Schoeffter P, Hoyer D (1990) 5-Hydroxytryptamine (5-HT)-induced endothelium-dependent relaxation of pig coronary arteries is mediated by 5-HT receptors similar to the 5-HT_{1D} receptor subtype. J Pharmacol Exp Ther 252(1):387–395

Selig MA, Bloomquist WM, Cohen ML, Fleisch JH (1988) Serotonin $(5HT)_2$ receptor mediated pulmonary vascular and airway constriction in isolated guinea pig lung. FASEB J2:A1560
Seong YH, Baba A, Matsuda T, Iwata H (1990) 5-Hydroxytryptamine modulation of electrically induced twitch responses of mouse vas deferens: involvement of multiple 5-hydroxytryptamine receptors. J Pharmacol Exp Ther 254(3): 1012–1016
Sheller JR, Holtzman MJ, Skoogh B-E, Nadel JA (1982) Interaction of serotonin with vagal- and ACh-induced bronchoconstriction in canine lungs. J Appl Physiol 52(4):964–966
Shen Y, Monsma FJ, Metcalf MA Jr, Jose PA, Hamblin MW, Sibley DR (1993) Molecular cloning and expression of a 5-hydroxytryptamine$_7$ serotonin receptor subtype. J Biol Chem 24:18200–18204
Shimokawa H, Flavahan NA, Vanhoutte PM (1989) Natural course of the impairment of endothelium-dependent relaxations after balloon endothelium removal in porcine coronary arteries. Possible dysfunction of a pertussis-sensitive G protein. Circ Res 65:740–753
Smith CFC, Bennett RT (1990) Characterization of the inhibitory 5-HT receptor in the rat vas deferens. Arch Int Pharmacodyn Ther 308:76–85
Steidle CP, Cohen ML, Hoover DM, Neubauer BL (1989) Comparative contractile responses among ventral, dorsal, and lateral lobes of the rat prostate. Prostate 15:53–63
Sumner MJ (1991) Characterization of the 5-HT receptor mediating endothelium-dependent relaxation in porcine vena cava. Br J Pharmacol 102:938–942
Sumner MJ, Feniuk W, Humphrey PPA (1989) Further characterization of the 5-HT receptor mediating vascular relaxation and elevation of cyclic AMP in porcine isolated vena cava. Br J Pharmacol 97:292–300
Szabo C, Hardebo JE, Owman C (1991) An amplifying effect of exogenous and neurally stored 5-hydroxytryptamine on the neurogenic contraction in rat tail artery. Br J Pharmacol 102:401–407
Szele FG, Murphy DL, Garrick NA (1988) Effects of fenfluramine, M-chlorophenylpiperazine, and other serotonin-related agonists and antagonists on penile erections in nonhuman primates. Life Sci 43:1297–1303
Taylor EW, Duckles SP, Nelson DL (1986) Dissociation constants of serotonin agonists in the canine basilar artery correlate to Ki values at the 5-HT1A binding site. J Pharmacol Exp Ther 236:118–125
Toda N, Okamura T (1990) Comparison of the response to 5-carboxamidotryptamine and serotonin in isolated human, monkey and dog coronary arteries. J Pharmacol Exp Ther 253(2):676–682
Tsuji T, Chilba S (1987) Vasoconstrictor mechanism of 5-hydroxytryptamine in isolated and perfused canine basilar arteries. Arch Int Pharmacodyn Ther 286: 111–122
Van Nueten JM, Leysen JE, Vanhoutte PM, Janssen PAJ (1982) Serotonergic responses in vascular and non-vascular tissues. Arch Int Pharmacodyn Ther 256:331–334
Vanhoutte PM (1991) Platelet-derived serotonin, the endothelium, and cardiovascular disease. J Cardiovasc Pharmacol 17(5):S6–S12
Verbeuren TJ, Jordaens FH, Bult H, Herman AG (1988) The endothelium inhibits the penetration of serotonin and norepinephrine in the isolated canine saphenous vein. J Pharmacol Exp Ther 244(1):276–282
Wang SS, Peroutka SJ (1988) Historical perspective. In: Sanders-Bush E (ed) The serotonin receptors. Humana, Clifton, pp 1–20
Wang Y, Baimbridge KG, Mathers DA (1991) Effect of serotonin on intracellular free calcium of rat cerebrovascular smooth muscle cells in culture. Can J Physiol Pharmacol 69:393–399

Watts SW, Cohen ML (1992) Characterization of the contractile serotonergic receptor in guinea pig trachea with agonists and antagonists. J Pharmacol Exp Ther 260:1101–1106

Weinshank RL, Zgombick JM, Macchi M, Branchek TA, Hartig PR (1993) The human serotonin 1D receptor is encoded by a subfamily of two distinct genes: 5-$HT_{1D\alpha}$ and 5-$HT_{1D\beta}$. Proc Natl Acad Sci USA (in press)

Yoshida S, Kuga T (1986) Probable pre- and postsynaptic modifications by 5-hydroxytryptamine of contractile responses to electrical stimulation of isolated guinea-pig vas deferens. Jpn J Pharmacol 41:315–323

Yu L, Nguyen H, Le H, Bloem LJ, Kozak CA, Hoffman BJ, Snutch TP, Lester HA, Davidson N, Lubbert H (1991) The mouse 5-HT_{1C} receptor contains eight hydrophobic domains and is X-linked. Mol Brain Res 11:143–149

Zhang I, Dyer DC (1991) 5-HT_2 receptor stimulated calcium influx in ovine uterine artery in late pregnancy. Arch Int Pharmacodyn Ther 310:46–55

Zmira N, Manpach M, Varon D, Maymon R, Bahary C (1991) Evaluation of ketanserin as a tocolytic agent in third trimester pregnant rats. Life Sci 48: 1809–1812

Section III
Pharmacological Agents and Smooth Muscle

CHAPTER 10

Excitation-Contraction Coupling in Gastrointestinal Smooth Muscles

K.M. Sanders and H. Ozaki

A. Introduction

The pharmacology of gastrointestinal (GI) muscles is complex. A multitude of naturally occurring and synthetic substances affect the electrical and contractile behaviors of these muscles. Previous reviews (e.g., see Daniel et al. 1989a,b) have capably and extensively documented many of the effects of endogenous and exogenous agents on GI motility and function. These reviews have also discussed the many families of receptors present in GI tissues. Pharmacological studies of GI motility in vitro and of intact muscles in vitro are complicated by the fact that many active compounds affect intrinsic nerves and other cells in addition to muscle cells. During the past 2 decades it has become possible to characterize electrical and mechanical responses of isolated smooth muscle cells to agonists. By comparing these responses to responses of intact muscles, it is possible to more accurately understand the direct effects of agonists on smooth muscle cells versus effects mediated by other cell types.

In the present review we have reviewed some of the basic cellular mechanisms of electrical rhythmicity and excitation-contraction coupling and discussed how various agonists affect these behaviors. Instead of an exhaustive listing of the actions of each class of active drug, we have concentrated the discussion on agonists that are known to provide neural and hormonal regulatory signals in the GI tract.

One of the complicating factors in understanding the excitation-contraction coupling in GI muscles is that the basic physiology of these muscles varies extensively in the different regions of the GI tract. For example, some muscles (such as those of the gastric antrum) gain little or no tone when stimulated by a slight elevation in external K^+ or moderate concentrations of acetylcholine (ACh) or pentagastrin. The response of these muscles is characterized by an increase in the amplitude of phasic contractions, and this increase correlates with an increase in the amplitude and duration of spontaneous electrical slow waves or the generation of action potentials at the peaks of slow waves (see Szurszewski 1987). Golenhofen (1976) classified these muscles as phasic muscles, and contrasted their activity with tonic muscles (such as muscles of the lower esophageal sphincter and many vascular and airway smooth muscles which are char-

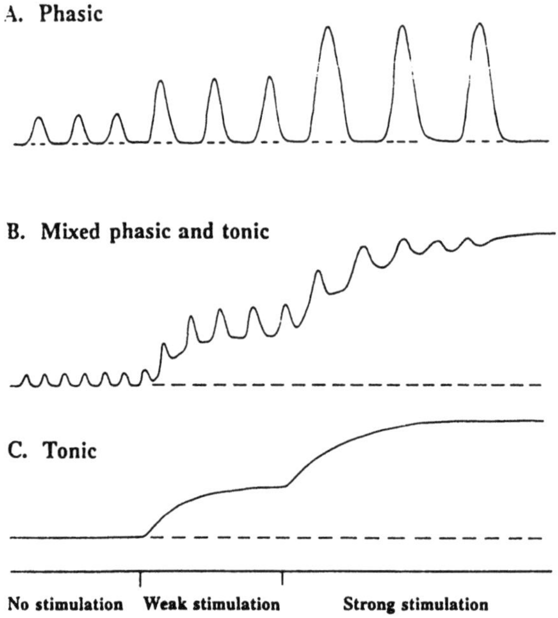

Fig. 1. Idealized examples of contractile responses from various regions of the GI tract. *Panel A* shows a typical response of a phasic muscle; *panel B* shows responses of a "mixed" type smooth muscle that displays both phasic and tonic properties; and *panel C* shows responses of a tonic muscle. (Redrawn from OZAKI et al. 1991b)

acterized by sustained contraction without rhythmic fluctuations (i.e., they generate tone; Fig. 1). The role of phasic and tonic muscles is quite different in GI motility, and the mechanisms that lead to these behaviors are also different. Thus the specific response elicited by a given pharmacological agent varies depending upon the region of the GI tract. A further complication arises from the fact that many GI muscles appear to exhibit a pattern of contraction that can best be described as a mixture of phasic and tonic activity. Muscles of the gastric fundus, guinea pig taenia caecum, longitudinal muscle of the ileum are examples of muscles that exhibit both phasic and tonic contractile behaviors. Until recently, the differences between smooth muscle types were considered to be due primarily to different mechanisms of electrical excitability. Recent work, however, suggests that sources of Ca^{2+} and possibly the mechanisms regulating the contractile apparatus differ in phasic and tonic muscles. Such mechanisms might be adaptations that favor rapid and strong transient contraction in phasic muscle and sustained contraction with less energy consumption in tonic muscles.

B. Mechanisms of Electrical Behavior of GI Smooth Muscles

Despite the diversity in electrical activity throughout the GI tract (see SZURSZEWSKI 1987), the predominant ionic currents expressed by smooth muscle cells of the various organs are similar. This suggests that the diversity in electrical activity results either from subtle differences in ionic channel structure, properties and/or expression, or from differences in regulation of the major ionic currents. Agonists have been shown to affect some of the major conductances, and these experiments will be discussed after a description of the basic ionic currents.

Voltage-clamp studies of isolated gastric muscle cells of Bufo marinus, which led the way in developing an understanding of the excitability mechanisms in smooth muscles, demonstrated three predominant phases of ionic current in response to depolarization (WALSH and SINGER 1981, 1987). The first phase is an inward current that reaches a peak within 10–20 ms. Shortly after depolarization the current reverses as an outward current develops. The initial phase of the outward current is also transient; after reaching a peak, the outward current relaxes to a steady-state level that persists for the duration of depolarization. The inward current activates at about −30 mV, reaches a peak at about +10 mV, and reverses at about +30 mV. The inward current is carried by Ca^{2+} ions: it was blocked by Mn^{2+}, and it was unaffected by low Na^{+}. The transient phase of the outward current depends upon the magnitude of Ca^{2+} influx, because: (a) the transient outward current was reduced when the inward current was inactivated by holding cells at positive potentials (the inward current displayed voltage-dependent inactivation over the range of −90 to −40 mV) and (b) Mn^{+} replacement of external Ca^{2+} blocked the transient outward current. The transient outward current was also reduced by TEA. These observations suggest that the transient outward current is due to activation of Ca^{2+}-activated K^{+} channels. The sustained phase of the outward current had properties similar to delayed rectifier-type K^{+} currents. Many studies of mammalian GI smooth muscle cells have confirmed the existence of these three major current components.

I. Major Ionic Currents Identified in GI Smooth Muscles

1. Properties of Voltage-Dependent Ca^{2+} Currents

Voltage-dependent Ca^{2+} channels are an extremely important aspect of excitation-contraction coupling in GI muscles. These channels are expressed by all of the GI muscles studied to date, and they are activated within a range of membrane potentials experienced by these cells. Activation of these channels appears to result in action potentials and slow waves in phasic GI muscles. Regulation of the open probability of voltage-dependent

Ca^{2+} channels also plays an important role in controlling contractile force in tonic muscles when membrane potentials of these muscles are altered by agonists. Regulation of voltage-dependent Ca^{2+} channels may occur as a primary response to agonists, or the open probability of these channels may be regulated by depolarizations or hyperpolarizations caused by other conductances.

Most studies of Ca^{2+} currents have utilized the patch clamp technique (see HAMILL et al. 1981). Pipette solutions in these studies typically contain Cs^+ replacement for K^+. After breaking the membrane patch under the tip of the patch pipette (i.e., establishing "whole-cell" recording conditions) the cells are dialyzed with the Cs^+ in the pipette solution. Cs^+ blocks most of the outward current and provides excellent resolution of the inward current. In one of the first reports MITRA and MORAD (1985) characterized Ca^{2+} currents in single rabbit and guinea pig gastric muscle cells that were isolated from the corpus region. In these experiments Ca^{2+} currents were resolved at potentials positive to −30 mV, reached a maximum of about 100–200 pA of inward current at about +10 mV, and reversed at about +40 mV. The current was blocked in Ca^{2+}-free solutions, and it was reduced by Co^{2+}, Cd^{2+} and dihydropyridines. These observations showed that the inward current was carried by Ca^{2+}. DROOGMANS and CALLEWAERT (1986) observed similar inward currents in isolated longitudinal cells of guinea pig ileum. The inward current in these cells was resolvable at potentials positive to −40 mV and reached a maximum at −10 mV. The inward current was transient and demonstrated voltage-dependent inactivation. Currents of up to 200 pA were obtained with 1.5 m*M* Ca^{2+} in the external solution.

The dominant Ca^{2+} current observed in GI muscles is an "L-type," dihydropyridine-sensitive current (Fig. 2; see BEAN 1989 for general description of Ca^{2+} channels; AARONSON and RUSSELL 1991; AKBARALI and GILES 1993; DROOGMANS and CALLEWAERT 1986; GANITKEVICH et al. 1986; INOUE et al. 1989; LANGTON et al. 1989a; LEE and SANDERS 1993; MAYER et al. 1990; SIMS 1992a; VIVAUDOU et al. 1991; VOGALIS et al. 1991, 1993; WARD et al. 1992a; YAMAMOTO et al. 1989). The channels that provide this class of Ca^{2+} current have also been termed "high-threshold" because they activate at relatively more depolarized levels (i.e., this current can usually be resolved with depolarization positive to about −40 mV). Single channel studies have also clearly demonstrated dihydropyridine-sensitive Ca^{2+} channels in several GI muscles (YOSHINO and YABU 1985; YOSHINO et al. 1988, 1989; VIVAUDOU et al. 1991; RICH et al. 1993). These channels are ideally suited for gastrointestinal muscles, many of which might be classified as phasic muscles, because they tend to activate periodically when cells are depolarized several mV above resting potential, and then they deactivate when cells repolarize. Although these channels inactivate (see below), inactivation is incomplete, and there is continual influx of Ca^{2+} through these channels during sustained depolarization. This pattern of activation and deactivation yields periodic influx of Ca^{2+} and phasic contractions, followed by periods of low Ca^{2+}

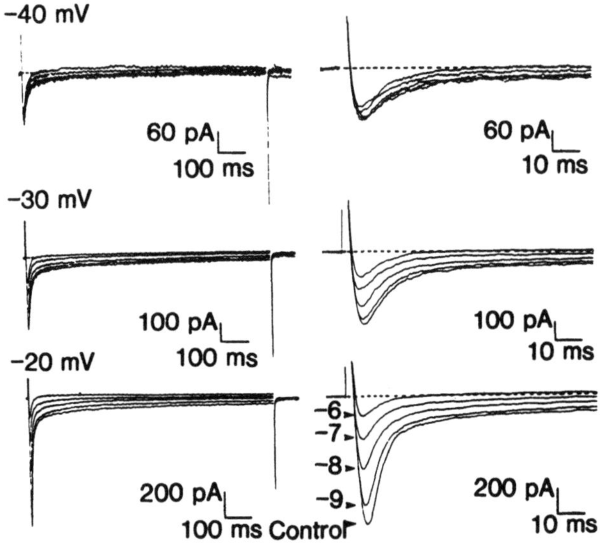

Fig. 2. The dominant Ca^{2+} current observed in GI muscles is a dihydropyridine-sensitive, "L-type" current. The figure shows whole-cell voltage-clamp records from an isolated colonic myocyte. The *left column* of traces shows inward currents generated by depolarization to −40 mV, −30 mV and −20 mV as denoted. Test pulses were 900 ms in duration; holding potential was −85 mV. Current initially activated and then inactivated to a long-lasting or "sustained" level. Multiple traces recorded from the same cell during responses to nifedipine (10^{-9}–10^{-6} *M*) are superimposed. The *right column* is a reproduction of the first 100 ms of the traces to the left, showing the dose-dependency of the block by nifedipine (concentrations noted in *bottom right panel*). Notice that the block increased with more depolarized test potentials, demonstrating the voltage-dependence of the block by nifedipine. There is still a small, initial component of inward current that was unblocked with 10^{-6} *M* nifedipine. The sustained current was completely blocked with this concentration of nifedipine. Pipette contained Cs^{+} to replace K^{+} and the experiment was conducted at 35°C. (From Ward and Sanders 1992b)

influx, restoration of resting cytosolic Ca^{2+} levels, and relaxation. This can all be accomplished simply by changes in membrane potential. Thus high-threshold Ca^{2+} channels are the ionic basis for several types of motility patterns observed in the GI tract, such as segmentation, in which periods of contraction are followed by relaxation and refilling, and peristalsis, in which sweeps of contraction over many centimeters are also followed by relaxation and refilling.

High-threshold Ca^{2+} channels do not appear to explain the generation of rhythmic pacemaker activity found in many GI muscles, however, because resting potentials of many rhythmic tissues lie far below the activation range of L-type Ca^{2+} channels. Other conductances, including "low-threshold" Ca^{2+} channels, may facilitate this behavior. Evidence for this hypothesis is discussed below in the section on "Pacemaker Currents."

Evidence also exists for "low-threshold" Ca^{2+} currents in some GI muscles. For example, in cat esophageal muscles a plot of peak inward current in response to depolarization steps has a "hump" in the I–V curve between −50 and −20 mV (SIMS et al. 1990a). At least two types of Ca^{2+} conductances can also be observed in single channel records from toad gastric muscle cells (VIVAUDOU et al. 1991), and there is also evidence in whole cell records for two types of Ca^{2+} current (VIVAUDOU et al. 1988). These authors found two conductance levels of Ca^{2+} channels, one averaging 11 pS and the other about 26 pS. The smaller conductance channels were present in the majority of patches, whereas the large conductance channels were difficult to resolve in the absence of the Ca^{2+} channel agonist, Bay K 8644. The larger conductance channels were considered to fit within the description of L-type Ca^{2+} channels, whereas the small conductance channels were not well described by standard definitions of either "N-" or "T-type" Ca^{2+} channels. Two levels of channel conductance were also noted in canine colonic cells (RICH et al. 1993). Although the conductance levels in these cells were similar to the levels reported in toad gastric cells, the smaller conductance channels were far more rare than the larger channels and both conductance levels were blocked by dihydropyridines. Northern blot analysis of α subunits revealed only one class of Ca^{2+} channels, and these corresponded to dihydropyridine-sensitive Ca^{2+} channels. At present it appears that there are tissue and species differences in the types of Ca^{2+} channels expressed, but every muscle studied to date displays dihydropyridine-sensitive Ca^{2+} currents. These fit well within the general description of L-type Ca^{2+} channels. More single channel studies that clearly demonstrate the existence of dihydropyridine-resistant currents are necessary before the conclusion that "T-type" Ca^{2+} channels are common to GI muscles can be reached.

Most studies of Ca^{2+} currents in isolated GI muscle cells have been performed at room temperatures (i.e., 21–25°C). Comparison of these currents at room temperature and physiological temperatures showed that the kinetics of activation and inactivation are increased at higher temperatures (WARD and SANDERS 1992b). The magnitude of the inward current was also greatly increased at physiological temperatures, but the sensitivity of Ca^{2+} currents to block by dihydropyridines appeared to decrease when temperature was raised. The increase in inward current at 35–37°C and the fact that currents initially "run-up" when cells are dialyzed with solutions containing ATP (LANGTON et al. 1989a) could mean that Ca^{2+} currents are metabolically regulated. But this hypothesis has not been investigated at present.

Ca^{2+} currents of GI muscle cells inactivate. Upon depolarization, inward current increases rapidly and then decreases despite constant depolarization (Fig. 2). Experiments have suggested that a portion of the inactivation is voltage-dependent (DROOGMANS and CALLEWAERT 1986; LANGTON et al. 1989a; GANITKEVICH et al. 1986). When cells are pre-conditioned by depolarization the availability of Ca^{2+} channels that can be activated upon depolarization decreases. Inactivation of Ca^{2+} currents is also Ca^{2+}-dependent (GANITKEVICH et al. 1987). When Ba^{2+} or Na^{+} are used to carry charge through Ca^{2+} channels, the rate of inactivation decreases (VOGALIS et al. 1992). These properties of inactivation affect the excitability of GI muscles. Voltage- and Ca^{2+}-dependent inactivation tend to provide negative feedback to regulate the frequency of excitable events and Ca^{2+} influx into GI muscle cells. For example, depolarized regions of muscle may have lower excitability (less ability to produce action potentials or slow waves) because of a reduction in the number of Ca^{2+} channels available for activation. Repolarization of action potentials or slow waves brings membrane potential back to the resting level rapidly, but GI muscles experience relatively long refractory periods (PUBLICOVER and SANDERS 1986). This refractoriness is at least partially due to the slow rate at which Ca^{2+} channels recover from inactivation. Stuides of the time course of recovery from inactivation have shown that full recovery requires at least 1 s in guinea pig taenia caeci cells (GANITKEVICH et al. 1986). In canine gastric muscles recovery from inactivation after depolarization is even slower, and the time constant for recovery matches the time constant for restoration of resting levels of intracellular Ca^{2+} ($[Ca^{2+}]_i$) (VOGALIS et al. 1992). Thus, depolarization and an increase in $[Ca^{2+}]_i$ reduce the availability of Ca^{2+} channels and provide negative feedback against further influx of Ca^{2+}.

As described above, Ca^{2+} currents appear to be transient in GI muscles. Although a transient inward current might explain "spikes" or action potentials, more stained inward currents would seem to be involved in electrical slow waves because these events have durations of several seconds. GANITKEVICH et al. (1986) found that the inward current of taenia caeci cell actually occurred in two phases: depolarization elicited an initial peak current that decayed within about 100–200 ms to a steady-state level. Others have also shown that inactivation is incomplete (e.g., AARONSON and RUSSELL 1991; AKBARALI and GILES 1993; DROOGMANS and CALLEWAERT 1986; LANGTON et al. 1989a, WARD and SANDERS 1992b). With depolarizations in the range of −40 to −30 mV, the magnitude of the sustained current can be 25% or more of the peak current (WARD et al. 1992). Characterization of the voltage dependence of activation and inactivation showed that these curves overlap (Fig. 3; LANGTON et al. 1989a; DROOGMANS and CALLEWAERT 1986). Thus the current activates over a range of potentials in which full inactivation has not been achieved. This phenomenon, known as "window current," can result in sustained inward current (COHEN and LEDERER 1987; LEE et al. 1984). It is important to note that window current occurs in GI muscle cells at potentials experienced by intact muscles during typical electrical events such as electrical slow waves (SZURSZEWSKI 1987; SANDERS 1989, 1992). The window current range also corresponds to the range of potentials referred to by some authors as the "mechanical threshold," where depolarization is coupled to the development of tension (e.g., BARAJAS-

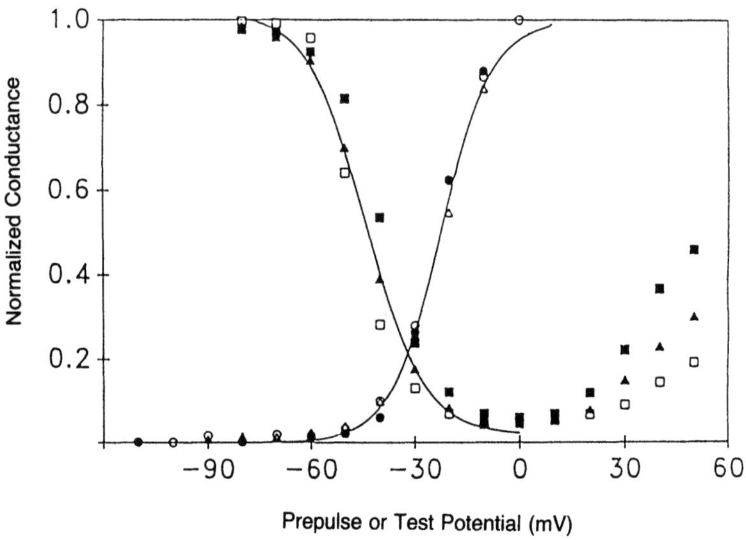

Fig. 3. Voltage-dependence of activation and inactivation of Ca^{2+} current recorded from three isolated colonic myocytes (*different symbol* for each cell). Voltage-clamp protocols are described in LANGTON et al. (1989a). Activation curve shows that current was resolvable with test depolarizations positive to about −50 mV. Maximal activation occurred at about 0 mV. The inactivation curve shows that the current was maximally available when cells were held at −80 mV before depolarization, and the current inactivated over the range of −70 to 0 mV. Data were fitted with Boltzman functions. The activation and inactivation curves overlap, demonstrating "window current" in the range of −50 to −10 mV. (From LANGTON et al. 1989a)

LOPEZ and HUIZINGA 1989; BAUER and SANDERS 1985; KURIYAMA et al. 1970; MORGAN and SZURSZEWSKI 1980; MORGAN et al. 1981; SZURSZEWSKI 1987; SANDERS and PUBLICOVER 1989).

Although the magnitude of the sustained Ca^{2+} current that occurs in the range of "window current" (i.e., about −50 to −20 mV) is small, the amount of Ca^{2+} influx relative to the volume of these cells is significant. For example, by integrating the inward current during depolarizations similar in duration to electrical slow waves, it was found that depolarizations in the range of −40 to −20 mV would increase intracellular Ca^{2+} in colonic muscle cells by several tens of μM (LANGTON et al. 1989a). Even with 100-fold buffering of intracellular Ca^{2+}, this influx is sufficient to increase $[Ca^{2+}]_i$ to levels necessary for contraction. Direct measurements of the changes in free $[Ca^{2+}]_i$ with fluorescent Ca^{2+} dyes indicate that the influx of Ca^{2+} is sufficient to increase $[Ca^{2+}]_i$ from the resting level to levels known to cause contraction (BECKER et al. 1989; VOGALIS et al. 1991). Thus Ca^{2+} influx is capable of explaining the coupling between slow waves and contractions in gastrointestinal muscles (BARAJAS-LOPEZ and HUIZINGA 1989; CHRISTENSEN et al. 1969; MORGAN and SZURSZEWSKI 1980; OZAKI et al. 1991c; SANDERS and SMITH 1986; SANDERS 1983). Small changes in the amplitude or duration of slow waves can greatly affect the magnitude of the Ca^{2+} influx (VOGALIS et al. 1991), and this explains the steep relationship between voltage and tension in GI muscles (BARAJAS-LOPEZ and HUIZINGA 1989; BAUER and SANDERS 1985; SZURSZEWSKI 1975; MORGAN and SZURSZEWSKI 1980; MORGAN et al. 1981).

Since Ca^{2+} influx is such an important determinant of the contractile performance of many GI muscles, the regulation of voltage-dependent Ca^{2+} channels is of fundamental importance in GI pharmacology. Agents that affect this conductance directly influence the force of contractions. The voltage-dependent Ca^{2+} channels of GI muscles have properties and pharmacology similar to Ca^{2+} channels of many other cells including the heart and vascular smooth muscle. Recent studies have also demonstrated molecular similarities between the Ca^{2+} channels of the GI tract and other tissues (RICH et al. 1993), making it difficult to selectively alter the Ca^{2+} permeability of GI muscles without possibly deleterious side effects. It is interesting to note that voltage-dependent Ca^{2+} channels are not typically the primary target for agonist regulation in many mammalian GI muscles, although examples to the contrary exist in mammals (e.g., MAYER et al. 1990) and lower vertebrates (CLAPP et al. 1987; VIVAUDOU et al. 1988). In many cells, the open probability of Ca^{2+} channels appears to be primarily regulated by other conductances that either depolarize membrane potential and increase the open probability of Ca^{2+} channels or hyperpolarize and decrease the open probability. Excitatory agonists primarily regulate K^+ conductances or activate a nonselective cation conductance, and inhibitory agonists appear to primarily activate K^+ conductances. Specific effects of agonists will be discussed below in the section "Regulation of Ionic Currents by Agonists." As specific channel agonists and antagonists are developed in the future to regulate GI motility, perhaps the same design philosophy should be followed: Ca^{2+} influx in GI muscles may be best manipulated by channel blocking drugs directed at specific conductances that regulate the open probability of Ca^{2+} channels by affecting membrane potential.

2. Properties of K^+ Currents

As is true for most excitable cells there is significant diversity in the K^+ currents of GI muscles. Since the first records of WALSH and SINGER (1981) from experiments on Bufo marinus gastric muscle cells, it has been recognized that there are at least two components of outward current. One component is Ca^{2+}-dependent, and the other appears to be a "delayed-rectifier"-type current. Other, less well characterized and less well understood, K^+ currents are also expressed by GI muscles. These are particularly interesting in terms of their regulation by agonists and their profound effects on the contractile activity of GI muscles.

a) Ca^{2+}-Dependent K^+ Currents

Ca^{2+}-dependent K^+ conductances are ubiquitous in muscles of the GI tract, and many studies have characterized the properties of the whole cell currents attributable to these channels and the properties of the channels responsible for these currents. The primary contributor to this class of conductances is the large conductance Ca^{2+}-activated K^+ channel known as "maxi K" or BK. SINGER and WALSH (1984, 1987) found large conductance Ca^{2+}-activated K^+ channels in isolated gastric cells of Bufo marinus. They studied excised, inside-out patches. By changing the solution that bathed the inner surface, they studied the selectivity and Ca^{2+} dependence of these channels. In symmetrical K^+ gradients (130 m*M*/130 m*M*) the channels were voltage and Ca^{2+} dependent and had a unitary conductance of approximately 250 pS. The open probability of these channels increased as the Ca^{2+} concentration at the inner surface of the patch was increased. The channels were highly selective for K^+, because reversal potentials in various solutions were close to those predicted by the Nernst relationship. Ca^{2+}-activated K^+ channels of the same type are also found in mammalian GI muscles. For example, BENHAM et al. (1986) characterized similar channels in longitudinal jejunal muscle cells of the rabbit. In physiological gradients these channels had unitary conductances of about 100 pS, and in symmetrical 126 m*M* K^+ solutions the channel conductance was about 200 pS. The relative permeability of the channels (with respect to K^+) was 1.0 K^+ : 0.7 Rb^+ : 0.05 Na^+ : 0.05 Cs^+. Studies on canine colonic and gastric muscle cells have demonstrated a steep relationship between Ca^{2+} and open probability (Fig. 4; CARL and SANDERS 1989; CARL et al. 1990). A change in $[Ca^{2+}]_i$ from 100 n*M* to 1 μ*M* was found to be equivalent to a depolarization of 130 mV in terms of the increase in open probability.

An important feature of Ca^{2+}-activated K^+ channels is that open probability increases as a function of $[Ca^{2+}]_i$ and voltage. This feature could provide negative feedback to repolarize membrane potential after periods of excitation and to limit the entry of Ca^{2+}. At resting membrane potentials and at resting levels of $[Ca^{2+}]_i$l (i.e., about 100 n*M*), the open probability of Ca^{2+}-activated K^+ channels is very low and this conductance may contribute very little to the determination of resting potential. Some GI muscles express a very large number (10 000–20 000) of these channels, so even with extremely low open probabilities a portion of the resting conductance could be due to these channels. Upon depolarization, the open probability of voltage-dependent Ca^{2+} channels increases and, as a result, $[Ca^{2+}]_i$ increases. These changes, depolarization and an increase in $[Ca^{2+}]_i$, increase the open probability of Ca^{2+}-activated K^+ channels. The resulting increase in outward current may serve to terminate excitable events and cause repolarization. Thus regulation of Ca^{2+}-activated K^+ channels could be an effective means of controlling the excitability of GI muscles (e.g., COLE et al. 1989; COLE and SANDERS 1989b).

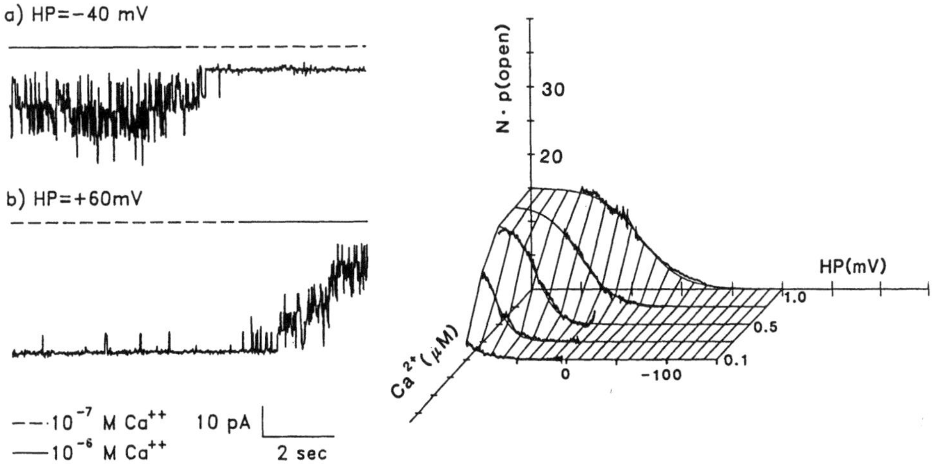

Fig. 4. Ca^{2+} and voltage dependent activation of K channels (canine prosdina colon). Ca^{2+}-activated K current depends upon voltage and Ca^{2+} concentration. Recording *at left* was taken from an isolated patch of membrane from a colonic myocyte. The patch contained several Ca^{2+}-activated K channels (inside-out configuration; see HAMILL et al. 1981). The recording was made in a symmetrical KCl gradient, so the reversal potential for the K current was 0 mV. In *trace A* the patch was held at −60 mV (inside surface of patch negative). In the beginning of the recording the inside surface was bathed with a solution containing 10^{-6} *M* Ca^{2+} (*black bar along top of trace*). Multiple channel openings were observed (at least three channels are apparent in this section of the trace; currents were inward at this potential). Then Ca^{2+} was reduced to 10^{-7} *M* (*dotted line along top of trace*). Open probability decreased such that no channel openings were noted in the remainder of the trace. Then the patch was stepped to +40 mV (*panel B*). The depolarization increased the open probability and a few brief openings were resolved at this potential (current was outward at this potential), even at this low concentration of Ca^{2+}. Switching back to 10^{-6} *M* Ca^{2+} caused a significant increase in open probability. Open probability was determined at potentials from −150 to +100 mV (using ramp potential protocols) and in Ca^{2+} concentrations ranging from 10^{-7} to 10^{-6} *M*. Data were fitted with Boltzman functions. A three-dimensional plot is shown *at the right of the figure* constructed by fitting the Boltzman curves with a surface. This surface describes the voltage and Ca^{2+} dependence of Ca^{2+}-activated K channels over the physiological range of these parameters. (Data redrawn from CARL and SANDERS 1989)

At the whole cell level, Ca^{2+} activated K^+ channels generate a current that reflects changes in $[Ca^{2+}]_i$. As described in the previous section, depolarization initiates inward current carried by Ca^{2+}. If $[Ca^{2+}]_i$ is not highly buffered, the inward current is rapidly obscured by the development of a large outward current that reaches a peak rapidly and then decays with continued depolarization. The peak of the outward current and a portion of the sustained outward current are reduced when external Ca^{2+} is reduced, Ca^{2+} channel blocking drugs are added, or when buffering of $[Ca^{2+}]_i$ is high. A portion of the outward current can also be blocked by low concentrations of tetraethylammonium ions and the scorpion toxin, charybdotoxin. Both of these agents block Ca^{2+}-activated K^+ channels. These observations all indicate that a portion of the outward current response to depolarization is due to the activation of Ca^{2+}-activated K^+ channels.

Another phenomenon attributable to Ca^{2+}-activated K^+ channels occurs in some isolated GI muscle cells. BENHAM and BOLTON (1986) observed spontaneous transient outward currents (STOCs) in longitudinal muscle cells of the rabbit jejunum and in some vascular smooth muscle cells. When cells were held positive to −50 mV, STOCs of about 100 ms duration with amplitudes as great as 250 pA were observed. If the pipette solution contained concentrations of EGTA above 1 m*M* or Cs^+, STOCs were blocked. STOCs are thought to be due to spontaneous and periodic release of Ca^{2+} from intracellular stores (BOLTON and LIM 1989; OHYA et al. 1987). The amplitude of STOCs suggests that up to 100 Ca^{2+}-activated K^+ channels might open simultaneously as a result of a localized release of Ca^{2+}. The occurrence of STOCs has been used as an assay of Ca^{2+} release in a number of studies (BOLTON and LIM 1989; BENHAM and BOLTON 1986). For example, addition of caffeine (5 m*M*) induced a burst of STOC activity, followed by a sustained period of quiescence and the inability to generate STOCs with other agents that normally release Ca^{2+} from intracellular stores. These experiments directly demonstrate that Ca^{2+}-activated K^+ channels could provide negative feedback in the case of stimulation by agents such as acetylcholine that depolarize membrane potential and release Ca^{2+} from internal stores (OZAKI et al. 1992a). The occurrence of STOCs also suggests that the release of Ca^{2+} from stores can be compartmentalized. Localized dumping of stores could generate a high Ca^+ concentration near the inner surface of the cell membrane, but this may not elevate $[Ca^{2+}]_i$ to a point where contraction is initiated. The localized increase in $[Ca^{2+}]_i$ appears to activate Ca^{2+}-activated K^+ channels and cause STOCs, and it probably inactivates voltage-dependent Ca^{2+} channels. Both factors would tend to lower the excitability of the cell.

b) Delayed Rectifier Currents

If the entry of Ca^{2+} into cell is blocked or the intracellular Ca^{2+} concentration is buffered, another component of voltage-dependent K^+ current can be resolved in response to depolarization. At room temperature this current activates slowly and is typically reduced by 4-aminopyridine or tetraethylammonium ions (BIELEFELD et al. 1990; COLE and SANDERS 1989a; THORNBURY et al. 1992a). Similar currents in other smooth muscles have been referred to as "delayed rectifiers" (BEECH and BOLTON 1989b; MURAKI et al. 1990; OKABE et al. 1987). The activation kinetics of this current at physiological levels of depolarization are relatively slow, and it was therefore difficult to see how this current could contribute to action potentials or the initial phase of electrical slow waves. Recent studies, however, have shown that delayed rectifiers are quite sensitive to temperature, and increasing temperature to

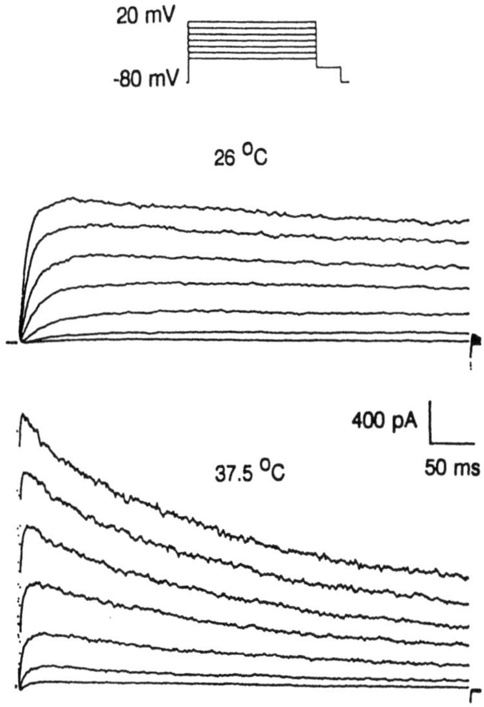

Fig. 5. Delayed rectifier currents of colonic myocytes. Effect of temperature on voltage-dependent outward (delayed rectifier) currents. Bath contained nifedipine (10^{-6} *M*) to block influx of Ca^{2+}. Family of currents were recorded from the same cell at 26°C (*upper traces*) and 37.5°C (*lower trace*). Holding potential was −80 mV and responses to test pulses of −40 to +20 mV are shown. Raising temperature increased the rate of activation, increased the magnitude of the delayed rectifier current, and increased the rate of inactivation. Maximal current approximately doubled in magnitude at +20 mV. (Redrawn from Thornbury et al. 1992a)

37°C greatly increases the magnitude of the current and the rates of activation and inactivation (Fig. 5; Thornbury et al. 1992a,b). At physiological temperatures this current reached a peak of activation within 20 ms with a depolarization to −20 mV, suggesting the current activates fast enough to influence excitable events in GI muscle cells.

Comparison of delayed rectifier currents in circular and longitudinal muscle cells of the colon suggests that different species of channels may contribute to electrical events in these cells (Thornbury et al. 1992a,b). The delayed rectifier current in longitudinal cells inactivated at more negative potentials (half-inactivation occurred at −63 mV) than in circular muscle cells (half inactivation at −36 mV). The current in longitudinal cells was also much less sensitive to 4-aminopyridine than the current in circular cells. Both currents were somewhat sensitive to extracellular TEA. It is possible that diversity of voltage-dependent K currents could help explain the diversity in electrical activity in various regions of the GI tract (see section "Mechanism of Electrical Rhythmicity in Phasic GI Smooth Muscles").

It is interesting to note that quite different species of delayed rectifiers appear to be expressed in different regions of the GI tract and in different species. For example, in comparison to the species of delayed rectifier expressed in canine colonic muscles, a different form appears in small intestinal cells of the same species. FARRGIA et al. (1993) have recently characterized a noninactivating, voltage-dependent outward current in jejunal cells that was resolved upon depolarizations positive to −65 mV. This current was blocked by quinidine, but it was insensitive to 4-aminopyridine. The current was activated by fenamates. Thus, the delayed rectifier current in the small intestine of dog is clearly different from the types of delayed rectifiers expressed in the proximal colon (i.e., THORNBURY et al. 1992a,b).

Delayed rectifiers have not been studied at the single channel level to the extent that Ca^{2+}-activated K^+ currents have been studied. This is because the conductance of delayed rectifiers is less than Ca^{2+}-activated K^+ channels, and they are frequently difficult to resolve in the presence of large-conductance Ca^{2+}-activated K^+ channels. A small channel of rabbit longitudinal jejunal muscle cells that was characterized by BENHAM and BOLTON (1983) may be a delayed rectifer. These channels had a unitary conductance of about 50 pS in a physiological potassium gradient. The channels were blocked reversibly by tetraethylammonium (5–10 m*M*), but were unaffected by Ca^{2+}-free EGTA solution. The density of channels varied from 1 to 12 per patch. The channels were also voltage-dependent. Very few channel openings were observed at potentials near the resting potential in situ, but depolarization into the range of −40 to +40 mV greatly increased the open probability. These channels showed little or no inactivation, but experiments were performed at room temperature which can de-emphasize the inactivation of delayed rectifier channels (THORNBURY et al. 1992a). Clearly the hypothesis that two or more species of delayed rectifier channels are expressed in GI muscles must be tested by single channel experiments.

Voltage-clamp studies of colonic circular muscle cells (THORNBURY et al. 1992a) suggest that more than one species of delayed rectifier contributes to non-Ca^{2+}-dependent outward currents, but the pharmacological tools available are inadequate to clearly separate these currents. Molecular studies have also supported the notion that there is diversity in the species of delayed rectifiers in GI muscles (ADLISH et al. 1991). Recently, a 4-aminopyridine-sensitive voltage-dependent K^+ current (CSMK1) has been cloned from colonic muscle and expressed in Xenopus oocytes (HART et al. 1993). These channels are 91% homologous to RAK, a $K_v1.2$ class K^+ channel cloned from rat heart (PAULMICHL et al. 1991), but the colonic channels are nearly an order of magnitude more sensitive to 4-aminopyridine than the cardiac species (HART et al. 1993). Northern analysis suggests that this channel is widely expressed in GI smooth muscles, but not in vascular and other visceral smooth muscles. The diversity of expression of voltage-dependent K^+ channels is likely to prove a very important area of study since these conductances contribute to the varying pharmacology and electrical activities from region to region within the GI tract.

c) "A Currents"

A current with many of the properties attributed to A currents has been identified in rat proximal colon smooth muscle cells (VOGALIS et al. 1993). This current activated at potentials positive to −60 mV and rapidly inctivated. The current was blocked by 3–5 m*M* 4-aminopyridine, and it was relatively insensitive to TEA. The activation and inactivation curves overlapped in the range of potentials from −60 to −40 mV, suggesting that a portion of the channels responsible for the current remain activated at these potentials even during sustained depolarization. This conductance may serve to retard the rising phase of the action potentials in rat colonic muscle cells, since this phase was accelerated in cells treated with 4-aminopyridine. The A currents in rat proximal colon cells may be molecularly related to other voltage-dependent K currents such as the "delayed rectifier" currents in proximal colon cells of the dog (see THORNBURY et al. 1992a,b). Although the activation/inactivation properties and pharmacologies of these channels differ from the A current in rat colonic cells, similar differences exist between channels in the Shaker family of K^+ channels due to relatively minor differences in molecular structure (see STRUMER et al. 1989).

d) "M Currents"

Another K^+ channel in GI muscles that appears to contribute to agonist responses is the "M" current (SIMS et al. 1985). Application of cholinergic agonists to gastric muscle cells of Bufo marinus resulted in a decrease in membrane conductance and depolarization (Fig. 6). This observation is not consistent with a conductance that is activated by cholinergic stimulation, but rather it must be explained by suppression of an outward conductance. This hypothesis was tested in voltage clamp experiments in which cells were held at relatively positive potentials (i.e., −20 to −35 mV) and then stepped periodically to more hyperpolarized levels. This allowed characterization of outward tail currents that represented deactivation of voltage-dependent outward currents. The outward current identified in this manner had properties similar to the "M" current of sympathetic neurons (ADAMS et al. 1982a,b): (a) the current remained activated at a constant level with sustained depolarization; (b) the current deactivated upon hyperpolarization although this occurred with a slower time course in the gastric muscle cells; and (c) the current was suppressed by cholinergic stimulation. The development of outward current upon depolarization and relaxation in the current upon repolarization were not affected by replacement of external Ca^{2+} with Mn^{2+}

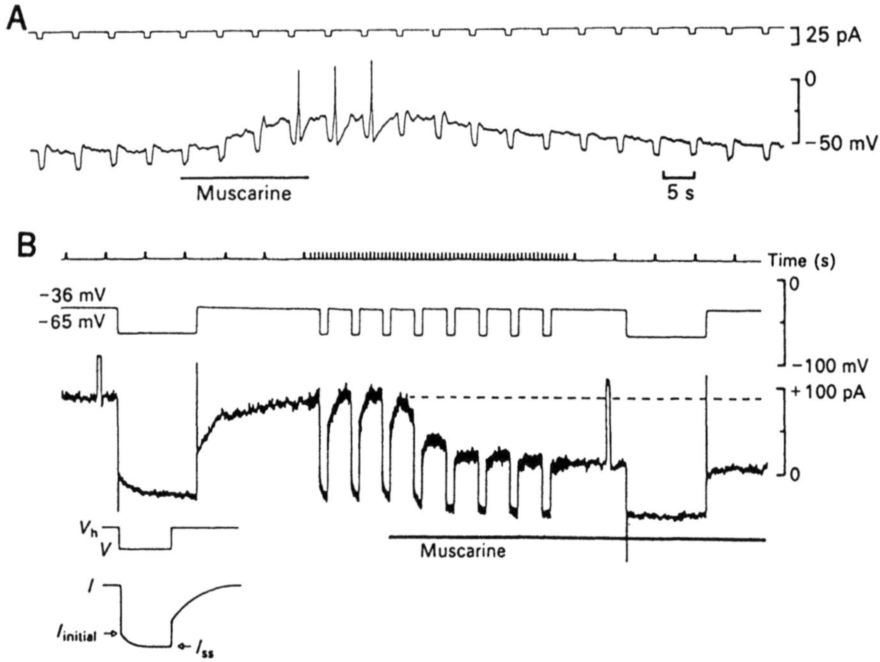

Fig. 6. Application of cholinergic agonist, muscarine (0.5 m*M*), to gastric muscle cells of Bufo marinus caused depolarization (*bottom trace in panel A*) and a decrease in conductance (as determined by the increase in the amplitude of electrotonic responses to constant current injections shown in *top trace in panel*). At peak of response action potentials were elicited at the braek of electrotonic potentials. *Panel B* shows cholinergic responses studied under voltage clamp. The membrane was held at −36 mV and periodically switched for 2 s to −65 mV (voltage-clamp protocol is shown in *second trace of panel*). Membrane currents are shown in *bottom trace*. Upon hyperpolarization there was a time-independent jump in current ($I_{initial}$ *in the inset at the bottom left*) and then a time-dependent current that developed to a steady-state (I_{SS} *in the inset*). ACh (0.5 m*M* in the application pipette) caused a net inward current to develop at the holding potential (current *below dotted line*), and current excursions upon hyperpolarizations were smaller. I_{SS}, which represents a K^+ current turning off upon hyperpolarization, was suppressed by ACh. This current is referred to as the M-current, and the suppression of this current was proposed as the mechanism for the depolarization response observed in *panel A*. (From SIMS et al. 1985)

suggesting that the current was not Ca^{2+} dependent. There does not appear to be an equivalent of an "M" current in mammalian GI muscles.

e) Inward Rectifiers

Although inward rectifiers might help explain the generation of spontaneous electrical rhythmicity of GI muscles, few examples of these currents have been found in smooth muscles. BENHAM et al. (1987) described an inwardly rectifying current in rabbit jejunal longitudinal muscle cells. These cells had resting potentials in the range of −30 to −60 mV. Under current clamp conditions, the electrotonic responses elicited by hyperpolarizing currents did not reach a steady-state. Instead, hyperpolarizations negative to −70 mV reached a peak and then decayed with time, suggesting activation of a time-dependent conductance. Under voltage clamp, hyperpolarization negative to −70 mV elicited a time-independent current followed by the development of a time-dependent current over a period of several seconds. Tests of the reversal potential showed that the current reversed at about −30 mV, suggesting that current was due to a nonselective conductance pathway that carried potassium and sodium ions. Cesium (1 m*M*) blocked the inward rectifier current, and Ba^{2+} (10 m*M*) partially blocked the current. TEA had no effect on the current.

3. Properties of Chloride Currents

Few studies have detected Cl^- currents in GI muscles. One example was found in anococcygeus muscle cells (Byrne and Large 1987). Application of norepinephrine induced large depolarizations that were associated with an increase in membrane conductance. The depolarization was associated with the activation of an inward current at holding potentials of −50 to −60 mV. The response was blocked by phentolamine suggesting that it was mediated by α-adrenoceptors.

Studies to determine the ionic mechanism of the norepinephrine-induced inward current in anococcygeus muscles showed that the current reversed at about 0 mV when solutions containing normal concentrations of K^+ were used. This finding suggsted that the current was either a nonselective cation current or a chloride current. When K^+ was removed from pipette and bath solutions and TEA was added to block K^+ channels, the norepinephrine-induced current reversed at 1 mV. Adjustments in the Na^+ and Cl^- gradients demonstrated that the current reversed near the equilibrium potential for Cl^- ions. These studies indicated that the current activated by norepinephrine was carried by Cl^- ions.

Cl^- current was also activated in anococcygeus muscle cells by the Ca^{2+} ionophore, A23187, and by caffeine. Removal of external Ca^{2+} decreased the Cl current, but did not block the current, suggesting that external Ca^{2+} is not essential for the adrenergic response. These authors concluded that adrenergic stimulation caused depolarization in anococcygeus muscles by activation of a Ca^{2+}-dependent Cl^- conductance.

Ca^{2+}-activated Cl^- currents have also been found in rabbit esophageal muscularis muscosae muscles (Akbarali and Giles 1993). Depolarizations of cells dialyzed with solutions containing CsCl and no EGTA resulted in initial inward Ca^{2+} currents followed by substantial outward currents. Upon repolarization, the outward currents reversed to long-lasting inward tail currents. Outward currents were activated at more negative potentials when the equilibrium potential for Cl^- ions was shifted to more negative potentials, suggesting that the outward current was carried by Cl^-. Niflumic acid, a Cl^- channel blocking compound, reduced the outward current initiated by depolarization and the inward tail currents that occurred upon repolarization. The inward tail currents were abolished when Ba^{2+} was used as the charge carrier or when cells were dialyzed with high concentrations of EGTA. These data suggest that a rise in $[Ca^{2+}]_i$ is necessary for activation of the tail currents. The inward tail currents were not due to Na/Ca exchange because replacement of external Na^+ with TEA did not significantly alter their amplitude.

4. Properties of Sodium Currents

There have been two types of sodium currents identified in GI muscles. One is attributed to a nonselective cation conductance that is activated by muscarinic stimulation (INOUE et al. 1987). Although this conductance is about equally selective to Na^+ and K^+ (VOGALIS and SANDERS 1990), it predominantly carries sodium over the range of physiological potentials. This conductance will be discussed in detail in the section "Regulation of Currents by Muscarinic Stimulation." Tetrodotoxin-sensitive sodium currents are a rarity in GI muscles, although examples have been reported in other smooth muscles (e.g., OHYA and SPERELAKIS 1989; OKABE et al. 1990; MIRONNEAU et al. 1990). One report has documented a TTX-sensitive conductance in muscle cells of the rat gastric fundus (MURAKI et al. 1991). At 21°C this current activated and inactivated within 10 ms. The current, which was highly sensitive to external Na^+ concentration, reached a peak at about −10 mV and had a half inactivation value of −74 mV. The participation of the Na^+ conductance in electrical activity of fundus cells was studied with TTX. Application of 10 n*M* TTX slowed, but did not block, the action potentials stimulated by depolarizing pulses from a holding potential of −70 mV. Action potentials were never generated when Ca^{2+} was replaced with Mn^{2+}. These data suggest that Na^+ current may contribute to action potentials in fundus cells, but Ca^{2+} current is dominant and necessary for action potentials.

Actually, the physiological role for Na^+ channels tends to be a bit unclear because the resting potentials for cells in the fundus region in situ are usually positive to −60 mV (SZURSZEWSKI 1987). At the potential, over 70% of the Na^+ channels found in rat fundus muscles were inactivated (MURAKI et al. 1991). Secondly, the membrane potentials of cells in the fundus of many species change slowly; stimulation by transmitters and hormones elicits slow depolarizations rather than spike potentials. These changes would tend to further inactivate Na^+ channels before these channels could contribution much inward current. Even in slow wave or spike producing muscles, the role for Na^+ channels would tend to be very minor based on their voltage-dependent properties. This may be why few GI muscle cells express a TTX-sensitive component of inward current.

II. Contribution of Ionic Channels to Electrical Activity and Electromechanical Coupling

The ionic currents described above generate a variety of electrical behavior in GI muscles. In some regions of the GI tract (e.g., the gastric fundus and the lower esophageal sphincter), membrane potential is relatively stable. Activation appears to be due to depolarization of resting potential, and inhibition is caused by hyperpolarization. These changes in membrane potential can be elicited by neurotransmitters, hormones or paracrine substances. These regions of the GI tract often display tonic contraction. The resting potentials of these muscles is close to or within the steep part of the activation curve for voltage-dependent Ca^{2+} channels (i.e., −55 to −45 mV; see SZURSZEWSKI 1987). Therefore subtle changes in membrane potential can increase or decrease the open probability of voltage-dependent Ca^{2+} channels and alter the influx of Ca^{2+}. At the resting potentials of these cells, the open probability of Ca^{2+} channels is low, and only a few picoamperes of Ca^{2+} current can be measured at these potentials. However, the current influx is continuous, and over time the amount of Ca^{2+} that enters cells is significant in terms of $[Ca^{2+}]_i$. Changes in Ca^{2+} influx, even subtle changes, can lead to net increases or decreases in $[Ca^{2+}]_i$ and alter the contractile state. It is likely that the small inward Ca^{2+} current is balanced by a K^+ current in the tonic muscles. Although this current(s) has not been specifically identified, it is likely to result from the delayed rectifier class of currents because these currents are activated and remain activated in the same range of potentials as the voltage-dependent Ca^{2+} currents (THORNBURY et al. 1992a). Therefore, one can hypothesize that the resting potential of "tonic" muscles in the GI tract may be partially regulated by delayed rectifier channels.

Many regions of the GI tract display autorhythmicity, that is, in the absense of external stimuli, the muscles are electrically active. The most common form of autorhythmicity is slow wave activity, and events of this type are commonly recorded in the stomach, small intestine and colon (for reviews see SANDERS and PUBLICOVER 1989; SANDERS and SMITH 1989; SZURSZEWSKI 1987). Recent studies on the slow waves of the canine colon have suggested the following hypothesis for the mechanism of electrical slow waves (see SANDERS 1992). Slow waves are composed of an initial upstroke depolarization that has a maximum velocity of about 1 V/s. The upstroke does not reach 0 mV, and in fact it rarely even reaches a maximum depolarization of −20 mV. After reaching the peak of the upstroke phase, membrane potential partially repolarizes, and then settles for up to several seconds into a "plateau phase" before repolarizing to the resting potential level. The upstroke phase is due to an increase in Ca^{2+} conductance which is activated at potentials positive to about −60 mV. A portion of this current appears to be insensitive to dihydropyridines at concentrations up to $10^{-6}\,M$, and slow waves continue in the presence of these "L-type" Ca^{2+} channel blocking agents (BARAJAS-LOPEZ et al. 1989a; WARD and SANDERS 1992a). The resistance to dihydropyridines is probably due to the voltage dependence of the block of Ca^{2+} channels by these agents (BEAN 1984; NELSON and WORLEY 1989). The upstroke depolarization does not continue on to more positive potentials because voltage-dependent K currents activate rapidly and serve as a "break." Activation of K currents causes the net current to reverse and become outward within a few milliseconds. This leads to the partial repolarization phase, but repolarization is incomplete, because as membrane potential becomes more negative, relatively more outward current deactivates, and a stable potential is reached where sustained inward Ca^{2+} current balances the outward current. This balance is reached between −45 and −35 mV creating the plateau phase of the slow wave. During the plateau phase Ca^{2+} continues to enter the cell and can lead to a build-up in $[Ca^{2+}]_i$ (see OZAKI et al. 1991c; VOGALIS et al. 1991). This is likely to explain the coupling between the plateau phase of slow waves and contractile activity (see SZURSZEWSKI 1987). Actually, the balance between inward and outward currents during the plateau phase may be quite analogous to the balance between these currents at the resting potential in tonic muscles. Complete repolarization of slow waves occurs when the balance is broken, either by reduction in the inward current, perhaps as a result of Ca^{2+}-dependent inactivation of Ca^{2+} current or an increase in the open probability of Ca^{2+}-dependent K current. As repolarization occurs, voltage-dependent inward and outward currents deactivate.

In most regions of the GI tract electrical slow waves are resistant to complete block by dihydropyridines (BARAJAS-LOPEZ et al. 1989; WARD et al. 1992a; and see SANDERS and PUBLICOVER 1989), yet the amplitude and rate-of-rise of slow waves is dependent upon the Ca^{2+} gradient (e.g., EL-SHARKAWAY et al. 1978; HUIZINGA et al. 1991a; WARD et al. 1992a). These, and similar, observations have led some investigators to suggest that slow wave generation is dependent upon two classes of Ca^{2+} channels, low-threshold ("L-type") channels and high-threshold ("L-type") channels. There is clearly diversity among smooth muscles, and some cells display a low threshold conductance (VIVAUDOU et al. 1991; YOSHINO et al. 1989), but in other cases where low-threshold Ca^{2+} channels have been postulated to explain the inability of dihydropyridines to block slow waves, only L-type Ca^{2+} currents and channels can be found (SIMS et al. 1992a; WARD et al. 1992b; RICH et al. 1993). It is likely that the lack of effects of dihydropyridines on these muscles can be explained by the voltage-dependence of the block of Ca^{2+} channels (WARD et al. 1992; NELSON and WORLEY 1989; BEAN 1984). At negative potentials dihydropyridines do not bind to Ca^{2+} channels with high affinity. Therefore, at the initiation of depolarization, the channels are available to carry current. As more channels pass into the inactivated state, dihydropyridine binding increases and channels become blocked. Thus some of the initial current through L-type Ca^{2+} channels is resistant to quite high levels of dihydropyridines (i.e., μM), but the sustained phase of the Ca^{2+} current is blocked. This explains why the upstroke depolarization of slow waves is not blocked by dihydropyridines at 1 μM, but the plateau phase is greatly reduced. Because of regional and species differences in the expression of Ca^{2+} channels, it is possible that other mechanisms, including other ion channels, might participate in slow waves elsewhere.

In addition to slow waves, some regions of the GI tract produce fast Ca^{2+} action potentials that are also referred to as "spikes." In some cases spikes occur during the plateau phase of slow waves, and in other cases small generator potentials or myenteric potential oscillations, as they have been termed in the colon (see SMITH et al. 1987b), appear to initiate spikes. Many of the same ionic channels found in nonspiking regions are also found in regions where action potentials occur. The diversity in electrical activity therefore must depend upon slight differences in channel density and/or the activation/inactivation properties of the species of channels present. For example, in the longitudinal muscle of the canine proximal colon where spiking is common, the same three major currents found in most GI muscles are apparent (i.e., voltage-dependent Ca^{2+} current, Ca^{2+}-dependent K^+ current, and delayed rectifier-type K^+ currents). In comparing the ionic currents of circular and longitudinal muscles it is apparent that there are differences in the voltage-dependent (delayed rectifier) K^+ currents. This species of current in longitudinal muscles inactivates at more negative potentials than the equivalent current in circular muscle cells (half inactivation in longitudinal cell was −63 mV vs. −36 mV in circular muscle cells; THORNBURY et al. 1992a,b). Thus during the slow depolarization of the myenteric potential oscillations (which serve as generator potentials in the longitudinal muscle layer) there is sufficient time for much of the delayed rectifier current to inactivate. When the action potential thresholds reached (at about −40 mV), the "breaking" action of the delayed rectifier has been inactivated, and membrane potential rapidly depolarizes to about 0 mV. Depolarization and Ca^{2+} entry during the upstroke of the spike may activate a large Ca^{2+}-dependent outward current and cause immediate repolarization.

1. Pacemaker Currents

Despite the progress in determining the ionic mechanisms of the electrical activities in GI muscles, the actual pacemaker mechanism that generates rhythmicity has remained elusive. After considerable work on this subject during the 1970s, three major hypotheses emerged as reviewed by TOMITA (1981): (a) slow waves are due to conductance changes (primarily attributed to DANIEL's group; e.g., EL-SHARKAWY and DANIEL 1975); (b) slow waves are due to oscillations in the activity of the Na/K ATPase (primarily attributed to Prosser's group; e.g., JOB 1969; LIU et al. 1969; CONNOR et al. 1974), and (c) slow waves occurred in two phases, the first phase or pacemaker activity was thought to be due to a non-voltage-dependent transporter and the second phase was due to conductance changes (this concept came from Tomita's laboratory; e.g., OHBA et al. 1975, 1977). Tomita discussed the evidence for and against these hypotheses in detail (1981). The actual pacemaker mechanism is still not understood, and mechanistic studies have not provided explanations for the observations that: (a) the frequency of slow waves is not very voltage-dependent (OHBA et al. 1975) and (b) under voltage clamp conditions (double sucrose gap voltage clamp), small inward currents can be recorded at about the same frequency as slow waves (OHBA et al. 1975; CONNOR et al. 1974).

Many isolated GI smooth muscle cells are capable of generating spontaneous electrical activity, usually in the form of action potentials (e.g., DROOGMANS and CALLEWAERT 1986; BENHAM et al. 1987). A recent study shows that the spontaneous excitability is enhanced by raising temperature to physiological levels (POST and HUME 1992). These observations clearly demonstrate that GI muscle cells are excitable and can generate spontaneous activity on their own. But cells in situ have more negative resting potentials than isolated cells, and the resting potentials is often 10–30 mV below the potential at which resolvable inward current can be elicited in voltage-clamp studies. Therefore, it is unclear how smooth muscle cells are depolarized to the threshold at which regenerative slow waves or action potentials are initiated.

During the past decade, several investigators have begun to believe that electrical activity in the GI tract originates in a type of cell referred to as interstitial cells of Cajal (IC; SUZUKI et al. 1986; THUNEBERG et al. 1983; DANIEL and BEREZIN et al. 1992; SANDERS 1992; KOBAYASHI et al. 1989a). Although there are several classes of these cells (see THUNEBERG 1982, 1989), networks of ICs are found in regions that have been shown to be pacemaker areas (BEREZIN et al. 1988; SUZUKI et al. 1986; SMITH et al. 1987a,b; WARD et al. 1991a; WARD and SANDERS 1990). These cells are usually loaded with mitochondria, and some types of ICs form gap junctions with each other and with surrounding smooth muscle cells. It has been very difficult to test the hypothesis that ICs generate rhythmicity. Attempts to do this have either employed dissection techniques to remove the ICs suspected of pacemaker activity (e.g., SUZUKI et al. 1986; HARA et al. 1986; SERIO et al. 1991) or used chemicals to selectively lesion ICs (THUNEBERG et al. 1983; WARD et al. 1990). With the former technique it is impossible to remove only the ICs or all of the ICs that may be involved in pacemaker activity (see WARD and SANDERS 1990). No chemical agent that is completely specific for ICs has been identified (e.g., see SANDERS et al. 1989). In one study the authors claimed that ICs within intact muscle strips were impaled with microelectrodes and slow waves were recorded from these cells (BARAJAS-LOPEZ et al. 1989). These authors have subsequently claimed that their study proved that slow waves originate in ICs (HUIZINGA et al. 1991b). This claim may be a bit too enthusiastic because: (a) the morphological techniques employed to localize the impaled cells (i.e., finding the hole left by the microelectrode) was less than convincing, and (b) even if slow waves were actually recorded from ICs, it would not be possible to conclude that slow waves originated from these cells with the techniques used (i.e., impalement of one cell with a microelectrode). ICs are electrically coupled to smooth muscle cells. Therefore, electrical signals similar to those recorded from adjacent smooth muscle cells would be expected to be observed in ICs whether these cells generated the signals or conducted the signals. Furthermore, it is impossible to determine the site of origin in a syncytium of cells when recording from a single site.

Another approach to study ICs has been to isolate the cells and study their electrical properties (LANGTON et al. 1989b). These experiments showed that ICs from the slow wave pacemaker region of the canine proximal colon were spontaneously active and generated electrical events from resting potentials similar to those measured in situ. The spontaneous events in ICs had amplitudes and characteristics similar to electrical slow waves recorded from muscle strips. The original study on ICs was conducted on dialyzed

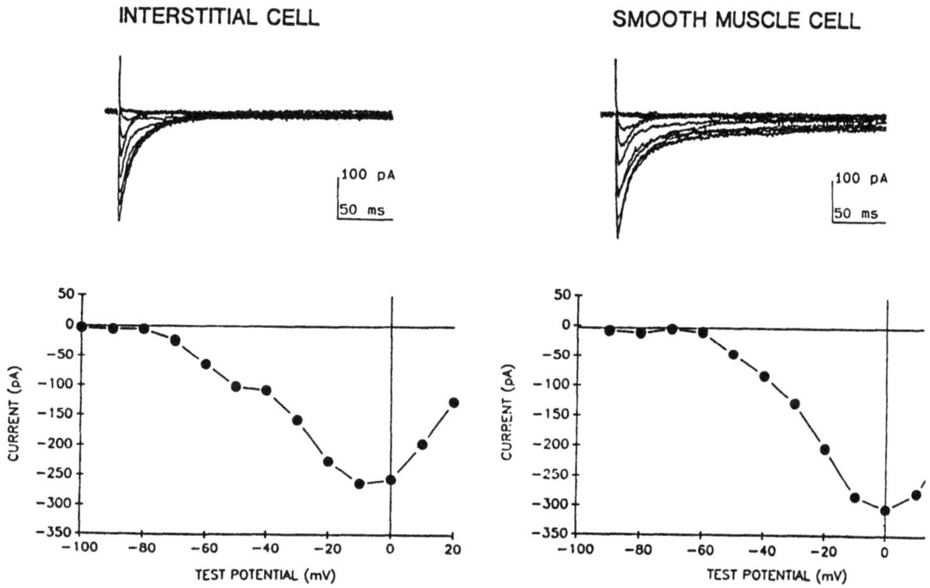

Fig. 7. Comparison of inward currents elicited in interstitial cells and smooth muscle cells of the circular muscle layer of the canine colon. Cells were studied with the "permeabilized" patch technique to preserve intracellular buffers and regulatory substances. Figure shows families of currents elicited by test potentials from −100 mV to +20 mV. Plotted below the current records are I–V curves. I–V curve for interstitial cell shows a distinctive hump at potentials ranging from −70 mV to about −40 mV. A similar hump is not seen in the circular muscle cell. This hump in the I–V curve is one of the signatures for a low-threshold (or T-type current). Inward current could be resolved at more negative potentials in interstitial cells. The low threshold current may facilitate pacemaker activity in these cells. (After LEE and SANDERS 1993)

cells at room temperature, and the ionic currents observed were similar to the currents observed in circular muscle cells from the pacemaker region. Other studies on freshly dispersed and cultured ICs have shown that these cells display Ca^{2+} oscillations that appear to be due to periodic influx of Ca^{2+} (PUBLICOVER et al. 1992). The cultured model of ICs is potentially a powerful tool for studies of pacemaker activity because these cells appear to retain the rhythmic phenotype under culture conditions through several passages.

In recent studies the ionic currents that may be responsible for initiation of slow wave activity have been studied in isolated ICs (LEE and SANDERS 1993). The currents of ICs were compared with those of circular muscle cells isolated from the same pacemaker region of the canine colon. Depolarization of both types of cells initiated transient inward, followed by slowly inactivating outward currents. When inward currents and Ca^{2+}-dependent outward currents were blocked, ICs displayed voltage-dependent outward currents that rapidly activated, reached a peak, and then inactivated. This current was relatively resistant to 4-aminopyridine as compared to the equivalent current in smooth muscle cells. The outward current of ICs also inactivated at more negative potentials (i.e., half inactivation at −53 mV in ICs, half inactivation at −20 mV in smooth muscle cells). With the perforated patch technique (using amphotericin B), a negatively activating inward current was observed in ICs that could be resolved at about −70 mV. Inward currents in smooth muscle cells were resolved at test potentials positive to −50 mV. The current-voltage relationship for ICs showed a distinct "hump," which is often the signature of a T-type current (Fig. 7). The availability of an inward current near the resting potential of ICs and the presence of a negatively inactivating outward current may facilitate spontaneous activity in these cells.

If slow waves originate in ICs, studies on colonic ICs point toward a voltage-dependent mechanism of slow wave generation (e.g., the low-threshold Ca^{2+} channels in ICs are clearly voltage-dependent). It is difficult to reconcile such a mechanism on the basis of Tomita's observation that slow wave frequency in the guinea stomach is not very voltage-dependent (see TOMITA 1981). Of course it is possible that different mechanisms are responsible for pacemaker activity in different regions of the GI tract or in different species. At any rate, cellular voltage clamp studies have as yet been unable to identify a voltage-independent mechanism that can generate a pacemaker current either in smooth muscle cells or in ICs.

C. Regulation of Ionic Currents by Agonists

The purpose of electrical activity in GI muscles is to regulate the influx of Ca^{2+} which is the excitatory trigger for contraction. Electrical activity is an ideal way in which to coordinate contractions in a large population of cells because electrical events spread from cell to cell via low resistance gap junctions. By regulating the frequency of electrical events at the pacemaker site or by locally influencing the membrane potential and/or amplitude of electrical events, it is possible to regulate the contractile behavior of GI muscles. In sections below we will discuss how agonists regulate contractile activity at steps beyond the plasma membrane, and in this section we will discuss how agonists affect electrical activity. Investigations have begun to make progress in determining the regulation of ionic currents by various agonists, but there is far more to know about the electrical mechanisms of wide variety of agonists that influence the contractile activities of GI muscles.

I. Regulation of Currents by Muscarinic Stimulation

Acetylcholine is the major excitatory neurotransmitter released from postganglionic motor neurons in the GI tract, and its importance in the regulation of motility is well recognized. BOLTON (1979, 1981) has extensively reviewed the literature regarding the excitatory actions of this compound in a variety of smooth muscles. Smooth muscle cells of the GI tract ubiquitously express muscarinic receptors, but the electrical responses to ACh vary dramatically from region to region suggesting that a variety of cellular effectors are coupled to these receptors (see SZURSZEWSKI 1987). Tonic smooth muscles are generally depolarized by ACh and this leads to an increase in tone in these muscles. ACh also depolarizes many phasic smooth muscles. However, in these muscles ACh also causes an increase in spiking in the case of taenia coli and other muscles, an increased likelihood of action potentials in the small intestine, an enhancement in the frequency of slow waves in gastric muscles, or an increase in the amplitude and duration of slow waves in muscles of the stomach, small bowel and colon. All of these responses can be viewed as excitatory because they each lead to an increase in the open probability of voltage-dependent Ca^{2+} channels and an increase in contraction. The motor output of a given muscle depends to a large extent upon the pattern of electrical activity. The actions of cholinergic agonists that explain some of the electrical responses are discussed below.

1. Nonselective Cation Currents

Application of muscarinic agonists evokes an inward current and depolarization in most GI muscles (e.g., Fig. 8; BENHAM et al. 1985; INOUE et al. 1987; INOUE and ISENBERG 1990a–c; INOUE 1991; PACAUD and BOLTON 1991; SIMS 1992b; VOGALIS and SANDERS 1990). The source of this inward current has been investigated in studies of several muscles. These experiments show that muscarinic receptor stimulation is coupled to the activation of a nonselective cation current (I_{ACh}). This current reverses close to 0 mV, and the channels responsible for I_{ACh} are thought to be about equally permeable to Na^+ and K^+ (VOGALIS and SANDERS 1990). At the negative resting potentials of GI muscle cells, therefore, the net current through g_{ACh} is inward and primarily carried by Na^+ ions, and this can explain the depolarization caused by muscarinic agonists in several muscles (BENHAM et al. 1985; INOUE and ISENBERG 1990a; PACAUD and BOLTON 1991; SIMS 1992b; VOGALIS and SANDERS 1990). In fact by using ramp potentials to fully characterize the current-voltage relationship of I_{ACh} in canine gastric cells, SIMS (1992b) found that the zero current level of the I–V curves (which corresponds to resting potential) shifted to more positive potentials. The depolarization caused by ACh was also associated with cell contraction (SIMS 1992b). I_{ACh}

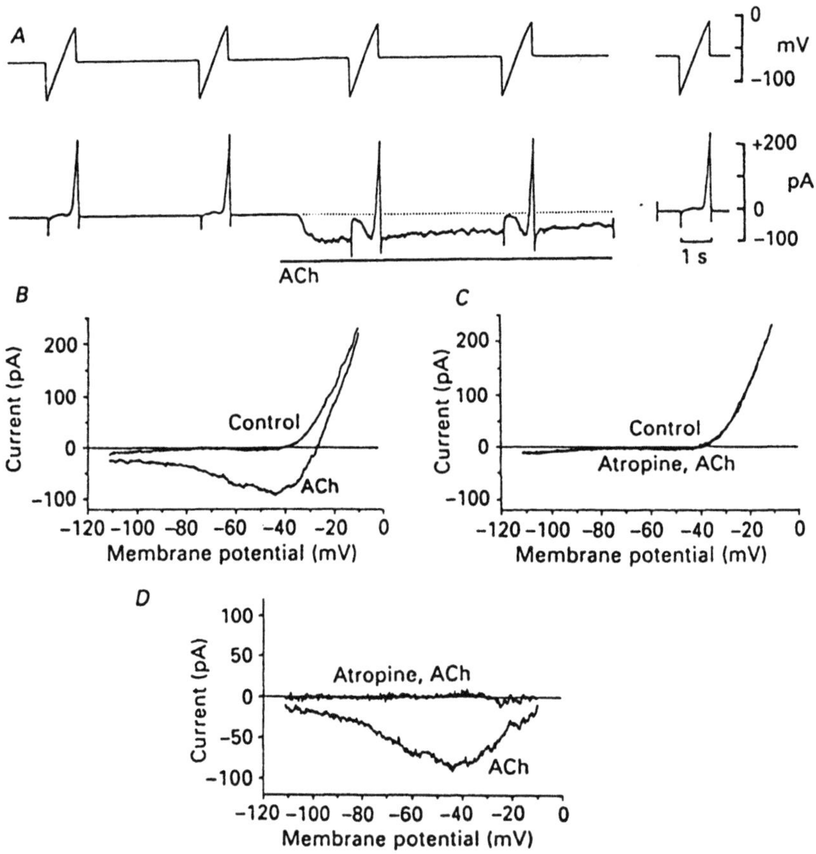

Fig. 8. Muscarinic stimulation activates a nonselective cation conductance in canine gastric muscle cells. *Panel A* shows voltage-clamp protocol (*top trace*) and current records (*bottom trace*) elicited before and after application of ACh (100 μM in application pipette). Ramp protocols were periodically applied to depolarize cell from −120 to −10 mV. Application of ACh evoked an inward current and changed the current response to the ramp depolarization. *Panel B* shows comparison of I–V curves elicited by ramps in control conditions and in the presence of ACh. The response to ACh was blocked by atropine (*panel C*). *Panel D* shows the difference currents (control response subtracted from response in presence of ACh). ACh elicited an inward current (I_{ACh}) that increased in magnitude from −100 mV to −40 mV and reversed at about −8 mV. (From SIMS 1992b)

has been found to be voltage-dependent in many cells. The conductance of these channels is low at negative potentials (i.e., −80 mV) but increases with depolarization. Because the reversal potential of I_{ACh} is near 0 mV, the current through I_{ACh} reaches a maximum near −40 mV and then begins to decrease with further depolarization (Fig. 8; SIMS 1992b). Under some circumstances I_{ACh} also conducts divalent ions (INOUE and ISENBERG 1990a), but the significance of Ca^{2+} entry through these channels in the presence of physiological concentrations of ions is debatable (see discussion below and PACAUD and BOLTON 1991).

I_{ACh} is dependent upon intracellular Ca^{2+} (INOUE and ISENBERG 1990b; PACAUD and BOLTON 1991; LIM and BOLTON 1988; SIMS 1992b). This is consistent with the observation that Ca^{2+}-free/EGTA solution blocks the depolarization response to muscarinic stimulation (BOLTON and KITAMURA 1983). Similarly, under current clamp conditions, application of ACh caused depolarization of guinea pig ileum cells (INOUE and ISENBERG 1990c), but this response was markedly attenuated when cells were placed in Ca^{2+}-free buffer. These authors also found that when activation of I_{ACh} was preceded by a depolarization that activated voltage-dependent Ca^{2+} current, I_{ACh} was greatly enhanced. The degree of enhancement of I_{ACh} was dependent upon the duration of the pre-conditioning depolarization and therefore the amount of Ca^{2+} that entered the cell. Ba^{2+} and Sr^{2+} were not able to substitute for Ca^{2+}. These data suggest that Ca^{2+} entry is an important facilitator of the response to ACh. INOUE and ISENBERG (1990c) referred to I_{ACh} as a "Ca^{2+} facilitated" current. This designation was suggested to make it clear that the current could not be activated directly by an increase in $[Ca^{2+}]_i$. In contrast, SIMS (1992b) reported that application of caffeine (which releases Ca^{2+} from intracellular stores) caused activation of I_{ACh} and contraction of cells in the absence of muscarinic stimulation. This observation suggests that in canine gastric cells I_{ACh} may be *activated* by Ca^{2+}. Other studies of canine colonic smooth muscle cells (LEE et al. 1993) are in agreement with a facilitating role for Ca^{2+} (i.e., Ca^{2+} alone could not activate the current; Ca^{2+} facilitated the current activated by muscarinic stimulation).

Influx of Ca^{2+} is not the only source of Ca^{2+} involved in facilitation of I_{ACh}. PACAUD and BOLTON (1991) showed that I_{ACh} activated by carbachol in guinea pig jejunal cells developed rapidly to a peak and relaxed to a "plateau phase" during sustained application of the drug. Low concentrations of heparin, which block inositol triphosphate (IP_3) receptors (KOBAYASHI et al. 1988; KOMORI and BOLTON 1990), inhibited the rapid phase of I_{ACh} and reduced, but did not block, the sustained phase. Heparin did not interfere with the potentiation of I_{ACh} caused by influx of Ca^{2+}. Dialysis of cells with ryanodine also reduced the response to carbachol. These results show that release of Ca^{2+} from internal stores facilitates I_{ACh}. The initial rise in $[Ca^{2+}]_i$ following muscarinic stimulation appears to be due to an increased production of IP_3. This increase in $[Ca^{2+}]_i$ causes the initial rapid phase in I_{ACh}. The sustained phase of the increase in Ca^{2+} and I_{ACh} appears to be due to Ca^{2+} entry. One of the important roles of Ca^{2+} stores in GI muscles may be to potentiate the initial enhancement of I_{ACh} in response to cholinergic stimulation.

As discussed above the typical response of GI muscles to muscarinic stimulation is depolarization. In some muscles this response appears to cause an increase in contractile tone, but in others (e.g., phasic muscles) there is little increase in tone between phasic contractions (see SZURSZEWSKI 1987). In the latter group of muscles the changes in slow wave amplitude and/or action potential frequency are the responses that lead to enhanced contraction. The facilitation of I_{ACh} by $[Ca^{2+}]_i$ is of critical importance in both types of muscle. In tonic muscles, depolarization leads to an increase in the open probability of voltage-dependent Ca^{2+} channels, enhanced influx of Ca^{2+}, and facilitation of I_{ACh}. In phasic muscles, Ca^{2+} entry is enhanced by the increase in depolarization during slow waves or an increase in action potential frequency. These events increase $[Ca^{2+}]_i$ and facilitate I_{ACh} in phasic muscles. Facilitation of I_{ACh} by an increase in $[Ca^{2+}]_i$ represents positive feedback because I_{ACh} contributes to depolarization and depolarization leads to enhanced influx of Ca^{2+} via voltage-dependent Ca^{2+} channels (INOUE and ISENBERG 1990c).

In muscles that generate electrical slow waves I_{ACh} appears to increase the amplitude and extend the duration of slow waves (LEE et al. 1993). Enhanced slow waves are associated with enhanced phasic contractions (see SZURSZEWSKI 1987; SANDERS and SMITH 1989). Because of its Ca^{2+}- and voltage-dependence, I_{ACh} is dynamic during the time course of slow waves. The voltage-dependence causes an increase in I_{ACh} over the voltage range of slow waves (i.e., approximately -80 mV to -40 mV; see SANDERS and SMITH 1989), and the Ca^{2+} dependence causes I_{ACh} to increase as Ca^{2+} enters cells during slow waves. The latter further illustrates the positive feedback produced by I_{ACh}: I_{ACh} would tend to sustain or even increase the amplitude of the plateau phase of slow waves by providing additional inward current. This results in an increase in the open probability of voltage-dependent Ca^{2+} channels and increases the period during which the open probability is elevated. Thus the total amount of Ca^{2+} influx increases, and the increase in $[Ca^{2+}]_i$ that occurs when the plateau phase is sustained results in further activation of I_{ACh}. We have used the term Ca^{2+}-induced Ca^{2+} entry (LEE et al. 1993) to describe the interdependence between voltage-dependent Ca^{2+} current and I_{ACh}.

In contrast to nicotinic ACh receptors, which are coupled directly to ion channels that these receptors activate, the I_{ACh} activated by muscarinic receptor stimulation in GI muscles appears to be activated via a G protein (INOUE and ISENBERG 1990c). Cells of the guinea pig ileum treated with pertussis toxin, which ADP ribosylates some G proteins (RIBEIRO-NETO et al. 1985), were no longer responsive to ACh. Addition of ADPβS, which competes with GTP for G protein binding sites, to the pipette solution also blocked the development of I_{ACh} in response to cholinergic stimulation.

However, when GTPγS was added to the pipette solution, a noisy inward current was induced as cells were dialyzed. This current had properties and sodium permeability similar to I_{ACh}, and during steady-state activation of I_{ACh} the effects of GTPγS and ACh were not additive. The current activated by GTPγS was also sensitive to $[Ca^{2+}]_i$, and entry of Ca^{2+} via voltage-dependent Ca^{2+} channels greatly augmented the current, similar to the facilitation of I_{ACh} by Ca^{2+} (INOUE and ISENBERG 1990b; PACAUD and BOLTON 1991).

Coupling of G proteins to ion channels can occur within the membrane or via a cytosolic second messenger. In guinea pig ileum the former mechanism was suggested because putative second messengers alone (i.e., IP_3, cAMP or Ca^{2+}) were incapable of activating I_{ACh} directly (INOUE and ISENBERG 1990c). Others have suggested that the G protein link in activation of I_{ACh} is due to release of Ca^{2+} from internal stores (KOMORI and BOLTON 1990). In these experiments, performed on rabbit jejunal muscle cells, GDPβS was less effective in blocking activation of I_{ACh} in response to carbachol than in experiments on guinea pig ileum. It appeared that the reduction in I_{ACh} in the presence of GDPβS was due to a reduction in release of Ca^{2+} from stores. Heparin, which blocks IP_3-dependent release of Ca^{2+} in smooth muscles (KOBAYASHI et al. 1988), had very similar effects to GDPβS, suggesting that the primary effect of GDPβS was to interfere with the G-protein coupling to IP_3 production and release of intracellular Ca^{2+}. As discussed above, Ca^{2+} release clearly can be an important determinant for activation of I_{ACh} (PACAUD and BOLTON 1991). In rabbit jejunal cells dialysis with GTPγS-containing solution induced a small, noisy inward current that reversed near 0 mV. This is similar to the findings reported by INOUE and ISENBERG (1990c) and tends to support some direct G-protein activation of I_{ACh}. It would appear that the G-protein coupling of muscarinic receptor occupation and activation of I_{ACh} is complicated and in need of further experiments to fully understand the pathways involved.

It is unclear whether the nonselective cation channels activated by muscarinic agonists conduct sufficient numbers of Ca^{2+} ions to affect $[Ca^{2+}]_i$. INOUE and ISENBERG (1990a) showed that I_{ACh} was not substantially reduced when external Na^+ ions were replaced with Ca^{2+} ions, suggesting the nonselective cation channels are permeable to Ca^{2+}. However, it is difficult to conclude from these experiments that these channels conduct Ca^{2+} when physiological ionic gradients are present. In many GI muscles contractions in response to cholinergic stimulation are essentially eliminated by blockade of voltage-dependent Ca^{2+} channels (BRADING and SNEDDON 1980; and see discussion in PACAUD and BOLTON 1991). This tends to suggest that the Ca^{2+} that may enter cells via the nonselective cation conductances is insufficient to activate the contractile apparatus. However, this reasoning neglects the interdependence of I_{ACh} and voltage-dependent Ca^{2+} current. Blockers of voltage-dependent Ca^{2+} channels must also reduce I_{ACh} because this channel is facilitated by the influx of Ca^{2+}. Therefore, the actual contribution of

I_{ACh} to Ca^{2+} influx cannot be evaluated when Ca^{2+} influx is affected. Additional work is needed to determine the contribution of Ca^{2+} influx is affected. Additional work is needed to determine the contribution of Ca^{2+} influx via nonselective cation channels to $[Ca^{2+}]_i$ during muscarinic stimulation.

One of the problems with evaluating the contribution of I_{ACh} to physiological responses has been the lack of selective blockers for this current. INOUE (1991) showed that low concentrations of divalent ions (i.e., Cd^{2+}, Ni^{2+} with K_d of about $100\,\mu M$) reduced the amplitude of I_{ACh} in guinea pig ileal muscle cells. Similar results have been found for Cd^{2+} and Ni^{2+} in canine colonic myocytes (LEE et al. 1993). These divalents are also well-known blockers of voltage-dependent Ca^{2+} channels, so it is not possible to use these agents to dissect the contribution of I_{ACh}.

2. Regulation of Calcium Currents

Although one of the ultimate purposes of excitatory and inhibitory agonists is to regulate the influx of Ca^{2+}, regulation of voltage-dependent Ca^{2+} currents does not appear to be a common phenomenon in GI muscles. Rather, the regulation of Ca^{2+} influx appears to be accomplished primarily by regulation of other currents that affect membrane potential and therefore the open probability of voltage-dependent Ca^{2+} channels. A few examples of agonist regulation of Ca^{2+} channels, however, have been reported.

In studies of gastric muscle cells of Bufo marinus, CLAPP et al. (1987) showed that acetylcholine application increased the amplitude of voltage-dependent Ca^{2+} currents elicited by depolarization and slowed the rate of inactivation of this current. Later studies attempted to dissect the particular component of voltage-dependent Ca^{2+} current that was specifically activated by muscarinic stimulation (VIVAUDOU et al. 1988). Toad gastric cells display both high and low threshold Ca^{2+} currents. The low threshold component activates at relatively negative potentials and inactivates rapidly. The other component activates at more depolarized potentials and inactivates more slowly. ACh increased the high threshold current, but did not affect the low threshold current. An analog of diacylglycerol and an activator of protein kinase C (PKC), sn-1,2-dioctanoyl glycerol, mimicked the effects of ACh on Ca^{2+} currents. These observations suggested that diacylglycerol may be the second messenger that mediates cholinergic effects in toad gastric muscles. The actions of sn-1,2-dioctanoyl glycerol were probably via activation of PKC since a structurally similar analog that was ineffective in activating PKC failed to enhance Ca^{2+} current. The results also suggested that muscarinic actions on Ca^{2+} current could be adequately explained by the action of diacylglycerol and did not require additional actions by other products of PIP_2 metabolism such as IP_3 and IP_4.

Although there is one report of potentiation of inward current in cat colonic cells by muscarinic stimulation (BIELEFELT et al. 1990), in most cases muscarinic regulation of voltage-dependent Ca^{2+} currents appears to be different in toad and mammalian cells. For example, INOUE and ISENBERG (1990a) found that voltage-dependent Ca^{2+} currents were decreased by ACh. This effect was observed under conditions in which buffering of $[Ca^{2+}]_i$ was very low, but it was not observed when high Ca^{2+} buffering conditions were used. These observations suggest that the decrease in I_{Ca} caused by ACh may have been due to Ca^{2+} inactivation and not a direct effect of the agonist on voltage-dependent Ca^{2+} channels. A decrease in Ca^{2+} current in response to muscarinic stimulation has also been noted in colonic myocytes (LEE et al. 1993). In these studies, membrane patches under the patch pipette were permeabilized with amphotericin B, and therefore it is unlikely that normal cellular Ca^{2+} buffering was disturbed. ACh reduced peak Ca^{2+} current, but whether this effect was due to Ca^{2+} inactivation is unclear. Other than Ca^{2+} inactivation, a second messenger pathway to explain the inhibitory effect of ACh on Ca^{2+} currents in mammalian GI smooth muscle cells has not been described.

RUSSELL and AARONSON (1990) also reported inhibition of voltage-dependent Ca^{2+} current by muscarinic agonists in studies of rabbit jejunal longitudinal muscle cells. High internal Ca^{2+} buffering was used in these experiments, and the effect was not blocked by treatment with ryanodine to pre-release internal stores of Ca^{2+} nor by use of Ba^{2+} as the charge carrier. KOMORI and BOLTON (1991) suggested that the inhibition caused by muscarinic stimulation was at least partially due to Ca^{2+}-dependent inactivation. They used "caged" inositol trisphosphate (IP_3) to show that rapid release of IP_3 and stored Ca^{2+} inhibited the Ca^{2+} current. This effect was blocked by heparin, an inhibitor of the IP_3 receptor on the sarcoplasmic reticulum. When these authors included 10 m*M* EGTA in the pipette solution, the inhibition of Ca^{2+} current by photolytic release of IP_3 was almost entirely prevented. It is not possible to determine from these experiments whether direct or indirect mechanisms are responsible for the decrease in Ca^{2+} current in response to muscarinic stimulation. Certainly it appears that a portion of this effect may be due to Ca^{2+}-dependent inactivation, but an additional, second messenger-dependent effect also appears to be possible because of the reduction observed by RUSSELL and AARONSON (1990) in the presence of high EGTA. It is possible that phosphorylation, perhaps via protein kinase C, could be involved in the inhibition of Ca^{2+} currents in mammalian cells. Although this would be opposite to the mechanism observed in amphibian cells (CLAPP et al. 1987; VIVAUDOU et al. 1988), there is some evidence that a phosphorylation-dependent mechanism decreases L-type Ca^{2+} currents in canine gastric and colonic muscle cells (e.g., WARD et al. 1991b). Regardless of the actual mechanism, it appears that cholinergic stimulation decreases Ca^{2+} current in mammalian cells from several species. This response occurs in amphotericin B-permeabilized cells with physiological Ca^{2+} concentrations and at physiological temperatures, suggesting that it is possibly an important mechanism involved in the regulation of Ca^{2+} entry (LEE et al. 1993).

3. Regulation of Outward Currents

Muscarinic stimulation also regulates outward currents in GI muscles. This mechanism is quite different from the traditional view that cholinergic stimulation causes depolarization and enhancement in action potential generation by activating a conductance and lowering membrane resistance (see BOLTON 1981).

The first report of cholinergic regulation of a K^+ current came from studies of toad gastric muscle cells (SIMS et al. 1985). These authors found that at least a portion of the depolarization response caused by acetylcholine or muscarine could be attributed to the suppression of a K^+ current that had properties similar to the "M-current" observed in bullfrog sympathetic nourons (ADAMS et al. 1982a,b). In gastric muscle cells ACh caused a decrease in conductance and depolarization (see Fig. 6). Muscarinic stimulation induced net inward currents when cells were held under voltage clamp positive to the equilibrium potential for K^+ ions (E_K), but net outward current when cells were held negative to E_K. The reversal potential for the current induced by ACh was shifted by changes in the K^+ gradient in a manner consistent with the idea that the current was carried by K^+ ions. The current displayed voltage-dependent activation with almost complete deactivation at -70 mV and a sigmoid activation curve positive to this level. Maximum conductance was activated at about -10 mV. The effects of muscarinic stimulation persisted in external solutions in which Ca^{2+} was replace by Mn^{2+}, suggesting the current was not Ca^{2+}-dependent. It was also noted that Ba^{2+} (2–5 m*M*) mimicked the suppression of the M-current caused by muscarine. Although atropine blocked the muscarinic suppression of the M-current, it did not affect the response to Ba^{2+}.

M-currents have not been found in mammalian smooth muscles, but in some mammalian cells it appears that muscarinic stimulation can suppress activation of Ca^{2+}-activated K^+ currents. BENHAM and BOLTON (1986) reported that muscarinic stimulation suppressed the occurrence of spontaneous transient outward currents (STOCs). STOCs were attributed to the spontaneous release of Ca^{2+} from internal stores and activation of Ca^{2+}-activated K^+ channels (see above). These authors suggested that the inhibitory effect of ACh on STOCs was due to emptying of internal stores of Ca^{2+}. After stores had emptied, no additional Ca^{2+} could be released to activate STOCs. COLE et al. (1989) attributed the effects of ACh on Ca^{2+}-dependent outward current to a more direct mechanism. These authors showed that ACh applied to colonic muscle cells shifted the activation curve for Ca^{2+}-activated K^+ channels to more positive potentials. Muscarinic stimulation decreased the open probability at any given potential or Ca^{2+} concentration. It also was found that suppression of Ca^{2+}-activated K^+ channels by ACh was reversible, blocked by atropine (COLE and SANDERS 1989b), and mediated by G proteins. The effect could not be demonstrated unless cells were dialyzed with GTP. In the presence of GTP, ACh decreased Ca^{2+}-dependent outward currents and the effect was reversible. In contrast, when nonhydrolyzable analogs of GTP were used (such as GTPγS), the suppression of K^+ current elicited by ACh was not reversible. The effects of ACh were blocked by dialysis of cells with pertussis toxin. Although the complete second messenger cascade that mediates this effect has not been elucidated in GI muscles, work on tracheal muscles suggests that part of the suppression of Ca^{2+}-activated K^+ current could be due to a direct effect of an α-subunit of G proteins on the channel (KUME et al. 1992). In these muscles there appears to be positive and negative regulation of the open probability of Ca^{2+}-activated K^+ channels because G_i-linked receptor agonists decreased P_o and G_s-linked receptor agonists increased P_o in the same membrane patches. These authors also showed enhancement in P_o of Ca^{2+}-activated K^+ channels upon direct application of activated α_o subunit to the interior surface of excised membrane patches.

Ca^{2+}-dependent currents are also suppressed by cholinergic stimulation in cat esophageal muscles (SIMS et al. 1990a). These authors found that the STOCs that occur in esophageal cells were suppressed by ACh. Several tests suggested that these events were due to Ca^{2+}-activated K^+ currents. In these cells ACh often transiently increased K^+ current, and this preceded the suppression of STOCs. The transient increase in K^+ current may have been due to initial stimulation of Ca^{2+}-activated K^+ current by release of Ca^{2+} from internal stores. Suppression of this current may have occurred via the G protein mechanism described above. Atropine, which had no effect on currents alone, blocked the initial transient increase in outward current and the suppression of STOCs.

II. Regulation of Currents by Adrenergic Stimulation

Direct adrenergic control of GI smooth muscles is probably not as important as the prejunctional control exerted by adrenergic nerves in these muscles. Nevertheless, GI smooth muscles express α- and β-adrenoceptors, and in most cases adrenergic agonists exert an inhibitory effect on GI muscles. The role of adrenergic pathways in the regulation of ionic conductances in GI muscles has not been as extensively studied as cholinergic effects, but some suggestions of the mechanisms by which these agonists work are available. At present, it appears that adrenergic effects on electrical activity are primarily mediated through regulation of K^+ channels.

Isoproterenol induced an M-current in single, voltage-clamped toad gastric muscle cells (SIMS et al. 1988). The current persisted for several minutes, but it was rapidly suppressed by further addition of ACh. This demonstrated dual regulation of the M-current, in which adrenergic agonists activate the current and cholinergic stimulation shuts it off. Since the effects of β-adrenergic stimulation are commonly mediated through enhanced production of cAMP, analogs of cAMP were tested for their ability to induce the M-current. Dibutyryl cAMP and 8-bromo-cAMP induced a current with identical properties to the current induced by isoproterenol. ACh also suppressed the current induced by the cAMP analogs, suggesting that regulation by ACh is downstream from the adenylate cyclase.

Several inhibitory agonists utilize cAMP-dependent pathways in mammalian muscle cells as well. Although few specific studies of these agonists exist, the effects of cAMP and cAMP-dependent protein kinase (protein kinase A) have been tested on Ca^{2+}-activated K channels of canine colonic muscles (CARL et al. 1991). Under "on cell" recording conditions, forskolin increases the open probability of Ca^{2+}-activated K^+ channels (Carl, personal communication), and this observation is consistent with the increase in open probability that is stimulated by the activated catalytic subunit of protein kinase A (Fig. 9; CARL et al. 1991). Calyculin A and okadaic acid,

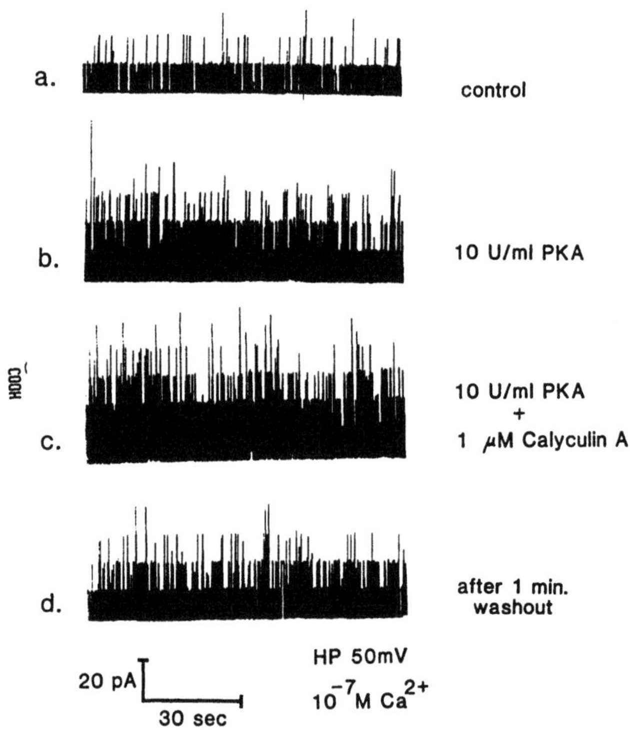

Fig. 9. Effects of activated catalytic subunit of protein kinase A on open probability of Ca^{2+}-activated K^+ channels. *Top trace* shows single channel currents recorded from a patch in the inside-out configuration held at +50 mV. K^+ gradient was symmetrical. Many single channel openings are seen, occasionally three channels open simultaneously. Addition of 10 U/ml protein kinase A catalytic subunit increased open probability (*panel B*; three channels open simultaneously more common, and even four or more channels can be seen to open simultaneously). Addition of calyculin A (1 μM), a phosphatase inhibitor, increased open probability even more. These effects reversed after removal of the drugs. Data illustrate regulation of Ca^{2+}-activated K channels by phosphorylation. (From CARL et al. 1991)

two phosphatase inhibitors, further increased the stimulation caused by protein kinase A. These results are similar to those from studies on tracheal smooth muscles (KUME et al. 1989), suggesting that phosphorylation via cAMP-dependent mechanisms may be a common mechanism of regulation of Ca^{2+}-activated K^+ channels in several smooth muscles.

Phosphorylation, although not necessarily via protein kinase A, also appears to regulate Ca^{2+} currents in some GI muscles. WARD et al. (1991b) reported that the phosphatase inhibitors calyculin A and okadaic acid reduced the amplitude of Ca^{2+} currents in gastric and colonic muscles. This response was also manifest at the level of muscle strips since these agents also inhibited electrical slow waves in a manner consistent with inhibitory effects on L-type Ca^{2+} currents. These data suggest that phosphorylation may either directly, or indirectly via changes in $[Ca^{2+}]_i$, regulate Ca^{2+} entry in these muscles. If the effect is direct (i.e., via phosphorylation of Ca^{2+} channels), this would be a mechanism opposite from the regulation of Ca^{2+} channels in the heart (TRAUTWEIN and HESCHELER 1990).

Several agonists that are important in the regulation of GI motility (i.e., β-adrenergic agents, VIP, and CGRP) have effects that are probably mediated via cAMP. Therefore, the regulation of channels by cAMP-dependent phosphorylation may be a common pathway for a number of inhibitory agonists. Actually, there are also other means by which cAMP-dependent effects can be manifest. Under some circumstances, cAMP may cross react and activate cGMP-dependent protein kinase (JIANG et al. 1992), and this effect may mediate some of the effects of cAMP in smooth muscles (LINCOLN et al. 1990). A recent report suggests that part of the inhibitory effect of high concentrations of isoproterenol in gastric muscles may be due to activation of cGMP-dependent protein kinase (JIN et al. 1993). Whether this cross-activation of cGMP-dependent protein kinase by cAMP affects ionic conductances in GI muscles is to be determined.

III. Regulation of Currents by Nitrosovasodilators

An exciting development recently has been the identification of nitric oxide (NO) as a noncholinergic, nonadrenergic (NANC) inhibitory neurotransmitter in GI muscles (for review see SANDERS and WARD 1992). Release of NO from enteric neurons is apparent because the enzyme that synthesizes NO is found in enteric inhibitory neurons (BREDT et al. 1991; YOUNG et al. 1992; WARD et al. 1992a), and TTX-sensitive release of NO occurs during electrical field stimulation (BOECKXSTAENS et al. 1991; CHAKDER and RATTAN 1993). Release of NO is implicated in the hyperpolarization response known as inhibitory junction potentials that occur in response to inhibitory nerve stimulation (DAZIEL et al. 1991; THORNBURY et al. 1991; STARK et al. 1991; CONKLIN and DU 1992).

The mechanism by which NO causes hyperpolarization has not been fully established. Most investigators would agree that inhibitory junction potentials are due to an increase in K^+ conductance (TOMITA 1972); however other mechanisms have been proposed (e.g., CRIST et al. 1991). THORNBURY et al. (1991) found that application of NO to isolated colonic muscle cells caused an enhancement in the open probability of Ca^{2+}-activated K^+ channels, and this action may play a role in some of the inhibitory effects caused by NO and other nitrosovasodilators, such as sodium nitroprusside. The mechanism of action of NO on Ca^{2+}-activated K^+ channels has not been determined. It is possible that cGMP or its degradation product, 5′-GMP, could directly influence the open probability of these channels (WILLIAMS et al. 1988), or it is possible that the increase in open probability could be mediated via cGMP-dependent phosphorylation as recently demonstrated in coronary artery smooth muscle cells (TANIGUCHI et al. 1992). It is unlikely that the large conductance Ca^{2+}-activated K^+ channels activated by NO are responsible for inhibitory junction potentials in GI muscles because these events are not blocked by rather high concentrations of TEA (10 m*M*; Bayguinov and Sanders, to be published), which is known to block these channels at much lower concentrations (CARL et al. 1993). Other K^+ conductances activated by NO have not been identified in GI muscles, but there may be several. For example, inhibitory junction potentials in the canine colon are not blocked by apamin (SMITH et al. 1989), but in many other regions such as the pyloric sphincter apamin reduces IJPs that appear to be due to NO (WARD et al. 1992; BAYGUINOV et al. 1993; BAYGUINOV and SANDERS 1993).

In vascular muscle sodium nitroprusside has been shown to inhibit voltage-dependent Ca^{2+} current and to increase the occurrence of spontaneous transient outward currents (STOCs). The latter was interpreted as a result of unloading of overloaded internal stores (CLAPP and GURNEY 1991). Regulation of Ca^{2+} channels in GI muscles by nitrosovasodilators has not yet been reported.

IV. Regulation of Currents by Peptides

Many biologically active peptides are important as regulators of GI motility. These peptides are released either as neurotransmitters or as hormones, and they have a wide variety of important effects in GI muscles (for review see DANIEL et al. 1989b), including potent effects on the electrical activities of these muscles (see SZURSZEWSKI 1987). At present there is very little known about the ionic channels affected by most of the peptides that are important in GI motility.

In toad gastric muscle cells substance P suppresses the same K^+ current affected by ACh (i.e., the M-current, see section above on "Regulation of currents by muscarinic stimulation"; SIMS et al. 1986). Substance P depolarized isolated gastric muscle cells and this effect was associated with a decrease in membrane conductance. Under voltage clamp conditions, substance P elicited a net inward current while net conductance decreased. These changes were attributed to suppression of an outward K^+ current that had similar properties to the cholinergically activated M-current (SIMS et al. 1985). When the M-current was fully suppressed by ACh, substance P had no further effect. Atropine blocked the effects of ACh, but did not affect the responses to substance P. These authors also noted that many of the cells studied were capable of responding in a similar manner to ACh and substance P, suggesting expression of multiple types of excitatory receptors on the same cells (see also LASSIGNAL et al. 1986).

At low doses, substance P was also found to activate Ca^{2+}-activated K^+ channels when it was applied to rabbit colonic myocytes (MAYER et al. 1990). At higher doses, substance P first activated Ca^{2+}-activated K^+ channels and then produced long lasting inhibition (MAYER et al. 1989). The stimulatory effect on these channels was dependent upon extracellular Ca^{2+} and blocked by dihydropyridines, suggesting that the activation seen with substance P was due to an increased influx of Ca^{2+} through L-type Ca^{2+} channels. The potentiation was blocked when $[Ca^{2+}]_i$ was buffered with 10 mM EGTA in the pipette solution. The effects of substance P were mimicked with low concentrations (10^{-12}–10^{-10} M) of substance P methylester (SPME), an NK-1 receptor agonist (REGOLI et al. 1987). SPME also significantly increased the amplitude of the voltage-dependent Ca^{2+} current, which was viewed to be a potentiation of L-type Ca^{2+} current because it was dihydropyridine sensitive. These interesting results contrast with the actions of acetylcholine on mammalian muscle cells (see section on cholinergic "Regulation of Calcium Currents" above), and suggest that different second messenger pathways mediate the effects of the two excitatory agonists.

V. Regulation of Currents by K^+ Channel Agonists

An interesting area of investigation has been the regulation of a class of K^+ channels that may correspond to ATP-dependent K^+ channels in other muscles (e.g., SANGUINETTI et al. 1988; QUASTHOFF et al. 1989; HAMILTON et al. 1986; see also WESTON 1989). A number of agents that open K^+ conductances in smooth muscles have been identified (WESTON 1989). These "K^+ channel agonists" also appear to activate K^+ channels in GI muscles, although this has not been as intensely studied as it has in vascular smooth muscle. Cromakalim and its (−)optical isomer, lemakalim, hyperpolarize membrane potential, shorten the duration of electrical slow waves, and decrease input resistance in canine colonic muscles (Fig. 10; POST et al. 1991; FARRAWAY and HUIZINGA 1991). These effects are consistent with the notion that these compounds increase K^+ conductance. Whole cell voltage clamp studies demonstrated that cromakalim and lemakalim increased a "time-independent" outward current and cromakalim also inhibited voltage-dependent Ca^{2+} current (POST et al. 1991). The nature of the outward current activated by these agents was not explored, but a number of K^+ channels have been linked to K^+ channel agonists in smooth muscles includ-

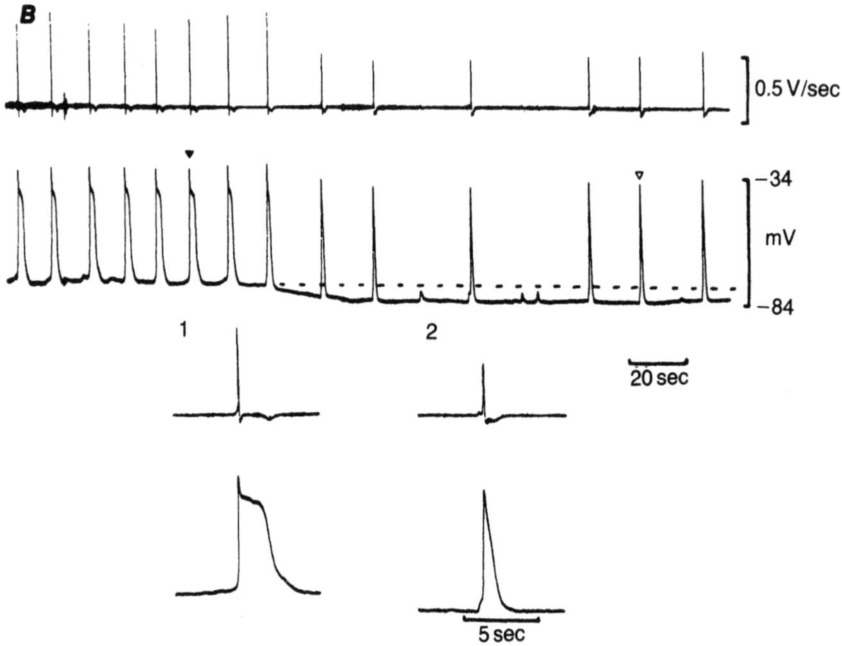

Fig. 10. Effects of lemakalim on electrical slow waves of canine colonic smooth muscle. *Top trace* shows first derivative of membrane potential, *second trace* shows intracellular recording of slow wave activity. Addition of lemakalim (*arrowhead*; 10 μM) caused hyperpolarization, decreased the duration of slow waves, and reduced the rise times of slow waves. *Below*, a control slow wave and a slow wave recorded during lemakalim are displayed at a faster sweep speed. Note loss of plateau phase during lemakalim. (From POST et al. 1991)

ing a large-conductance ATP-sensitive K^+ conductance (STANDEN et al. 1989); large conductance Ca^{2+}-activated K^+ channels (GELBAND et al. 1989); delayed rectifiers (BEECH and BOLTON 1989a); and small conductance Ca^{2+}-activated K^+ channels (OKABE et al. 1990). In colonic muscles cromakalim and lemakalim activated large conductance Ca^{2+}-activated K^+ channels (CARL et al. 1992). For example, in inside out patches lemakalim reversibly increased open probability of these channels, and glyburide prevented this effect. Under conditions where Ca^{2+}-activated K^+ channels were maximally activated lemakalim failed to activate a second type of K^+ conductance in these experiments, and when TEA was present in the pipette solution to partially block Ca^{2+}-activated K^+ channels a second type of K^+ channel activated by lemakalim was not observed. The effects of lemakalim on ensemble average currents obtained from multiple openings of Ca^{2+}-activated K^+ channels mimicked the effects of this agent on whole cell currents. Although these data suggest that part of the hyperpolarization and inhibition of electrical slow waves caused by lemakalim and cromakalim might be mediated by an increase in the open probability of Ca^{2+}-activated K^+

channels, another conductance is also likely to contribute to these effects since hyperpolarization responses of colonic muscles to lemakalim were not blocked by charybdotoxin (K. Keef, personal communication). The expression and characteristics of other conductances in GI muscles that are affected by K^+ channel agonists is in need of further investigation. But this may be a highly fruitful avenue of investigation because K^+ channel agonists may find eventual therapeutic use in relieving spasmodic conditions in the GI tract.

D. Coupling Between Electrical Excitation and Contraction

The electrical activity described above facilitates the entry of Ca^{2+} into GI smooth muscle cells. In tonic muscles small depolarizations may be sufficient to increase the open probability of voltage-dependent Ca^{2+} channels and increase influx. In phasic muscles electrical events, slow waves and action potentials, result in Ca^{2+} influx because these events are associated with an increase in voltage-dependent Ca^{2+} currents. Although the magnitude of the currents associated with tonic depolarization and electrical excitability in GI muscles seem small (i.e., only a few picoamps in the case of a small tonic depolarization to a few hundred picoamps in the case of an action potential), these currents are sufficient to change cytoplasmic Ca^{2+} in a meaningful way. If fact, as discussed below, the changes in cytoplasmic free Ca^{2+} ($[Ca^{2+}]_i$) observed with physiological depolarizations are far less than the actual changes in total cell Ca^{2+} that result from Ca^{2+} influx because Ca^{2+} buffering is high in smooth muscle cells.

The study of the coupling between electrical activity and $[Ca^{2+}]_i$ has been greatly enhanced by the introduction of Ca^{2+}-sensitive indicators, such as fura-2 and indo-1 (see GRYNKIEWICZ et al. 1985). These indicators can be delivered directly to the cytoplasm of cells via patch clamp electrodes, or they can be delivered in a membrane permeable form that is taken up by cells and then metabolized to produce a "trapped" active form of the indicator. By ratioing emissions elicited by two different excitation wavelengths (fura-2) or ratioing two emission wavelengths excited by a single excitation wavelength (indo-1), quantitative information about $[Ca^{2+}]_i$ can be obtained. Dynamic changes in $[Ca^{2+}]_i$ can be followed as a function of time (Ca^{2+} transients) while other measurements, such as contraction, membrane potential, or membrane current are simultaneously collected.

I. Relationship Between Ca^{2+} Current and $[Ca^{2+}]_i$

For several years $[Ca^{2+}]_i$ has been recognized as the primary regulator of contractile activity in smooth muscles (KAMM and STULL 1985). Several investigators have studied the relationship between $[Ca^{2+}]_i$ and force in skinned smooth muscle strips (IINO 1981), and YAGI et al. (1988) demonstrated the relationship between $[Ca^{2+}]_i$ and contractile force in isolated toad gastric smooth muscle cells. Later this group also investigated the regulation of $[Ca^{2+}]_i$ by characterizing changes in $[Ca^{2+}]_i$ in fura-2 loaded cells while monitoring electrical activity with intracellular microelectrodes (BECKER et al. 1989). Basal $[Ca^{2+}]_i$ under the conditions of this study averaged 226 n*M*. Transient increases in $[Ca^{2+}]_i$ occurred during action potentials or in response to depolarizing command potentials under voltage clamp. Integration of the Ca^{2+} current suggested that total cell Ca^{2+} increased far more than free Ca^{2+} (as measured with fura-2). This observation demonstrated that most of the Ca^{2+} that enters cells is buffered, and these authors calculated that Ca^{2+} buffering exceeded 40-fold. Upon repetitive depolarizations, Ca^{2+} transients summed and $[Ca^{2+}]_i$ reached a maximal level of about 850 n*M*. A "ceiling" in $[Ca^{2+}]_i$ was probably reached because: (a) peak Ca^{2+} current decreased with each depolarization, presumably due to inactivation of Ca^{2+} channels and (b) the process for removing Ca^{2+} was accelerated by higher levels of $[Ca^{2+}]_i$. The latter was thought to be due to Ca^{2+}-dependent facilitation of Ca^{2+} extrusion mechanisms such as the sarcolemmal Ca^{2+} pumps.

The relationship between Ca^{2+} current and $[Ca^{2+}]_i$ has also been studied in mammalian GI muscle cells (VOGALIS et al. 1991). Canine gastric muscle cells had an average basal $[Ca^{2+}]_i$ of 144 n*M*. Depolarization into the voltage range in which Ca^{2+} current could be resolved caused an increase in $[Ca^{2+}]_i$ (Fig. 11). The change in $[Ca^{2+}]_i$ mirrored the current-voltage relationship for Ca^{2+} current. Nifedipine blocked most of the Ca^{2+} current and abolished Ca^{2+} transients. Integration of the inward current (shown to be exclusively Ca^{2+} current in these cells) showed that total cell Ca^{2+} actually increased to nearly 100 μM during a 2-s pulse to 0 mV, but the change in $[Ca^{2+}]_i$ was more than 100-fold less than this suggesting that the majority of the Ca^{2+} entering these cells was buffered, taken up by internal stores, or rapidly extruded. After repolarization, Ca^{2+} transients relaxed, but the rate of relaxation was surprisingly slow (time constant about 2 s). The maximum rate was dependent upon the level of $[Ca^{2+}]_i$ reached during the depolarization, as noted by BECKER et al. (1989).

In studies of a variety of smooth muscles investigators have noted a "mechanical threshold," which is the voltage that must be obtained before a

⟶

Fig. 11. Isolated gastric muscle cells were voltage-clamped with protocols that simulated slow waves. Voltage-clamp protocols (shown in *top panel*) consisted of an initial upstroke depolarization to −10 mV, partial repolarization to a stable plateau potential for 6 s, and repolarization to the holding potential of −60 mV. The plateau potential was varied from −45 mV to −25 mV to simulate the changes in the plateau that occur in intact muscles stimulated with excitatory and inhibitory agonists. The *bottom traces* show Ca^{2+} currents elicited by the slow wave-like depolarization. Current responses (responses to −25 mV and −45 mV depolarization are denoted) consisted of a transient inward current corresponding to the upstroke depolarization and a small sustained inward current corresponding to the plateau

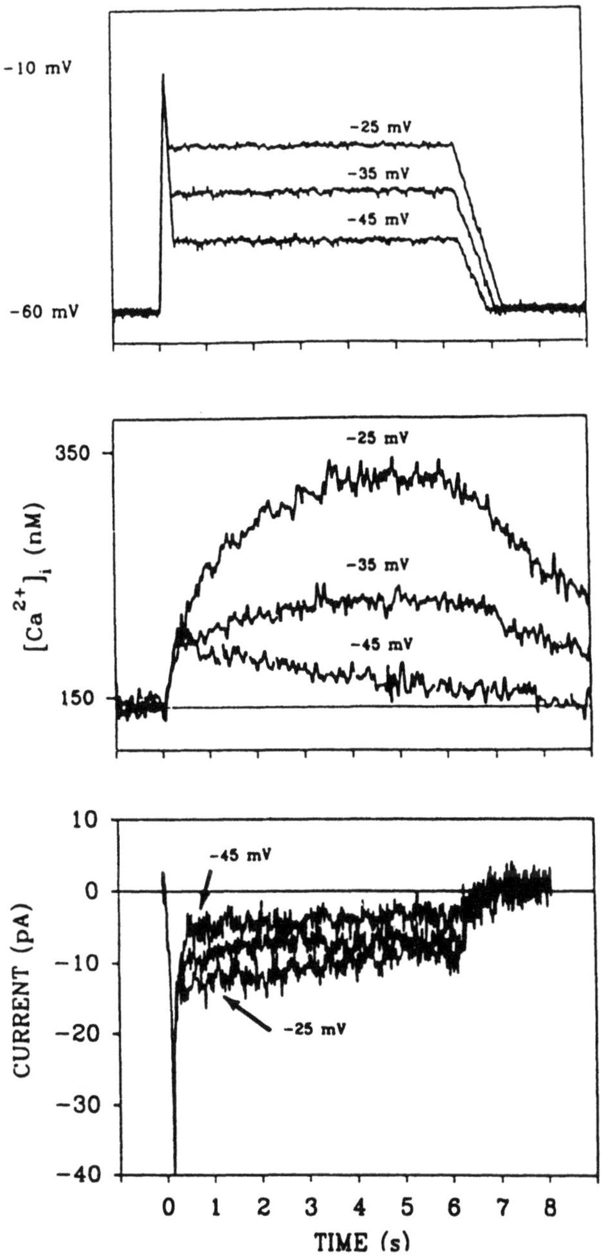

depolarization. The sustained currents varied by only a few pA in amplitude. The *middle traces* shows changes in $[Ca^{2+}]_i$ in response to the influx of Ca^{2+}. Between −45 mV and −35 mV, a transition occurred in which $[Ca^{2+}]_i$ either fell during the plateau phase or continued to rise. The continual increase in $[Ca^{2+}]_i$ leads to a second phase of contraction in intact muscles, and this is determined by subtle changes in the magnitude of the Ca^{2+} current. This is the same range of potentials referred to as the "mechanical threshold" (see SZURSZEWSKI 1987) in these cells. (From VOGALIS et al. 1991) depolarization elicits contraction. The source of the mechanical threshold was investigated by VOGALIS et al. (1991). As described above, the rise in $[Ca^{2+}]$ $_{w}$as closely associated with the Ca^{2+} current. Voltage-clamp protocols that simulated the voltage changes observed in intact gastric muscles showed that depolarization of cells over the range of potentials that occur during slow waves (i.e., −45 to −25 mV during plateau phase; see Fig. 11) caused an increase in the magnitude of Ca^{2+} transients. In gastric muscles contractile transients are often biphasic, and previous work has suggested that the first phase is associated with the upstroke depolarization (which is relatively constant from event to event) and the second phase is associated with the magnitude of the plateau phase of the slow wave (see SZURSZEWSKI 1987). The plateau phase is highly variable and it is regulated by many excitatory and inhibitory agonists. In voltage-clamp studies, when the amplitude of the "plateau" phase of the test potentials was adjusted over the range of −45 to −25 mV, Ca^{2+} transients changed from reaching a peak soon after initiation of the depolarization (at −45 mV) to constantly increasing over the 6-s depolarization (−35 to −25 mV). The latter would be expected to initiate a small contraction soon after the onset of depolarization and then the amplitude of contraction would be expected to continue to increase during the sustained increase in $[Ca^{2+}]_i$. Thus, Ca^{2+} entry during these simulated slow waves could predict the contractile events observed in situ and explain why the voltage level of about −40 mV appears as the "mechanical threshold" in these muscles. One could also predict from these experiments that agonists that slightly reduce the amplitude of the plateau (perhaps by reduction of Ca^{2+} current or by an increase in K^+ current during the plateau) could greatly affect the mechanical output of these muscles. This is exactly the observation that others have made in many studies of the electromechanical coupling in gastric muscles (see MORGAN and SZURSZEWSKI 1980; MORGAN et al. 1981; SZURSZEWSKI 1987).

II. Regulation of Ca^{2+} Influx by $[Ca^{2+}]_i$

One of the important means by which Ca^{2+} influx is regulated occurs by Ca^{2+}-dependent inactivation of Ca^{2+} channels. This is a common phenomenon that occurs in many cells (see LEE et al. 1985; PLANT et al. 1983; ECKERT and TILLOTSON 1981; ECKERT and CHAD 1984; PLANT 1988), and it appears to be a very important mechanism in mammalian GI smooth muscles (GANITKEVICH et al. 1987; and see section "Properties of Voltage-Dependent Ca^{2+} Currents" above). The regulation of Ca^{2+} current by $[Ca^{2+}]_i$ was investigated in canine gastric muscle cells in studies in which these parameters were simultaneously measured (VOGALIS et al. 1992). Basal $[Ca^{2+}]_i$ appeared to exert some influence on the magnitude of Ca^{2+} current because buffering of $[Ca^{2+}]_i$ to low levels with EGTA increased the average amplitude of the Ca^{2+} current. When Ca^{2+} flows through Ca^{2+} channels, the current rapidly inactivates. Inactivation in canine gastric cells can be fit with

2 exponentials (time constants of 7.4 and 42 ms). When Ba^{2+} or Na^{+} were used as charge carriers, the fast component of inactivation was abolished and the current inactivated more slowly. The fast component of inactivation appeared to be rather independent of $[Ca^{2+}]_i$ which, as measured with indo-1, is a measurement of cytoplasmic free Ca^{2+}. The data suggested that Ca^{2+} within the channel or very near the mouth of the channel is probably responsible for Ca^{2+} inactivation. Theoretical calculations suggest that the Ca^{2+} concentration can be very high near the mouth of the channel just after the channel opens (SALA and HERNANDEZ-CRUZ 1990; SHERMAN et al. 1990). Changes in $[Ca^{2+}]_i$ were not observed for at least 20 ms following the onset of Ca^{2+} influx. This is well into the period during which Ca^{2+} inactivation occurs.

Following repolarization of cells, several seconds are required before Ca^{2+} current fully recovers from inactivation. The time course is probably too slow to be accounted for by a purely voltage-dependent phenomenon, so recovery from inactivation studies were performed to test the hypothesis that the slow washout of Ca^{2+} following repolarization may be responsible for the slow recovery from inactivation. These studies demonstrated that $[Ca^{2+}]_i$ decays with approximately the same time constant as the recovery of Ca^{2+} current, suggesting that the long recovery time is due to prolonged Ca^{2+} inactivation. The slow recovery from inactivation may partially explain the long refractory periods following slow wave repolarization in muscle strips (PUBLICOVER and SANDERS 1986). It is interesting to note that recovery from inactivation was hardly faster when Ba^{2+} was used as a charge carrier. It is possible that divalents other than Ca^{2+} can inactivate Ca^{2+} channels just as observed in Helix neurons (CHAD and ECKERT 1986), but this process may occur more slowly than the inactivation produced by Ca^{2+}.

III. Relationship Between Electrical Activity and $[Ca^{2+}]_i$ in Muscle Strips

At one time, and still present in many textbooks of gastroenterology, a "division of labor" was thought to exist between electrical slow waves and "spikes." Most investigators believed that slow waves timed contractions, and spikes, which occurred at the peak of slow waves, initiated contractions. This led, in part, to the terminology of electrical control activity (ECA) for slow waves and electrical response activity (ERA) for action potentials (SARNA 1975). Now it is recognized that slow waves in the stomach, small bowel and colon can, in the absence of spiking, elicit contractions (SZURSZEWSKI 1975; MORGAN and SZURSZEWSKI 1980; SANDERS 1983; HARA et al. 1986; CHRISTENSEN et al. 1969; SANDERS and SMITH 1986; and see SZURSZEWSKI 1987; SANDERS and PUBLICOVER 1989 for reviews). Certainly spiking is an important mechanism for E-C coupling in muscles of the small intestine, the longitudinal layer of the colon, and pyloric sphincter. But contractions also result from slow waves because these events result, in part, from an

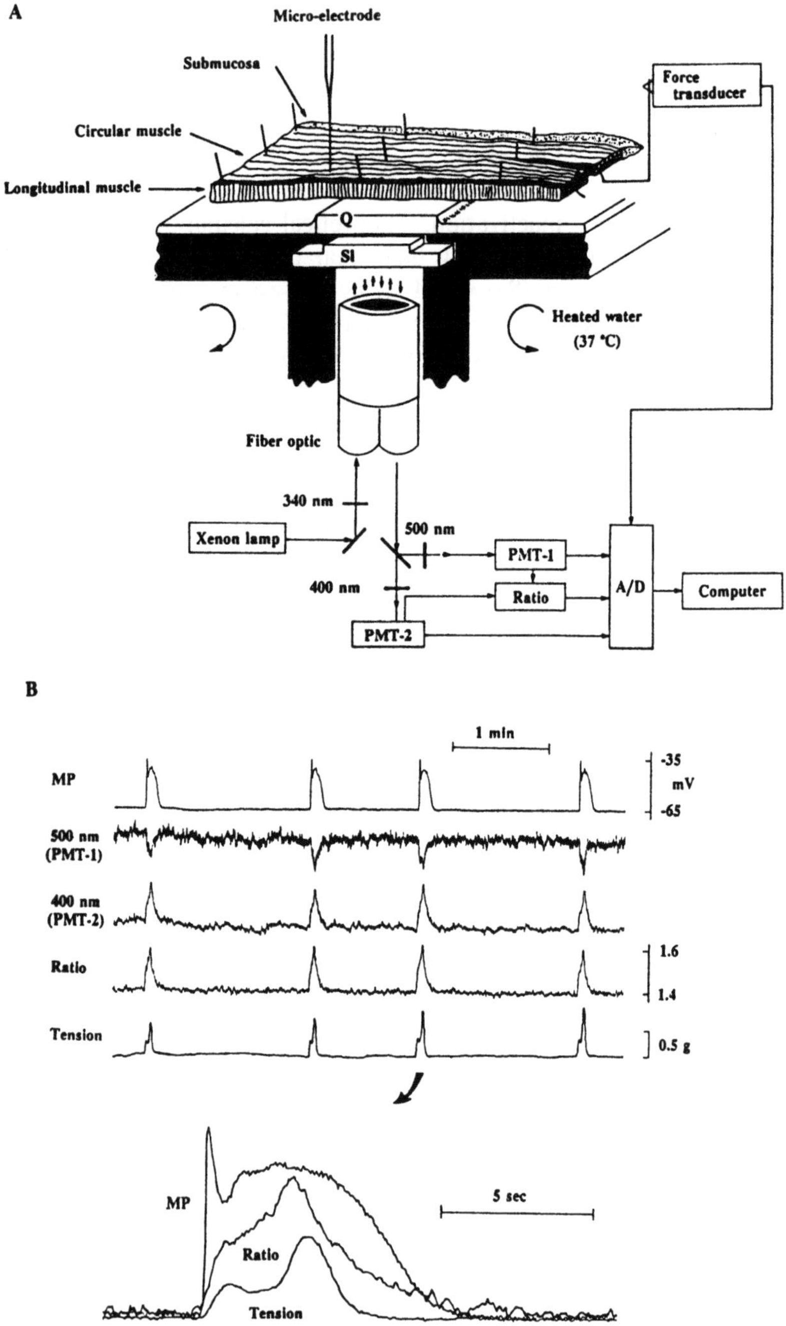
A
Micro-electrode
Submucosa
Circular muscle
Longitudinal muscle
Force transducer
Q
SI
Heated water (37 °C)
Fiber optic
340 nm
Xenon lamp
500 nm
PMT-1
400 nm
Ratio
A/D
Computer
PMT-2
B
1 min
MP
-35
mV
-65
500 nm (PMT-1)
400 nm (PMT-2)
Ratio
1.6
1.4
Tension
0.5 g
MP
5 sec
Ratio
Tension

influx of Ca^{2+} through dihydropyridine-sensitive Ca^{2+} channels (see section above on "Contribution of Ionic Channels to Electrical Activity and Electromechanical Coupling" and review by SANDERS 1992). If the influx of Ca^{2+} causes $[Ca^{2+}]_i$ to reach a level where myosin light chain kinase is activated, then contractile responses will result. There appears to be a sufficient increase in the open probability of voltage-dependent Ca^{2+} channels during the plateau phase of slow waves in most muscles (i.e., plateau occurs at −50 to −35 mV) to increase $[Ca^{2+}]_i$ to levels in which slow waves are coupled to contractions.

The ability of slow waves to initiate Ca^{2+} transients has been directly studied in canine gastric smooth muscle strips using a technique in which changes in membrane potential, $[Ca^{2+}]_i$, and tension could be simultaneously monitored (see OZAKI et al. 1991c; Fig. 12). Each electrical slow wave initiated a rise in $[Ca^{2+}]_i$ and a small phasic contraction in gastric muscles. Contractions in the canine gastric antrum are often biphasic (MORGAN et al. 1978). The upstroke phase of the slow wave is fairly constant from event to event and this initiates a small rise in $[Ca^{2+}]_i$ and a small contraction. The plateau phase of the slow wave is quite variable. If it is of sufficient amplitude, $[Ca^{2+}]_i$ rises during this phase and a second phase of contraction is initiated. ACh increased the amplitude of the plateau phase, increased $[Ca^{2+}]_i$, and increased the force of phasic contractions. Similar responses were observed with the dihydropyridine Ca^{2+}-channel agonist Bay K 8644. Nicardipine antagonized the enhancement in slow waves, Ca^{2+} transients and contractions. These data directly demonstrate that quite small changes in slow wave amplitude and duration can result in significant changes in $[Ca^{2+}]_i$ and contractile force. They also demonstrated that entry of Ca^{2+} through dihydropyridine-sensitive (L-type) Ca^{2+} channels is critical for excitation-contraction coupling in phasic GI muscles. It is interesting to note that while responses to ACh were mimicked by Bay K 8644, it is unlikely that ACh

←

Fig. 12. Measurement of membrane potential, fluorescence and contraction from smooth muscle strips. *Panel A* shows a schematic of the preparation and apparatus used. Muscles were mounted in a specially designed chamber. Cross-sectional orientation of the muscle strip allowed impalement of cells at any point through muscularis externa. Muscle was pinned over quartz window (*Q*) in bottom of chamber. Excitation and emission light were guided to and from muscle strip by a fiber optic light guide. Fluorescence from specific regions of the muscle could be specifically monitored by a movable slit (*Sl*) mounted beneath the quartz window. The fiber optic was connected to a fluorimeter in which a xenon lamp produced excitation light (340 nm) and two photomultiplier tubes (PMT) collected emissions at 400 nm and 500 nm. After analog determination of the F400/F500 ratio, signals were digitized along with tension and membrane potentials and analyzed by the computer. *Panel B* shows representative records showing spontaneous slow wave activity (*MP*), 500 nm fluorescence (F500, detected by PMT-1), 400 nm fluorescence (F400, detected by PMT-2), F400/F500 ratio which is proportional to $[Ca^{2+}]_i$, and tension. *Traces at bottom* show superposition of membrane potential, F400/F500 ratio, and tension. (From OZAKI et al. 1991c)

directly increases the open probability of Ca^{2+} channels (see section on "Regulation of Calcium Channels" by cholinergic agonists). Instead other mechanisms, such as activation of nonselective cation current and/or suppression of K^+ currents, serve to maintain or increase the level of depolarization achieved during the slow wave plateau. The more depolarized level and the longer period of depolarization increases the open probability of Ca^{2+} channels, increases the period of Ca^{2+} influx, and causes enhanced $[Ca^{2+}]_i$. Similar enhancements in $[Ca^{2+}]_i$ undoubtedly occur in spiking muscles since action potentials in GI muscles are due to activation of L-type Ca^{2+} channels, but equivalent experiments in which the Ca^{2+} transients that occur in response to action potentials have not yet been performed.

IV. Ca^{2+} and Metabolism

Glycolysis (glycogenolysis) is stimulated not only by inorganic phosphate or ADP which activates phosphofructokinase, but also by Ca^{2+} and calmodulin which activates phosphorylase b kinase. Reduced pyridine nucleotides are fluorescent substances which are located both in mitochondria and in the cytosol. Oxidized flavoproteins are also fluorescent substances which are located specifically in the inner mitochondrial membrane. Utilizing these characteristics, redox states of pyridine nucleotides and flavoproteins have been recorded fluorometrically in a variety of isolated tissues or cells. By choosing the experimental conditions (anoxia, glucose depletion and a combination of these), it is possible to determine the metabolic regulation in cytosol and mitochondria in intact smooth muscle. As shown in Fig. 13, reduced pyridine nucleotides and oxidized flavoproteins increase in response to spontaneous mechanical activities in guinea pig taenia caecum (OZAKI et al. 1988) indicating that large oxidation-reduction potentials are generated across the mitochondrial membrane during contractions. The amount of reduced pyridine nucleotides is closely correlated with force of contractions in guinea pig ileum (SHIMIZU et al. 1991). Interestingly, the change in flavoproteins fluorescence starts to increase 0.5–1 s before the initiation of contraction and this time course corresponds to the change in $[Ca^{2+}]_i$. Further, Ca^{2+} dependency of contraction, pyridine nucleotide fluorescence and flavoprotein fluorescence indicates that the Ca^{2+} sensitivity is in the order of: flavoproteins fluorescence > pyridine nucleotide fluorescence > muscle contraction. CHANCE (1965) observed that Ca^{2+} increases rate of respiration and electron transport in mitochondria. Further, key, intramitochondrial enzymes for oxidative metabolism, such as dehydrogenases, are activated by micromolar concentrations of Ca^{2+} (see BALABAN 1990). From these findings, it is suggested that the increase in $[Ca^{2+}]_i$ may directly and independently activate cytoplasmic glycolysis, mitochondrial oxidation of flavoproteins and contraction in intact smooth muscle tissue. ATP may be generated in smooth muscle cells before its consumption due to Ca^{2+}-dependence of mitochondrial respiration.

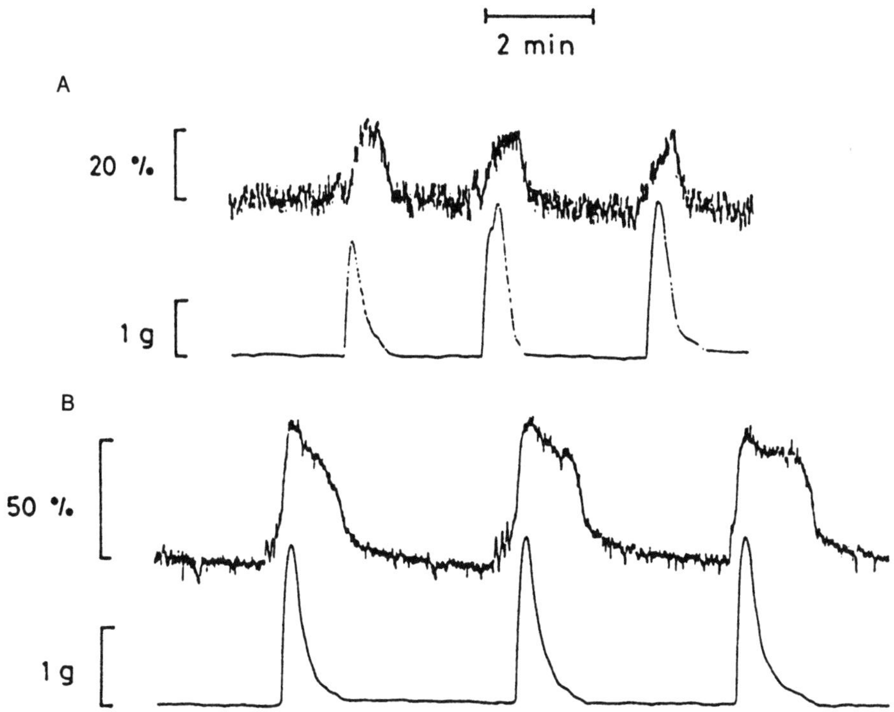

Fig. 13. Changes in reduced pyridine nucleotides (*PN_{red}*; *panel A*) and oxidized flavoproteins (*FP_{ox}*; *panel B*) measured fluorometrically during spontaneous contractions of taenia caeci. When excited by light at 340 nm PN_{red} emits blue fluorescence (emission peak at 470), and FP_{ox} emits yellow-green fluorescence (emission peak at 530 nm) when excited by 430–470 nm light. Data show that PN_{red} and FP_{ox} change during contractions. (From OZAKI et al. 1988)

V. Relationship Between $[Ca^{2+}]_i$ and Force

Depolarization of many tonic smooth muscles with elevated external K^+ ($[K^+]_o$) causes development of sustained contraction (tone) without visible rhythmic fluctuations. Generation of tone appears to be associated with sustained elevations in $[Ca^{2+}]_i$ over basal levels in vascular smooth muscle (SATO et al. 1988a; ABE et al. 1990; HIMPENS et al. 1990). Elevated $[K^+]_o$ also induces a sustained increase in $[Ca^{2+}]_i$, contraction and myosin phosphorylation in tracheal smooth muscles (GERTHOFFER et al. 1989; OZAKI et al. 1990a). Contractions and elevated $[Ca^{2+}]_i$ are completely blocked by organic Ca^{2+} antagonists, such as verapamil and nifedipine in all tissues tested (KARAKI and WEISS 1984). These results suggest that the contraction caused by elevated $[K^+]_o$ in tonic muscles is mediated by an increase in Ca^{2+} influx though voltage-dependent Ca^{2+} channels and the resulting increase in myosin light chain (MLC) phosphorylation.

In contrast, agonists such as norepinephrine in vascular smooth muscles and muscarinic agents in airway smooth muscles increase $[Ca^{2+}]_i$ and induce contraction without membrane depolarization. SOMLYO and SOMLYO (1968) called this phenomenon "pharmaco-mechanical coupling." Recent results obtained in rat aorta have suggested that L-type Ca^{2+} channel blockers strongly inhibit the rise in $[Ca^{2+}]_i$ stimulated by norepinephrine (OZAKI et al. 1990b; KARAKI et al. 1991; SAKATA et al. 1989; SATO et al. 1992), but these inhibitors only partially inhibit contraction. NELSON et al. (1988) suggested that in rabbit mesenteric artery norepinephrine increases open probability of voltage-dependent Ca^{2+} channels and the influx of Ca^{2+} induces contraction. These results suggest that, although norepinephrine increases $[Ca^{2+}]_i$ by activation of L-type Ca^{2+} channels, this may not be the only mechanism responsible for inducing tonic contraction.

The relationship between $[Ca^{2+}]_i$ and force depends upon the method of stimulation in tonic smooth muscles (see KARAKI 1989, 1990; SOMLYO and HIMPENS 1989). For a given increase in $[Ca^{2+}]_i$, agonists such as norepinephrine in ferret portal vein (MORGAN and MORGAN 1984; DEFEO and MORGAN 1985), norepinephrine, prostaglandins and endothelin in rat aorta (SATO et al. 1988a; SAKATA et al. 1989; OZAKI et al. 1990b), and carbachol in canine trachea (GERTHOFFER et al. 1990; OZAKI et al. 1990a) induce much greater contractions than elevated $[K^+]_o$. These observations suggest that the Ca^{2+}-sensitivity of the contractile apparatus may be increased by certain agonists, and this may explain how L-type Ca^{2+} channel blockers can dissociate $[Ca^{2+}]_i$ and force.

Permeabilization of smooth muscles with detergents, such as Triton ×100 or saponin, has been a valuable technique for understanding the regulation of contraction. However, soluble regulatory components may be lost during the process of permeabilization (RUEGG and PAUL 1982). A significant advance came with the discovery of α-toxin as a means to permeabilize muscles (NISHIMURA et al. 1988; KITAZAWA et al. 1989). With this agent intracellular regulatory proteins are not lost since the pore size is only 2–3 nm in diameter. Membrane signal transduction systems also appear to be preserved in α-toxin permeabilized muscle (NISHIMURA et al. 1988; KITAZAWA et al. 1989). These two studies showed that α-adrenergic stimulation increased the force generated by a given pCa in several arterial smooth muscles and demonstrated the phenomenon of regulation of Ca^{2+} sensitivity. Ca^{2+} sensitization by α-adrenergic stimulation requires the presence of GTP and GTPγS mimics the effects of α-adrenergic stimulation. Activation of protein kinase C (PKC) with phorbol esters also increases the Ca^{2+} sensitivity of the contractile apparatus. Studies of this mechanism have led to the following concept: (a) Binding of agonist to receptor stimulates phospholipase C which is coupled with GTP-binding proteins. (b) Activation of PKC following receptor activation phosphorylates some, as yet unidentified, regulatory proteins which accelerate myosin-actin cross bridge cycling. Many functional proteins can be phosphorylated by PKC (i.e., myosin, myosin light chain kinase, caldesmon, calponin, and others). Therefore, it is difficult to define the target protein(s) for PKC that are responsible for Ca^{2+} sensitization. Caldesmon or calponin, which are actin binding proteins and inhibit the interaction with myosin, may be good candidates for PKC phosphorylation, since PKC phosphorylation of these proteins reduces their inhibitory potency (CLARK et al. 1986; ABE et al. 1990). However, at present, there is no clear evidence showing a correlation between the phosphorylation of these actin binding proteins and Ca^{2+} sensitization. It is also possible that myosin phosphatase could be involved in regulation of Ca^{2+} sensitization. Recent studies have suggested that stimulation of GTP proteins inhibits phosphatase activity in α-toxin permeabilized muscle or muscle homogenates (KITAZAWA et al. 1991; KUBOTA et al. 1992). Somlyo and his colleagues (GONG et al. 1992) have also reported that arachidonic acid, which may also be generated in response to receptor stimulation via activation of phospholipase A_2, inhibits myosin phosphatase. This is also a mechanism that could be involved in G-protein mediated Ca^{2+} sensitization (GONG et al. 1992).

Receptors that are coupled to an increase in phospholipase C activity generate inositol 1,4,5-trisphosphate (IP_3). IP_3 is known to release Ca^{2+} from the sarcoplasmic reticulum (SR). Thus part of contractile responses that are due to pharmaco-mechanical coupling may be explained by an increase in $[Ca^{2+}]_i$ through release of Ca^{2+} stored in the SR (KARAKI and WEISS 1988; SATO et al. 1988a,b). IP_3 production, however, is usually transient, persisting only a few minutes. Receptor-mediated contraction, on the other hand, can be maintained for several tens of minutes or sometimes over several hours. Therefore, release of Ca^{2+} from stores may trigger contraction and initiate the upstroke phase of tonic contraction in these muscles.

In some tonic muscles, such as the swine carotid artery, force increases to a constant level in response to stimulation. Shortening velocity increases as contraction increases, but then shortening velocity decreases while tension remains constant (DILLON et al. 1981). In the presence of agonists, myosin light chain (MLC) phosphorylation rapidly decreases to a level slightly above the resting level. ATPase activity is extremely low in this state (GUTH and JUNG 1982). Murphy and his collaborators have called this phenomena the "Latch-state" (HAI and MURPHY 1989). Latch-bridges have now been observed in skinned smooth muscle fibers (CHATTAJEE and MURPHY 1983; MORELAND et al. 1986). This mechanism may be important in some, but perhaps not all, tonic muscles (see MARSTON 1989).

The rhythmic electrical activity in most GI muscles leads to phasic mechanical activity. Stimulation by moderately elevated $[K^+]_o$ or acetylcholine (ACh) increases the frequency and/or amplitude of electrical events and this increases contractile frequency and amplitude (see SZURSZEWSKI 1987). Measurements of changes in $[Ca^{2+}]_i$ have been performed on a number of GI muscles and cells, for example see studies on guinea pig taenia caecum (OZAKI et al. 1988; MITSUI and KARAKI 1990), guinea pig ileum (HIMPENS and SOMLYO 1988) and gastric smooth muscle cells of Bufo marinus (YAGI et al. 1988). In muscles such as the guinea pig taenia caecum that generate tone with phasic contractions superimposed (i.e., "mixed" type smooth muscles), phasic contractions correlate strongly with rhythmic increases in $[Ca^{2+}]_i$, and tone occurs as a function of sustained increases in $[Ca^{2+}]_i$ (OZAKI et al. 1988; MITSUI and KARAKI 1990). In the taenia, the $[Ca^{2+}]_i$-force relationships were similar when muscles were stimulated with either $[K^+]_o$ or carbachol. Spontaneous rhythmic contractions were also closely associated with rhythmic changes in $[Ca^{2+}]_i$ in the guinea pig ileum, another "mixed" type smooth muscle (HIMPENS and SOMLYO 1988). These results support the hypothesis that cytosolic Ca^{2+} is a primary regulator of contraction in GI muscles. However, regulation of Ca^{2+} sensitivity also appears to be an important secondary mechanism that regulates contractile responses in the GI tract. Both stimulatory (Ca^{2+}-sensitization) and inhibitory (Ca^{2+}-desensitization) forms of regulation have been found to modulate the $[Ca^{2+}]_i$-force relationship.

During the contraction-relaxation cycle of isolated toad stomach muscle cells, YAGI et al. (1988) found that the relationship between $[Ca^{2+}]_i$ and force was represented by a clockwise hysteresis loop. This indicates that Ca^{2+} sensitivity decreases with time. HIMPENS and CASTEELS (1990) also found that Ca^{2+}-sensitivity decreased during contractions of the guinea pig ileum that were elicited by elevated external K^+ solutions. These data suggest that Ca^{2+}-sensitivity is regulated in the absence of agonists. This type of regulation may be mediated by changes in $[Ca^{2+}]_i$. In canine antrum (a typical phasic muscle), similar Ca^{2+} desensitization occurs within the normal contractile cycle of phasic contractions. This appears to be an important mechanism for maintaining the phasic nature of these muscles. Figure 14 shows an example of the changes in $[Ca^{2+}]_i$ and muscle tension that occur in circular muscles of the canine antrum. Ca^{2+} transients and contractions consisted of two phases. The first phase of the Ca^{2+} transient and contraction corresponds to the upstroke depolarization, and the second phase is regulated by the plateau depolarization (see SZURSZEWSKI 1978; SANDERS and PUBLICOVER 1989; OZAKI et al. 1991c). Although the magnitude of the second phase of the Ca^{2+} transient was equal to, or greater than, the first phase, the second contraction was smaller than the first contraction. Further, in both cases muscle tension decayed more rapidly than the decrease in $[Ca^{2+}]_i$ at the termination of phasic contractions. The rate of MLC dephosphorylation during relaxation was also significantly greater than the decrease in $[Ca^{2+}]_i$. These results suggest that the Ca^{2+}-sensitivity of the contractile element decreases as a function of time during spontaneous phasic contractions (OZAKI et al. 1991a,b).

ACh raised basal or "resting" $[Ca^{2+}]_i$, increased the amplitude of Ca^{2+} transients, and increased the force of phasic contractions in canine antrum (Fig. 15). Despite the increase in the level of resting $[Ca^{2+}]_i$, ACh did not increase the level of tone between phasic contractions. When muscles were

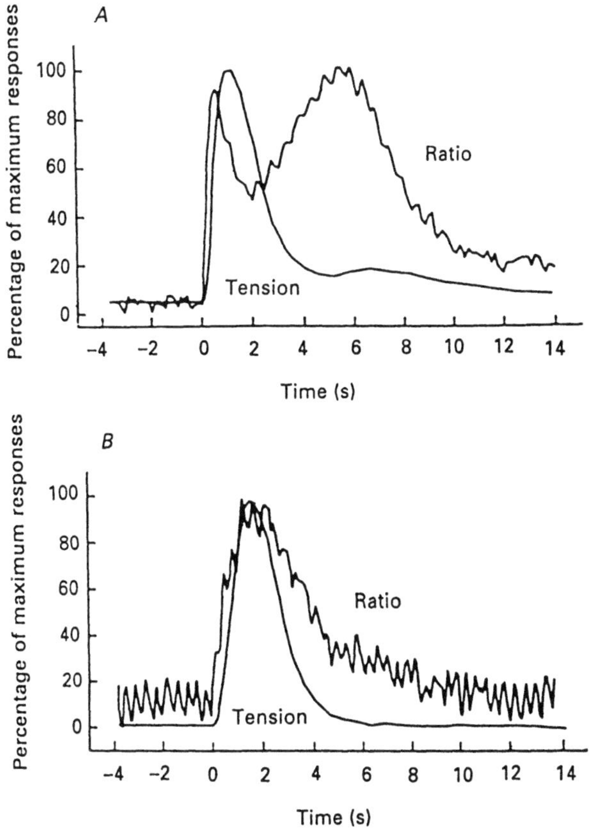

Fig. 14. Spontaneous contractions and Ca^{2+} transients in antral circular muscle. The initiation of the increase in $[Ca^{2+}]_i$ preceded the increase in force. In the population of muscles studied, two types of activity were observed. *Panel A* shows responses of a muscle in which Ca^{2+} transients and contractions consisted of two phases. Although the peak of the second phase of the Ca^{2+} transient was similar in magnitude to the first, the amplitude of the contractile response was usually far less than the first (decrease in Ca^{2+} sensitivity). *Panel B* shows a typical response of a group of muscles that generated single-phase Ca^{2+} transients. These muscles all displayed single-phase contractions. In both groups of muscles tension fell during relaxation more rapidly than $[Ca^{2+}]_i$. Data in these traces are normalized against the maximum of the Ca^{2+} transient and the contractile maximum. (From Ozaki et al. 1991a)

depolarized with elevated $[K^+]_o$, the increase in baseline $[Ca^{2+}]_i$ was small in relation to the $[Ca^{2+}]_i$-force relationship obtained during phasic responses (Ozaki et al. 1991a). Thus, these muscles appear to be "protected" from tonic contractions, perhaps by the decrease in sensitivity of the contractile apparatus to Ca^{2+} discussed above.

At present it is unclear what mediates the decrease in Ca^{2+}-sensitivity, but it is possible that MLC phosphorylation (activation) and/or dephosphorylation (inactivation) may both be regulated by elevation of $[Ca^{2+}]_i$.

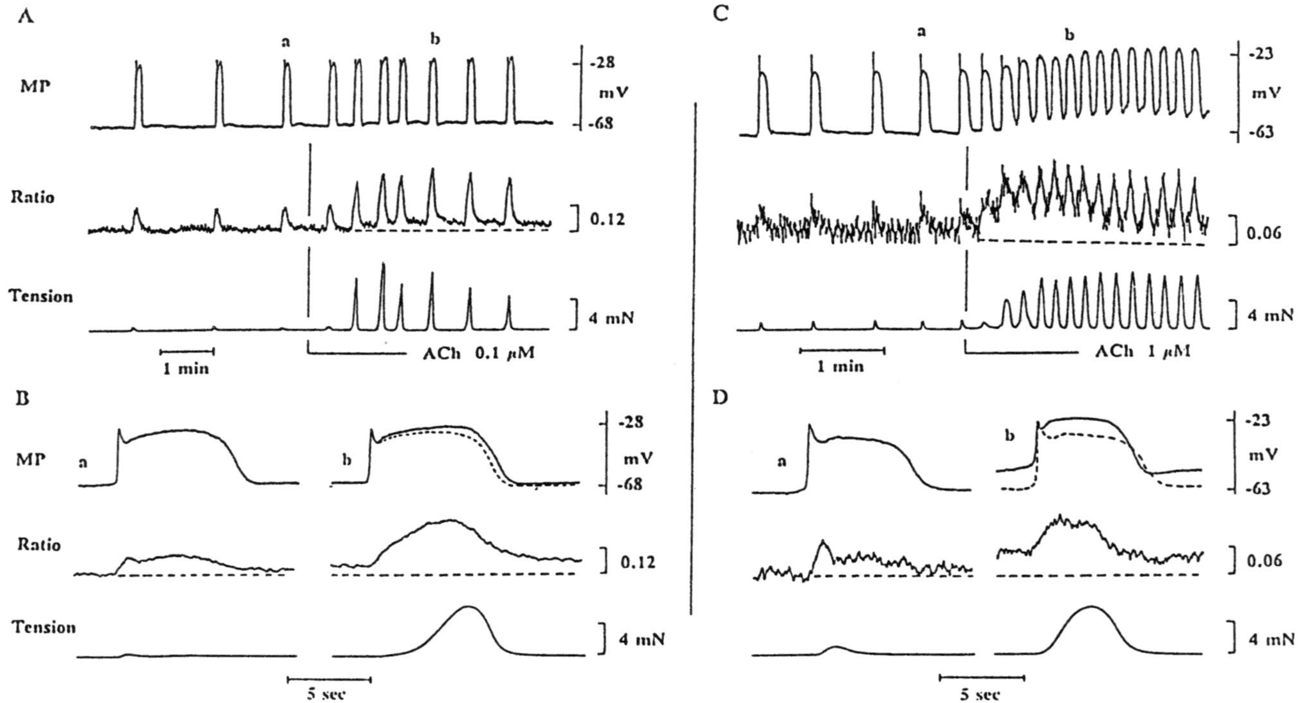

Fig. 15. Effect of ACh on electrical activity, $[Ca^{2+}]_i$ (measured with indo-1), and muscle tension in circular muscle of canine antrum. Addition of $0.1\,\mu M$ ACh (*panel A*) increased the amplitude and duration of slow waves. ACh also increased Ca^{2+} transients and elevated basal $[Ca^{2+}]_i$. The increases in Ca^{2+} transients were associated with an increase in the amplitude of phasic contractions, but the elevation in $[Ca^{2+}]_i$ between Ca^{2+} transients did not cause tone. *Panel B* shows expanded traces of electrical slow waves, Ca^{2+} transients and contractions before (denoted by *a*) and during $0.1\,\mu M$ ACh application (denoted by *b*). ACh ($1\,\mu M$; *panel C*) depolarized membrane potential and increased the amplitude and frequency of slow waves. ACh also increased basal $[Ca^{2+}]_i$ but produced only a transient rise in tone in a few preparations. ACh also increased the maximal level of Ca^{2+}-transients and phasic contractions. *Panel D* shows expanded traces of electrical slow waves, Ca^{2+} transients and contractions before (denoted by *a*) and during $1\,\mu M$ ACh application (denoted by *b*). All parameters were simultaneously monitored. (From OZAKI et al. 1992a)

When $[Ca^{2+}]_i$ increases, the initial response may be activation of MLC kinase, MLC phosphorylation, and contraction. With time, however, elevated $[Ca^{2+}]_i$ might lead to an increase in the activity of the phosphatases that dephosphorylate myosin. The relatively rapid increase in phosphorylation and delay in the acceleration of dephosphorylation would tend to produce a time-dependent decrease in the sensitivity of the contractile element to Ca^{2+} as the "activation-inactivation balance" shifts from an initially high level of MLC phosphorylation to an increasingly higher rate of dephosphorylation. This shift in the "activation-inactivation balance" may explain why Ca^{2+} sensitivity decreases with time and why muscle tension decreases more rapidly than $[Ca^{2+}]_i$ during relaxation. Differences in the time-dependence and Ca^{2+}-dependence of the activation and inactivation processes may determine the amplitude of contractions in phasic muscles.

In bovine tracheal smooth muscle, Ca^{2+}-calmodulin-dependent protein kinase II is capable of phosphorylating MLCK at "site A" (Stull et al. 1990) during responses to carbachol and elevated $[K^+]_o$. Similar results are obtained in swine carotid artery (Gilbert et al. 1991), suggesting that this may function as a Ca^{2+}-dependent negative-feedback control. This mechanism should also be explored in GI muscles.

A time-dependent change in the "activation-inactivation balance" might also explain why the slow increase in baseline $[Ca^{2+}]_i$ in response to ACh, and elevated $[K^+]_o$ does not induce tonic contraction in phasic muscles. Since the reduction in Ca^{2+}-sensitivity occurs within a few seconds (i.e., within the time-course of phasic contractions), a slow increase in $[Ca^{2+}]_i$ may not be fast enough to initiate the activation process (MLC phosphorylation) before the inactivation process (MLC dephosphorylation) occurs. Although the increase in $[Ca^{2+}]_i$ may increase the rate of MLC phosphorylation, an increase in the rate of dephosphorylation would tend to negate the increase in MLC phosphorylation. A slow increase in $[Ca^{2+}]_i$ could even shift the threshold level of $[Ca^{2+}]_i$ necessary to elicit the development of force (see Ozaki et al. 1991b). This again may protect phasic muscles from developing tone, even when basal Ca^{2+} rises.

Another mechanism that regulates contractile responses is an increase in Ca^{2+} sensitivity that occurs during agonist stimulation in many smooth muscles. For exmaple, Kitazawa et al. (1989) showed that acetylcholine increased the Ca^{2+} sensitivity of the contractile apparatus in guinea pig ileal muscles. Similar effects could be induced in permeabilized muscles by GTPγS. Studies in canine antral smooth muscles have also demonstrated this phenomenon. ACh increased the amplitudes of Ca^{2+} transients by 2–3 fold, and this was associated with a 15–20 fold increase in the force of phasic contractions (Ozaki et al. 1993). The digestive hormone analogs, pentagastrin and cholecystokinin-octapeptide, had similar effects on Ca^{2+} transients and phasic contractions. These results demonstrate the increase in the steepness of the $[Ca^{2+}]_i$-force relationship (or the nonlinearity of this relationship) in response to agonists. However, comparison of the $[Ca^{2+}]_i$-force relationships in muscles stimulated with BAY K 8644 or TEA (nonreceptor agonists) also showed that these agents, which would be unlikely to stimulate phosphatidyl inositol turnover or a G-protein dependent mechanism, also enhanced Ca^{2+} sensitivity. Thus multiple pathways might contribute to regulation of Ca^{2+} sensitivity.

The mechanisms responsible for the very steep $[Ca^{2+}]_i$-force relationship are of great interest. Hyperbolic $[Ca^{2+}]_i$-force relationships and $[Ca^{2+}]_i$-MLC phosphorylation-force relationships (KAMM and STULL 1986; REMBOLD and MURPHY 1988; OZAKI et al. 1990a; GERTHOFFER et al. 1989) have also been demonstrated in tonic smooth muscles, but these relationships were not as steep as the apparent dependence in antral muscles. This is possibly an important distinction between phasic and tonic smooth muscles. In canine gastric muscle, the level of MLC phosphorylation increased from 0.1 to 0.3 (mol P_i/mol myosin light chain) during stimulation with ACh (OZAKI et al. 1991a). The steep $[Ca^{2+}]_i$-force ratio seen in antral muscles may be explained by a type of "cooperativity," in which a small degree of MLC phosphorylation induces substantial contraction, such as observed in permeabilized venous smooth muscles (see VYAS et al. 1992). It is possible that thin filament regulation by actin binding proteins such as caldesmon (SUTHERLAND and WALSH 1989) and calponin (TAKAHASHI et al. 1986; TAKAHASHI and NADAL-GINARD 1991) could be involved in regulating cross bridge cycling rates in phasic muscles. Unphosphorylated forms of caldesmon or calponin inhibit actin-myosin interactions (CLARK et al. 1986; ABE et al. 1990), and the inhibition is relieved by phosphorylation by some Ca^{2+}-dependent protein kinases, such as calmodulin dependent protein kinase II and C kinase (ADAM et al. 1989; SUTHERLAND and WALSH 1989; WINDER and WALSH 1990). This may increase cross-bridge cycling in parallel with MLC phosphorylation, thus producing "cooperativity," such that modest levels of MLC phosphorylation can lead to high rates of cross-bridge cycling in phasic muscles. HAEBERLE et al. (1991) have shown that caldesmon content in "phasic" muscles (such as ileum, taenia coli and uterus) is higher than levels of this protein in "tonic" muscles (such as aorta, carotid artery and trachea). This is consistent with an important role for thin filament regulation in phasic smooth muscle of canine colon (GERTHOFFER et al. 1991).

Figure 16 shows the comparison of Ca^{2+} sensitivity of contractile elements in various smooth muscles (canine antrum, guinea pig ileal longitudinal smooth muscle, rat uterus, rabbit mesenteric artery and bovine trachea) permeabilized with α-toxin. The rank order of the Ca^{2+} sensitivity is bovine trachea > rabbit mesenteric artery > rat uterus > guinea pig ileum > canine antrum. It would appear from this analysis that the Ca^{2+} sensitivities of phasic muscles are lower than tonic muscles. This may explain the shift in mechanical thresholds observed in the fundus, corpus and antrum of the stomach (see SZURSZEWSKI 1987). The difference in Ca^{2+} sensitivities may be partly due to a higher level of phosphatase activity in phasic muscles (KITAZAWA et al. 1991b). Tonic muscles may be activated for

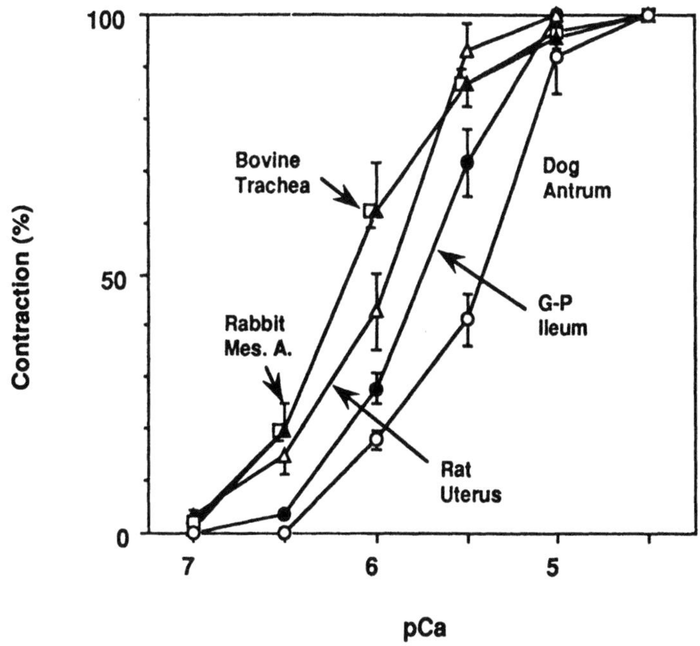

Fig. 16. Comparison of Ca^{2+} sensitivities of contractile elements of various smooth muscles (canine antrum, guinea pig ileal longitudinal smooth muscle, rat uterus, rabbit mesenteric artery and bovine trachea) permeabilized with α-toxin. Contraction is plotted as a function of pCa. Bovine trachea and rabbit mesenteric artery are more sensitive to Ca^{2+} than canine antrum

longer duration by a lower concentration of Ca^{2+}, while phasic muscles may be activated periodically with higher Ca^{2+} concentration. Higher Ca^{2+} sensitivity seen in tonic muscles could be a protective mechanism against the cytotoxic effects of Ca^{2+}.

E. Role of Phosphatidyl Inositol Turnover

In most smooth muscle cells, activation of certain receptors leads to an increase in phosphatidyl inositol (PI) turnover and generation of inositol-1,4,5-trisphosphate (IP_3) and diacylglycerol (DG) (for reviews, see NISHIZUKA 1986; ABDEL-LATIF 1986). IP_3 releasep Ca^{2+} from sarcoslasmic reticulum (SR) and increases $[Ca^{2+}]_i$. DG activates protein kinase C (PKC), which phosphorylates many functional proteins including ion channels (for review, see SHEARMAN et al. 1989).

I. Regulation Via IP_3

In most smooth muscles activation of PI turnover stimulates contraction. This occurs because the release of stored Ca^{2+} causes a significant rise in

$[Ca^{2+}]_i$ and activation of myosin light chain kinase. This response occurs even when Ca^{2+} entry is restricted or when it is blocked by removing extracellular Ca^{2+}. For example, addition of norepinephrine to vascular smooth muscle in Ca^{2+} free solution causes a transient increase in $[Ca^{2+}]_i$ and contraction (see KARAKI 1990; SATO et al. 1988a). Application of IP_3 induces Ca^{2+} release from smooth muscle microsomes (CARSTEN and MILLER 1985), and it causes a transient contraction in vascular smooth muscles permeabilized with saponin (SUEMATSU et al. 1984; KOBAYASHI et al. 1989b). In GI muscles cholinergic stimulation also stimulates PI turnover via activation of muscarinic receptors (JAFFERJI and MICHELL 1976; BEST et al. 1985; GARDNER et al. 1988; OZAKI et al. 1992a). In guinea pig ileum (mixed type muscle), carbachol induced a transient increase in $[Ca^{2+}]_i$ and force in Ca^{2+} free solutions (HIMPENS and SOMLYO 1988). In canine antrum, ACh raised basal $[Ca^{2+}]_i$ with very little depolarization, and membrane potentials of atral cells are far negative to the activation range for voltage-dependent Ca^{2+} channels (OZAKI et al. 1991a). These results are consistent with the concept that the increase in $[Ca^{2+}]_i$ could be due to the release of Ca^{2+} from internal stores via an increase in IP_3.

Agonists besides cholinergic agents also increase PI turnover. An interesting example of this is ATP. ATP is an inhibitory substance in most GI muscles that may be released from intrinsic nerves as a co-transmitter (BURNSTOCK and KENNEDY 1985; GORDON 1986; HOYLE and BURNSTOCK 1989). The effects of ATP are likely to be mediated via P_2 receptors, which are classified into two subtypes: P_{2X} and P_{2Y} (BURNSTOCK and KENNEDY 1985). P_{2X} receptors are usually linked to ion channels (such as a nonselective cation channel) and P_{2Y} receptors are linked to phospholipase C (KENNEDY 1990). REMBOLD et al. (1991) demonstrated that ATP transiently increases $[Ca^{2+}]_i$, myosin light chain phosphorylation and force in swine carotid artery. These authors demonstrated that ATP induces contraction in these muscles by an increase in transmembrane Ca^{2+} influx through the activation of P_{2X} receptor. In other vascular smooth muscle, ATP induces transient increases in $[Ca^{2+}]_i$ and contraction (TAWADA et al. 1987). In canine antral muscles ATP stimulated PI turnover and generated IP_3 and diacylglycerol (OZAKI et al. 1992a). ATP also transiently increased $[Ca^{2+}]_i$ without membrane depolarization. During the increase in basal $[Ca^{2+}]_i$ by ATP electrical slow waves and phasic contractions were suppressed. Other agents that tonically increased $[Ca^{2+}]_i$ also decreased the amplitude of the plateau potential and the amplitude of phasic contractions. These results suggest the hypothesis that IP_3 could provide negative feedback to contractile activity in some phasic muscles. This might occur because Ca^{2+} release may cause a local increase in Ca^{2+} concentration that affects ion channels (i.e., by causing inactivation of voltage-dependent Ca^{2+} channels and activation of Ca^{2+}-dependent K^+ channels), but IP_3-dependent release may not be sufficient to raise $[Ca^{2+}]_i$ to the point where the contractile apparatus is activated (OZAKI et al. 1992a). An example of this has been demonstrated with photolytic release of Ca^{2+} in antral muscle. This apparently did not increase $[Ca^{2+}]_i$

sufficiently to generate a contractile response, but the increase in Ca^{2+} was sufficient to inhibit slow waves and associated phasic contractions (CARL et al. 1990).

II. Role of Activators of Protein Kinase C

Another product of phosphatidyl inositol turnover is diacylglycerol, which activates protein kinase C. Although PKC activation induces contraction or increases the Ca^{2+} sensitivity of the contractile apparatus as described above, the full role of PKC in excitation-contraction coupling is not yet well understood in GI muscles. In canine antrum, phorbol ester selectively inhibited the plateau potential of slow waves, the associated increase in $[Ca^{2+}]_i$, and phasic contraction. Thus, in phasic smooth muscles, the activation of PKC seems to be inhibitory and suggests that the activation of PKC during muscarinic stimulation may serve as negative feedback to regulate Ca^{2+} influx. A discussion of the effects of cholinergic agonists and activators of PKC is given in the section on "Regulation of Currents by Muscarinic Stimulation."

PKC phosphorylates several functional proteins in smooth muscle cells and these might regulate certain steps in excitation-contraction coupling (NISHIZUKA 1986). It is possible that the reduction in Ca^{2+} transients caused by phorbol esters could be related to an enhancement in Ca^{2+} uptake and/or extrusion, because, for example, the plasma membrane Ca^{2+}-ATPase has been shown to be activated by a PKC-dependent mechanism (FURUKAWA et al. 1989). Repeated or prolonged application of receptor agonists frequently causes a decrease in responsiveness in many smooth muscles. This is known as "tachyphylaxis" or "desensitization." Extensive studies on nonmuscle cells suggest that a major function of PKC may be related to negative feedback regulation of proteins involved in receptor-activation (see NISHIZUKA 1986; ABDEL-LATIF 1986).

F. Regulation of the Contractile Apparatus by Cyclic Nucleotides

There are several inhibitory agonists that have similar effects on the electrical and mechanical activities of GI muscles. It is possible that the effects of several of these compounds are mediated by common mechanisms. For example, β-adrenergic stimulation causes relaxation of most smooth muscles, and this is partially mediated by hyperpolarization (reviews, see STILES et al. 1984; BULBRING and TOMITA 1987) and reduction of $[Ca^{2+}]_i$ (KWON et al. 1993). These effects are thought to be mediated by stimulation of adenylate cyclase and an increase in synthesis of cAMP. In the phasic smooth muscles, such as the canine antrum and colon, β-adrenergic stimulation reduces the amplitude and duration of slow waves in addition to causing hyperpolarization (EL-SHARKAWY and SZURSZEWSKI 1978; SMITH et al. 1993). Vasoactive intestinal peptide (VIP), which has been considered as a candidate for a nonadrenergic, noncholinergic (NANC) inhibitory transmitter in gastrointestinal muscles (FAHRENKRUG et al. 1978; ITO et al. 1990), is also known to suppress contractile activity by stimulating adenylate cyclase in gastric and other smooth muscles (BITAR and MAKHLOUF 1982; MORGAN et al. 1978). It has also been reported that calcitonin gene-related peptide (CGRP), which could also be released from nerves in some tissue, also inhibits via a cAMP-dependent mechanism (MATON et al. 1988). Nerves that contain CGRP have been found in circular muscle layers throughout the gut, with the highest concentrations found in the stomach and proximal intestine (STERNINI et al. 1987).

In canine antral smooth muscle, isoprenaline, VIP and CGRP selectively inhibit the plateau potential of slow waves, reduce associated Ca^{2+} transients, and inhibit contractions (OZAKI et al. 1992b). Forskolin, an adenylate cyclase activator, causes similar results in the antrum and colon (SMITH et al. 1993; HUIZINGA et al. 1991b). These results suggest that elevation of cAMP directly or indirectly reduces the open probability of Ca^{2+} channels. Cyclic AMP could indirectly regulate Ca^{2+} influx by increasing the activity of K channels (e.g., see SIMS et al. 1990; CARL et al. 1991; and see section on "Regulation of Currents by Adrenergic Stimulation").

In addition to affecting ion channels, the Ca^{2+} sensitivity of the contractile apparatus is regulated by cAMP. For example, in rat aorta elevation of cAMP by forskolin or addition of dibutyryl cAMP reduces $[Ca^{2+}]_i$ and contraction stimulated by norepinephrine. These agents also reduce contractions induced by elevated $[K^+]_o$ without affecting $[Ca^{2+}]_i$ (ABE and KARAKI 1989, 1992). Similar responses occur in canine trachea and guinea pig taenia coli (OZAKI et al. 1991; KWON et al. 1993). These results suggest that cAMP inhibits smooth muscle contraction by at least two mechanisms: reduction in Ca^{2+} influx and decrease in the Ca^{2+} sensitivity of the contractile apparatus. In agreement with these observations, isoprenaline, VIP, CGRP and forskolin inhibit more strongly contraction than Ca^{2+} transients in canine antral muscles (OZAKI et al. 1992b). One possible explanation for the Ca^{2+} desensitization caused by elevations in cAMP is that MLCK can be desensitized to Ca^{2+} by cAMP-dependent phosphorylation at a specific site (site A) (ADELSTEIN et al. 1978; STULL et al. 1990). During the inhibition of contraction by isoprenaline in bovine trachea, MLC phosphorylation was greatly decreased. In permeabilized muscle of rabbit mesenteric artery (NISHIMURA and VAN BREEMEN 1989), coronary artery (PFITZER et al. 1984), and canine antrum (OZAKI et al. 1992b), addition of cAMP (with or without adding catalytic subunit of A kinase) shifts the pCa-force curve to the right.

The inhibitory effects of cAMP do not appear to be limited to a decrease in MLC kinase activity. Elevation of cAMP could inhibit receptor-mediated signal transduction resulting in inhibition of $[Ca^{2+}]_i$ (see ABDEL-LATIF 1991). The inhibition by forskolin and dibutyryl cyclic AMP of phosphatidyl-inositol hydrolysis has been demonstrated in tracheal (HALL et al. 1989) and vascular smooth muscles (AHN et al. 1992). Activation of protein kinase A by forskolin in canine colonic smooth muscle desensitizes muscarinic receptors and inhibits phosphatidyl-inositol turnover (ZHANG and BUXTON 1993).

It is also possible that the reduction in Ca^{2+} transients caused by cAMP elevation could be related to an enhancement in Ca^{2+} uptake and/or extrusion. For example, activation of plasma membrane and/or sarcoplasmic reticulum Ca^{2+}-ATPase has been shown to occur via cAMP-dependent mechanism (ANDERSON and NILSSON 1972; HISAYAMA and TAKAYANAGI 1983). Others have also suggested that isoprenaline could activate an electrogenic Ca^{2+} extrusion pump which could partially account for membrane hyperpolarization and reduction in slow wave activity in some muscles (BÜLBRING and DEN HERTOG 1980).

Regulation by cGMP is also an important and growing area of investigation in GI muscles since it has been recognized that nitric oxide is an important inhibitory neurotransmitter in these GI tissues (see SANDERS and WARD 1992). In vascular smooth muscle, elevation of cGMP by sodium nitroprusside (SNP), nitroglycerin or endothelium derived relaxing factor reduces $[Ca^{2+}]_i$ and contraction stimulated by norepinephrine, prostaglandin $F_{2\alpha}$ or histamine. In contrast SNP reduces contractions elicited by elevated $[K^+]_o$ without affecting $[Ca^{2+}]_i$ (KARAKI et al. 1988; YANAGISAWA et al. 1989; ABE et al. 1990; SATO et al. 1990; BALWIERCZAK 1991). In canine gastric muscle, sodium nitroprusside inhibits slow waves and the Ca^{2+} transients, but it inhibits mechanical transients more strongly than electrical and Ca^{2+} transients (OZAKI et al. 1991d). In α-toxin permeabilized muscles, cGMP has been shown to inhibit contraction when Ca^{2+} is clamped to specific levels (NISHIMURA and VAN BREEMEN 1989; OZAKI et al. 1991d). These results suggest that, like cAMP, cGMP-dependent mechanisms regulate the sensitivity of the contractile apparatus to Ca^{2+}.

FRANCIS et al. (1988) showed, in studies of porcine coronary and guinea pig tracheal smooth muscle, that cGMP-dependent protein kinase specific cyclic nucleotide analogs, but not cAMP-specific analogs, cause relaxation. This result tends to suggest that since elevation of cAMP in these tissue causes relaxation, cAMP-dependent protein kinase may not be the mediator of the cAMP effects. LINCOLN et al. (1990) demonstrated that forskolin reduces vasopressin-stimulated $[Ca^{2+}]_i$ in primary cultured vascular smooth muscle cells, but it failed to reduce Ca^{2+} levels after repetitive passages of these cells. Both primary and passaged cells contain cAMP-kinase, although cGMP-kinase is greatly reduced in passaged cells. The introduction of cGMP kinase into the passaged cells restored the capacity of forskolin to reduce vasopressin-stimulated Ca^{2+} concentration. These results suggest that activation of cGMP-kinase by cAMP leads to the reduction of intracellular Ca^{2+}. Such cross-activation of cGMP-kinase by cAMP has been shown in porcine coronary arterial smooth muscle tissue (JIANG et al. 1992).

G. Summary

Since ultimately Ca^{2+} triggers contraction in GI smooth muscles, the first line of regulation of contractile activity is organized to control $[Ca^{2+}]_i$. This is first accomplished by a plasma membrane that restricts Ca^{2+} entry unless specific voltage-dependent Ca^{2+} channels are activated. The open probability of these channels can be controlled either directly by processes such as Ca^{2+}-dependent inactivation or phosphorylation, or these channels can be controlled by regulating membrane potential via other conductances (primarily K conductances and the nonspecific cation conductance). In addition to regulation at the plasma membrane, the sarcoplasmic reticulum provides storage of Ca^{2+} within cells, and release and uptake of Ca^{2+} by the SR is regulated by several factors. The cell also contains a large capacity to buffer Ca^{2+}, so that the vast majority of Ca^{2+} ions that enter the cytoplasmic compartment are bound in a nonspecific manner.

It is quite clear now that Ca^{2+} triggers contraction, but after initial activation several downstream biochemical events can dissociate $[Ca^{2+}]_i$ from force. These biochemical pathways including, but probably not restricted to, protein kinase C-, Ca^{2+}-, cAMP- and cGMP-dependent mechanisms, regulate the Ca^{2+} sensitivity of the contractile apparatus. Activation of these pathways shifts the $[Ca^{2+}]_i$-force relationship and changes contractile responses for given levels of $[Ca^{2+}]_i$. The compexity of regulatory pathways in GI muscles provides ample opportunity for therapeutic intervention, but the fact that many of these mechanisms are shared by other cells and smooth muscles of other organs means that careful characterization and a thorough understanding of these processes will be necessary before drugs that are specific for GI motility can be developed.

Acknowledgements. This review was supported by a grant from the National Institutes of Health, DK 41315. The authors thank Ms. Susan Bowen for editorial assistance.

References

Aaronson PI, Russell SN (1991) Rabbit intestinal smooth muscle calcium current in solutions with physiological calcium concentrations. Exp Physiol 76:539–551
Abdel-Latif AA (1986) Calcium-mobilizing receptors, polyphosphoinositides and the generation of second messengers. Pharmacol Rev 38:227–272
Abdel-Latif AA (1991) Biochemical and functional interactions between the inositol 1,4,5-trisphosphate-Ca^{2+} and cyclic AMP signalling systems in smooth muscle. Cell Signal 3:371–385
Abe A, Karaki H (1989) Effect of forskolin on cytosolic Ca^{2+} level and contraction in vascular smooth muscle. J Pharmacol Exp Ther 249:895–900
Abe A, Karaki H (1992) Mechanisms underlying the inhibitory effect of dibutyryl cyclic AMP in vascular smooth muscle. Eur J Pharmacol 211:305–311
Abe M, Takahashi K, Hiwada K (1990) Effect of calponin on actin-activated myosin ATPase activity. J Biochem (Tokyo) 108:835–838
Abe S, Kanaide H, Nakamura M (1990) Front-surface fluorometry with fura-2 and effects of nitroglycerin on cytosolic calcium concentrations and on tension in the coronary artery of the pig. Br J Pharmacol 101:545–552

Adam LP, Haeberle JR, Hathaway DR (1989) Phosphorylation of caldesmon in arterial smooth muscle. J Biol Chem 264:7698–7703
Adams PR, Brown DA, Constanti A (1982a) M-currents and other potassium currents in bullfrog sympathetic neurons. J Physiol (Lond) 330:537–572
Adams PR, Brown DA, Constanti A (1982b) Pharmacological inhibition of the M-current. J Physiol (Lond) 332:223–262
Adelstein RS, Conti MA, Hathaway DR, Klee CB (1978) Phosphorylation of smooth muscle myosin light chain kinase by catalytic subunitof adenosin 3′:5′-monophosphate-dependent protein kinase. J Biol Chem 253:8347–8350
Adlish JD, Overturf KE, Hart P, Duval D, Sanders KM, Horowitz B (1991) Molecular cloning and differential expression of potassium and calcium channels expressed in canine colonic circular muscles. Biophys J 59:454a
Ahn HY, Kang SE, Chang KC, Karaki H (1992) Dibutyryl cyclic AMP and forskolin inhibit phosphatidylinositol hydrolysis, Ca^{2+} influx and contraction in vascular smooth muscle. Jpn J Pharmacol 59:263–265
Akbarali HI, Giles WR (1993) Ca^{2+} and Ca^{2+}-activated Cl^- currents in rabbit oesophageal smooth muscle. J Physiol (Lond) 460:117–133
Andersson R, Nilsson K (1972) Cyclic AMP and calcium in relaxation in intestinal smooth muscle. Nature [New Biol] 238:19–120
Balaban RS (1990) Regulation of oxidataive phosphorylation in the mammalian cell. Am J Physiol 258:C377–C389
Balwierczak JL (1991) The relationship of KCl- and prostaglandin $F_{2\alpha}$-mediated increases in tension of the pocine coronary artery with changes in intracellular Ca^{2+} measured with fura-2. Br J Pharmacol 104:373–378
Barajas-Lopez C, Huizinga JD (1989) Different mechanisms of contraction generation in circular muscle of canine colon. Am J Physiol 256:G570–G580
Barajas-Lopez C, Den Hertog A, Huizinga JD (1989a) Ionic basis of pacemaker generation in dog colonic smooth muscle. J Physiol (Lond) 416:385–402
Barajas-Lopez C, Berezin I, Daniel EE, Huizinga JD (1989b) Pacemaker activity recorded in interstitial cells of Cajal of the gastrointestinal tract. Am J Physiol 257:C830–C835
Bauer AJ, Sanders KM (1985) Gradient in excitation-contraction coupling in canine gastric antral circular muscle. J Physiol (Lond) 369:283–294
Bayguinov O, Sanders KM (1993) Role of nitric oxide as an inhibitory neurotransmitter in the canine pyloric sphincter. Am J Physiol (in press)
Bayguinov O, Vogalis F, Morris B, Sanders KM (1992) Patterns of electrical activity and neural responses of canine proximal duodenum. Am J Physiol 263:G887–G894
Bean BP (1984) Nifedipine block of cardiac calcium channels: high-affinity binding to the inactivated state. Proc Natl Acad Sci USA 81:6388–6392
Bean BP (1989) Classes of calcium channels in vertebrate cells. Annu Rev Physiol (Lond) 51:367–384
Becker PL, Singer JJ, Walsh JV Jr, Fay FS (1989) Regulation of calcium concentration in voltage-clamped smooth muscle cells. Science 244:211–214
Beech DJ, Bolton TB (1989a) Properties of the cromakalim-induced potassium conductance in smooth muscle cells isolated from the rabbit portal vein. Br J Pharmacol 98:851–864
Beech DJ, Bolton TB (1989b) Two components of potassium current activated by depolarization of single smooth cells from the rabbit portal vein. J Physiol (Lond) 418:293–309
Benham CD, Bolton TB (1983) Patch-clamp studies of slow potential-sensitive potassium channels in longitudinal smooth muscle cells of rabbit jejunum. J Physiol (Lond) 340:469–486
Benham CD, Bolton TB (1986) Spontaneous transient outward currents in single visceral and vascular smooth muscle cells of the rabbit. J Physiol (Lond) 381:385–406

Benham CD, Bolton TB, Lang RJ (1985) Acetylcholine activates an inward current in single mammalian smooth muscle cells. Nature 316:345–347

Benham CD, Bolton TB, Lang RJ, Takewaki T (1986) Calcium-activated potassium channels in single smooth muscle cells of rabbit jejunum and guinea-pig mesenteric artery. J Physiol (Lond) 371:45–67

Benham CD, Bolton TB, Denbigh JS, Lang RJ (1987) Inward rectification in freshly isolated single smooth muscle cells of the rabbit jejunum. J Physiol (Lond) 383:461–476

Berezin I, Huizinga JD, Daniel EE (1988) Interstitial cells of Cajal in the canine colon: a special communication network at the inner border of the circular muscle. J Comp Neurol 273:42–51

Best L, Brooks KJ, Bolton TB (1985) Relationship between stimulated inositol lipid hydrolysis and contractility in guinea-pig visceral longitudinal smooth muscle. Biochem Pharmacol 34:2297–2301

Bielefeld DR, Hume JR, Krier J (1990) Action potentials and membrane currents of isolated single smooth muscle cells of cat and rabbit colon. Pflugers Arch 415:678–687

Bitar KN, Makhlouf GM (1982) Relaxation of isolated gastric smooth muscle cells by vasoactive intestinal peptide. Science 216:531–533

Boeckxstaens GE, Pelckmans PA, Ruytjens IF, Bult H, De Man JG, Herman AG, Van Maercke YM (1991) Bioassay of nitric oxide released upon stimulation of non-adrenergic non-cholinergic nerves in the canine ileocolonic junction. Br J Pharmacol 103:1085–1091

Bolton TB (1979) Mechanisms of action of transmitters and other substances on smooth muscle. Physiol Rev 59:606–718

Bolton TB (1981) Action of acetylcholine on the smooth muscle membrane. In: Bulbring E, Brading AF, Jones AW, Tomita T (eds) Smooth muscle: an assessment of current knowledge. Arnold, London, p 119

Bolton TB, Kitamura K (1983) Evidence that ionic channels associated with the muscarinic receptor of smooth muscle may admit calcium. Br J Pharmacol 78(2):405–416

Bolton TB, Lim SP (1989) Properties of Ca^{2+} stores and transient outward currents in single smooth muscle cells of rabbit intestine. J Physiol (Lond) 409:385–401

Brading AF, Sneddon P (1980) Evidence for multiple sources of calcium for activation of the contractile mechanism of guinea-pig taenia coli on stimulation with carbachol. Br J Pharmacol 70:229–240

Bredt DS, Hwang PM, Snyder SH (1991) Localization of nitric oxide synthase indicating a neural role for nitric oxide. Nature 347:768–770

Bulbring E, Hertog A (1980) The action of isoprenaline on the smooth muscle of the guinea-pig taenia coli. J Physiol (Lond) 304:277–296

Bulbring E, Tomita T (1987) Catecholamine action on smooth muscle. Pharmacol Rev 39:49–96

Burnstock G, Kennedy C (1985) Is there a basis for distinguishing two types of P_2-purinoceptor? Gen Pharmacol 16:433–440

Byrne NG, Large WA (1987) Action of noradrenaline on single smooth muscle cells freshly dispersed from the rat anococcygeus muscle. J Physiol (Lond) 389: 513–525

Carl A, Sanders KM (1989) Ca^{2+}-activated K^+ channels of canine colonic myocytes. Am J Physiol 257:C470–C480

Carl A, McHale NG, Publicover NG, Sanders KM (1990) Participation of Ca^{2+}-activated K^+ channels in electrical activity of canine gastric smooth muscle. J Physiol (Lond) 429:205–221

Carl A, Kenyon JL, Uemura D, Fusetani N, Sanders KM (1991) Regulation of Ca^{2+}-activated K^+ channels by protein kinase A and phosphatase inhibitors. Am J Physiol 261:C387–C392

Carl A, Bowen S, Gelband CH, Sanders KM, Hume JR (1992) Cromakalim and lemakalim activated Ca^{2+}-dependent K^+ channels in canine colon. Pflugers Arch 421:67–76
Carl A, Frey BW, Ward SM, Sanders KM, Kenyon JL (1993) Inhibition of slow wave repolarization and Ca^{2+}-activated K^+ channels by quaternary ammonium ions. Am J Physiol 264:C625–C631
Carsten ME, Miller JD (1985) Ca^{2+} release by inositol trisphosphate from Ca^{2+} transporting microsomes derived from uterine sarcoplasmic reticulum. Biophys Biochem Res Commun 130:1027–1031
Chad JE, Eckert R (1986) An enzymatic mechanism for calcium current inactivation in dialyzed *Helix* neurones. J Physiol (Lond) 378:31–51
Chakder S, Rattan S (1993) Release of nitric oxide by activation of nonadrenergic, noncholinergic neurons of internal anal sphincter. Am J Physiol 264:G7–G12
Chance B (1965) The energy-linked reaction of calcium with mitochondria. J Biol Chem 240:2729–2748
Chatterjee M, Murphy RA (1983) Ca^{2+}-dependent tension maintenance without myosin phosphorylation in skinned smooth muscle. Science 211:464–466
Christensen J, Caprilli R, Lund GF (1969) Electrical slow waves in the circular muscle of cat colon. Am J Physiol 234:771–776
Clapp LH, Gurney AM (1991) Modulation of calcium movements by nitroprusside in isolated vascular smooth muscle cells. Pflugers Arch 418:462–470
Clapp LH, Vivaudou MB, Walsh JV Jr, Singer JJ (1987) Acetylcholine increases voltage-activated Ca^{2+} current in freshly dissociated smooth muscle cells. Proc Natl Acad Sci USA 84:2092–2096
Clark T, Ngai PK, Sutherland C, Groschel-Stewart U, Walsh MP (1986) Vascular smooth muscle caldesmon. J Biol Chem 261:8028–8035
Cohen NM, Lederer WJ (1987) Calcium current in isolated neonatal rat ventricular myocytes J Physiol (Lond) 391:169–191
Cole WC, Sanders KM (1989a) Characterization of macroscopic outward currents of canine colonic myocytes. Am J Physiol 257:C461–C469
Cole WC, Sanders KM (1989b) G proteins mediate suppression of Ca^{2+}-activated K^+ current by acetylcholine in smooth muscle cells. Am J Physiol 257:C596–C600
Cole WC, Carl A, Sanders KM (1989) Muscarinic suppression of Ca^{2+}-dependent K^+ current in colonic smooth muscle. Am J Physiol 257:C481–C487
Conklin J, Du C (1992) Guanylate cyclase inhibitors: effect on inhibitory junction potentials in esophageal smooth muscle. Am J Physiol 263:G87–G90
Connor JA, Prosser CL, Weems WA (1974) A study of pacemaker activity in intestinal smooth muscle. J Physiol (Lond) 240:671–701
Crist JR, He XD, Goyal RK (1991) Chloride-mediated inhibitory junction potentials in opossum esophageal circular smooth muscle. Am J Physiol 261:G752–G762
Daniel EE, Berezin I (1992) Interstitial cells of Cajal: are they major players in control of gastrointestinal motility? J Gastrointest Motil 4:1–24
Daniel EE, Collins SM, Fox JET, Huizinga JD (1989a) Pharmacology of drugs acting on gastrointestinal motility. In: Wood JD, Schultz S (eds) The gastrointestinal system. American Physiological Society, Bethesda, p 715 (Handbook of physiology)
Daniel EE, Collins SM, Fox JET, Huizinga JD (1989b) Pharmacology of neuroendocrine peptides. In: Wood JD, Schultz S (eds) The gastrointestinal system. American Physiological Society, Bethesda, p 759 (Handbook of physiology)
Daziel HH, Thornbury KD, Ward SM, Sanders KM (1991) Involvement of nitric oxide synthetic pathway in inhibitory junction potentials in canine proximal colon. Am J Physiol 260:G789–G792
DeFeo TT, Morgan KG (1985) Calcium-force relationship as detected by aequorin in two different vascular smooth muscle of the ferret. J Physiol (Lond) 369: 269–282

Dillon PF, Aksoy MO, Driska SP, Murphy RA (1981) Myosin phosphorylation and cross-bridge cycle in smooth muscle. Science 211:495–497

Droogmans G, Callewaert G (1986) Ca^{2+}-channel current and its modification by the dihydropyridine agonist BAY K 8644 in isolated smooth muscle cells. Pflugers Arch 406:259–265

Eckert R, Chad JE (1984) Inactivation of Ca^{2+} channels. Prog Biophys Mol Biol 44:215–267

Eckert R, Tillotson DL (1981) Calcium-mediated inactivation of the calcium conductance in calcium-loaded giant neurons of *Aplysia Californica*. J Physiol (Lond) 314:265–280

El-Sharkawy TY, Daniel EE (1975) Ionic mechanisms of the intestinal electrical control activity. Am J Physiol 229:1287–1298

El-Sharkawy TY, Szurszewski JH (1978) Modulation of canine antral circular muscle by acetylcholine, noradrenaline and pentagastrin. J Physiol (Lond) 279:309–320

El-Sharkawy TY, Morgan KG, Szurszewski JH (1978) Intracellular electrical activity of canine and human gastric smooth muscle. J Physiol (Lond) 279:291–307

Fahrenkrug J, Haglund U, Jodal M, Lundgren O, Olbe L, Shaffalitzky De Muckadell OB (1978) Nervous release of vasoactive intestinal polypeptide in the gastrointestinal tract of the cat; possible physiological implications. J Physiol (Lond) 284:291–305

Farraway L, Huizinga JD (1991) Potassium channel activation by cromakalim affects the slow wave type action potential of colonic smooth muscle. Am J Pharmacol Exp Ther 257:35–41

Farrugia G, Rae JL, Szurszewski JH (1993) Characterization of an outward potassium current in canine jejunal circular muscle and its activation by fenamates. J Physiol (Lond) (in press)

Fish RD, Sperti GW, Colucci WS, Clapham DE (1988) Phorbol ester increases the dihydropyridine-sensitive calcium conductance in a vascular smooth muscle cell line. Circ Res 62:1049–1054

Francis SH, Noblett BD, Todd BW, Wells JN, Corbin JD (1988) Relaxation of vascular and tracheal smooth muscle by cyclic nucleotide analogs that preferentially activate purified cGMP-dependent protein kinase. Mol Pharmacol 34:506–517

Furukawa K, Tawada Y, Shigekawa M (1989) Protein kinase C activation stimulates plasma membrane Ca^{2+} pump in cultured vascular smooth muscle cells. J Biol Chem 264:4844–4849

Ganitkevich VY, Shuba MF, Smirnov SV (1986) Potential-dependent calcium inward current in a single isolated smooth muscle cell of the guinea-pig taenia caeci. J Physiol (Lond) 380:1–16

Ganitkevich VY, Shuba MF, Smirnov SV (1987) Calcium-dependent inactivation of potential-dependent calcium inward current in an isolated guinea-pig smooth muscle cell. J Physiol (Lond) 392:431–449

Cardner AL, Choo LK, Mitchelson F (1988) Comparison of the effects of some muscarinic agonists on smooth muscle function and phosphatidylinositol turnover in the guinea-pig taenia caeci. Br J Pharmacol 94:199–211

Gelband CH, Lodge NJ, van Breemen C (1989) A Ca^{2+}-activated K^+ channel from rabbit aorta: modulation by cromakalim. Eur J Pharmacol 167:201–210

Gerthoffer WT, Murphey KA, Gunst SJ (1989) Aequorin luminescence, myosin phosphorylation and active stress in tracheal smooth muscle. Am J Physiol 257:C1062–C1068

Gerthoffer WT, Murphey KA, Mangini J, Boman S, Lattanzio FA Jr (1991) Myosin phosphorylation and calcium in tonic and phasic contractions of colonic smooth muscle. Am J Physiol 260:G958–G964

Gilbert EK, Weaver BA, Rembold CM (1991) Depolarization decreases the $[Ca^{2+}]_i$ sensitivity of myosin light chain kinase in arterial smooth muscle; comparison of aequorin and fura 2 $[Ca^{2+}]_i$ estimates. FASEB J 5:2593–2599

Golenhofen K (1976) Theory of P and T systems for calcium activation in smooth muscle. In: Bulbring E, Shuba MF (eds) Physiology of smooth muscle. Raven, New York, p 197

Golenhofen K (1981) Differentiation of calcium activation processes in smooth muscle using selective antagonists. In: Bulbring E, Brading AF, Jones AW, Tomita T (eds) Smooth muscle: an assessment of current knowledge. Arnold, London, p 157

Gong MC, Fuglsang A, Alessi D, Kobayashi S, Cohen P, Somlyo AV, Somlyo AP (1992) Arachidonic acid inhibits myosin light chain phosphatase and sensitizes smooth muscle to calcium. J Biol Chem 267:21492–21498

Gordon JL (1986) Extracellular ATP: effects, sources and fate. Biochem J 233: 309–319

Grynkiewicz G, Poenie M, Tsien RY (1985) A new generation of Ca^{2+} indicators with greatly improved fluorescence properties. J Biol Chem 260:3440–3450

Guth K, Jung J (1982) Low Ca^{2+} impedes crossbridge detachment in chemically skinned taenia coli. Nature 300:775–776

Haeberle JR, Hathaway DR, Smith CL (1991) Caldesmon content of mammalian smooth muscles. J Muscle Res Cell Motil 13:81–89

Hai CM, Murphy RA (1989) Ca^{2+}, crossbridge phosphorylation, and contraction. Annu Rev Physiol 51:285–298

Hall IP, Donaldson J, Hill SJ (1989) Inhibition of histamine-stimulated inositol phospholipid hydrolysis by agents which increase cyclic AMP levels in bovine tracheal smooth muscle. Br J Pharmacol 97:603–613

Hamill OP, Marty A, Neher E, Sakmann B, Sigworth FJ (1981) Improved patch-clamp techniques for high-resolution current recording from cells and cell-free membrane patches. Pflugers Arch 391:85–100

Hamilton TC, Weir SW, Weston AH (1986) Comparison of the effects of BRL 34915 and verapamil on electrical and mechanical activity in rat portal vein. Br J Pharmacol 88:103–111

Hara Y, Kubota M, Szurszewski JH (1986) Electrophysiology of smooth muscle of the small intestine of some mammals. J Physiol (Lond) 372:501–520

Hart P, Overturf KE, Russell SN, Hume J, Sanders KM, Horowitz B (1993) Cloning and expression of a delayed rectifier K^+ channel from canine colonic smooth muscle. J Physiol (Lond) (in press)

Himpens B, Casteels R (1990) Different effects of depolarization and muscarinic stimulation on the Ca^{2+}-force relationship during the contraction-relaxation cycle in the guinea-pig ileum. Pflugers Arch 416:28–35

Himpens B, Somlyo AP (1988) Free-calcium and force transients during depolarization and pharmacomechanical coupling in guinea-pig smooth muscle. J Physiol (Lond) 395:507–530

Himpens B, Matthjis G, Somlyo AP (1989) Desensitization to cytosolic Ca^{2+} and Ca^{2+} sensitivity in guinea-pig ileum and rabbit pulmonary artery. J Physiol (Lond) 413:489–503

Himpens B, Kitazawa T, Somlyo AP (1990) Agonist-dependent modulation of Ca^{2+} sensitivity in rabbit pulmonary artery smooth muscle. Pflugers Arch 417:21–28

Hisayama T, Takayanagi I (1983) Effects of cyclic AMP and protein kinase on calcium uptake in a microsomal fraction from guinea-pig taeni caecum. Biochem Pharmacol 32:3197–3202

Hoyle CHV Burnstock G (1989) Neuromuscular transmission in the gastrointestinal tract. In: Schultz SG, Wood JD (eds) The gastrointestinal system I. American Physiological Society, Bethesda, pp 435–464 (Handbook of physiology)

Huizinga JD, Farraway L, Den Hertog A (1991a) Generation of slow-wave-type action potentials in canine colon smooth muscle involves a non-L-type Ca^{2+} conductance. J Physiol (Lond) 442:15–29

Huizinga JD, Farraway L, Den Hertog A (1991b) Effect of voltage and cyclic AMP on frequency of slow wave-type action potentials in canine colon smooth muscle. J Physiol (Lond) 442:31–45

Iino M (1981) Tension responses of chemically skinned fibre bundles of the guinea-pig taenia caeci under varied ionic environments. J Physiol (Lond) 320:449–467
Inoue R (1991) Effect of external of Cd^{2+} and other divalent cations on carbachol-activated non-selective cation channels in guinea-pig ileum. J Physiol (Lond) 442:447–463
Inoue R, Isenberg G (1990a) Effect of membrane potential on acetylcholine-induced inward current in guinea-pig ileum. J Physiol (Lond) 424:57–71
Inoue R, Isenberg G (1990b) Intracellular calcium ions modulate acetylcholine-induced inward current in guinea-pig ileum. J Physiol (Lond) 424:73–92
Inoue R, Isenberg G (1990c) Acetylcholine activates nonselective cation channels in guinea-pig ileum through a G protein. Am J Physiol 258:C1173–C1178
Inoue R, Kitamura K, Kuriyama H (1987) Acetylcholine activates single sodium channels in smooth muscle cells. Pflugers Arch 410:69–74
Inoue R, Xiong Z, Kitamura K, Kuriyama H (1989) Modulation produced by nifedipine of the unitrary Ba current of dispersed smooth muscle cells of the rabbit ileum. Pflugers Arch 414:534–542
Ito S, Kurokawa A, Ohga A, Ohta T, Sawabe K (1990) Mechanical, electrical and cyclic nucleotide responses to peptide VIP and inhibitory nerve stimulation in rat stomach. J Physiol (Lond) 430:337–353
Jafferji SS, Michell RH (1976) Stimulation of phosphatidylinositol turnover by histamine, 5-hydroxytryptamine and adrenaline in longitudinal smooth muscle of guinea-pig ileum. Biochem Pharmacol 25:1429–1430
Jiang H, Colbran JL, Francis SH, Corbin JD (1992) Direct evidence for cross-activation of cGMP-dependent protein kinase by cAMP in guinea-pig coronary arteries. J Biol Chem 267:1015–1019
Jin J-G, Murthy KS, Grider JR, Makhlouf GM (1993) Activation of distinct cAMP- and cGMP-dependent pathways by relaxant agents in isolated gastric muscle cells. Am J Physiol 264:G470–G477
Job DD (1969) Ionic basis of intestinal electrical activity. Am J Physiol 217:1534–1541
Kamm KE, Stull JT (1985) The function of myosin and myosin light chain kinase phosphorylation in smooth muscle. Annu Rev Pharmacol Toxicol 25:593–620
Kamm KE, Stull JT (1986) Activation of smooth muscle contraction: relation between myosin phosphorylation and stiffness. Science 232:80–82
Karaki H, Weiss GB (1984) Calcium channels in smooth muscle. Gastroenterology 87:960–970
Karaki H, Weiss GB (1988) Calcium release in smooth muscle. Life Sci 42:111–122
Karaki H (1989) Ca^{2+} localization and sensitivity in vascular smooth muscle. Trends Pharmacol Sci 10:320–325
Karaki H (1990) The intracellular calcium-force relationship in vascular smooth muscle. Am J Hypertens 3:253S–256S
Karaki H, Sato K, Ozaki H, Murakami K (1988) Effects of sodium nitroprusside on cytosolic calcium level in vascular smooth muscle. Eur J Phamracol 156:259–266
Karaki H, Sato K, Ozaki H (1991) Different effects of verapamil on cytosolic Ca^{2+} and contraction in norepinephrine-stimulated vascular smooth muscle. Jpn J Pharmacol 55:35–42
Kennedy C (1990) P_1- and P_2-purinoceptor subtype: an update. Arch Int Pharmacodyn Ther 303:30–50
Kitazawa T, Kobayashi S, Horiuchi K, Somlyo AV, Somlyo AP (1989) Receptor-coupled, permeabilized smooth muscle. J Biol Chem 264:5339–5342
Kitazawa T, Masuo M, Somlyo AP (1991a) G proteins-mediated inhibition of myosin light-chain phosphatase in vascular smooth muscle. Proc Natl Acad Sci USA 88:9307–9310
Kitazawa T, Gaylinn BD, Denney GH, Somlyo AP (1991b) G-protein-mediated Ca^{2+} sensitization of smooth muscle contraction through myosin light chain phosphorylation. J Biol Chem 266:1708–1715
Kobayashi S, Somlyo AV, Şomlyo AP (1988) Heparin inhibits the inostiol 1,4,5-triphosphatedependent, but not the independent, calcium release induced by

guanine nucleotide in vascular smooth muscle. Biochem Biophys Res Commun 153:625–631
Kobayashi S, Furness JB, Smith TK, Pompolo S (1989a) Histological identification of the interstitial cells of Cajal in the guinea-pig small intestine. Arch Histol Cytol 52:267–286
Kobayashi S, Kitazawa T, Somlyo AV, Somlyo AP (1989b) Cytosolic heparin inhibits muscarinic and α-adrenergic Ca^{2+} release in smooth muscle. J Biol Chem 264:17997–18004
Komori S, Bolton TB (1990) Role of G-proteins in muscarinic receptor inward and outward currents in rabbit jejunal smooth muscle. J Physiol (Lond) 427:395–419
Komori S, Bolton TB (1991) Inositol trisphosphate releases stored calcium to block voltage-dependent calcium channels in single smooth muscle cells. Pflugers Arch 418:437–441
Kubota Y, Nomura M, Stull JT (1992) GTP-γ-S-dependent regulation of smooth muscle contractile elements. Am J Physiol 262:C405–C410
Kume H, Takai A, Tokuno H, Tomita T (1989) Regulation of Ca^{2+}-dependent K^+-channel activity in tracheal myocytes by phosphorylation. Nature 341:152–154
Kume H, Graziano MP, Kotlikoff MI (1992) Stimulatory and inhibitory regulation of calcium-activated potassium channels by guanine nucleotide-binding proteins. Proc Natl Acad Sci USA 89:11051–11055
Kuriyama H, Osa T, Tasaki H (1970) Electrophysiological studies of the antrum muscle fibers of the guinea-pig stomach. J Gen Physiol 55:48–62
Kwon SC, Ozaki H, Karaki H (1993) Isoproterenol changes the relationship between cytosolic Ca^{2+} and contraction in guinea-pig taenia caecum. Jpn J Pharmacol 61:57–64
Langton PD, Burke EP, Sanders KM (1989a) Participation of Ca^{2+} currents in colonic electrical activity. Am J Physiol 257:C451–C460
Langton PD, Ward SM, Carl A, Norell MA, Sanders KM (1989b) Spontaneous electrical activity of interstitial cells of Cajal isolated from canine proximal colon. Proc Natl Acad Sci USA 86:7280–7284
Lassignal NL, Singer JJ, Walsh JV Jr (1986) Multiple neuropeptides exert a direct effect on the same isolated smooth muscle cells. Am J Physiol 250:C792–C798
Lee HK, Sanders KM (1993) Comparison of ionic currents from interstitial cells and smooth muscle cells of canine colon. J Physiol (Lond) 460:135–152
Lee HK, Bayguinov O, Sanders KM (1993) Role of non-selective cation current in muscarinic responses of canine colonic muscle. Am J Physiol (in press)
Lee KS, Nobel D, Lee E, Spindler AJ (1984) A new calcium current underlying the plateau of the cardiac action potential. Proc R Soc Lond [Biol] 223:35–48
Lee KS, Marban E, Tsien RW (1985) Inactivation of calcium channels in mammalian heart cells: joint dependence on membrane potential and intracellular calcium. J Physiol (Lond) 364:395–411
Lim SP, Bolton TB (1988) A calcium dependent rather than a G-protein mechanism is involved in the inward current evoked by muscarinic receptor stimulation in dialysed single smooth muscle cells of small intestine. Br J Pharmacol 95:325–327
Lincoln TM, Cornwell TL, Taylor AE (1990) cGMP-dependent protein kinase mediates the reduction of Ca^{2+} by cAMP in vascular smooth muscle cells. Am J Physiol 258:C399–C407
Liu J, Prosser CL, Job DD (1969) Ionic dependence of slow waves and spikes in intestinal muscle. Am J Physiol 217:1542–1547
Lorland B, Lembeck F, Holzer P (1987) Calcitonin gene-related peptide is a potent relaxant of intestinal muscle. Eur J Pharmacol 135:449–451
Marston SB (1989) What is latch? New ideas about tonic contraction in smooth muscle. J Muscle Res Cell Motil 10:97–100
Maton PN, Sutliff VE, Zhao ZC, Collins SM, Gardener JD, Jensen RT (1988) Characterization of receptors for calcitonin gene-related peptide on gastric smooth muscle cells. Am J Physiol 254:G789–G794

Mayer EA, Loo DDF, Kodner A, Reddy SN (1989) Differential modulation of Ca^{2+}-activated K^+ channels by substance P. Am J Physiol 257:G887–G897
Mayer EA, Loo DDF, Snape WJ Jr, Sachs G (1990) The activation of calcium and calcium-activated potassium channels in mammalian colonic smooth muscle by substance P. J Physiol (Lond) 420:47–71
Mironneau J, Martin C, Arnaudeau S, Jmari K, Rakotoarisoa L, Sayet I, Mironneau C (1990) High-affinity binding sites for [^{3}H]saxitoxin are associated with voltage-dependent sodium channels in portal vein smooth muscle. Eur J Pharmacol 184:315–319
Mitra R, Morad M (1985) Ca^{2+} and Ca^{2+}-activated K^+ currents in mammalian gastric smooth muscle cells. Science 229:269–272
Mitsui M, Karaki H (1990) Dual effects of carbachol on cytosolic Ca^{2+} and contraction in the intestinal smooth muscle. Am J Physiol 258:C787–C793
Moreland S, Moreland RS, Singer HA (1986) Apparent dissociation between myosin phosphorylation and maximal velocity of shortening in KCL-depolarized swine carotid artery: effect of temperature and KCl concentration. Pflugers Arch 408:139–145
Morgan JP, Morgan KG (1984) Stimulus-specific patterns of intracellular calcium levels in smooth muscle of ferret portal vein. J Physiol (Lond) 351:312–317
Morgan KG, Szurszewski JH (1980) Mechanism of phasic and tonic actions of pentagastrin on canine gastric smooth muscle. J Physiol (Lond) 301:229–242
Morgan KG, Schmalz PF, Szurszewski JH (1978) The inhibitory effects of vasoactive intestinal polypeptide on the mechanical and electrical activity of canine antral smooth muscle. J Physiol (Lond) 282:437–450
Morgan KG, Muir TC, Szurszewski JH (1981) The electrical basis for contraction and relaxation in canine fundal smooth muscle. J Physiol (Lond) 311:475–488
Muraki K, Imaizumi Y, Kojima T, Kawai T, Watanabe M (1990) Effects of tetraethylammonium and 4-aminopyridine on outward currents and excitability in canine tracheal smooth muscle cells. Br J Pharmacol 100:507–515
Muraki K, Imaizumi Y, Watanabe M (1991) Sodium currents in smooth muscle cells freshly isolated from stomach fundus of the rat and ureter of the guinea-pig. J Physiol (Lond) 442:351–375
Nelson MT, Worley JF (1989) Dihydropyridine inhibition of single calcium channels and contractions in rabbit mesenteric artery depends on voltage. J Physiol (Lond) 412:65–91
Nelson MT, Standen NB, Brayden JE, Worley JF (1988) Noradrenaline contracts arteries by activating voltage-dependent calcium channels. Nature 336:382–385
Nishimura J, Van Breemen C (1989) Direct regulation of smooth muscle contractile elements by second messengers. Biochem Biophys Res Commun 163:929–935
Nishimura J, Kolber M, van Breemen C (1988) Norepinephrine and GTP-γ-S increase myofilament Ca^{2+} sensitivity in α-toxin permeabilized arterial smooth muscle. Biochem Biophys Res Commun 157:677–683
Nishizuka Y (1986) Studies and perspectives of protein kinase C. Science 233: 305–312
Ohba M, Sakamoto Y, Tomita T (1975) The slow wave in the circular muscle of the guinea-pig stomach. J Physiol (Lond) 253:505–516
Ohba M, Sakamoto Y, Tomita T (1977) Effects of sodium, potassium and calcium ions on the slow wave in the circular muscle of the guinea-pig stomach. J Physiol (Lond) 267:167–180
Ohya Y, Sperelakis N (1989) Fast Na^+ and slow Ca^{2+} channels in single uterine muscle cells from pregnant rats. Am J Physiol 257:C408–C412
Ohya Y, Kitamura K, Kuriyama K (1987) Cellular calcium regulates outward currents in rabbit intestinal smooth muscle cell. Am J Physiol 252:C401–C410
Okabe K, Kitamura K, Kuriyama H (1987) Features of 4-aminopyridine sensitive outward current observed in single smooth muscle cells from the rabbit pulmonary artery. Pflugers Arch 409:561–568

Okabe K, Kajioka S, Nakao K, Kitamura K, Kuriyama H, Weston AH (1990) Actions of cromakalim on ionic currents recorded from single smooth muscle cells of the rat portal vein. J Pharmacol Exp Ther 252:832–839
Ozaki H, Satoh T, Karaki H, Ishida Y (1988) Regulation of metabolism and contraction by cytoplasmic calcium in the intestinal smooth muscle. J Biol Chem 263:14074–14079
Ozaki H, Kwon S-C, Tajimi M, Karaki H (1990a) Changes in cytosolic Ca^{2+} and contraction induced by various stimulants and relaxants in canine tracheal smooth muscle. Pflugers Arch 416:351–359
Ozaki H, Ohyama T, Sato K, Karaki H (1990b) Ca^{2+} dependent and independent mechanism of sustained contraction in vascular smooth muscle of rat aorta. Jpn J Pharmacol 52:509–512
Ozaki H, Gerthoffer WT, Publicover NG, Fusetani N, Sanders KM (1991a) Time-dependent changes in Ca^{2+} sensitivity during phasic contraction of canine antral smooth muscle. J Physiol (Lond) 440:207–224
Ozaki H, Gerthoffer WT, Publicover NG, Sanders KM (1991b) Changes in the Ca^{2+} sensitivity in antral smooth muscle: possible role of a Ca^{2+}-dependent phosphatase
in rhythmic contractions. In: Moreland RS (ed) Regulation of smooth muscle. Plenum, New York, p 481
Ozaki H, Blondfield DP, Stevens RJ, Publicover NG, Sanders KM (1991c) Simultaneous measurement of membrane potential, cytosolic calcium and muscle tension in smooth muscle tissue. Am J Physiol 260:C917–C925
Ozaki H, Blondfield DP, Hori M, Publiscover NG, Kato I, Sanders KM (1991d) Spontaneous release of nitric oxide inhibits electrical, Ca^{2+} and mechanical transients in canine gastric smooth muscle. J Physiol (Lond) 445:231–247
Ozaki H, Zhang L, Buxton ILO, Sanders KM, Publicover NG (1992a) Negative-feedback regulation of excitation-contraction coupling in gastric smooth muscle. Am J Physiol 263:C1160–C1171
Ozaki H, Blondfield DP, Hori M, Sanders KM, Publicover NG (1992b) Cyclic AMP mediated regulation of excitation-contraction coupling in gastric smooth muscle. J Physiol (Lond) 447:351–372
Ozaki H, Gerthoffer WT, Hori M, Karaki H, Sanders KM, Publicover NG (1993) Ca^{2+}-regulation of the contractile apparatus in canine gastric smooth muscle. J Physiol (Lond) 460:33–50
Pacaud P, Bolton TB (1991) Relation between muscarinic receptor cationic current and internal calcium in guinea-pig jejunal smooth muscle cells. J Physiol (Lond) 441:447–499
Paulmichl M, Nasmith P, Hellmiss R, Reed K, Boyle WA, Nerbonne JM, Peralta EG, Clapham DE (1991) Cloning and expression of a rat cardiac delayed rectifier potassium channel. Proc Natl Acad Sci USA 88:7892–7895
Pfitzer G, Hofmann F, DiSalvo J, Ruegg JC (1984) CGMP and CAMP inhibit tension development in skinned coronary artery. Pflugers Arch 401:277–280
Plant TD (1988) Properties and calcium-dependent inactivation of calcium currents in cultured mouse pancreatic β-cells. J Physiol (Lond) 404:731–747
Plant TD, Standen NB, Ward TA (1983) The effects of injection of calcium ions and calcium chelators on calcium channel inactivation in *Helix* neurons. J Physiol (Lond) 334:189–212
Post JM, Hume JR (1992) Ionic basis for spontaneous depolarizations in isolated smooth muscle of canine colon. Am J Physiol 263:C691–C699
Post JM, Stevens R, Sanders KM, Hume JR (1991) Effect of cromakalim and lemakalim on K^+ and Ca^+ currents in colonic smooth muscle. Am J Physiol 260:C375–C382
Publicover NG, Sanders KM (1986) Effects of frequency on the waveform of propagated slow waves in canine gastric antral muscle. J Physiol (Lond) 371:179–189

Publicover NG, Horowitz NN, Sanders KM (1992) Calcium oscillations in freshly dispersed and cultured interstitial cells from canine colon. Am J Physiol 262:C589–C597

Quasthoff S, Spuler A, Horn-Lehman F, Grafe P (1989) Cromakalim, pinacidil and RP 49356 activate a tolbutamide-sensitive K^+ conductance in human skeletal muscle fibers. Pflugers Arch 414:S179–S180

Regoli D, Drapeau G, Dion S, D'Orleans-Juste P (1987) Pharmacological receptors for substance P and neurokinins. Life Sci 40:109–117

Rembold CM, Murphy RA (1988) $[Ca^{2+}]$-dependent myosin phosphorylation in phorbol ester stimulated smooth muscle contraction. Am J Physiol 255:C719–C723

Rembold CM, Weaver BA, Linden J (1991) Adenosine triphosphate induces a low $[Ca^{2+}]_i$ sensitivity of phosphorylation and an unusual form of receptor desensitization in smooth muscle. J Biol Chem 266:5407–5411

Ribeiro-Neto FAP, Mattera R, Hildebrandt JD, Codina J, Field JB, Birnbaumer L, Sekura R (1985) ADP-ribosylation of membrane components by pertussis and cholera toxins. Methods Enzymol 109:566–577

Rich A, Kenyon JL, Hume JR, Overturf K, Horowitz B, Sanders KM (1993) Dihydropyridine-sensitive calcium channels expressed in canine colonic smooth muscle cells. Am J Physiol 264:C745–C754

Ruegg JC, Paul RJ (1982) Vascular smooth muscle: calmodulin and cyclic AMP-dependent protein kinase alter calcium sensitivity in porcine carotid skinned fibers. Circ Res 50:394–399

Russell SN, Aaronson PI (1990) Carbachol inhibits the voltage-gated calcium current in smooth muscle cells isolated from the longitudinal muscle of the rabbit jejunum. J Physiol (Lond) 426:23P

Sakata K, Ozaki H, Kwon S-C, Karaki H (1989) Effects of endothelin on the mechanical activity and cytosolic calcium levels of various types of smooth muscle. Br J Pharmacol 98:483–492

Sala F, Hernandez-Cruz A (1990) Calcium diffusion modelling in a spherical neuron: relevance of buffering properties. Biophys J 57:313–324

Sanders KM (1983) Excitation-contraction coupling without Ca^{2+} action potentials in small intestine. Am J Physiol 244:C356–C361

Sanders KM (1989) Electrophysiology of dissociated gastrointestinal muscle cells. In: Schultz SG, Wood JD (eds) The gastrointestinal system. American Physiological Society, Bethesda, p 163 (Handbook of physiology)

Sanders KM (1992) Ionic mechanisms of electrical rhythmicity in gastrointestinal smooth muscles. Annu Rev Physiol 54:439–453

Sanders KM, Publicover NG (1989) Electrophysiology of the gastric musculature. In: Schultz SG, Wood JD (eds) The gastrointestinal system. American Physiological Society, Bethesda, p 187 (Handbook of physiology)

Sanders KM, Smith TK (1986) Motorneurons of the submucous plexus regulate electrical activity of the circular muscle of the canine proximal colon. J Physiol (Lond) 380:293–310

Sanders KM, Smith TK (1989) Electrophysiology of colonic smooth muscle. In: Schultz SG, Wood JD (eds) The gastrointestinal system. American Physiological Society, Bethesda, p 251 (Handbook of physiology)

Sanders KM, Ward SM (1992) Nitric oxide as a mediator of non-adrenergic, non-cholinergic neurotransmission. Am J Physiol 262:G379–G392

Sanders KM, Burke EP, Stevens RJ (1989) Effects of methylene blue on electrical rhythmicity of the canine colon. Am J Physiol 256:G779–G784

Sanguinetti MC, Scott AL, Zingaro GJ, Siegl PKS (1988) BRL 34915 (cromakalim) activates ATP-sensitive K^+ current in cardiac muscle. Proc Natl Acad Sci USA 85:8360–8364

Sarna SK (1975) Gastrointestinal electrical activity: terminology. Gastroenterology 68:1631–1635

Sato K, Ozaki H, Karaki H (1988a) Changes in cytosolic calcium level in vascular smooth muscle strip measured simultaneously with contraction using fluorescent calcium indicator fura-2. J Pharmacol Exp Ther 246:294–300
Sato K, Ozaki H, Karaki H (1988b) Multiple effects of caffeine on contraction and cytosolic free levels in vascular smooth muscle of rat aorta. Naunyn Schmiedebergs Arch Pharmacol 338:443–448
Sato K, Ozaki H, Karaki H (1990) Differential effects of carbachol on cytosolic calcium levels in vascular endothelium and smooth muscle. J Pharmacol Exp Ther 255:114–119
Sato K, Hori M, Ozaki H, Takano-Ohmuro H, Tsuchiya T, Sugi H, Karaki H (1992) Myosin phosphorylation-independent contraction induced by phorbol ester in vascular smooth muscle. J Pharmacol Exp Ther 261:497–505
Serio R, Barajas-Lopez C, Daniel EE, Berezin I, Huizinga JD (1991) Slow-wave activity in colon: role of network of submucosal interstitial cells of Cajal. Am J Physiol 260:G636–G645
Shearman MS, Sekiguchi K, Nishizuka Y (1989) Modulation of ion channel activity: key function of the protein kinase C enzyme family. Pharmacol Rev 41:211–237
Sherman A, Keizer J, Rinzel J (1990) Domain model for Ca^{2+}-inactivation of Ca^{2+} channels at low channel density. Biophys J 58:985–995
Shimizu K, Kaburagi T, Nakajo S, Urakawa N (1991) Decrease in muscle tension and reduced pyridine nuclortides of the guinea-pig ileal longitudinal smooth muscle in high K^+, Na^+-deficient solution. Jpn J Pharmacol 56:53–59
Sims SM (1992a) Calcium and potassium currents in canine gastric smooth muscle cells. Am J Physiol 262:G859–G867
Sims SM (1992b) Cholinergic activation of a non-selective cation current in canine gastric smooth muscle is associated with contraction. J Physiol (Lond) 449: 377–398
Sims SM, Singer JJ, Walsh JV Jr (1985) Cholinergic agonists suppress a potassium current in freshly dissociated smooth muscle cells of the toad. J Physiol (Lond) 367:503–529
Sims SM, Walsh JV Jr, Singer JJ (1986) Substance P and acetylcholine both suppress the same K^+ current in dissociated smooth muscle cells. Am J Physiol 251: C580–C587
Sims SM, Singer JJ, Walsh JV Jr (1988) Antagonistic adrenergic-muscarinic regulation of M current in smooth muscle cells. Science 239:190–193
Sims SM, Vivaudou MB, Hillemeier C, Biancani P, Walsh JV Jr, Singer JJ (1990a) Membrane currents and cholinergic regulation of K^+ current in esophageal smooth muscle cells. Am J Physiol 258:G794–G802
Sims SM, Clapp LH, Walsh JV, Singer JJ (1990b) Dual regulation of M current in gastric smooth muscle cells: β-adrenergic-muscarinic antagonism. Pflugers Arch 417:291–302
Singer JJ, Walsh JV Jr (1984) Large conductance Ca^{2+}-activated K^+ channels in smooth muscle cell membrane. Biophys J 45:68–70
Singer JJ, Walsh JV Jr (1987) Characterization of calcium-activated potassium channels in single smooth muscle cells using the patch-clamp technique. Pflugers Arch 408:98–111
Smith TK, Reed JB, Sanders KM (1987a) Origin and propagation of electrical slow waves in circular muscle of canine proximal colon. Am J Physiol 252:C215–C224
Smith TK, Reed JB, Sanders KM (1987b) Interaction of two electrical pacemakers in muscularis of canine proximal colon. Am J Physiol 252:C290–C299
Smith TK, Reed JB, Sanders KM (1989) Electrical pacemakers of canine proximal colon are functionally innervated by inhibitory motor neurons. Am J Physiol 252:C215–C224
Smith TK, Ward SM, Zhang L, Buxton ILO, Gerthoffer WT, Sanders KM, Keef KD (1993) β-adrenergic inhibition of electrical and mechanical activity in canine colon: Role of cyclic AMP. Am J Physiol 264:G708–G717

Somlyo AP, Himpens B (1989) Cell calcium and its regulation in smooth muscle. FASEB J 3:2266–2276
Somlyo AP, Somlyo AV (1968) Vascular smooth muscle: normal structure, pathology, biochemistry and biophysics. Pharmacol Rev 20:197–272
Standen NB, Qualye JM, Davis NW, Brayden JE, Huang Y, Nelson MT (1989) Hyperpolarizing vasodilators activate ATP-sensitive K^+ channels in arterial smooth muscle. Science 245:177–180
Stark ME, Bauer AJ, Szurszewski JH (1991) Effect of nitric oxide on circular muscle of the canine small intestine. J Physiol (Lond) 444:743–761
Sternini C, Reeve JR, Brecha N (1987) Distribution and characterization of calcitonin gene-related peptide immunoreactivity in the digestive system of normal and capsaicin-treated rats. Gastroenterology 93:852–862
Stiles GL, Caron MG, Lefkowitz RJ (1984) β-adrenergic receptors: biochemical mechanisms of physiological regulation. Physiol Rev 64:661–743
Strumer W, Ruppersberg JP, Schröter KH, Sakmann B, Stocker M, Giese KP, Perschke A, Baumann A, Pongs O (1989) Molecular basis of functional diversity of voltage-gated potassium channels in mammalian brain. EMBO J 8:3235–3244
Stull JT, Hsu L-C, Tansey M, Kamm KE (1990) Myosin light chain phosphorylation in tracheal smooth muscle. J Biol Chem 265:16683–16690
Suematsu E, Hirata M, Hashimoto T, Kuriyama H (1988) Inositol 1,4,5-trisphosphate releases Ca^{2+} from intracellular store sites in single cells of porcine coronary artery. Biochem Biophys Res Commun 120:481–485
Sutherland C, Walsh MP (1989) Phosphorylation of caldesmon prevents its interaction with smooth muscle myosin. J Biol Chem 264:578–583
Suzuki N, Prosser CL, Dahms V (1986) Boundary cells between longitudinal and circular layers: essential for electrical slow waves in cat intestine. Am J Physiol 280:G287–G294
Szurszewski JH (1975) Mechanism of action of pentagastrin and acetylcholine on the longitudinal muscle of the canine antrum. J Physiol (Lond) 252:335–361
Szurszewski JH (1987) Electrical basis for gastrointestinal motility. In: Johnson LR (ed) Physiology of the gastrointestinal tract. Raven, New York, p 1435
Takahashi K, Nadal-Ginard B (1991) Molecular cloning and sequence analysis of smooth muscle calponin. J Biol Chem 266:13284–13288
Takahashi K, Hiwada K, Kokubu T (1986) Isolation and characterization of a 34000-dalton calmodulin- and F-actin-binding protein from chicken gizzard smooth muscle. Biochem Biophys Res Commun 26:20–26
Taniguchi J, Furukawa K-I, Shigekawa M (1992) Maximum K^+ channel activity stimulated by G kinase in the canine coronary arterial smooth muscle. Jpn J Pharmacol 58(Suppl 2):397P
Tawada Y, Furukawa K, Shigekawa M (1987) ATP-induced calcium transient in cultured rat aortic smooth muscle cells. J Biochem (Tokyo) 102:1499–1509
Thornbury KD, Ward SM, Dalziel HH, Carl A, Westfall DP, Sanders KM (1991) Nitric oxide and nitrosocysteine mimics nonadrenergic, noncholinergic hyperpolarization in canine proximal colon. Am J Physiol 261:G553–G557
Thornbury KD, Ward SM, Sanders KM (1992a) Participation of fast-activating, voltage-dependent potassium currents in electrical slow waves of colonic muscle. Am J Physiol 263:C226–C236
Thornbury KD, Ward SM, Sanders KM (1992b) Outward currents in longitudinal colonic muscle cells contribute to spiking electrical behavior. Am J Physiol 263:C237–C245
Thuneberg L (1982) Interstitial cells of Cajal: intestinal pacemaker cells. Adv Anat Embryol Cell Biol 71:1–130
Thuneberg L (1989) Interstitial cells of Cajal. In: Wood JD (ed) The gastrointestinal system. American Physiological Society, Bethesda, p 349 (Handbook of physiology)

Thuneberg L, Johansen V, Rasmenssen JJ, Andersen BG (1983) Interstitial cells of Cajal (ICC): selective uptake of methylene blue inhibits slow wave activity. In: Roman C (ed) Gastrointestinal motility. MTP Press, Lancaster, p 495
Tomita T (1972) Conductance change during the inhibitory junction potential in the guinea-pig taenia coli. J Physiol (Lond) 225:693–703
Tomita T (1981) Electrical activity (spikes and slow waves) in gastrointestinal smooth muscle. In: Bulbring E, Brading AF, Jones Aw, Tomita T (eds) Smooth muscle: an assessment of current knowledge. University of Texas Press, Austin, p 127
Trautwein W, Hescheler J (1990) Regulation of cardiac L-type calcium current by phosphorylation and G proteins. Annu Rev Physiol 52:257–274
Van Breemen C, Aaronson P, Loutzenhiser R (1979) Na^+, Ca^{2+} interaction in mammalian smooth muscle. Pharmacol Rev 30:167–208
Vivaudou MB, Clapp LH, Walsh JV Jr, Singer JJ (1988) Regulation of one type of Ca^{2+} current in smooth muscle cells by diacylglycerol and acetylcholine. FASEB J 2:2497–2504
Vivaudou MB, Singer JJ, Walsh JV Jr (1991) Multiple types of Ca^{2+} channels in visceral smooth muscle cells. Pflugers Arch 418:144–152
Vogalis F, Sanders KM (1990) Cholinergic stimulation activates a non-selective cation current in canine pyloric circular muscle cells. J Physiol (Lond) 429: 223–236
Vogalis F, Publicover NG, Hume JR, Sanders KM (1991) Relatinship between calcium current and cytosolic calcium concentration in canine gastric smooth muscle cells. Am J Physiol 260:C1012–C1018
Vogalis F, Publicover NG, Sanders KM (1992) Regulation of calcium current by voltage and cytoplasmic calcium in canine gastric smooth muscle. Am J Physiol 262:C691–C700
Vogalis F, Lang RJ, Bywater RAR, Taylor GS (1993) Voltage-gated ionic currents in smooth muscle cells of the guinea-pig proximal colon. Am J Physiol 264:C527–C536
Vyas TB, Moores SU, Narayan SR, Witherell JC, Siegman MJ, Butler TM (1992) Cooperative activation of myosin by light chain phsophorylation in permeabilized muscle. Am J Physiol 263:C210–C219
Walsh JV Jr, Singer JJ (1981) Voltage clamp of single freshly dissociated smooth muscle cells: current-voltage relationships for three currents. Pflugers Arch 390:207–210
Walsh JV Jr, Singer JJ (1987) Identification and characterization of major ionic currents in isolated smooth msucle cells using the voltage-clamp technique. Pflugers Arch 408:83–97
Ward SM, Burke EP, Sanders KM (1990) Use of rhodamine 123 to label and lesion interstitial cells of Cajal in canine colonic circular muscle. Anat Embryol (Biol) 182:215–224
Ward SM, Sanders KM (1990) Pacemaker activity in septal structures of canine colonic circular muscle. Am J Physiol 259:G264–G273
Ward SM, Sanders KM (1992a) Dependence of electrical slow waves of canine colonic smooth muscle on calcium current. J Physiol (Lond) 455:307–319
Ward SM, Sanders KM (1992b) Upstroke component of electrical slow waves in canine colonic smooth muscle due to nifedipine-resistant Ca^{2+} current. J Physiol (Lond) 455:321–337
Ward SM, Vogalis F, Bondfield DP, Ozaki H, Fusetani N, Uemura D, Publicover NG, Sanders KM (1991a) Inhibition of electrical slow waves and Ca^{2+} currents of gastric and colonic smooth muscle by phosphatase inhibitors. Am J Physiol 261:C64–C70
Ward SM, Keller RG, Sanders KM (1991b) Structure and organization of electrical activity of canine distal colon. Am J Physiol 260:G724–G735
Ward SM, Vogalis F, Blondfield DP, Ozaki H, Fusetani N, Uemura D, Publicover NG, Sanders KM (1991c) Inhibition of electrical slow waves and calcium

currents of gastric and colonic smooth muscles by phosphatase inhibitors. Am J Physiol 261:C64–C70
Ward SM, McKeen ES, Sanders KM (1992a) Role of nitric oxide in non-adrenergic, non-cholinergic inhibitory junction potentials in canine ileocolonic sphincter. Br J Pharmacol 105:776–782
Ward SM, Xue C, Shuttleworth CWR, Bredt DS, Snyder SH, Sanders KM (1992b) NADPH diaphorase and nitric oxide synthase colocalization in enteric neurons of canine proximal colon. Am J Physiol 25:G277–G284
Weston AH (1989) Smooth muscle K^+ channel openers: their pharmacology and clinical potential. Pflugers Arch 414:S99–S105
Williams DL, Katz GM, Roy-Contancin L, Reuben JP (1988) Guanosine 5′-monophosphate modulates gating of high conductance Ca^{2+}-activated K^+ channels in vascular smooth muscle cells. Proc Natl Acad Sci USA 85:9360–9364
Winder SJ, Walsh MP (1990) Smooth muscle calponin: inhibition of actomyosin MgATPase and regulation by phosphorylation. J Biol Chem 265:10148–10155
Yagi S, Becker PL, Fay FS (1988) Relationship between force and Ca^{2+} concentration in smooth muscle as revealed by measurements on single cells. Proc Natl Acad Sci USA 85:4109–4113
Yamamoto Y, Hu SI, Kao CY (1989) Inward current in single smooth muscle cells of the guineapig taenia coli. J Gen Physiol 93:521–550
Yanagisawa T, Kawada M, Taira N (1989) Nitroglycerin relaxes canine coronary arterial smooth muscle without reducing intracellular Ca^{2+} concentrations measured with fura-2. Br J Pharmacol 98:469–482
Yoshino M, Yabu H (1985) Single Ca^{2+} channel currents in mammalian visceral smooth muscle cells. Pflugers Arch 404:285–286
Yoshino M, Someya T, Nishio A, Yabu H (1988) Whole-cell and unitary Ca^{2+} channel currents in mammalian intestinal smooth muscle cells: evidence for the existence of two types of Ca^{2+} channels. Pflugers Arch 411:229–231
Yoshino M, Someya T, Nishio A, Yazawa K, Usuki T, Yabu H (1989) Multiple types of Ca^{2+} channels in mammalian intestinal muscle cells. Pflugers Arch 414:401–409
Young HM, Furness JB, Shuttleworth CWR, Bredt DS, Snyder SH (1992) Co-localization of nitric oxide synthase immunoreactivity and NADPH diaphorase staining in neurons of the guineapig intestine. Histochemistry 97:375–378
Zhang L, Buxton ILO (1993) Protein kinase regulation of muscarinic receptor signaling in colonic smooth muscle. Br J Pharmacol 108:613–621

CHAPTER 11

Airway Smooth Muscle

A.J. KNOX and A.E. TATTERSFIELD

A. Introduction

Smooth muscle relaxant drugs have provided the mainstay of treatment for obstructive lung disease for several decades. Although the emphasis for treatment of asthma has been changing to anti-inflammatory agents bronchodilators still provide useful symptomatic treatment for asthma and are the first line drugs for the treatment of non-asthmatic obstructive lung disease. A greater understanding of the various parts of the contractile and relaxant signal transduction systems has enabled several advances to be made in recent years although the main therapeutic relaxant agents remain the β-adrenergic agonist groups of drugs. It is hoped that in the next 10 years greater understanding of airway smooth muscle receptor pharmacology and post-receptor signalling systems will lead to therapeutic advances. In this review we will first give an overview of the contractile and relaxant pathways in airway smooth muscle. Understanding these mechanisms is a prerequisite to understanding the sites and actions of the wide range of pharmacological agents which act on this effector organ.

B. Contractile Mechanisms

Contractile mechanisms in airway smooth muscle can be divided broadly into receptor operated and depolarisation-induced contraction mechanisms.

I. Receptor-Operated Contraction

Receptor-operated contraction of airway smooth muscle results from binding of a contractile agonist to its surface receptor and activation of the inositol phospholipid second messenger system (HALL and CHILVERS 1989). This form of contraction has been termed pharmacomechanical coupling (SOMYLO and SOMYLO 1968). Most of the contractile agents which have been studied (e.g., histamine, 5-hydroxytryptamine, cholinergic agonists) activate this pathway (MADISON and BROWN 1988; HALL and HILL 1988; GRANDORDY et al. 1986).

II. Initiation

Ligand-receptor binding results in activation of a G protein (G_s) coupled to phospholipase C causing cleavage of inositol bisphosphate to form a series of different inositol phosphate derivatives, the most important of which is inositol 1,4,5 trisphosphate ($InsP_3$) (BERRIDGE and IRVINE 1989). $InsP_3$ binds

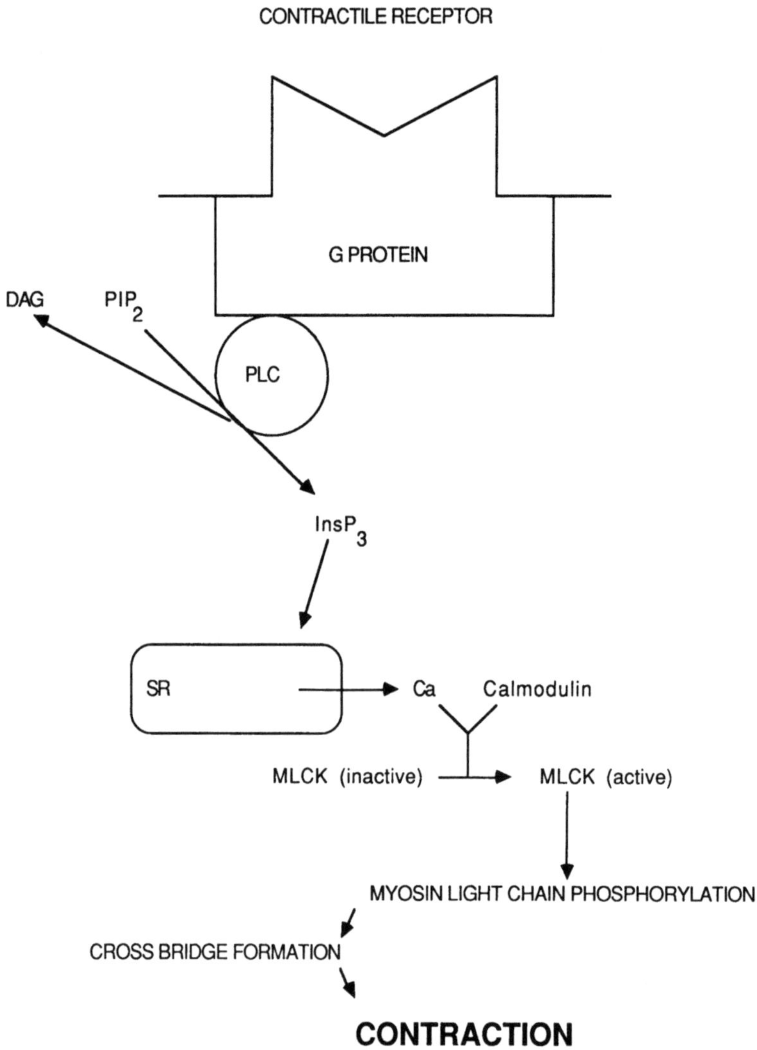

Fig. 1. Sequence of events whereby contraction is initiated. Binding of a contractile agonist receptor is coupled through a G protein to phospholipase C leading to cleavage of membrane PIP_2 to form $InsP_3$. $InsP_3$ releases calcium from the sarcoplasmic reticulum (*SR*) which then binds to calmodulin to form activated calcium calmodulin complex. This converts myosin light chain kinase (*MLCK*) to its active form. Myosin light chain is then phosphorylated, cross bridge formation begins and contraction takes place

to a receptor on the sarcoplasmic reticulum causing release of calcium from internal stores. The subsequent rise in intracellular calcium, from basal levels of 0.1 nM to 10–100 μM, is responsible for stimulating a series of reactions leading to phosphorylation of myosin light chain and the initiation of contraction (STULL 1980; ADELSTEIN and EISENBERG 1980). Calcium binds to the four binding sites on calmodulin to form activated calcium-calmodulin complex and this in turn activates myosin light chain kinase. Myosin light chain kinase then phosphorylates myosin light chain, resulting in myosin ATPase activation by actin, cross bridge formation and contraction (Fig. 1).

III. Maintenance

Although the sequence of events leading to the initiation of a contractile response are now well understood, the events responsible for maintaining the response are not. The rise in intracellular calcium concentration following agonist-receptor binding is short lived, calcium levels returning to values only slightly above baseline within minutes (MURRAY and KOTLIKOFF 1991) despite continued agonist binding. It is not clear how contraction is subsequently maintained in the presence of a calcium level only slightly above baseline. Most evidence points to protein kinase C being responsible since a system whereby $InsP_3$ initiates a cellular response which is subsequently maintained by protein kinase C is widespread in cellular signal transduction systems in different tissues and species (BERRIDGE and IRVINE 1989; NISHIZUKA 1984). Cleavage of membrane inositol bisphosphate by phospholipase C leads to formation of $InsP_3$ and diacylglycerol (DAG). DAG remains membrane associated and in the presence of phosphatidyl serine residues and calcium activates protein kinase C (NISHIZUKA 1988). Protein kinase C may maintain the contractile response by phosphorylation of certain contractile proteins in addition to effects on membrane ion transport (PARK and RASMUSSEN 1985; SCHRAMM and GRUNSTEIN 1989; SOUHRADA and SOUHRADA 1989a; KNOX et al. 1989a). The contractile protein caldesmon is a substrate for protein kinase C in smooth muscle in vitro (UMEKAWA and HIDAKA 1985) and several late phase contractile proteins such as synemin, caldesmon and 5 cytosolic proteins are phosphorylated after application of phorbol esters to bovine trachealis in vitro (PARK and RASMUSSEN 1986). As phorbol esters also feedback to block the inositol phospholipid response to spasmogens (KOTLIKOFF et al. 1987), protein kinase C may have a dual role in switching off the events which initiate contraction while switching on the processes responsible for the maintenance phase (Fig. 2).

The role of inositol phosphates other than $InsP_3$ in airway smooth muscle is unknown. Cyclic inositol bisphosphate can cause calcium release in other tissues although the concentrations necessary to do this are generally an order of magnitude greater than those for $InsP_3$; in some tissues $InsP_4$ is thought to cause calcium entry and refilling of intracellular calcium stores (BERRIDGE and IRVINE 1989). In bovine tracheal smooth muscle, however,

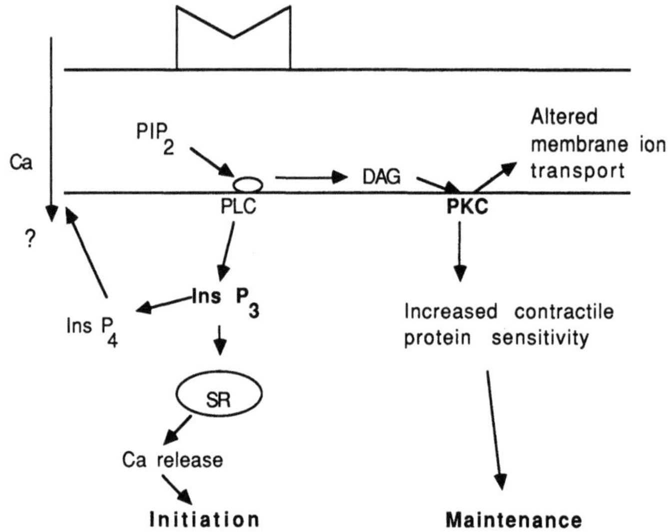

Fig. 2. Complementary roles of inositol trisphosphate (*InsP₃*) and diacylglycerol (*DAG*) activated protein kinase C (*PKC*) in bringing about the initiation and maintenance of the contractile response

$InsP_4$ concentrations are increased only slightly following stimulation with carbachol (CHILVERS and NAHORSKI 1990).

IV. Depolarisation-Induced Contraction

This is the main mode of contraction produced by agents such as KCl or tetraethylammonium (TEA). Contraction is due to membrane depolarisation with influx of calcium via the voltage-dependent calcium channels and is often referred to as electro-mechanical coupling (SOMYLO and SOMYLO 1968). Influx of calcium causes activation of calmodulin and subsequent myosin light chain phosphorylation (COBURN and YAMAGUCHI 1977; FARLEY and MILES 1977; FOSTER et al. 1983, 1984). A similar sequence of events can be produced by ouabain (an Na/K ATPase inhibitor) which induces airway smooth muscle contraction by membrane depolarisation and influx of calcium through voltage-dependent calcium channels and possibly stimulation of calcium entry via Na/Ca exchange (SOUHRADA et al. 1981; KNOX et al. 1990a).

V. Sources of Calcium Utilised During Contraction

Airway smooth muscle relies upon mobilisation of calcium ions from both extracellular and intracellular sources to support contraction. The main intracellular sources or calcium are the sarcoplasmic reticulum and calcium bound to invaginations of the internal surface of the plasmalemma (TRIGGLE

1985). Calcium may also come from several external sources including a leak pathway, receptor-operated channels or voltage-dependent channels. The leak pathway occurs due to inward diffusion down the electrochemical gradient. Receptor operated channels are activated by a ligand-receptor interaction and may admit the entry of calcium, possibly along with other ions. Until recently good electrophysiological evidence for the existence of receptor-operated channels in airways smooth muscle was lacking. Voltage dependent channels are activated by depolarising stimuli such as potassium chloride or tetraethylammonium (TEA) and are sensitive to the dihydropyridine group of calcium channel antagonists such as nifedipine. Initial mobilisation of calcium from any of these sources may lead to calcium induced calcium release, whereby a small rise in intracellular calcium mobilises a larger intracellular calcium pool (SAIDA and VAN BREEMAN 1984).

Several of these processes may be involved in the initiation and maintenance phases of airway smooth muscle contraction though studies measuring calcium influx in airway smooth muscle cells during the initiation phase have failed to demonstrate any increase in calcium influx (AHMED et al. 1984; RAEBURN et al. 1984) suggesting that the main sources of activator calcium are the intracellular calcium stores. One study in guinea pig airways, however, demonstrated increased calcium influx during the maintenance phase of receptor-operated contraction suggesting that extracellular calcium may be important at this time (WEISS et al. 1985). In a more recent study of cultured human airway smooth muscle calcium entered during the maintenance phase of contraction via a novel receptor operated calcium channel (MURRAY and KOTLIKOFF 1991). Confirmation of these studies and the precise elucidation of their role is eagerly awaited. In contrast to agents which act by receptor operated mechanisms, stimuli such as KCl or TEA which contract airway smooth muscle through membrane depolarisation have been shown to cause and influx of radiolabelled calcium in guinea pig trachealis (FOSTER et al. 1983).

C. Relaxant Mechanisms

The two second messengers responsible for airway smooth muscle relaxation are cyclic adenosine monophosphate (cAMP) and cyclic guanosine monophosphate (cGMP).

I. Elevation of Cyclic AMP

The most important relaxant signal transduction pathway in airway smooth muscle involves cyclic adenosine monophosphate (cAMP). The main receptor which has been studied with respect to this pathway is the β_2-adrenoceptor which is coupled via a G protein to adenylate cyclase. Activation of adenylate cyclase leads to conversion of ATP to cAMP, cAMP then activates a group of cyclic AMP-dependent protein kinases which in turn phosphorylate a number of different sites to cause relaxation. Other agents which act via cAMP include prostaglandin E_2 (Tauber et al. 1973), vasoactive intestinal peptide (VIP) (Robberecht et al. 1981) and forskolin. The sites of action of cAMP-dependent protein kinases may include inhibition of inositol phosphate hydrolysis (Madison and Brown 1988; Hall et al. 1989), increased calcium uptake by internal stores (Mueller and Van Breeman 1979), inactivation of myosin light chain kinase (Silver and Stull 1982) and activation of cell membrane ion channels and transporters such as K^+ channels (Kume et al. 1989) and Na/K ATPase (Gunst and Stropp 1988; Knox and Brown 1991) (Fig. 3). The net effect of these processes would be to reduce intracellular calcium concentration and to reduce contractile protein phosphorylation thereby producing relaxation. The relative importance of each of these mechanisms in mediating the relaxant effects of cAMP-dependent protein kinase is as yet unclear.

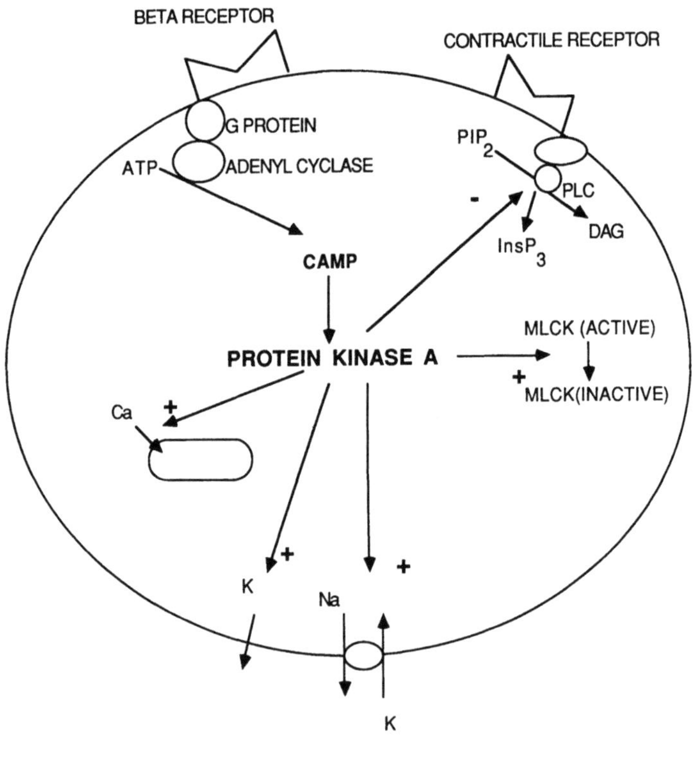

Fig. 3. Schematic representation of the sites of phosphorylation for protein kinase A. The combined effect of these processes is to lead to airway smooth muscle relaxation

II. Elevation of Cyclic GMP

It has recently become clear that cyclic guanosine monophosphate (cGMP) is also a relaxant second messenger in smooth muscles in various sites including airway smooth muscle (SUZUKI et al. 1986). Agents which cause relaxation of airway smooth muscle by increasing cGMP include the nitrates, atrial natriuretic peptide (ISHII and MURAD 1989) and the cGMP-specific phosphodiesterase inhibitors (TORPHY and UNDEM 1991). In contrast to cAMP-dependent protein kinases which are the major intracellular effector system for cAMP, cGMP-dependent protein kinases do not mediate all the intracellular effects of cGMP, at least in non-muscle cells. In the retina a light inhibited sodium channel can be directly regulated by cGMP without the participation of protein kinases (FESENKO et al. 1985). Whether or not ion channels in airway smooth muscle can be regulated by cGMP directly has not been studied to date. The structure and regulation of guanylate cyclase and of cGMP-dependent protein kinases are very different from adenylate cyclase and the cAMP-dependent protein kinases (WALTER 1989). Guanylate cyclase exists in two forms, a soluble form and a particulate form, both of which have been purified and characterised (WALTER 1989). Soluble guanylate cyclase exists as a heterodimer composed of two subunits. It is a haem protein which can be activated by nitric oxide and the nitro group of vaso/bronchodilators. Particulate guanylate cyclase has been purified in the lung, but its precise structure has not yet been elucidated. It is one of the receptors mediating the effects of atrial natriuretic peptide. Guanylate cyclase catalyses the conversion of GTP to cGMP which then activates a group of distinct cGMP-dependent protein kinases. Considerable sequence homology exists between cGMP-dependent protein kinases and other kinases, but particularly cAMP-dependent protein kinases (TAKIO et al. 1984). Various sites of action have been suggested for cGMP-dependent protein kinase in airway smooth muscle as with cAMP-dependent protein kinases (HOGABOOM et al. 1982; FELBEL et al. 1988; PFITZER et al. 1982). The relative importance of the different intracellular sites of action of cGMP dependent protein kinases have still to be established.

III. Breakdown of cAMP and cGMP

cAMP and cGMP are broken down by a family of enzymes called phosphodiesterases. Hydrolysis of the 3′ phosphoesterbond on cAMP and cGMP by phosphodiesterases converts them to their inactive 5′ nucleotide metabolites 5′ AMP and 5′ GMP. Several phosphodiesterase isoenzymes each differing in its kinetic and physical characteristics and substrate selectivities have been identified using molecular biological techniques (TORPHY and UNDEM 1991). The role of specific inhibitors of these isoenzymes as relaxant agents in airway smooth muscle is discussed later.

D. Calcium Removal Mechanisms

Three calcium removal mechanisms are thought to operate in airway smooth muscle, a calcium efflux pump, uptake of calcium into intracellular organelles such as sarcoplasmic reticulum and mitochondria, and sodium calcium exchange (RODGER 1988).

I. Calcium Efflux Pump

The calcium efflux pump was first described in erythrocytes (SCHATZMANN 1966) and a similar enzyme has since been demonstrated in the plasma membrane of most cells including smooth muscle (HOGABOOM and FEDAN 1981). It is a magnesium and ATP-dependent enzyme which extrudes one calcium ion for two hydrogen ions and is thus electroneutral. Several regulatory mechanisms control the activity of this enzyme. It is stimulated by calcium-calmodulin complex and, as this is formed in response to a rise in intracellular calcium, it may act as a feedback mechanism to limit the rise in intracellular calcium caused by a given stimulus (HOGABOOM and FEDAN 1981; VINCENZI et al. 1980; WAISMAN et al. 1981). The activity of the calcium efflux pump depends on the activity of cyclic AMP dependent protein kinases in several types of smooth muscle (Daniel et al. 1983) but the only study to date in airway smooth muscle failed to show this (SANDS and MASCALI 1978).

II. Sodium-Calcium Exchange

The sodium-calcium exchanger transports one calcium ion for three sodium ions and is driven by the sodium gradient across the membrane, which in turn is maintained by the activity of Na/K ATPase (BLAUSTEIN 1974). There is good evidence for the existence of sodium-calcium exchange in cardiac muscle (SMITH et al. 1984), but the presence of Na-Ca exchange in some smooth muscle is controversial (BRADING and LATEGAN 1985) and relatively little is known of this process in airway smooth muscle. Such a mechanism probably exists in the airways since reducing external sodium causes contraction of canine (ITO and INOUE 1989), bovine (BULLOCK et al. 1981; KNOX et al. 1990b), and human (CHIDECKEL et al. 1987) airway smooth muscle in vitro. A study in bovine tracheal smooth muscle membranes has suggested that Na/Ca exchange sites are abundant in airway smooth muscle (SLAUGHTER et al. 1987). The relative contribution of the calcium efflux pump and the Na-Ca exchanger to calcium efflux in airway smooth muscle has still to be established.

III. Intracellular Calcium Removal Mechanisms

The intracellular organelles which are thought to be important for the removal of calcium are the sarcoplasmic reticulum and the mitochondria.

Studies in microsomal fractions of airway smooth muscle suggest that the sarcoplasmic reticulum accumulates calcium via a cAMP dependent mechanism (GROVER et al. 1980; HOGABOOM and FEDAN 1981; SANDS and MASCALI 1978). The mitochondria take up calcium and store it as a non-ionic calcium-phosphate complex and release it back into the cytosol via a calcium efflux pathway (BORLE 1981). The mitochondrial calcium store does not appear to have a physiological role but may protect the cell against calcium overload in pathological conditions (RODGER 1988).

IV. Calcium During Relaxation

The relationship between free intracellular calcium concentration and airway smooth muscle relaxation has not been studied as extensively as the relationship with contraction. The studies which have been done have concentrated on cAMP mediated pathways and have produced conflicting results. In canine trachealis contracted with 5-hydroxytryptamine or cholinergic agonists, the relaxation which accompanied isoproterenol was preceded by a fall in intracellular calcium (measured by aequorin luminescence) (GUNST and BANDYOPADHYAY 1989). The fall in intracellular calcium correlated strongly with the reduction in tension suggesting that it was responsible for the reduction. In bovine tracheal smooth muscle however there was no fall in aequorin luminescence in response to a β-agonist (TAKUWA et al. 1988) but this may be because the experiments were carried out at 22°C when many intracellular processes including Na/K ATPase would be inhibited.

E. Effect of Drugs on Airway Smooth Muscle Function

There are several possible sites where pharmacological agents can act to modify airway smooth muscle function (Fig. 4).

1. Surface receptors.
2. G proteins, intracellular second messengers and contractile proteins.
3. Cell membrane ion transport.

I. Drugs Acting Via Smooth Muscle Receptors

1. Adrenergic Agents

a) α-Adrenergic Agonists/Antagonists

Sympathetic innervation to the airways is very much less than cholinergic innervation in man and provides only a small percentage of the total nerve supply (PARTENAN et al. 1982). Despite this adrenergic receptors are abundant as shown by autoradiographic studies of human lungs (BARNES et al. 1980).

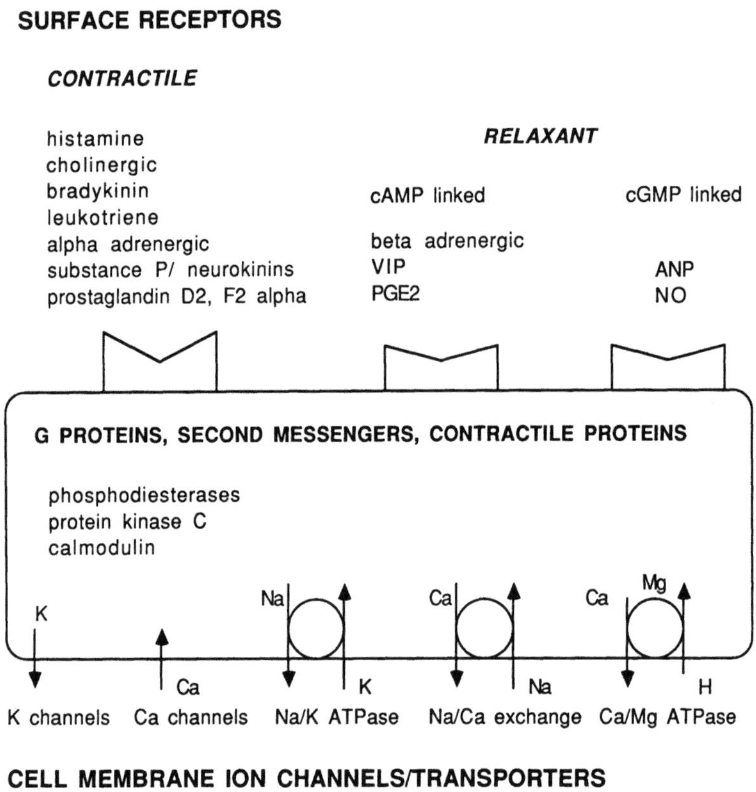

Fig. 4. Diagram showing the different sites that pharamacological agents can act on to modify the contractile process in airway smooth muscle

α-Adrenergic receptors are relatively sparse compared to β-receptors and more abundant in bronchioles than in large airways (BARNES et al. 1983a,b). They have been classified as α_1-and α_2-receptors based on the relative potencies of a series of agonists and antagonists (HOFFMAN and LEFKOWITZ 1980) and subtypes of both classes of receptors have been identified more recently (HARRISON et al. 1991). Although α_2-receptors predominate over α_1-receptors in canine trachealis in a ratio of 5:1 (BARNES et al. 1983a) noradrenaline-induced contraction is inhibited by both α_1- and α_2-receptor antagonists suggesting a functional role for both receptor subtypes (LEFF and MUNOZ 1981).

When studied on human airway smooth muscle in vitro non-selective adrenoceptor agonists such as adrenaline, noradrenaline and phenylephrine cause contraction when the β-receptors are blocked with a drug such as propranolol (MATHE et al. 1971; SIMONNSON et al. 1972; KNEUSSL and RICHARDSON 1978) and phentolamine, an α-blocker, causes relaxation (MATHE et al. 1971). The contraction produced by α-agonists is relatively weak, however, being only 10%–20% of the maximum response elicited by acetylcholine or histamine. The mechanism of α-adrenergic contraction has not been studied extensively, though α-receptors may be coupled to inositol phospholipid hydrolysis since noradrenaline produces a small increase in $InsP_3$ in bovine tracheal smooth muscle which can be blocked by α-blocking drugs (HALL and HILL 1988).

In patients with asthma α-agonists have usually produced no change in airway calibre even with doses which have marked cardiovascular effects (LARSSON et al. 1986). Bronchoconstriction occurred with methoxamine, but this may be due to stimulation of α-receptors on cells other than airway smooth muscle since it is inhibited by ipratropium bromide and sodium cromoglycate (BLACK et al. 1985). α_2-Receptors are present on sympathetic nerve endings where they have a negative feedback function.

Several drugs with α-blocking activity have been studied in man including drugs with predominantly α_1-blocking activity, phenoxybenzamine, phentolamine, indoramin, thymoxamine and prazosin, and the α_2-blockers midaglizole and clonidine. Some caution is needed in interpreting these studies since some α-blocking drugs lack specificity and they often have other effects such as antihistamine activity. The older α_1-antagonists caused slight bronchodilatation in some studies and some protection against histamine, exercise and allergen-induced bronchoconstriction (BIANCO et al. 1974; KERR et al. 1970; PATEL and KERR 1975). They had little or, more often, no beneficial effect when given regularly to subjects with asthma (DYSON et al. 1980) and the same is true for the more selective α_1-antagonist prazosin after both short term (BARNES et al. 1981a) and long term (BAUDOUIN et al. 1988) treatment. Midaglizole, an α_2-selective antagonist, caused some bronchodilatation in patients with moderate to severe asthma although the changes were relatively small (YOSHIE et al. 1989). Clonidine caused some reduction in the response to allergen when given by inhalation (LINDGREN et al. 1986) but the response to histamine was increased following oral administration (DINH XUAN et al. 1988). Some of the effects of α-adrenergic agonists and antagonists in man are likely to be due to central nervous system effects or an effect on cells other than airway smooth muscle but the extent to which this occurs is difficult to determine. Side effects have been a problem with α-blocking drugs, particularly postural hypotension and drowsiness. It seems unlikely that α-blockers will develop a therapeutic place in the treatment of airflow obstruction.

b) β-Adrenergic Agonists/Antagonists

β-Receptors were classified into β_1- and β_2-receptors by LANDS et al. in 1967. It has been shown more recently that many cells possess an atypical β-receptor which has been classified as β_3 (EMORINE et al. 1989; ZAAGSMA and NAHORSKI 1990). The distribution of β-adrenoceptors within the airways has been studied in several species including man. Smooth muscle contains mainly β_2-receptors, whereas the terminal airways and alveoli contain a mixture of β_2 and β_1 (CARSWELL and NAHORSKI 1983; GOLDIE et al. 1986a). The density of β_2-receptors in airway smooth muscle increases progressively from the trachea to small bronchioles (CARSTAIRS et al. 1985). There are no reports as yet of β_3-receptors in lungs. Data on β-receptor localisation in asthmatic airways are limited but numbers and distribution appear to be similar to non-asthmatic airways (SPINA et al. 1989; SHARMA and JEFFERY 1990). β_2-Receptors are also found on parasympathetic ganglia in animals and probably in man and may modulate cholinergic transmission to the airways (RHODEN et al. 1988).

β_2-Receptors are the most important receptors for pharmacological relaxation of airway smooth muscle. Despite their large number, the role of β-adrenoceptors in normal man appears to be small. Large doses of β-adrenoceptor antagonists such as propranolol cause no detectable change in airway function or airway responsiveness to constrictor stimuli in most normal subjects (TATTERSFIELD et al. 1973; ZAID and BEALL 1966). In contrast to non-asthmatic subjects β-adrenoceptor antagonists can cause marked bronchoconstriction in patients with asthma, implying that asthmatic airways are being dilated tonically by β-adrenoceptor stimulation. Since sympathetic innervation of airway smooth muscle is relatively sparse the stimulus is usually presumed to be circulating catecholamines, though levels of circulating adrenaline are not increased in patients with asthma (IND et al. 1985) nor have asthmatic airways been shown to be more sensitive to β-agonists than non-asthmatic airways (TATTERSFIELD et al. 1983). An alternative suggestion is that β-blocking drugs block inhibitory β_2-receptors on parasympathetic ganglia, thus increasing cholinergic traffic and stimulation of airway smooth muscle (BARNES 1989). There is some evidence for this since anti-muscarinic drugs inhibit the bronchoconstrictor response to β-blocking drugs in animals and man (RHODEN et al. 1988; IND et al. 1989).

The β_2-receptor is coupled through an intermediary stimulatory G protein (G_s) to adenylate cyclase. Adenylate cyclase can be inhibited through muscarinic M_2 receptors via an inhibitory G protein (G_i). Adenylate cyclase catalyses the intracellular conversion of ATP to cyclic 3′5′AMP. Intracellular [cAMP] causes relaxation of airway smooth muscle by various mechanisms including activation of cAMP-dependent protein kinases. These kinases phosphorylate intracellular proteins to activate several relaxant mechanisms (see relaxant mechanisms).

β-Agonists are potent relaxants of airway smooth muscle in several species including man. Isoprenaline relaxes bronchial smooth muscle with EC_{50} values in the range 10^{-9}–$10^{-7}\,M$. When tissue is preconstricted with contractile agents such as histamine or acetylcholine isoprenaline is less potent with EC_{50} values ranging from 10^{-7} to $10^{-3}\,M$ (GOLDIE et al. 1986b). Some β_2-selective agonists such as salbutamol are partial agonists when compared to isoprenaline in vitro. Receptor desensitisation occurs with prolonged exposure to high concentrations of β-receptors via several mechanisms including the relatively slow process of β-receptor down-regulation and the more rapid process of β-adrenoceptor phosphorylation by β-adrenoceptor kinase (βARK) and protein kinase A (HAUSDORFF et al. 1990). Receptor desensitisation is associated with reduced intracellular second messenger in response to β-agonist stimulation.

Two new β_2-adrenoceptor agonists formoterol and salmeterol produce more potent and prolonged relaxation of guinea pig trachea and human bronchial smooth muscle in vitro than drugs such as salbutamol (JOHNSON 1990). The longer action of salmeterol may be due to its large non-polar N-substituent which is thought to interact with non-polar groups on the receptor protein; formoterol has a pyridine nucleus which gives it a higher receptor affinity. Both drugs are more lipophilic than salbutamol and non-specific binding to the cell membrane may also prolong their action. Salmeterol is a partial agonist compared to both isoprenaline and salbutamol (BALL et al.1991).

In vivo, β_2-agonists cause bronchodilatation in normal man and to a greater extent in subjects with airflow obstruction. Since β_2-agonists have actions other than those on airway smooth muscle (reduction of mediator release from inflammatory cells, reduced cholinergic transmission, increased mucociliary clearance, decreased vascular permeability) their effects in vivo cannot be attributed solely to an effect on airway smooth muscle, though this is likely to be largely responsible for the acute effects. The duration of bronchodilatation varies according to the type of β-agonist but lasts for up to 12 h with salmeterol and formoterol (ULLMAN and SVEDMYR 1988; CAMPOS-GONGORA et al. 1991).

In addition to causing bronchodilatation β_2-agonists cause a reduction in the response to a wide range of constrictor stimuli, including exercise and inhaled histamine, methacholine and dry air. For example, the acute administration of a β-agonist by inhalation will allow between two and four more doubling doses of histamine to be given before the FEV_1 falls by 20% (TATTERSFIELD 1987). This action of β-agonists is analagous to the functional antagonism seen in vitro and is presumably a direct effect on airway smooth muscle since it occurs rapidly. The inhibitory effect of β_2-agonists on the early response to allergen is probably due to actions on inflammatory cells rather than airway smooth muscle, whereas functional antagonism probably accounts for most of the inhibition of the late response seen with the long acting β_2-agonists and with high doses of the shorter acting drugs (TWENTYMAN et al. 1990, 1991, 1992).

Following prolonged treatment with β-agonists there is some evidence that tolerance may develop. The bronchodilator response to β-agonists is usually maintained but the protection afforded by β-agonists against bronchoconstrictor stimuli is reduced following regular treatment with both short-acting and long-acting β_2-agonists (VATHENEN et al. 1988; CHEUNG et al. 1992). Bronchial reactivity is also increased after cessation of β-agonist treatment (KERREBIJN et al. 1987; VATHENEN et al. 1988; WAHEDNA et al.

1992) as would then be expected with receptor desensitisation since endogenous sympathetic stimulation would be less effective. These effects have been associated with a fall in FEV_1 (WAHEDNA et al. 1993) and slight clinical deterioration in some studies (SEARS et al. 1990). Despite these concerns about the long term use of β-agonists regularly the drugs are very widely used to treat asthma and chronic obstructive lung disease and are the most effective bronchodilators for acute attacks of asthma.

2. Muscarinic Agents

In contrast to the sparse sympathetic innervation of airway smooth muscle, parasympathetic nerves are abundant (RICHARDSON 1979; BARNES 1986). The main neural constrictor fibres to airway smooth muscle come from the parasympathetic nervous system via the vagus nerve and parasympathetic ganglia. Post-ganglionic fibres pass to airway smooth muscle where release of acetylcholine from the nerve endings causes airway smooth muscle contraction via post-junctional muscarinic cholinergic receptors (BARNES 1987). It has long been recognised that muscarinic receptor agonists such as acetylcholine, carbachol and methacholine constrict airway smooth muscle in vitro and in vivo (WIDDICOMBE 1963) and anticholinergic drugs such as the atropine derivatives have been used to treat asthma and obstructive lung diseases for several centuries (GANDEVIA 1975).

Five distinct subtypes of muscarinic receptors, termed M_1 to M_5, have been sequenced, cloned and expressed recently, and selective inhibitors of M_1, M_2 and M_3 receptors are now available (BARNES 1989). Pirenzepine selectively inhibits M_1 receptors, gallamine, AF-DX116 and methoctramine selectively inhibit M_2 receptors and 4 diphenylacetoxy-*N*-methylpiperidine methiodide (4-DAMP) and hexahydrosiladifenidol selectively inhibit M_3 receptors (MITCHELSON 1988). Muscarinic agonists with high selectivity for specific muscarinic receptors are not yet available though McN-A-343 and pilocarpine are relatively selective for M_1 and M_2 receptors respectively (BARNES 1989) (Table 1).

Autoradiographic, molecular biological and functional studies have been used to study muscarinic receptor subtypes in airway smooth muscle.

Table 1. Muscarinic receptor subtypes

Type	Site	Inhibitors
M_1	Ganglionic	Pirenzipine
M_2	Cholinergic nerves	Gallamine AF-DX 116 Methoctramine
M_3	Airway smooth muscle	4-DAMP Hexahydrosiladifenidol

Receptor binding studies using ^{3}H-quinclidinyl benzilate (QNB) have shown marked differences in muscarinic receptor distribution in different species, the receptors being equally prominent in peripheral and central human airways (MAK and BARNES 1990), but showing predominant binding to central airways in small animals such as the ferret (BARNES et al. 1983a). Autoradiographic studies in human and guinea pig lung have looked at ^{3}H-QNB labelling in the presence of unlabelled M_1, M_2 and M_3 antagonists. ^{3}H-QNB binding was inhibited in guinea pig airways by both methoctramine and 4-DAMP, indicating the presence of both M_2 and M_3 receptors. Only 4-DAMP prevented ^{3}H-QNB binding in human airways, however, suggesting that human airway smooth muscle contains only M_3 receptors (MAK and BARNES 1990). However, mRNA for both M_2 and M_3 receptors has been demonstrated in both porcine and human airway smooth muscle using cDNA probes (MAEDA et al. 1988; MAK et al. 1992). In functional studies the M_3 antagonists have a larger effect than the M_2 antagonists AF-DX 116 and methoctramine on contraction in bovine trachea (ROFFEL et al. 1988) and the M_3 receptor antagonists 4-DAMP and hexahydrosiladifenidol are potent relaxants in human airway smooth muscle (ROFFEL et al. 1990). Taken together these studies suggest that the receptor mediating contraction in human airway smooth muscle is predominantly M_3 although M_2 receptors are also present on airway smooth muscle from several animals. Studies looking at signal transduction pathways in airway smooth muscle for the different muscarinic receptor subtypes suggest that the M_3 receptor is coupled to phosphoinositide hydrolysis and is thus the receptor responsible for the contractile effects of acetylcholine. In canine trachealis an M_2 receptor inhibits adenylate cyclase via an inhibitory protein G_i (SANKARY et al. 1988; SCHAEFER et al. 1992).

Muscarinic receptors are found on other airway structures in addition to airway smooth muscle and these may affect the response to muscarinic agents in vivo. They include M_1 receptors in parasympathetic ganglia and alveolar walls and auto-inhibitory M_2 receptors on parasympathetic nerve endings, stimulation of which reduces the output of acetylcholine from parasympathetic nerve endings (MAK and BARNES 1990).

The antimuscarinic drugs used to treat obstructive lung disease have included atropine, ipratropium bromide and oxitropium bromide. When inhaled they cause bronchodilatation in normal subjects and in subjects with chronic obstructive airways disease and asthma (GROSS 1988). Their onset of action when given by inhalation is slower than that of the β-adrenergic agonists such as salbutamol (the peak effect occurring after 30–90 min), but they have a longer duration of action. There is some evidence that intravenous atropine may produce more complete cholinergic blockade than inhaled ipratropium bromide (MORRISON and PEARSON 1989). Although antimuscarinic drugs cause some protection against bronchoconstrictor stimuli such as exercise, histamine and cold air (GROSS 1988), the degree of protection is small (TATTERSFIELD 1987) and much smaller than that seen with β-adrenergic agonists when given in doses that produce the same amount of bronchodilatation (BRITTON et al. 1988).

Atropine, ipratropium and oxitropium bromide are non-selective muscarinic antagonists, blocking all three muscarinic receptor subtypes in the lung. Specific M_3 receptor antagonists would theoretically be more useful for the treatment of asthma since they would not have the unwanted effect of increasing release of acetylcholine from cholinergic nerve terminals (an M_2 mediated effect which could potentiate bronchoconstriction). Few studies have looked at the effect of more specific antagonists of muscarinic receptors subtypes in vivo. Pirenzipine an M_1 antagonist caused bronchodilatation in asthmatic subjects in one study (MORRISON and PEARSON 1988). High doses of pirenzipine will antagonise M_3 receptors, however, but there is some evidence that it can inhibit vagal activity in man when given in doses that do not cause M_3 receptor inhibition (LAMMERS et al. 1989).

Following regular treatment with ipratropium bromide the response to methacholine was increased in a study by NEWCOMB et al. (1985), in keeping with up-regulation of muscarinic receptors following chronic receptor blockade. If responsiveness to endogenous acetylcholine is increased following regular treatment with antimuscarinic drugs asthma would be expected to deteriorate; we have recently shown a small fall in FEV_1 following treatment with ipratropium bromide (P. Wilding, personal communication).

3. NANC System Neurotransmitters

Non-adrenergic, non-cholinergic inhibitory fibres which cause airway smooth muscle relaxation have been demonstrated in vitro using electrical field stimulation after adrenergic and cholinergic blockade in a variety of species including man (RICHARDSON 1981). This has been confirmed in animals in vivo (IRVIN et al. 1980) and appears to be present in man since mechanical and chemical stimulation of the larynx causes non-adrenergic bronchodilatation (LAMMERS et al. 1988). Whether or not the NANC inhibitory system has a physiological role has not been established. Some species, notably the guinea pig, have a NANC excitatory component. Electrical field stimulation in guinea pig bronchi and trachea in vitro and stimulation of the vagus in vivo causes bronchoconstriction, part of which is not inhibited by atropine (ANDERSON and GRUNDSTROM 1983).

a) Inhibitory

The main candidates proposed for the neurotransmitter mediating NANC relaxation are vasoactive intestinal peptides (VIP), the related peptide histine isoleucine (PHI) and more recently nitric oxide. Airway smooth muscle from several species including man is innervated by nerve fibres which show VIP immunoreactivity (DEY et al. 1981) and VIP receptors have also been identified in airway smooth muscle using ^{125}I-VIP (ROBBERECHT et al. 1981). Autoradiography has shown that VIP receptors are predominantly in large rather than small airways (Carstairs and Barnes 1986a). VIP relaxes airway smooth muscle from several species in vitro including human airways and is the most potent endogenous bronchodilator, being 100 times more potent than isoprenaline on human bronchi (Palmer et al. 1986). Studies using VIP in vivo have been less promising. Inhaled VIP has no bronchodilator effect in man and provides only small protection against histamine induced constriction (Barnes and Dixon 1984). The lack of effect of inhaled VIP may be due to enzymatic degradation of the peptide or diffusion problems (Palmer et al. 1986). Intravenous VIP also had little effect in normal and asthmatic subjects (Morice et al. 1983), but the dose which can be given by this route is limited by cardiovascular effects (Barnes 1988b).

Peptide histidine isoleucine (PHI) and its human equivalent peptide histidine methionine (PHM) share structural similarities with VIP and also share immunocytochemical distribution in the lung with VIP. PHI is a potent relaxant of human bronchi in vitro (Palmer et al. 1986).

More recently studies in pig and human airways have suggested that nitric oxide is a mediator of NANC responses (Kannan and Johnson 1992; Belvisi et al. 1992a) as NO synthetase inhibitors markedly inhibit NANC induced relaxant responses.

b) Excitatory

Although the evidence for an excitatory NANC system in the guinea pig is fairly compelling, a role in man has not been established. Substance P and the neurokinins are thought to be the neurotransmitters of the non-cholinergic excitatory nerves. Substance P has been shown to be localised to nerves in the airways of several species including humans (Lundberg et al. 1984) although one study failed to demonstrate substance P in human airways (Laitinen et al. 1983). Substance P and the other related neurokinins, neurokinin A and B, act on specific tachykinin receptors. These are found in high density in airway smooth muscle from trachea down to small bronchioles (Carstairs and Barnes 1986b). There are at least three types of tachykinin receptors NK1, NK2 and NK3.

Substance P and the neurokinins contract airway smooth muscle in vitro in several species including man (Casale 1991). Neurokinin A is more potent than substance P in guinea pig and human airways in vitro, whereas neurokinin B has no effect in human airways. Substance P causes bronchoconstriction in animals in vivo but has little effect in man (Casale 1991). The lack of effect in man may be due to the fact that cardiovascular actions of the peptide limit the dose that can be given. In contrast neurokinin A causes bronchoconstriction in asthmatic subjects when given either by inhalation (Joos et al. 1987) or by intravenous infusion (Clarke et al. 1987). Calcitonin gene related peptide (CGRP) is another constrictor neuropeptide which is localised to human airway nerves (Palmer et al. 1987) and is often co-localised with substance P. It contracts human airways in vitro, being more potent than substance P and probably acts through specific CGRP receptors. Several other neuropeptides have been identified in lung such as neuropeptide Y, gallolin and gastrin releasing peptide, but their functions are as yet uncertain.

The tachykinins are broken down by angiotensin converting enzyme and neutral endopeptidase (NEP). NEP inhibitors potentiate the contractile effects of the tachykinins in airway smooth muscle in vitro (BLACK et al. 1988). Neurogenic inflammation may play a role in asthma, since loss of airway epithelium could expose sensory nerve endings, stimulation of which could cause bronchoconstriction via an axon reflex.

Analogues of vasoactive intestinal peptide could potentially act as bronchodilators, but none are available as yet. Tachykinin receptor antagonists might play a possible role as bronchodilators, but those developed so far have lacked specificity, have only weak activity or being peptides, may be rapidly degraded (BARNES et al. 1990). Whether more effective and specific drugs will have significant bronchodilator activity in man depends on the importance of the NANC system in mediating bronchoconstriction in asthma. This is undetermined as yet.

4. Agents Acting on Guanylate Cyclase

The relaxant effects of agents which elevate cGMP have been studied on airway smooth muscle in vitro. Nitrates and sodium nitroprusside relax canine and bovine and atrial natriuretic peptide relaxes guinea pig and bovine airway smooth muscle (KATSUKI and MURAD 1976; GRUETTER et al. 1989; O'DONNELL et al. 1985; ISHII and MURAD 1989). Studies looking at the effects of the nitrates in vivo have been disappointing (MILLER and SHULTZ 1979), although recent studies with atrial natriuretic peptide have shown it to be a bronchodilator and protect against induced bronchoconstriction in asthmatic subjects (HULKS et al. 1991). These studies, when taken in conjunction with recent evidence suggesting that nitric oxide may be the neurotransmitter of NANC inhibitory nerves, may stimulate renewed interest in cGMP as a relaxant mediator. Whether agents which act through cGMP can act in a synergistic fashion to increase the maximal bronchodilator effect of agents acting through cAMP such as the β-adrenergic agonists has not been established.

5. Inflammatory Mediators/Epithelial Products

a) Histamine

Histamine is a potent contractor of airway smooth muscle from several species including man in vitro (WIDDICOMBE 1963) and an important constrictor agent in vivo (COCKROFT 1985). The functional response to histamine varies in different parts in the tracheobronchial tree. In guinea pig airways histamine and cholinergic agonists are approximately equipotent in the trachea but histamine is much more active in parenchymal strips (DRAZEN and SCHNEIDER 1978). In isolated human bronchioles, histamine and muscarinic agents are approximately equipotent (FINNEY et al. 1985). Histamine receptors can be divided into H_1, H_2 and H_3 receptors (ARRANG et al. 1983). The H_1 antagonists diphenhydramine and pyrilamine inhibit histamine induced contraction in human and guinea pig airway smooth muscle, suggesting that the receptor mediating the contractile effect of histamine on airway smooth muscle is the H_1 receptor. The mechanism of action of histamine on airway smooth muscle is via coupling of the H_1 receptor through a G protein to phospholipase C with stimulation of inositol phospholipid hydrolysis (HALL and HILL 1988). H_2 antagonists are without effect on contractile tone. Studies in vitro in human airways have shown that tachyphylaxis occurs to histamine and that this is mediated by an H_2 receptor resident on airway epithelium and is probably prostaglandin dependent (KNIGHT et al. 1992). H_3 receptors may modulate cholinergic and NANC neurotransmission in guinea pig and human airways (ICHINOSE and BARNES 1989a,b).

The results with H_1 and H_2 blockers in vitro have been mirrored in vivo in normal and asthmatic subjects (CASTERLINE and EVANS 1977; NOGRADY and BEVAN 1981). Antihistamines have proved relatively disappointing in the treatment of asthma, however (WHITE et al. 1987). Many of the older preparations caused sedation and systemic side effects. Recently terfenadine and astemazole have been developed which have less CNS sedative effect. The small effects of histamine antagonists are presumably due to the fact that histamine is only one of many inflammatory mediators in the asthmatic process.

b) Bradykinin

Three distinct kinins have been isolated in human tissue, namely bradykinin, kallidin and Met-Lys-bradykinin (REGOLI and BARABE 1980). Bradykinin has different effects on airway smooth muscle in different species. It has no effect on rabbit, rat or dog airways, but causes constriction on isolated guinea pig tracheal rings and isolated rat lung (COLLIER 1963; REGOLI and BARABE 1980). In human bronchial rings bradykinin has inconsistent effects causing small or no contractions at high concentrations (SIMONSSON et al. 1973; FULLER et al. 1987). Inhaled bradykinin causes bronchoconstriction in asthmatics but not normal subjects in vivo (FULLER et al. 1987). Its mode of action as a bronchoconstrictor is uncertain, but its lack of effects in vitro suggests that it is not acting directly on smooth muscle.

c) Prostaglandins

In guinea pig airways, prostaglandins are responsible for basal tone and they modulate the response to a variety of contractile agonists (OREHEK et al. 1975). However, prostanoids are not responsible for resting tone of isolated human bronchioles as indomethacin, a prostaglandin synthetase inhibitor, has little effect on resting tone (BRINK et al. 1980). Studies looking at the effect of specific prostaglandins on isolated smooth muscle preparations have shown that PGD_2, $PGF_{2\alpha}$ and thromboxane A_2 are contractile agents whilst PGE_1 and prostacyclin are relaxants (BRINK 1988). In isolated human airways PGE_2 has variable effects, causing relaxation in some and contraction in others (GARDINER 1975; SWEATMAN and COLLIER 1968).

Studies looking at the effects of prostaglandins in vivo have often produced complex findings. This may be due to the balance of the effects of prostaglandins on different cell types in the airway. WALTERS and DAVIES (1982) demonstrated a biphasic response to aerosolised PGE_2 in normal subjects where an initial bronchoconstriction was followed by bronchodilatation. Both prostaglandin D_2 and prostaglandin $F_{2\alpha}$ cause bronchoconstriction in normal and in asthmatic subjects (HARDY et al. 1984; SZCZEKLIK et al. 1977). More recently it has been postulated that PGE_2 may have an important role as a negative modulator of inflammatory processes in the airway. PGE_2 inhibits inflammatory cell function and neurogenic bronchoconstriction in vitro (ITO et al. 1990). Inhaled PGE_2 provides marked protection against indirect bronchoconstrictor challenges in asthma (PAVORD et al. 1991a). In addition, evidence is accumulating to suggest that the diuretic agent frusemide, which has recently been shown to protect against several indirect constrictor challenges in asthma, may be acting in part by release of PGE_2 (PAVORD et al. 1991b). These studies have renewed interest in the protective role of prostaglandins in the airways.

Attempts have been made to develop prostaglandin analogues. Oral misoprostol, a PGE_1 analogue, has been shown to protect against methacholine induced constriction but has not been studied extensively. Thromboxane synthetase inhibitors and thromoxane A_2 receptor antagonists have small effects of bronchoconstrictor responsiveness but do not alter airway calibre (FUJIMARA et al. 1986, 1991).

d) Leukotrienes

Leukotriene B_4 is a potent chemotactic factor for leukocytes whereas the sulphidopeptide leukotrienes (LT) C_4, D_4 and E_4 are potent constrictors of airway smooth muscle from a wide range of species including man (DAHLEN et al. 1980; KRELL et al. 1981) and like other spasmogens act via inositol phospholipid pathways. Leukotrienes C_4, D_4 and E_4 are of similar potency in human airways in vitro and all are potent bronchoconstrictors in normal and asthmatic subjects following inhalation (DRAZEN and AUSTEN 1987). Considerable effort has been expended to develop sulphidopeptide leukotriene antagonists and lipoxygenase inhibitors for the treatment of asthma, the lipoxygenase inhibitors having the potential advantage of inhibiting the formation of leukotriene B_4 in addition to leukotrienes C_4, D_4 and E_4 (for review see KRELL 1989). The early oral leukotriene antagonists had relatively weak activity in inhibiting the bronchoconstrictor response to inhaled leukotrienes in man (BARNES et al. 1987) and proved to be disappointing when studied in patients with asthma, on antigen challenge (BRITTON et al. 1987) or in clinical measures of asthma (CLOUD et al. 1989). More active, orally available, leukotriene antagonists have been developed in the last few years including MK-571, ICI 204219, SK&F 104353 and RG 12525. These drugs cause much greater inhibition of the response to inhaled leukotrienes (SMITH et al. 1990; KIPS et al. 1991; WAHEDNA et al. 1991), reduce exercise-induced bronchoconstriction (MANNING et al. 1990; FINNERTY et al. 1992), the response to allergen (TAYLOR et al. 1991) and aspirin induced asthma (CHRISTIE et al. 1991). Several have shown evidence of clinical efficacy including bronchodilatation (GADDY et al. 1992; HUI et al. 1991). The 5-lipoxygenase inhibitors that have been developed include A-64077 and MK-886. A-64077 has been shown to reduce the response to cold air (ISRAEL et al. 1990) though the response to allergen was not inhibited significantly (HUI et al. 1991). Although more effective than the early leukotriene antagonists the question still remains as to whether a single mediator antagonist or inhibitor will prove to be as effective as drugs such as β-agonists or inhaled steroids.

e) Platelet Activating Factor

Platelet activating factor (PAF) is a potent bronchoconstrictor agent in animals and man in vivo and it produced a prolonged increased in airway responsiveness in vivo (CUSS et al. 1986; BARNES 1988a), though others have been unable to confirm this. The bronchoconstrictor response rapidly develops tachyphylaxis. The mechanism of bronchoconstriction is thought to be due to release of other spasmogens from the airway cells rather than a direct effect on airway smooth muscle (BARNES 1988). The broncho-constrictor response in man was inhibited by leukotriene antagonist (SPENCER et al. 1991). Recent studies with PAF antagonists in asthma have been disappointing.

II. Drugs Acting on Post-receptor Mechanisms

1. Phosphodiesterase Inhibitors

Theophylline and other xanthine derivatives have been used to treat obstructive lung disease for several years. These agents cause airway smooth muscle relaxation in vitro and bronchodilatation in vivo and, although they may have several cellular mechanisms of action, phosphodiesterase (PDE) inhibition would seem to be important in their smooth muscle relaxant effects (TORPHY and UNDEM 1991). Theophyllines have been shown to relax smooth muscle contracted with a wide range of spasmogens in vitro from several species including man (PERSSON and KARLSSON 1987). Theophylline

Table 2. Phosphodiesterase isoenzymes in airway smooth muscle

Isoenzymes	Species
I Ca/Calmodulin stimulated	Canine, bovine, human
II cGMP stimulated	Canine, bovine, human
III cGMP inhibited	Canine, guinea pig, human
IV cAMP specific	Canine, guinea pig, bovine, human
V cGMP specific	Canine, human

is about 1000 times less potent than isoprenaline in relaxing contraction due to moderate doses of carbachol in human bronchi (Svedmyr 1977). The value of theophylline is limited in vivo by its narrow therapeutic index which reflects the fact that it is a rather non-selective inhibitor of tissue phosphodiesterases.

Recent studies have identified five classes of phosphodiesterase isoenzymes which vary in their tissue distribution (Beavo and Reifsnyder 1990). Drugs which inhibit specific isoenzymes may therefore be relatively tissue specific in their effects. This has led to a search for selective isoenzyme inhibitors and has renewed interest in the potential role of phosphodiesterase inhibitors as bronchodilators. Studies of the distribution of the different phosphodiesterase isoenzymes (PDE) in airway smooth muscle in different animal species has shown several isoenzymes in all types of airway smooth muscle (Table 2). Drugs which selectively hydrolyse PDE types III, IV and V have been studies in airway smooth muscle in several species. Inhibitors of type III PDE (SK&F 94836, SK&F 94120) relax canine (Torphy et al. 1991), guinea pig (Bryson and Rodger 1987) and bovine trachealis (Hall et al. 1990). Inhibitors of the cAMP specific phosphodiesterase, PDE IV (RO20 – 1724, rolipram), have been shown to produce concentration dependent relaxation of canine (Torphy et al. 1991), bovine (Hall et al. 1990) and guinea pig (Harris et al. 1989) trachealis. Inhibitors of type V cGMP-specific phosphodiesterase (Zaprinast and SK&F 96231) potentiate nitroprusside-induced relaxation in canine (Torphy et al. 1991), but not bovine trachealis (Nicholson et al. 1990). These studies suggested that PDE type III, IV and V are important in canine and guinea pig airway smooth muscle but only types III and IV are important in bovine airway smooth muscle. Recent studies have shown that human airway smooth muscle contains PDE types II, III, IV and V and that inhibitors of type III, IV and V were all relaxants (Giembycz et al. 1992; Belvisi et al. 1992b).

In vivo, type III, IV and V phosphodiesterase inhibitors have bronchodilator activity in the guinea pig, but few studies have been performed with isoenzyme-selective inhibitors in man. Oral Zaprinast (a type V inhibitor) has little bronchodilator activity in asthma and there have been limited reports using type IV PDE inhibitors which have had small and variable effects (for review see Torphy and Undem 1991). Recently zardaverine, a type III and IV inhibitor, was shown to have bronchodilator properties (BRUNNEE et al. 1992).

2. Miscellaneous

There are several other post-receptor sites which drugs can interact with to modify the contractile and relaxant processes in airway smooth muscle. These include guanosine nucleotide regulatory proteins (G proteins), intracellular calcium stores and enzymes regulating the contractile process such as calmodulin and protein kinase C. Few agents which affect these processes have been studied. Several calmodulin antagonists are available but they generally lack specificity. Calmodulin antagonists have been shown to inhibit contraction in response to antigen in guinea pig trachea (BURKA 1984; NAGAI et al. 1985). The putative protein kinase C inhibitors H7 and staurosporin inhibit contractile responses to spasmogens in guinea pig (SOUHRADA and SOUHRADA 1989b) and bovine (KNOX et al. 1989b) airway smooth muscle but these drugs are rather non-specific and may have other effects such as inhibition of myosin light chain kinase.

Lithium, which inhibits myoinositol phosphate leading to depletion of membrane phospholipid pools, causes relaxation and protects against histamine induced contractions in guinea pig airways in vitro (VENUGOPALAN et al. 1985). In a recent study in subjects with mild asthma oral lithium protected against histamine induced bronchoconstriction (KNOX et al. 1992). This suggests that modification of inositol phospholipid turnover may modify airway responsiveness.

III. Drugs Acting on Cell Membrane Ion Transport

1. Calcium Channel Antagonists

The calcium channel antagonists which have been studied most extensively in airway smooth muscle are the inhibitors of voltage dependent calcium channels. Whole cell patch clamp studies have identified voltage-dependent calcium channels in canine (KOTLIKOFF 1987), guinea pig (HISADA et al. 1990) and human (MARTHAN et al. 1989) airway smooth muscle and single channel studies have characterised their properties in rat and human airway smooth muscle cells (WORLEY and KOTLIKOFF 1990).

The voltage-dependent calcium channel antagonists can be divided into the dihydropyridines (nifedipine, nicardipine, nisoldipine), the phenylalkylamines (verapamil, gallopamil) and the benzothiazepines (diltiazem). Verapamil, gallopamil, nicardipine and nifedipine have been shown to inhibit contractile responses to a variety of spasmogens activating receptor-operated pathways (histamine, cholinergic agonists, leukotreine D4) in addition to agents acting by depolarisation such as KCL and TEA, in airway smooth muscle from several species including man (FOSTER et al. 1983; BLACK et al.

1986; DRAZEN et al. 1983; WEICHMAN et al. 1983; ADVENIER et al. 1984). These drugs also reduce the response of sensitised airway smooth muscle to antigen in guinea pig (BURKA 1984) and human airway smooth muscle in vitro (HENDERSON et al. 1983). The order of potency of these drugs in airway smooth muscle is nifedipine > verapamil > diltiazem. Generally, the effects of these drugs have been modest and have required high drug concentrations. This may reflect the fact that the initiation of airway smooth muscle contraction is due to release of calcium from internal stores rather than calcium influx. Several studies have looked at the effects of calcium channel antagonists on airway calibre and responsiveness in normal and asthmatic subjects in vivo and these have been comprehensively reviewed by BARNES (1985). Calcium antagonists have shown relatively small or no effects on resting bronchomotor tone, small and inconsistent effects on the airway response to cold air, histamine, methacholine and antigen and a fairly small but consistent inhibitory effect on exercise induced bronchoconstriction. The maximum benefits seem to occur when drugs are given sublingually or by inhalation. All the available drugs are more potent relaxants of vascular than airway smooth muscle and vascular side effects would be limiting. More recently a novel receptor operated calcium channel has been shown to be responsible for calcium entry during the maintenance phase of the contractile response in cultured human airway cells (MURRAY and KOTLIKOFF 1991).

2. Potassium Channel Activators

Airway smooth muscle is electrically stable and does not normally exhibit slow wave activity or action potentials. Early studies showed that tetraethylammonium (TEA), a potassium channel blocker, could induce action potentials, slow wave activity and spontaneous contractions in airway smooth muscle suggesting that potassium channels might be responsible for its electrical stability (KIRKPATRICK 1975; STEPHENS et al. 1975; KROEGER and

Table 3. K^+ channels in airway smooth muscle

Type	Species identified	Inhibitors	Activators
Voltage sensitive	Dog, guinea pig, man	TEA Cs Ba	
Calcium activated	Dog, guinea pig, man, rabbit	TEA Charybdotoxin Ba	β-Agonists
Receptor operated	None identified to date		
ATP sensitive	Guinea pig, man	Glibenclamide and sulphonylureas	Cromakalim, pinacidil, nicorandil, diazoxide

STEPHENS 1975; SUZUKI et al. 1976). A diverse range of potassium channels including voltage sensitive channels, calcium activated channels, receptor operated channels and metabolically activated channels have been described in a wide variety of tissues (Table 3). The role of potassium channels in airway smooth muscle has only recently been studied.

a) Voltage Sensitive Channels

Voltage sensitive K^+ channels which have been identified in non-airway smooth muscle cells include delayed (outward) rectifier (BENHAM and BOLTON 1983), the "A" current (OKABE et al. 1987) and the inward rectifier (BENHAM et al. 1987; EDWARDS et al. 1988). These channels are activated by changes in membrane potential; cell depolarisation is accompanied by increased activity of the outward rectifier and cell hyperpolarisation activates the inward rectifier. The main function of voltage-sensitive K channels would seem to be to provide membrane stability. Recent studies using the whole cell patch clamp technique have identified voltage-sensitive delayed rectifier currents in canine, guinea pig and human airway smooth muscle (KOTLIKOFF 1987; HISADA et al. 1990; MARTHAN et al. 1989).

b) Calcium Activated Channels

Several calcium activated K^+ channels (K[Ca]) have been described in non-airway smooth muscle cells including the maxi-K^+ channel (BENHAM et al. 1985), and a small (INOUE et al. 1985) and medium conductance K^+ channel (INOUE et al. 1986). Large conductance calcium activated K^+ channels similar to the maxi-K^+ channels have been described in airway smooth muscle from several species including man (MCCANN and WELSH 1986; KOTLIKOFF 1990; MARTHAN et al. 1989; HISADA et al. 1990). These channels can be blocked by TEA, barium and charybdotoxin. Whole cell recordings in guinea pig airway smooth muscle have suggested two discrete calcium activated components of the outward K current which differ in their calcium sensitivity (HISADA et al. 1990). Calcium activated K^+ channels in isolated canine myocytes can be activated by methacholine probably in response to calcium release from internal stores (KOTLIKOFF 1990). β_2-Adrenergic agonists activate a K^+ (Ca) channel in rabbit trachea via cAMP-dependent protein kinase (KUME et al. 1989). Charybdotoxin, and inhibitor of the maxi-K^+ channels, blocks cAMP mediated relaxation of guinea pig and human airway smooth muscle (MURRAY et al. 1991; MIURA et al. 1992), confirming a role for these channels in cAMP mediated airway smooth muscle relaxation.

c) Receptor Operated Channels

Receptor operated channels which open or close in response to agonists such as acetylcholine have been described mainly in non-smooth muscle cells, and not in airway smooth muscle as yet. These channels can be linked to the receptor directly or via a G protein.

d) Metabolically Controlled Channels

There has been a great deal of interest in drugs which open the ATP-sensitive K^+ channel in airway smooth muscle. These channels open in response to a reduction in ATP and are blocked by the sulphonylurea group of drugs (glibenclamide, tolbutamide, chlorpropamide). Three K^+ channel activators, cromakalim, nicorandil and pinacidil, have been studied in airway smooth muscle in vitro. These agents increase 86-Rb efflux, cause cell membrane hyperpolarisation and relax spontaneous tone in guinea pig airways. They also reduce tone induced by histamine, prostaglandin D_2 and leukotriene C_4 in guinea pig airways and cholinergic tone in human but not guinea pig airways (TAYLOR et al. 1988; HALL and MACLAGAN 1988; BLACK et al. 1990). K^+ channel openers are more effective than β-adrenergic agonists at reducing spontaneous tone but less effective at reversing receptor mediated contractile tone.

K^+ channel activators relax airway smooth muscle by several mechanisms which includes membrane hyperpolarisation with reduction of calcium influx via voltage dependent calcium channels, activation of Na/K ATPase causing increased intracellular Na and Ca efflux via Na/Ca exchange and possibly by regulating intracellular calcium release (BLACK and BARNES 1990). They may therefore have greater therapeutic potential than the calcium antagonists which have so far proved disappointing in the management of airways disease.

Potassium channel activators have been studied in vivo in both guinea pig and man. Pinacidil and cromakalim given intravenously protect against 5-hydroxytryptamine induced bronchocontriction in guinea pigs and cromakalim provides weak protection against cholinergic bronchoconstriction in guinea pigs. Oral cromakalim has been shown to protect against histamine induced bronchoconstriction in normal subjects (BAIRD et al. 1988)·and asthmatic subjects (WILLIAMS et al. 1988) and to reduce nocturnal bronchoconstriction in asthmatic subjects (WILLIAMS et al. 1990). Side effects are a problem, however, particularly headache and postural hypotension. Increased specificity for airway smooth muscle is needed if the drugs are to find a role in clinical practice.

3. Others

Drugs which modify sodium transport can alter the contractile state of airway smooth muscle. Ouabain, a Na/K ATPase inhibitor, has been shown to cause contraction of airway smooth muscle from several species including man (CHIDECKEL et al. 1987; KNOX et al. 1990a). Inhibition of Na/K ATPase would increase intracellular sodium causing an increase in calcium influx via Na/Ca exchange. Ouabain induced contraction can be inhibited by nifedipine suggesting that it is partly due to influx of calcium via voltage dependent calcium channels possibly in addition to inhibition of Ca efflux via Na/Ca exchange (KNOX et al. 1990a). Amiloride, an inhibitor of several sodium transport processes, relaxes and inhibits receptor-operated contraction in canine and bovine trachealis and inhibits antigen induced responses in guinea pig airways in vitro (KRAMPETZ and BOSA 1988; KNOX et al. 1990a; SOUHRADA et al. 1988). Ouabain causes bronchoconstriction in guinea pigs, but neither ouabain nor amiloride alter airway calibre or histamine reactivity in asthmatic subjects when given by inhalation (KNOX et al. 1988, 1990b). Activators of Ca ATPase or Na/Ca exchange would be expected to act as airway smooth muscle relaxants by stimulating calcium extrusion across the cell membranes. No such agents are currently available.

F. Conclusion

The pharmacology of airway smooth muscle is complex due to its varied innervation, the wide range of receptors present on airway smooth muscle cells, and the complex post-receptor mechanisms imvolved in the contractile and relaxant responses. This provides a wide range of potential sites for pharmacological agents to modify smooth airway muscle contractility. Although the established relaxant drugs, β-adrenergic agonists and anticholinergic drugs, are currently the more useful, advances are being made in several areas. Modifications of the structures of β-adrenergic agonists have produced agents with a longer duration of action. Advances in molecular biology have enabled the identification of several receptor subtypes for many receptors. This will allow more precise targeting of pharmacological agents to the more important receptor subtypes in the future. In the case of muscarinic cholinergic receptors, the development of M_3 receptor antagonists would target the muscarinic receptor responsible for airway smooth muscle contraction. Identification of different isoenzymes of phosphodiesterase and the development of selective isoenzyme inhibitors may produce drugs which can exert tissue selectivity by acting on isoenzymes present in airway smooth muscle tissue but not other tissues. This might prevent some of the undesirable side effects of non-selective phosphodiesterase inhibitors such as theophyllines. There have also been several advances made in the development of specific mediator antagonists, in particular the leukotriene antagonists. It would seem unlikely, however, that one specific mediator antagonist would provide an important therapeutic advance in asthma where several mediators play an important pathophysiological role. Ion transport inhibitors have proved rather disappointing therapeutic agents to date. Although the calcium channel antagonists have relaxant properties in vitro they have not proved useful in vivo. The potassium channel activators have additional properties to inhibition of calcium entry via voltage-dependent calcium channels but greater specificity for airway smooth muscle is needed if they are to be useful relaxant agents. A number of other steps in the contractile and relaxant responses are open to targeting by pharmacological agents, including modulation of G proteins, the inositol phospholipid cascade and calcium release mechanisms, calmodulin and protein kinase C. There is also potential for drugs which facilitate calcium extrusion across the cell membrane by activating sodium-calcium exchange, sodium potassium ATPase or calcium magnesium ATPase. Airway smooth muscle pharmacology is an exciting and developing area for which the future holds much promise.

Acknowledgements. We thank Hilary Alexander for typing the manuscript.

References

Adelstein RS, Eisenberg E (1980) Regulation and kinetics of the actin-myosin-ATP interaction. Annu Rev Biochem 49:921–956

Advenier C, Cerrina J, Duroux P, Floch A, Renier A (1984) Effects of five different calcium antagonists on guinea pig isolated trachea. Br J Pharmacol 82:727–733

Ahmed F, Foster RW, Small RC, Weston AH (1984) Some features of the spasmogenic actions of acetylcholine and histamine in guinea-pig isolated trachealis. Br J Pharmacol 83:227–233

Anderson RGG, Grundstrom N (1983) The excitatory non-cholingergic, non-adrenergic nervous system of guinea pig airways. Eur J Respir Dis 64:141–157

Arrang JM, Garbarg M, Schwartz J-C (1983) Autoinhibition of brain histamine release mediated by a novel (H_3) class of histamine receptor. Nature 302:832–837

Baird A, Hamilton T, Richards D, Tasher T, Williams AJ (1988) Cromakalim, a potassium channel activator, inhibits histamine induced bronchoconstriction in healthy volunteers. Br J Clin Pharmacol 25:114

Ball DI, Brittain RT, Coleman RA et al. (1991) Salmeterol, a novel, long-acting β_2-adrenoceptor agonist: characterisation of pharmacological activity in vitro and in vivo. Br J Pharmacol 104:665–671

Barnes N, Piper PJ, Costello J (1987) The effect of an oral leukotriene antagonist L-649,923 on histamine and leukotriene D_4-induced bronchoconstriction in normal man. J Allergy Clin Immunol 79:816–821

Barnes PJ (1985) Clinical studies with calcium antagonists in asthma. Br J Clin Pharmacol 20:2895–2985

Barnes PJ (1986) Neural control of human airways in health and disease. Am Rev Respir Dis 134:1289–1314

Barnes PJ (1987) Cholinergic control of airway smooth muscle. Am Rev Respir Dis 136:S42–S45

Barnes PJ (1988a) Platelet-activating factor and asthma. J Allergy Clin Immunol 81:152–159

Barnes PJ (1988b) Neuropeptides and airway smooth muscle. Pharmacol Ther 36:119–129

Barnes PJ (1989) Muscarinic receptor subtypes: implications for lung disease: (Editorial). Thorax 44:161–167

Barnes PJ, Basbaum CB (1983) Mapping of adrenoceptors in the trachea by autoradiography. Exp Lung Res 4:183–192

Barnes PJ, Dixon CM (1984) The effect of inhaled vasoactive intestinal peptide on bronchial reactivity to histimine in humans. Am Rev Respir Dis 130:120–126

Barnes PJ, Karliner JS, Dollery CT (1980) Human lung adrenoceptors studied by radioligand binding. Clin Sci 50:457–461

Barnes PJ, Ind PW, Dollergy CT (1981a) Inhaled prazosin in asthma. Thorax 36:378–381

Barnes PJ, Wilson NM, Vickers H (1981b) Prazosin, an α_1-adenoceptor antagonist, partially inhibits exercise induced asthma. J Allergy Clin Immunol 68:411–415

Barnes PJ, Basbaum CB, Nadel JA (1983a) Autoradiographic localisation of autonomic receptors in airways smooth muscle: marked differences between large and small airways. Am Rev Respir Dis 127:758–762
Barnes PJ, Basbaum CB, Nadel JA, Roberts JM (1983b) Pulmonary alpha adrenoceptors: autoradiographic localisation using [^{3}H] Prozosin. Eur J Pharmacol 88:57–62
Barnes PJ, Belvisi MG, Rogers DF (1990) Modulation of neurogenic inflammation: novel approaches to inflammatory disease. TIPS: 11:185–189
Baudouin SV, Aitman TJ, Johnson AJ (1988) Prazosin in the treatment of chronic asthma. Thorax 43:385–387
Beavo JA, Reifsnyder DH (1990) Primary sequence of cyclic nucleotide phosphodiesterase isozymes and the design of selective inhibitors. TIPS 11:150–155
Belvisi MG, Stretton CD, Barnes PJ (1992a) Evidence that nitric oxide is the neurotransmitter of inhibitor NANC nerves in human airways. Eur J Pharmacol 210:221
Belvisi MG, Miura M, Peters MJ, Ward JK, Tadjkarimi S, Yacoub MH, Giembycz MA, Barnes PJ (1992b) Effect of isoenzyme-selective cyclic nucleotide phosphodiesterase inhibitors on human tracheal smooth muscle tone. Br J Pharmacol 107:53P
Benham CD, Bolton TB (1983) Patch clamp studies of slow potential sensitive potassium channels in longtitudinal smooth muscle of rabbit jejunum. J Physiol (Lond) 340:469–486
Benham CD, Bolton TB, Lang R, Takewaki T (1985) The mechanism of action of Ba and TEA on single Ca activated K channels in arterial and intestinal cell membranes. Pflugers Arch 403:120–127
Benham CD, Bolton TB, Denbigh JS, Lang R (1987) Inward rectification in freshly isolated single smooth muscle cells of the rabbit. J Physiol (Lond) 383:461–476
Berridge MJ, Irvine RF (1989) Inositol phosphates and cell signalling. Nature 341: 197–203
Bianco S, Griffin JP, Kamburoff PL, Prime FJ (1974) Prevention of exercise induced asthma by indoramin. Br Med J 4:18–20
Black JL, Barnes PJ (1990) Potassium channels and airway function: new therapeutic prospects. Thorax 45:213–218
Black JL, Vincenc K, Salome C (1985) Inhibition of methoxamine-induced bronchoconstriction by ipratropium bromide and disodium cromoglycate in asthmatic subjects. Br J Clin Pharmacol 20:41–46
Black JL, Armour C, Johnson P, Vincenc K (1986) The calcium dependence of histamine, carbachol and potassium chloride-induced contraction in human airways in vitro. Eur J Pharmacol 125:159–168
Black JL, Johnson PRA, Armour CL (1988) Potentiation of the contractile effects of neuropeptides in human bronchus by an enkephalinase inhibitor. Pulm Pharmacol 1:21–23
Black JL, Armour CL, Johnson PRA, Alouan LA, Barnes PJ (1990) The action of a potassium channel activator BRL 38227 (lemakalim) on human airway smooth muscle. Am Rev Respir Dis 142:1384–1389
Blaustein MP (1974) The interrelationship between sodium and calcium fluxes across cell membranes. Rev Physiol Biochem Pharmacol 70:33–82
Borle AB (1981) Control, modulation and regulation of cell calcium. Rev Physiol Biochem Pharmacol 90:13–153
Brading Af, Lategan TW (1985) Na-Ca exchange in vascular smooth musCle. J Hypertens 3:109–116
Brink C (1988) Prostaglandins. In: Barnes PJ, Rodger IW, Thomson NC (eds) Asthma: basic mechanisms and clinical management. Academic, London, p 203
Brink C, Grimand C, Guillot C, Orehek J (1980) The interaction between indomethacin and contractile agents on human isolated smooth muscle. Br J Pharmacol 69: 383–388

Britton JR, Hanley SP, Tattersfield AE (1987) The effect of an oral leukotriene D4 antagonist L-649,923 on the response to inhaled antigen in asthma. J Allergy Clin Immunol 79:811–816
Britton JR, Hanley SP, Garrett HV, Hadfield JW, Tattersfield AE (1988) Dose related effects of salbutamol and ipratropium bromide on airway calibre and reactivity in subjects with asthma. Thorax 43:300–305
Brunnee T, Engelstatter, Steinijans VW, Kunkel G (1992) Bronchodilatory effect of inhaled zardaverine, a phosphodiesterase III and IV inhibitor, in patients with asthma. Eur Respir J 5:982–985
Bryson SE, Rodger IW (1987) Effects of phosphodiesterase inhibitors on normal and chemically skinned isolated smooth muscle. Br J Pharmacol 92:673–681
Bullock CG, Fettes JJF, Kirkpatric CT (1981) Tracheal smooth muscle: second thoughts on sodium-calcium exchange. J Physiol (Lond) 318:46P
Burka JF (1984) Effects of calcium channel blockers and a calmodulin antagonist on contractions of guinea pig airways. Eur J Pharmacol 99:257–268
Campos-Gongora H, Wisniewski A, Tattersfield AE (1991) A single dose comparison of inhaled albuterol and two formulations of salmeterol on airway reactivity in asthmatic subjects. Am Rev Respir Dis 144:626–629
Carstairs JR, Barnes PJ (1986a) Visualisation of vasoactive intestinal peptide receptors in human and guinea-pig lung. J Pharmacol Exp Ther 239:249–255
Carstairs JR, Barnes PJ (1986b) Autoradiographic mapping of substance P receptors in lung. Eur J Pharmacol 127:295–296
Carstairs JR, Nimmo AJ, Barnes PJ (1985) Autoradiographic visualisation of β-adrenoceptor subtypes in human lung. Am Rev Respir Dis 132:541–547
Carswell H, Nahorski SR (1983) Beta-adrenoceptors in guinea-pig airways: comparison of functional and receptor labelling studies. Br J Pharmacol 79:965–971
Casale TB (1991) Neuropeptides and the lung. J Allergy Clin Immunol 88:1–14
Casterline CL, Evans R (1977) Further studies on the mechanism of human histamine-induced asthma. J Allergy Clin Immunol 58:607–612
Cheung D, Timmers MC, Zwinderman AH, Bel EH, Dijkman JH, Sterk PJ (1992) Long-term effects of a long-acting β_2-adrenoceptor agonist, salmeterol, on airway hyperresponsiveness in patients with mild asthma. N Engl J Med 327:1198–1203
Chideckel EW, Frost JL, Mike P, Fedan JS (1987) The effect of ouabain on tension in isolated respiratory tract smooth muscle of humans and other species. Br J Pharmacol 92:609–614
Chilvers ER, Nahorski SR (1990) Phosphoinositide metabolism in airway smooth muscle. Am Rev Respir Dis 141:S137–S140
Christie PE, Smith CM, Lee TH (1991) The potent and selective sulfidopeptide leukotriene antagonist, SK&F 104353, inhibits asprin-induced asthma. Am Rev Respir Dis 144:957–958
Clarke B, Evans TW, Dixon CMS, Conradson T-B, Barnes PJ (1987) Comparison of the cardiovascular and respiratory effects of substance P and neurokinin A in man (Abstr). Clin Sci 72:41P
Cloud ML, Enas GC, Kemp J et al. (1989) A specific LTD4/LTE4-receptor antagonist improves pulmonary function in patients with mild, chronic asthma. Am Rev Respir Dis 140:1336–1339
Coburn RF (1979) Electromechanical coupling in canine trachealis muscle: acetylcholine contractions. Am J Physiol 236:C177–C184
Coburn RF, Yamaguchi T (1977) Membrane potential-dependent and independent tension in canine trachealis muscle. J Pharmacol Exp Ther 201:276–284
Cockcroft DW (1985) Bronchial inhalation tests. I. Measurement of nonallergic bronchial responsiveness. Ann Allergy 55:527–534
Collier HOJ (1963) The action and antagonism of kinines on bronchioles. Ann NY Acad Sci 104:290–298
Cuss FM, Dixon CMS, Barnes PJ (1986) Effects of inhaled platelet-activating factor on pulmonary function and bronchial responsiveness in man. Lancet 2:189–192

Dahlen SE, Hedqvist P, Hammarstrom S, Samuelsson B (1980) Leukotriene are potent constrictors of human bronchi. Nature 229:484–486
Daniel EE, Grover AK, Kwan CY (1983) Calcium. In: Stephens NL (ed) Biochemistry of smooth muscle, vol 111. CRC, Boca Raton, pp 1–88
Dey Rd, Shannon WA, Said SI (1981) Localisation of VIP-immunoreactive nerves in airways and pulmonary vessels of dogs, cats and human subjects. Cell Tissue Res 220:231–238
Dinh Xuan AT, Regnard J, Matran R, Mantrand P, Advenior C, Lockhart A (1988) Effects of clonidine of bronchial responses to histamine in normal and asthmatic subjects. Eur Respir J 1:345–350
Dinh Xuan AT, Regnard J, Similowski T, Rey J, Marsac J, Lockhart A (1990) Effects of SK&F 104353, a leukotriene receptor antagonist, on the bronchial responses to histamine in subjects with asthma: A comparative study with terfenadine. J Allergy Clin Immunol 85:865–871
Drazen JM, Austen KF (1987) Leukotrienes and airway responses. Am Rev Respir Dis 136:985–998
Drazen JM, Schneider MW (1978) Comparative responses of tracheal spirals and parenchymal strips to histamine and carbachol in vitro. J Clin Invest 61: 1441–1447
Drazen JM, Fanta CH, Lacontre PG (1983) Effect of nifedipine on constriction of human tracheal strips in vitro. Br J Pharmacol 78:687–691
Dyson AJ, Hills EA, Mackay AD, Woods JB (1980) Is indoramin useful in the treatment of bronchial asthma? Br J Dis Chest 74:403–404
Edwards FR, Hirst GDS, Silverberg GD (1988) Inward rectification in rat cerebral arterioles; involvement of potassium ions in autoregulation. J Physiol (Lond) 404:455–466
Emorine LJ, Marullo S, Briend-Sutren MM et al. (1989) Molecular characterisation and the human beta 3-adenergic receptor. Science 245:1118–1121
Farley JM, Miles PR (1977) Role of depolarisation in acetyl-choline induced contractions of dog trachealis muscle. J Pharmacol Exp Ther 201:199–205
Felbel J, Trockur B, Ecker T, Landgraf W, Hofmann F (1988) Regulation of cytosolic calcium by cAMP and cGMP in freshly isolated smooth muscle cells from bovine trachea. J Biol Chem 263:16764–16771
Fesenko EE, Kolesnikov SS, Lyubarsky AL (1985) Induction by cyclic GMP of cationic conductance in plasma membrane of retinal rod outer segment. Nature 313:310–313
Finnerty JP, Wood-Baker R, Thomson H, Holgate ST (1992) Role of leukotrienes in exercise-induced asthma. Am Rev Respir Dis 145:746–749
Finney MJB, Karlsson JA, Persson CGA (1985) Effects of bronchoconstrictors and bronchodilators on a novel human small airway preparation. Br J Pharmacol 85:29–36
Foster RW, Small RC, Weston AH (1983) The spasmogenic action of potassium chloride on guinea-pig trachealis. Br J Pharmacol 80:553–559
Foster RW, Okpalugo MI, Small RC (1984) Antagonism of Ca and other actions of verapamil in guinea pig isolated trachea. Br J Pharmacol 81:499–507
Fujimura M, Sasaki F, Nakatsumi Y, Takahashi Y, Hifumi S, Tag K, Mifune J-I, Tanaka T, Matsuda T (1986) Effects of a thromboxane synthetase inhibitor (OXY-046) and a lipoxygenase inhibitor (AA-861) on bronchial responsiveness to acetylcholine in asthmatic subjects. Thorax 41:955–959
Fujimura M, Sakamoto S, Saito M, Miyake Y, Matsuda T (1991) Effect of a thromboxane A_2 receptor antagonist (AA-2414) on bronchial hyperresponsiveness to methacholine in subjects with asthma. J Allergy Clin Immunol 87:23–27
Fuller RW, Dixon CMS, Cuss FMC, Barnes PJ (1987) Bradykinin-induced bronchoconstriction in man: mode of action. Am Rev Respir Dis 135:176–181
Gaddy JS, Margolskee DJ, Bush RK, Williams VC, Busse WW (1992) Bronchodilatation with a potent and selective leukotriene D_4 (LTD_4) receptor antagonist (MK-571) in patients with asthma. Am Rev Respir Dis 146:358–363

Gandevia B (1975) Historical review of the use of parasympatholytic agents in the treatment of respiratory disorders. Postgrad Med J 51 [Suppl 7]:13–20
Gardiner PJ (1975) The effects of some natural prostaglandins on isolated human circular bronchial muscle. Prostaglandins 10:607–616
Giembycz MA, Belvisi MG, Miura M, Perkins RS, Kelly J, Tadjkarimi S, Yacoub MH, Barnes PJ (1992) Soluble cyclic nucleotide phosphodiesterase isoenzymes from human tracheal smooth muscle. Br J Pharmacol 107:52P
Goldie RG, Papidimitriou SM, Paterson SW et al. (1986a) Autoradiographic localisation of beta-adrenoceptors in pig lung using [125I]-idocyanopindolol. Br J Pharmacol 88:621–628
Goldie RG, Spina D, Henry PJ, Lulich KM, Paterson JW (1986b) In vitro responsiveness of human asthmatic bronchus to carbachol, histamine, b-adrenoceptor agonists and theophylline. Br J Clin Pharmacol 22:669
Grandordy BM, Cuss FM, Sampson AS, Palmer JB, Barnes PJ (1986) Phosphatidcholinergic agonists in airway smooth muscle: relationship to contraction and muscarinic receptor occupancy. J Pharmacol Exp Ther 238:273–279
Gross NJ (1988) Medical intelligence: ipratropium bromide. N Engl J Med 319: 486–494
Grover AK, Kannan MS, Daniel EE (1980) Canine trachealis membrane fractionation and characterisation. Cell Calcium 1:135–146
Gruetter CA, Childers CE, Bosserman MK, Lemke SM, Ball JG, Valentovic MA (1989) Comparison of relaxation induced by glyceryl trinitrate, isosorbide dinitrate, and sodium nitroprusside in bovine airways. Am Rev Respir Dis 139: 1192–1197
Gunst SS, Bandyopadhyay S (1989) Contractile force and intracellular Ca^{2+} during relaxation of canine tracheal smooth muscle. Am J Physiol 257:C355–C364
Gunst SS, Stropp JQ (1988) Effect of Na-K adenosinetriphosphatase activity on relaxation of canine tracheal smooth muscle. J Appl Physiol 64:635–641
Hall AK, MacLagan J (1988) Effect of cromakalim on cholinergic neurotransmission in the guinea-pig trachea (Abstr). 96:792P
Hall IP, Chilvers ER (1989) Inositol phosphates and airway smooth muscle. Pulm Pharmacol 2:113–120
Hall IP, Hill SJ (1988) B-adrenoceptor stimulation inhibits histamine-stimulated inositol phospholipid hydrolysis in bovine tracheal smooth muscle. Br J Pharmacol 95:1204–1212
Hall IP, Donaldson J, Hill SJ (1989) Inhibition of histamine stimulated inositol phospholipid hydrolysis by agents which increase cyclic AMP levels in bovine tracheal smooth muscle. Br J Pharmacol 97:603–613
Hall IP, Walker D, Hill SJ, Tattersfield AE (1990) Effect of isozyme selective phosphodiesterase inhibitors on bovine tracheal smooth muscle tone. Eur J Pharmacol 183:1096–1099
Hardy CC, Robinson C, Tattersfield AE, Holgate ST (1984) The bronchoconstrictor effect of inhaled prostaglandin D_2 in normal and asthmatic men. N Engl J Med 311:209–213
Harris AL, Connell MJ, Ferguson EW, Wallace AM, Gordon RJ, Pagani ED, Silver PJ (1989) Role of low Km cyclic AMP phospodiesterase inhibitors in tracheal relaxation and bronchodilatation in the guinea pig. J Pharmacol Exp Ther 251:199–206
Harrison JK, Pearson WR, Lynch KR (1991) Molecular characterisation of α_1- and α_2-adrenoceptors. Trends Pharmacol Sci 12:62–67
Hausdorff WP, Caron MG, Lefkowitz RJ (1990) Turning off the signal: desensitisation of β-adrenergic receptor function. FASEB J 4:2881–2889
Henderson AF, Heaton RW, Dunlop LS, Costello JF (1983) Effects of nifedipine on antigen-induced bronchoconstriction. Am Rev Respir Dis 127:549–553
Hisada T, Yoshihisa K, Tsuneaki S (1990) Properties of membrane current from guinea-pig trachea. Pflugers Arch 416:151–161

Hoffman BB, Lefkowitz RJ (1980) Alpha-adrenergic receptor subtypes. N Engl J Med 302:1390–1396
Hogaboom GK, Fedan JS (1981) Calmodulin stimulation of Ca uptake and Ca-Mg ATPase activities in microsomes from canine tracheal smooth muscle. Biochem Biophys Res Commun 99:737–744
Hogaboom GK, Emler CA, Butcher FR, Fedan JS (1982) Concerted phosphorylation of endogenous tracheal smooth muscle membrane proteins by Ca^{2+}-calmodulin-, cyclic GMP-, and cyclic AMP-dependent protein kinases. FEBS Lett 139:309–312
Hui KP, Barnes NC (1991) Lung function improvement in asthma with a cysteinyl-leukotriene receptor antagonist. Lancet 337:1062–1063
Hui KP, Taylor IK, Taylor GW, Rubin P, Kesterson J, Barnes NC, Barnes PJ (1991) Effect of a 5-lipoxygenase inhibitor on leukotriene generation and airway responses after allergen challenge in asthmatic patients. Thorax 46:184–189
Hulks G, Jardine AG, Connell JMC, Thomson NC (1991) Influence of elevated plasma levels of atrial natiuretic factor or bronchial reactivity in asthma. Am Rev Respir Dis 143:778–782
Ichinose M, Barnes PJ (1989a) Histamine H3-receptors modulate nonadrenergic noncholinergic neural bronchoconstriction in guinea pig in vivo. Eur J Pharmacol 174:49–55
Ichinose M, Barnes PJ (1989b) Inhibitory histamine H3-receptors on cholinergic nerves in human airways. Eur J Pharmacol 163:383–386
Ind PW, Causon RC, Brown MJ, Barnes PJ (1985) Circulating catecholamines in acute asthma. Br Med J 290:267–269
Ind PW, Dixon CMS, Fuller RW, Barnes PJ (1989) Anticholinergic blockade of β-blocker-induced bronchoconstriction. Am Rev Respir Dis 139:1390–1394
Inoue R, Kitamura K, Kuriyama H (1985) Two Ca-dependent K-channels by the application of tetraettyl ammonium dibutyrate to smooth muscle membranes of the rabbit portal vein. Pflugers Arch 405:173–179
Inoue R, Okabe K, Kitamura K, Kurilyama H (1986) A newly identified Ca dependent K channels in the smooth muscle membrane of single cells dispersed from the rabbit portal vein. Pflugers Arch 406:138–143
Irvin CG, Boileau R, Tremblay J, Martin RR, Macklem PT (1980) Bronchodilatation: Noncholingergic, nonadrenergic mediation demonstrated in vivo in the cat. Science 207:791–792
Ishii K, Murad F (1989) ANP relaxes bovine tracheal smooth muscle and increases cGMP. Am J Physiol 256:C495–C500
Israel E, Dermarkarian R, Rosenberg M, Sperling R, Taylor G, Rubin P, Drazen JM (1990) The effects of a 5-lipoxygenase inhibitor on asthma induced by cold, dry air. N Engl J Med 323:1740–1744
Ito I, Inoue T (1989) Contracture and change in membrane potential produced by sodium removal in the dog trachea and broniole. J Appl Physiol 67:2078–2086
Ito I, Suzuki H, Aizawa H, Hirose T, Hakoda H (1990) Pre-junctional inhibitory action of prostaglandin E_2 on excitatory neuro-effector transmission in the human bronchus. Prostaglandins 39:639–655
Johnson M (1990) The pharmacology of salmeterol. Lung Suppl:115–119
Joos G, Pauwels R, van der Straeten M (1987) Effect of inhaled substance P and neurokinin A on the airways of normal and asthmatic subjects. Thorax 42:779
Kannan MS, Johnson DE (1992) Nitric oxide mediates the neural nonadrenergic, noncholinergic relaxation of pig tracheal smooth muscle. Am J Physiol 262: L511–L514
Katsuki S, Murad F (1976) Regulation of adenosine cyclic 3′,5′-monophosphate and guanosine cyclic 3′,5′-monophosphate levels and contractility in bovine tracheal smooth muscle. Mol Pharmacol 13:330–341
Kerr KW, Gavindoraj M, Patel KR (1970) Effect of alpha receptor blocking drugs and disodium chromoglycate on histamine hypersensitivity in bronchial asthma. Br Med J 22:139–141

Kerrebijn KF, van Essen-Zandvliet EEM, Neijens HJ (1987) Effect of long term treatment with inhaled corticosteroids and beta agonists on the bronchial responsiveness in children with asthma. J Allergy Clin Immunol 79:653–659
Kips JC, Joos GF, de Lepeleire I, Margolskee DJ, Buntinx A, Pauwels RA, van der Straeten ME (1991) MK-571, a potent antagonist of leukotriene D_4-induced bronchoconstriction in the human. Am Rev Respir Dis 144:617–621
Kirkpatrick CT (1975) Excitation and contraction in bovine tracheal smooth muscle. J Physiol (Lond) 244:263–281
Kneussl MP, Richardson JB (1978) Alpha-adrenergic receptors in human and canine tracheal and bronchial smooth muscle. J Appl Physiol 45 (2):307–311
Knight DA, Stewart GA, Thompson PJ (1992) Histamine tachyphylaxis in human airway smooth muscle. Am Rev Respir Dis 146:137–140
Knox AJ, Brown JK (1991) Differential regulation of Na/K ATPase by the cyclic nucleotide dependent protein kinases in airway smooth muscle. Clin Res 39: 331A
Knox AJ, Tattersfield AE, Britton JR (1988) The effect of inhaled ouabain on bronchial reactivity to histamine in man. Br J Clin Pharmacol 25:758–760
Knox AJ, Ajao P, Cragoe EJ, Tattersfield AE (1989a) Sodium and calcium dependence of phorbol ester induced contraction of bovine airways. Eur Respir J 2 [Suppl 5]:295S
Knox AJ, Clark J, Tattersfield AE (1989b) The effect of protein kinase C inhibition on histamine and methacholine induced constriction of airway smooth muscle. Am Rev Respir Dis 139:A76
Knox AJ, Ajao P, Tattersfield AE (1989c) The effect of removal of external sodium on airway smooth muscle tone. Am Rev Respir Dis 139:A353
Knox AJ, Ajao P, Britton JR, Tattersfield AE (1990a) The effect of sodium transport inhibitors on airway smooth muscle contractility in vitro. Clin Sci 79: 315–323
Knox AJ, Britton JR, Tattersfield AE (1990b) The effect of sodium transport inhibitors on bronchial reactivity in vivo. Clin Sci 79:325–330
Knox AJ, Higgins BG, Hall IP, Tattersfield AE (1992) The effect of lithium on bronchial reactivity in asthma. Clin Sci 82:407–412
Kotlikoff MI (1987) Potassium currents in isolated airway smooth muscle cells. Am Rev Respir Dis 135:A74
Kotlikoff MI (1988) Calcium currents in isolated airway smooth muscle cells. Am J Physiol 254:C793–C801
Kotlikoff MI (1990) Potassium currents in canine airway smooth muscle cells. Am J Physiol 259:L394–L395
Kotlifoff MI, Murray RK, Reynolds EE (1987) Histamine-induced calcium release and phorbol antagonism in cultured airway smooth muscle cells. Am J Physiol 245:C561–C566
Krampetz IK, Bose R (1988) Relaxant effect of amiloride on canine tracheal smooth muscle. J Pharmacol Exp Ther 246:641–648
Krell RD (1989) The emergence of potent and selective peptide leukotriene receptor antagonists. Pulm Pharmacol 2:27–31
Krell RD, Osborn R, Vickery L, Falcone K, Gleason J (1981) Contraction of isolated airway smooth muscle by synthetic leukotrienes C_4 and D_4. Prostaglandins 22:387–409
Krell RD, Aharony D, Buckner CK, Keith RA, Kusner EJ, Snyder DW, Bernstein PR, Matassa VG, Yee YK, Brown FJ, Hesp B, Giles RE (1990) The preclinical pharmacology of ICI 204,219. Am Rev Respir Dis 141:978–987
Kroeger EA, Stephens NL (1975) Effect of tetraethylammonium on tonic airway smooth muscle: Initiation of phasic electrical activity. Am J Physiol 228:633–636
Kume H, Takai A, Tokuno H, Tomita T (1989) Regulation of Ca-dependent K channel activity in tracheal myocytes by phosphorylation. Nature 341:152–154
Kumlin M, Dahlen B, Bjorck T, Zetterstrom O, Granstrom E, Dahlen S-E (1992) Urinary excretion of leukotriene E_4 and 11-dehydro-thromboxane B_2 in response

to bronchial provocations with allergen, aspirin, leukotriene D_4, and histamine in asthmatics. Am Rev Respir Dis 146:96–103
Laitinen LA, Laitinen A, Panula P, Partenun M, Terrok, Teroo T (1983) Immunohistochemical demonstration of substance P in the lower respiratory tract of the rabbit and not of man. Thorax 38:531–536
Lammers J-WJ, Minette P, McCusker MT, Chung KF, Barnes PJ (1988) Nonadrenergic bronchodilator mechanisms in normal human subjects in vivo. J Appl Physiol 64:1817–1822
Lammers J-WJ, Minette P, McCusker M, Barnes PJ (1989) The role of pirenzepine-sensitive (M_1) muscarinic receptors in vagally mediated bronchoconstriction in human. Am Rev Respir Dis 139:446–449
Lands AM, Arnold A, McAuliff JP et al. (1967) Differentiation of receptor systems activated by sympathomimetric amines. Nature 214:597–598
Larsson K, Martinsson A, Hjemdahl P (1986) Influence of circulating α-adrenoceptor agonists on lung function in patients with exercise-induced asthma in healthy subjects. Thorax 41:552–558
Leff AR, Munoz NI (1981) Evidence for two subtypes of alpha-adrenergic receptors in canine airway smooth muscle. J Pharmacol Exp Ther 217:530–535
Lindgren BR, Ekstrom T, Andersson RGG (1986) The effect of inhaled clonidine in patients with asthma. Am Rev Respir Dis 134:266–269
Lindgren BR, Ekstrom T, Andersson RG (1986) The effect of inhaled clonidine in patients with asthma. Am Rev Respir Dis 134:266–269
Lofdahl C-G, Svedmyr N (1989) Formoterol fumarate, a new β_2-adrenoceptor agonist. Allergy 44:264–271
Lundberg JM, Hokfelt T, Martling C-R, Saria A, Cuelloh (1984) Substance P-immunoreactive sensory nerves in the lower respiratory tract of various mammals including man. Cell Tissue Res 235:251–261
Madison JM, Brown JK (1988) Differential effects of forskolin, isoproterenol, and dibutyryl-cyclic adenosine monophosphate on phosphoinositide hydrolysis in canine tracheal smooth muscle. J Clin Invest 82:1462–1465
Maeda A, Kubo T, Mishina M, Numa S (1988) Tissue distribution of mRNAs encoding muscarinic acetylcholine receptor subtyes. FEBS Lett 239:339–342
Mak JCW, Barnes PJ (1990) Autoradiological visualisation of muscarinic receptor subtyes in human and guinea pig lung. Am Rev Respir Dis 141:1559–1568
Mak JCW, Baraniuk JN, Barnes PJ (1992) Localization of muscarinic receptor subtype mRNAs in human lung. Am J Resp Cell Mol Biol 7:344–348
Manning PJ, Watson RM, Margolskee DJ, Williams VC, Schwartz JI, O'Byrne PM (1990) Inhibition of exercise-induced bronchoconstriction by MK-571, a potent leukotriene D_4-receptor antagonist. 323:1736–1739
Marthan R, Martin C, Thierry A, Mironneau J (1989) Calcium channels in isolated smooth muscle cells from human bronchus. J Appl Physiol 66(4):1706–1714
Mathe AA, Astrom A, Persson NA (1971) Some bronchoconstricting and bronchodilating responses of human isolated bronchi: evidence for the existence of α-adrenoceptors. J Pharm Pharmacol 23:905–910
McCann JD, Welsh MJ (1986) Calcium-activated potassium channels in canine airway smooth muscle. J Physiol (Lond) 372:113–127
Miller WC, Shultz TF (1979) Failure of nitroglycerin as a bronchodilator. Am Rev Respir Dis 120:471–472
Mitchelson F (1988) Muscarinic receptor differentiation. Pharmacol Ther 37:357–423
Miura M, Belvisi MG, Stretton CD, Yacoub MH, Barnes PJ (1992) Role of potassium channels in bronchodilator responses in human airways. Am Rev Respir Dis 146:132–136
Morrison JFJ, Pearson SB (1988) Type I muscarinic receptors in asthma. Am Rev Respir Dis 137:A197
Morrison JFJ, Pearson SB (1989) The effect of the circadian rhythm of vagal activity on bronchomotor tone in asthma. Br J Clin Pharmacol 28:545–549

Mueller E, Van Breeman C (1979) Role of intracellular Ca2 sequestration in beta adrenergic relaxation of airway smooth muscle. Nature 281:682–683
Murray MA, Kotlikoff MI (1991) Receptor-activated calcium influx in human airway smooth muscle cells. J Physiol (Lond) 435:123–144
Murray MA, Berry JL, Cook SJ, Foster RW, Green KA, Small RC (1991) Guinea-pig isolated trachealis: the effects of charybdotoxin on mechanical activity, membrane potential changes and the activity of plasmalemmal K+ channel. J Biol Chem 261:14607–14613
Nagai H, Yamada H, Goto S, Inagaki N, Koda A (1985) Reduction of antigen-induced contraction of sensitised guinea-pig tracheal smooth muscle in vitro by calmodulin inhibitors. Int Arch Allergy Appl Immunol 76:251–255
Newcomb R, Tashkin DP, Hui KK, Conolly ME, Lee E, Dauphinee B (1985) Rebound hyperresponsiveness to muscarinic stimulation after chronic therapy with an inhaled muscarinic antagonist. Am Rev Respir Dis 132:12–15
Nicholson SD, Shalid M, van Amsterdam RGM, Zaagsima J (1990) Cyclic nucleotide phosphodiesterase (PDE) isoenzymes in bovine trachea smooth muscle, their inhibition and the ability of isoenzyme inhibitors to relax precontracted preparations. Eur J Pharmacol 183:1097–1098
Nishizuka Y (1984) The role of protein kinase C in cell surface signal transduction and tumour promotion. Nature 308:693–697
Nishizuka Y (1988) The molecular heterogenity of protein kinase C and its implications for cellular regulation. Nature 334:661–665
Nogrady SG, Bevan C (1981) H2-receptor blockade and bronchial hyperreactivity to histamine in asthma. Thorax 36:268–271
O'Donnell M, Garippa R, Welton AF (1985) Relaxant effect of atriopeptins in isolated guinea-pig airway and vascular smooth muscle. Peptides 6:597–601
Okabe K, Kitamura K, Kuriyama H (1987) Features of 4-aminopyridine sensitive outwards current observed in single smooth muscle cells from the rabbit pulmonary artery. Pflugers Arch 409:561–568
Ollerenshaw S, Jarvis D, Woolcock A, Sullivan C, Scheibner T (1989) Absence of immunoreactive vasoactive intestinal polypeptide in tissue from the lungs of patients with asthma. N Engl J Med 320:1244–1248
Orehek J, Douglas JS, Bouhuyj A (1975) Contractile responses of the guinea pig trachea in vitro. Modification by prostaglandin synthesis inhibiting drugs. J Pharmacol Exp Ther 194:554–564
Palmer JBD, Cuss FMC, Barnes PJ (1986) VIP and PHM and their role in non-adrenergic inhibitory responses in isolated human airways. J Appl Physiol 61: C115–C119
Palmer JBD, Cuss FMC, Mulderry PK et al. (1987) Calcitonin gene-related peptide is localised to human airway nerves and potently constricts human airway smooth muscle. Br J Pharmacol 91:95–101
Park S, Rasmussen H (1985) Activation of tracheal smooth muscle contraction: syndergism between Ca and C kinase activators. Proc Natl Acad Sci USA 82:8835–8839
Park S, Rasmussen H (1986) Carbachol-induced protein phosphorylation changes in bovine trachial smooth muscle. J Biol Chem 261:15734–15739
Partenan M, Laitinen A, Hervonen A, Toivanen M, Laitenen LA (1982) Catecholamine and acetylcholinesterase containing nerves in human lower respiratory tract. Histochemistry 76:175–188
Patel KR, Kerr JW (1975) Effect of alpha receptor blocking drug, thymoxamine, on allergen induced bronchoconstriction in extrinsic asthma. Clin Allergy 5: 311–316
Pavord ID, Knox AJ, Cole A, Tattersfield AE (1991a) Effect of frusemide on release of PGE_2 on bovine tracheal mucosa. Thorax 46:715P
Pavord ID, Wisniewski A, Mathur R, Wahedna I, Knox AJ, Tattersfield AE (1991b) Effect of inhaled prostaglandin E_2 on bronchial reactivity to sodium metabisulphite and methacholine in patients with asthma. Thorax 46:633–637

Persson CGA, Karlsson JA (1987) In vitro responses to bronchodilator drugs. In: Jenne JW, Murphy S (eds) Drug therapy for asthma. Dekker, New York, pp 129–176

Pfitzer G, Ruegg JC, Flockerzi V, Hofmann F (1982) cGMP-dependent protein kinase decreases calcium sensitivity of skinned cardiac fibres. FEBS Lett 149: 171–175

Raeburn D, Rodger IW (1984) Lack of effect of leukotriene D_4 on ^{45}Ca uptake in airway smooth muscle. Br J Pharmacol 83:499–504

Regoli D, Barabe J (1988) Pharmacology of bradykinin and related peptides. Pharmacol Rev 32:1–46

Rhoden KJ, Meldrum LA, Barnes PJ (1988) Inhibition of cholinergic neurotransmission in human airways by β-adrenoceptors. J Appl Physiol 65:700–705

Richardson JB (1979) Nerve supply to the lungs. Am Rev Respir Dis 119:785–802

Richardson JB (1981) Non-adrenergic inhibitory innervation of the lung. Lung 159:315–322

Robberecht P, Chatelain P, De Neef P, Camus J-C, Waelbroeck M, Christophe J (1981) Presence of vasoactive intestinal peptide receptors coupled to adenylate cyclase in rat lung membranes. Biochem Biophys Acta 678:76–82

Rodger I (1988) Biochemistry of airway smooth muscle contraction. In: Barnes PJ, Rodger I, Thomson N (eds) Asthma: basic mechanisms and clinical management. Academic, London, pp 57–79

Roffel AF, Elzinga CRS, van Amsterdam RGM, de Zeeuw RA, Zaagsma J (1988) Muscarinic M_2 receptors in bovine tracheal smooth muscle: discrepancies between binding and function. Eur J Pharmacol 153:72–83

Roffel AF, Elzinga CRS, Zaagsma J (1990) Muscarinic M3-receptors mediate contraction of human central and peripheral airway smooth muscle. Pulm Pharmacol 3:47–52

Saida K, Van Breeman C (1984) Characteristics of the noradrenaline sensitive Ca store in vascular smooth muscle. Blood Vessels 21:43–52

Sands H, Mascali J (1978) Effects of cyclic AMP and of protein kinase on the calcium uptake by various tracheal smooth muscle organelles. Arch Int Pharmacodyn Ther 236:180–191

Sankary RM, Jones CA, Madison M, Brown JK (1988) Muscarinic cholinergic inhibition of cyclic AMP accumulation in airway smooth muscle. Am Rev Respir Dis 138:145–150

Schaefer OP, Ethier MF, Madison JM (1992) Characterisation of the muscarinic receptor subtype mediating inhibition of cyclic AMP accumulation inbovine trachealis cells. Am Rev Respir Dis 145:A438

Schatzmann HJ (1966) ATP dependent Ca extrusion from human red cells. J Physiol (Lond) 235:551–569

Schramm CM, Grunstein MM (1989) Mechanisms of protein kinase C regulation of airway contractility. J Appl Physiol 66:1935–1941

Sears MR, Taylor DR, Print CG, Lake DC, Li Q, Falnnery EM, Yates DM, Lucas MK, Herbison GP (1990) Regular inhaled beta-agonist treatment in bronchial asthma. Lancet 336:1391–1396

Sharma RK, Jeffery PK (1990) Airway β-adrenoceptor number in cystic fibrosis and asthma. Clin Sci 78:409–417

Silver P, Stull JT (1982) Phosphorylation of myosin light chain and phosphorylase in tracheal smooth muscle in response to KCl and carbachol. Mol Pharmacol 25:267–274

Simonnson BG, Svedmyr N, Skoogh BE, Anderson R, Bergh NB (1972) In vitro studies on alpha-adrenoceptors in human airways potentiation with bacterial endotoxin. Scand J Respir Dis 53:227–236

Simmonnson BG, Skoogh B-E, Bergh NP, Anderson R, Svedmyr N (1973) In vivo and in vitro effects of bradykinin on bronchial motor tone in normal subjects and in patients with airflow obstruction. Respiration 30:378–388

Slaughter RS, Welton AF, Morgan DW (1987) Sodium-calcium exchange in sarcolemmal vesicles from tracheal smooth muscle. Biochim Biophys Acta 904: 92–104
Smith LJ, Geller S, Ebright L, Glass M, Thyrum PT (1990) Inhibition of leukotriene D_4-induced bronchoconstriction in normal subjects by the oral LTD_4 receptor antagonist ICI 204, 219. Am Rev Respir Dis 141:988–992
Smith TW, Antman EM, Fridman PL, Blatt CM, Marsh JD (1984) Digitalis glycosides: mechanisms and manifestations of toxicity. Prog Cardiovasc Dis 26(6)
Somylo AV, Somylo AP (1968) Electromechanical and pharmacomechanical coupling in vascular smooth muscle. J Pharmacol Exp Ther 259:129–145
Souhrada M, Souhrada JF (1989a) Sodium and calcium influx induced by phorbol esters in airway smooth muscle cells. Am Rev Respir Dis 139:927–932
Souhrada M, Souhrada JF (1989b) The role of protein kinase C in sensitisation and antigen response of airway smooth muscle. Am Rev Respir Dis 140:1567–1572
Souhrada M, Souhrada JF, Cherniack RM (1981) Evidence for a sodium electrogenic pump in airway smooth muscle. J Appl Physiol 51:346–352
Souhrada M, Souhrada MH, Souhrada JF (1988) The inhibition of sodium influx attenuates airway response to a specific antigen challenge. Br J Pharmacol 93:884–892
Spencer DA, Evans JM, Green SE, Piper PJ, Costello JF (1991) Participation of the cysteinyl leukotrienes in the acute bronchoconstrictor response to inhaled platelet activating factor in man. Thorax 46:441–445
Spina D, Rigby PJ, Paterson JW, Goldie RG (1989) Autoradiographic localisation of β-adrenoceptors in asthmatic human lung. Am Rev Respir Dis 140:1410–1415
Stephens NL, Kroeger EA, Kromer U (1975) Induction of a myogenic response in tonic airway smooth muscle by tetraethylammonium. Am J Physiol 228:628–632
Stull JT (1980) Phosphorylation of contractil proteins in relation to muscle function. Adv Cyclic Nucleotide Res 13:39–93
Suzuki H, Morita K, Kuriyama H (1976) Innervation and properties of the smooth muscle of the dog trachea. Jpn J Physiol 26:303–320
Suzuki K, Takagi K, Satake T, Sugiyama S, Osawa T (1986) The relationship between tissue levels of cyclic GMP and tracheal smooth muscle relaxation in the guinea pig. Clin Exp Pharmacol Physiol 13:39–46
Svedmyr N (1977) Treatment with Ã-adrenostimulants. Scand J Respir Dis 101: 59–68
Sweatman WJF, Collier HOJ (1968) Effects of prostaglandins on human bronchial muscle. Nature 217:69
Sykes AP, Ayres JG (1990) A study of the duration of the bronchodilator effect of 12 ug and 24 ug of inhaled formoterol and 200 mcgs inhaled salbutamol in asthma. Respir Med 84:135–138
Szczeklik A, Nizankowska E, Nizankowska R (1977) Bronchial reactivity to prostaglandins F2 alpha, E2 and histamine in different types of asthma. Respiration 34:323–331
Takio K, Wado RD, Smith SB, Krebs EG, Walsh KA, Titani K (1984) Guanosine cyclic 3′-5′-phosphate protein kinase, a chimeric protein homologous with two separate protein familes. Biochemistry 23:4207–4218
Takuwa Y, Takuwa N, Rasmussen H (1988) The effects of isoproterenol on intracellular calcium concentration. J Biol Chem 263:762–768
Tattersfield AE (1987) Beta agonists and anticholinergic drugs – effect on bronchial reactivity. Am Rev Respir Dis 136:564–568
Tattersfield AE, Leaver DG, Pride NB (1973) Effects of β-adrenergic blockade and stimulation on normal human airways. J Appl Physiol 35:613–619
Tattersfield AE, Holgate ST, Harvey JE, Gribbin HR (1983) Is asthma due to partial β-blockade of airways? In: Morley J, Rainsford KD (eds) Pharmacology of asthma; agents and actions. Birkhäuser, Basel

Tauber AI, Kaliner M, Stechschulte DJ, Austen KF (1973) Immunological release of histamine and slow reacting substance of anaphylaxis from human lung. J Immunol 111:27–32
Taylor IK, O'Shaughnessy KM, Fuller RW, Dollery CT (1991) Effect of cysteinyl-leukotriene receptor antagonist ICI 204.219 on allergen-induced bronchoconstriction and airway hyperreactivity in atopic subjects. Lancet 337:690–694
Taylor SG, Bumstead J, Morris JEJ, Shaw DJ, Taylor JF (1988) Cromakalim inhibits cholinergic-mediated responses in human isolated bronchioles but not in guinea-pig airways (Abstr). Br J Pharmacol 96:795P
Torphy TJ, Undem BJ (1991) Phosphodiesterase inhibitors: new opportunities for the treatment of asthma. Thorax 46:512–523
Torphy TJ, Zhou H-L, Burman M, Huang LBF (1991) Role of cyclic nucleotide phosphodiesterase isozymes in intact canine trachealis. Mol Pharmacol 39: 376–384
Triggle DJ (1985) Calcium ions and respiratory muscle function. Br J Clin Pharmacol 20:213S–219S
Twentyman OP, Finnerty JP, Harris A, Palmer J, Holgate ST (1990) Protection against allergen-induced asthma by salmeterol. Lancet 336:1338–1342
Twentyman OP, Finerty JP, Holgate S (1991) The inhibitory effect of nebulised albuterol on the early and late asthmatic reactions and increase in airway responsiveness provoked by inhaled allergen in asthma. Am Rev Respir Dis 144:782–787
Ullman A, Svedmyr N (1988) Salmeterol, a new long activing inhaled β_2 adrenoceptor agonist: comparison with salbutamol in adult asthmatic patients. Thorax 43: 674–678
Umekawa H, Hidaka H (1985) Phosphorylation of caldesman by protein kinase C. Biochem Biophys Res Commun 132:56–62
Vathenen AS, Knox AJ, Higgins BG, Britton JR, Tattersfield AE (1988) Rebound increase in bronchial responsiveness after inhaled terbutaline. Lancet 1:554–557
Venugopalan CS, O'Rourke YM, Tucker TA (1985) Bronchorelaxing activity of lithium in vitro. J Pharmacol Sci 74:1120–1122
Vincenzi FF, Hines TR, Raess BV (1980) Calmodulin and the plasma membrane calcium pump. Ann NY Acad Sci 256:233–244
Wahedna I, Wisniewski AS, Tattersfield AE (1991) Effect of RG 12525, an oral leukotriene D_4 antagonist, on the airway response to inhaled leukotrine D_4 in subjects with mild asthma. Br J Clin Pharmacol 32:512–515
Wahedna I, Wong CS, Wisniewski AFZ, Pavord ID, Tattersfield AE (1992) Asthma control during and after cessation of regular beta$_2$ agonist treatment. Am Rev Respir Dis 148:702–712
Waisman DM, Gimble J, Goodman DPB, Rasmussen H (1981) Studies of the Ca transport mechanism of human inside-out plasma membrane vesicles. I. Regulation of the Ca pump by calmodulin. J Biol Chem 256:409–414
Walter U (1989) Physiological role of cGMP and cGMP-dependent protein kinase in the cardiovascular system. Rev Physiol Biochem Pharmacol 113:41–88
Walters EH, Davies BH (1982) Dual effects of prostaglandin E_2 on normal airways smooth muscle in vivo. Thorax 37:918–922
Weichman BM, Muccitelli RM, Tucker SS, Wasserman MA (1983) Effect of calcium antagonists on leukotrine D_4-induced contractions of the guinea-pig trachea and lung parenchyma. J Pharmacol Exp Ther 225:310–315
Weiss GB, Pang I-H, Goodman FR (1985) Relationship between ^{45}Ca movements, different calcium components and responses to acetylcholine and potassium in tracheal smooth muscle. J Pharmacol Exp Ther 233:389–394
White JP, Mills J, Eiser NM (1987) Comparison of the effects of histamine H1- and H2-receptor agonists on large and small airways in normal and asthmatic subjects. Br J Dis Chest 81:155–169

Widdicombe JG (1963) Regulation of tracheobronhial smooth muscle. Physiol Rev 43:1–37
Williams AJ, Vyse T, Richards DH, Lee T (1988) Cromakalim, a potassium channel activator, inhibits histamine induced bronchoconstriction and nocturnal bronchoconstriction in patients with asthma (Abstr). NEJ Allergy Proc 9:249
Williams AJ, Lee TH, Cochrane GM, Hopkirk A, Vyse T, Chiew F, Lavender E, Richards DH, Owen S, Stone P, Church S, Woodwook AA (1990) Attenuation of nocturnal asthma by cromakalim. Lancet 336:334–336
Worley JF, Kotlikoff MI (1990) Dihydropyridine-sensitive single calcium channels in airway smooth muscle cells. Am J Physiol 259:L468–L480
Yoshie Y, Iizuka K, Nakazawa T (1989) The inhibitory effect of a selective alpha 2 adrenergic receptor antagonist on moderate to severe type asthma. J Allergy Clin Immunol 84:747–752
Zaagsma J, Nahorski SR (1990) Is the adipocyte β-adrenoceptor a prototype for the recent cloned atypical "β_3-adrenoceptor"? TIPS 11:3–7
Zaid G, Beall GN (1966) Bronchial response to beta-adrenergic blockade. N Engl J Med 275:580–584

CHAPTER 12

Uterine Smooth Muscle: Electrophysiology and Pharmacology

J. MIRONNEAU

A. Introduction

In the myometrium, an increase in the concentration of the free calcium in the cytoplasm is essential for the initiation of contraction. Calcium can enter the cell through plasma membrane channels activated either by membrane depolarization or by specific binding of a ligand to its receptor. In addition, calcium may be released from stores within the cell in response to activation of membrane receptors. The link between these two events is the breakdown of plasma membrane phosphoinositides with the generation of second messengers. This chapter focuses on the different ion channels present in myometrial cells and how they are modulated during gestation and in response to endogenous hormones that are known to influence uterine motility.

B. Electrical Activity of Myometrium

I. Relationship Between Membrane Potential and Contraction

In the myometrium, as in other smooth muscles, contraction is initiated when the concentration of free calcium in the cytoplasm increases from $10^{-7}\,M$ to $10^{-6}\,M$. This calcium may be released from stores within the cell, but the most important source of calcium for durable and repeated contractions is that which enters the cell from the extracellular space (MIRONNEAU 1973, 1974). At the resting membrane potential the influx of calcium is very small but increases considerably as the membrane becomes depolarized. Thus, membrane electrical events play an important role in the regulation of calcium influx and thereby contraction.

Changes in resting membrane potential occur during the course of pregnancy as a consequence of the changing hormonal environment and the stretch imposed on the myometrium by the developing foetus. In the longitudinal myometrium of the rat, the resting potential is about −60 mV at midterm and falls to near −45 mV at term. These changes in resting potential are accompanied by alterations in the pattern of action potential discharge (MARSHALL 1962).

In the longitudinal myometrial layer spontaneous electrical activity is currently recorded. This activity is characterized by slow depolarizations of the membrane followed by repolarization back to the resting level. These can take the form of small events up to 5–10 mV in amplitude and several seconds in duration but they are not associated with changes in tension unless the threshold for an action potential is reached. Although all myometrial cells appear capable of generating slow depolarizations, their occurrence is probably restricted to a few areas within the myometrium; however, the pacemaker areas are mobile rather than fixed (LODGE and SPROAT 1981). Slow depolarizations are myogenic in nature but their ionic mechanisms are not clear. In the longitudinal myometrium the slow depolarization disappears when external sodium is reduced, suggesting that the slow depolarization may be dependent on an increase in sodium permeability (REINER and MARSHALL 1975). When the slow depolarization reaches a critical threshold, a second much faster depolarization occurs: the pacemaker potential. The pacemaker potential generates the action potential. The rate of depolarization dictates the frequency of action potential within a burst.

In contrast to the longitudinal myometrial layer, the electrical activity of the circular layer consists of single, plateau-type action potentials during most of gestation (PARKINGTON and COLEMAN 1990).

Simultaneous recordings of contraction and action potentials are abolished in calcium-free solution and in the presence of calcium channel blockers (MIRONNEAU 1973), suggesting that the calcium carried inward during the action potential could be sufficient to initiate a contraction. The amount of calcium entering the cell during a typical action potential (70 mV in amplitude, 50 ms in duration) has been calculated from the calcium current density and the estimated cell volume. It is estimated that cytoplasmic calcium increases by $10\,\mu M$ during such an action potential and this value is similar to that calculated in other smooth muscles (KLÖCKNER and ISENBERG 1985). Although much of the calcium entering the cell during an action potential may become buffered by uptake into intracellular storage sites, a sustained increase of calcium to $1–2\,\mu M$ inside the cell has been shown to cause near-maximal contraction in skinned myometrium (SAVINEAU et al. 1988).

II. Ion Channels: Identification and Modulation

The ion channels involved in the generation of electrical signals can be separated into at least two different groups.

Ligand-gated ion channels are opened in response to activation of an associated receptor. Typical channels of this type include the nonspecific channels that are opened by activation of muscarinic receptors (BENHAM et al. 1985) or ATP receptors (BENHAM and TSIEN 1987; FRIEL 1988; HONORÉ et al. 1989b). Activation of ligand-gated ion channels mediates local increase in ion conductance, producing depolarization or hyperpolarization of the membrane.

In contrast, voltage ion channels mediate rapid, voltage-gated changes in ion conductance during action potential. In myometrium, the action potential is largely dependent on an increase in calcium conductance that initiates excitation-contraction coupling (MIRONNEAU 1973) and multiple calcium-activated biochemical processes. In addition, calcium ion channels can be regulated by receptor-dependent processes, including protein phosphorylation and interaction with GTP-binding proteins.

1. Voltage-Gated Calcium Channels

The general properties of calcium channels have been elucidated using the double-sucrose gap technique to voltage-clamp strips of longitudinal myometrium from pregnant rats and, more recently, using both single microelectrode or patch electrode to voltage clamp single cells from this preparation in short-term primary culture.

a) Channel Type

Recent studies have indicated that vascular and visceral smooth muscle cells contain two or three different kinds of calcium channels (LOIRAND et al. 1986; STUREK and HERMSMEYER 1986; BENHAM et al. 1987a; YOSHINO et al. 1989). Two of these channels most likely correspond to the calcium channels termed T (transient) and L (long-lasting), which have been studied in cardiac and neuronal tissues (NILIUS et al. 1985; FOX et al. 1987). The two types of calcium channels have different selectivity, single-channel conductance, and pharmacology. Furthermore, T-type channels are inactivated by steady-state holding potentials to about −40 mV, while L-type channels need much more positive holding potentials for steady-state inactivation.

Calcium current in single rat myometrial cells might be composed of two components, demonstrated by applying depolarizing test pulses from two different holding potentials (−40 and −80 mV). In 89% of cells used, neither change in kinetics of inactivation nor shift in the peak current was observed between the two families of calcium currents (n = 125). Much evidence suggests that the calcium channel current appears to be mainly an L-type current: (a) Equimolar replacement of barium for calcium induces an increase in the peak current and a decrease in the inactivation rate. (b) Residual inward currents are recorded at the end of the pulses (400 ms in duration). (c) Membrane potential for half-inactivation was about −40 mV. However, in 11% of cells tested, two components of calcium channel currents can be separated by variations in holding potential (HONORÉ et al. 1989a). In human myometrium, evidence for two different calcium channels has been obtained at the single-channel level (INOUE et al. 1990).

b) Inactivation and Selectivity of L-Type Calcium Channels

Calcium channels are also time dependent such that, under voltage clamp, a depolarizing step in the membrane sufficient to activate the channels results in an inward current that peaks within 10 ms and then declines (AMÉDÉE et al. 1987). Part of the decline is caused by a voltage-dependent inactivation of the calcium channels (JMARI et al. 1986). At small depolarizations only a small number of calcium channels enter the inactivated state, and it requires a positive depolarization before the maximum number of calcium channels become inactivated. There is an additional part of inactivation caused by an accumulation of calcium ions on the inside surface of the membrane. It is assumed that free calcium ions react reversibly with calcium channel binding sites. This may lead to an enzymatic dephosphorylation of a channel component, resulting in inactivation of the channels (ECKERT and CHAD 1984). There is no evidence that accumulation of calcium ions in a restricted extracellular or intracellular space may change the membrane potential in isolated myometrial cells (AMÉDÉE et al. 1987).

Sodium ions can permeate these calcium channels but only when the external calcium concentration is less than $10^{-6} M$ (JMARI et al. 1987). The channels are also permeable to the divalent cations barium and strontium, which pass preferentially over calcium. A key element in the movement of cations through channels is the interaction between the charged ion and the electric field within the membrane. The force acting on the cation to move it into the cell is countered by reactive groups and steric configurations, i.e., energy barriers, that function to impede the flow of cations through the channel. This energy barrier or selectivity filter is structured in such a way as to permit certain types of cations to pass through the channel while excluding others (HILLE and SCHWARZ 1978; HESS and TSIEN 1984). In myometrial cells, the relative selectivity of the calcium channel for the divalent cations followed the sequence $Ca^{2+} > Sr^{2+} > Ba^{2+}$. The selectivity for calcium depends on the presence of external calcium as shown for calcium channels of different excitable cells (TSIEN et al. 1987). Other divalent cations, particularly cobalt and manganese but also nickel and cadmium, do not readily permeate the channels but bind to sites in the channel so strongly that they block the channel. Using large organic cations as probes, the pore size of the Ca^{2+} channel was estimated to be approximately 6 Å (TSIEN et al. 1987). This is three times the diameter of the calcium ion and approximately the same size as a hydrated calcium ion.

Functional voltage-gated calcium channels have been recently expressed in *Xenopus* oocytes injected with poly (A^{+})mRNA obtained from pregnant rat myometrium (FOURNIER et al. 1989).

c) Calcium Channel Blockers

A diverse group of compounds have the capacity to alter calcium influx across the cell membrane. These can be divided into two groups: inorganic inhibitors such as cobalt, manganese, nickel, and cadmium; and organic compounds. The latter agents are generally referred to as calcium entry blockers (FLECKENSTEIN 1977).

A number of calcium entry blockers, including nifedipine, nicardipine, diltiazem, verapamil, and gallopamil, have been shown to inhibit uterine contractions in various species including the rat, rabbit, sheep, monkey, and human (HOLLINGSWORTH and DOWNING 1988). The dihydropyridines are the most potent and selective inhibitors of uterine tension and therefore are of considerable interest for both therapeutic and experimental purposes. In cardiac cells the concentrations of dihydropyridines which give rise to half-maximal electrophysiological responses are several orders of magnitude greater than the dissociation constant values determined from ligand experiments (KOKUBUN et al. 1987; HAMILTON et al. 1987). Recent electrophysiological experiments in smooth muscle cells have demonstrated a voltage dependence of dihydropyridine antagonist action, and this has been postulated to account for the difference between the binding affinities and the dissociation constants determind from the pharmacological effects of these drugs (DACQUET et al. 1988; LOIRAND et al. 1989).

In the simplest model, calcium channels are in one of three configurations: closed, open, or inactivated. Transitions usually occur between the three configurations as random events, with each transition having a different probability. Under resting conditions the probability of most channels being open is small as most channels are closed. The effect of the membrane depolarization is to increase the frequency of opening. The channels remain open for a brief but random period. During maintained depolarization calcium channels make the transition to the inactivated state and become nonconducting. This state, however, differs from being closed in that the channel usually requires the membrane to be repolarized before it can be opened again (reactivated). The probability of leaving the inactivated state is small at depolarized potentials and becomes much greater at more negative potentials. Binding of calcium entry blockers to one of the calcium channel states produces inhibition of the calcium current.

In myometrium, L-type calcium channels are largely blocked by the dihydropyridine derivatives nifedipine at $10^{-7}\,M$ and (+)isradipine at $10^{-8}\,M$. To study the mechanism of blockade of the calcium channel current with (+)isradipine in more detail, HONORÉ et al. (1989a) used different protocols to assess the relative contribution of initial, conditioned, and tonic blockade of calcium channels. First, when voltage-clamp depolarization is applied at a frequency of 0.05 Hz, the inhibition induced by $2 \times 10^{-8}\,M$ (+)isradipine reaches 50% ± 5% ($n = 7$) within 5 min. Second, (+)isradipine (2×10^{-8} M) is applied during a rest period of 5 min and blockade is assessed as the difference between peak current in the control and the first pulse after drug exposure. Under these conditions, the calcium channel current is inhibited by 48% ± 6% ($n = 4$), indicating that the blockade is not dependent on the number of voltage depolarizations applied at 0.05 Hz (absence of use-dependent inhibition). Initial blockade is defined as the blockade that can be removed by application of a long-lasting hyperpolarization. As hyperpolarizing the membrane to $-90\,mV$ for 1 min restores the inward current it can be assumed that isradipine blockade is largely initial. Furthermore, (+)isradipine does not bind to the open state of the calcium channel. Third, in order to investigate whether (+)isradipine binds with a higher affinity to the inactivated state of the calcium channel, the effects of isradipine have been determined on the voltage-dependent steady-state inactivation of the calcium channel current. From the shift in the inactivation curve against voltage, it is possible to estimate the dissociation constant for (+)isradipine binding in the resting and inactivated state by using an approach described by BEAN (1984), assuming one-to-one binding of drug to the resting and inactivated states. The dissociation constant for binding to the inactivated state (K_i) can be calculated using the equation:

$\Delta Vh = k\,ln(1 + [\mathrm{Isr}]/K_i)/(1 + [\mathrm{Isr}]/K_r)$, where ΔVh is the shift of the midpoint of the steady-state inactivation curve, k is the slope factor of the inactivation curve, [Isr] is the (+)isradipine concentration used, and K_r is determined as the potency of (+)isradipine for the resting calcium channel. With $\Delta Vh = 24\,mV$, $k = 5.7\,mV$, $K_r = 23\,nM$ and $[\mathrm{Isr}] = 20\,nM$, K_i is estimated to be $0.13\,nM$.

These findings support the idea that (+)isradipine not only binds to the resting state of calcium channels but also that it has a higher affinity for the inactivated state (voltage-dependent inhibition).

d) Characterization of (+)-[^{3}H]Isradipine Binding Sites

As the potency of (+)isradipine blockade changes with the holding potential in a manner consistent with a higher affinity to the inactivated state of the calcium channel compared to the resting state, it is of interest to verify whether the K_i that is determined by electrophysiological data is similar to the dissociation constant of the high-affinity binding site for (+)-[^{3}H]isradipine determined with radioligand binding studies in microsomal fractions of myometrium. Scatchard analysis of the specific binding of (+)-[^{3}H]isradipine results in a linear plot, thereby indicating specific binding to a single class of sites. The dissociation constant, K_D, is $0.10 \pm 0.01\,nM$, and the maximal binding capacity, B_{max}, is 95 ± 5 fmol/mg protein ($n = 8$). These results indicate that the K_D value from binding experiments is similar to the apparent dissociation constant obtained from electrophysiological studies.

e) Modulation of Calcium Channels by Oxytocin

It is well known that oxytocin increases the frequency and the force of spontaneous contractions of the uterus. This effect has been correlated with an increase in calcium current (MIRONNEAU 1976) and a release of calcium from the intracellular stores. The concentration of oxytocin receptors in myometrium depends on the steroid concentration, because estrogens are capable of causing an increase in oxytocin receptor concentration and progesterone antagonizes the estrogen effect (SOLOFF 1984).

In the absence of estrogens in the culture medium, oxytocin (2×10^{-10} M) produces a very small increase in the maximal calcium current. In contrast, when isolated cells from pregnant rats are cultured with $10^{-6} M$ diethylstilbestrol (DES for 12 h) the increase in calcium current induced by $2 \times 10^{-10} M$ oxytocin reaches 55% ± 8% (n = 5). As oxytocin does not affect the apparent reversal potential, the analysis may be confined to possible changes in calcium conductance or inactivation parameters. Since the steady-state inactivation curve of the calcium current is unchanged in the presence of oxytocin, it is assumed that oxytocin acts by increasing the calcium conductance of myometrial cells. Further experiments with recordings from single channels are needed to propose a mechanism of action for oxytocin in myometrium.

f) Changes in Calcium Channels During Gestation

Calcium channel current density, normalized by cell capacitance, significantly increases during gestation. The averaged density of calcium channel current at day 10 of gestation (3.1 ± 0.7 μA/μF, n = 4) is similar to that of nonpregnant rat myometrium (2.5 ± 0.3 μA/μF, n = 9). In contrast, the averaged density of calcium channel current is maximal at day 18 of gestation (9.2 ± 1.1 μA/μF, n = 5) and remains elevated near term (7.3 ± 2.1 μA/μF, n = 8) at day 21 of gestation. The electrophysiological properties of calcium channels are not modified during gestation. Threshold for calcium channel current (about −40 mV), the voltage at the peak inward current (about +10 mV), and extrapolated reversal potential (about +65 mV) are virtually identical in nonpregnant and pregnant rat myometrial cells between days 10 and 21 of gestation. Sensitivity of calcium current to (+)isradipine is not affected during gestation as the concentration producing half-inhibition is about $2 \times 10^{-8} M$. These results differ from those obtained by Inoue and Sperelakis (1991), which showed no change in calcium channel density during pregnancy. In order to confirm the increase in calcium channel density near term, (+)-[^{3}H]isradipine specific binding is determined in intact strips of myometrium at day 10 and 18 of gestation. The dissociation constant, K_D, is unchanged (0.23 nM, n = 8) while the maximal binding capacity, B_{max}, increases from 21.2 fmol/mg wet wt. at day 10 to 32.5 ± 0.5 fmol/mg wet wt. at day 18 of gestation. The difference is highly significant ($P < 0.001$, n = 12). These results suggest that the calcium channel protein may be synthesized and incorporated in a greater amount into myometrial cells near pregnancy. The mechanisms responsible for the increase in calcium channel density are not known. However, it can be postulated that changes in hormonal status or stretching of the cells caused by fetuses may act as signals for synthesis of calcium channel proteins.

2. Voltage-Gated Sodium Channels

Characterization of sodium channels has been advanced through the use of the selective sodium blockers saxitoxin (STX) and tetrodotoxin (TTX). It is now clear that mammalian sodium channels are encoded by a multigene family (ROGART et al. 1989) and show different functional properties in different tissues. The most widely known difference is that TTX-sensitive sodium channels in nerve and skeletal muscles are blocked by nanomolar concentrations of the neurotoxins (PAPONNE 1980; BENOIT et al. 1985), whereas TTX-resistant sodium channels in cardiac muscle and noninnervated skeletal muscle are blocked only by micromolar concentrations of the neurotoxins (RENAUD et al. 1983; GUO et al. 1987).

In myometrial smooth muscle cells, sodium-dependent action potentials were first reported using the microelectrode technique (AMÉDÉE et al. 1986). These action potentials are totally blocked by micromolar concentrations of TTX. More recently, TTX-sensitive sodium currents have been described in dissociated cells from pulmonary artery (OKABE et al. 1988), portal vein (MIRONNEAU et al. 1990), and pregnant myometrium of rats (OHYA and SPERELAKIS 1989; MARTIN et al. 1990), and humans (YOUNG and HERNDON-SMITH 1991).

a) Binding Sites for [^{3}H]Saxitoxin in Myometrial Membranes

A Scatchard plot of the specific binding component for [^{3}H]STX indicates the existence of both high- and low-affinity binding sites. The dissociation constants, K_D, and maximal binding capacities, B_{max}, are 0.53 ± 0.11 nM and 39 ± 5 fmol/mg protein for the high-affinity site, and 27 ± 6 nM and 350 ± 45 fmol/mg protein for the low-affinity site ($n = 6$).

Therefore, the properties of the high-affinity site are examined using 0.4 nM [^{3}H]STX. Under these conditions about 80% of the signal originates from binding to the high-affinity sites and 20% from binding to the low-affinity sites. Increasing concentrations of unlabeled neurotoxins gradually inhibit [^{3}H]STX binding. The inhibition constant values, K_i, for STX and TTX are 0.18 ± 0.02 and 0.94 ± 0.04 nM, respectively ($n = 4$). The Hill coefficients are close to 1, suggesting the drugs bind to a single family of binding sites.

The properties of the low-affinity binding site are investigated using 15 nM [^{3}H]STX as 80% of the total binding is associated with the low-affinity binding site. Increasing concentrations of STX and TTX inhibit [^{3}H]STX binding to low-affinity sites. The K_i values for STX and TTX are 55 ± 9 nM ($n = 3$) and 1.2 ± 0.4 μM ($n = 3$), respectively.

Binding of [^{3}H] STX to dispersed myometrial cells gives similar dissociation constant values for both high- and low-affinity binding sites, indicating that both binding sites for [^{3}H]STX are well located in the plasma membrane of myometrial cells.

b) Electrophysiological Properties of Sodium Channels

Because the fast sodium current peaks within 2–3 ms, capacitive transients have to be reduced by using small cells and electrodes with a relatively low resistance. As myometrial cells are long cylinders, ranging from 80 to 100 μm in length and from 6 to 12 μm in width (AMÉDÉE et al. 1986), the first 1–3 ms are not properly clamped and the peak current amplitude may be underestimated, especially for the high depolarizing pulses. Neurotoxins such as veratridine and sea anemone toxins have been shown to reveal the voltage-dependent sodium current by stabilizing an open form of the sodium channel (CATTERALL 1980). When $10^{-4}\,M$ veratridine is applied in the bath solution, the inward calcium channel current is progressively inhibited within 2–3 min (ROMEY and LAZDUNSKI 1982). After the pulse, there is a standing inward tail current representing a population of veratridine-modified channels that does not close at −80 mV. Veratridine seems to bind to resting sodium channels, without preferential binding to open or inactivated states, as previously shown in nerve (ULBRICHT 1972).

The voltage-gating of the sodium current is determined by studying both activation and inactivation curves. Sodium current starts to activate at about −40 mV and reaches a maximum at +20 mV. Half-maximal activation is obtained near −15 mV. The sodium current is fully available at very negative membrane potentials (−110 mV), presents a half-inactivation potential near −70 mV, and is completely inactivated at −40 mV.

c) Sodium Channel Blockers

Both STX and TTX applied for 5 min reduce the amplitude of the sodium current in a concentration-dependent manner. The concentrations of STX and TTX inhibiting 50% of the maximal sodium current (IC_{50}) are estimated to be 1.4 ± 0.3 and 8.8 ± 0.2 nM, respectively ($n = 5$). The slope factor of the concentration-response curves is about 0.6, suggesting that the neurotoxins have multiple sites of interaction in myometrial membranes.

Interestingly, the dissociation constant of STX measured by electrophysiology (1.4 nM) does not correspond to the K_D of [^{3}H]STX for both high- (0.53 nM) and low-affinity sites (27 nM). These observations suggest that the high-and low-affinity receptors for STX identified by binding experiments could reflect the existence of two different subtypes of STX-sensitive sodium channels, which could correspond to different membrane proteins. Recently, PIDOPLICHKO (1986) has reported that two types of sodium currents recorded from ventricular cardiac cells can be distinguished by their sensitivity to TTX (K_D = 80 nM and 7 μM). Unequivocal answers to these molecular questions will require more detailed electrophysiological and biochemical experiments in myometrium. However, it is now clear that mammalian sodium channels are encoded by a multigene family and that separate genes encode for high- and low-affinity STX receptors (ROGART et al. 1989).

In isolated myometrial smooth muscle cells, the dissociation constants for TTX against the sodium current range between 8 nM (MARTIN et al. 1990) and 27 nM (OHYA and SPERELAKIS 1989), suggesting that the inhibitory effects of TTX may depend largely on an interaction with the high-affinity binding sites for neurotoxins. Thus, the low-affinity binding sites may be associated with a nonfunctional subtype of sodium channel.

d) Changes in Sodium Channels During Gestation

There are some reports showing that sodium ions may be important to the generation of action potentials in the myometrium (ANDERSON et al. 1971; KAO and MCCULLOUGH 1975). Recently, changes in sodium channel density have been studied during gestation (INOUE and SPERELAKIS 1991). During gestation, there is a quasilinear increase in averaged density of sodium current which is maximal near term. This effect is not produced by an increase in sodium current density per cell, but from an increase in the fraction of cells which possess fast sodium channels. The role of the fast sodium channel is not known as yet. However, the insertion of fast sodium channels into the cell membrane may cause an elevation of intracellular calcium concentration by the sodium-calcium exchange system functioning in the reverse mode (SAVINEAU et al. 1987), which can, in turn, potentiate the myometrial contraction.

It has been reported, using multicellular preparations of pregnant rat myometrium, that in midpregnancy the ratio of calcium current to sodium current is 0.57 whereas at term the ratio decreases to 0.31, indicating that the importance of sodium current increases near term (NAKAI and KAO 1983). However, it has to be noted that at term the resting membrane potential is rather depolarized (−45 mV) and the contribution of sodium channels to excitation-contraction is expected to be limited. In contrast, at midpregnancy, the resting membrane potential ranges between −60 and −70 mV. In this voltage range, sodium channels are largely activatable and may contribute to spike generation and cell-to-cell conduction. Further experiments are necessary to clarify the physiological role of the sodium channels in pregnant myometrial smooth muscle cells.

3. Ligand-Gated Nonspecific Channels

Adenosine trisphosphate (ATP) has received increasing attention, because it has similar effects to nonadrenergic and noncholinergic nerves in various smooth muscles, including myometrium (BURNSTOCK 1972). Evidence has accumulated indicating that ATP is involved in physiological regulation of nonvascular smooth muscle tone through liberation from "purinergic" nerve endings.

In artery and vas deferens, ATP has been shown to activate a conductance which is cation selective with little discrimination between small monovalent cations (BENHAM et al. 1987b; FRIEL 1988), but with a 3:1 selectivity for calcium over sodium (BENHAM and TSIEN 1987).

a) Electrophysiological Properties of ATP-Activated Channels

In myometrial cells the ATP-activated conductance is a nonspecific monovalent cation conductance. A pressure ejection pulse of 1 m*M* ATP (for 2 s) on a single myometrial cell maintained at a holding potential of −70 mV produces an inward current that is fast in onset (less than 100 ms) and has a short time to peak (less than 1 s). Furthermore, it is possible to obtain several similar responses to ATP in the cell with no decline in the amplitude of the response. The absence of desensitization in myometrium clearly differs from the high desensitization observed in other smooth muscles. In fact, a facilitation process for the ATP responses of myometrium has been reported (HONORÉ et al. 1989b).

The concentration-response curve can be fitted well by assuming that a receptor needs to bind one ATP molecule at a single site to activate current. From the fitted curve, the interpolated ATP concentration for half-activation of the conductance is about 0.6 m*M*. However, increasing the divalent cation concentration from 2.5 to 7.5 m*M* results in a shift of the concentration-response curve to higher concentrations of ATP, as expected from an ATP-chelation mechanism. Therefore, it is likely that ATP^{4-} is the only effective form of ATP acting in myometrium, and the concentration for half-activation of the conductance is then decreased to 0.09 m*M* ATP^{4-}.

By changing the ionic composition of the internal and external solutions it has been established that the ATP-activated channel is cation selective but with little discrimination between at least sodium, potassium, and cesium ions. It is interesting that there is a measurable, although very small, inward current remaining when sodium ions are substituted with choline (in the presence of atropine), suggesting that the channel could be permeable to divalent cations. However, increasing the external barium concentration decreases and suppresses the ATP-activated current. This observation does not imply the channel activated by ATP in myometrial cells is not permeable to calcium ions, since, owing to calcium chelation of ATP, the free acid form is reduced to subthreshold levels (with $10^{-3}\,M$ ATP under standard conditions, the ATP-free acid form is reduced to $<1.2 \times 10^{-6}\,M$). Further work is needed to resolve the question of whether calcium is permeant through the ATP-activated channel in myometrium.

b) Pharmacology of ATP-Activated Channels

The ligand specificity and pharmacology of the ATP-activated channel in myometrium is most similar to the P_2 type of receptors (BURNSTOCK and KENNEDY 1985). Like other responses mediated by P_2-receptors in ear artery and vas deferens, the ATP-activated response in myometrium is absolutely specific for ATP over adenosine, AMP, and ADP. The finding that ATP is by far the most effective phosphorylated adenosine derivative in activating the channels raises the possibility that ATP might actually be used as a substrate in some enzymatic reactions. ATP-γ-S and α,β-methylene ATP are found to be capable of mimicking the action of ATP, although with a lower potency. These results are consistent with the idea that the channel pathway is governed by a ligand receptor with very high specificity for ATP but not necessarily involving hydrolysis of ATP.

The cellular mechanism underlying ATP-activated channels is calcium independent, as ATP-activated current is obtained in the presence of both external barium (replacing calcium) and internal EGTA. The two other possibilities would be either a direct control of a channel closely linked to a ligand receptor or the generation of a chemical second messenger. It is quite possible that activation of the ATP-activated channel involves a chemical second messenger as onset kinetics of ATP-activated current are relatively fast (<100 ms) and, furthermore, ATP-activated current does not undergo any desensitization. These possibilities remain, however, to be explored in more detail.

c) Changes in ATP-Activated Channels During Gestation

The response of myometrial cells to ATP is dependent on the stage of pregnancy. In rat myometrial cells isolated at day 21 of gestation, the amplitude of the ATP-activated current is tenfold smaller than the responses elicited in cells isolated at day 10 and 19 of gestation. Furthermore, pretreatment of cultured cells isolated at day 19 of gestation with $10^{-5}\,M$ diethylstilbestrol for 15 h depresses the amplitude of the ATP response, which becomes similar to that obtained at day 21 of gestation. Therefore, the ATP-induced response appears to be strongly reduced near term in response to estrogen synthesis.

4. Potassium Channels

The repolarization phase of the action potential in myometrium is mediated by an outward current carried by potassium ions (MIRONNEAU 1974; KAO and MCCULLOUGH 1975). As with other smooth muscle cells, it has been suggested that more than one set of channels may be involved in longitudinal myometrium. From studies involving sucrose gap voltage clamp, it appears that one component of the current is partly blocked by tetraethylammonium ions, and it has been suggested that the current is calcium dependent (MIRONNEAU and SAVINAU 1980). Single-channel recordings in pregnant guinea pig myometrium have described a large conductance potassium channel sensitive to calcium ions (COLEMAN and PARKINGTON 1987). Another component of the potassium current is most prominent during midpregnancy. It is quickly activated, voltage dependent, and blocked by both 4-aminopyridine and tetraethylammonium ions (MIRONNEAU et al. 1981). This current is largely inactivated at the resting membrane potential in myometrium near term and can only be maximally activated by depolarization from membrane potentials more negative than −80 mV. In midgestation the resting potential is more negative than near term. As a result, this fast potassium current could be more available and could be therefore activated to a greater extent during the action potential. This might explain the more rapid rate of repolarization and the afterhyperpolarization observed in midgestation. A very slowly activated potassium current has been expressed in *Xenopus* oocytes injected with rat uterine RNA (BOYLE et al. 1987). The expression of this potassium channel appears to be increased by estrogen treatment (BOYLE and KACZMAREK 1989).

5. Other Channels

Chloride channels, activated by both release of calcium from the intracellular store and calcium influx through voltage-dependent calcium channels, have been identified in visceral and vascular smooth muscles (BYRNE and LARGE 1987; PACAUD et al. 1989). In isolated patches of myometrial membrane, single-channel currents with large conductance have been recorded (COLEMAN and PARKINGTON 1987). Acetylcholine and oxytocin activate a calcium-dependent chloride current, which tends to depolarize the cell membrane toward threshold for activation of calcium channels (Arnaudeau, Martin, and Mironneau, unpublished results).

Nonspecific cation channels activated by stretch have been identified in smooth muscle cells dissociated from the stomach of amphibia (KIRBER et al. 1988). When the membrane is stretched, the activity of the channel increases, and this would tend to depolarize the membrane. This channel may therefore underlie stretch-induced contraction.

Hyperpolarization-activated channels in myometrium have a reversal potential between −30 and −15 mV. The current is thought to be carried by a mixture of sodium and potassium ions (COLEMAN and PARKINGTON 1990) and would tend to depolarize the membrane. It may be speculated that this channel gives rise to pacemaker potentials.

C. Contractile Activity of Myometrium

For contraction to occur, the intracellular calcium concentration must be changed from $10^{-7} M$ to $10^{-6} M$. This may occur from either or both of two sources. Calcium can enter the cell (from the extracellular medium) or be released from membrane-bound intracellular storage areas. Relaxation would occur when the calcium is removed from the cytoplasm across the same membranes. Thus, in myometrium the source of calcium may be extracellular or intracellular.

I. Calcium Entry from the Extracellular Medium

A well-established mechanism for calcium entry into the myometrial cell is calcium influx through voltage-gated calcium channels. Receptor-activated channels permeable to calcium channels have also been proposed but have not yet been characterized (see Sect. B.II, this chapter).

II. Calcium Release from the Intracellular Calcium Store

Although it has been much more difficult to demonstrate calcium release than calcium uptake in sarcoplasmic reticulum preparations, several lines of evidence suggest that the calcium stored in the sarcoplasmic reticulum can be released upon stimulation and is of sufficient quantity to activate a maximal contraction of smooth muscle.

1. Agonist Effects on the Intracellular Calcium Store

When uterine smooth muscle is bathed in a calcium-free solution (2 m*M* EGTA), there is a single transient phasic contraction to agonists such as acetylcholine, angiotensin II or oxytocin which cannot be repeated until intracellular calcium stores are replenished (SAKAI et al. 1982; LALANNE et al. 1984; MIRONNEAU et al. 1984; ASHOORI et al. 1985). In isolated smooth muscle cells from rat myometrium, maximal contractions can be obtained in response to acetylcholine, angiotensin II, high-potassium solution, and action potential (AMÉDÉE et al. 1986). Interestingly, a maximal transient contraction is induced by acetylcholine after extracellular calcium has been removed or following blockade of calcium channels with nitrendipine, but the contraction declines exponentially with time in calcium-free solution (KYOZYKA et al. 1987).

Although the amplitude of contraction following extracellular calcium removal does correlate with the volume of the sarcoplasmic reticulum in different smooth muscles (JOHANSSON and SOMLYO 1980), it is likely that, in addition to the volume of the sarcoplasmic reticulum, other factors such as functional properties of excitation-contraction coupling in calcium-free solutions or the rate of passive calcium leak from the sarcoplasmic reticulum contribute to the varying maintenance of contractility in the absence of extracellular calcium.

2. Characteristics of the Intracellular Calcium Store

The properties of contractile proteins and the involvement of the intracellular calcium store in the contraction have been more directly studied in muscle strips where cells have been made hyperpermeable by membrane skinning (ENDO et al. 1977). Saponin-skinned strips prepared from pregnant rat myometrium in 4 m*M* EGTA exhibit a high calcium sensitivity as 80% of the maximal contraction is achieved in the presence of 10^{-6} *M* calcium (SAVINEAU et al. 1988). Other divalent cations induce concentration-dependent contractions and the rank order of potency is $Ca^{2+} > Mn^{2+} > Sr^{2+} > Ba^{2+}$. All these divalent cation-activated contractions are antagonized by trifluoperazine, suggesting the involvement of a calcium-calmodulin complex in the myometrial contraction. This complex allows the activation of myosin light chain kinase and it has been clearly shown that phosphorylation of 20000-dalton myosin light chains (LC20) via this enzyme is a prerequisite for the contraction of uterine smooth muscle (HAEBERLE et al. 1985). Thus, a common mechanism would be implied in these divalent cation-activated contractions. Moreover, pretreatment of skinned myometrial strips with cyclic AMP results in a reduction of the calcium-activated contraction, suggesting a direct effect of cyclic AMP on the contractile process, e.g., phosphorylation of myosin light chain kinase reducing its affinity for the calcium-calmodulin complex, as previously proposed (ADELSTEIN and HATHAWAY 1979). The cyclic AMP-induced relaxation of rat myometrium is dependent on the calcium concentration. It has been shown that at low calcium concentrations two sites of myosin light chain kinase are phosphorylated by the cyclic AMP-dependent protein kinase, while at higher concentrations only one site is phosphorylated and the myosin light chain kinase is not fully inhibited (CONTI and ADELSTEIN 1981).

When saponin-treated myometrial strips are exposed to low concentrations of EGTA (0.1 m*M*), calcium ions can accumulate into the intracellular store. Several lines of evidence support this proposal. After calcium loading, replacement of potassium ions by choline or addition of ionophore A23187 or D-myo-inositol trisphosphate ($InsP_3$) induce contractions, suggesting release of calcium from a store which has been filled. Replacement of potassium by a large cation such as choline produces a net positive charge on the outside (cytoplasmic side) of the sarcoplasmic reticulum, and thus generates an electrical state comparable to that occurring with membrane depolarization in the intact muscle. Thus, the internal calcium store of myometrium appears to be sensitive to electrical phenomena. A similar observation is obtained by applying large depolarizations in intact strips perfused in calcium-free solution or in the presence of calcium entry blockers (MIRONNEAU 1973).

$InsP_3$, at concentrations between 2 and 20 μM, produces contractions similar to those induced by agonists (SAVINEAU et al. 1988; KANMURA et al. 1988). The contractile response is concentration dependent and correlates with the increase in free calcium. No data are available about the value of $InsP_3$ concentration in rat myometrium in response to agonists, but the range of $InsP_3$ concentration is in good agreement with that required for ^{45}Ca-release from isolated sarcoplasmic reticulum of pregnant uterus (CARSTEN and MILLER 1985). For a calcium loading of 3–4 min with $10^{-6} M$ calcium (in the presence of $10^{-4} M$ EGTA) the amplitude of the $InsP_3$-induced contraction is maximal. However, for calcium loading longer than 6 min, the $InsP_3$-induced contraction appears to be reduced, indicating a decrease in the amount of stored calcium and the possible existence of a calcium-induced calcium release mechanism. The identification of the calcium store as sarcoplasmic reticulum is supported by the fact that the $InsP_3$-induced contraction is inhibited by ryanodine and procaine (two drugs which alter the calcium-release mechanism from the sarcoplasmic reticulum; Ito et al. 1986), but remains unaffected by inhibitors of mitochondrial oxidative phosphorylation such as sodium azide or oligomycin.

Skinned myometrial strips exhibit repeated contractile responses for about 1 h, without exposure to calcium-containing solution between each application of agonist, a phenomenon also reported in intact muscle strips bathed in calcium-free solution (Mironneau et al. 1984). Reduction below 4 min of the time interval separating two successive $InsP_3$-induced contractions depresses the amplitude of the second contraction. Moreover, addition of vanadate, a potent inhibitor of the calcium-ATPase of sarcoplasmic reticulum (Missiaen et al. 1989), slows the relaxation of the $InsP_3$-induced contraction and reduces the subsequent response. Therefore, these results suggest that the internal calcium store of myometrium may be able to requester, by an active process, the major part of the released calcium.

The capacity of the internal calcium store is increased during gestation (Batra 1986), indicating that the involvement of an intracellular calcium source in the contractile processes can thus be more efficient in the last stages of gestation.

3. Relaxant Effects of Caffeine in Myometrium

In uterine smooth muscle, caffeine fails to evoke a contraction although the existence of an internal calcium store that is sensitive to several substances has clearly been demonstrated. The absence of caffeine-induced contraction in rat myometrium seems to be a characteristic of this tissue, since caffeine causes contraction of other smooth muscles from the same animal (e.g., portal vein and colon). As it is widely accepted that caffeine-induced contraction is due to the release of calcium from sarcoplasmic reticulum (Fabiato and Fabiato 1977), the lack of contractile effect of caffeine in rat myometrium could be due to the absence of a specific calcium release mechanism activated by caffeine, as is the case in the sarcoplasmic reticulum of canine aorta (Ehrlich and Watras 1988). In this respect, the sarcoplasmic reticulum of myometrium could be of the sβ-type according to the classification proposed by Iino et al. (1988), i.e., with a calcium-release mechanism activated by $InsP_3$ but not by caffeine.

More interestingly, caffeine exhibits a potent inhibitory effect on myometrium in both intact and skinned strips (Savineau and Mironneau 1990a). In saponin-skinned strips, application of caffeine (5–10 mM) during loading of the calcium store increases the subsequent contraction induced by $InsP_3$, showing that more calcium ions have been stored. As application of cyclic AMP produces the same effect, it is suggested that caffeine (as theophylline and isobutyl methylxanthine) may increase the cyclic AMP concentration by inhibiting some phosphodiesterases. In addition, caffeine decreases calcium-activated contractions in skinned strips lacking a functional internal calcium store. This effect is reduced by the cyclic AMP-dependent protein kinase inhibitor (PKA inhibitor), indicating that caffeine might decrease the calcium sensitivity of the contractile proteins through the cyclic AMP-dependent phosphorylation of the myosin light chain kinase (ADELSTEIN and HATHAWAY 1979).

Another matter of debate is the effect of caffeine on the voltage-dependent calcium channels (MARTIN et al. 1989). In isolated cells from pregnant rat myometrium caffeine inhibits the L-type calcium channel current in a concentration-dependent manner. The concentration producing half-inhibition is estimated to be about 30 m*M*. The caffeine inhibition is not enhanced when calcium channels are opened by a conditioning depolarizing pulse sequence or when the number of inactivated calcium channels is increased at depolarized potentials. Thus, caffeine may interact with all states of the calcium channels. This assumption is supported by the caffeine inhibition of specific (+)-[^{3}H]isradipine binding for myometrial membranes which are, of course, depolarized. It is shown that caffeine behaves as a noncompetitive inhibitor of (+)-[^{3}H]isradipine binding, reducing the maximal binding capacity without change in the dissociation constant. Thus, at concentrations higher than 1 m*M*, caffeine interacts with a site closely associated with the L-type calcium channels and, in turn, inhibits calcium influx.

III. Phosphoinositide Cycle

It is now well established that a variety of agonists acting at cell surface receptors can alter cellular function by activating phospholipase C isoforms (via a guanine nucleotide exchange protein). The major target of this enzyme is phosphatidylinositol 4,5-bisphosphate (PIP_2), and its immediate products [$InsP_3$ and (1,2)diacylglycerol, DAG] are important second messengers (BERRIDGE 1984; ABDEL-LATIF 1986).

1. Action of $InsP_3$

An increase in $InsP_3$ concentration of myometrium occurs in the presence of carbachol, oxytocin, angiotensin II, α-adrenergic agonists, and platelet-activating factor (MARC et al. 1986; SCHREY et al. 1986; VAROL et al. 1989). In human myometrium carbachol is ineffective, contrary to findings in animals (SCHREY et al. 1988).

$InsP_3$ rapidly releases calcium from a microsomal fraction of myometrium. At the highest concentration used (5 μM), $InsP_3$ releases about 40% of the calcium releasable by an ionophore (CARSTEN and MILLER 1985). Specific binding sites for $InsP_3$ have been demonstrated and analyzed by

Scatchard plots in adrenal cortex (GUILLEMETTE et al. 1987). Interestingly, there is approximately a 1000-fold difference between the nanomolar binding affinity of $InsP_3$ and the micromolar concentration needed for calcium mobilization. Using permeabilized cells, it has been possible to compare the time of onset of contraction evoked by an agonist and that of $InsP_3$ (SOMLYO et al. 1985). The $InsP_3$ latency is 0.5 s compared to 1.5 s for α-agonist-induced contraction. The time difference is expected to be long enough to allow for release of $InsP_3$ by the agonist.

Application of $InsP_3$ to saponin-skinned pregnant myometrium leads to a contractile response (SAVINEAU et al. 1988; KANMURA et al. 1988).

2. Action of Phorbol Esters

Phorbol esters are known to mimic the action of DAG, the natural activator of protein kinase C in various cellular types (NISHIZUKA 1986). In pregnant rat myometrium, phorbol esters have multiple and opposing effects on the mechanical activity, according to the concentrations used and the duration of application (SAVINEAU and MIRONNEAU 1990b).

At low concentrations (10^{-9} to $10^{-8}\,M$), phorbol dibutyrate (PDB) increases the amplitude of phasic contractions, but has no effect in the presence of Ca-entry blockers and after removal of external calcium. PDB stimulates L-type calcium channel of pregnant rat myometrium (Arnaudeau, Martin and Mironneau, unpublished data), as previously described in vascular smooth muscle (FISH et al. 1988; LOIRAND et al. 1990). The increase in calcium current may account, at least in part, for the early increase in amplitude of both potassium and oxytocin-induced contractions.

Furthermore, direct effects on the contractile proteins have been described and may play a role in the relaxant effect of phorbol esters. In skinned myometrial strips, high concentrations of PDB ($10^{-6}\,M$) decrease the calcium-activated contraction (in the absence of functional intracellular calcium store) but have no effect on the internal calcium store. Therefore, it is likely that the PDB-induced changes are due to a direct action of PDB on the contractile proteins. A possible mechanism is that protein kinase C can phosphorylate myosin light chain kinase and hence reduce the affinity of phosphorylated myosin light chain for the calcium-calmodulin complex, leading to a decrease in tension, by a mechanism similar to that induced by cyclic AMP-dependent protein kinase (IKEBE et al. 1985).

Finally, the possibility that protein kinase C may inhibit the hydrolysis of phosphoinositides and the contraction induced by the release of $InsP_3$ by several agonists has been proposed (MCMILLAN et al. 1986; ITOH et al. 1988). In myometrium, it has been reported that PDB may antagonize phosphoinositide hydrolysis (VAROL et al. 1989) by negative feedback control on phospholipase C activity.

D. Summary and Conclusions

Control of myometrial contractility during gestation can be discussed in terms of ion channel functioning. Many of these channels are primarily voltage dependent and the probability of their being in the open, conducting state is determined by membrane potential. Although some channel types can be directly activated by ligands, it is accepted that membrane potential during pregnancy is important in determining myometrial contractility.

Calcium channels are the fundamental units regulating calcium influx. Other channels exist that provide specific conductance pathways for other ions, such as sodium, potassium, and chloride, or that are permeable for cations without strong discrimination between mono- and divalent cations. Although it is not well established that the nonspecific cation channels are permeable to calcium ions, all these channels control the membrane potential and thus regulate activity of the calcium channels because of their voltage dependence. Therefore, it appears that it is the totality of channel activity in the myometrial cell membrane which governs action potential discharge and contraction. Ion channels are involved in regulation of the membrane potential during gestation and are modulated by hormones and neuromediators.

Among the calcium entry blockers, the dihydropyridines appear to be the most effective in inhibiting both calcium channels and contractility. Because of deleterious effects on the fetus, the use of calcium entry blockers in controlling gestation will be dependent on the development of agents with greater specificity for the myometrium and reduced side effects.

Myometrial contraction occurs as a result of both calcium influx and calcium release from the intracellular store. One of the links between membrane receptor occupation and calcium mobilization has been identified in the action of phospholipase C, which hydrolyzes membrane phosphoinositides, leading to the generation of $InsP_3$ and diacylglycerol. It is presently thought that $InsP_3$ releases calcium from the sarcoplasmic reticulum while diacylglycerol activates protein kinase C. These are some arguments supporting the fact that activation of protein kinase C by phorbol esters increases calcium channel current, presumably through some phosphorylation processes. Relaxant effects of phorbol esters have also been obtained which seem to involve interactions with contractile proteins.

Although the present understanding of myometrial contractility has been largely clarified by using electrophysiological techniques, the greatest advances in the future will certainly depend on an integrative approach involving biochemistry, immunology, molecular biology, and electrophysiology applied to human myometrial preparations.

References

Abdel-Latif AA (1986) Calcium-mobilizing receptors, phosphoinositides, and the generation of second messengers. Pharmacol Rev 38:227–272

Adelstein RS, Hathaway DR (1979) Role of calcium and cyclic nucleotide 3:5′ monophosphate in regulating smooth muscle contraction. Am J Cardiol 44: 783–787

Amédée T, Mironneau C, Mironneau J (1986) Isolation and contractile responses of single pregnant rat myometrial cells in short-term primary culture and the effects of pharmacological and electrical stimuli. Br J Pharmacol 88:873–880

Amédée T, Mironneau C, Mironneau J (1987) The calcium channel current of pregnant rat single myometrial cells in short-term primary culture. J Physiol (Lond) 392:253–272

Anderson NC, Ramon F, Snyder A (1971) Studies on calcium and sodium in uterine smooth muscle excitation under current-clamp and voltage-clamp conditions. J Gen Physiol 58:322–339

Ashoori F, Takai A, Tomita T (1985) The response of non-pregnant rat myometrium to oxytocin in Ca-free solution. Br J Pharmacol 84:175–183

Batra S (1986) Effect of estrogen and progesterone on calcium uptake by the myometrium and smooth muscle of the lower urinary tract. Eur J Pharmacol 127:37–42

Bean B (1984) Nitrendipine block of cardiac calcium channels: high affinity binding to the inactivated state. Proc Natl Acad Sci USA 81:1–30

Benham CD, Tsien RW (1987) A novel receptor-operated Ca^{2+}-permeable channel activated by ATP in smooth muscle. Nature 328:275–278

Benham CD, Bolton TB, Lang RJ (1985) Acetylcholine activates an inward current in single mammalian smooth muscle cells. Nature 316:345–347

Benham CD, Hess T, Tsien R (1987a) Two types of calcium channels in single smooth muscle cells from rabbit ear artery studied with whole-cell and single-channel recordings. Circ Res 61:I10–I16

Benham CD, Bolton TB, Byrne NG, Large WA (1987b) Action of externally applied adenosine trisphosphate on single smooth muscle cells dispersed from rabbit ear artery. J Physiol (Lond) 387:473–488

Benoit E, Corbier A, Dubois JM (1985) Evidence for two transient sodium currents in the frog node of Ranvier. J Physiol (Lond) 361:339–360

Berridge MJ (1984) Inositol trisphosphate and diacylglycerol as second messengers. Biochem J 220:345–360

Boyle MB, Kaczmarek LK (1989) Regulation of potassium channel mRNA by estrogen in uterine smooth muscle. Neurol Neurobiol 50:167–181

Boyle MB, Azhderian EM, McLusky NJ, Naftolin F, Kaczmarek LK (1987) Xenopus oocytes injected with rat uterine RNA express very slowly activating potassium currents. Science 235:1221–114

Burnstock G (1972) Purinergic nerves. Pharmacol Rev 24:509–581

Burnstock G, Kennedy C (1985) Is there a basis for distinguishing two types of P_2-purinoceptors? Gen Pharmacol 16:433–440

Byrne NG, Large WA (1987) Action of noradrenaline on single smooth muscle cells freshly dispersed from the rat anococcygeus muscle. J Physiol (Lond) 389:513–525

Carsten ME, Miller JD (1985) Ca^{2+} release by inositol trisphosphate from Ca^{2+}-transporting microsomes derived from uterine sarcoplasmic reticulum. Biochem Biophys Res Commun 130:1027–1031

Catterall WA (1980) Neurotoxins that act on voltage-sensitive sodium channels in excitable membranes. Annu Rev Pharmacol Toxicol 20:15–43

Coleman HA, Parkington HC (1987) Single channel Cl^- and K^+ currents from cells of uterus not treated with enzymes. Pflugers Arch 410:560–562

Coleman HA, Parkington HC (1990) Hyperpolarization-activated channels in myometrium: a patch clamp study. In: Speralakis N, Wood JD (eds) Frontiers in smooth muscle research. Liss, New York, pp 665–672
Conti MA, Adelstein RS (1981) The relationship between calmodulin binding and phosphorylation of smooth muscle kinase by the catalytic subunit 3,5-cAMP-dependent protein kinase. J Biol Chem 256:3178–3181
Dacquet C, Pacaud P, Loirand G, Mironneau C, Mironneau J (1988) Comparison of binding affinities and calcium current inhibitory effects of a 1,4-dihydropyridine derivative (PN 200–110) in vascular smooth muscle. Biochem Biophys Res Commun 152:1165–1172
Eckert R, Chad JD (1984) Inactivation of Ca channels. Prog Biophys Mol Biol 44:215–267
Ehrlich BE, Watras J (1988) Inositol 1,4,5-trisphosphate activates a channel from smooth muscle sarcoplasmic reticulum. Nature 336:583–586
Endo M, Kitazawa T, Yagi S, Iino M, Kakuta Y (1977) Some properties of chemically skinned smooth muscle. In: Casteels R, Godfraind T, Rüegg JC (eds) Excitation-contraction coupling in smooth muscle. Elsevier/North-Holland, Amsterdam, pp 199–209
Fabiato A, Fabiato F (1977) Calcium release from the sarcoplasmic reticulum. Circ Res 40:119–129
Fish RD, Sperti G, Colucci WS, Lapham DE (1988) Phorbol ester increases the dihydropyridine-sensitive calcium conductance in a vascular smooth muscle cell line. Circ Res 62:1049–1054
Fleckenstein A (1977) Specific pharmacology of calcium ion in myocardium, cardiac pacemakers, and vascular smooth muscle. Annu Rev Pharmacol Toxicol 17:149–166
Fournier F, Honoré E, Brûlé G, Mironneau J, Guilbault P (1989) Expression of Ba currents in Xenopus oocyte injected with pregnant rat myometrium mRNA. Pflugers Arch 413:682–684
Fox A, Nowycky M, Tsien R (1987) Kinetic and pharmacological properties distinguish three types of calcium currents in chick sensory neurones. J Physiol (Lond) 394:149–172
Friel D (1988) An ATP-sensitive conductance in single smooth muscle cells from the rat vas deferens. J Physiol (Lond) 401:361–380
Guillemette G, Balla T, Baukal AJ, Catt AJ (1987) Inositol 1,4,5-trisphosphate binds to a specific receptor and releases microsomal calcium in the anterior pituitary gland. Proc Natl Acad Sci USA 84:8195–8199
Guo X, Uehara A, Ravindran A, Bryant SH, Hall S, Moczydlowski E (1987) Kinetic basis for insensitivity to tetrodotoxin and saxitoxin in sodium channels of canine heart and denervated rat skeletal muscle. Biochemistry 26:7546–7556
Haeberle JR, Hott JW, Hathaway DR (1985) Regulation of isometric force and isotonic shortening velocity by phosphorylation of the 20000 dalton myosin light chain of rat uterine smooth muscle. Pflugers Arch 403:215–219
Hamilton S, Yatani A, Brush K, Schwartz A, Brown A (1987) A comparison between the binding and electrophysiological effects of dihydropyridines on cardiac membranes. Mol Pharmacol 31:221–231
Hess P, Tsien RW (1984) Mechanism of ion permeation through calcium channels. Nature 309:453–455
Hille B, Schwarz W (1978) Potassium channels as multi-ion single-file pores. J Gen Physiol 72:409–442
Hollingsworth M, Downing S (1988) Calcium entry blockers and the uterus. Med Sci Res 16:1–16
Honoré E, Amédée T, Martin C, Dacquet C, Mironneau J (1989a) Calcium channel current and its sensitivity to (+)isradipine in cultured pregnant rat myometrial cells. Pflugers Arch 414:477–483

Honoré E, Martin C, Mironneau C, Mironneau J (1989b) An ATP-sensitive conductance in cultured smooth muscle cells from pregnant rat myometrium. Am J Physiol 257:C297–C309

Iino M, Kobayashi T, Endo M (1988) Use of ryanodine for functional removal of the calcium store in smooth muscle cells of the guinea pig. Biochem Biophys Res Commun 152:417–422

Ikebe M, Inagaki M, Kanmura K, Hidaka H (1985) Phosphorylation of smooth muscle myosin light chain kinase by Ca^{2+}-activated phospholipid-dependent protein kinase. J Biol Chem 260:4547–4550

Inoue Y, Sperelakis N (1991) Gestational change in Na^{+} and Ca^{2+} channel current densities in rat myometrial smooth muscle cells. Am J Physiol 260:C658–C663

Inoue Y, Nakao K, Okabe K, Izumi H, Kanda S, Kitamura K, Kuriyama H (1990) Some electrical properties of human pregnant myometrium. Am J Obstet Gynecol 162:1090–1098

Ito K, Takakura S, Sato K, Sutko JL (1986) Ryanodine inhibits the release of calcium from intracellular stores in guinea pig aortic smooth muscle. Circ Res 58:730–734

Itoh T, Kubota Y, Kuriyama H (1988) Effects of a phorbol ester on acetylcholine-induced Ca^{2+}-mobilization and contraction in the porcine coronary artery. J Physiol (Lond) 397:401–419

Jmari K, Mironneau C, Mironneau J (1986) Inactivation of calcium channel current in rat uterine smooth muscle: evidence for calcium and voltage-mediated mechanisms. J Physiol (Lond) 380:111–126

Jmari K, Mironneau C, Mironneau J (1987) Selectivity of calcium channels in rat uterine smooth muscle: interactions between sodium, calcium and barium ions. J Physiol (Lond) 384:247–261

Johansson B, Somlyo AP (1980) Electrophysiology and excitation-contraction coupling. In: Bohr DF, Somlyo AP, Sparks HV (eds) Vascular smooth muscle. American Physiological Society, Bethesda, pp 301–323 (Handbook of physiology, sect 2/II)

Kanmura Y, Missiaen L, Casteels R (1988) Properties of intracellular calcium stores in pregnant rat myometrium. Br J Pharmacol 95:284–290

Kao CY, McCullough JR (1975) Ionic currents in the uterine smooth muscle. J Physiol (Lond) 246:1–36

Kirber MT, Walsh JV Jr, Singer JJ (1988) Stretch-activated ion channels in smooth muscle. A mechanism for the initiation of stretch-induced contraction. Pflugers Arch 4121:339–345

Klöckner M, Isenberg G (1985) Calcium currents of cesium loaded isolated smooth muscle cells (urinary bladder of the guinea pig). Pflugers Arch 405:340–348

Kokubun S, Prod'Hom B, Becker C, Porzig H, Reuter H (1987) Studies on Ca channels in intact cardiac cells: voltage-dependent effects and cooperative interactions of dihydropyridine enantiomers. Mol Pharmacol 30:571–584

Kyozuka M, Crankshaw J, Berezin I, Collins SM, Daniel EE (1987) Calcium and contractions of isolated smooth muscle cells from rat myometrium. Can J Physiol Pharmacol 65:1966–1975

Lalanne C, Mironneau C, Mironneau J, Savineau JP (1984) Contraction of rat uterine smooth muscle induced by acetylcholine and angiotensin II in Ca-free' medium. Br J Pharmacol 81:317–326

Lodge S, Sproat JE (1981) Resting membrane potentials of pacemaker and non-pacemaker areas in rat uterus. Life Sci 28:2251–2256

Loirand G, Pacaud P, Mironneau C, Mironneau J (1986) Evidence for two distinct calcium channels in rat vascular smooth muscle cells in short-term primary culture. Pflugers Arch 407:566–568

Loirand G, Mironneau C, Mironneau J, Pacaued P (1989) Two types of calcium currents in single smooth muscle cells from rat portal vein. J Physiol (Lond) 412:333–349

Loirand G, Pacaud P, Mironneau C, Mironneau J (1990) GTP-binding proteins mediate noradrenaline modulation of calcium and chloride conductances in cultured rat portal vein myocytes. J Physiol (Lond) 428:517–529
Marc S, Leiber D, Harbon S (1986) Carbachol and oxytocin stimulate the generation of inositol phosphates in the guinea pig myometrium. FEBS Lett 201:9–14
Marshall JM (1962) Regulation of activity in uterine smooth muscle. Physiol Rev 42:213–227
Martin C, Dacquet C, Mironneau C, Mironneau J (1989) Caffeine-induced inhibition of calcium channel current in cultured smooth muscle cells from pregnant rat myometrium. Br J Pharmacol 98:493–498
Martin C, Arnaudeau S, Jmari K, Rakotoarisoa L, Sayet I, Dacquet C, Mironneau C, Mironneau J (1990) Identification and properties of voltage-sensitive sodium channels in smooth muscle cells from pregnant rat myometrium. Mol Pharmacol 38:667–673
McMillan M, Chernow B, Roth BL (1986) Phorbol esters inhibit alpha-adrenergic receptor-stimulated phosphoinositide hydrolysis and contraction in rat aorta: evidence for a link between vascular contraction and phosphoinositide turnover. Biochem Biophys Res Commun 134:970–974
Mironneau J (1973) Excitation-contraction coupling in voltage-clamped uterine smooth muscle. J Physiol (Lond) 233:127–141
Mironneau J (1974) Voltage clamp analysis of the ionic currents in uterine smooth muscle using the double sucrose gap method. Pflugers Arch 352:197–210
Mironneau J (1976) Effects of oxytocin on ionic currents underlying rhythmic activity and contraction in uterine smooth muscle. Pflugers Arch 363:113–116
Mironneau J, Savineau JP (1980) Effects of calcium ions on outward membrane currents in rat uterine smooth muscle. J Physiol (Lond) 302:411–425
Mironneau J, Savineau JP, Mironneau C (1981) Fast outward current controlling electrical activity in rat uterine smooth muscle during gestation. J Physiol (Paris) 77:851–858
Mironneau J, Mironneau C, Savineau JP (1984) Maintained contractions of rat uterine smooth muscle incubated in Ca^{2+}-free solution. Br J Pharmacol 82: 735–743
Mironneau J, Martin C, Arnaudeau S, Jmari K, Rakotoarisao L, Sayet I, Mironneau C (1990) High affinity binding sites for [^{3}H]saxitoxin are associated with voltage-dependent sodium channels in portal vein smooth muscle. Eur J Pharmacol 184: 315–319
Missiaen L, Vrolix M, Raemaekers L, Casteels R (1989) Regulation of the Ca^{2+} transport ATPase of smooth muscle. Adv Protein Phosphatases 5:239–260
Nakai Y, Kao CY (1983) Changing properties of Na^+ and Ca^{2+} components of the early inward current in the rat myometrium during pregnancy. Fed Proc 42:313
Nilius B, Hess P, Lansman JB, Tsien RW (1985) A novel type of cardiac calcium channels in ventricular cells. Nature 316:443–446
Nishizuka Y (1986) Studies and perspectives of protein kinase C. Science 233:305–312
Ohya Y, Sperelakis N (1989) Fast Na^+ and slow Ca^{2+} channels in single uterine muscle cells from pregnant rats. Am J Physiol 257:C408–C412
Okabe K, Kitamura K, Kuriyama H (1988) The existence of a highly tetrodotoxin sensitive Na channel in freshly dispersed smooth muscle cells of the rabbit main pulmonary artery. Pflugers Arch 411:423–428
Pacaud P, Loirand G, Mironneau C, Mironneau J (1989) Noradrenaline activates a calcium-activated chloride conductance and increases the voltage dependent calcium current in cultured single cells of rat portal vein. Br J Pharmacol 97: 139–146
Paponne P (1980) Voltage-clamp experiments in normal and denervated mammalian skeletal muscle fibres. J Physiol (Lond) 306:377–410
Parkington HC, Coleman HA (1990) The role of membrane potential in the control of uterine motility. In: Carsten ME, Miller JD (eds) Uterine function: molecular and cellular aspects. Plenum, New York, pp 195–248

Pidoplichko VI (1986) Two different tetrodotoxin-separable inward sodium currents in the membrane of isolated cardiomyocytes. Gen Physiol Biophys 6:593–604
Reiner O, Marshall JM (1975) Action of D600 on spontaneous and electrically stimulated activity of the parturient rat uterus. Naunyn Schmiedebergs Arch Pharmacol 290:21–22
Renaud JF, Kazaoglou T, Lombet A, Chicheportiche R, Jaimovich E, Romey G, Lazdunski M (1983) The Na^+ channel in mammalian cardiac cells. J Biol Chem 258:8799–8805
Rogart RB, Cribbs LL, Muglia LK, Kephart DD, Kaiser MW (1989) Molecular cloning of a putative tetrodotoxin-resistant rat heart Na^+ channel isoform. Proc Natl Acad Sci USA 86:8170–8174
Romey G, Lazdunski M (1982) Lipid-soluble toxins thought to be specific for Na^+ channels block Ca^{2+} channels in neuronal cells. Nature 297:79–80
Sakai K, Higuchi K, Yamaguchi T, Uchida M (1982) Oxytocin-induced Ca-free contractions of rat uterine smooth muscle: effects of preincubation with EGTA and drugs. Gen Pharmacol 13:393–400
Savineau JP, Mironneau J (1990a) Caffeine acting on pregnant rat myometrium: analysis of its relaxant action and its failure to release Ca^{2+} from intracellular stores. Br J Pharmacol 99:261–266
Savineau JP, Mironneau J (1990b) An analysis of the action of phorbol 12,13-dibutyrate on mechanical activity in rat uterine smooth muscle. J Pharmacol Exp Ther 255:133–139
Savineau JP, Mironneau J, Mironneau C (1987) Influence of the sodium gradient on contractile activity in pregnant rat myometrium. Gen Physiol Biophys 6:535–560
Savineau JP, Mironneau C, Mironneau J (1988) Contractile properties of chemically skinned fibers from pregnant rat myometrium: existence of an internal Ca-store. Pflugers Arch 411:296–303
Schrey MP, Read AN, Steer PJ (1986) Oxytocin and vasopressin stimulate inositol phosphate production in human gestational myometrium and decidua cells. Biosci Rep 6:613–619
Schrey MP, Cornford PA, Read AN, Steer PJ (1988) A role for phosphoinositide hydrolysis in human uterine smooth muscle during parturition. Am J Obstet Gynecol 159:964–970
Soloff MS (1984) Regulation of oxytocin action at the receptor level. In: Bottari S, Thomas JP, Vokaer A, Vokaer R (eds) Uterine contractility. Masson, Paris, pp 261–264
Somlyo AV, Bond M, Somlyo AP, Scarpa A (1985) Inositol trisphosphate-induced calcium release and contraction in vascular smooth muscle. Proc Natl Acad Sci USA 82:5231–5235
Sturek M, Hermsmeyer K (1986) Calcium and sodium channels in spontaneously contracting vascular muscle cells. Science 233:475–478
Tsien RW, Hess P, McCleskey EW, Rosenberg RL (1987) Calcium channels: mechanisms of selectivity, permeation and block. Annu Rev Biophys Chem 16:265–290
Ulbricht W (1972) Rate of veratridine action on the nodal membrane. I. Fast phase determined during sustained depolarization in the voltage clamp. Pflugers Arch 336:187–199
Varol FG, Hadjiconstantinou M, Zuspan FP, Neff NH (1989) Pharmacological characterization of the muscarinic receptors mediating phosphoinositide hydrolysis in rat myometrium. J Pharmacol Exp Ther 249:11–15
Yoshino M, Someya T, Nishio A, Yazawa K, Usuki T, Yabu H (1989) Multiple types of voltage-dependent Ca channels in mammalian intestinal smooth muscle cells. Pflugers Arch 414:401–409
Young RC, Herndon-Smith L (1991) Characterization of sodium channels in cultured human uterine smooth muscle cells. Am J Obstet Gynecol 164:175–181

CHAPTER 13

Effect of Potassium Channel Modulating Drugs on Isolated Smooth Muscle

G. EDWARDS and A.H. WESTON

A. Introduction

I. Modulation of Potassium Channels: General Principles

The intracellular potassium (K) concentration $[K^+]_i$ of smooth muscle cells is approximately 150 m*M* and this contrasts markedly with the much lower extracellular concentration of this ion $[K^+]_o$ which lies in the range 3–5 m*M* (JONES 1980). This disequilibrium arises because the smooth muscle plasma membrane exhibits very differing permeabilities to Na^+, K^+, Ca^{2+} and Cl^-. This fact, together with the activity of electrogenic ion pumps, especially Na^+/K^+ ATPase, and the intracellular synthesis of large non-diffusible ions gives rise to the observed 25:1 ratio of $[K^+]_i:[K^+]_o$ and to the basal membrane potential which is typically in the range −50 mV to −60 mV.

At rest, the smooth muscle cell is relatively permeable to K^+, probably because a number of (as yet) poorly defined K-channels are in a conducting state (see NOACK 1992). If additional and sufficient numbers of K-channels can be opened so that the ensuing K-conductance dominates all other ion conductances, the cell will hyperpolarize and the membrane potential will lie at the K-equilibrium potential, E_K (approximately −90 mV). Such an effect can be achieved using the drugs known as the K-channel openers (Sect. C.I.). In their presence, not only does the intracellular electrical potential move towards E_K, but the membrane becomes essentially voltage-clamped at E_K. Thus any tendency for the membrane potential to move in a positive (depolarizing) direction immediately results in further K^+ loss from the cell and the membrane remains "clamped" at or near E_K. Conversely, blockade or inhibition of K-channels (Sect. C.III.) when the cell is at rest results in membrane depolarization. This is due to the reduction of the "hyperpolarizing" effect which basal K-conductances exert on the resting membrane potential.

II. Mechanical Consequences of K-Channel Modulation

1. Plasmalemmal Effects

Smooth muscle tension changes result from fluctuations in the intracellular calcium concentration $[Ca^{2+}]_i$ which modifies the activity of contractile proteins. An increase in the concentration of this cation may be due to agonist–receptor interactions which induce inositol trisphosphate ($InsP_3$) production and stimulate Ca^{2+} release from intracellular stores. In addition, the opening of voltage-sensitive Ca-channels allows Ca^{2+} influx, which stimulates Ca^{2+}-induced Ca^{2+} release from intracellular stores (GANITKEVICH and ISENBERG 1992). The resultant rise in $[Ca^{2+}]_i$ enhances myosin light chain phosphorylation and generates force development (for reviews see MAYER et al. 1992; BERRIDGE 1993). Thus the modulation of K-channels indirectly exerts a marked inhibitory effect on the gating of depolarization-dependent Ca-channels. The membrane hyperpolarization which follows from K-channel opening increases the probability of closure of L-type Ca-channels and relaxation ensues. Conversely, K-channel closure generates depolarization and an increase in contractile activity.

2. Other Intracellular Effects

The K-channel openers may also exert additional intracellular effects which modify the contractile state of the smooth muscle cell. These include a modification of Ca^{2+}-release and refilling processes at intracellular Ca stores and the direct interference with contractile proteins. For details, see Sect. C.I.2.b)γ).

B. K-Channels in Smooth Muscle

Many types of K-channel exist. Each differs in its calcium, voltage or agonist sensitivity and makes a unique contribution to the complex modulation of membrane potential and tissue excitability. In this section, smooth muscle K-channels have been categorised according to their voltage, calcium or agonist sensitivities. Such an approach has certain advantages since the role of a particular K-channel under specific (quasi) physiological conditions can readily be appreciated. However, it overlooks possible structural similarities between channels of different sensitivities and the fact that some channels are modulated by multiple factors. In the future, an alternative classification based on structural families (see JAN and JAN 1992) should be possible, but this must await further advances using the techniques of molecular biology.

I. Voltage-Sensitive Channels

1. General Features of Activation and Inactivation

With the exception of the inward rectifier ("anomalous") K-channel (see Sect. B.I.3.c), voltage-sensitive channels open on depolarization of the plasma membrane. Some voltage-sensitive K-channels also demonstrate inactivation, i.e., current through the channel declines despite maintained depolarization. During recovery from inactivation the channel again passes through the open state (RUPPERSBERG et al. 1991). ARMSTRONG and BEZANILLA (1977) proposed that the amino-terminal region of sodium channel subunits could be responsible for channel inactivation. They suggested that the voltage-induced changes in transmembrane charge distribution, which allow channel opening, would permit binding of the charged amino terminal to a site associated with the channel, thus occluding the pore and resulting in channel inactivation. The model envisaged a "ball and chain" structure for the channel protein N-terminal. Mutation studies have now indicated that such a mechanism is likely to be responsible for inactivation of voltage-dependent K_A-channels and that the rate of inactivation is dependent upon the length of the "chain" tethering the "ball" (HOSHI et al. 1990; ZAGOTTA et al. 1990; see also HILLE 1992).

2. Channel Substructure

Much of the work relating to the protein structure of voltage-sensitive K-channels is derived from studies of fly genes. *Shaker* mutants of the fruit fly *Drosophila* have abnormal neuronal "A"-currents, a finding which led to the identification of a gene for the A-type K-channel which was present in the normal *Drosophila*, but not possessed by the *Shaker* mutant (SALKOFF 1983). Hydropathy plots of the proteins forming certain voltage-dependent K-channels suggest that these are formed from subunits consisting of six membrane-spanning regions, five of which (S1, S2, S3, S5, S6) are hydrophobic, and with N-terminal and C-terminal regions lying at the cytoplasmic side of the membrane (PONGS et al. 1988; Fig. 1). The S4 region is highly positively charged (every third amino acid is an arginine or lysine) and is thought to constitute the voltage sensor of the channel (STÜHMER et al. 1989; see also HILLE 1992). A further hydrophobic region between S5 and S6, termed H5, is thought to lie within the membrane (in the form of two antiparallel β-sheets) and to line the channel pore (MACKINNON and YELLEN 1990; HARTMAN et al. 1991; YOOL and SCHWARZ 1991). Four six-component subunits are thought to combine to form the K-channel, with the β-sheets of the H5 regions forming an eight-staved β-barrel which surrounds the aqueous pore of the channel.

Subsequently, other genes which encode for the voltage-sensitive K-channels, *Shab*, *Shal* and *Shaw*, have been isolated, firstly from *Drosophila* and later from mouse cDNA libraries (see ADAMS et al. 1992 for review). The sensitivity to blockers, voltage sensitivity and kinetics of activation and inactivation of the cloned homomultimeric channels differ. However, when *Xenopus* oocytes are injected with mixed mRNA, the heteromultimeric channels thus formed have properties intermediate between those of the homomultimers (ALDRICH 1990; ISACOFF et al. 1990). Thus, the apparent diversity of K-channels in different tissues could be due not only to alternative splicing or to post-translational modifications such as phosphorylation, but also to the formation of heteromultimeric channels (JAN and JAN 1990; RUDY et al. 1991).

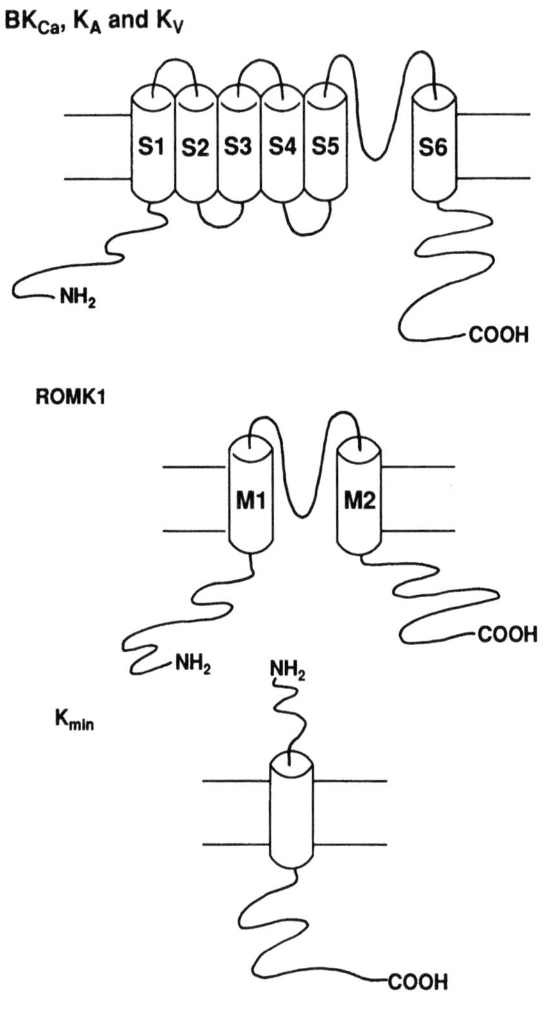

Fig. 1. Diversity of K-channel general structure. The channels (K_A, K_V and BK_{Ca}) responsible for carrying the A-current, delayed rectifier current and the large conductance Ca-dependent K-current, respectively, have similar basic structures (*upper panel*). These K-channels seem to be multimers consisting of four subunits, each containing six transmembrane segements S1–S6, with the carboxylic- and amino-termini in the cytoplasm. The ROMK1-channel subunit (*middle panel*) contains two membrane-spanning segments (*M1*, *M2*), again with carboxylic- and amino-termini orientated intracellularly. The electrophysiological characteristics of ROMK1 are typical of an inwardly rectifying K-channel, but the number of such units (possibly four?) required to assemble a complete channel is not known. The basic repeat unit of the K_{min} channel (*lower panel*) consists of a single transmembrane-spanning segment with the amino and carboxylic-termini orientated extra- and intracellularly, respectively. This channel, like the more complex structure shown in the *upper panel*, also exhibits the electrophysiological characteristics of a delayed rectifier, and has been described in uterine smooth muscle. For further details, see text

3. Voltage-Sensitive K-Channel Subtypes

a) The "A"-Channel (K_A)

Currents ($I_{K(A)}$) which are similar to the neuronal "A"-current (e.g., GUSTAFSSON et al. 1982; SEGAL et al. 1984; BELLUZZI et al. 1985) have been described in several smooth muscle cell types (BILBAUT et al. 1988; IMAIZUMI et al. 1989, 1990; BEECH and BOLTON 1989a; LANG 1989; NOACK et al. 1990, 1992b). The channels (K_A; unitary conductance 14 pS, quasi-physiological K^+ gradient; IMAIZUMI et al. 1990) which underlie this calcium-independent transient current activate and inactivate rapidly on depolarization after a period of hyperpolarization. K_A are partially inactivated at resting membrane potentials; in the rabbit portal vein, 50% are inactivated at −79 mV, and the channels are only fully activated on depolarization from potentials negative to −90 mV (BEECH and BOLTON 1989a). Since the channel underlying the "A"-current inactivates so rapidly on depolarization, it is unlikely that it contributes significantly to reversal of action potentials. However, particularly in cells which hyperpolarize significantly after action potential generation (e.g., guinea pig ileum; NAKAO et al. 1986), K_A probably acts to slow the rate of depolarization to the threshold for action potential firing with consequent reduction in action potential frequency. Such an effect is consistent with the finding of HARA et al. (1980) that 0.5 m*M* 4-aminopyridine (a relatively selective blocker of K_A in low millimolar concentrations) increases smooth muscle spike frequency (see Sect. C.III.2.).

b) The Delayed Rectifiers (K_V)

In tissues other than smooth muscle, the delayed rectifier K-current ($I_{K(V)}$) is considered to be carried by two or three subtypes of a calcium-insensitive K-selective channel (K_V), all of which open after a short delay on depolarization (DUBOIS 1981, 1983; BENOIT and DUBOIS 1986; SANGUINETTI and JURKIEWICZ 1990). The unitary conductance of K_V is approximately 9 pS under quasi-physiological conditions (BEECH and BOLTON 1989b; BOYLE et al. 1992; VOLK and SHIBATA 1993). Like the big conductance Ca-sensitive K-channel (BK_{Ca}; see Sect. B.II.1.), K_V channels play an important role in reversing cellular depolarization. In some smooth muscle cells, these channels inactivate in a time-dependent manner, fully inactivating within a few seconds (rabbit pulmonary artery, OKABE et al. 1987; rabbit portal vein, BEECH and BOLTON 1989b; rabbit coronary artery, VOLK et al. 1991; canine trachealis, MURAKI et al. 1990; pregnant rat myometrium, MIYOSHI et al. 1991; rat anococcygeus, MCFADZEAN and ENGLAND 1992). In others, e.g. rabbit jejunum, the channels appear to be non-inactivating (BENHAM and BOLTON 1983). Thus, most smooth muscle cells appear to possess delayed rectifier K-channels which may be of at least two different types. However IMAIZUMI et al. (1989) were unable to detect a delayed rectifier-type K-current in cells isolated from guinea pig ureter, and K_V may also be absent from the guinea pig bladder (KLÖCKNER and ISENBERG 1985).

In the heart, the channel underlying the delayed rectifier K-current is phosphorylation dependent, being enhanced by protein kinases A and C and inhibited by phosphatases (TOHSE et al. 1987; WALSH and KASS 1988; WALSH et al. 1991). In addition, the myocardial delayed rectifier current is extremely sensitive to the intracellular magnesium concentration, leading DUCHATELLE-GOURDON and coworkers (1991) to suggest that magnesium could modulate intracellular phosphatases thus modifying the phosphorylation and hence the activity of K_V. It is not yet known whether K_V in smooth muscle is phosphorylation dependent, although the finding that $I_{K(V)}$ runs down under conditions in which the intracellular concentration of ATP ($[ATP]_i$) is reduced (NOACK et al. 1992c; EDWARDS et al. 1993) suggests that it probably is. Indeed, recent studies suggest that modification of the phosphorylation of K_V may result in the conversion of this channel to a form currently known as the ATP-sensitive K-channel, K_{ATP}; this possibility is further discussed in Sect. C.I.3.a).

c) The Inward Rectifier (K_{IR})

In general, the inward rectifier K-channels (K_{IR}) conduct well at potentials negative to E_K, but pass relatively little (outward) current at potentials positive to this value. It is thought that interference with the outward passage of K^+ is mainly due to blockade of the channel by intracellular magnesium (Mg^{2+}), whereas inwardly moving K^+ is able to dislodge this blocking ion. In addition, inward rectification may also be an intrinsic property of the channel, the open probability of which increases with hyperpolarization (see HILLE 1992). In cardiac muscle, K_{IR} plays an important role in the termination of the action potential (ARENA et al. 1990). This channel may also be present in smooth muscle, and EDWARDS et al. (1988; EDWARDS and HIRST 1988) have described a K-conductance in the submucosal arterioles of the guinea pig ileum and rat middle cerebral arteries which is increased by hyperpolarisation. The underlying channel is partly activated at the resting membrane potential of the cell, passing outward current and thus probably contributing to the resting membrane potential (EDWARDS and HIRST 1988).

Two independent laboratories have recently described the cDNA sequence of inwardly rectifying K-channels (HO et al. 1993; KUBO et al. 1993). Typical of inwardly rectifying K-channels, the channel designated ROMK1 (renal outer medulla) described by HO and coworkers shows marked inward rectification and is blocked by barium (Ba^{2+}). Atypically, however, the channel open probability *increases* with depolarization and millimolar concentrations of MgATP are required for channel opening. Northern blot analysis of total RNA from a variety of tissues indicates the presence of ROMK1 in kidney, spleen, eye, thalamus, pituitary, hypothalamus and lung with no evidence for its presence in cardiac, skeletal or vascular smooth muscle (Ho et al. 1993). Similarly, membrane depolarization increases the open probability of the inwardly rectifying K-channel (IRK1) cloned by KUBO et al. (1993) from mouse macrophage cells. This channel shows considerable homology with the ROMK1 channel. Rectification is probably due to channel blockade by intracellular Mg^{2+}, and the channel is blocked both by caesium (Cs^{+}) and by Ba^{2+} (KUBO et al. 1993).

Although the mRNA for IRK1 is present in mouse forebrain and cerebellum as well as in skeletal and cardiac muscle, it remains to be determined whether smooth muscle expresses such a channel. A very interesting feature of the IRK1 and ROMK1 channels is that each channel subunit is relatively small and is thought to comprise only two membrane-spanning segments linked by a sequence which shows extensive similarity to the H5 regions of the voltage-dependent *Shaker*, *Shal*, *Shaw*, and *Shab* K-channels and the calcium-dependent *Slo* K-channels (Ho et al. 1993; KUBO et al. 1993; see Fig. 1). The structure of the IRK1 and ROMK1 channels thus differs markedly from that of those channels which comprise subunits consisting of six putative membrane-spanning regions with an additional hydrophobic pore-lining domain. This basic structure is apparently conserved by voltage-sensitive Ca- and Na-channels as well as by voltage- and by Ca-sensitive K-channels (CATTERALL 1988). Thus IRK1 and ROMK1 may belong to a new K-channel superfamily, members of which are related to, but which differ from, those of the voltage- or Ca-dependent K-channel family (KUBO et al. 1993; see Fig. 1).

d) K_{min}

This channel was first cloned from rat kidney by TAKUMI et al. (1988) and was originally named I_{SK}. Perhaps unkindly, other workers have referred to the channel as IsK?, since its protein structure (Fig. 1) is much simpler than that of any other known ion channel, and it was thought unlikely that the encoded protein was of sufficient complexity to form a channel. However, mutation studies designed to eliminate the possibility that the protein was merely activating a silent K-channel in oocytes have clearly indicated that K_{min} forms an integral part of the channel (TAKUMI et al. 1991). (Note that for the purpose of this review we have decided to use the term K_{min}, rather than I_{SK}, to avoid confusion with abbreviations for "intermediate" or "current"). It appears that the channel protein spans the membrane with the amino- and carboxy-terminals on the extracellular and intracellular sides, respectively (SUGIMOTO et al. 1990).

K_{min} mRNA is present in renal epithelial cells (SUGIMOTO et al. 1990), neonatal rat and mouse heart (FOLANDER et al. 1990; HONORE et al. 1991; LESAGE et al. 1992), rat myometrium (FOLANDER et al. 1990; PRAGNELL et al. 1990) and human T lymphocytes (ATTALI et al. 1992). When expressed in *Xenopus* oocytes, K_{min} is voltage sensitive, slowly activating on depolarisation. However, it shows no inactivation and is blocked by Ba^{2+}, Cs^{+} and tetraethylammonium (TEA) ions (TAKUMI et al. 1988; MURAI et al. 1989; HAUSDORFF et al. 1991). It is thought that K_{min} may be responsible for carrying the delayed rectifier current in myometrium (FOLANDER et al. 1990). Expression of the myometrial delayed rectifier channel is dependent upon the presence of oestrogen (BOYLE et al. 1987), and the dramatic changes in myometrial excitability during oestrus and pregnancy can be attributed to associated changes in the magnitude of the delayed rectifier K-current (BOYLE et al. 1987).

II. Calcium-Sensitive K-Channels

Mutant *Drosophila* which are homozygous for *slow-poke* (*Slo*) alleles lack a calcium-activated K-current. Comparison of amino acid sequences encoded by the normal and mutant *Drosophila* mRNA has allowed identification of the missing Ca-activated K-channel gene and its expression in *Xenopus* oocytes (ADELMAN et al. 1992). Like the voltage-sensitive K-channels described above, this channel is thought to be formed from four basic subunits each with six membrane-spanning regions. Although it shows little homology with other cloned K-channels, the H5 region of *Slo* is well conserved (ADELMAN et al. 1992).

Ca-sensitive K-channels have been divided into three groups based primarily on their unitary conductance – the "big" conductance K-channel (BK_{Ca}), the "intermediate" conductance K-channel (IK_{Ca}) and the "small" conductance K-channel (SK_{Ca}; ROMEY and LAZDUNSKI 1984; PENNEFATHER et al. 1985; BLATZ and MAGLEBY 1986, 1987). Of these, BK_{Ca} appears to be the most important channel in the regulation of smooth muscle activity, although IK_{Ca} and SK_{Ca} are probably also present.

1. BK_{Ca}

BK_{Ca} is both voltage and Ca sensitive, it has a conductance of 150–250 pS in a symmetrical (100–140 m*M*) K^{+} gradient, and has been described in uterine, bladder, airways, gastrointestinal tract and vascular smooth muscle (BENHAM et al. 1985; INOUE et al. 1985; BENHAM et al. 1986; MCCANN and WELSH 1986; CECCHI et al. 1986; SINGER and WALSH 1987; SADOSHIMA et al. 1988a; CARL and SANDERS 1989; KUME et al. 1989; MAYER et al. 1990; MURAKI et al. 1990). The density of the channel is high and has been calculated to be approximately per 10000 cell (BENHAM and BOLTON 1986). This, together with its large conductance, means that a small change in BK_{Ca} open probability can exert a large effect on the membrane potential of a smooth muscle cell. In the presence of a low calcium concentration (100 n*M*), BK_{Ca} only opens at potentials positive to −20 mV in most mammalian smooth muscle tissues (BENHAM et al. 1985; MCCANN and WELSH 1986, 1987; SADOSHIMA et al. 1988a) and in some tissues potentials more positive than +40 mV are required (CARL and SANDERS 1989; STUENKEL 1989; TORO et al. 1990). Nevertheless, the channel can open at depolarized potentials even in the absence of calcium (BLATZ and MAGLEBY 1987; LATORRE et al. 1989). Ca^{2+} is thought to bind to a site on the channel and to enhance channel opening, almost in the manner of an agonist. In the presence of higher calcium concentrations (e.g., 1 μM) BK_{Ca} opens at more hyperpolarized potentials (around −40 mV; see BOLTON and BEECH 1992). Typically, BK_{Ca} shows no inactivation (BENHAM et al. 1986; BEECH and BOLTON 1989b).

In both rat aorta and rabbit trachea, channel phosphorylation, catalysed by a cAMP-dependent kinase, modifies BK_{Ca} activity (KUME et al. 1989; SADOSHIMA et al. 1988b; MINAMI et al. 1993), possibly by increasing the sensitivity of BK_{Ca} to $[Ca^{2+}]_i$ (SADOSHIMA et al. 1988b). Since most patch-clamp experiments are performed using solutions which lack ATP or substrates for the tricarboxylic acid cycle, the Ca^{2+} sensitivity of BK_{Ca} determined using such techniques may be an underestimate of its actual sensitivity. In porcine coronary artery cells, phosphorylation of BK_{Ca} by protein kinase C paradoxically *inhibits* channel opening, suggesting that protein kinases A and C have different target phosphorylation sites (MINAMI et al. 1993). Such a finding has also been described by WANG and GIEBISCH (1991), who found similar (i.e., opposite) effects of protein kinases C and A on the activity of a renal ATP-sensitive K-channel (K_{ATP}). Other factors which may influence the calcium sensitivity of BK_{Ca} include pH (KUME et al. 1990), G proteins (COLE and Sanders 1989; TORO et al. 1990) and GMP (EWALD et al. 1985; WILLIAMS et al. 1988).

In spontaneously contracting tissues, membrane depolarization associated with action potentials enhances the intracellular calcium concentration ($[Ca^{2+}]_i$) and induces tissue contraction. In response both to the rise in $[Ca^{2+}]_i$ and to the depolarization, BK_{Ca} opens, thus terminating the contraction. In quiescent tissue, BK_{Ca} may contribute to the stability of the membrane, opposing any depolarizing influences. Thus, in airway smooth muscle, BK_{Ca} is thought to be responsible for the strong outward rectification of the membrane which prevents the generation of action potentials in normal physiological salt solutions (McCANN and WELSH 1986; see reviews by KOTLIKOFF 1993 and SMALL et al. 1993). BK_{Ca} also carries the spontaneous transient outward currents (STOCs) which are thought to be induced by transient release of Ca^{2+} from $InsP_3$-sensitive Ca stores (BEECH and BOLTON 1989a,b; STRONG et al. 1989; MURAKI et al. 1990).

a) BK: A Variant of BK_{Ca}

A large conductance calcium-insensitive channel has been described in both bovine trachealis and human myometrium (GREEN et al. 1991; KHAN et al. 1993). In each of these tissues, the unitary conductance of the calcium-insensitive K-channel (BK) is identical to that of BK_{Ca} (GREEN et al. 1991; KHAN et al. 1993). During labour, myometrial cells appear to possess almost exclusively the calcium-insensitive large conductance K-channels, whereas during pregnancy BK is not evident and only BK_{Ca} is found (KHAN et al. 1993). BK_{Ca} differs from BK in its sensitivity both to barium and to TEA (GREEN et al. 1991; KHAN et al. 1993). KHAN et al. (1993) proposed that BK may be a modification of BK_{Ca} which has lost its calcium sensitivity. Since its barium sensitivity is also reduced, the binding sites for calcium and barium may be closely associated (KHAN et al. 1993).

2. SK_{Ca}

Apamin, a toxin derived from the venom of the honey bee, *Apis mellifera*, is a highly selective inhibitor of a small conductance, highly calcium-dependent K-channel, SK_{Ca}, in rat skeletal muscle (BLATZ and MAGLEBY 1986). In a variety of tissue types, SK_{Ca} (unitary conductance 10–20 pS), unlike BK_{Ca}, has no intrinsic voltage sensitivity. Instead, its apparent voltage sensitivity arises from voltage-induced changes in $[Ca^{2+}]_i$ (HILLE 1992). Patch-clamp experiments have largely failed to identify this type of channel in smooth muscle, although COLLIER et al. (1990) found that apamin (1 μM) inhibits a K-current in bovine trachealis cells. In addition, apamin is capable of inhibiting the contractile effects of noradrenaline in the guinea pig taenia caecum, and it increases the frequency of spontaneous contractions in the guinea pig ileum and bladder, consistent with inhibition of a K-channel (WEIR and WESTON 1986a; BAUER and KURIYAMA 1982; ZOGRAFOS et al. 1992). Since SK_{Ca} does not have voltage sensitivity, it may be activated by increases in $[Ca^{2+}]_i$ which occur at potentials close to the resting membrane potential and thus play a role in the regulation of action potential frequency (McMANUS 1991). The SK_{Ca} which INOUE et al. (1985) described in rabbit portal vein cells apparently differs from that identified in other tissues, since its opening was voltage dependent, although these workers did not specifically examine its apamin sensitivity.

3. IK_{Ca}

Although an IK_{Ca} has been described in smooth muscle (INOUE et al. 1985; WANG and MATHERS 1991), data concerning this channel are sparse. In the rabbit portal vein the channel has a unitary conductance of 92 pS (symmetrical K^+ gradient) and is less sensitive to TEA than BK_{Ca} (INOUE et al. 1985). Channels with a conductance of 91 pS or 100 pS, and with a similar sensitivity to TEA, but which lack calcium sensitivity have been detected in rat pancreatic arterioles (STUENKEL 1989), frog and toad stomach (BERGER et al. 1984) and rabbit jejunum (BOLTON et al. 1986). In contrast, the channel found in rat cerebral arteries (92 pS) was very sensitive to TEA (K_d, 0.31 mM; WANG and MATHERS 1991). It thus seems possible that the intermediate-conductance K-channel described by INOUE et al. (1985) is the same as that found in other smooth muscle preparations, but that the experimental conditions employed by the Japanese group may have affected its apparent calcium sensitivity.

4. $K_{Ca(o)}$

There are few reports of K-channels sensitive to extracellular calcium. However, INOUE et al. (1986) described the presence of such a channel, with a large unitary conductance (180 pS), in rabbit portal vein cells. A smaller conductance (30 pS) K-channel which was sensitive to extracellular calcium, but which was opened by nicorandil (20–200 μM) and by 4-acetamide-4′-isothiocyanostilbene-2,2′disulfonic acid (SITS; 1–10 μM) in the presence of a low extracellular calcium concentration, has also been described in porcine coronary artery (INOUE et al. 1989). The significance of $K_{Ca(o)}$ channels has yet to be determined.

III. ATP-Sensitive K-Channels

The existence of K-channels which are inhibited by physiological concentrations of intracellular ATP ($[ATP]_i$) and which open as $[ATP]_i$ falls has been reported in numerous cell types (ASHCROFT 1988; ASHCROFT and ASHCROFT 1990; EDWARDS and WESTON 1993a). These channels, currently designated K_{ATP}, appear to be the site of action of the K-channel openers (Sect. C.I.1.a–g). Until recently, a single report (STANDEN et al. 1989) showed that such a channel might be also present in smooth muscle cells, although the unitary conductance of this channel (approximately 100 pS, recalculated for quasi-physiological conditions) is much higher than that typically reported for K_{ATP} (approximately 20 pS in quasi-physiological conditions; see ASHCROFT 1988).

Recently, there have been several reports of a non-inactivating K-current ($I_{K(ATP)}$) in smooth muscle which is stimulated by reducing $[ATP]_i$ (CLAPP and GURNEY 1992; NOACK et al. 1992c; SILBERBERG and VAN BREEMEN 1992; BONEV and NELSON 1993). The underlying channel has a unitary conductance of 10–20 pS when measured under quasi-physiological conditions (NOACK et al. 1992c), consistent with the conductance of so-called type 1 ATP-sensitive K-channels in other tissues (ASHCROFT 1988; ASHCROFT and ASHCROFT 1990). This is similar to the conductance of the smooth muscle K-channel opened by several K-channel openers (NOACK et al. 1992a; IBBOTSON et al. 1993a; BEECH et al. 1993b; BONEV and NELSON 1993; CRIDDLE et al. 1994).

Recent studies have cast doubt on whether K_{ATP} is a discrete channel in its own right or whether it is a partially dephosphorylated state of the delayed rectifier channel K_V (EDWARDS et al. 1993; EDWARDS and WESTON 1993b; IBBOTSON et al. 1993b). In both vascular smooth muscle and insulinoma cells, these workers showed that K_V and K_{ATP} shared a common pharmacology and that dephosphorylating conditions or exposure to K-channel openers, both of which induced $I_{K(ATP)}$, was always associated with a parallel reduction in $I_{K(V)}$. Furthermore, the unitary conductances of these two channels are essentially identical (both in the range 6–20 pS; see EDWARDS et al. 1993 for further details).

Irrespective of whether K_{ATP} is a state of K_V or a distinct entity, its physiological importance in smooth muscle is uncertain, although it may play a role under conditions of ischaemia (DAUT et al. 1990). It remains to be determined whether changes in $[ATP]_i$ are primarily responsible for modulating the opening of K_{ATP} (or conversion of K_V) in vivo or whether agonists such as adenosine (DAUT et al. 1990; KIRSCH et al. 1990; ORITO et al. 1993), second messenger systems or nucleotide phosphates other than ATP (TUNG and KURACHI 1991; BEECH et al. 1993a; PFRÜNDER et al. 1993) are more important in the modulation/conversion process. In other tissues, the regulation of the K_{ATP} state is complex (see EDWARDS and WESTON 1993a for review) and may additionally involve nucleotide diphosphates (TUNG and KURACHI 1991; SHEN et al. 1991), G proteins (ITO et al. 1992) and phosphorylation by protein kinases A and C (WOLLHEIM et al. 1988; DE WEILLE et al. 1989; RIBALET et al. 1989).

C. Modulators of Smooth Muscle K-Channels

I. Synthetic K-Channel Openers

1. Chemical Classification

The term "K-channel opener" was first used to describe the smooth muscle relaxant actions of the synthetic agent cromakalim (HAMILTON et al. 1985, 1986). A few years earlier, the hyperpolarizing effect of nicorandil on vascular smooth muscle had been described (FURUKAWA et al. 1981). However, the importance of this finding was somewhat overshadowed by the subsequent observation that nicorandil was an activator of guanylate cyclase (HOLZMANN 1983), and it was some time before it was realised that this agent was both a K-channel opener and a nitrovasodilator (WEIR and WESTON 1986a,b). Since these initial observations, several apparently distinct chemical classes of synthetic K-channel opener have been recognised (EDWARDS and WESTON 1990a; ROBERTSON and STEINBERG 1990), and details of the background to the synthesis of cromakalim have been published (STEMP and EVANS 1993). Claims for the existence of a pharmacophore common to all classes of K-channel opener have recently been published (KOGA et al. 1993), although it seems doubtful that such a structure can account for the highly potent thioformamide derivatives (BROWN et al. 1993).

Attempts to classify the synthetic K-channel openers are still rather arbitrary. Although distinct chemical nuclei (e.g., benzopyran) can be recognised as the basic feature of several molecules, others such as the so-called cyanoguanidines and thioformamides often possess a pyridine moiety. This feature would allow them to be classed along with the pyridine nicorandil. However, because of potency and other differences, they are usually categorised separately. Although the claim that a pharmacophore exists which is common to many of the K-channel openers (KOGA et al. 1993), it is by no means certain that all openers of, say, K_{ATP} (see Sect. C.I.3.a) interact with a common site to modulate this channel.

a) Benzopyrans

The prototype substance in this series can be regarded as the racemate, cromakalim. This is a mixture of two *trans* enantiomers, the relatively inactive molecule BRL 38226 and the more active (−)-enantiomer, levcromakalim (formerly BRL 38227; Fig. 2). The presence of two chiral carbon atoms at positions 3 and 4 of the benzopyran nucleus means that a total of four enantiomers exists, but virtually all published work has been performed with the *trans* configured molecules BRL 38226 and levcromakalim. Recently, however, details of the pharmacology of the two *cis* enantiomers have been published and, as expected, neither was as potent as levcromakalim in vivo or in vitro (QUAST and VILLHAUER 1993).

More variants of the benzopyran nucleus have been synthesised than of any other chemical class of K-channel opener. Such molecules include Ro31-6930, bimakalim, emakalim, EMD 57283, BRL 55834 and the pro-drug Y27152. These were produced in an attempt to increase both potency and duration of action and to introduce tissue selectivity (Figs. 2, 3; see also EDWARDS and WESTON 1990b). Evidence for selective actions on individual tissues is based largely on in vivo evaluation and may be the consequence of selective localisation of the agent in a particular organ. Rilmakalim (Fig. 3; formerly HOE234; see MIURA et al. 1993) and BRL55834 (Figs. 3, 4; BOWRING et al. 1993) are under investigation as selective bronchodilators, while bimakalim and emakalim (Fig. 3) exhibit a potentially favourable anti-anginal profiles (GROSS 1991).

Molecules which can be regarded as hybrids of the benzopyrans and the cyanoguanidines (see Sect. C.I.1.b) have been synthesised. Some of these, including the pyranylcyanoguanidine BMS 180448 (Fig. 3) are relatively impotent smooth muscle relaxants in vitro. However, in vivo they exhibit cardiac anti-ischaemic properties without lowering systemic blood pressure (ATWAL et al. 1993).

b) Cyanoguanidines

The prototype agent in this class is the racemate pinacidil (Fig. 2), an agent synthesised in the early 1970s but not recognised as a K-channel opener until much later (BRAY et al. 1987a,b). Evidence that actions additional to K-channel opening may contribute to its pharmacological profile (COOK et al.

Benzopyrans

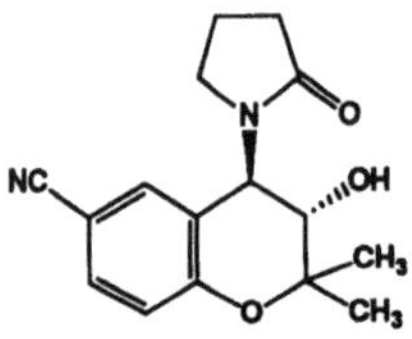

Levcromakalim

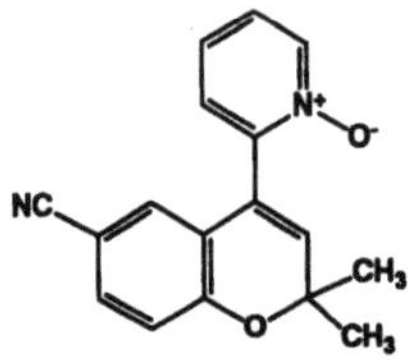

Ro 31-6930

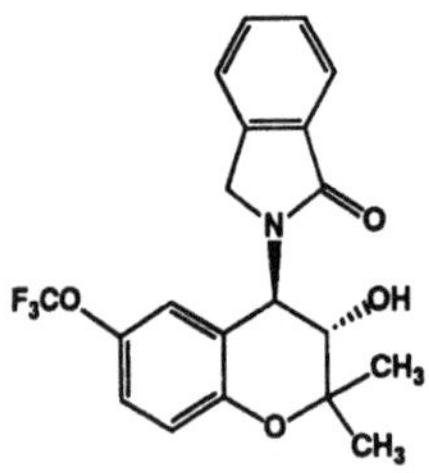

Celikalim

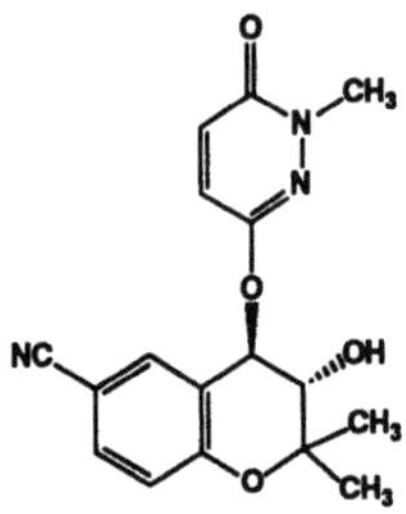

EMD 57283

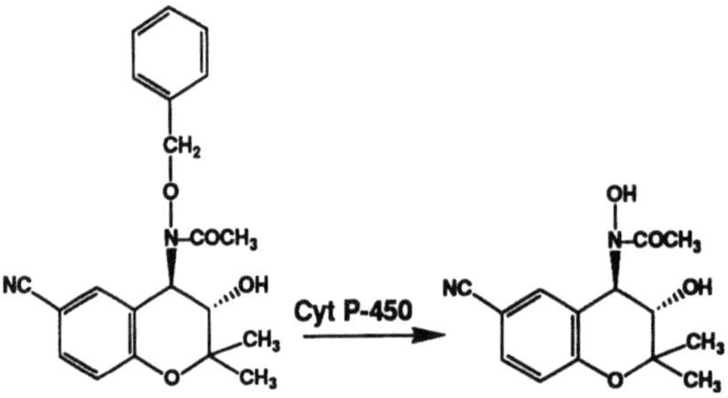

Y-27152
(inactive)

Y-26763
(active)

Cyanoguanidines

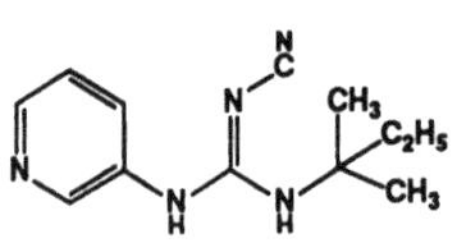

P1075

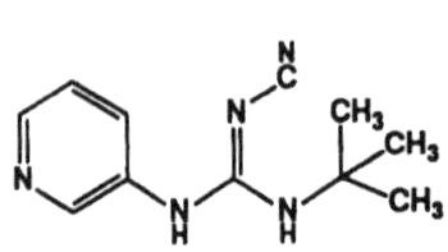

P1060

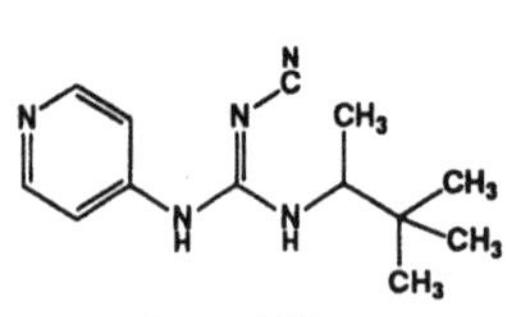

Pinacidil

Pyridines

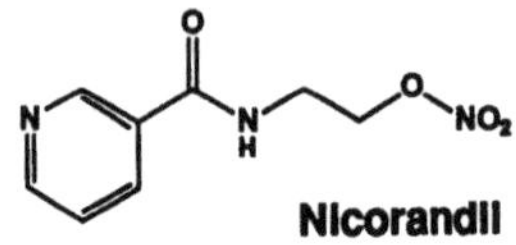

Nicorandil

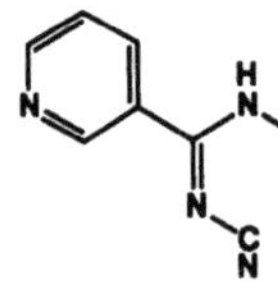

KRN 2391

Thioformamides

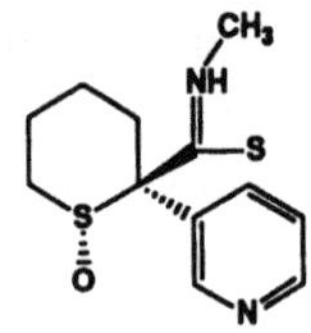

Aprikalim

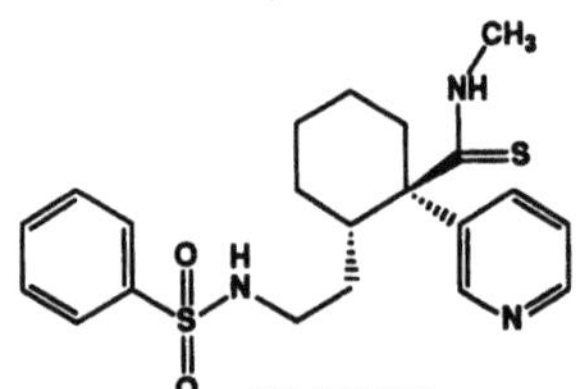

RP 66784

Miscellaneous

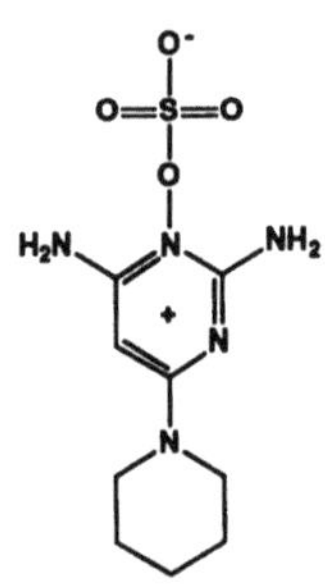

Minoxidil sulphate

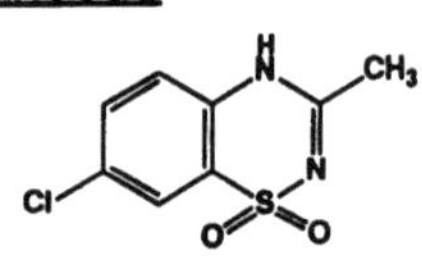

Diazoxide

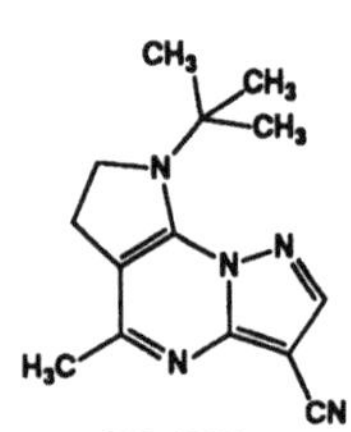

LP-805

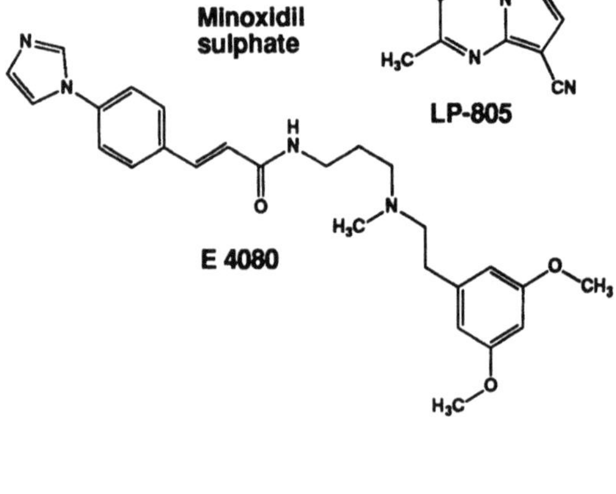

E 4080

1989) has prompted the use of related molecules, including P1060 (IBBOTSON et al. 1993a) and P1075 (BRAY and QUAST 1992a).

c) Thioformamides

These K-channel openers, the prototype molecule of which is the racemate RP49356, were derived from agents which were originally developed as inhibitors of K^+/H^+ ATPase (MONDOT et al. 1989). The active enantiomer of this racemate has been designated aprikalim (formerly RP52891). Within the series of thioformamide (or carbothioamide) K-channel openers exist the most potent molecules so far developed based on in vitro tests. Typical of these is RP66784 (Fig. 2), with a reported IC90 value of 0.3 n*M* against a 20 m*M* KCl-induced contraction in rat aorta (BROWN et al. 1993).

d) Pyridines

Nicorandil, the K-channel opener with additional guanylate cyclase-stimulating properties, is the protoype of this class. The use of the term "pyridine" illustrates the rather arbitrary nature of the chemical classification of K-channel openers since the cyanoguanidines (Sect. C.I.1.b) and the thioformamides (Sect. C.I.1.c) could also be placed in this category (see EDWARDS and WESTON 1993a). The pharmacology of only a few structural variants of nicorandil has been described (MARUYAMA et al. 1982; INOUE et al. 1984) and the structural features of nicorandil appear optimal for maximum potency. A closely related molecule, KRN 2391 (Fig. 2), with reported potency greater than that of nicorandil, is also under development (KANETA et al. 1990; MIWA et al. 1993).

e) Pyrimidines

Until recently the only member of this group was minoxidil sulphate. Based on the protocols employed to show that cromakalim was a K-channel opener (HAMILTON et al. 1985, 1986), some evidence was obtained that this agent, the active metabolite of minoxidil (MCCALL et al. 1983), was also capable of opening smooth muscle K-channels (MEISHERI et al. 1988). This view was subsequently confirmed using a variety of organ bath, ion flux and micro-

Fig. 2. Chemical structures of typical synthetic K-channel openers believed to exert their effects by opening the ATP-sensitive K-channel, K_{ATP}. The molecules shown have been grouped according to their chemical class and are those currently under clinical evaluation or which have been relatively widely investigated. The miscellaneous group comprises the pyrimidines (minoxidil sulphate and LP-805), the benzothiadiazine, diazoxide and the butenoic acid derivative, E4080. Key: *E*, Eisai; *EMD*, E. Merck; *KRN*, Kirin Brewery; *LP*, Pola; *P*, Leo, Denmark; *Ro*, Roche; *RP*, Rhône-Poulenc Rorer; *Y*, Yoshitomi. For further details, see text

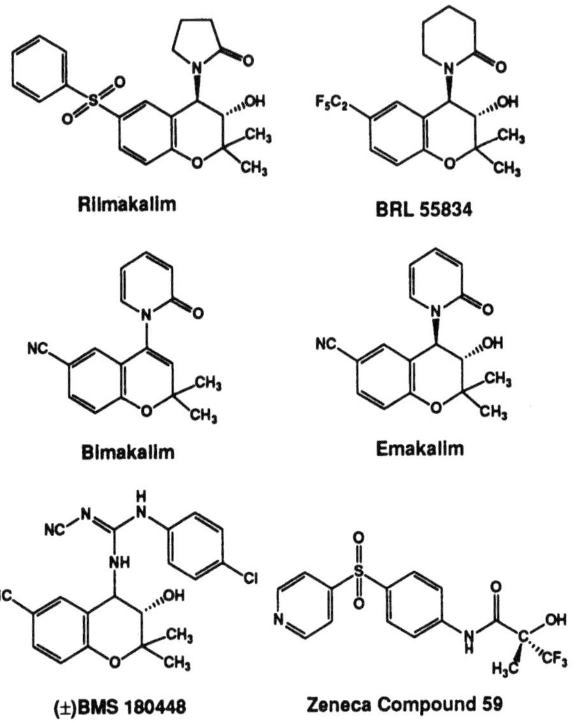

Fig. 3. Chemical structures of synthetic K-channel openers which exhibit some tissue selectivity, especially in vivo. Rilmakalim (formerly HOE 234) and BRL 55834 are airway selective, while bimakalim and emakalim (formerly EMD 52692 and EMD 58431, respectively) are active on coronary blood vessels. The racemic pyranylcyanoguanidine BMS 180448 and Zeneca compound 59 selectively exhibit anti-ischaemic and bladder-relaxant properties, respectively. Key: *BMS*, Bristol Myers Squibb; *EMD*, E. Merck; *HOE*, Hoechst. For further details, see text

electrode techniques (WINQUIST et al. 1989; NEWGREEN et al. 1990; BRAY and QUAST 1991a). More recently, another pyrimidine-based smooth muscle relaxant, LP-805 (Fig. 2), has been described with properties typical of those of a K-channel opener (KISHII et al. 1992a,b; KAMOUCHI et al. 1993).

f) Benzothiadiazines

Diazoxide is essentially the only example of this type of K-channel opener. Originally developed as an antihypertensive agent, its use has been restricted by its prominent hyperglycaemic side effect (EDWARDS and WESTON 1990b). This action is associated with its ability to open K_{ATP} channels in pancreatic β-cells (ZÜNKLER et al. 1988). Subsequently, its glibenclamide-sensitive vasodilator activity was reported (QUAST and COOK 1989a; NEWGREEN et al. 1990).

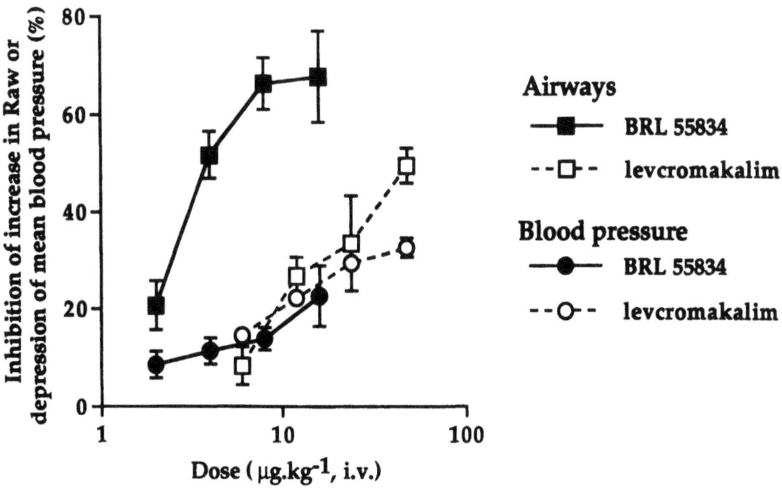

Fig. 4. In vivo comparison of the ability of levcromakalim and of BRL 55834 to inhibit the increase in airways resistance (R_{aw} □, ■ respectively) produced by a histamine aerosol and to depress mean blood pressure (○, ●, respectively) in the guinea pig. (Adapted from BOWRING et al. 1993)

Diazoxide is a close chemical relative of the thiazide diuretics. The acute action of hydrochlorthiazide on blood vessels is unaffected by glibenclamide, but inhibited by charybdotoxin, suggesting the possible involvement of Ca-dependent K-channels in the antihypertensive action of this agent (CALDER et al. 1992).

g) Miscellaneous Agents

α) E4080. The butenoic acid derivative E4080 (Fig. 2) exhibits bradycardic and coronary vasodilating properties (KAWAMURA et al. 1990). In vitro studies have indicated that this molecule can open a glibenclamide-sensitive K-channel in smooth muscle, although effects on Ca-currents and on intracellular Ca-handling have also been reported (KAMOUCHI et al. 1991; OKADA et al. 1992).

β) SCA40. SCA40 is a novel imidazopyrazine derivative with smooth muscle relaxant properties (BONNET et al. 1992; Fig. 5). In a detailed analysis of its bronchodilator action, the relaxant effects of this agent on guinea pig trachealis were unaffected by glibenclamide, but were inhibited by charybdotoxin (LAURENT et al. 1993). These observations led to the conclusion that SCA40 exerts its bronchodilator effects by opening BK_{Ca} (LAURENT et al. 1993). SCA40 is chemically related to the xanthine smooth muscle relaxants such as aminophylline and theophylline and, like these, SCA40 is a phosphodiesterase inhibitor (BONNET et al. 1992). Since the relaxant action of theophylline is also charybdotoxin sensitive (MURRAY et

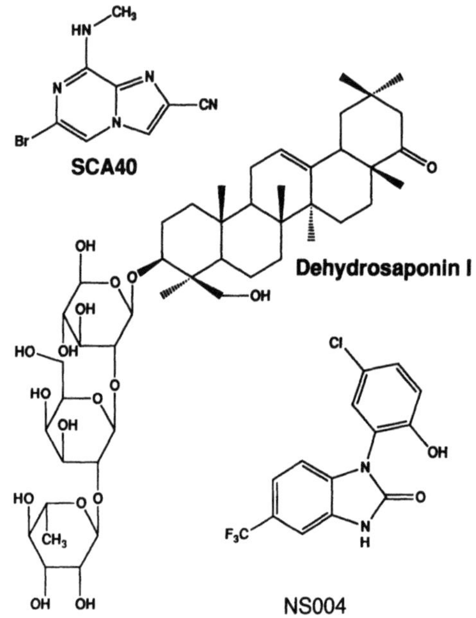

Fig. 5. Chemical structures of K-channel openers believed to be active at the large, calcium-activated K-channel, BK_{Ca}, in smooth muscle. Dehydrosaponin 1 is a modification of the active principle found in the medicinal herb *Desmodium adscendens* while SCA40 and NS004 are synthetic molecules. For further details, see text

al. 1991), it seems probable that any effects of SCA40 on a charybdotoxin-sensitive K-channel are the consequence of phosphodiesterase inhibition. To test this, studies on the specific phosphodiesterase isoenzymes associated with relaxation in the trachealis muscle (ELLIOTT et al. 1991) should be performed.

As with the β-adrenoceptor agonists (see Sect. C.II.7.b), the importance of K-channel opening and especially that of BK_{Ca} in the relaxant actions of agents which effectively increase cAMP concentrations in smooth muscle is a matter of conjecture. In high K^+ physiological salt solutions, both the xanthines (SMALL et al. 1989) and β-adrenoceptor agonists such as isoprenaline (ALLEN et al. 1985) are still able to relax bronchial smooth muscle. Such observations are a possible indication that K-channel opening may not be of primary importance in the bronchodilator effects of these agents.

γ) Benzimidazoles. A single brief report has described the opening effects of some benzimidazole derivatives (for a typical structure, see Fig. 4) on BK_{Ca} in cultured bovine aortic and guinea pig tracheal smooth muscle cells (OLESEN and WÄRJEN 1992). Some of the structural features of this class of agent are similar to those of SCA40 (see Sect. C.I.1.g)β); Fig. 5) and of LP 805 (Sect. C.I.1.e); Fig. 2). Further studies are necessary to establish whether benzimidazoles represent an important new class of K-channel opener, active at BK_{Ca}.

δ) *Triterpenoid Glycosides.* In Ghana, crude extracts of the herb *Desmodium adscendens* have long been used in the treatment of bronchial asthma. Crude extracts of this plant were found to inhibit the binding of ^{125}I-charybdotoxin to BK_{Ca} in bovine trachea (McManus et al. 1993a,b). Further analysis and purification revealed that these extracts contained three active principles, the most potent of which was the triterpenoid glycoside, dehydrosaponin 1 (Fig. 4). When applied to the intracellular, but not extracellular, side of the membrane, this agent reversibly increased the open probability of BK_{Ca} channels from bovine trachea when these were incorporated into planar lipid bilayers (McManus et al. 1993a,b).

ε) *Methylpropanamides.* A single report concerning the pharmacological characteristics of a series of chemically novel methylpropanamides with K-

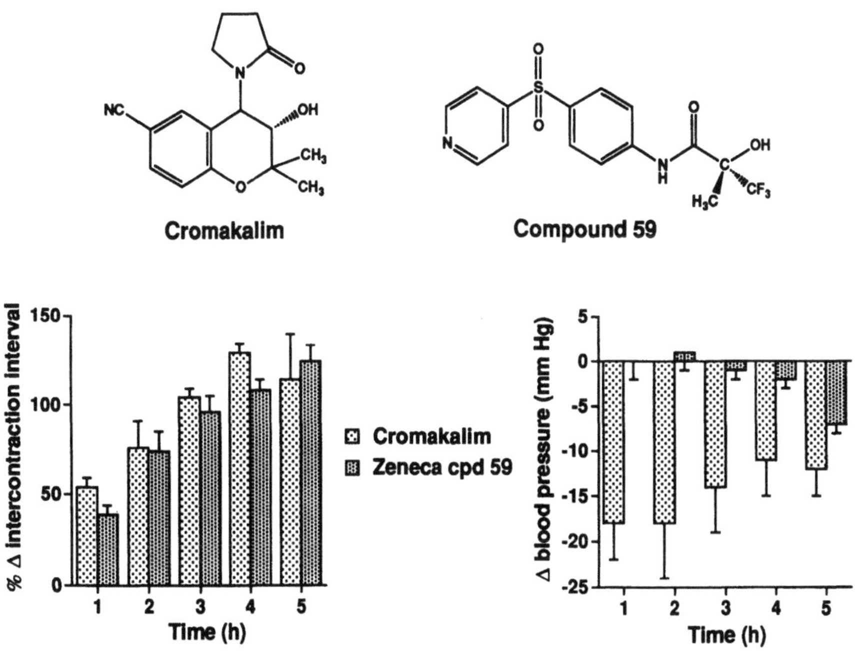

Fig. 6. In vivo comparison of the effects of cromakalim (1 mg/kg, po) and Zeneca compound 59 (3 mg/kg, po) in the conscious rat. The *lower left panel* shows that both agents produce a similar time-dependent increase (Δ) in the intercontraction interval of rat bladder whereas compound 59 produces relatively little fall in blood pressure (*right panel*) compared with cromakalim over the same time interval. Each histogram is derived from four experiments; *Vertical bars* show s.e. mean values. (Figure drawn from data given in Russell et al. 1993)

channel opening properties has so far been published (RUSSELL et al. 1993). These workers describe the in vitro mechano-inhibitory actions of several variants on the propanamide structure, a typical example of which is the agent designated Zeneca compound 59 (Fig. 3). Using rat isolated bladder as the test tissue, the in vitro potency of the methylpropanamides is similar to that of cromakalim. However, in vivo, evidence of bladder-selective relaxation with little effect on systemic blood pressure has been obtained (RUSSELL et al. 1993; Fig. 6). Whether the basis of this selectivity is pharmacokinetic remains to be established, but agents with tissue selectivity have enormous clinical potential (see Sect. D.).

2. Effects on Smooth Muscle In Vitro

As a group, the synthetic openers of K_{ATP} (see Sect. C.I.3.a) exert a series of characteristic effects on smooth muscle-containing tissues. These include inhibition of spontaneous mechanical activity, relaxation of agonist- and KCl ($\leq$20–25 m*M*)-induced contractions, an increase in the rate of $^{42}K/^{86}Rb$ exchange and membrane hyperpolarisation. Together these constitute a unique profile which distinguishes these agents from other smooth muscle relaxants.

a) Spontaneous Mechanical Activity

The spontaneous mechanical activity of all spontaneously active smooth muscles so far tested is inhibited by the K-channel openers. In rat portal vein, HAMILTON et al. (1985, 1986) first demonstrated that cromakalim reduced the amplitude and frequency of spontaneous contractions which were abolished at sufficiently high concentrations. Subsequent studies in the rat and guinea pig subsequently showed that this was a general feature of the K-channel openers as a group (BRAY et al. 1987a; SOUTHERTON et al. 1988; NEWGREEN et al. 1990). In guinea pig trachealis muscle, spontaneous tone is greatly reduced (ALLEN et al. 1985; BRAY et al. 1987a; PACIOREK et al. 1990; RAEBURN and BROWN 1991), although the K-channel openers are not as efficacious as isoprenaline or aminophylline (ALLEN et al. 1986; MURRAY et al. 1989; BERRY et al. 1991). Inhibition of spontaneous uterine activity was first demonstrated for pinacidil (COHEN and COLBERT 1986), an observation which was extended to cromakalim by HOLLINGSWORTH et al. (1987) and to other K-channel openers by PIPER et al. (1990). Inhibition of spontaneous contractions occurs in the normal and hypertrophied bladders of various species (ANDERSSON et al. 1988; FOSTER et al. 1989a,b; MALMGREN et al. 1990).

b) Agonist-Induced Contractions

α) KCl. Early studies with cromakalim showed that this agent could relax the vascular contractile effects of low ($\leq$20–25 m*M*) KCl concentrations, whereas the effects of higher (>40 mM) KCl concentrations were little affected (HAMILTON et al. 1986; WEIR and WESTON 1986b). Since then, this characteristic action of the K-channel openers has become established as a key distinguishing feature of this class of agent. It was originally argued that the potassium equilibrium potential (E_K) which would obtain when the physiological salt solution [K^+] totalled 25 mM was approximately −50 mV. Thus an established 20-mM KCl-induced contraction would be relaxed by a K-channel opener since, in its presence, the membrane potential would move towards the new E_K which was sufficiently negative to close any open voltage-dependent Ca-channels (see HAMILTON et al. 1986). With higher $[K^+]_o$, the membrane potential in the presence of a K-channel opener would be more positive (for example, E_K is approximately −20 mV with added 80 mM KCl), and the open probability of voltage-dependent Ca-channels would remain high with no tissue relaxation.

Physiological salt solutions containing a raised KCl concentration have been widely employed to contract smooth muscle by stimulating Ca^{2+} influx (see HOF and VUORELA, 1983). Furthermore, the ability of K-channel openers to inhibit contractions induced by low (≤20–25 mM) KCl concentrations has seemed entirely consistent with a plasmalemmal K-channel opening action (see above). However, a study of the action of K-channel openers in Rb-containing physiological salt solutions has shown that these agents can relax contractions produced by RbCl without any apparent plasmalemmal K-channel opening (GREENWOOD and WESTON 1993). Since membrane depolarization (and by inference the ensuing contraction) seems to involve significant Ca^{2+} influx-induced Ca^{2+} release from intracellular stores (GANITKEVICH and ISENBERG 1992), it seems possible that some of the ability of K-channel openers to relax 20 mM KCl-induced contractions stems from an inhibitory effect on such stores (see GREENWOOD and WESTON 1993; Sect. C.I.2.b)γ).

β) *Other Agonists.* Agonist-induced contractions in all tissues are inhibited by the K-channel openers as a group and such an action would be predicted from the ability of these agents to open plasmalemmal K-channels (Sect. A.I.1.). If this were the sole mechanism by which these agents exert their effects, the relaxant profile of a K-channel opener should be very similar to that of a Ca entry-blocking agent such as nifedipine. This, however, is not the case. Several studies have highlighted the greater "antispasmogenic" action of the K-channel openers compared with that of the Ca entry blockers (COOK et al. 1988; CAIN and NICHOLSON 1989; TURNER et al. 1989; BRAY et al. 1991). Furthermore, the ability of K-channel openers to retain much of their relaxant profile in Rb-containing physiological salt solutions has emphasised the additional non-plasmalemmal actions of these agents (GREENWOOD and WESTON 1993).

γ) *Intracellular Effects.* Based largely on work with prototype K-channel openers such as levcromakalim (see Sect. C.I.1.a), noradrenaline-induced

Ca^{2+} release from intracellular Ca stores is inhibited (BRAY et al. 1991; ITO et al. 1991; QUAST and BAUMLIN 1991). Such an effect may be generated via a decrease in $InsP_3$ production (ITO et al. 1991). In addition, the K-channel openers inhibit the refilling of agonist-sensitive Ca stores (aorta, BRAY et al. 1991; airways muscle, CHOPRA et al. 1992) and reduce resting $[Ca^{2+}]_i$ in a variety of smooth muscle-containing tissues (YANAGISAWA et al. 1990; ITO et al. 1991; ITOH et al. 1992).

Both nicorandil and pinacidil shift the $[Ca^{2+}]_i$-force relationship to the right whereas cromakalim is without effect. This is a further indication that some K-channel openers can exert significant effects on contractile mechanisms (YANAGISAWA et al. 1990). In skinned vascular muscle, pinacidil inhibits Ca-induced contractions possibly at a site between Ca^{2+}-calmodulin complex formation and myosin light chain phosphorylation (ITOH et al. 1991).

These changes could be indirect consequences of the membrane hyperpolarization following plasmalemmal K-channel opening. Alternatively, they could be reflections of the mechanism by which the K-channel openers modulate plasmalemmal K-channels but exerted at loci other than the K-channel itself (see Sect. C.I.3.a)γ).

c) $^{42}K/^{86}Rb$ Efflux

Measurements of drug-induced changes in the efflux of ^{86}Rb or ^{42}K from smooth muscle-containing tissues has become one of the diagnostic tests for this group of agents. In spite of the reduction in electrical driving force on the outward movement of K^+ which accompanies membrane hyperpolarization, concentration-dependent increases in $^{42}K/^{86}Rb$ efflux can routinely be detected (blood vessels, HAMILTON et al. 1986; WEIR and WESTON 1986b; airways, LONGMORE et al. 1990, 1991; bladder, EDWARDS et al. 1991). For reasons still not explained, little or no such increase can be observed in uterine tissues (HOLLINGSWORTH et al. 1987, 1989).

Studies of the absolute changes in ^{42}K and ^{86}Rb efflux have revealed that any K-channel opener-induced increase in ^{42}K exchange significantly exceeds that when ^{86}Rb is employed (QUAST and BAUMLIN 1988; NEWGREEN et al. 1990). This reflects the lower general permeability of K-channels to Rb^+ (SMITH et al. 1986) and may also include a small component of K-channel blockade by the Rb^+ itself (see GREENWOOD and WESTON 1993). In general, the relaxant potencies of K-channel openers are also reflected in their ability to increase $^{42}K/^{86}Rb$ exchange, although minoxidil sulphate exhibits relatively low efficacy in stimulating ^{86}Rb efflux (NEWGREEN et al. 1990). Such findings prompted the suggestion that K-channel openers might open two types of K-channel, one which was permeable to both K^+ and Rb^+ while the other was Rb-impermeable (NEWGREEN et al. 1990). Subsequent whole-cell clamp studies have, however, failed to confirm this view (NOACK et al. 1992a; IBBOTSON et al. 1993a).

d) Membrane Potential in Whole Tissues

Initial studies with microelectrodes in guinea pig portal vein showed that nicorandil abolished spontaneous electrical complexes and hyperpolarized the membrane (KARASHIMA et al. 1982). This work was extended to rat portal vein with similar results being obtained using cromakalim, pinacidil, diazoxide and minoxidil sulphate (HAMILTON et al. 1986; SOUTHERTON et al. 1987; NEWGREEN et al. 1990). At the upper end of their effective concentration range the K-channel openers produce a remarkable membrane hyperpolarization which is usually maintained in the continuing presence of the agent and the membrane is held at or close to E_K (HAMILTON et al. 1986; SOUTHERTON et al. 1988). In electrically quiescent vessels, a concentration-dependent hyperpolarization is also observed (aorta, TAYLOR et al. 1988; BRAY et al. 1991; pulmonary artery, KREYE et al. 1987; mesenteric artery, NAKAO et al. 1988; VIDEBAEK et al. 1988; MCHARG et al. 1990; NAKASHIMA et al. 1990).

In other smooth muscles, electrical changes which are qualitatively similar to those observed in blood vessels are usually observed. Thus, in guinea pig trachealis, spontaneous electrical slow-waves are suppressed and the membrane hyperpolarizes to a level close to the calculated E_K (ALLEN et al. 1986; MURRAY et al. 1989; BERRY et al. 1991). In bovine trachealis, in which there are essentially no spontaneous fluctuations of membrane potential, cromakalim, RP49356 and diazoxide each produce a concentration-dependent hyperpolarization (LONGMORE et al. 1990, 1991). The smooth muscle of the urinary bladder of both the guinea pig and pig is spontaneously active with spike potentials occurring in bursts or as a continuous discharge (FOSTER et al. 1989a,b; FUJII et al. 1990a). Exposure to cromakalim reduces spike frequency and hyperpolarizes the membrane close to E_K.

Relatively few membrane potential measurements have been made in the smooth muscles of other genito-urinary tissues. In the uterus of the term-pregnant rat, cromakalim had relatively little effect on basal membrane potential, although spike discharges were rapidly inhibited (HOLLINGSWORTH et al. 1987). In a preliminary study on guinea pig ureter, cromakalim and the benzopyran derivative S0121 each produced membrane hyperpolarization (KLAUS et al. 1990a,b).

3. Effects on K-Channels

a) K_{ATP}

Glibenclamide, the inhibitor of ATP-sensitive K-channels (K_{ATP}) in insulinoma and in pancreatic β-cells (STURGESS et al. 1985, 1988) antagonises both the in vivo and the in vitro actions of the K-channel openers (BUCKINGHAM et al. 1989). This finding, made simultaneously by several laboratories, led to the proposal that K_{ATP} was the target channel for the K-channel openers in smooth muscle (see QUAST and COOK 1989b). However, in rabbit intestinal muscle, OHYA et al. (1987) were unable to show that raising $[ATP]_i$ had any effect on K-currents. Furthermore, the rank order of potency for inhibition of spontaneous tension changes in smooth muscle (portal vein) is cromakalim > pinacidil > diazoxide (NEWGREEN et al. 1990) which is quite different from that for the inhibition of insulin release (diazoxide > pinacidil > cromakalim; GARRINO et al. 1989).

α) *Isolated Patch Experiments*. Patch-clamp experiments have so far yielded indecisive results concerning the nature of the K-channel which is modulated by the K-channel openers. In the first published report, STANDEN et al. (1989) described the characteristics of a relatively high conductance (approximately 100 pS, conductance recalculated for quasi-physiological conditions), K-selective channel which was opened by cromakalim in a glibenclamide-sensitive manner. The channel was also closed by raising the [ATP] on the inner surface of the patch. Subsequently, KAJIOKA et al. (1990, 1991) reported that nicorandil (rat portal vein) and pinacidil (rabbit portal vein) opened a relatively small conductance (10–30 pS: quasi-physiological K^+ gradient) ATP-sensitive K-channel. In the isolated patches from rabbit portal vein, GDP was required before effects of pinacidil could be obtained (KAJIOKA et al. 1991).

More recently, BEECH et al. (1993a,b) have described the characteristics of a 17-pS (conductance recalculated for quasi-physiological conditions) K-channel in rabbit portal vein. This channel was essentially inactive in the absence of nucleoside diphosphates, but in the presence of these (and especially of GDP), channel openings occurred. These openings could sometimes be enhanced by levcromakalim in a glibenclamide-sensitive fashion and inhibited by ATP provided that nucleoside diphosphates were present. Such data are thus analogous to those described by TUNG and KURACHI (1991) in guinea pig ventricular myocytes in which nucleoside diphosphates could stimulate K_{ATP} opening after rundown of this channel.

The interpretation of these isolated patch data needs careful reconsideration in the light of the finding that K_{ATP} could be a partially dephosphorylated form of K_V, a possibility which is further discussed in Sect. C.I.3.a)γ).

β) *Whole-Cell Experiments*. The use of freshly isolated smooth muscle cells in this configuration has provided some of the best data on the action of the K-channel openers. In portal veins, early recordings of current noise indicated that the target channel for the K-channel openers was of relatively low conductance, although specific unitary values were not calculated (rabbit, BEECH and BOLTON 1989c; rat, OKABE et al. 1990). More recently, whole-cell voltage clamp experiments in rat portal vein have shown that not only levcromakalim, but also the chemically distinct openers P1060 and aprikalim open a small conductance (10–20 pS in quasi-physiological conditions), glibenclamide-sensitive K-channel (NOACK et al. 1992a; IBBOTSON et al. 1993a; Fig. 7). A similar K-channel is activated by levcromakalim in cells separated

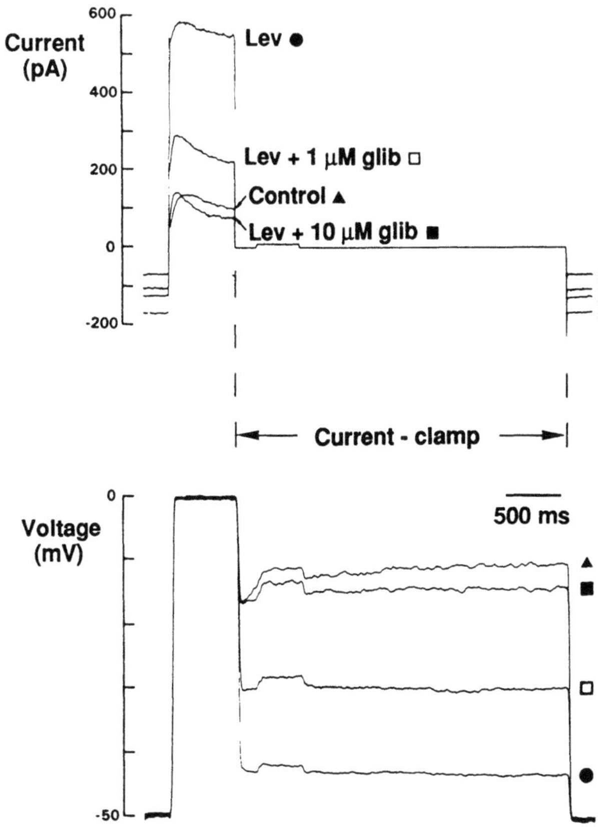

Fig. 7. Effect of levcromakalim (*Lev*) on membrane currents and membrane potential in an isolated, single cell from rat portal vein and reversal of the effect by glibenclamide (*glib*). In the *upper panel* the superimposed current traces obtained under voltage-clamp conditions show that the outward current obtained on depolarization from the holding potential of −50 mV to 0 mV (control, △) was markedly increased in the presence of levcromakalim (●). In the *lower panel*, the voltage recorded on switching to current-clamp (*0 pA*) shows that levcromakalim (●) hyperpolarized the membrane by approximately 30 mV from control levels (▲). The addition of glibenclamide (1 μM, □ or 10 μM, ■) in the presence of levcromakalim reversed the increase in current produced by levcromakalim and repolarized the membrane in a concentration-dependent manner (□, ■). Current injection of 10 pA immediately after switching from voltage- to current-clamp was used to assess membrane resistance. Each current and voltage trace is the computer-derived average derived from ten identical and successive voltage/current-clamp protocols in the same isolated cell. (From NOACK et al. 1992c)

from fourth generation mesenteric arterioles (CRIDDLE et al. 1994). In current-clamp mode, NOACK et al. (1992c) observed that levcromakalim generated a marked membrane hyperpolarization which was reversed by glibenclamide (NOACK et al. 1992c; Fig. 7). Depletion of $[ATP]_i$ by removal of metabolic substrates from the pipette and bath solutions produced a substantial glibenclamide-sensitive K-current, an event which was accompanied by marked membrane hyperpolarization (NOACK et al. 1992c). Based on current noise analysis, the K-channel responsible for carrying this "metabolic" current had a unitary conductance in the range 10–20 pS (quasi-physiological conditions). After rundown of this current, effects of levcromakalim could no longer be observed (NOACK et al. 1992c).

γ) Mechanism of Opening. An important feature of the action of agents such as levcromakalim, aprikalim and P1060 is their ability not only to induce the current known as $I_{K(ATP)}$, but also simultaneously to reduce the delayed rectifier current $I_{K(V)}$ in smooth muscle (NOACK et al. 1992a,c; IBBOTSON et al. 1993a; CRIDDLE et al. 1994). Since the channels (K_V) which carry $I_{K(V)}$ must be phosphorylated before they can be opened (PEROZO and BEZANILLA 1990), the K-channel openers could reduce the phosphorylation of K_V by inhibiting the actions of protein kinases or enhancing the activity of protein phosphatases. Indeed, recent experiments have suggested that the K-channel openers may be dephosphorylating agents, the primary site of action of which is the delayed rectifier K_V. This becomes converted into a voltage-insensitive state known as K_{ATP} (EDWARDS et al. 1993). On prolonged exposure to dephosphorylating conditions or to K-channel openers, K_{ATP} is further dephosphorylated to an inactive state. Such an action (dephosphorylation) would account for the ability of the K-channel openers to inhibit K_{ATP} in the absence of ATP (KOZLOWSKI et al. 1989; DUNNE 1990). Furthermore, it is entirely consistent with the findings that K-channel openers can inhibit not only ATP-sensitive chloride channels (SHEPPARD and WELSH 1992), but also L-type calcium channels (OKABE et al. 1990), the opening of both of which is phosphorylation dependent (see FULLER and BENOS 1992; ONO and FOZZARD 1993). Preliminary experiments (EDWARDS et al. 1993) suggest that the apparent dephosphorylation brought about by the K-channel openers is associated with stimulation of membrane phosphatases.

b) BK_{Ca}

BK_{Ca} in smooth muscle was originally reported to be ATP sensitive (SILBERBERG and VAN BREEMEN 1990), thus making this channel a potential target for the K-channel openers. However, this finding is now known to be largely an artefact due to the chelation of Ca^{2+} by the disodium salts used in the study (see KLÖCKNER and ISENBERG 1992). Nevertheless, several groups have reported that benzopyran K-channel openers can open BK_{Ca} in smooth muscle (GELBAND et al. 1989; CARL et al. 1992; KLÖCKNER and ISENBERG 1992). Since the relaxant effects of cromakalim and minoxidil sulphate in whole smooth muscles are not inhibited by charybdotoxin (WINQUIST et al. 1989; MURRAY et al. 1989; WICKENDEN et al. 1991), it seems possible that the properties of the channels in the isolated cells utilised by GELBAND et al. (1989), CARL et al. (1992) and KLÖCKNER and ISENBERG (1992) had become changed by cell separation procedures or by the recording conditions.

The large conductance of BK_{Ca}, together with its high density in smooth muscle (BENHAM and BOLTON 1986), means that openers of this channel would exert powerful inhibitory effects. Development of such agents is in its early stages (see Sects. C.I.1.g)β–δ), C.II.7.b); Fig. 5).

c) SK_{Ca}

The relaxant effects of noradrenaline and of exogenously applied ATP are apamin sensitive in certain smooth muscles. Based on this finding, it is assumed that these mechanical changes result from the opening of SK_{Ca}, although this channel has not been described in smooth muscle. For further details, see Sects. B.II.2., C.II.7.a) and C.III.5.e).

II. Endogenous K-Channel Openers

1. Adenosine

The effects of adenosine on the cardiovascular system have recently been reviewed and both A_1 and A_2 receptor subtypes are involved (OLSSON and PEARSON 1990). In cultured rat ventricular myocytes, adenosine-induced activation of A_1-receptors produces opening of K_{ATP} channels via a G_i protein (KIRSCH et al. 1990) and evidence of adenosine-induced opening of K_{ATP} in smooth muscle has also been reported. Thus in guinea pig isolated hearts, the coronary vasodilator effect of adenosine is glibenclamide sensitive (DAUT et al. 1990; VON BECKERATH et al. 1991), an action which seems to involve the A_2-receptor subtype (UEEDA et al. 1991). In contrast, glibenclamide-sensitive relaxations in porcine coronary artery are mediated by the stable A_1-selective adenosine analogue, N^6-cyclopentyladenosine whereas the A_2 agonist N^6-[2-(3,5-dimethoxyphenyl)-2-(2-methylphenyl)-ethyl]adenosine (DPMA) induces glibenclamide-insensitive relaxations (MERKEL et al. 1992). Electrophysiological studies in pig isolated coronary myocytes have confirmed the existence of an A_1-receptor-generated K-current which is glibenclamide sensitive (DART and STANDEN 1993). Evidence for the complex nature of the cardiovascular effects mediated by adenosine receptor subtypes is provided by studies in the pithed rat. In this model no involvement of glibenclamide-sensitive K-channels in responses mediated by either A_1- or A_2-receptor subtypes has been detected (FOZARD and CARRUTHERS 1993).

Thus in some vessels the relaxant actions of adenosine seem linked to the opening of glibenclamide-sensitive K-channels and evidence which implicates either A_1- or A_2-receptor subtypes exists. One possible source of confusion could be the involvement of the vascular endothelium and the release of an endothelium-derived hyperpolarising factor (EDHF), the relaxant actions of which in most tissues do not involve glibenclamide-sensitive K-channels (see Sect. C.I.2.c).

2. Calcitonin Gene-Related Peptide

There is no clear view concerning the role of K-channel opening in the vasodilator actions of calcitonin gene-related peptide, CGRP. In some tissues, the actions of this agent are endothelium dependent (rat coronary artery: PRIETO et al. 1991; rat aorta: BRAIN et al. 1985; GRACE et al. 1987). In rat aorta, early studies showed that CGRP did not increase cGMP levels, suggesting that the endothelium-derived relaxant factor (EDRF, thought to be nitric oxide) was not involved (GRACE et al. 1987). Later, however, L-NMMA was shown to reduce CGRP-induced relaxations (GRAY and MARSHALL 1990), indicating the involvement of nitric oxide and conflicting with the earlier investigation (GRACE et al. 1987). A more recent report has shown that some of the effects of the peptide are inhibited by glibenclamide, while others are sensitive to L-NAME (ANDERSSON 1992).

A study in rabbit mesenteric resistance vessels involving several techniques concluded that CGRP opened a glibenclamide-sensitive K-channel and that this event caused the ensuing relaxation (NELSON et al. 1990). However, other workers have either failed to detect a significant membrane hyperpolarization following exposure to CGRP or have shown relaxant effects of this peptide which are glibenclamide insensitive (feline cerebral arteries, SAITO et al. 1989; pig coronary arteries, BENY et al. 1989; rat coronary arteries, PRIETO et al. 1991). A recent study in the rabbit ophthalmic artery showed that CGRP hyperpolarized the tissue, an action which could be reversed by application of glibenclamide (ZSCHAUER et al. 1992). However, the apparent antagonistic effect of this sulphonylurea could have been of a functional nature since no measurement of the effect of glibenclamide was made in the absence of CGRP. Furthermore, in the same study, glibenclamide failed to antagonise the *relaxant* actions of CGRP in the opthalmic artery. Thus, CGRP can hyperpolarize a variety of smooth muscles, but the glibenclamide sensitivity of this event is equivocal. CGRP-induced relaxations are in general *not* glibenclamide sensitive and thus the role of K-channels, and especially those designated K_{ATP}, remains to be established.

3. Endothelium-Derived Hyperpolarising Factor

Experiments using conventional microelectrodes and $^{86}Rb/^{42}K$ efflux techniques have shown that acetylcholine-induced hyperpolarization of endothelium-intact blood vessels is associated with the release of a K-channel opening factor from the vascular endothelium (SOUTHERTON et al. 1987; CHEN et al. 1988; FELETOU and VANHOUTTE 1988; TAYLOR et al. 1988; BRAY and QUAST 1991b; CHEN et al. 1991; RAND and GARLAND 1992). Although nitric oxide (FURCHGOTT 1988) or a closely related molecule (MYERS et al. 1990) is probably the major endothelium-derived relaxing factor, the use of NO-synthase inhibitors and of the guanylate-cyclase modifying agents haemoglobin and methylene blue strongly indicates that the endogenous K-channel opening factor is different from EDRF (CHEN et al. 1988, 1991;

NAGAO and VANHOUTTE 1992). This endogenous K-channel opener has been designated "endothelium-derived hyperpolarizing factor" (EDHF; CHEN et al. 1988).

Some studies have indicated that nitric oxide can itself hyperpolarize vascular smooth muscles. This phenomenon has been described in uterine arteries (TARE et al. 1990), in rat small mesenteric arteries (GARLAND and MCPHERSON 1992) and in rabbit basilar artery (RAND and GARLAND 1992). However, these effects are essentially only observed close to the maximally effective relaxant concentrations of nitric oxide and they contrast with the results of other vascular studies involving exposure to nitric oxide in which no hyperpolarization was detected (BENY and BRUNET 1988; KOMORI et al. 1988; BRAYDEN 1990). Furthermore, relaxant agents such as sodium nitroprusside which can be regarded as nitric oxide carriers have little effect on membrane potential in vascular smooth muscle (ITO et al. 1978; CHEUNG and MACKAY 1985; HUANG et al. 1988; SOUTHERTON et al. 1988).

The K-channel opened by EDHF has not been unequivocally identified. Early studies in rabbit middle cerebral artery (STANDEN et al. 1989; BRAYDEN 1990) found that acetylcholine-induced hyperpolarizations were glibenclamide sensitive, suggesting that an ATP-sensitive K-channel was the target channel. However, in other vessels such as the guinea pig coronary artery and rat small mesenteric artery with intact endothelial layers, glibenclamide failed to inhibit acetylcholine-induced hyperpolarizations (FUJII et al. 1990b; CHEN et al. 1991; MCPHERSON and ANGUS 1991; ECKMAN et al. 1992; GARLAND and MCPHERSON 1992; VAN DE VOORDE et al. 1992). Additionally, BRAY and QUAST (1991b) showed that the ACh-induced increase in ^{42}K efflux from rat aorta was not reduced by glibenclamide. Furthermore, the endothelium-dependent hyperpolarization and nitro-L-arginine-resistant relaxation produced by bradykinin in canine coronary artery was inhibited by calmidazolium (ILLIANO et al. 1992; NAGAO et al. 1992). This agent is described as a calmodulin antagonist (VAN BELLE 1981) and other molecules with this property inhibit K_{ATP} and antagonise the effects of K-channel openers in *Xenopus* oocytes and in insulinoma cells (MÜLLER et al. 1991; SAKUTA et al. 1992). However, in the concentrations employed in canine coronary artery, calmidazolium antagonised the endothelium-dependent effects of bradykinin without inhibiting the hyperpolarization induced by levcromakalim (ILLIANO et al. 1992; NAGAO et al. 1992).

These conflicting observations could be explained if EDHF opens glibenclamide-sensitive K-channels (presumably a variant of K_{ATP}) in rabbit middle cerebral artery (STANDEN et al. 1989; BRAYDEN 1990), while a glibenclamide-insensitive and, as yet, unidentified K-channel is involved in EDHF-induced changes in guinea pig and dog coronary arteries, in rat small mesenteric artery and in rat aorta (FUJII et al. 1990b; CHEN et al. 1991; MCPHERSON and ANGUS 1991; BRAY and QUAST 1991b; ECKMAN et al. 1992; GARLAND and MCPHERSON 1992; NAGAO et al. 1992; VAN DE VOORDE et al.

1992). Alternatively, more than one "EDHF" could exist. What is clear, however, is that there is little support for the view (NELSON et al. 1990; STANDEN 1992) that the effects of EDHF in *most* vessels are mediated via K_{ATP}.

Although the majority of studies on EDHF have utilised acetylcholine as the releasing agent, other substances such as adenosine diphosphate (BRAYDEN 1991), bradykinin (NAGAO et al. 1992), carbachol and substance P (BOLTON et al. 1984; BOLTON and CLAPP 1986; BRAY and QUAST 1991b) and histamine (CHEN and SUZUKI 1989) also seem capable of releasing an endothelium-derived hyperpolarising factor. However, as already stated, it is not clear whether this factor is a single entity or whether a family of EDHFs exists.

4. Fatty Acids

In smooth muscle cells from toad stomach, pulmonary artery and human aorta, a variety of fatty acids has been found to open a small conductance, Ca-dependent K-channel and a large conductance Ca-dependent K-channel (reviewed by ORDWAY et al. 1991). The significance of these observations is not clear, but they suggest that fatty acids could have a widespread role in modulating K-channels in vivo.

5. Prostacyclin

In canine artery, the stable prostacyclin analogue, iloprost, increases the K-permeability of the component smooth muscle cells (SIEGEL et al. 1990). In isolated, perfused rabbit hearts, recent studies have suggested that prostacyclin itself, as well as iloprost, produces vasodilatation of the coronary circulation by opening glibenclamide-sensitive K-channels (JACKSON et al. 1993). Furthermore, iloprost-induced relaxation of bovine and rabbit isolated coronary vessels was also glibenclamide sensitive (JACKSON et al. 1993). Previous studies have suggested that prostacyclin-induced relaxations are associated with an increase in cAMP levels (GRYGLEWSKI et al. 1991), a change sometimes associated with the opening of charybdotoxin-sensitive K-channels (Sects. C.I.1.g)β) and C.II.7.). Surprisingly, iloprost still induced a concentration-dependent increase in cAMP levels even in the presence of a concentration of glibenclamide which prevented iloprost-induced relaxation of bovine coronary artery (JACKSON et al. 1993).

Further studies are clearly necessary to establish the importance of K-channel opening as a general phenomenon in the action of prostacyclin and its derivatives.

6. Vasoactive Intestinal Polypeptide

Vasoactive intestinal polypeptide (VIP) produces relaxation of a variety of smooth muscle-containing tissues. In some blood vessels, the dilator effects of this agent are largely endothelium dependent (rat aorta, DAVIES and WILLIAMS 1984; bovine intrapulmonary artery, IGNARRO et al. 1987), effects which are probably associated with EDRF release (see MOORE et al. 1990).

In other blood vessels, VIP was shown to hyperpolarize the smooth muscle, an effect which was glibenclamide sensitive, consistent with the opening of K_{ATP} (STANDEN et al. 1989). However, in rabbit mesenteric artery, none of the effects of VIP was glibenclamide sensitive (HATTORI et al. 1992), and in canine colon and intestine and in opossum oesophagus, VIP had little effect on membrane potential (CHRISTINCK et al. 1991; HUIZINGA et al. 1992). In rat stomach and in rabbit mesenteric artery, VIP-induced relaxations were associated with increases in both cAMP and cGMP concentrations (ITOH et al. 1985; ITO et al. 1990).

7. Adrenoceptor Agonists

a) Role of α-Adrenoceptors

In gastrointestinal muscle, adrenaline-induced relaxations and hyperpolarization are inhibited by apamin (MAAS et al. 1980). This constituent of bee venom is a selective inhibitor of a small conductance, Ca-sensitive K-channel designated SK_{Ca} (see Sect. C.III.5.e). Furthermore, relaxations induced by α-adrenoceptor agonists in guinea pig taenia caeci, together with the associated ^{86}Rb efflux, are also apamin sensitive (GATER et al. 1985; WEIR and WESTON 1986a). Such results clearly implicate SK_{Ca} as the mediator of the observed relaxations.

b) Role of β-Adrenoceptors

Several pieces of evidence suggest that the opening of K-channels and particularly of BK_{Ca} contributes to the relaxant effects of β_2-adrenoceptor agonists, especially in trachealis muscle. Early studies showed that catecholamines relaxed and hyperpolarized airway smooth muscle, effects which were propranolol sensitive and associated with a decrease in membrane conductance (ITO and TAJIMA 1982). More recently, KUME et al. (1989) showed that isoprenaline increased the open probability of BK_{Ca} channels in tracheal smooth muscle, an action probably associated with activation of protein kinase A and subsequent channel phosphorylation. The ability of charybdotoxin to antagonise the effects of β-adrenoceptor agonists in both guinea pig trachealis and human bronchus is also consistent with the opening of BK_{Ca} (JONES et al. 1990; MURRAY et al. 1991; MIURA et al. 1992).

Collectively, these data support the idea that β_2-adrenoceptor-mediated relaxations are causally linked to the opening of BK_{Ca}. However, the most recent studies (CHIU et al. 1993; COOK et al. 1993) cast doubt on this view and also suggest that both β_1- and β_2-adrenoceptors are involved in isoprenaline-induced relaxations, at least in guinea pig trachealis. In a comprehensive series of experiments, mechanical relaxation by isoprenaline, procaterol, salbutamol and salmeterol was only associated with β_2-adrenoceptor stimulation. However, the membrane hyperpolarization and increased ^{86}Rb efflux produced by isoprenaline, procaterol and salbutamol were antagonised by very selective inhibitors of either β_1 (CGP 20712A) or β_2 (ICI 118551) adrenoceptors. Surprisingly, the long-acting β_2-agonist salmeterol failed to hyperpolarize the trachealis cells or to increase the rate of ^{86}Rb exchange.

These recent findings, together with earlier observations that the relaxant effects of both isoprenaline and salbutamol are still present in high K^+ (80–120 mM) physiological salt solution (ALLEN et al. 1985; SMALL et al. 1989), suggest that the opening of BK_{Ca} channels is not the sole mechanism which mediates β-adrenoceptor-induced relaxations, at least in guinea pig trachealis.

8. Endogenous Relaxants: Involvement of K-Channels

Data which support the view that K-channel opening is the underlying mode of action of the endogenous relaxants considered in Sect. C.II. are far from unequivocal. Convincing evidence relates to the action of EDHF in blood vessels and there seems little doubt that this agent, which remains to be identified, can be regarded as an endogenous K-channel opener. The results of most workers indicate that EDHF is a different entity from EDRF and that the target K-channel for EDHF is not K_{ATP}. In the gastrointestinal tract, apamin-sensitive relaxations induced via α-adrenoceptors indicate the involvement of SK_{Ca}. However, the opening of BK_{Ca} channels seems to play only a partial role in the relaxations associated with stimulation of β-adrenoceptors, at least in guinea pig trachealis.

In the case of adenosine, CGRP, prostacyclin and VIP, the picture is uncertain. One of the problems in interpreting the available data is whether the observed changes involve the direct interaction of the agent with smooth muscle or whether the effects are indirect via release of an endogenous mediator such as EDRF or EDHF. What does seem clear, however, is that the view that these endogenous vasodilators owe their relaxant actions to the opening of a glibenclamide-sensitive K_{ATP} channel (STANDEN et al. 1989; NELSON et al. 1990; STANDEN 1992) is not supported by most of the available data.

III. K-Channel Inhibitors

1. Inorganic Cations

A low concentration of barium (Ba^{2+}; 40 μM) is a relatively selective blocker of the inwardly rectifying K-channel (KAMEYAMA et al. 1983; GILES and IMAIZUMI 1988). Millimolar concentrations of Ba^{2+} also block the cardiac ATP-sensitive K-channel and depress the delayed rectifier and transient outward K-currents (KAKEI and NOMA 1984; HUME and LEBLANC 1989; ARENA and KASS 1989). Ba^{2+} (1–10 mM) has little or no effect on the single channel current carried by BK_{Ca}, but reduces the channel open probability at positive potentials (BENHAM et al. 1985; MCCANN and WELSH 1986; GREEN et al. 1991; KHAN et al. 1993). Since the atomic radius of Ba^{2+} is similar to that of K^+ and since the blocking ability of Ba^{2+} is sensitive to the K^+ concentration, it is thought that Ba^{2+} inhibits the channel by binding to a site within the channel pore and is displaced from this site by K^+ (NEYTON and MILLER 1988a,b). NEYTON and MILLER (1988a) suggest that BK_{Ca} may have gates on both sides of the pore which close to trap the blocking ion Ba^{2+}. Interestingly, in patch-clamp experiments in which single cells separated from canine and bovine trachealis were used, intracellular Ba^{2+} was found to enhance the opening of BK_{Ca} at negative potentials, presumably by substituting for Ca^{2+} (MCCANN and WELSH 1986; GREEN et al. 1991).

In cells isolated from rabbit jejunum or guinea pig small mesenteric artery, extracellular caesium (Cs^+, 10 mM) blocks the inward K-current through BK_{Ca}, although intracellular application of the same concentration has little effect on the outward K-current (BENHAM et al. 1986; GREEN et al. 1991). Cs^+ is also capable to inhibiting K_{ATP} in frog skeletal muscle (QUAYLE et al. 1988), although its effect on the smooth muscle K_{ATP} remains to be determined. Other cations which were found to have inhibitory effects on a calcium-sensitive K-channel (although not defined, its TEA sensitivity suggests that it was probably BK_{Ca}) but to have no effect on the delayed rectifier K-channel in canine renal arterial cells are nickel (500 μM) and cadmium (300 μM; GELBAND and HUME 1992).

2. Aminopyridines, TEA

In this review, these K-channel inhibitors (Fig. 8) are grouped together for historical rather than pharmacological reasons. Until relatively recently, these two drugs were the most widely used tools in the identification of K-channel types. Although neither is selective for a single channel type, K-channels differ in their sensitivity to inhibition and thus, by use of low concentrations, some channel selectivity can be achieved.

The smooth muscle K-channel which is most sensitive to the blocking action of TEA is BK_{Ca}, which is blocked by 0.1–1 mM external TEA, but, compared to K_V it is less sensitive to 4-aminopyridine. In contrast, in a variety of smooth muscle preparations, K_V channels are inhibited by 4-aminopyridine (500 μM–5 mM) but insensitive to TEA (1–4 mM; BEECH and BOLTON 1989b; IMAIZUMI et al. 1990; GREEN et al. 1991; VOLK et al. 1991; BOYLE et al. 1992; GELBAND and HUME 1992). Both aminopyridines and TEA (0.3–10 mM) are capable of antagonising the effects of the K-channel openers (ALLEN et al. 1986; COLDWELL and HOWLETT 1987; QUAST 1987; SOUTHERTON et al. 1988; WINQUIST et al. 1989; WILSON et al. 1988b; LEBRUN et al. 1990).

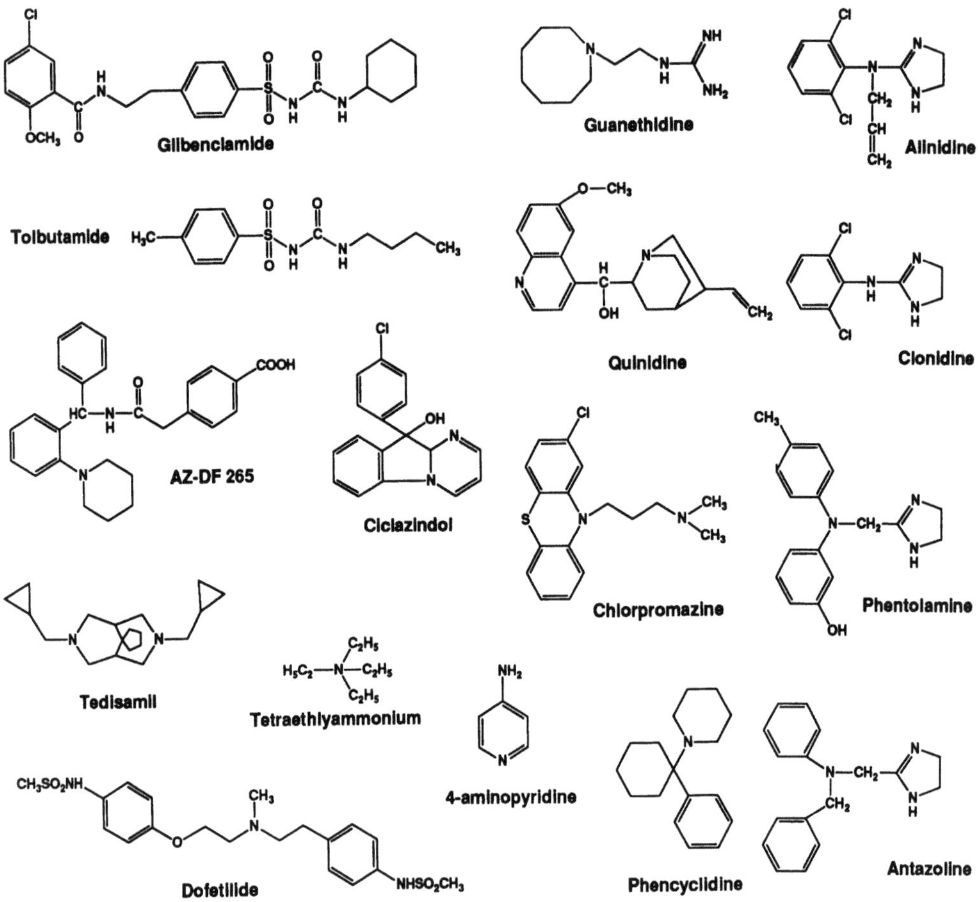

Fig. 8. Structural diversity of synthetic inhibitors of smooth muscle K-channels. Simple molecules such as tetraethylammonium and 4-aminopyridine probably block the pore of various K-channel types and are correctly described as K-channel blockers. However, the mode of action of the other agents is not known and they are more correctly described as channel inhibitors. For further details, see text

Site-directed mutagenesis experiments have demonstrated that the binding site for TEA is within the S5–S6 region (Fig. 1; Stocker et al. 1991; Yellen et al. 1991). Since this region is well conserved in several types of K-channel, it is perhaps not surprising that TEA is a relatively non-selective K-channel inhibitor. TEA is thought to bind to a site within the pore of the K-channel and to produce physical block of the channel (see Hille 1992).

In trachealis isolated from a variety of species, TEA produces a marked depolarization and induces action potentials which can generate rhythmic contractions (Kirkpatrick 1975; Stephens et al. 1975; Murray et al. 1989). These effects are also associated with a rise in basal tension (Murray et al.

1989). In high concentrations (10–100 mM), TEA is capable of inducing phasic contractions in normally electrically quiescent arterial smooth muscle (CASTEELS et al. 1977; DROOGMANS et al. 1977; HARDER and SPERELAKIS 1979; FUJIWARA and KURIYAMA 1983). However, in a concentration which is more selective for BK_{Ca} ($\leq$10 mM), TEA has been found to have no effect on rat aorta or on the frequency or amplitude of spontaneous contractions in the guinea pig mesotubarium (COOK 1989; LYDRUP 1991). In concentrations in which the aminopyridines are relatively selective for K_V, 4-aminopyridine (1 mM) and 3,4-diaminopyridine (0.3–1 mM) induce membrane depolarization in the guinea pig pulmonary artery and portal vein and stimulate contractions in the rat portal vein and rabbit aorta, respectively (HARA et al. 1980; SOUTHERTON et al. 1988; COOK 1989). However, in the guinea pig urinary bladder there appears to be no 4-aminopyridine-sensitive K-current (KLÖCKNER and ISENBERG 1985).

In neurones, K_A is more sensitive than Ca^{2+}-activated or delayed rectifier currents to 4-aminopyridine, with, in general, a K_d between 1 and 5 mM (see RUDY 1988). Similarly, in vascular and ureteral smooth muscle, $I_{K(A)}$ is very sensitive to 4-aminopyridine (BEECH and BOLTON 1989a; IMAIZUMI et al. 1989; LANG 1989) and is fully inhibited by 5 mM 4-aminopyridine (BEECH and BOLTON 1989a).

3. Class III Antiarrhythmic Agents

The potent class III antiarrhythmic agents E-4031, UK-66,914 and dofetilide (UK-68,798) appear to be selective inhibitors of K_V in the heart (SANGUINETTI and JURKIEWICZ 1990; GWILT et al. 1991). There are no data for the effects of these compounds on smooth muscle K-channels. However, class III agents are capable of enhancing spontaneous activity in the rat isolated portal vein (BASKIN et al. 1992), consistent with an inhibitory effect on K-channels in this tissue.

Tedisamil (Fig. 8) is a bradycardic agent which inhibits the delayed rectifier and transient K-currents in rat cardiac myocytes (DUKES et al. 1990; DUKES and MORAD 1991). In the guinea pig portal vein, this agent produces a concentration-dependent inhibition (1 μM–100 μM) of the opening of BK_{Ca} by reducing the mean channel open time and is without effect on the mean channel closed time (PFRÜNDER and KREYE 1991). Like certain other K-channel inhibitors, (see Sect. C.III.4.), tedisamil reduces both the delayed rectifier K-current (PFRÜNDER and KREYE 1992) and the non-inactivating K-current stimulated by the K-channel openers (BRAY and QUAST 1992b). BRAY and QUAST (1992b) found that cromakalim-induced K^+ efflux was markedly inhibited by concentrations of tedisamil which had only a minor effect on the relaxant activity of cromakalim in rat isolated aorta. In contrast, the mechano-inhibitory and K^+ efflux-stimulating effects of minoxidil sulphate were inhibited by similar concentrations of tedisamil. This might be an indication that the relaxant effects of cromakalim are not due entirely to the opening of plasmalemmal K-channels in rat aorta (see GREENWOOD and WESTON 1993). A similar differential sensitivity of cromakalim-stimulated Rb^+ efflux and cromakalim-induced relaxation to tedisamil was not observed in rabbit isolated aorta (KREYE et al. 1992).

4. Imidazoline-Guanidino Receptor Site Ligands

Several compounds which possess an imidazoline or guanidine moiety (Fig. 8) inhibit the relaxant effects of the K-channel openers in smooth muscle (McPHERSON and ANGUS 1989; MURRAY et al. 1989; BANG and NIELSEN-KUDSK 1992; NOACK et al. 1992b; SCHWIETERT et al. 1992; IBBOTSON et al. 1993b). These compounds, which at lower concentrations are pharmacologically active at diverse receptor sites, may antagonise the K-channel openers via an interaction with imidazoline-guanidino receptor sites (IGRS; also known as non-adrenoceptor imidazoline binding sites, NAIBS) in vascular smooth muscle. In addition, phentolamine (an α-adrenoceptor antagonist), antazoline (an H_1 antagonist) and guanabenz (an α_2-adrenoceptor agonist) all enhance spontaneous activity in the rat portal vein (SCHWIETERT et al. 1992; IBBOTSON et al. 1993b). This stimulatory effect on smooth muscle may be due to inhibition of delayed rectifier currents since patch-clamp studies have shown that phentolamine (30–100 μM), antazoline (30–100 μM) and guanabenz (30 μM) each inhibit delayed rectifier K-currents in pancreatic β-cells and/or in vascular smooth muscle (JONAS et al. 1992; NOACK et al. 1992b; RUSSELL et al. 1992; IBBOTSON et al. 1993a,b).

The effects of the putative IGRS ligands on spontaneous activity are particularly marked when the extracellular calcium concentration is low (SCHWIETERT et al. 1992), possibly because calcium-sensitive K-channels are less active and the repolarizing effect of K_V becomes relatively more important. In contrast, in guinea pig isolated trachealis, phentolamine (1–100 μM) has no effect on spontaneous tone (BERRY et al. 1991). This probably reflects the relatively small contribution of the K_V, in comparison to BK_{Ca}, towards repolarization in airways smooth muscle. In patch-clamp experiments on single cells isolated from bovine trachea, BERRY et al. (1991) found no effect of phentolamine (1–100 μM) on BK_{Ca}.

Direct evidence that several putative IGRS ligands (antazoline, cirazoline and phentolamine) are also capable of inhibiting the K-current stimulated by the K-channel openers has been derived from whole-cell voltage-clamp recordings (NOACK et al. 1992b; RUSSELL et al. 1992; IBBOTSON et al. 1993a,b). In addition, several inhibitors of K_V (phentolamine, antazoline, midaglizole and tolazoline) inhibit the mechano-inhibitory effects of K-channel openers on rat isolated portal veins (OKUMURA et al. 1992; SCHWIETERT et al. 1992; IBBOTSON et al. 1993b).

The ability to inhibit K_{ATP} appears to be a feature of almost all inhibitors of delayed rectifier K-channels and may provide further evidence that these channel types are related (see Sect. C.I.3.a)γ). In addition to the IGRS ligands, other inhibitors of K_V, for example phencyclidine (1–100 μM; a selective inhibitor of the delayed rectifier K-current in cultured hippocampal neurones; FRENCH-MULLEN et al. 1988), ciclazindol (structurally related to the appetite suppressant mazindol), 4-aminopyridine and quinidine, also inhibit the delayed rectifier current and the K-channel opener-induced currents in portal vein cells (BEECH and BOLTON 1989c; NOACK et al. 1992b). The reported exception, E-4031 (SCHWIETERT et al. 1992), is a highly selective inhibitor of one component of cardiac delayed rectifier currents (the rapid transient outward current). In contrast, tedisamil, which inhibits the effect of K-channel openers in vascular smooth muscle (BRAY and QUAST 1992b), is a less selective inhibitor of cardiac transient outward currents (see COLATSKY 1992). This may be an indication that the E-4031-induced increase in spontaneous mechanical activity in rat portal vein (SCHWIETERT et al. 1992) is not exerted via K-channel inhibition. Alternatively, if a K-channel is involved, it may not be K_V.

5. Toxins

The toxins have proved to be invaluable in the identification and characterisation of ion channels. The use of highly selective toxins to inhibit different K-channel types has been reviewed by CASTLE et al. (1989), DREYER (1990) and GARCIA et al. (1991).

a) Charybdotoxin

Charybdotoxin is a potent, relatively selective, reversible inhibitor of large conductance calcium-activated K-channels in a variety of tissues (MILLER et al. 1985; ANDERSON et al. 1988; GUGGINO et al. 1987; GIMENEZ-GALLEGO et al. 1988; REINHART et al. 1989). However, low concentrations of charybdotoxin are capable of blocking calcium-insensitive K-channels in lymphocytes (PRICE et al. 1989; SANDS et al. 1989) and calcium-insensitive, dendrotoxin-sensitive K-channels in rat dorsal root ganglion cells (SCHWEITZ et al. 1989). Charybdotoxin was also reported to be capable of inhibiting the *Shaker* gene encoded A-current (MACKINNON et al. 1988) and an intermediate conductance calcium-sensitive K-channel (IK_{Ca}) in a variety of tissues (FARLEY and RUDY 1988; REINHART et al. 1989; HERMANN and ERXLEBEN 1987; BEECH et al. 1987; STRONG et al. 1989; GRINSTEIN and SMITH 1989). In addition, charybdotoxin inhibits the binding of [^{125}I]-labelled α-dendrotoxin, β-dendrotoxin and mast cell degranulating peptide (MCDP) to high-affinity binding sites for these compounds in the rat brain (SCHWEITZ et al. 1989; SORENSEN et al. 1990). In neuronal tissue, charybdotoxin was found to block voltage-sensitive K-channels (SCHWEITZ et al. 1989; SORENSEN et al. 1990), although the delayed rectifier channel in the rabbit portal vein is not inhibited by charybdotoxin (100 nM; BEECH and BOLTON 1989a). Charybdotoxin has no effect on the relaxant response of the rat portal vein to K-channel openers (WINQUIST et al. 1989) and appears only functionally to antagonise the effects of cromakalim on the guinea pig detrusor (GRANT and ZUZACK 1991).

Thus a variety of different K-channel types may possess a common binding site for charybdotoxin, although many of the inhibitory effects attributed to charybdotoxin are now thought to have been due to impurities and many of the earlier observations require confirmation (G.J. Kaczorowski, personal communication). Nevertheless, site-directed mutation studies (MACKINNON and MILLER 1989) have located the charybdotoxin binding site to the pore-lining H5 region of the channel protein. This region is well-conserved in a variety of calcium- and/or voltage-sensitive K-channels (see Sects. B.I. and B.II.) and could explain any lack of K-channel selectivity by charybdotoxin.

b) Leiurotoxin

Charybdotoxin is separated from the venom of the scorpion *Leiurus quinquestriatus hebraeus* (GIMENEZ-GALLEGO et al. 1988). A second toxin isolated from this venom (and later termed leiurotoxin I) was found to inhibit apamin binding to rat brain synaptosomal membranes and, in addition, to block small conductance, apamin- and calcium-sensitive K-channels in guinea pig hepatocytes (CHICCHI et al. 1988; CASTLE and STRONG 1986). Leiurotoxin I has no effect on spontaneous activity in the rabbit portal vein, but, like apamin, it inhibits the adrenaline-induced relaxation of the guinea pig taenia caeci (CHICCHI et al. 1988). An inhibitory effect of the crude scorpion toxin on cromakalim-stimulated ^{86}Rb efflux (QUAST and COOK 1988) was later attributed to leiurotoxin I (STRONG et al. 1989). However, in the guinea pig detrusor, which apparently possesses SK_{Ca}, leiurotoxin I had no effect on the relaxant response to cromakalim, suggesting that, in this tissue at least, SK_{Ca} does not contribute to the effect of cromakalim (GRANT and ZUZACK 1991).

c) Noxiustoxin

Noxiustoxin inhibits BK_{Ca} in skeletal muscle (IC_{50} 450 n*M*; VALDIVIA et al. 1988) at a concentration lower than that required to inhibit K_V in squid giant axons (IC_{50}, 1.5 μM; CARBONE et al. 1987). Despite this, its ability to increase spontaneous activity in the guinea pig isolated portal vein has been taken as evidence of inhibition of a delayed rectifier K-channel (WICKENDEN et al. 1991). Noxiustoxin had no effect on the spontaneous activity of rat detrusor in vitro (GRANT and ZUZACK 1991). The inability of noxiustoxin to inhibit the relaxant effects of cromakalim in either the guinea-pig portal vein or rat bladder preparations was assumed to indicate that the site of action of the K-channel openers was not the delayed rectifier K-channel (GRANT and ZUZACK 1991; WICKENDEN et al. 1991), although definitive studies on the effects of this toxin on K_V in smooth muscle have not been carried out.

d) Dendrotoxin

Dendrotoxin is a potent and selective inhibitor of the delayed rectifying K-current in myelinated frog nerve fibres and node of Ranvier (WELLER et al. 1985; BENOIT and DUBOIS 1986; JONAS et al. 1989) and dorsal root ganglion cells of the rat or guinea pig (STANSFELD and FELTZ 1988; PENNER et al. 1986). There are no reports of the effects of dendrotoxin in smooth muscle, and it remains to be determined whether the delayed rectifier K-channel in this tissue is inhibited by dendrotoxin.

e) Apamin

Apamin, isolated from the venom of the honey bee, *Apis mellifera*, is a selective inhibitor of a small conductance, TEA-insensitive, K-channel in skeletal muscle which is very sensitive to intracellular calcium (SK_{Ca}; BLATZ and MAGLEBY 1986; HUGUES et al. 1982). As already described, the effects produced by apamin on some whole smooth muscle-containing tissues are consistent with inhibition of K-channels. Thus, in vitro, apamin has no effect on spontaneous activity in the rat portal vein or on tone in the guinea pig trachealis (CHICCHI et al. 1988; WINQUIST et al. 1989), whereas it increases spontaneous activity in guinea pig urinary bladder, and antagonises both α-adrenoceptor-induced K^+ efflux and adrenaline-induced relaxation in guinea pig taenia caeci (WEIR and WESTON 1986a; CHICCHI et al. 1988). This may indicate that SK_{Ca} is present in only some smooth muscle-containing tissues. However, the presence of an apamin-sensitive small conductance, calcium-dependent K-channel in smooth muscle remains to be demonstrated using patch-clamp techniques. The relaxant response to the openers of K_{ATP} is not antagonised by apamin in the rat portal vein, guinea pig trachealis, guinea pig taenia caecum or guinea pig detrusor (ALLEN et al. 1986; WEIR and WESTON 1986a; WINQUIST et al. 1989; GRANT and ZUZACK 1991).

f) Iberiotoxin

Iberiotoxin, extracted from scorpion venom, appears to be a selective inhibitor of BK_{Ca} in bovine aortic smooth muscle (IC_{50} approximately 250 pM; GALVEZ et al. 1990; GIANGIACOMO et al. 1992). Charybdotoxin (in nanomolar concentrations) also inhibits BK_{Ca} in smooth muscle (TALVENHEIMO et al. 1988; STRONG et al. 1989; VÁZQUEZ et al. 1989), although it may be less selective than iberiotoxin for this channel (see Sect. C.III.5.a). Spontaneously active tissues derived from the guinea pig were found to differ in their sensitivity to inhibitors of BK_{Ca}. Of the tissues examined (bladder, portal vein, taenia coli and uterus), the spontaneous activity of the bladder was most markedly increased by charybdotoxin (10–100 nM), iberiotoxin (10–100 nM) and TEA (0.3–3 mM), whereas that of the portal vein was less affected, although still markedly enhanced (SUAREZ-KURTZ et al. 1991;

WICKENDEN et al. 1991). In the rat portal vein, charybdotoxin (1–100 nM) produces a concentration-dependent increase in contractile activity (WINQUIST et al. 1989). Spontaneous activity in the rat portal vein or rat bladder is also enhanced by inhibitors of delayed rectifier currents (see Sects. C.III.3. and C.III.4.). However, using guinea pig bladder, GRANT and ZUZACK (1991) were unable to demonstrate any effect of noxiustoxin (up to 300 nM) on either basal mechanical activity or on that stimulated by depolarisation (15 mM KCl). Surprisingly, charybdotoxin and iberiotoxin each stimulated contraction of guinea pig aorta and trachea, both quiescent tissues (SUAREZ-KURTZ et al. 1991), suggesting that BK_{Ca} contributes significantly to the resting membrane potentials in these tissues.

6. Hypoglycaemic Agents

The mechanism by which many hypoglycaemic agents stimulate insulin secretion from the pancreatic β-cell is believed to involve selective inhibition of the β-cell K_{ATP} (STURGESS et al. 1988; ASHCROFT and RORSMAN 1990; EDWARDS and WESTON 1993a). Therapeutically, the sulphonylureas, such as glibenclamide and tolbutamide, form the most important group of hypoglycaemic agents. These compounds have also been investigated in smooth muscle in which K_{ATP} is also thought to be their site of action (EDWARDS and WESTON 1993a). However, in contrast to K_{ATP} in the pancreatic β-cell, which is inhibited by low nanomolar concentrations of glibenclamide, the vascular smooth muscle K_{ATP} requires micromolar concentrations of glibenclamide (EDWARDS and WESTON 1993a). Nevertheless, even in these concentrations (1–20 μM), glibenclamide has no effect on K_V in airways or vascular smooth muscle (BOYLE et al. 1992; IBBOTSON et al. 1993a), although some inhibition is seen at 50 μM (BEECH et al. 1993a).

Glibenclamide (1–10 μM) has no effect on mechanical activity in a variety of smooth muscle preparations (MURRAY et al. 1989; GRANT and ZUZACK 1991; ECKMAN et al. 1992; NOACK et al. 1992b; SCHWIETERT et al. 1992), although in a similar concentration range (0.1–35 μM) it is reported to produce a depolarization of several millivolts in the guinea pig trachealis and coronary artery and in rabbit mesenteric artery (MURRAY et al. 1989; MCPHERSON and ANGUS 1991; ECKMAN et al. 1992). In general, it is thought that under physiological conditions the channel inhibited by glibenclamide in smooth muscle, K_{ATP}, is predominantly closed (or in its phosphorylated form, K_V, see Sects. B.III. and C.I.3.a)γ). Early studies established that glibenclamide can inhibit the vascular smooth muscle mechano-inhibitory effects of the K-channel openers in vitro and their hypotensive effects in vivo (WILSON et al. 1988a; WINQUIST et al. 1989; BUCKINGHAM et al. 1989; CAVERO et al. 1989; QUAST and COOK 1989a). These findings have now been extended to a range of smooth muscle-containing tissue types, including guinea pig and rat detrusor (FUJII et al. 1990a; EDWARDS et al. 1991; GRANT and ZUZACK 1991), guinea pig trachealis (NIELSEN-KUDSK and BANG 1991) and rat uterus (PIPER et al. 1990). Tolbutamide, although less potent than glibenclamide, is also able to antagonise the effects of the K-channel openers in rabbit isolated mesenteric artery (WILSON 1989), although no inhibitory effect of tolbutamide ($10\,\mu M$) on the relaxant responses to cromakalim, pinacidil or nicorandil in rat isolated basilar artery was observed (KSOLL et al. 1991). More recently, direct evidence of the ability of glibenclamide to inhibit the K-channel opened by the K-channel openers (and thought to be K_{ATP}) has been derived from whole-cell voltage clamp recordings (NOACK et al. 1992a; RUSSELL et al. 1992; IBBOTSON et al. 1993a).

Studies using membrane fractions have failed to detect the binding of labelled benzopyran, cyanoguanidine or thioformamide K-channel openers. However, GOPALAKRISHNAN et al. (1991) have demonstrated the ability of [^{3}H]glibenclamide to bind to both high- and low-affinity binding sites in a smooth muscle membrane fractions. Although [^{3}H]glibenclamide binding was slightly inhibited by diazoxide, there was no inhibition by cromakalim, nicorandil, pinacidil or minoxidil sulphate. Using whole segments of rat aorta, BRAY and QUAST (1992, 1993) detected a binding site for the ligand [^{3}H]P1075. Similarly, HOWLETT and LONGMAN (1992) were only able to measure [^{3}H]cromakalim binding in whole, cultured airway smooth muscle cells. In all three studies, the binding of these ligands was inhibited by other K-channel openers, but only BRAY and QUAST (1992, 1993) reported inhibition of binding by sulphonylureas. The relationship between these findings and the mode and site of action of the K-channel openers and of the sulphonylureas remains to be determined. However, as discussed by EDWARDS and WESTON (1993a), it now seems unlikely that these two groups of agent interact at a common site.

The binding of [^{3}H]glibenclamide to membrane fractions is increased under dephosphorylating conditions (SCHWANSTECHER et al. 1991). Such a result is consistent with the view that K_{ATP} may be a dephosphorylated state of the delayed rectifier channel, K_V (see Sects. B.III. and C.I.3.a)γ).

7. Calmodulin Antagonists

Several calmodulin antagonists (trifluoperazine, thioridazine and chlorpromazine) inhibit BK_{Ca} in airway smooth muscle with K_d values in the range $1-5\,\mu M$ (MCCANN and WELSH 1987). These and related agents (W-7, trifluoperazine, trifluopromazine, fluphenazine and chlor-promazine) also inhibit K_{ATP} in *Xenopus* oocytes or HIT-T15 insulinoma cells (MÜLLER et al. 1991; SAKUTA et al. 1992). In addition, endothelium-dependent hyperpolarization of the canine coronary artery induced by bradykinin or the calcium ionophore A23187 is inhibited by the calmodulin antagonists calmidazolium and fendiline (ILLIANO et al. 1992; NAGAO et al. 1992; see Sect. C.II.3.). These observations have been taken as evidence for a role of calmodulin or calmodulin-dependent processes in the regulation of K-channels (SAKUTA et al. 1992; NAGAO et al. 1992). Alternatively, the calmodulin antagonists could have a direct effect on the channel (McCANN and WELSH 1987). The uses and limitations of calmodulin antagonists have been detailed (VEIGL et al. 1989) and these agents are also inhibitors of protein kinase C (see NORMAN 1991). Since several K-channel types appear to be inhibited by the calmodulin antagonists and at least two of these (BK_{Ca} and K_{ATP}) are phosphorylation dependent (see Sects. B.II.1. and B.III.), it seems possible that inhibition of protein kinase C, and consequently channel dephosphorylation, rather than a direct effect on the channel, may also contribute to the actions of the calmodulin antagonists.

D. Selectivity of Action

A detailed discussion of the possible basis for the selective action of K-channel modulators on individual channel subtypes or specific tissues is premature. The main purpose of this review has been to detail the great variety of agents which is now available for the modulation of K-channels. However, from the therapeutic viewpoint, the development of selective molecules is of enormous importance, since the very ubiquity of K-channels in all tissues could result in a variety of side effects following the administration of a non-selective K-channel modulator.

Molecules with some degree of *channel* selectivity are already available. Perhaps the best examples of these are the naturally occurring toxins, some of which exhibit a remarkable degree of selectivity for individual K-channel subtypes. Certain synthetic agents are also channel selective and over recent years the development of modulators of K_{ATP} has proceeded rapidly. For example, glibenclamide seems to have few inhibitory effects on channels other than K_{ATP}. Furthermore, the effects of the K-channel openers appear to be largely confined to the channel known as K_{ATP} (although this may be due to a dephosphorylating action on K_V).

The development of *tissue* selectivity is now a critically important research objective if K-channel modulators are to become established in the clinic. Although glibenclamide apparently inhibits type 1 K_{ATP} channels in all cells so far tested, its effects on the pancreatic β-cell predominate in vivo. A factor here may be that in most tissues K_{ATP} is normally closed (or in the form of the delayed rectifier K_V) and that only in the β-cell do fluctuations in ATP levels probably play a critical physiological role in the modulation of K_{ATP}/K_V. In respect of the K-channel openers, some progress towards tissue selectivity has been made (see Figs. 3, 5, 6). It remains to be established whether selective molecules derive this property from pharmacokinetic factors or whether they are able to exploit any differences which may exist between similar K-channel subtypes in different tissues.

Acknowledgement. Gillian Edwards was supported by Pfizer Central Research.

References

Adams LA, Houamed KM, Tempel BL (1992) Potassium channel genes: genomic complexity, molecular properties and differential expression. In: Weston AH, Hamilton TC (eds) Potassium channel modulators: pharmacological, molecular and clinical aspects. Blackwell, Oxford, pp 14–33

Adelman JP, Shen K-Z, Kavanaugh MP, Lagrutta A, Bond CT, North RA (1992) Calcium-activated potassium channels expressed from cloned complementary DNAs. Neuron 9:209–216

Aldrich RW (1990) Mixing and matching. Nature 345:475–476

Allen SL, Beech DJ, Foster RW, Morgan GP, Small RC (1985) Electrophysiological and other aspects of the relaxant action of isoprenaline in guinea-pig isolated trachealis. Br J Pharmacol 86:843–854

Allen SL, Boyle JP, Cortijo J, Foster RW, Morgan GP, Small RC (1986) Electrical and mechanical effects of BRL 34915 in guinea-pig isolated trachealis. Br J Pharmacol 89:395–405

Anderson C, MacKinnon R, Smith C, Miller C (1988) Charybdotoxin block of single Ca^{2+}-activated K^+ channels. Effects of channel gating, voltage, ionic strength. J Gen Physiol 91:317–333

Andersson K-E, Andersson P-O, Fovaeus M, Hedlund H, Malmgren A, Sjögren C (1988) Effects of pinacidil on bladder muscle. Drugs 36 [Suppl 7]:41–49

Andersson SE (1992) Glibenclamide and L-N^G-nitro-arginine methyl ester modulate the ocular and hypotensive effects of calcitonin gene-related peptide. Eur J Pharmacol 224:89–91

Arena JP, Kass RS (1989) Enhancement of potassium-sensitive current in heart cells by pinacidil: evidence for modulation of the ATP-sensitive potassium channel. Circ Res 65:436–445

Arena JP, Walsh KB, Kass RS (1990) Measurement, block, and modulation of potassium channel currents in the heart. In: Colatsky T (ed) Potassium channels: basic function and therapeutic aspects. Liss, New York, pp 43–63

Armstrong CM, Bezanilla F (1977) Inactivation of the sodium channel. II. Gating current experiments. J Gen Physiol 70:567–590

Ashcroft FM (1988) Adenosine 5′-triphosphate-sensitive potassium channels. Annu Rev Neurosci 11:97–118

Ashcroft FM, Rorsman P (1990) ATP-sensitive K^+ channels: a link between β-cell metabolism and insulin secretion. Biochem Soc Trans 18:109–111

Ashcroft SJH, Ashcroft FM (1990) Properties and functions of ATP-sensitive K-channels. Cell Signal 2:197–214

Attali B, Romey G, Honore E, Schmidalliana A, Mattei MG, Lesage F, Ricard P, Barhanin J, Lazdunski M (1992) Cloning, functional expression, regulation of 2 K^+ channels in human lymphocytes-T. J Biol Chem 267:8650–8657

Atwal K, Grover GJ, Ahmed SZ, Ferrara FN, Harper TW, Kim KS, Sleph PG, Dzwonczyk S, Russell AD, Moreland S, McCullough JR, Normandin DE (1993) Cardioselective anti-ischaemic ATP-sensitive potassium channel openers. J Med Chem 36:3971–3974

Bang L, Nielsen-Kudsk JE (1992) Smooth muscle relaxation and inhibition of responses to pinacidil and cromakalim induced by phentolamine in guinea-pig isolated trachea. Eur J Pharmacol 211:235–241

Baskin E, Serik C, Wallace A, Jurkiewicz N, Winquist R, Lynch J (1992) Vascular effects of class-III antiarrhythmic agents. Drug Dev Res 26:481–488

Bauer V, Kuriyama H (1982) The nature of non-cholinergic, non-adrenergic transmission in longitudinal and circular muscles of the guinea-pig ileum. J Physiol (Lond) 332:375–391

Beech DJ, Bolton TB (1989a) A voltage-dependent outward current with fast kinetics in single smooth muscle cells isolated from rabbit portal vein. J Physiol (Lond) 412:397–414

Beech DJ, Bolton TB (1989b) Two components of potassium current activated by depolarization of single smooth muscle cells from the rabbit portal vein. J Physiol (Lond) 418:293–309
Beech DJ, Bolton TB (1989c) Properties of the cromakalim-induced potassium conductance in smooth muscle cells isolated from the rabbit portal vein. Br J Pharmacol 98:851–864
Beech DJ, Bolton TB, Castle NA, Strong PN (1987) Characterisation of a toxin from scorpion (*Leiurus quinquestriatus*) venom that blocks in vitro both large (BK) K^+ channels in rabbit vascular smooth muscle and intermediate (IK) conductance Ca^{2+}-activated K^+ channels in human red cells. J Physiol (Lond) 387:32P
Beech DJ, Zhang H, Nakao K, Bolton TB (1993a) K channel activation by nucleotide diphosphates and its inhibition by glibenclamide in vascular smooth muscle cells. Br J Pharmacol 110:573–582
Beech DJ, Zhang H, Nakao K, Bolton TB (1993b) Single channel and whole-cell K-currents evoked by levcromakalim in smooth muscle cells from the rabbit portal vein. Br J Pharmacol 110:583–590
Belluzzi O, Sacchi O, Wanke E (1985) A fast transient outward current in the rat sympathetic neurone studied under voltage-clamp. J Physiol (Lond) 358:91–108
Benham CD, Bolton TB (1983) Patch-clamp studies of slow potential-sensitive potassium channels in longitudinal smooth muscle cells of rabbit jejunum. J Physiol (Lond) 340:469–486
Benham CD, Bolton TB (1986) Spontaneous transient outward currents in single visceral and vascular smooth muscle cells of the rabbit. J Physiol (Lond) 381: 385–406
Benham CD, Bolton TB, Lang RJ, Takewaki T (1985) The mechanism of action of Ba^{2+} and TEA on single Ca^{2+}-activated K^+ channels in arterial and intestinal smooth muscle cell membranes. Pflügers Arch 403:120–127
Benham CD, Bolton TB, Lang RJ, Takewaki T (1986) Calcium-activated potassium channels in single smooth muscle cells of rabbit jejunum and guinea-pig mesenteric artery. J Physiol (Lond) 371:45–67
Benoit E, Dubois J-M (1986) Toxin I from the snake *Dendroaspis polylepsis*: a highly specific blocker of one type of potassium channel in myelinated nerve fiber. Brain Res 377:374–377
Bény JL, Brunet PC (1988) Neither nitric oxide nor nitroglycerin accounts for all the characteristics of endothelially-mediated vasodilatation of pig coronary arteries. Blood Vessels 25:308–311
Bény JL, Brunet PC, Huggel H (1989) Effects of substance P, calcitonin gene related peptide and capsaicin on tension and membrane potential of pig coronary artery in vitro. Regul Pept 25:25–36
Berger W, Grygorcyk R, Schwarz W (1984) Single K^+ channels in membrane evaginations of smooth muscle cells. Pflügers Arch 402:18–23
Berridge MJ (1993) Inositol trisphosphate and calcium signalling. Nature 361: 315–325
Berry JL, Elliott KRF, Foster RW, Green KA, Murray MA, Small RC (1991) Mechanical, biochemical and electrophysiological studies of RP-49356 and cromakalim in guinea-pig and bovine trachealis muscle. Pulm Pharmacol 4: 91–98
Bilbaut A, Hernandez-Nicaise M-L, Leech CA, Meech RW (1988) Membrane currents that govern smooth muscle contraction in a ctenophore. Nature 331: 533–535
Blatz AL, Magleby KL (1986) Single apamin-blocked Ca-activated K^+ channels of small conductance in cultured rat skeletal muscle. Nature 323:718–720
Blatz AL, Magleby KL (1987) Calcium-activated potassium channels. Trends Neurosci 10:463–467
Bolton TB, Beech DJ (1992) Smooth muscle potassium channels: their electrophysiology and function. In: Weston AH, Hamilton TC (eds) Potassium channel

modulators: pharmacological, molecular and clinical aspects. Blackwell, Oxford, pp 144–180
Bolton TB, Clapp LH (1986) Endothelial-dependent relaxant actions of carbachol and substance P in arterial smooth muscle. Br J Pharmacol 87:713–723
Bolton TB, Lang RJ, Takewaki T (1984) Mechanism of action of noradrenaline and carbachol on smooth muscle of guinea-pig anterior mesenteric artery. J Physiol (Lond) 351:549–572
Bolton TB, Lang RJ, Takewaki T, Benham CD (1986) Patch and whole-cell voltage-clamp studies on single smooth muscle cells. J Cardiovasc Pharmacol 8:S20–S24
Bonev AD, Nelson MT (1993) ATP-sensitive potassium channels in smooth muscle cells from guinea pig urinary bladder. Am J Physiol 264:C1190–C1200
Bonnet PA, Michel A, Laurent F, Sablayrolles C, Rechlencq E, Mani JC, Boucard M, Chapat JP (1992) Synthesis and antibronchospastic activity of 8-alkoxy- and 8-(alkylamino)imidazo[1,2α]pyrazines. J Med Chem 35:3353–3358
Bowring NE, Arch JRS, Buckle DR, Taylor JF (1993) Comparison of the airways relaxant and hypotensive potencies of the potassium channel activators BRL 55834 and levcromakalim (BRL 38227) in vivo in guinea-pigs and rats. Br J Pharmacol 109:1133–1139
Boyle JP, Tomasic M, Kotlikoff MI (1992) Delayed rectifier potassium channels in canine and porcine airway smooth muscle cells. J Physiol (Lond) 447:329–350
Boyle MB, MacLusky NJ, Naftolin F, Kaczmarek LK (1987) Hormonal regulation of K^+-channel messenger RNA in rat myometrium during oestrus cycle and in pregnancy. Nature 330:373–375
Brain SD, Williams TJ, Tippins JR, Morris HR, Macintyre I (1985) Calcitonin gene related peptide is a potent vasodilator. Nature 313:54–56
Bray KM, Newgreen DT, Small RC, Southerton JS, Taylor SG, Weir SW, Weston AH (1987a) Evidence that the mechanism of the inhibitory action of pinacidil in rat and guinea-pig smooth muscle differs from that of glyceryl trinitrate. Br J Pharmacol 91:421–429
Bray KM, Newgreen DT, Weston AH (1987b) Some effects of the enantiomers of the potassium channel openers BRL 34915 and pinacidil on rat blood vessels. Br J Pharmacol 191:357P
Bray KM, Quast U (1991a) Some degree of overlap exists between the K^+-channels opened by cromakalim and those opened by minoxidil sulphate in rat isolated aorta. Naunyn Schmiedebergs Arch Pharmacol 344:351–359
Bray KM, Quast U (1991b) Differences in the K^+-channels opened by cromakalim, acetylcholine and substance-P in rat aorta and porcine coronary artery. Br J Pharmacol 102:585–594
Bray KM, Quast U (1992a) A specific binding site for K^+ channel openers in rat aorta. J Biol Chem 267:11689–11692
Bray KM, Quast U (1992b) Differential inhibition by tedisamil (KC-8857) and glibenclamide of the responses to cromakalim and minoxidil sulphate in rat isolated aorta. Naunyn Schmiedebergs Arch Pharmacol 345:244–250
Bray KM, Quast U (1993) Binding of the K^+ channel opener [^{3}H]P1075 in rat isolated aorta: relationship to functional effects of openers and blockers. Mol Pharmacol 43:474–481
Bray KM, Weston AH, Duty S, Newgreen DT, Longmore J, Edwards G, Brown TJ (1991) Differences between the effects of cromakalim and nifedipine on agonist-induced responses in rabbit aorta. Br J Pharmacol 102:337–344
Brayden JE (1990) Membrane hyperpolarization is a mechanism of endothelium-dependent cerebral vasodilation. Am J Physiol 259:H668–H673
Brayden JE (1991) Hyperpolarization and relaxation of resistance arteries in response to adenosine diphosphate. Distribution and mechanism of action. Circ Res 69:1415–1420
Brown TJ, Chapman RF, Mason JS, Palfreyman MN, Vicker N, Walsh RJA (1993) The synthesis and biological activities of potent potassium channel openers

derived from (±)-2-oxo-1-pyridin-3-yl-cyclohexane carbothioic acid methylamide: new potassium channel openers. J Med Chem 36:1604–1612
Buckingham RE, Hamilton TC, Howlett DR, Mootoo S, Wilson C (1989) Inhibition by glibenclamide of the vasorelaxant action of cromakalim in the rat. Br J Pharmacol 97:57–64
Cain CR, Nicholson CD (1989) Comparison of the effects of cromakalim, a potassium conductance enhancer, and nimodipine, a calcium antagonist, on 5-hydroxytryptamine responses in a variety of vascular smooth muscles. Naunyn Schmiedebergs Arch Pharmacol 340:293–299
Calder JA, Schachter M, Sever PS (1992) Direct vascular actions of hydrochlorothiazide and indapamide in isolated small vessels. Eur J Pharmacol 220: 19–26
Carbone E, Prestipino G, Spadavecchia L, Franciolini F, Possani LD (1987) Blocking of the squid axon K^+ channel by noxiustoxin: a toxin from the venom of the scorpion *Centruroides noxius*. Pflügers Arch 408:423–431
Carl A, Bowen S, Gelband CH, Sanders KM, Hulme JR (1992) Cromakalim and levcromakalim activate Ca^{2+}-dependent K^+ channels in canine colon. Pflügers Arch 421:67–76
Carl A, Sanders KM (1989) Ca^{2+}-activated K channels of canine colonic myocytes. Am J Physiol 257:C470–C480
Casteels R, Kitamura K, Kuriyama H, Suzuki H (1977) Excitation-coupling in the smooth muscle cells of the rabbit pulmonary artery. J Physiol (Lond) 271:63–79
Castle NA, Haylett DG, Jenkinson DH (1989) Toxins in the characterization of potassium channels. Trends Neurosci 12:59–65
Castle NA, Strong PN (1986) Identification of two toxins from scorpion (*Leiurus quinquestriatus*) venom which block distinct classes of calcium-activated potassium channel. FEBS Lett 209:117–121
Catterall WA (1988) Structure and function of voltage-sensitive ion channels. Science 242:50–61
Cavero I, Mondot S, Mestre M (1989) Vasorelaxant effects of cromakalim in rats are mediated by glibenclamide-sensitive potassium channels. J Pharmacol Exp Ther 248:1261–1268
Cecchi X, Alvarez O, Wolff D (1986) Characterization of a calcium-activated potassium channel from rabbit intestinal smooth muscle incorporated into planar bilayers. J Membr Biol 91:11–18
Chen G, Suzuki H (1989) Some electrical properties of the endothelium-dependent hyperpolarisation recorded from rat arterial smooth muscle cells. J Physiol (Lond) 410:91–106
Chen G, Suzuki H, Weston AH (1988) Acetylcholine releases endothelium-derived hyperpolarising factors and EDRF from rat blood vessels. Br J Pharmacol 95:1165–1174
Chen G, Yamamoto Y, Miwa K, Suzuki H (1991) Hyperpolarization of arterial smooth muscle induced by endothelial humoral substances. Am J Physiol 260:H1888–H1892
Cheung DW, MacKay MJ (1985) The effects of sodium nitroprusside on smooth muscle cells of rabbit pulmonary artery and portal vein. Br J Pharmacol 86: 117–124
Chicchi GG, Gimenez-Gallego G, Ber E, Garcia ML, Winquist R, Cascieri MA (1988) Purification and characterization of a unique, potent inhibitor of apamin binding from *Leiurus quinquestriatus hebraeus* venom. J Biol Chem 263: 10192–10197
Chiu P, Cook SJ, Small RC, Berry JL, Carpenter JR, Downing SJ, Foster RW, Miller AJ, Small AM (1993) β-Adrenoceptor subtypes and the opening of plasmalemmal K^+-channels in bovine trachealis muscle: studies of mechanical activity and ion fluxes. Br J Pharmacol 109:1149–1156
Chopra LC, Twort CHC, Ward JPT (1992) Direct action of BRL 38227 and glibenclamide on intracellular calcium stores in cultured airway smooth muscle of rabbit. Br J Pharmacol 105:259–260

Christinck F, Jury J, Cayabyab F, Daniel EE (1991) Nitric oxide may be the final mediator of nonadrenergic, noncholinergic inhibitory junction potentials in the gut. Can J Physiol Pharmacol 69:1448–1458
Clapp LH, Gurney AM (1992) ATP-sensitive K^+ channels regulate resting potential of pulmonary arterial smooth muscle cells. Am J Physiol 262:H916–H920
Cohen ML, Colbert WE (1986) Comparison of the effects of pinacidil and its metabolite pinacidil N-oxide in isolated smooth and cardiac muscle. Drug Dev Res 7:111–124
Colatsky TJ (1992) Potassium channel blockers: synthetic agents and their antiarrhythmic potential. In: Weston AH, Hamilton TC (eds) Potassium channel modulators: pharmacological, molecular and clinical aspects. Blackwell, Oxford, pp 304–340
Coldwell MC, Howlett DR (1987) Specificity of action of the novel antihypertensive agent BRL 34915 as a potassium channel activator: Comparison with nicorandil. Biochem Pharmacol 36:3663–3669
Cole WC, Sanders KM (1989) G proteins mediate suppression of Ca^{2+}-activated K^+ current by acetylcholine in smooth muscle cells. Am J Physiol 257:C596–C600
Collier ML, Twort CHC, Cameron IR, Ward JPT (1990) BRL 38227 and ATP-dependent potassium channels in airway smooth muscle cells. Proceedings of the XIth International Congress of Pharmacology, New Drugs for Asthma Symposium, P4
Cook NS (1989) Effect of some potassium channel blockers on contractile responses of the rabbit aorta. J Cardiovasc Pharmacol 13:299–306
Cook NS, Quast U, Manley PW (1989) K^+ channel opening does not alone explain the vasodilator activity of pinacidil and its enantiomers. Br J Pharmacol 96:181P
Cook NS, Weir SW, Danzeisen MC (1988) Anti-vasoconstrictor effects of the K^+ channel opener cromakalim on the rabbit aorta-comparison with the Ca^{2+}-antagonist isradipine. Br J Pharmacol 95:741–752
Cook SJ, Small RC, Berry JL, Chiu P, Downing SJ, Foster RW (1993) β-Adrenoceptor subtypes and the opening of plasmalemmal K^+-channels in trachealis muscle: electrophysiological and mechanical studies in guinea-pig tissue. Br J Pharmacol 109:1140–1148
Criddle DN, Greenwood IA, Weston AH (1994) Levcromakalim-induced modulation of membrane potassium currents, intracellular calcium and mechanical activity in rat mesenteric artery. Naunyn Schmiedebergs Arch Pharmacol 349:422–430
Dart C, Standen NB (1993) Adenosine-activated potassium current in smooth muscle cells isolated from the pig coronary artery. J Physiol 471:767–786
Daut J, Maier-Rudolph W, von-Beckerath N, Mehrke G, Gunther K, Goedel-Meinen L (1990) Hypoxic dilation of coronary arteries is mediated by ATP-sensitive potassium channels. Science 247:1341–1344
Davies JM, Williams KI (1984) Endothelial-dependent relaxant effects of vaso-active intestinal polypeptide and arachidonic acid in rat aortic strips. Prostaglandins 27:195–200
De Weille JR, Schmid-Antomarchi H, Fosset M, Lazdunski M (1989) Regulation of ATP-sensitive K^+ channel in insulinoma cells: activation by somatostatin and protein kinase C and the role of cAMP. Proc Natl Acad Sci USA 86:2971–2975
Dreyer F (1990) Peptide toxins and potassium channels. Rev Physiol Biochem Pharmacol 115:93–136
Droogmans G, Raeymaekers L, Casteels R (1977) Electro- and pharmacomechanical coupling in the smooth muscle cells of the rabbit ear artery. J Gen Physiol 70:129–148
Dubois JM (1981) Evidence for the existence of three types of potassium channels in the frog Ranvier node membrane. J Physiol (Lond) 318:297–316
Dubois JM (1983) Potassium currents in the frog node of Ranvier. Prog Biophys Mol Biol 42:1–20
Duchatelle-Gourdon I, Lagrutta AA, Hartzell HC (1991) Effects of Mg^{2+} on basal and β-adrenergic-stimulated delayed rectifier potassium current in frog atrial myocytes. J Physiol (Lond) 435:333–347

Dukes ID, Cleemann L, Morad M (1990) Tedisamil blocks the transient and delayed rectifier-K^+ currents in mammalian cardiac and glial cells. J Pharmacol Exp Ther 254:560–589
Dukes ID, Morad M (1991) The transient K^+ current in rat ventricular myocytes-evaluation of its Ca^{2+} and Na^+ dependence. J Physiol (Lond) 435:395–420
Dunne MJ (1990) Effects of pinacidil, RP 49356 and nicorandil on ATP-sensitive potassium channels in insulin-secreting cells. Br J Pharmacol 99:487–492
Eckman DM, Frankovich JD, Keef KD (1992) Comparison of the actions of acetylcholine and BRL 38227 in the guinea pig coronary artery. Br J Pharmacol 106:9–16
Edwards FR, Hirst GDS (1988) Inward rectification in submucosal arterioles of guinea-pig ileum. J Physiol (Lond) 404:437–454
Edwards FR, Hirst GDS, Silverberg GD (1988) Inward rectification in rat cerebral arterioles; involvement of potassium ions in autoregulation. J Physiol (Lond) 404:455–466
Edwards G, Henshaw M, Miller M, Weston AH (1991) Comparison of the effects of several potassium-channel openers on rat bladder and rat portal vein in vitro. Br J Pharmacol 102:679–686
Edwards G, Ibbotson T, Weston AH (1993) Levcromakalim may induce a voltage-independent K-current in rat portal veins by modifying the gating properties of the delayed rectifier. Br J Pharmacol 110:1037–1048
Edwards G, Weston AH (1990a) Structure activity relationships of K^+ channel openers. Trends Pharmacol Sci 11:417–422
Edwards G, Weston AH (1990b) Potassium channel openers and vascular smooth muscle relaxation. Pharmacol Ther 48:237–258
Edwards G, Weston AH (1993a) The pharmacology of ATP-sensitive potassium channels. Annu Rev Pharmacol Toxicol 33:597–637
Edwards G, Weston AH (1993b) Induction of a glibenclamide-sensitive K-current by modification of a delayed rectifier channel in rat portal vein and insulinoma cells. Br J Pharmacol 110:1280–1281
Elliott KRF, Berry JL, Bate AJ, Foster RW, Small RC (1991) The isoenzyme selectivity of AH 21-132 as an inhibitor of cyclic nucleotide phosphodiesterase activity. J Enzym Inhib 4:245–251
Ewald DA, Williams A, Levitan IB (1985) Modulation of single Ca^{2+}-dependent K^+ channel activity by protein phosphorylation. Nature 315:503–506
Farley J, Rudy B (1988) Multiple types of voltage-dependent Ca^{2+}-activated K^+ channels of large conductance in rat brain synaptosomal membranes. Biophys J 53:919–934
Feletou M, Vanhoutte PM (1988) Endothelium-dependent hyperpolarization of canine coronary artery muscle. Br J Pharmacol 93:512–524
Ffrench-Mullen JMH, Rogawski MA, Barker JL (1988) Phencyclidine at low concentrations selectively blocks the sustained but not the transient voltage-dependent potassium current in cultured hippocampal neurons. Neurosci Lett 88:325–330
Folander K, Smith JS, Antanavage J, Bennet C, Stein RB, Swanson R (1990) Cloning and expression of the delayed rectifier IsK channel from neonatal rat heart and diethylstilbestrol-primed rat uterus. Proc Natl Acad Sci USA 87: 2975–2979
Foster CD, Fujii K, Kingdon J, Brading AF (1989a) The effects of cromakalim on the smooth muscle of the guinea-pig urinary bladder. Br J Pharmacol 97:281–291
Foster CD, Speakman MJ, Fujii K, Brading AF (1989b) The effects of cromakalim on the detrusor muscle of human and pig urinary bladder. Br J Urol 63:284–294
Fozard JR, Carruthers AM (1993) The cardiovascular effects of selective adenosine A_1 and A_2 receptor agonists in the pithed rat: no role for glibenclamide-sensitive potassium channels. Naunyn Schmiedebergs Arch Pharmacol 347:192–196

Fujii K, Foster CD, Brading AF, Parekh AB (1990a) Potassium channel blockers and the effects of cromakalim on the smooth muscle of the guinea-pig bladder. Br J Pharmacol 99:779–785
Fujii K, Kobayashi K, Koga T, Fujishima M (1990b) Decreased endothelium-dependent hyperpolarization in spontaneously hypertensive rats (SHR). Circulation 82 [Suppl 4]:III-344
Fujiwara S, Kuriyama H (1983) Effects of agents that modulate potassium permeability on smooth muscle cells of the guinea-pig basilar artery. Br J Pharmacol 79:23–25
Fuller CM, Benos DJ (1992) CFTR! Am J Physiol 263:C267–C286
Furchgott RF (1988) Studies on relaxation of rabbit aorta by sodium nitrate: the basis for the proposal that the acid-activatable inhibitory factor from bovine retractor penis is inorganic nitrate and the endothelium-derived relaxing factor is nitric oxide. In: Vanhoutte PM (ed) Mechanisms of vasodilatation. Raven, New York, pp 401–414
Furukawa K, Itoh T, Kajiwara M, Kitamura K, Suzuki H, Ito Y, Kuriyama H (1981) Vasodilating actions of 2-nicotinamidoethyl nitrate on porcine and guinea-pig coronary arteries. J Pharmacol Exp Ther 218:248–259
Galvez A, Gimenez-Gallego G, Reuben JP, Roy-Contancin L, Feigenbaum P, Kaczorowski GJ, Garcia ML (1990) Purification and characterisation of a unique, potent, peptidyl probe for the high conductance, calcium-activated potassium channel from venom of the scorpion *Buthus tamulus*. J Biol Chem 265:11083–11099
Ganitkevich VY, Isenberg G (1992) Contribution of Ca^{2+}-induced Ca^{2+} release to the $[Ca^{2+}\text{-}]_i$ transients in myocytes from guinea-pig urinary bladder. J Physiol (Lond) 458:119–137
Garcia ML, Galvez A, Garciacalvo M, Kin VF, Vazquez J, Kaczorowski GJ (1991) Use of toxins to study potassium channels. J Bioenerg Biomembr 23:615–646
Garland CJ, McPherson GA (1992) Evidence that nitric oxide does not mediate the hyperpolarization and relaxation to acetylcholine in the rat small mesenteric artery. Br J Pharmacol 105:429–435
Garrino MG, Plant TD, Henquin JC (1989) Effects of putative activators of K^+ channels in mouse pancreatic β-cells. Br J Pharmacol 98:957–965
Gater PR, Haylett DG, Jenkinson DH (1985) Neuro-muscular blocking agents inhibit receptor-mediated increases in the potassium permeability of intestinal smooth muscle. Br J Pharmacol 86:861–868
Gelband CH, Hume JR (1992) Ionic currents in single smooth muscle cells of the canine renal artery. Circ Res 71:745–758
Gelband CH, Lodge NJ, van Breemen C (1989) A Ca^{2+}-activated K^+ channel from rabbit aorta: modulation by cromakalim. Eur J Pharmacol 167:201–210
Giangiacomo KM, Garcia ML, McManus OB (1992) Mechanism of iberiotoxin block of the large-conductance calcium-activated potassium channel from bovine aortic smooth muscle. Biochem 31:6712–6727
Giles WR, Imaizumi Y (1988) Comparison of potassium currents in rabbit atrial and ventricular cells. J Physiol (Lond) 405:123–145
Gimenez-Gallego G, Navia MA, Reuben JP, Katz GM, Kaczorowski GL, Garcia ML (1988) Purification, sequence, model structure of charybdotoxin, a potent selective inhibitor of calcium-activated potassium channels. Proc Natl Acad Sci USA 85:3329–3333
Gopalakrishnan M, Johnson DE, Janis RA, Triggle DJ (1991) Characterization of binding of the ATP-sensitive potassium channel ligand, $[^3H]$-glyburide, to neuronal and muscle preparations. J Pharmacol Exp Ther 257:1162–1171
Grace GC, Dusting GJ, Kemp BE, Martin TJ (1987) Endothelium and the vasodilator action of rat calcitonin gene-related peptide (CGRP). Br J Pharmacol 91:729–733

Grant TL, Zuzack JS (1991) Effects of K^+ channel blockers and cromakalim (BRL 34915) on the mechanical activity of guinea pig detrusor smooth muscle. J Pharmac Exp Ther 259:1158–1164
Gray DW, Marshall I (1990) Calcitonin gene-related peptide (CGRP) endothelium-dependent relaxation in rat aorta is inhibited by L-NMMA. Br J Pharmacol 99: 104P
Green KA, Foster RW, Small RC (1991) A patch-clamp study of K^+-channel activity in bovine isolated tracheal smooth muscle cells. Br J Pharmacol 102: 871–878
Greenwood IA, Weston AH (1993) Effects of rubidium on responses to potassium channel openers in rat isolated aorta. Br J Pharmacol 109:925–932
Grinstein S, Smith JD (1989) Ca^{2+} induces charybdotoxin-sensitive membrane potential changes in rat lymphocytes. Am J Physiol 257:C197–C206
Gross GJ (1991) Coronary blood flow studies with potassium channel openers. Curr Drugs Potassium Channel Modulators 1:KCMB82–KCMB92
Gryglewski RJ, Botting RM, Vane JR (1991) Prostacyclin: from discovery to clinical application. In: Rubanyi GM (ed) Cardiovascular significance of endothelium-derived vasoactive factors. Futura, Mount Kisco, pp 3–37
Guggino SE, Guggino WB, Green N, Sacktor B (1987) Blocking agents of Ca^{2+}-activated K^+ channels in cultured medullary thick ascending limb cells. Am J Physiol 252:C128–C137
Gustafsson B, Galvan M, Grafe P, Wigstrom H (1982) A transient outward current in mammalian central neurone blocked by 4-aminopyridine. Nature 299: 252–254
Gwilt M, Arrowsmith JE, Blackburn RA, Burges PE, Cross HW, Dalrymple HW, Higgins AJ (1991) UK-68,798: a novel, potent and highly selective class III antiarrhythmic agent which blocks potassium channels in cardiac cells. J Pharmacol Exp Ther 256:318–324
Hamilton TC, Weir SW, Weston AH (1985) Comparison of the effects of BRL34915 and verapamil on rat portal vein. Br J Pharmacol 86:443P
Hamilton TC, Weir SW, Weston AH (1986) Comparison of the effects of BRL 34915 and verapamil on electrical and mechanical activity in rat portal vein. Br J Pharmacol 88:103–111
Hara Y, Kitamura K, Kuriyama H (1980) Actions of 4-aminopyridine on vascular smooth muscle tissues of the guinea-pig. Br J Pharmacol 68:99–106
Harder DR, Sperelakis N (1979) Action potentials induced in guinea-pig arterial smooth muscle by tetraethylammonium. Am J Physiol 237:C75–C80
Hartman HA, Kirsch GE, Drewe JA, Taglialatela M, Joho RH, Brown AM (1991) Exchange of conduction pathways between two related K^+ channels. Science 251:942–944
Hattori Y, Nagashima M, Endo Y, Kanno M (1992) Glibenclamide does not block arterial relaxation caused by vasoactive intestinal polypetide. Eur J Pharmacol 213:147–150
Hausdorff SF, Goldstein SA, Rushin EE, Miller C (1991) Functional characterization of a minimal K^+ channel expressed from a synthetic gene. Biochemistry 30: 3341–3346
Hermann A, Erxleben C (1987) Charybdotoxin selectively blocks small Ca-activated K channels in aplysia neurones. J Gen Physiol 90:24–47
Hille B (1992) Ionic channels of excitable membranes. Sinauer, Sunderland, pp 193–197
Ho K, Nichols CG, Lederer WJ, Lytton J, Vassilev PM, Kanazirska MV, Hebert SC (1993) Cloning and expression of an inwardly rectifying ATP-regulated potassium channel. Nature 362:31–38
Hof RP, Vuorela HJ (1983) Assessing calcium antagonism on vascular smooth muscle: a comparison of three methods. J Pharmacol Methods 91:41–52
Hollingsworth M, Amédée T, Edwards D, Mironneau J, Savineau J, Small RC, Weston AH (1987) The relaxant action of BRL34915 in rat uterus. Br J Pharmacol 91:803–813

Hollingsworth M, Edwards D, Miller M, Rankin JR, Weston AH (1989) Potassium channels in isolated rat uterus and the action of cromakalim. Med Sci Res 17:461–463
Holzmann S (1983) Cyclic GMP a possible mediator of coronary arterial relaxation by nicorandil (SG-75). J Cardiovasc Pharmacol 5:364–370
Honoré E, Attali B, Romey G, Heurteaux C, Ricard P, Lesage F, Lazdunski M, Barhanin J (1991) Cloning, expression, pharmacology and regulation of a delayed rectifier K^+ channel in mouse heart. EMBO J 10:2805–2811
Hoshi T, Zagotta WN, Aldrich RW (1990) Biophysical and molecular mechanisms of *Shaker* potassium channel inactivation. Science 250:533–538
Howlett DR, Longman SD (1992) Identification of a binding site for [^{3}H]-cromakalim in vascular and bronchial smooth muscle cells. Br J Pharmacol 107:396P
Huang AH, Busse R, Bassenge E (1988) Endothelium-dependent hyperpolarization of smooth muscle cells in rabbit femoral arteries is not mediated by EDRF (nitric oxide). Naunyn Schmiedebergs Arch Pharmacol 338:438–442
Hugues M, Romey G, Duval D, Vincent JP, Lazdunski M (1982) Apamin as a selective bocker of the calcium-dependent potassium channel in neuroblastoma cells: voltage clamp, biochemical characterization of the toxin receptor. Proc Natl Acad Sci USA 79:1308–1312
Huizinga JD, Tomlinson J, Pintin-Quezada J (1992) Involvement of nitric oxide in nerve-mediated inhibition and action of vasoactive intestinal peptide in colonic smooth muscle. J Pharmacol Exp Ther 260:803–808
Hume JR, Leblanc N (1989) Macroscopic K^+ currents in single smooth muscle cells of the rabbit portal vein. J Physiol (Lond) 413:49–73
Ibbotson T, Edwards G, Noack T, Weston AH (1993a) Effects of P1060 and aprikalim on whole-cell currents in rat portal vein; inhibition by glibenclamide and phentolamine. Br J Pharmacol 108:991–998
Ibbotson T, Edwards G, Weston AH (1993b) Antagonism of levcromakalim by imidazoline- and guanidine-derivatives in rat portal vein cell: involvement of the delayed rectifier. Br J Pharmacol 110:1556–1564
Ignarro LJ, Byrns RE, Buga GM, Wood KS (1987) Mechanisms of endothelium-dependent vascular smooth muscle relaxation elicited by bradykinin and VIP. Am J Physiol 253:H1074–H1082
Illiano S, Nagao T, Vanhoutte PM (1992) Calmidazolium, a calmodulin inhibitor, inhibits endothelium-dependent relaxations resistant to nitro-L-arginine in the canine coronary artery. Br J Pharmacol 107:387–392
Imaizumi Y, Muraki K, Watanabe M (1989) Ionic currents in single smooth muscle cells from the ureter of the guinea-pig. J Physiol (Lond) 413:49–73
Imaizumi Y, Muraki K, Watanabe M (1990) Characteristics of transient outward currents in single smooth muscle cells from the ureter of the guinea-pig. J Physiol (Lond) 427:301–324
Inoue I, Nakaya Y, Nakaya S, Mori H (1989) Extracellular Ca^{2+}-activated K channel in coronary artery smooth muscle cells and its role in vasodilation. FEBS Lett 255:281–284
Inoue R, Kitamura K, Kuriyama H (1985) Two Ca-dependent K-channels classified by the application of tetraethylammonium distributes to smooth muscle membranes of the rabbit portal vein. Pflügers Arch 405:173–179
Inoue R, Okabe K, Kitamura K, Kuriyama H (1986) A newly identified Ca^{2+} dependent K^+ channel in the smooth muscle membrane of single cells dispersed from the rabbit portal vein. Pflügers Arch 406:138–141
Inoue T, Kanmura Y, Fujisawa K, Itoh T, Kuriyama H (1984) Effects of 2-nicotinamidoethyl nitrate (nicorandil; SG-75) and its derivatives on smooth muscle cells of the canine mesenteric artery. J Pharmacol Exp Ther 229:793–802
Isacoff EY, Jan YN, Jan LY (1990) Evidence for the formation of heteromultimeric potassium channels in *Xenopus* oocytes. Nature 345:530–534

Ito H, Tung RT, Sugimoto T, Kobayashi I, Takahashi K, Katada T, Ui M, Kurachi Y (1992) On the mechanism of G protein beta gamma subunit activation of the muscarinic K^+ channel in guinea pig atrial cell membrane. Comparison with the ATP-sensitive K^+ channel. J Gen Physiol 99:961–983

Ito S, Kurokawa A, Ohga A, Ohta T, Sawabe K (1990) Mechanical, electrical and cyclic nucleotide responses to peptide VIP and inhibitory nerve stimulation in rat stomach. J Physiol (Lond) 430:337–353

Ito S, Kajikuri J, Itoh T, Kuriyama H (1991) Effects of lemakalim on changes in Ca^{2+} concentration and mechanical activity induced by noradrenaline in the rabbit mesenteric artery. Br J Pharmacol 104:227–233

Ito Y, Suzuki H, Kuriyama K (1978) Effects of sodium nitroprusside on smooth muscle cells of rabbit pulmonary artery and portal vein. J Pharmacol Exp Ther 207:1022–1031

Ito Y, Tajima K (1982) Dual effects of catecholamines on pre- and postjunctional membranes in the dog trachea. Br J Pharmacol 75:433–440

Itoh T, Sasaguri T, Makita Y, Kanmura Y, Kuriyama H (1985) Mechanisms of vasodilation induced by vasoactive intestinal polypeptide in rabbit mesenteric artery. Am J Physiol 249:H231–H240

Itoh T, Seki N, Suzuki S, Ito S, Kajikuri J, Kuriyama H (1992) Membrane hyperpolarization inhibits agonist-induced synthesis of inositol 1,4,5-trisphosphate in rabbit mesenteric artery. J Physiol (Lond) 541:307–328

Itoh T, Suzuki S, Kuriyama H (1991) Effects of pinacidil on contractile proteins in high K^+-treated intact and in β-escin-treated skinned smooth muscle of the rabbit mesenteric artery. Br J Pharmacol 103:1697–1702

Jackson WF, König A, Dambracher TD, Busse R (1993) Prostacyclin-induced vasodilation in rabbit heart is mediated by ATP-sensitive potassium channels. Am J Physiol 264:H238–H243

Jan LY, Jan YN (1990) How might the diversity of potassium channels be generated. Trends Neurosci 13:415–419

Jan LY, Jan YN (1992) Tracing the roots of ion channels. Cell 69:715–718

Jonas JC, Plant TD, Henquin JC (1992) Imidazoline antagonists of α_2-adrenoceptors increase insulin release in vitro by inhibiting ATP-sensitive K^+ channels in pancreatic β-cells. Br J Pharmacol 107:8–14

Jonas P, Bräu ME, Hermsteiner M, Vogel W, (1989) Single-channel recording in myelinated nerve fibres reveals one type of Na channel but different K channels. Proc Natl Acad Sci USA 86:17238–17243

Jones AW (1980) Content and fluxes of electrolytes. In: Bohr DF, Somlyo AP, Sparks HV (eds) Handbook of Physiology. Section 2: The cardiovascular system. Volume II. Vascular smooth muscle American Physiological Society, Bethesda, pp 253–299

Jones TR, Charette L, Garcia ML, Kaczorowski GJ (1990) Selective inhibition of relaxation of guinea-pig trachea by charybdotoxin, a potent Ca^{2+}-activated K^+-channel inhibitor. J Pharmacol Exp Ther 255:697–706

Kajioka S, Kitamura K, Kuriyama H (1991) Guanosine diphosphate activates an adenosine 5′-triphosphate-sensitive K^+ channel in the rabbit portal vein. J Physiol (Lond) 444:397–418

Kajioka S, Oike M, Kitamura K (1990) Nicorandil opens a calcium-dependent potassium channel in smooth muscle cells of the rat portal vein. J Pharmacol Exp Ther 254:905–913

Kakei M, Noma A (1984) Adenosine-5′-triphosphate-sensitive single potassium channel in the atrioventricular node cell of the rabbit heart. J Physiol (Lond) 352:265–284

Kameyama K, Kiyosue T, Soejima M (1983) Single channel analysis of the inward rectifier K current in the rabbit ventricular cells. Jpn J Physiol 33:1039–1056

Kamouchi M, Xiong Z, Teramoto N, Kajioka S, Okabe K, Kitamura K (1991) Ionic currents involved in vasodilating actions of E4080, a newly synthesized

bradycardia-inducing agent, in dispersed smooth muscle cells of the rabbit portal vein. J Pharmacol Exp Ther 259:1396–1403
Kamouchi M, Kajioka S, Sakai T, Kitamura K, Kuriyama H (1993) A target K^+ channel for the LP-805-induced hyperpolarization in smooth muscle cells of the rabbit portal vein. Naunyn Schmiedebergs Arch Pharmacol 347:329–335
Kaneta S, Jinno Y, Harada K, Ohta H, Ogawa N, Nishikori K (1990) Cardiohaemodynamic effect of KRN 2391, a novel vasodilator, in anaesthetised dogs. Jpn J Pharmacol 52:374P
Karashima T, Itoh T, Kuriyama H (1982) Effects of 2-nicotinamidoethyl nitrate on smooth muscle cells of the guinea-pig mesenteric and portal veins. J Pharmacol Exp Ther 221:472–480
Kawamura T, Ogawa T, Adachi H (1990) Cardiovascular effects of E4080, a novel bradycardic agent with a coronary vasodilating property, in dogs: a comparison with alinidine. Eur J Pharmacol 183:794–795
Khan RN, Smith SK, Morrison JJ, Ashford MLJ (1993) Properties of large-conductance K^+ channels in human myometrium during pregnancy and labour. Proc R Soc Lond [Biol] 251:9–15
Kirkpatrick CT (1975) Excitation and contraction in bovine tracheal smooth muscle. J Physiol (Lond) 244:263–281
Kirsch GE, Codina J, Birnbaumer L, Brown AM (1990) Coupling of ATP-sensitive K^+ channels to A1-receptors by G-proteins in rat ventricular myocytes. Am J Physiol 259:H820–H826
Kishii K-I, Inazu M, Morimoto T, Tsujitani M, Takayanagi I (1992a) Effects of LP-805, a new vasodilating agent, on cytosolic Ca^{2+} and contraction in vascular smooth muscle of rat aorta. Gen Pharmacol 23(3):355–363
Kishii K-I, Morimoto T, Nakajima N, Yamazaki K, Tsujitani M, Takayanagi I (1992b) Effects of LP-805, a novel vasorelaxant agent, a potassium channel opener, on rat thoracic aorta. Gen Pharmacol 23(3):347–353
Klaus E, Englert HC, Hropot M, Mania D, Zwergel U (1990a) Inhibition of the rhythmic contractions of ureters by K^+-channel openers. Naunyn Schmiedebergs Arch Pharmacol 340:R59
Klaus E, Englert HC, Hropot M, Mania D, Zwergel U (1990b) K^+-channel openers inhibit the KCI-induced phasic-rhythmic contractions in the upper urinary tract. Eur J Pharmacol 183:673
Klöckner U, Isenberg G (1985) Action potentials and net membrane currents of isolated smooth muscle cells (urinary bladder of the guinea-pig). Pflügers Arch 405:329–339
Klöckner U, Isenberg G (1992) ATP suppresses activity of Ca^{2+}-activated K^+ channels by Ca^{2+} chelation. Pflügers Arch 420:101–105
Koga H, Ohta M, Sato H, Ishizawa T, Nabata H (1993) Design of potent K^+ channel openers by pharmacophore model. Bioorg Med Chem Lett 3(4): 625–631
Komori K, Lorenz RR, Vanhoutte PM (1988) Nitric oxide ACh and electrical and mechanical properties of canine arterial smooth muscle. Am J Physiol 255: H207–H212
Kotlikoff MI (1993) Potassium channels in airway smooth muscle – a tale of 2 channels. Pharmacol Ther 58:1–12
Kozlowski RZ, Hales CN, Ashford MLJ (1989) Dual effects of diazoxide on ATP-K^+ currents recorded from an insulin-secreting cell line. Br J Pharmacol 197: 11039–11050
Kreye VAW, Gerstheimer F, Weston AH (1987) Effects of BRL 34915 on resting membrane potential and ^{86}Rb efflux in rabbit tonic vascular smooth muscle. Naunyn Schmiedebergs Arch Pharmacol 335:R64
Kreye VAW, Pfründer D, Theiss U (1992) Effects of the K^+ channel blocker tedisamil on ^{86}Rb efflux induced by cromakalim, high potassium and noradrenaline, on mechanical tension in rabbit isolated vascular smooth muscle. Naunyn Schmiedebergs Arch Pharmacol 345:238–243

Ksoll E, Parsons AA, Mackert JR, Schilling L, Wahl M (1991) Analysis of cromakalim-, pinacidil-, and nicorandil-induced relaxation of the 5-hydroxytryptamine precontracted rat isolated basilar artery. Naunyn Schmiedebergs Arch Pharmacol 343:377–383
Kubo Y, Baldwin TJ, Jan YN; Jan LY (1993) Primary structure and functional expression of a mouse inward rectifier potassium channel. Nature 362:127–133
Kume H, Takagi K, Satake T, Tokuno H, Tomita T (1990) Effects of intracellular pH on calcium-activated potassium channels in rabbit tracheal smooth muscle. J Physiol (Lond) 424:445–457
Kume H, Takai A, Tokuno H, Tomita T (1989) Regulation of Ca^{2+}-dependent K^+-channel activity in tracheal myoctyes by phosphorylation. Nature 341:152–154
Lang RJ (1989) Identification of the major membrane currents in freshly dispersed single smooth muscle cells of guinea-pig ureter. J Physiol (Lond) 412:375–395
Latorre R, Oberhauser A, Labarca P, Alvarez O (1989) Varieties of calcium-activated potassium channels. Annu Rev Physiol 51:385–399
Laurent F, Michel A, Bonnet PA, Chapat JP, Boucard M (1993) Evaluation of the relaxant effects of SCA40, a novel charybdotoxin-sensitive potassium channel opener, in guinea-pig isolated trachealis. Br J Pharmacol 108:622–626
Lebrun P, Fang ZY, Antoine MH, Herchuelz A, Hermann M, Berkenboom G, Fontaine J (1990) Hypoglycemic sulfonylureas antagonize the effects of cromakalim and pinacidil on ^{86}Rb fluxes and contractile activity in the rat aorta. Pharmacology 41:36–48
Lesage F, Attali B, Lazdunski M, Barhanin J (1992) ISK, a slowly activating voltage-sensitive K^+ channel-characterization of multiple cDNAs and gene organization in the mouse. FEBS Lett 301:168–172
Longmore J, Bray KM, Weston AH (1991) The contribution of Rb-permeable potassium channels to the relaxant and membrane hyperpolarizing actions of cromakalim, RP49356 and diazoxide in bovine tracheal smooth muscle. Br J Pharmacol 102:979–985
Longmore J, Newgreen DT, Weston AH (1990) Effects of cromakalim, RP49356, diazoxide and galanin in rat portal vein. Eur J Pharmacol 190:75–84
Lydrup ML (1991) Role of K^+ channels in spontaneous electrical and mechanical activity of smooth muscle in the guinea-pig mesotubarium. J Physiol (Lond) 433:327–340
Maas AJJ, Den Hertog A, Van der Akker J (1980) The action of apamin on guinea-pig taenia caeci. Eur J Pharmacol 67:265–274
MacKinnon R, Miller C (1989) Mutant potassium channels with altered binding of charybdotoxin, a pore-blocking peptide inhibitor. Science 245:1382–1385
MacKinnon R, Reinhart PH, White MM (1988) Charybdotoxin block of Shaker K^+ channels expressed in *Xenopus* oocytes suggests that functionally different K^+ channels share common structural features. Neuron 1:997–1001
MacKinnon R, Yellen G (1990) Mutations affecting TEA blockade and ion permeation in voltage-activated K^+-channels. Science 250:276–279
Malmgren A, Andersson K-E, Andersson P-O, Fovaeus M, Sjögren C (1990) Effects of cromakalim (BRL34915) and pinacidil on normal and hypertrophied rat detrusor in vitro. J Urol 143:828–834
Maruyama M, Satoh K, Taira N (1982) Effects of nicorandil and its congeners on musculature and vasculature of the dog trachea in situ. Arch Int Pharmacodyn Ther 258:260–266
Mayer EA, Loo DDF, Snape WJ, Sachs G (1990) The activation of calcium and calcium-activated potassium channels in mammalian colonic smooth muscle by substance-P. J Physiol (Lond) 420:47–71
Mayer EA, Sun XP, Willenbucher RF (1992) Contraction coupling in colonic smooth muscle. Annu Rev Physiol 54:395–414
McCall JM, Aiken JW, Chidchester CG, Ducharme DW, Wendling MG (1983) Pyrimidine and triazine 3-oxide sulphates: a new family of vasodilators. J Med Chem 26:1791–1793

McCann JD, Welsh MJ (1986) Calcium-activated potassium channels in canine airway smooth muscle. J Physiol (Lond) 372:113–127
McCann JD, Welsh MJ (1987) Neuroleptics antagonize a calcium-activated potassium channel in airway smooth muscle. J Gen Physiol 89:339–352
McFadzean I, England S (1992) Properties of the inactivating outward current in single smooth muscle cells isolated from the rat anococcygeus. Pflügers Arch 421:117–124
McHarg AD, Southerton JS, Weston AH (1990) A comparison of the actions of cromakalim and nifedipine on rabbit isolated mesenteric artery. Eur J Pharmacol 185:137–146
McManus OB (1991) Calcium-activated potassium channels-regulation by calcium. J Bioenerg Biomembr 23:537–560
McManus OB, Giangiacomo KL, Harris GH, Addy ME, Reuben JP, Kaczorowski GJ, Garcia ML (1993a) A maxi-K channel agonist isolated from a medicinal herb. Biophys J 64:3a
McManus OB, Harris GH, Giangiacomo KM, Feigenbaum P, Reuben JP, Addy ME, Burka JF, Kaczorowski GJ, Garcia ML (1993b) An activator of calcium-dependent potassium channels isolated from a medicinal herb. J Biol Chem 32:6128–6133
McPherson GA, Angus JA (1989) Phentolamine and structurally related compounds selectively antagonise the vascular actions of the K^+ channel opener cromakalim. Br J Pharmacol 97:941–949
McPherson GA, Angus JA (1991) Evidence that acetylcholine-mediated hyperpolarization of the rat small mesenteric artery does not involve the K^+ channel opened by cromakalim. Br J Pharmacol 103:1184–1190
Meisheri KD, Cipkus LA, Taylor CJ (1988) Mechanism of action of minoxidil sulfate-induced vasodilation: a role for increased K^+ permeability. J Pharmacol Exp Ther 245:751–760
Merkel LA, Lappe RW, Rivera LM, Cox BF, Perrone MH (1992) Demonstration of vasorelaxant activity with an A_1-selective adenosine agonist in porcine coronary artery – involvement of potassium channels. J Pharmacol Exp Ther 260:437–443
Miller C, Moczydlowski E, Latorre R, Phillios M (1985) Charybdotoxin, a protein inhibitor of single Ca^{++}-activated K^+ channels from mammalian skeletal muscle. Nature 313:316–318
Minami K, Fukuzawa K, Nakaya Y (1993) Protein kinase-C inhibits the Ca^{2+}-activated K^+ channel of cultured porcine coronary artery smooth muscle cells. Biochem Biophys Res Commun 190:263–269
Miura M, Belvisi MG, Stretton CD, Yacoub M, Barnes PJ (1992) Role of potassium channels in bronchodilator responses in human airways. Am Rev Respir Dis 146:132–135
Miura M, Belvisi MG, Ward KJ, Tadjkarimi S, Yacoub M, Barnes PJ (1993) Bronchodilating effects of the novel potassium channel opener HOE 234 in human airways in vitro. Br J Clin Pharmacol 35:318–320
Miwa A, Kaneta S, Motoki K, Jinno Y, Kasai H, Okada Y, Fukushima H, Ogawa N (1993) Vasorelaxant mechanism of KRN2391 and nicorandil in porcine coronary arteries of different sizes. Br J Pharmacol 109:632–636
Miyoshi H, Urabe T, Fujiwara A (1991) Electrophysiological properties of membrane currents in single myometrial cells isolated from pregnant rats. Pflügers Arch 419:386–393
Mondot S, Cailland CG, James C, Aloup J-C, Cavero I (1989) Antihypertensive and haemodynamic profile of 49356RP, a novel vasorelaxant agent with potassium channel activating properties. Fundam Clin Pharmacol 3:158
Moore PK, al-Swayeh OA, Chong NWS, Evans RA, Gibson A (1990) L-N^G-nitro arginine-reversible inhibitor of endothelium-dependent vasodilatation in vitro. Br J Pharmacol 99:408–412
Müller M, de Weille JR, Lazdunski M (1991) Chlorpromazine and related phenothiazines inhibit the ATP-sensitive K^+ channel. Eur J Pharmacol 198:101–104

Murai T, Kakizuka A, Takumi T, Ohkubo H, Nakanishi S (1989) Molecular cloning and sequence analysis of human genomic DNA encoding a novel membrane protein which exhibits a slowly activating potassium channel activity. Biochem Biophys Res Commun 161:176–181

Muraki K, Imaizumi Y, Kojima T, Kawai T, Watanabe M (1990) Effects of tetraethylammonium and 4-aminopyridine on outward currents and excitability in canine tracheal smooth muscle cells. Br J Pharmacol 100:507–515

Murray MA, Berry JL, Cook SJ, Foster RW, Green KA, Small RC (1991) Guinea-pig isolated trachealis: the effects of charybdotoxin on mechanical activity, membrane potential changes and the activity of plasmalemmal K^+-channels. Br J Pharmacol 103:1814–1818

Murray MA, Boyle JP, Small RC (1989) Cromakalim-induced relaxation of guinea-pig isolated trachealis: antagonism by glibenclamide and by phentolamine. Br J Pharmacol 98:865–874

Myers PR, Minor RL, Guerra R, Bates JN, Harrison DG (1990) Vasorelaxant properties of the endothelium-derived relaxing factor more closely resemble S-nitrosocysteine than nitric oxide. Nature 345:161–163

Nagao T, Illiano S, Vanhoutte PM (1992) Calmodulin antagonists inhibit endothelium-dependent hyperpolarization in the canine coronary artery. Br J Pharmacol 107:382–386

Nagao T, Vanhoutte PM (1992) Hyperpolarization as a mechanism for endothelium-dependent relaxations in the porcine coronary artery. J Physiol (Lond) 445: 355–367

Nakao K, Inoue R, Yamanaka K, Kitamura K (1986) Actions of quinidine and apamin on after-hyperpolarization of the spike in circular smooth muscle cells of the guinea-pig ileum. Naunyn Schmiedebergs Arch Pharmacol 334:508–513

Nakao K, Okabe K, Kitamura K, Kuriyama H, Weston AH (1988) Characteristics of cromakalim-induced relaxations in the smooth muscle cells of guinea-pig mesenteric artery and vein. Br J Pharmacol 95:795–804

Nakashima M, Li YJ, Seki N, Kuriyama H (1990) Pinacidil inhibits neuromuscular transmission indirectly in the guinea-pig and rabbit mesenteric arteries. Br J Pharmacol 101:581–586

Nelson MT, Patlak JB, Worley JF, Standen NB (1990) Calcium channels, potassium channels and voltage dependence of arterial smooth muscle tone. Am J Physiol 259:C3–C18

Newgreen DT, Bray KM, McHarg AD, Weston AH, Duty S, Brown BS, Kay PB, Edwards G, Longmore J, Southerton JS (1990) The action of diazoxide and minoxidil sulphate on rat blood vessels: a comparison with cromakalim. Br J Pharmacol 100:605–613

Neyton J, Miller C (1988a) Potassium blocks barium permeation through a calcium-activated potassium channel. J Gen Physiol 92:549–567

Neyton J, Miller C (1988b) Discrete Ba^{2+} block as a probe of ion occupancy and pore structure in the high conductance Ca^{2+}-activated K+ channel. J Gen Physiol 92:569–586

Nielsen-Kudsk JE, Bang L (1991) Effects of pinacidil and other cyanoguanidine derivatives on guinea-pig isolated trachea, aorta and pulmonary artery. Eur J Pharmacol 201:97–102

Noack T (1992) Potassium channels in excitable cells: a synopsis. In: Weston AH, Hamilton TC (eds) Potassium channel modulators: pharmacological, molecular and clinical aspects. Blackwell, Oxford, pp 1–13

Noack T, Deitmer P, Golenhofen K (1990) Features of a calcium independent, caffeine sensitive outward current in single smooth muscle cells from guinea pig portal vein. Pflügers Arch 416:467–469

Noack T, Deitmer P, Edwards G, Weston AH (1992a) Characterization of potassium currents modulated by BRL 38227 in rat portal vein. Br J Pharmacol 106: 717–726

Noack T, Edwards G, Deitmer P, Greengrass P, Morita T, Andersson P-O, Criddle D, Wyllie MG, Weston AH (1992b) The involvement of potassium channels in the action of ciclazindol in rat portal vein. Br J Pharmacol 106: 17–24

Noack T, Edwards G, Deitmer P, Weston AH (1992c) Potassium channel modulation in rat portal vein by ATP depletion – a comparison with the effects of levcromakalim (BRL-38227). Br J Pharmacol 107:945–955

Norman J (1991) Calmodulin antagonists. In: Doods HN, van Meel JCA (eds) Receptor data for biological experiments: a guide to drug selectivity. Horwood, Chichester, pp 169–174

Ohya Y, Kitamura K, Kuriyama H (1987) Modulation of ionic currents in smooth muscle balls of the rabbit intestine by intracellularly perfused ATP and cyclic AMP. Pflügers Arch 408:465–473

Okabe K, Kajioka S, Nakao K, Kitamura K, Kuriyama H, Weston AH (1990) Actions of cromakalim on ionic currents recorded from single smooth muscle cells of the rat portal vein. J Pharmacol Exp Ther 252:832–839

Okabe K, Kitamura K, Kuriyama H (1987) Features of 4-aminopyridine sensitive outward current observed in single smooth muscle cells from the rabbit pulmonary artery. Pflugers Arch 409:561–568

Okada Y, Yanagisawa T, Taira N (1992) E4080 has a dual action, as a K^+ channel opener and a Ca^{2+} channel blocker, in canine coronary artery smooth muscle. Eur J Pharmacol 218:259–264

Okumura K, Ichihara K, Nagasaka M (1992) Effects of imidazoline-related compounds on the mechanical response to nicorandil in the rat portal vein. Eur J Pharmacol 215:253–257

Olesen SP, Wärjen F (1992) Benzimidazole derivatives, their preparation and use. European Patent Application EP-477819-A

Olsson RA, Pearson JD (1990) Cardiovascular purinoceptors. Physiol Rev 70: 761–845

Ono K, Fozzard HA (1993) Two phosphatase sites on the Ca^{2+} channel affecting different kinetic functions. J Physiol 470:73–84

Ordway RW, Singer JJ, Walsh JV Jr (1991) Direct regulation of ion channels by fatty acids. Trends Pharmacol Sci 14:96–100

Orito K, Satoh K, Taira N (1993) Involvement of ATP-sensitive K^+ channels in the sustained coronary vasodilator response to adenosine in dogs. Eur J Pharmacol 231:183–189

Paciorek PM, Burden DT, Burke YM, Cowlrick IS, Perkins RS, Taylor JC, Waterfall JF (1990) Preclinical pharmacology of Ro 31-6930 a new potassium channel opener. J Cardiovasc Pharmacol 15:188–197

Pennefather P, Lancaster B, Adams PR, Nicoll RA (1985) Two distinct Ca-dependent K-currents in bullfrog sympathetic ganglion cells. Proc Natl Acad Sci USA 82:3040–3044

Penner R, Petersen M, Pierau FK, Dreyer F (1986) Dendrotoxin: a selective blocker of a non-inactivating potassium current in guinea-pig dorsal root ganglion neurones. Pflugers Arch 407:365–369

Perozo E, Bezanilla F (1990) Phosphorylation affects voltage-gating of the delayed rectifier K^+ channel by electrostatic interactions. Neuron 5:685–690

Pfründer D, Kreye VAW (1991) Tedisamil blocks single large-conductance Ca^{2+}-activated K^+ channels in membrane patches from smooth muscle cells of the guinea-pig portal vein. Pflügers Arch 418:308–312

Pfründer D, Kreye VAW (1992) Tedisamil inhibits the delayed rectifier K^+ current in single smooth muscle cells of the guinea-pig portal vein. Pflügers Arch 421:22–25

Pfründer D, Anghelescu I, Kreye VAW (1993) Intracellular ADP activates ATP-sensitive K^+ channels in vascular smooth muscle cells of the guinea pig portal vein. Pflügers Arch 423:149–151

Piper I, Minshall E, Downing SJ, Hollingsworth M, Sadraei H (1990) Effects of several potassium channel openers and glibenclamide on the uterus of the rat. Br J Pharmacol 101:901–907

Pongs O, Kecskemethy N, Müller R, Krah-Jentgens I, Bauman A, Kiltz HH, Canal I, Llamazares S, Ferrus A (1988) *Shaker* encodes a family of putative potassium channel proteins in the nervous system of *Drosophila*. EMBO J 7: 1087–1096

Pragnell M, Snay KJ, Trimmer JS, MacLusky NJ, Naftolin F, Kaczmarek LK, Boyle MB (1990) Estrogen induction of a small, putative K^+ channel mRNA in rat uterus. Neuron 4:807–812

Price M, Lee SC, Deutsch C (1989) Charybdotoxin inhibits proliferation and interleukin 2 production in human peripheral blood lymphocytes. Proc Natl Acad Sci USA 86:10171–10175

Prieto D, Benedito S, Nielsen PJ, Nyborg NCB (1991) Calcitonin gene-related peptide is a potent vasodilator of bovine retinal arteries in vitro. Exp Eye Res 53:399–405

Quast U (1987) Effect of the K^+ efflux stimulating vasodilator BRL 34915 on $^{86}Rb^+$ efflux and spontaneous activity in guinea-pig portal vein. Br J Pharmacol 91:569–578

Quast U, Baumlin Y (1988) Comparison of the effluxes of $^{42}K^+$ and $^{86}Rb^+$ elicited by cromakalim (BRL34915) in tonic and phasic vascular tissue. Naunyn Schmiedebergs Arch Pharmacol 338:319–326

Quast U, Baumlin Y (1991) Cromakalim inhibits contractions of the rat isolated mesenteric bed induced by noradrenaline but not caffeine in Ca^{2+}-free medium: evidence for interference with receptor-mediated Ca^{2+} mobilization. Eur J Pharmacol 200:239–249

Quast U, Cook, NS (1988) *Leiurus quinquestriatus* venom inhibits BRL34915-induced $^{86}Rb^+$ efflux from the rat portal vein. Life Sci 42:805–810

Quast U, Cook NS (1989a) In vitro and in vivo comparison of two K^+ channel openers diazoxide and cromakalim and their inhibition by glibenclamide. J Pharmacol Exp Ther 250:261–271

Quast U, Cook NS (1989b) Moving together: K^+ channel openers and ATP-sensitive K^+ channels. Trends Pharmacol Sci 10:431–435

Quast U, Villhauer EB (1993) The individual enantiomers of cis-cromakalim possess K^+ channel opening activity. Eur J Pharmacol 245:165–171

Quayle JM, Standen NB, Stanfield PR (1988) The voltage-dependent block of ATP-sensitive potassium channels of frog skeletal muscle by caesium and barium ions. J Physiol (Lond) 405:677–697

Raeburn DM, Brown TJ (1991) RP49356 and cromakalim relax airway smooth muscle in vitro by opening a sulphonylurea-sensitive K^+ channel: a comparison with nifedipine. J Pharmacol Exp Ther 256:492–499

Rand VE, Garland GJ (1992) Endothelium-dependent relaxation to acetylcholine in the rabbit basilar artery: importance of membrane hyperpolarization. Br J Pharmacol 106:143–150

Reinhart PH, Chung S, Levitan IB (1989) A family of calcium-dependent potassium channels from rat brain. Neuron 2:1031–1041

Ribalet B, Ciani S, Eddlestone GT (1989) ATP mediates both activation and inhibition of K(ATP) channel activity via cAMP-dependent protein kinase in insulin-secreting cell lines. J Gen Physiol 94:693–717

Robertson DW, Steinberg MI (1990) Potassium channel modulators: scientific applications and therapeutic promise. J Med Chem 33:1529–1541

Romey G, Lazdunski M (1984) The coexistence in rat muscle cells of two distinct classes of Ca^{2+}-dependent K^+ channels with different pharmacological properties and different physiological functions. Biochem Biophys Res Commun 118:669–674

Rudy B (1988) Diversity and ubiquity of K channels. Neuroscience 25:729–749

Rudy B, Kentros C, Vega-Saenz de Miera E (1991) Families of potassium channel genes in mammals: toward an understanding of the molecular basis of potassium channel diversity. Mol Cell Neurosci 2:89–102
Ruppersberg JP, Frank R, Pongs O, Stockre M (1991) Cloned neuronal IK(A) channels reopen during recovery from inactivation. Nature 353:657–660
Russell K, Ohnmacht CJ, Gibson KH (1993) Therapeutic amides. European Patent Application 0524781A1
Russell SN, Smirnov SV, Aaronson PI (1992) Effects of BRL-38227 on potassium currents in smooth muscle cells Isolated from rabbit portal vein and human mesenteric artery. Br J Pharmacol 105:549–556
Sadoshima J, Akaike N, Kanaide H, Nakamura M (1988a) Cyclic AMP modulates Ca-activated K channel in cultured smooth muscle cells of rat aortas. Am J Physiol 255:H754–H759
Sadoshima J, Akaike N, Tomoike H, Kanalde H, Nakamura M (1988b) Ca-activated K channel in cultured smooth muscle cells of rat aortic media. Am J Physiol 255:H410–H418
Saito A, Masaki T, Uchiyama Y, Lee TJ-F, Goto K (1989) Calcitonin gene-related peptide and vasodilator nerves in large cerebral arteries of cats. J Pharmacol Exp Ther 248:455–462
Sakuta H, Sekiguchi M, Okamoto K, Sakai Y (1992) Inactivation of glibenclamide-sensitive K^+ channels in Xenopus oocytes by various calmodulin antagonists. Eur J Pharmacol 226:199–207
Salkoff L (1983) Genetic and voltage-clamp analysis of a Drosophila potassium channel. Cold Spring Harbor Symposium on Molecular Biology 48:221–231
Sands SB, Lewis RS, Cahalan MD (1989) Charybdotoxin blocks voltage-gated K^+ channels in human and murine T lymphocytes. J Gen Physiol 93:1061–1074
Sanguinetti MC, Jurkiewicz NK (1990) Two components of cardiac delayed rectifier K^+ current: differential sensitivity to block by class III antiarrhythmic agents. J Gen Physiol 96:195–215
Schwanstecher M, Löset S, Rietze I, Panten U (1991) Phosphate and thiophosphate group donating adenine and guanine nucleotides inhibit glibenclamide binding to membranes from pancreatic islets. Naunyn-Schmiedebergs Arch Pharmacol 343:83–89
Schweitz H, Stansfeld CE, Bidard J-N, Fagni L, Maes P, Lazdunski M (1989) Charybdotoxin blocks dendrotoxin-sensitive voltage-activated K^+ channels. FEBS Lett 250:519–522
Schwietert R, Wilhelm D, Wilffert B, van Zwieten PA (1992) The effect of some α-adrenoceptor antagonists on spontaneous myogenic activity in the rat portal vein and the putative involvement of ATP-sensitive K^+ channels. Eur J Pharmacol 211:87–95
Segal M, Rogawski MA, Barker JL (1984) A transient potassium conductance regulates the excitability of cultured hippocampal and spinal neurones. J Neurosci 4:604–609
Shen WK, Tung RT, Machulda MM, Kurachi Y (1991) Essential role of nucleotide diphosphates in nicorandil-mediated activation of cardiac ATP-sensitive K^+-channel – a comparison with pinacidil and lemakalim. Circ Res 69:1152–1158
Sheppard DN, Welsh MJ (1992) Effect of ATP-sensitive K^+ channel regulators on cystic fibrosis transmembrane conductance regulator chloride currents. J Gen Physiol 100:573–591
Siegel G, Mironneau J, Schnalke F, Schröder G, Schulz B-G, Grote J (1990) Vasodilation evoked by K^+ channel opening. Prog Clin Biol Res 327:299–306
Silberberg SD, van Breemen C (1990) An ATP, calcium and voltage sensitive potassium channel in porcine coronary artery smooth muscle cells. Biochem Biophys Res Commun 172:517–522
Silberberg SD, van Breemen C (1992) A potassium current activated by lemakalim and metabolic inhibition in rabbit mesenteric artery. Pflügers Arch 420:118–120

Singer JJ, Walsh JV (1987) Characterization of calcium-activated potassium channels in single smooth muscle cells using the patch-clamp technique. Pflügers Arch 408:98–111
Small RC, Berry JL, Cook SJ, Foster RW, Green KA, Murray MA (1993) Potassium channels in airways. In: Chung KF, Barnes PJ (eds) Pharmacology of the Respiratory Tract: Experimental and Clinical Research. Marcel Dekker, Inc., New York, 137–176
Small RC, Boyle JP, Duty S, Elliott KRF, Foster RW, Watt AJ (1989) Analysis of the relaxant effects of AH 21-132 in guinea-pig isolated trachealis. Br J Pharmacol 97:1165–1173
Smith JM, Sanchez AA, Jones AW (1986) Comparison of rubidium-86 and potassium-42 fluxes in rat aorta. Blood Vessels 23:297–309
Sorensen RG, Schneider MJ, Togowski RS, Blaustein MP (1990) Snake and scorpion neurotoxins as probes of rat brain synaptosomal potassium channels. In: Colatsky TJ (ed) Potassium channels: basic function and therapeutic aspects. Liss, New York, pp 279–301
Southerton JS, Taylor SG, Weston AH (1987) Comparison of the effects of BRL 34915 and of acetylcholine-liberated EDRF on rat isolated aorta. J Physiol (Lond) 391:77P
Southerton JS, Weston AH, Bray KM, Newgreen DT, Taylor SG (1988) The potassium channel opening action of pinacidil; studies using biochemical, ion flux and microelectrode techniques. Naunyn Schmiedebergs Arch Pharmacol 338:310–318
Standen NB (1992) Potassium channels, metabolism and muscle. Exp Physiol 77:1–25
Standen NB, Quayle JM, Davies NW, Brayden JE, Huang Y, Nelson MT (1989) Hyperpolarizing vasodilators activate ATP-sensitive K^+ channels in arterial smooth muscle. Science 245:177–180
Stansfeld CE, Feltz A (1988) Dendrotoxin-sensitive K^+ channels in dorsal root ganglion cells. Neurosci Lett 93:49–55
Stemp G, Evans JM (1993) Discovery and development of cromakalim and related potassium channel activators. In: Ganellin CR, Roberts SM (eds) Medicinal chemistry, 2nd edn. Academic, New York, pp 141–162
Stephens NL, Kroeger EA, Kromer U (1975) Induction of myogenic response in tonic airway smooth muscle by tetraethylammonium. Am J Physiol 228:628–632
Stocker M, Pongs O, Hoth M, Heinemann S, Stühmer W, Schröter KH, Ruppersberg JP (1991) Swapping of functional domains in voltage-gated K^+ channels. Proc R Soc Lond [Biol] 245:101–107
Strong PN, Weir SW, Beech DJ, Hiestand P, Kocher HP (1989) Effects of potassium channel toxins from *Leiurus quinquestriatus hebraeus* venom on responses to cromakalim in rabbit blood vessels. Br J Pharmacol 98:817–826
Stuenkel EL (1989) Single potassium channels recorded from vascular smooth muscle cells. Am J Physiol 257:H760–H769
Stühmer W, Conti F, Suzuki H, Wang X, Noda N, Yahagi N, Kubo H, Numa S (1989) Structural parts involved in activation and inactivation of the sodium channel. Nature 339:597–603
Sturgess NC, Ashford MLJ, Cook DL, Hales CN (1985) The sulphonylurea receptor may be an ATP sensitive K^+ channel. Lancet ii:474–475
Sturgess NC, Kozlowski RZ, Carrington CA, Hales CN, Ashford MLJ (1988) Effects of sulphonylureas and diazoxide on insulin secretion and nucleotide-sensitive channels in an insulin-secreting cell line. Br J Pharmacol 95:83–94
Suarez-Kurtz G, Garcia ML, Kaczorowski GJ (1991) Effects of charybdotoxin and iberiotoxin on the spontaneous motility and tonus of different guinea pig smooth muscle tissues. J Pharmacol Exp Ther 259:439–443
Sugimoto T, Tanabe Y, Shigemoto R, Iwai M, Takumi T, Ohkubo H, Nakanishi S (1990) Immunohistochemical study of a rat membrane protein which induces a

selective potassium permeation: its localization in the apical membrane portion of epithelial cells. J Membr Biol 113:39–47
Takumi T, Moriyoshi K, Aramori I, Ishii T, Oiki S, Okada Y, Ohkubo H, Nakanishi S (1991) Alteration of channel activities and gating by mutations of slow-ISK potassium channel. J Biol Chem 266:22192–22198
Takumi T, Ohkubo H, Nakanishi S (1988) Cloning of a membrane protein that induces a slow voltage-gated potassium current. Science 242:1042–1045
Talvenheimo JA, Lam G, Gelband C (1988) Charybdotoxin inhibits the 250 pS Ca2+-activated K^+ channel in aorta and contracts aorta smooth muscle. Biophys J 53:258a
Tare M, Parkington HC, Coleman HA, Nield TO, Dusting GJ (1990) Hyperpolarization and relaxation of arterial smooth muscle caused by nitric oxide derived from the endothelium. Nature 346:69–71
Taylor SG, Southerton JS, Weston AH, Baker JRJ (1988) Endothelium-dependent effects of acetylcholine in rat aorta: a comparison with sodium nitroprusside and cromakalim. Br J Pharmacol 94:853–863
Tohse N, Kameyama M, Irisawa H (1987) Intracellular Ca^{2+} and protein kinase C modulate K^+ current in single cells from frog atrium. J Gen Physiol 253: H1321–H1324
Toro L, Ramos-Franco J, Stefani E (1990) GTP-dependent regulation of myometrial K_{Ca} channel incorporated into lipid bilayers. J Gen Physiol 96:373–394
Tung RT, Kurachi Y (1991) On the mechanism of nucleotide diphosphate activation of the ATP-sensitive K^+ channel in ventricular cell of guinea-pig. J Physiol (Lond) 437:239–256
Turner NC, Dollery CT, Williams AJ (1989) Endothelin-I-induced contractions of vascular and tracheal smooth muscle: effects of nicardipine and BRL 34915. J Cardiovasc Pharmacol 135:S180–S182
Ueeda M, Thompson RD, Arroyo LH, Olsson RA (1991) 2-Alkoxy-adenosines: potent and selective agonists at the coronary artery A_2 adenosine receptor. J Med Chem 34:1334–1339
Valdivia HH, Smith JS, Martin BM, Coronado R, Possani LD (1988) Charybdotoxin and noxiustoxin, two homologous peptide inhibitors of the $K^+(Ca^{2+})$ channel. FEBS Lett 226:280–284
Van Belle H (1981) R 24 571: a potent inhibitor of calmodulin-activated enzymes. Cell Calcium 2:483–494
Van de Voorde J, Vanheel B, Leusen I (1992) Endothelium-dependent relaxation and hyperpolarization in aorta from control and renal hypertensive rats. Circ Res 70:1–8
Vázquez J, Feigenbaum P, Katz G, King VF, Reuben JP, Roy-Contancin L, Slaughter RS, Kaczorowski GJ, Garcia ML (1989) Characterization of high affinity binding sites for charybdotoxin in sarcolemmal membranes from bovine aortic smooth muscle. Evidence for a direct association with the high conductance calcium-activated potassium channel. J Biol Chem 264:20902–20909
Veigl ML, Klevit RE, Sedwick WD (1989) The uses and limitations of calmodulin antagonists. Pharmac Ther 44:181–239
Videbaek LM, Aalkjaer C, Mulvany MJ (1988) Pinacidil opens K^+-selective channels causing hyperpolarisation and relaxation of noradrenaline contractions in rat mesenteric resistance vessels. Br J Pharmacol 95:103–108
Volk KA, Shibata EF (1993) Single delayed rectifier potassium channels from rabbit coronary artery myocytes. Am J Physiol 264:H1146–H1153
Volk KA, Matsuda JJ, Shibata EF (1991) A voltage-dependent potassium current in rabbit coronary artery smooth muscle cells. J Physiol (Lond) 439:751–768
Von Beckerath N, Cyrys S, Dischner A, Daut J (1991) Hypoxic vasodilatation in isolated, perfused guinea-pig heart: an analysis of the underlying mechanism. J Physiol (Lond) 442:297–319
Walsh KB, Kass RS (1988) Regulation of a heart potassium channel by protein kinase A and C. Pflügers Arch 411:232–234

Walsh KB, Arena JP, Kwok W-M, Freeman L, Kass RS (1991) Delayed-rectifier potassium channel activity in isolated membrane patches of guinea pig ventricular myocytes. Am J Physiol 260:H1390–H1393

Wang W, Giebisch G (1991) Dual effect of adenosine triphosphate on the apical small conductance K^+ channel of the rat cortical collecting duct. J Gen Physiol 98:35–61

Wang YH, Mathers DA (1991) High sensitivity to internal tetraethylammonium in K(Ca) channels of cerebrovascular smooth muscle cells. Neurosci Lett 132: 222–224

Weir SW, Weston AH (1986a) Effect of apamin on responses to BRL 34915 nicorandil and other relaxants in guinea-pig taenia caeci. Br J Pharmacol 88:113–120

Weir SW, Weston AH (1986b) The effects of BRL34915 and nicorandil on electrical and mechanical activity and on $^{86}Rb^+$ efflux in rat blood vessels. Br J Pharmacol 88:121–128

Weller U, Bernhardt U, Siemen D, Dreyer F, Vogel W, Habermann E (1985) Electrophysiological and neurobiochemical evidence for the blockade of a potassium channel by dendrotoxin. Naunyn Schmiedebergs Arch Pharmacol 330:77–83

Wickenden AD, Grimwood S, Grant TL, Todd MH (1991) Comparison of the effects of the K^+-channel openers cromakalim and minoxidil sulphate on vascular smooth muscle. Br J Pharmacol 103:1148–1152

Williams DL, Katz GM, Roy-Contancin L, Reuben JP (1988) Guanosine 5′-monophosphate modulates gating of high conductance Ca^{2+}-activated K^+ channels in vascular smooth muscle. Proc Natl Acad Sci USA 85:9360–9364

Wilson C (1989) Inhibition by sulphonylureas of vasorelaxation induced by K^+ channel activators in vitro. J Auton Pharmacol 9:71–78

Wilson C, Buckingham RE, Mootoo S, Parrott LS, Hamilton TC, Pratt SC, Cawthorne MA (1988a) In vivo and in vitro studies of cromakalim (BRL 34915) and glibenclamide in the rat. Br J Pharmacol 93:126P

Wilson C, Coldwell MC, Howlett DR, Cooper SM, Hamilton TC (1988b) Comparative effects of K^+ channel blockade on the vasorelaxant activity of cromakalim pinacidil and nicorandil. Eur J Pharmacol 152:331–339

Winquist RJ, Heaney LA, Wallace AA, Baskin EP, Stein RB, Garcia ML, Kaczorowski GJ (1989) Glyburide blocks the relaxation response to BRL 34915 (cromakalim) minoxidil sulphate and diazoxide in vascular smooth muscle. J Pharmacol Exp Ther 248:149–156

Wollheim CB, Dunne MJ, Peter-Riesch B, Bruzzone R, Pozzan T, Petersen OH (1988) Activators of protein kinase C depolarize insulin-secreting cells by closing K^+ channels. EMBO J 7:2443–2449

Yanagisawa T, Teshigawara T, Taira N (1990) Cytoplasmic calcium and the relaxation of canine coronary arterial smooth muscle produced by cromakalim, pinacidil and nicorandil. Br J Pharmacol 101:157–165

Yellen G, Jurman ME, Abramson T, MacKinnon R (1991) Mutations affecting internal TEA blockade identify the probable pore-forming region of a K^+ channel. Science 251:939–942

Yool AJ, Schwarz TL (1991) Alteration of ionic selectivity of a K^+ channel by mutation of the H5 region. Nature 349:700–704

Zagotta WN, Hoshi T, Aldrich RW (1990) Restoration of inactivation in mutants of *Shaker* potassium channels by a peptide derived from *ShB*. Science 250:568–571

Zografos P, Li JH, Kau ST (1992) Comparison of the in vitro effects of K^+ channel modulators on detrusor and portal vein strips from guinea pigs. Pharmacology 45:216–230

Zschauer A, Uusitalo H, Brayden JE (1992) Role of endothelium and hyperpolarization in CGRP-induced vasodilation of rabbit ophthalmic artery. Am J Physiol 263:H359–H3650

Zünkler BJ, Lenzen S, Manner K, Panten U, Trube G (1988) Concentration-dependent effects of tolbutamide, meglitinide, glipizide, glibenclamide and diazoxide on ATP-regulated K^+ currents in pancreatic β-cells. Naunyn Schmiedebergs Arch Pharmacol 337:225–230

CHAPTER 14

Smooth Muscle of the Male Reproductive Tract

E. KLINGE and N.O. SJÖSTRAND

A. Introduction

The research into the smooth muscle of various organ systems which has been carried out for more than a century has largely been aimed at factors regulating the muscular tone. The nervous mechanisms have usually been the object of greater interest than the relevant properties of the smooth muscle cells themselves. This also holds true for research into the smooth muscles of the male reproductive tract, some of which have been used as a model of neurotransmission, e.g., the vas deferens for more than 30 years. Some of the trends characterizing the research in this field from the late 1970s onwards are the work done in order to isolate and identify the non-adrenergic non-cholinergic (NANC) inhibitory neurotransmitter from the bovine retractor penis muscle, and the increasing use of isolated human penile tissues for pharmacological experiments. Also the growing number of morphological studies, mainly performed with immunohistochemical techniques, have played an important role in the localization of several peptides and other compounds.

An earlier review focused on the smooth muscle of the internal reproductive organs (SJÖSTRAND 1981). This review will primarily deal with the innervation of smooth muscles in the external reproductive organs, i.e., the penile artery, the corpus cavernosum penis (in the following called the corpus cavernosum), the corpus spongiosum and the retractor penis. The anococcygeus muscle will not be dealt with because anatomically it is not included in the reproductive organs, and because its innervation has recently been reviewed (GILLESPIE et al. 1990; MARTIN and GILLESPIE 1991). The retractor penis muscle (in the following called the retractor penis) is considered important because its innervation is very similar to or identical with that of the penile artery (KLINGE and SJÖSTRAND 1974). The aspects of innervation dealt with will be confined to the extrinsic nerves, adrenergic and cholinergic mechanisms, and vasoactive intestinal polypeptide as a neurotransmitter candidate, because these essential issues are still unclear and disputed although they have been extensively studied. The L-arginine nitric oxide synthase pathway is included because it may provide an answer to the crucial question of the nature of the NANC inhibitory neurotransmitter in the external reproductive organs. Many endogenous compounds, e.g., prostanoids and peptides, that may influence the tone of the smooth muscles concerned have been omitted. To become acquainted with the role of such compounds, and to complete the knowledge of the subjects treated in this presentation, the reader is referred to the reviews of DE GROAT and STEERS (1988), ANDERSSON and HOLMQUIST (1990), STEERS (1990), GILLESPIE et al. (1990), MARTIN and GILLESPIE (1991), ANDERSSON (1993), DAIL (1993) and DE GROAT and BOOTH (1993).

B. Extrinsic Nerves

A rather detailed historical survey of the research into the innervation of the external reproductive organs has already been reported (KLINGE and SJÖSTRAND 1974). Some additional data were reported later (SJÖSTRAND and KLINGE 1979a). A similar historical survey of the studies of the innervation of the internal male accessory genital organs has also been published (SJÖSTRAND 1965). New techniques, especially tracing with horse radish peroxidase (e.g., DALSGAARD and ELFVIN 1979), have consolidated these studies. Good overviews of the innervation of the external genital organs have recently been written by DE GROAT and STEERS (1988, 1990), DAIL (1993) and DE GROAT and BOOTH (1993). In short, the sympathetic (thoracolumbar) outflow derives from the lower thoracic and the lumbar segments and runs to the sympathetic chains. From the chains the fibres reach the pelvic organs via three paths: (1) Via the inferior mesenteric ganglia and the hypogastric nerves. This is the path for the excitatory innervation of the internal male accessory reproductive organs and the sympathetic erectile fibres, both of which have their synaptic relay peripherally close to the smooth muscle. Also some of the excitatory adrenergic fibres to the penile smooth muscle follow the hypogastric nerve. (2) The main portion of the excitatory adrenergic innervation of the penile smooth muscle follows the pudic nerve. (3) Some of the excitatory adrenergic fibres to the penile smooth muscle follow the pelvic nerve.

The parasympathetic (sacral) outflow derives from the sacral segments and reaches the genital organs via the pelvic nerves, i.e., the nervi erigentes of Eckhard. The erectile fibres of the sacral parasympathetic outflow have their synaptic relay peripherally. Part of the postganglionic nerves seem also to have a preganglionic supply from the hypogastric nerves (SJÖSTRAND and KLINGE 1979a; DAIL et al. 1986). By definition these postganglionic fibres belong to both the sympathetic and the parasympathetic system.

There are certain species and individual variations concerning the level and number of spinal segments from which each division of the autonomic nervous system derives, the number of different strands in the hypogastric and pelvic nerves, the number and development of inferior mesenteric ganglia, the relative importance of pelvic versus hypogastric erectile fibres, the relative importance of the three paths for the excitatory nerves to the penis, and the structure of the pelvic plexus. These are dealt with in the papers already cited and in the papers cited under the following subheadings.

C. External Reproductive Organs

I. General Properties of the Smooth Muscle

Early studies (see KLINGE and SJÖSTRAND 1974) indicated that the retractor penis is spontaneously active. The automaticity and unitary character of the retractor penis, cavernous tissue and penile artery of all species studied has by now been well established in both in vivo and in vitro studies (KLINGE and SJÖSTRAND 1974, 1977a; ADAIKAN and KARIM 1978; SJÖSTRAND and KLINGE 1979a; SAMUELSON et al. 1983; BYRNE and MUIR 1984; SJÖSTRAND et al. 1993). Despite its general characteristics of single unit smooth muscle the retractor penis is provided with a dense plexus of nerve terminals (see below). Both excitatory and inhibitory junction potentials are seen on field stimulation of the bovine retractor penis (SAMUELSON et al. 1983; BYRNE and MUIR 1984, 1985). The former are abolished by guanethidine and reduced by phentolamine, while the latter seem to be related to NANC nerve stimulation, which, however, also seems to involve a component of relaxation without an electrical correlate.

II. Adrenergic Mechanisms

The presence of considerable amounts of noradrenaline has been demonstrated, e.g., in the corpus spongiosum and the corpus cavernosum of rabbit (PENTTILÄ and VARTIAINEN 1964), the corpus cavernosum of monkey (BAUMGARTEN et al. 1969) and man (MELMAN et al. 1980), the retractor penis of dog, horse, swine, bull, ram and goat (KLINGE and SJÖSTRAND 1977b; SJÖSTRAND et al. 1993) and the bovine penile artery (KLINGE et al. 1978). It has also been microscopically confirmed that, e.g., in the corpus spongiosum and the corpus cavernosum of rabbit (KLINGE and PENTTILÄ 1969; MCCONNELL et al. 1979), guinea pig (BHARGAVA et al. 1974), rat (MCCONNELL et al. 1979), cat, monkey (BAUMGARTEN et al. 1969; MCCONNELL et al. 1979) and man (SHIRAI et al. 1972; MCCONNELL et al. 1979), the retractor penis of dog (BELL and MCLEAN 1970), bull (KLINGE et al. 1970) and goat (SJÖSTRAND et al. 1993), and the penile artery of bull (KLINGE and POHTO 1971) and man (MCCONNELL and BENSON 1982) the smooth muscles have a rich supply of nerves which exhibit typical catecholamine fluorescence after treatment with formaldehyde or glyoxylic acid. Furthermore, electron microscopy has, after fixation in potassium permanganate or dichromate, revealed axons with typical small granular vesicles in the bovine retractor penis and penile artery (ERÄNKÖ et al. 1976; KLINGE et al. 1978) and human penile blood vessels (MCCONNELL and

BENSON 1982). The muscles are invariably and dose dependently contracted by exogenous noradrenaline (KLINGE and SJÖSTRAND 1977a; ADAIKAN and KARIM 1978; ANDERSSON et al. 1983), and the release of noradrenaline upon stimulation of the sympathetic chains has been demonstrated in dogs (DIEDERICHS et al. 1990).

The concept that noradrenaline is the principal excitatory neurotransmitter for the smooth muscles of the external male genital organs has not been challenged, and clear differences in this respect between various mammalian species have not been reported. There are, however, species differences when the contractions of the retractor penis muscle elicited by transmural stimulation of the motor nerves are studied. In the pig and sheep the contractions are more resistant to guanethidine than in other species (KLINGE and SJÖSTRAND 1977a; AMBACHE and KILLICK 1978), but they can be effectively abolished by a combined treatment with guanethidine and 6-hydroxydopamine (KLINGE and SJÖSTRAND 1977a).

The characterization of the pre- and postjunctional adrenoceptors has mainly been performed by functional studies using specific agonists and antagonists.

1. α-Adrenoceptors

a) Subclassification

In the corpus cavernosum of dog and rabbit the postjunctional α-receptors have been shown to be predominantly of the α_1-subtype (KIMURA et al. 1989; SAENZ DE TEJADA et al. 1989). The same situation prevails in the corpus spongiosum (HEDLUND et al. 1984; HOLMQUIST et al. 1990b) as well as in the corpus cavernosum of man (HEDLUND and ANDERSSON 1985a; SAENZ DE TEJADA et al. 1989; CHRIST et al. 1990; HOLMQUIST et al. 1990b). On the contrary, in the cavernous artery of man there seems to be a postjunctional predominance of α_2-receptors (HEDLUND and ANDERSSON 1985a) and the situation seems to be similar in the penile artery of dog (KIMOTO and ITO 1987). In the deep dorsal vein of the human penis there is a predominance of α_1-receptors (FONTAINE et al. 1986b), whereas in the penile circumflex vein both α_1- and α_2-receptors occur postjunctionally without a clear predominance of either (KIRKEBY et al. 1989).

The subclassification of the prejunctional α-receptors has been performed by studying the effect of specific ligands on the release of ^3H from tissues preloaded with ^3H-noradrenaline. These studies consistently indicate that in both the corpus spongiosum and the corpus cavernosum of man the prejunctional α-receptors are of the α_2-subtype (HEDLUND et al. 1984; MOLDERINGS et al. 1989; SAENZ DE TEJADA et al. 1989).

b) Physiological Significance

LANGLEY and ANDERSON (1895) showed that stimulation of the lumbar sympathetic chains caused in rabbits, cats and dogs strong contraction of the penile arteries as well as strong contraction of the retractor penis in the two last-mentioned species. Although the transmitter role of noradrenaline and the concept of α-receptors have been established much later, their studies can still be regarded as a basis of the present view of the essential role of postjunctional α-receptors in the detumescence, i.e., in the shrinkage of the penis after erection. This view has not been questioned but rather reinforced by more recent animal experiments (SJÖSTRAND and KLINGE 1979a; DE GROAT and STEERS 1988; DIEDERICHS et al. 1990). The neuroeffector junction constituted by adrenergic terminals and postjunctional α-receptors is characterized by a very high efficiency as demonstrated in animal tissues in vitro and in vivo (KLINGE and SJÖSTRAND 1974, 1977a; SJÖSTRAND and KLINGE 1979a; SJÖSTRAND et al. 1993). In corpus cavernosum strips from older impotent men, the efficiency is greater than in strips from younger impotent men (CHRIST et al. 1992).

For a long time it has been known that in man even the oral use of α-adrenoceptor antagonists can occasionally lead to priapism, and at present this occurs most commonly with trazodone (THOMPSON et al. 1990). On the other hand, there is a good theoretical basis for the treatment of impotence with intracavernosally injected phentolamine. The effectiveness of the therapy, but possibly also the risk of priapism, will increase on the combined use of a drug known to relax smooth muscles by a mechanism other than blockade of α-receptors, e.g., papaverine. Likewise, there are logical grounds for the use of intracavernosally injected α-adrenoceptor agonists for the pharmacological treatment of priapism.

Although there is much evidence supporting the crucial role of postjunctional α-receptors in the process of detumescence, their importance in the maintenance of the flaccid state of the penis is less well understood. It has been reported that intravenous phentolamine or phenoxybenzamine causes erection in anaesthetized cats (DOMER et al. 1978), but it is not known whether this occurs in conscious cats, and it does not occur in anaesthetized dogs (CARATI et al. 1987) or in conscious man. In rabbits, sectioning of the sympathetic chains causes protrusion of the penis, but other contractile mechanisms including autoregulation soon take over and lead to its hyperinvolution (SJÖSTRAND and KLINGE 1979a). Neither is lumbar sympathectomy reported to cause persistent protrusion of the penis in man (WHITELAW and SMITHWICK 1951). Finally, there are no data which would allow an evaluation of the physiological importance of the prejunctional α_2-receptors.

2. β-Adrenoceptors

a) Occurrence and Subclassification

The presence of postjunctional β-receptors, although unclassified, has been demonstrated by means of functional in vitro studies in the retractor penis of nine species, in the corpus spongiosum of six species including the macaque, and in the bovine penile artery (KLINGE 1970; KLINGE and SJÖSTRAND 1977a).

Similar studies indicate the presence of β_2-receptors in the corpus cavernosum of dog (CARATI et al. 1985), and in vivo studies indicate their presence in this tissue of cat (DOMER et al. 1978). Functional in vitro studies suggest that in the corpus cavernosum of man there are β_2- but possibly also β_1-receptors (ADAIKAN and KARIM 1981; HEDLUND and ANDERSSON 1985a), whereas no β-receptors have been detected in the penile artery of man (HEDLUND and ANDERSSON 1985a). In the corpus spongiosum of man the postjunctional β-receptors are mainly of the β_2-subtype (HEDLUND et al. 1984). Radioligand binding studies have shown that in the corpus cavernosum of man β-receptors are outnumbered by α-receptors (LEVIN and WEIN 1980) and have indicated that they are mainly of the β_2-subtype (DHABUWALA et al. 1985). Thus, postjunctional β-receptors occur in most tissues. The presence of prejunctional β-receptors has been shown only in the corpus spongiosum of man and here they are mainly of the β_2-subtype (HEDLUND et al. 1984).

b) Physiological Significance

Mainly owing to the dense adrenergic innervation of the penis and to the effect of intravenously administered β_2-adrenoceptor agonists it has been suggested that, at least in the cat, a substance released from adrenergic nerves and acting on β_2-receptors plays a predominant role in the induction of erection (DOMER et al. 1978; MCCONNELL and BENSON 1982). This theory has been developed further by suggesting that also in other species, e.g., the rat and dog and possibly also man, the final neurotransmitter of erection is noradrenaline released from short adrenergic neurons (KRANE and SIROKY 1981). The involvement of β-receptors in penile erection of man has also been suggested by ADAIKAN et al. (1986).

There are, however, several findings that argue against an essential role of β-adrenoceptors in erection. The penis is not innervated by short adrenergic neurons (DAIL 1987), but how could noradrenaline released from even long adrenergic neurons circumvent simultaneous activation of postjunctional α-receptors? In the corpus spongiosum and retractor penis of several species pretreatment with guanethidine does not prevent but on the contrary enhances the neurogenic relaxation induced by field stimulation (KLINGE and SJÖSTRAND 1977a). It seems therefore most unlikely that noradrenaline or another substance released from adrenergic nerves could effectively relax penile smooth muscles. Concerning other nerves, the transmitter(s) released from the NANC nerves in the bovine retractor penis does (do) not relax the muscle via β-receptors because guanylate cyclase, but not adenylate cyclase, is activated (BOWMAN and DRUMMOND 1984), and the relaxation is blocked to the same extent by both of the optical isomers of propranolol (KLINGE et al. 1992). The response to pelvic nerve stimulation is unaffected by propranolol in dogs (THÜROFF et al. 1986; CARATI et al. 1987), rabbits (SJÖSTRAND and KLINGE 1979a) and goats (SJÖSTRAND et al. 1993). Neither does propranolol affect the erectile response to hypogastric nerve stimulation in rabbits (SJÖSTRAND and KLINGE 1979a) or goats (SJÖSTRAND et al. 1993). There is thus no evidence to support the idea that erection could be produced by a substance which activates β-receptors after being released from any local nerves.

It is further unlikely that penile β-receptors could be selectively activated by a substance of humoral origin. In rabbits injection of adrenaline or noradrenaline into the penile artery did not affect the volume of the flaccid penis even after α-adrenoceptor blockade, and these amines invariably caused shrinkage of the protruded penis which also holds true for stimulation of the splanchnic nerves (SJÖSTRAND and KLINGE 1979a). Erection is not produced in man by intracavernous injection of salbutamol (BRINDLEY 1986) or in dogs by intracavernous injection of cyclic AMP (TRIGO-ROCHA et al. 1993a). If β-receptors were of crucial importance in the mediation of the erectile response in man, the incidence of impotence caused by continuous use of β-adrenoceptor blocking drugs should be significantly higher than it is. Taken together, it seems that, although species variations may exist, β-receptors play at most a minor contributory role in the induction of erection.

III. Cholinergic Mechanisms

HENDERSON and ROEPKE (1933) and BACQ (1935) showed that when penile erection was produced in dogs by stimulation of the pelvic nerves the effect of stimulation was remarkably intensified by physostigmine. The latter author also showed that the same enhancement occurred when erection was produced by stimulation of the hypogastric nerves. These results strongly suggest the involvement of an acetylcholine-mediated mechanism. In the experiments of HENDERSON and ROEPKE (1933) the effect of stimulation of the pelvic nerves was but slightly or not at all decreased by atropine, whereas in the experiments of BACQ (1935) atropine clearly curtailed the effect of stimulation of both sets of nerves. In view of the fact that all stimulations were preganglionic and the observation that in dogs the erectile response to pelvic nerve stimulation is abolished by hexamethonium (DORR and BRODY 1967), it is likely that part of the effect of physostigmine was due to facilitation of ganglionic transmission. There are, however, certain reasons, predominantly based on in vitro experiments, which suggest that in addition postganglionic cholinergic effects are involved in the induction and possibly also in the maintenance of erection.

The occurrence of considerable amounts of acetylcholine has been demonstrated in the corpus cavernosum of rabbit and the corpus spongiosum of rabbit and bull (PENTTLLÄ and VARTIAINEN 1964), the retractor penis of dog, horse, swine, bull and ram (KLINGE and SJÖSTRAND 1977b) and goat (SJÖSTRAND et al. 1993), and the bovine penile artery (KLINGE et al. 1978). It has also been shown that the corpus cavernosum of man and rat are able to synthesize acetylcholine (BLANCO et al. 1988, 1990; DAIL and HAMILL 1989). Further, by means of electron microscopy presumably cholinergic small empty vesicles have been identified in axon profiles of the bovine retractor penis and penile artery (ERÄNKÖ et al. 1976; KLINGE et al. 1978) and corpus cavernosum of man and monkey (McCONNELL and BENSON 1982; GU et al. 1983; STEERS et al. 1984). Finally, acetylcholinesterase-positive nerves have been identified in several tissues. This does not prove, although it suggests, the cholinergic nature of the nerves. Such tissues are e.g., the corpus cavernosum of guinea pig (GRIETEN and GEREBTZOFF 1957; BHARGAVA et al. 1974), rat (McCONNELL et al. 1979; DAIL et al. 1986), cat (McCONNELL et al. 1979), rabbit (KLINGE and PENTTILÄ 1969; McCONNELL et al. 1979), dog (STIEF et al. 1989a), monkey (McCONNELL et al. 1979; STIEF et al. 1989b) and man (SHIRAI et al. 1972; BENSON et al. 1980; GU et al. 1983), the corpus spongiosum of rat (McCONNELL et al. 1979), rabbit and bull (KLINGE and PENTTILÄ 1969; McCONNELL et al. 1979), cat, monkey and man (McCONNELL et al. 1979), the retractor penis of dog (BELL and McLEAN 1970), bull (KLINGE et al. 1970) and goat (SJÖSTRAND et al. 1993), and the bovine penile artery (KLINGE and POHTO 1971).

The above-mentioned data indicate that in several mammalian species the smooth muscles in the external male genitalia are innervated by cholinergic nerves in spite of the fact that so far there has been no successful demonstration of the release of unlabelled acetylcholine upon stimulation of the nerves. It is, however, difficult to characterize more exactly the cholinergic nervous mechanisms and to evaluate their physiological importance. In the following some possible muscarinic and nicotinic effects will be discussed.

1. Muscarinic Effects

a) Suppression of Excitatory Adrenergic Neurotransmission

More than 25 years ago German pharmacologists showed that in the isolated rabbit heart the release of noradrenaline evoked by stimulation of the sympathetic nerves is inhibited by exogenous muscarinic agonists as well as by simultaneous stimulation of vagal nerve fibres (see MUSCHOLL 1979). Mostly by using conventional in vitro techniques this presynaptic muscarinic inhibition of the release of noradrenaline has later been reported to occur in a great variety of tissues. A morphological background for the occurrence of this phenomenon in the genital organs is provided by the fact that in the bovine retractor penis and penile artery adrenergic and presumably cholinergic fibres run in close juxtaposition to each other (ERÄNKÖ et al. 1976; KLINGE et al. 1978). Pharmacological evidence has been obtained by showing that in the retractor penis of dog, cat, horse, boar, bull, elk, ram and goat as well as in the canine corpus spongiosum the contractions elicited by field stimulation of the adrenergic nerves are suppressed by acetylcholine or merely by physostigmine and that this suppression is abolished by scopolamine (KLINGE and SJÖSTRAND 1977b, 1982; SJÖSTRAND et al. 1993). Likewise,

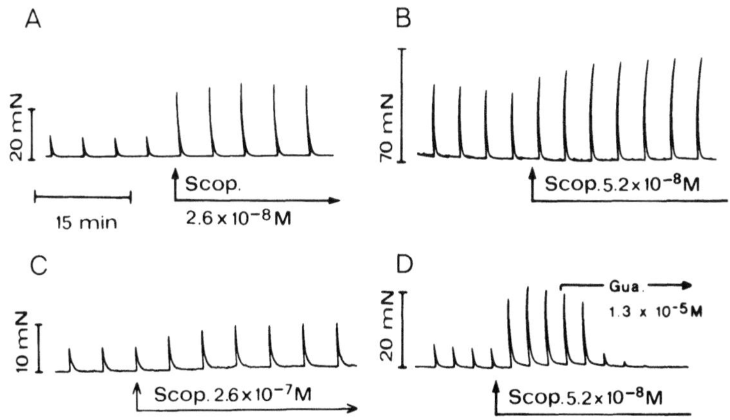

Fig. 1A–D. A low concentration of only scopalamine enhances the contractions induced by field stimulation (at regular intervals of 4 or 5 min) of the adrenergic excitatory nerves in untreated strips of the retractor penis muscle of dog (**A**) and stallion (**B**), the corpus spongiosum of man (**C**) and the ampullary part of the vas deferens of dog (**D**). The contractions induced by exogenous noradrenaline were unaffected by scopolamine. This phenomenon therefore suggests the existence of an endogenous acetylcholine-mediated muscarinic mechanism that counteracts the development of the contractions by interfering with the adrenergic neurotransmission. **A** and **B** are reproduced with permission from KLINGE and SJÖSTRAND (1977b) and **D** from ARVER and SJÖSTRAND (1982), respectively. In **C** the stimulation was performed with 8 Hz for 10 s and a pulse duration of 0.2 ms. *Gua*, guanethidine

in the corpus cavernosum of rat the contractions were suppressed by acetylcholine and the suppression was reversed by atropine (DAIL et al. 1987). In the retractor penis of stallion and dog the same contractions are enhanced merely by scopolamine (Fig. 1), and in the canine retractor penis the field stimulation evoked release of ^{3}H-noradrenaline is inhibited by physostigmine and enhanced by scopolamine (KLINGE and SJÖSTRAND 1977b). Further, in the bovine and dog retractor penis the excitatory junction potentials elicited by field stimulation of adrenergic nerves are attenuated by physostigmine, and this effect is reversed by scopolamine or atropine (SAMUELSON et al. 1983; BYRNE and MUIR 1984; KINEKAWA et al. 1984). Also, a low concentration of scopolamine or atropine enhances these potentials in the bovine retractor penis (SAMUELSON et al. 1983; BYRNE and MUIR 1984). Finally, it has been demonstrated that in strips of the corpus spongiosum of man and the corpus cavernosum of rabbit and man the field stimulation induced release of ^{3}H-noradrenaline is inhibited by muscarinic agonists and enhanced by muscarinic antagonists (HEDLUND et al. 1984; SAENZ DE TEJADA et al. 1989).

There is thus substantial evidence indicating that presynaptic muscarinic inhibition of adrenergic neurotransmission may operate in vitro, whereas attempts to demonstrate the functioning of this mechanism in vivo have been less numerous. It has been shown that in the anaesthetized rabbit the detumescence induced by low frequency stimulation of the sympathetic chains was clearly counteracted by simultaneous stimulation of the partially cholinergic hypogastric nerves (SJÖSTRAND and KLINGE 1979a). This was also true when the sympathetic tone was raised by means of carotid occlusion (SJÖSTRAND and KLINGE 1979b). The same pattern was seen in the goat when contraction of the retractor penis was induced by stimulation of the sympathetic chains (SJÖSTRAND et al. 1993).

b) Possible Endothelium-Mediated Relaxant Effects of Exogenous Muscarinic Agonists

HENDERSON and ROEPKE (1933) showed that in dogs atropine abolished the increase in penile blood flow induced by intraarterial administration of acetylcholine. By using intracavernous administration of drugs their observation was confirmed by STIEF et al. (1989a). When the penile artery was perfused in rabbits scopolamine totally blocked the acetylcholine-induced decrease in perfusion pressure (SJÖSTRAND and KLINGE 1979a). In monkeys, however, the increase in intracavernous pressure effected only by small doses of acetylcholine was abolished by atropine, whereas that induced by large doses required the additional use of a ganglion-blocking agent (STIEF et al. 1989b). This is in agreement with the general view that when mixed muscarinic and nicotinic agonists, such as acetylcholine or carbachol, are used low concentrations induce only muscarinic effects that strengthen along with the dose, whereas the threshold concentration for the nicotinic effects is significantly higher. It has further been shown that in the isolated perfused bovine penis, atropine abolished the vasodilatation induced by intra-arterially injected acetylcholine (PENTTILÄ 1966) and that both nerve–muscle preparations of rabbit corpus cavernosum and strips of the canine glans penis are effectively relaxed by rather low concentrations of acetylcholine (HUKOVIC and BUBIC 1967; KIMOTO and ITO 1988).

ADAIKAN et al. (1983) were the first to show that atropine blocks the acetylcholine-induced relaxation of strips of the human corpus cavernosum. Subsequently this observation was confirmed and it was also reported that scopolamine blocks the acetylcholine- and carbachol-induced relaxations of the isolated human corpus spongiosum and cavernous artery (HEDLUND et al. 1984; HEDLUND and ANDERSSON 1985a). More recently it was noticed that the acetylcholine-induced relaxations were virtually abolished in human corpus cavernosum strips as well as in strips of the deep penile artery denuded of endothelium (SAENZ DE TEJADA et al. 1988; SJÖSTRAND et al. 1988). Receptor binding studies have demonstrated the presence of muscarinic acetylcholine receptors in endothelial cells cultured from human corpus cavernosum tissue and suggested that they are of the M_2 or M_3 subtype (TRAISH et al. 1990).

It is conceivable that all the above-mentioned relaxations caused by exogenous muscarinic agonists are endothelium-mediated, and they could be produced via the synthesis and release of nitric oxide, although the release of other endothelium derived relaxing factors cannot be excluded (FURCHGOTT 1988; AZADZOI et al. 1990). The physiological role of such relaxations is, however, unclear because the origin of the acetylcholine that would activate the endothelial receptors is unknown. In the blood-stream acetylcholine is rapidly inactivated by the non-specific cholinesterase, and this may be one of the reasons why no signs of erection were observed in dogs or cats after intravenous administration of this choline ester (DORR and BRODY 1967; DOMER et al. 1978). Further, it has not been shown that the endothelium can be reached by acetylcholine released from local cholinergic nerves, but this possibility cannot be excluded either.

c) Unclassified Nerve-Mediated Muscarinic Relaxant Effects

Not all but several more recent authors (e.g., DORR and BRODY 1967; ANDERSSON et al. 1984; STIEF et al. 1989a; TRIGO-ROCHA et al. 1993a) have arrived at the conclusion that in dogs the erectile response to pelvic (or cavernous) nerve stimulation is in part, but never totally, blocked by intravenously or intracavernosally administered atropine. The same situation was shown to prevail in cats (ANDERSSON et al. 1987) and monkeys (STIEF et al. 1989b), whereas SJÖSTRAND and KLINGE (1979a) failed to demonstrate any effect of scopolamine on the response to pelvic nerve stimulation in rabbits. The studies with rabbits are, however, hampered by the frequent occurrence of plasma atropine esterase. Also in rats 1 mg/kg atropine failed to affect the vasodilator response produced by stimulation of the cavernous nerve (DAIL et al. 1989). On the other hand, SJÖSTRAND and KLINGE (1979a) were able to show that the erectile response to hypogastric nerve stimulation was curtailed by scopolamine in the rabbit and the same was shown for atropine in cats (BESSOU and LAPORTE 1961; ANDERSSON et al. 1987).

Also the results of in vitro studies are inconsistent. In the retractor penis of nine species and the corpus spongiosum of six species including the macaque KLINGE and SJÖSTRAND (1974, 1977a) were unable to show a blocking effect of atropine or scopolamine or an enhancing effect of physostigmine on the relaxations elicited by field stimulation of the inhibitory nerves. Likewise, in the penile artery of bull and dog such relaxations were unaffected by muscarinic blockade (KLINGE and SJÖSTRAND 1974; BOWMAN and GILLESPIE 1983). In the feline retractor penis physostigmine, on the contrary, curtailed the relaxations (KLINGE and SJÖSTRAND 1977b). It has also been shown that in strips of corpus spongiosum of man the relaxations were unaffected by a high concentration of atropine (ANDERSSON et al. 1983). It has, however, been reported from one laboratory that in corpus cavernosum strips of man these relaxations were slightly but significantly decreased by atropine and enhanced by physostigmine (SAENZ DE TEJADA et al. 1988).

Taken together, these functional studies support the concept that, at least in some species, there may be a cholinergic muscarinic component involved in the neurogenic relaxation of the smooth muscles that remain contracted during the flaccid state of the penis. It is possible that suppression of excitatory adrenergic neurotransmission and endothelium-mediated relaxations take place, but other possibilities cannot be excluded. It has been reported that the pure muscarinic agonist methacholine induces prostacyclin production in human corpus cavernosum tissue in vitro in an atropine-sensitive way (JEREMY et al. 1986), but the significance of this observation is unclear. It is, however, unlikely that acetylcholine relaxes the muscle cells by acting directly upon them, although some of them have been shown to be equipped with muscarinic receptors (KLINGE and SJÖSTRAND 1974, 1977a; TRAISH et al. 1990), because so far such receptors are known to mediate only weak contractile effects.

d) Physiological Significance of Muscarinic Effects

To the authors' knowledge there are no reports in the literature indicating that moderate doses of atropine would impair normal penile erection in healthy conscious animals of any mammalian species (for references see KLINGE and SJÖSTRAND 1974). Man has used atropine for more than 100 years, but evidence indicating that clinical doses of this drug would cause impotence has not been presented, which is in disagreement with the previously common opinion that erection is a cholinergic phenomenon. Recently it has been reported that in man penile erection is unaffected by a single i.v. injection of 0.035 mg/kg atropine (WAGNER and BRINDLEY 1980). This observation should be extended to include also the result of a more chronic administration of this drug, although the so-called centrally acting anticholinergics that are used for Parkinson's disease are not known to cause impotence (EADIE 1987; SEGRAVES 1989). The situation seems to be different with the quaternary anticholinergics used to treat certain gastrointestinal disorders in case that their ganglion-blocking activity is strong enough (ALARANTA et al. 1990).

The data available do not allow a more precise evaluation of the importance of cholinergic muscarinic mechanisms in the initiation and maintenance of penile erection. But they suggest that such mechanisms play a contributory role rather than being of crucial importance. The existence of unique atropine- and hexamethonium-resistant cholinergic mechanisms cannot, however, be excluded.

2. Nicotinic Effects

a) Ganglionic and Possible Postganglionic Nicotinic Receptors

The erectile response to stimulation of the pelvic nerves proximal to the pelvic plexus has been shown to be abolished by ganglion-blocking drugs in the cat (GOLDENBERG 1965), dog (DORR and BRODY 1967), rabbit (SJÖSTRAND and KLINGE 1979a) and rat (STEERS et al. 1988). In the rabbit and rat the response to stimulation of the hypogastric nerves and the sympathetic chains is identically blocked (SJÖSTRAND AND KLINGE 1979a; STEERS et al. 1988). It has not been questioned that the impulse traffic via the pelvic plexus or the sympathetic chains to the penis or the retractor penis is mediated by the type of cholinergic nicotinic receptors usually encountered in autonomic ganglia regardless of what type of neurotransmitter is released from the postganglionic neuron. But one of the questions that remain open is whether or not there are in the erectile pathway nerve cell bodies also distal to the somewhat diffuse pelvic plexus. Morphological evidence indicates that in the proximal part of the corpus cavernosum of man there are sparsely scattered nerve cell bodies (GU et al. 1983). The situation is the same in the retractor penis and the penile artery of the bull (ALARANTA et al. 1989) and in the retractor penis of the rat and goat (DAIL et al. 1990; SJÖSTRAND et al. 1993).

Strips of corpus cavernosum of man are relaxed by DMPP (1,1-dimethyl-4-phenylpiperazinium), which is a specific agonist for the type of nicotinic receptors occurring in ganglia, and the relaxations are blocked by hexamethonium and tetrodotoxin (ADAIKAN et al. 1983). It has also been shown that when the isolated corpus cavernosum of man is relaxed by carbachol only the relaxations elicited by low enough concentrations are blocked by atropine (HOLMQUIST et al. 1990a). This indirectly supports the idea that the relaxations induced by higher concentrations were due to a nicotinic mechanism. Similarly, in monkeys the erectile response to intracavernous injection of acetylcholine was blocked by atropine when small enough doses were used, while larger doses required the additional use of a ganglion-blocking agent (STIEF et al. 1989b).

Both the retractor penis and the corpus spongiosum of several animals are relaxed by nicotine (KLINGE and SJÖSTRAND 1977a) and this also applies to the erectile tissue of man (ADAIKAN et al. 1983; SJÖSTRAND et al. 1988). The nicotinic relaxations of the bovine retractor penis have been studied in more detail (KLINGE et al. 1988). The relaxations are, provided that appropriate concentrations of nicotinic agonists are used, highly susceptible to ganglion-blocking agents, local anaesthetics and hypoxia (Fig. 2). If too high concentrations are used the relaxations will lose their susceptibility to these blocking factors. It has not been shown that the site of action of exogenous nicotinic agonists is primarily at the cell bodies of the inhibitory nerves. But this seems likely as far as suitable concentrations are used, while higher concentrations may also affect the nerve endings.

b) Physiological Significance of Nicotinic Effects

The physiological and clinical significance of cholinergic nicotinic mechanisms in the nervous regulation of the tone of the penile smooth muscles that are actively relaxed only for penile erection is indisputable in view of the fact that ganglion-blocking drugs are well known to cause impotence (e.g.,

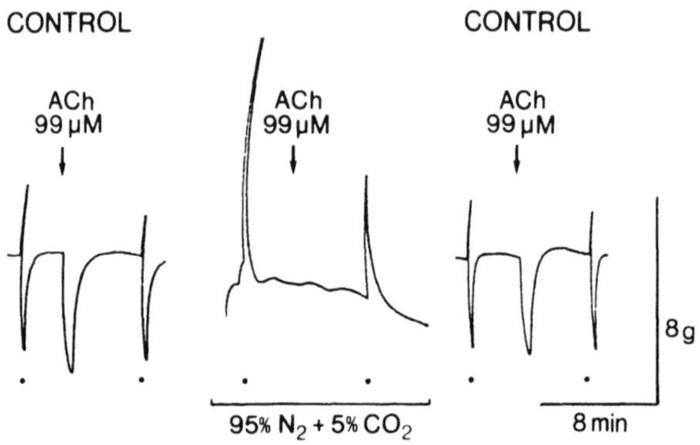

Fig. 2. Hypoxia totally but reversibly blocks the acetylcholine-induced nicotinic relaxation of the bovine retractor penis muscle (in the presence of 1.0 μM scopolamine and 1.1 μM physostigmine) as well as the relaxation elicited by field stimulation of the inhibitory nerves (3 Hz, 1 ms, 10 s, *at dots*) when the normal ventilation of the organ bath with 95% O_2 + 5% CO_2 is replaced by 95% N_2 + 5% CO_2. (From KLINGE et al. 1988)

VOLLE and KOELLE 1970; ERINA 1980). It is, however, not known whether the nicotinic receptors are confined to the nerve cell bodies or whether they also have important functions more distally in the postganglionic neurons. It has been reported that in paraplegics physostigmine makes erection easier to obtain and maintain in 40%–80% of cases (CHAPELLE et al. 1983), but it cannot be judged to what extent this is due to the strengthening of nicotinic or muscarinic functions.

c) Use of Nicotinic Relaxation of Bovine Retractor Penis Muscle as a Model

The relaxation of the isolated bovine retractor penis by appropriate concentrations of nicotinic agonists including acetylcholine involves a more selective and maybe also more physiological activation of the inhibitory nerves than that effected by field stimulation (KLINGE et al. 1988). It is conceivable that such a relaxation of the muscle offers a useful model for further characterization of the nervous transmission which leads to penile erection in the bull and possibly also in other mammals. In addition to serving as a neurophysiological and neuropharmacological model there is evidence indicating that the nicotine-induced relaxation of the muscle can be used for quantitative assessment of the ganglion-blocking activity of various drugs (ALARANTA et al. 1990; KLINGE et al. 1993).

IV. Peptides

Since the 1970s the involvement of several peptides, mostly neuropeptides, in the regulation of the tone of the vascular and non-vascular smooth muscles in the external genital organs has been the object of growing interest. Vasoactive intestinal polypeptide (VIP) has been most extensively studied, and it is the only peptide for which the role of an inhibitory neurotransmitter has been suggested in several studies. Accordingly, in this context only VIP will be dealt with. Concerning other peptides of interest the reader is referred to the reviews of OTTESEN et al. (1988), ANDERSSON and HOLMQUIST (1990) and ANDERSSON (1993).

1. VIP

a) Occurrence and Distribution

By means of radioimmunoassay, significant amounts of VIP immunoreactivity have been found in the penis of, e.g., cat, rabbit and man (LARSEN et al. 1981; WILLIS et al. 1983). In the human penis, which is more rich in VIP than the rabbit penis, the highest concentration was found in the deep artery and the corpus cavernosum, but in all species the concentrations in the penis were severalfold lower than in the vas deferens. Light microscopic studies have shown that in the corpus cavernosum of, e.g., rat, rabbit, monkey, man and dog VIP-immunofluorescent nerve fibres are mostly located in the adventitial layer of small and large arteries and within the cavernosal smooth muscle (DAIL et al. 1983; WILLIS et al. 1983; GU et al. 1983; BENSON 1983; STEERS et al. 1984; JÜNEMANN et al. 1987). The localization is similar in

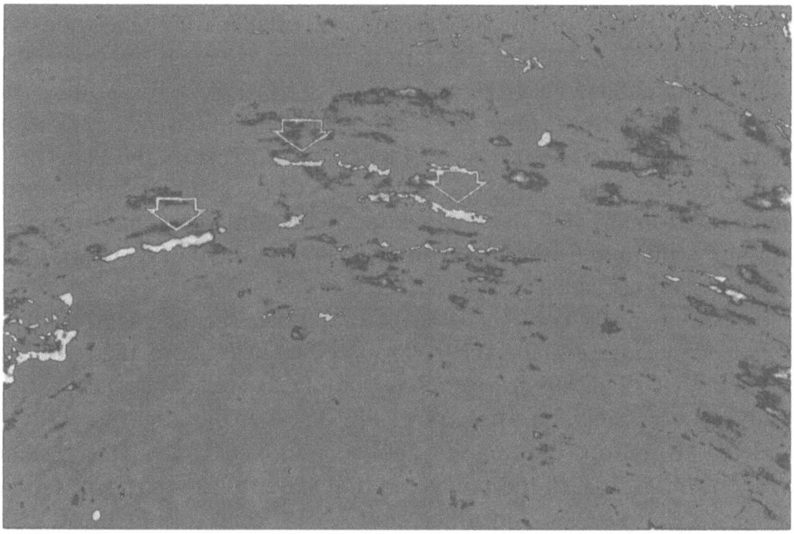

Fig. 3. VIP-immunoreactive nerve fibres (*arrowheads*) in the outer layer of the muscular wall of the bovine penile artery. (From PANULA et al. 1991). In spite of the presence of such nerve fibres the artery in vitro is only weakly and inconsistently relaxed by exogenous VIP (SJÖSTRAND et al. 1981; BOWMAN and GILLESPIE 1983)

the corpus spongiosum and penile circumflex vein of man (GU et al. 1983; WILLIS et al. 1983; ANDERSSON et al. 1983; KIRKEBY et al. 1992). Also in the bovine penile artery (Fig. 3) and retractor penis there are numerous VIP-immunofluorescent nerve fibres (PANULA et al. 1991). In the corpus cavernosum of man and rat there is a similar distribution between VIP-immunoreactive and acetylcholinesterase-positive nerves (GU et al. 1983; DAIL et al. 1986). Ultrastructural studies have revealed that in the corpus cavernosum of man and rhesus monkey VIP immunoreactivity is localized in large dense-cored vesicles coexisting in some axon terminals with small agranular vesicles similar to those found in classic cholinergic terminals (GU et al. 1983; STEERS et al. 1984). Contrary to what has been observed in other species, it has been reported that in the corpus cavernosum of green monkeys VIP is colocalized in the same axons with neuropeptide Y (SCHMALBRUCH and WAGNER 1989). In the root of the corpus cavernosum of man there are scattered VIP-immunoreactive neuronal cell bodies (GU et al. 1983). In the rat such cell bodies seem to be localized predominantly in the pelvic plexus (DAIL 1987).

b) Pharmacological Effects of Exogenous VIP

α) In Vitro. The relaxant effect of VIP has usually been studied in preparations precontracted with noradrenaline, phenylephrine, $PGF_{2\alpha}$ or some other agent. VIP relaxes with varying efficacy corpus cavernosum strips from e.g., rabbit, monkey and man, and the relaxations are unaffected by atropine, propranolol, naloxone and tetrodotoxin (WILLIS et al. 1983; STEERS et al. 1984). In corpus cavernosum of man the relaxations were suppressed by VIP antiserum contrary to the neurogenic relaxations elicited by field stimulation with 10 Hz (ADAIKAN et al. 1986). In another study, α-chymotrypsin did not affect the neurogenic relaxations, whereas it abolished the VIP-induced relaxations (PICKARD et al. 1993). In the same study it was further shown that the nitric oxide synthase inhibitor N^G-nitro-L-arginine abolished, and the guanylate cyclase inhibitor methylene blue significantly inhibited, the neurogenic relaxations contrary to the VIP-evoked relaxations which were unaffected in both cases.

In spontaneously contracted strips of human corpus cavernosum, the threshold concentration for the relaxant effect was more than 100 times lower for VIP than for acetylcholine and PGE_1, whereas the maximum relaxations were similar (ADAIKAN et al. 1986). In pharmacologically precontracted strips from any of the above-mentioned tissues, the maximum relaxations did not reach the precontraction level. In such strips it is difficult to evaluate the potency and efficacy of VIP, because both the threshold concentration and the degree of relaxation depend on the type and concentration of the contracting agent.

In addition, the corpus spongiosum of man is relaxed, although less effectively than the corpus cavernosum (ANDERSSON et al. 1983; HEDLUND and ANDERSSON 1985b). WILLIS et al. (1983) reported that VIP clearly relaxed the corpus spongiosum of rabbit, while SJÖSTRAND et al. (1981) found only weak and inconsistent effects on this tissue of cat, dog, rabbit and guinea pig. VIP relaxes, although inconsistently, the perfused canine and bovine penile artery (BOWMAN and GILLESPIE 1983). It also relaxes the cavernous artery and dorsal and circumflex veins of the human penis (HEDLUND and ANDERSSON 1985b; ADAIKAN et al. 1986; KIRKEBY et al. 1992) as well as the retractor penis of cat, dog and bull (SJÖSTRAND et al. 1981). The relaxant effect on the retractor penis exhibited considerable tachyphylaxis. Consistent contractile effects of VIP on the smooth muscle of the external genital organs have not been reported.

In corpus spongiosum of man VIP did not affect the rhythmic activity provoked by $PGF_{2\alpha}$ (ANDERSSON et al. 1983), whereas in corpus cavernosum strips from rabbit, monkey and man it abolished rhythmic contractions provoked by noradrenaline or oxytocin (WILLIS et al. 1983). In the corpus cavernosum, corpus spongiosum and cavernous artery of man, VIP suppressed contractions evoked by field stimulation of the excitatory nerves with frequencies up to 30 Hz (HEDLUND and ANDERSSON 1985b; ADAIKAN et al. 1986). This effect may have been directly muscular, because in human corpus cavernosum strips VIP did not affect the release of noradrenaline (SAENZ DE TEJADA et al. 1989).

β) *In Vivo*. Infusion of VIP into the two dorsal arteries of the canine penis produced a fivefold increase in venous outflow and a submaximal erectile response (ANDERSSON et al. 1984). It has also been reported that intracavernous injection of VIP induced erection in dogs mainly due to active venous outflow restriction and that VIP antibody blocked venous outflow restriction during neurostimulation-induced erection (JÜNEMANN et al. 1987), but it was not shown that the veins involved were contracted by VIP in vitro. In the monkey intracorporal injection of VIP had no effect, and administration of it into the internal iliac artery, while having no effect on the flaccid penis, caused detumescence of the penis erected by cavernous nerve stimulation (STEERS et al. 1984). An initial study on humans showed that intracavernous injection of 200 pmol VIP may cause full erection (OTTESEN et al. 1984). Later, more comprehensive studies have, however, demonstrated that intracavernous injection of even much larger doses of VIP only, i.e., 1–4 μg, does in most cases induce only tumescence but not the erection and rigidity adequate for intromission (ADAIKAN et al. 1986; KIELY et al. 1989; ROY et al. 1990).

c) Release of Endogenous VIP into Cavernous Blood

In a detailed study with dogs pelvic nerve stimulation caused a substantial output of VIP, which was correlated in onset and duration to the initial vasodilator response preceding the erectile response proper (ANDERSSON et al. 1984). Similarly, in cats selective stimulation of the pelvic or the hypogastric nerves was in either case found to cause about a fivefold increase in VIP output from the penis that was coordinated in time with the local blood flow increase during erection (ANDERSSON et al. 1987). When erection was induced by intracavernous injection of papaverine or by visual sexual stimulation in men examined for erectile dysfunction the concentration of VIP in the cavernous blood could increase up to 20-fold (VIRAG et al. 1982; OTTESEN et al. 1984). In the papaverine-induced erections the source of VIP is not fully clear. Conflicting results have been obtained, because in the study of KIELY et al. (1987) the mean VIP concentration did not alter significantly in either cavernosal or peripheral venous blood during similarly induced erections.

d) Physiological Significance

A more exact estimation of the physiological role of VIP in the external male genital organs has so far not been possible. Its occurrence and localization are not against a role in the functioning of the NANC inhibitory nerves, but there are several findings indicating that VIP is not the principal neurotransmitter of these nerves. To these belong the above-mentioned results of PICKARD et al. (1993), obtained with human tissues, as well as a lot of other observations, some of which have been summarized in the review of GILLESPIE et al. (1990). For example, there is a considerable discrepancy between the time course of the mechanical effect of exogenous VIP and that of field stimulation of the NANC nerves, and only the former has been shown to be susceptible to VIP antiserum. In several tissues the relaxant effect of VIP is only weak or moderate, and the in vitro and in vivo effects are not always consistent. In addition, the VIP-induced electrophysiological effects are inconsistent (SAMUELSON et al. 1983). Further, in the bovine retractor penis the neurogenic relaxation is mediated via cGMP without a change in the cAMP level (BOWMAN and DRUMMOND 1984), whereas VIP-induced relaxations are usually associated with an increase in the cAMP level (SCHOEFFTER and STOCLET 1985; FONTAINE et al. 1986a; KAMATA et al. 1988). The neurogenic relaxations of the bovine retractor penis are blocked by ethanol, oxyhaemoglobin, hypoxia and *N*-methylhydroxylamine, while those elicited by exogenous VIP are unaffected (BOWMAN et al. 1982; BOWMAN and DRUMMOND 1985; BOWMAN and MCGRATH 1985; Fig. 4).

The release of VIP on stimulation of the erectile nerves has been demonstrated in only a few species and even in them it alone did not account for the induction of full erection. This further consolidates the view that the importance of VIP in the neurogenic relaxation of the smooth muscles in the external reproductive organs is quite restricted, although certain species differences may exist. However, a cotransmitter or a neuromodulatory function of VIP cannot be excluded. An interaction with acetylcholine on both the prejunctional level (BARTFAI et al. 1988) and the postjunctional level (e.g., LUNDBERG et al. 1982; EVA et al. 1985) is possible,

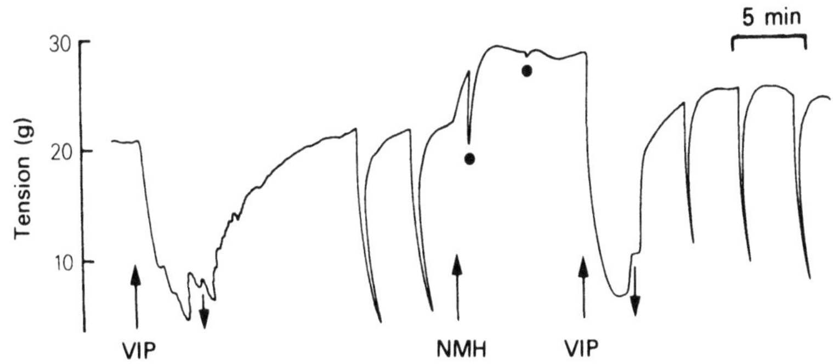

Fig. 4. Relaxation of the bovine retractor penis muscle induced by field stimulation of the inhibitory nerves (1 Hz for 10 s) is totally blocked by 2 m*M* *N*-methylhydroxylamine (*NMH*), an inhibitor of guanylate cyclase, whereas the VIP-induced relaxation remains unaffected, indicating that in this tissue the nerve-mediated relaxation involves the formation of cGMP while VIP uses a second messenger other than cGMP. (From Bowman and Drummond 1984)

although an attempt to demonstrate a synergism on the postjunctional level has failed (Kawanishi et al. 1990). A putative cotransmitter role with nitric oxide remains to be studied.

It has been suggested that a decrease of VIP in penises from impotent men supports its importance in the induction of erection (Gu et al. 1984; Lincoln et al. 1987). It is, however, unknown whether the decrease reflects a cause or rather a consequence of the condition. Synthesis of specific VIP receptor antagonists may further elucidate the physiological significance of this peptide.

V. L-Arginine Nitric Oxide Synthase Pathway

1. Background Constituted by the Inhibitory Factor Extracted from the Bovine Retractor Penis Muscle

The fact that the relaxation of the bovine retractor penis elicited by transmural stimulation of the inhibitory nerves was resistant to all autonomic blocking drugs (Klinge and Sjöstrand 1974) hinted at the possibility that the inhibitory neurotransmitter is an unusual compound. The first account of an attempt to isolate this transmitter was given by Ambache et al. (1975) who showed that a crude acid thermolabile extract yielded relaxations similar to those induced by stimulation of the inhibitory nerves. But this group published no further reports on the topic. Instead, the challenging and obviously difficult task was taken over by Gillespie and coworkers in Glasgow (Gillespie and Martin 1978).

In the following some of the essential findings of the Glasgow group will be mentioned. The observations of Ambache et al. (1975) were confirmed and it was shown that acid activation constituted a prerequisite for pharmacological activity of the extract (GILLESPIE and MARTIN 1980). It was also shown that the active principle was soluble in water and methanol but insoluble in ether, acetone or chloroform, and that it was not an adenine nucleotide or a peptide (BOWMAN et al. 1979; GILLESPIE et al. 1981). It relaxed arterial strips from various species, but the activity was abolished by an erythrocyte-associated material, possibly haemoglobin, in the same way as the response to stimulation of the inhibitory nerves in the bovine retractor penis (BOWMAN et al. 1981; BOWMAN and GILLESPIE 1981a,b). The active principle was called the inhibitory factor (IF); its susceptibility to haemoglobin was confirmed (BOWMAN et al. 1982; BOWMAN and GILLESPIE 1982) and its neuronal origin was supported by ultracentrifugation studies (GILLESPIE and HUNTER 1982). It was shown that it mimicked the inhibitory nerves also in the sense that both of them activated guanylate cyclase and used cyclic GMP, but not cyclic AMP, as a second messenger (BOWMAN and DRUMMOND 1984). However, hypoxia did not affect the relaxant effect of the inhibitory factor, whereas it effectively but reversibly blocked the neurogenic relaxation, suggesting a prejunctional site of the high demand for oxygen (BOWMAN and MCGRATH 1985).

The discovery of the endothelium-derived relaxing factor (EDRF; FURCHGOTT and ZAWADSKI 1980) had opened a pathway independent of that characterizing the inhibitory factor from the bovine retractor penis. It turned out, however, that the EDRF also exerted its action by activating guanylate cyclase and that its synthesis, release or effect was blocked by haemoglobin and hypoxia (FURCHGOTT 1984). Further, also the EDRF was shown to be a labile compound (ANGUS and COCKS 1987), the effect of which was blocked by borohydride (GRIFFITH et al. 1984) like that of the inhibitory factor (GILLESPIE et al. 1981). Thus, it became increasingly evident that the EDRF had several properties in common with the inhibitory factor and/or the NANC neurotransmitter in the bovine retractor penis. The comprehensive work done to identify the chemical structure of the EDRF eventually led to the conclusion that the principal EDRF is identical with nitric oxide (NO; PALMER et al. 1987; IGNARRO et al. 1987) or *S*-nitrosocysteine (MYERS et al. 1990). This was accompanied by the suggestion that also the inhibitory factor is NO, the precursor of which is nitrite (MARTIN et al. 1988; FURCHGOTT 1988). It was also suggested that the activated inhibitory factor consists of a nitroso compound and NO (YUI et al. 1989) and more recently that it is a *S*-nitrosothiol (KERR et al. 1992). A detailed description of the extraction, purification and characterization of the inhibitory factor is given in the reviews of GILLESPIE et al. (1990) and MARTIN and GILLESPIE (1991).

2. Nitric Oxide Synthase and the Use of Its Inhibitors

In endothelial cells and macrophages L-arginine, but not D-arginine, is a physiological precursor of NO (PALMER et al. 1988; SCHMIDT et al. 1988;

HIBBS et al. 1987; IYENGAR et al. 1987). One of the terminal guanidino nitrogen atoms of this amino acid is incorporated into NO by the enzyme nitric oxide synthase, L-citrulline being another reaction product (MARLETTA et al. 1988). Consequently, e.g., the following L-arginine analogues have been introduced to act as competitive and stereoselective inhibitors of the NO synthase: N^G-monomethyl-L-arginine (L-NMMA), N^G-nitro-L-arginine methyl ester (L-NAME), and N^G-nitro-L-arginine (L-NOARG, later also abbreviated L-NNA or L-NA; REES et al. 1989; PALACIOS et al. 1989; MOORE et al. 1990; MÜLSCH and BUSSE 1990).

The NO synthase (NOS; EC 1.14.23) has been purified from rat cerebellum, kinetically characterized, and cloned (BREDT and SNYDER 1990; SCHMIDT et al. 1991; BREDT et al. 1991). Immunohistochemical studies first demonstrated in the brain its neuronal localization, thus providing conclusive evidence for the synthesis of NO in nerves (BREDT et al. 1990; HOPE et al. 1991). Thereafter, by using immunohistochemistry and/or NADPH diaphorase staining, the presence of NOS has been demonstrated in the nerves of the bovine and rat retractor penis (SHENG et al. 1992; DAIL et al. 1993) as well as in the cell bodies and the peripheral parts of the nerves innervating the penile arteries and cavernosal smooth muscles of rat (BURNETT et al. 1992; KEAST 1992; ALM et al. 1993), dog (BURNETT et al. 1992) and man (BURNETT et al. 1993). The isoform of NOS found in the nerves is virtually absent from the smooth muscle cells (SHENG et al. 1992), whereas it is present in the endothelium of the arteries and the cavernosal sinusoids, although its staining is less intense there (BURNETT et al. 1992).

It has been shown that the enzymatic activity of the NOS purified from rabbit corpus cavernosum is, like that in rat cerebellum, completely dependent on NADPH, calmodulin and calcium and that the maximal rate of NO synthesis also requires tetrahydrobiopterin (BUSH et al. 1992a). The authors further showed that most of the activity of this constitutive isoform of NOS was present in the cytosolic fraction, which also, although indirectly, supports the concept that it is mainly of neuronal origin (SCHMIDT et al. 1991; FÖRSTERMANN et al. 1991). The same cofactors are required by the NOS isolated from the bovine retractor penis, which should be exclusively of neuronal origin (SHENG et al. 1991, 1992).

a) In Vitro Studies

Initially it was shown that L-NMMA inhibited the response to electrical field stimulation of the NANC inhibitory nerves in the rat anococcygeus (GILLESPIE et al. 1989), whereas it failed to do so in both the bovine retractor penis (GILLESPIE and XIAORONG 1989) and the corpus cavernosum of rabbit (KIM et al. 1990) and man (SJÖSTRAND et al. 1990). These results led to the idea that the L-arginine NOS pathway could be of importance in the NANC neurotransmission in the rat anococcygeus, but not in that of the other above-mentioned tissues. A re-evaluation of this concept became necessary, however, when HOBBS and GIBSON (1990) demonstrated that both L-NOARG and L-NAME are more potent inhibitors of the NOS than L-NMMA, which may in part be due to the fact that L-NMMA is metabolized by NOS to L-citrulline (HECKER et al. 1990). It turned out that L-NOARG blocked the relaxation elicited by electrical or pharmacological activation of the NANC nerves in the bovine retractor penis and penile artery, the IC_{50} value being about $2.3\,\mu M$ in both tissues (MARTIN et al. 1991; LIU et al. 1991; KOSTIAINEN et al. 1991). More recently it has been shown that in the bovine retractor penis L-NMMA inhibits the effect of L-NOARG and L-NAME with a greater potency than L-arginine (MARTIN et al. 1993).

L-NOARG and/or L-NAME was also effective in blocking the NANC relaxation in the corpus cavernosum of rabbit (IGNARRO et al. 1990; HOLMQUIST et al. 1992) and man (ADAIKAN et al. 1991; HOLMQUIST et al. 1991a, 1992; KIM et al. 1991; PICKARD et al. 1991; RAJFER et al. 1992). In the corpus cavernosum of man, $200\,\mu M$ L-NOARG reduced the size of the relaxations induced by electrical field stimulation of the inhibitory nerves to 0.8% and in the corpus cavernosum of rabbit to 16.2%, respectively, while it abolished in both species the relaxations elicited by exogenous acetylcholine (KNISPEL et al. 1992). In the corpora cavernosa of rabbit and man, the relaxations evoked by exogenous NO or compounds known to generate NO closely mimicked those produced by electrical field stimulation of the inhibitory nerves (BUSH et al. 1992b; KNISPEL et al. 1992). In addition, the fact that the neurogenic relaxation of the bovine retractor penis was inhibited by the superoxide anion generator LY 83583 in a superoxide dismutase-sensitive way (SHENG 1991) supports the possibility that NO plays a role in the relaxation (FURCHGOTT 1990). Finally, indirect evidence has been presented suggesting a basal release of NO in the corpus cavernosum of rabbit (KNISPEL et al. 1991).

b) In Vivo Studies

It has been shown that in pithed rats intravenously administered L-NAME significantly inhibits the erectile response to stimulation of the pelvic parasympathetic outflow (FINBERG and VARDI 1991; FINBERG et al. 1993). It has also been shown that the erectile response induced by direct electrical stimulation of the cavernous nerve is blocked in rabbits and dogs by intracavernosal administration of L-NOARG (HOLMQUIST et al. 1991b; TRIGO-ROCHA et al. 1993b) and in rats by L-NAME or L-NMMA injected into the jugular vein (BURNETT et al. 1992). In dogs about 75% of the erectile response to electrical stimulation of the pelvic nerve persisted after destruction of the sinusoidal endothelium with a detergent (TRIGO-ROCHA et al. 1993a).

Topical application of glyseryltrinitrate may improve erection in impotent men (HEATON et al. 1990; MEYHOFF et al. 1992). This also applies to intracavernosal administration of other NO donors, i.e. linsidomide chlorhydrate (SIN-1) and sodium nitroprusside, but their usefulness is still undergoing evaluation (STIEF et al. 1992; PORST 1993; BROCK et al. 1993).

3. Physiological Significance

Our present knowledge indicates that the L-arginine nitric oxide synthase pathway is involved in the initiation of penile erection in man and all other mammals studied. The two probable sources of nitric oxide are the inhibitory nerves responsible for the relaxation of the smooth muscles that normally maintain the flaccid state of the penis and the endothelium of the inflow resistance vessels and the cavernous sinusoids. No evidence of the involvement of an inducible NOS present in the smooth muscle cells has been presented. The identification as NO or NO only, is, however, still somewhat controversial in the case of the inhibitory neurotransmitter and the EDRF (for references see KERR et al. 1992).

As far as the NO of neuronal origin is concerned, the evidence obtained from morphological and biochemical studies, together with that obtained from functional in vitro and in vivo experiments, strongly suggests, and in the opinion of several authors even establishes, that it is the major, if not the sole, inhibitory neurotransmitter released from the nerves subserving erection. However, little is so far known, e.g., about the factors involved in vivo in the regulation of the synthesis, release and inactivation of this compound.

The physiological significance of the NO of endothelial origin is at present less well understood than that of neuronal origin. The importance of the former is well demonstrated in vitro, because the relaxation of strips from penile arteries and cavernosal tissue elicited by low concentrations of exogenous acetylcholine is abolished by atropine, inhibitors of the NOS or destruction of the endothelium. It is, however, not clear how the endothelium is reached in vivo by acetylcholine or other endogenous substances known to bring about the release of NO. There is, for example, so far no convincing evidence showing that in this vascular bed the endothelium is reached by acetylcholine released from nerves. On the other hand, it has not been excluded that the endothelial production of NO is triggered or augmented by the shear stress caused by the initial nerve-mediated increase in blood flow.

The functioning of the inhibitory nerves in the bovine retractor penis, whether activated by electrical or pharmacological means, is abruptly, but reversibly blocked by hypoxia (BOWMAN and MCGRATH 1985; KLINGE et al. 1988; Fig. 2). It is also known that the synthesis of NO accomplished by NOS requires, in addition to L-arginine, molecular oxygen as a substrate (e.g., LEONE et al. 1991). Strips of human and rabbit corpus cavernosum are, when exposed to arterial-like PO_2, relaxed by acetylcholine and electrical stimulation of the inhibitory nerves, but the relaxations are progressively inhibited as a function of decreasing PO_2 (KIM et al. 1993). The authors further showed that reinstitution of normoxic conditions restored the endothelium-dependent and neurogenic relaxation and that low PO_2 in -hibited NOS activity in corpus cavernosum cytosol. Taken together, these data suggest that the oxygen tension modulates penile erection by regulating the local synthesis of NO.

D. Internal Accessory Reproductive Organs

The smooth muscles of the internal male genital organs have been the subject of rather detailed earlier reviews (SJÖSTRAND 1965, 1976, 1981). A more recent review focused on results obtained from human material (WAGNER and SJÖSTRAND 1988).

I. General Properties of the Smooth Muscle

There is great variance with respect to the degree of electric coupling between the smooth muscle cells. Thus the guinea pig vas deferens behaves like an electric syncytium but in the rat and mouse vasa deferentia there is no or little longitudinal spread of electrotonus. The variance seems to be related to the degree of innervation of the smooth muscle and the influence of androgens. Denervated vasa deferentia and vasa deferentia from castrated specimens are inclined towards automaticity. Vasa deferentia with comparatively sparse innervation, as those of man and dog, are more inclined towards spontaneous activity than those of rodents. Furthermore, there is a tendency towards greater automaticity in the testicular end of the excretory ducts than in the prostatic end; the density of innervation follows the opposite direction (see SJÖSTRAND 1981). In a recent report it was shown that 28 h after neurotomy of the rat vas deferens the neuroeffector cleft was widened and the number of nexal contacts was doubled (SJÖSTRAND and ANDERSSON-FORSMAN 1991). This further substantiates the view that the intimacy of the neuromuscular contacts determines, at least in part, the intimacy between the smooth muscle cells.

Upon stimulation of the hypogastric nerves or field stimulation of intramural nerves excitatory junction potentials are seen (see SJÖSTRAND 1981; STJÄRNE 1989). Apart from one recent study no study has shown the presence of inhibitory junction potentials in the smooth muscle of the internal male reproductive organs. SEKI and SUZUKI (1989), however, demonstrated such potentials in the rabbit prostatic smooth muscles where they seem to be due to NANC nerves.

II. Adrenergic Mechamisms

The smooth muscles of the accessory male genital organs have in general a very dense supply of adrenergic terminals deriving from so called short adrenergic neurons and often have very high concentrations of noradrenaline (SJÖSTRAND 1965, 1981). The vas deferens of guinea pig, rat and mouse has become a favourite model organ for studies of adrenergic mechanisms. The recent review of STJÄRNE (1989) covers these aspects. Therefore the matter will be only briefly commented upon here.

1. α-Adrenoceptors

The smooth muscle is supplied with α_1-adrenoceptors, while α_2-adrenoceptors seem to be preferentially located in the nerve terminals (e.g., CONNAUGHTON and DOCHERTY 1990). Activation of postjunctional α_1-adrenoceptors seems to be the principal effect of the adrenergic component of the response to excitatory nerve stimulation (see below). The main function ascribed to the α_2-adrenoceptors is prejunctional inhibition of transmitter secretion (see STJÄRNE 1989).

2. β-Adrenoceptors and Their Significance

Early studies showed the presence of inhibitory β_2-receptors in the smooth muscle (see SJÖSTRAND 1981). β-Adrenoceptor agonists hyperpolarize the vas deferens of guinea pig (SJÖSTRAND 1973b). ARVER and SJÖSTRAND (1982) found that in the dog vas deferens, ampulla ductus deferentis and prostate, terbutaline and isoprenaline inhibited the response to field stimulation of nerves or direct stimulation of the smooth muscle. Sotalol blocked these effects. If α-adrenoceptors were blocked and the muscle brought into rhythmic activity by barium chloride, field stimulation of intramural nerves performed with high frequencies and long duration of the train produced suppression of the rhythmic activity. Also this effect was blocked by sotalol. Hence the β-adrenoceptors could be activated by intense nerve stimulation. But the procedures necessary for the demonstration of this phenomenon make its significance doubtful. We still lack data showing a physiological significance of the β-adrenoceptors present in the smooth muscle of the internal accessory male reproductive organs.

III. Cholinergic Mechanisms

In the early 1960s there was a certain interest in cholinergic mechanisms in the vas deferens. This was due to the by that time much discussed hypothesis of BURN and RAND suggesting that there was a cholinergic link in peripheral adrenergic neurotransmission (see SJÖSTRAND 1962, 1965). Later several light microscopic studies showed the presence of acetylcholinesterase-positive nerves in many internal male genital organs and the presence of nerve terminals with small agranular vesicles was demonstrated by means of electron microscopy (see SJÖSTRAND 1981). The presence of nicotinic receptors was indicated by the finding that contractions of the guinea pig vas deferens produced by high concentrations of acetycholine were partially blocked by hexamethonium (SJÖSTRAND 1962). More recently, WAKUI and FUKUSHI (1986) found that after muscarinic blockade high concentrations of nicotinic agonists produced a depolarization of the guinea pig vas deferens that seemed to be related to transmitter release from excitatory nerves. Detailed studies of nicotinic receptors in the internal reproductive genital organs are, however, still lacking.

SJÖSTRAND (1981) concluded that the cholinergic nerves in the smooth muscles might have the following functions:

1. They may just pass the muscle layers on their way to secretory cells.
2. They could constitute a small part of the excitatory innervation.
3. They may act by prejunctional inhibition of the excitatory adrenergic neurotransmission.

1. Excitatory Muscarinic Effects

SJÖSTRAND (1961) reported that approximately 1000 times smaller concentrations of acetylcholine than those producing direct contraction of the vas deferens caused an at least threefold increase in the motor response to hypogastric nerve stimulation. Atropine blocked this effect. It was, however, not specific to acetylcholine but shared by various other agonists, e.g., catecholamines, histamine, serotonin and substance P. The common base of this effect seems to be hypopolarization of the smooth muscle (SJÖSTRAND 1973a; WAKUI and FUKUSHI 1986). Physostigmine produces an effect similar to that of low concentrations of acetylcholine (FUKUSHI and WAKUI 1986a,b). According to a recent study (ELTZE 1988), the postjunctional muscarinic receptors in the smooth muscle of the rabbit vas deferens are of the M_2 subtype.

2. Inhibitory Muscarinic Effects

Since muscarinic suppression of excitatory adrenergic neurotransmission in vitro is very clear in the external male genital smooth muscle it was logical to search for this mechanism in the internal male genital smooth muscle. A major problem is the strong contractile action of muscarinic agonists that is generally seen in the latter muscles. ARVER and SJÖSTRAND (1982) investigated both circular and longitudinal strips of the ampulla ductus deferentis and various parts of the vas deferens and prostate of the dog. The choice of experimental animal was determined by the fact that in dog vas deferens there are comparatively few adrenergic nerves, but many acetylcholinesterase-positive ones (BELL and MCLEAN 1970). Low concentrations of scopolamine enhanced the responses to excitatory nerve stimulation (Fig. 1D), while physostigmine or acetylcholine suppressed them. It was concluded that in the smooth muscle of canine effectors of seminal emission prejunctional inhibition of the excitatory adrenergic neurotransmission was the likely function of cholinergic nerves. According to ELTZE (1988), the prejunctional muscarinic receptors in the rabbit vas deferens are of the M_1 subtype.

3. Physiological Significance of Cholinergic Nerves

Recent data suggest that cholinergic nerves may via muscarinic receptors act either as direct excitatory nerves or as indirect inhibitory nerves. It is at present not possible to evaluate precisely which is the main function. Concerning the role of cholinergic nerves as direct excitatory nerves of the smooth muscle it should, however, be emphasized that the neurogenic contractile response of the vas deferens and other internal male genital organs is totally blocked by guanethidine. Thus, if cholinergic nerves constitute a minor portion of the excitatory innervation, the presence of a simultaneous neurotransmission from the dominant adrenergic excitatory nerves is necessary for their action. Accordingly, it seems reasonable to assume that the possible function, if any, of cholinergic nerves in internal male genital smooth muscle depends on the spatial relationship between them and their two presumptive targets, i.e., the adrenergic nerves and the smooth muscle cells, and the relative sensitivity to muscarinic stimulation of these targets. Finally one should not neglect the possibility of acetylcholine-mediated nicotinic stimulation of nervous elements in the thick muscular walls of the internal male genital organs.

IV. Peptides

In the internal accessory genital organs many studies have demonstrated immunoreactivity against antisera to various neuropeptides (see review of OTTESEN et al. 1988, and original reports of FEHLER and BURNSTOCK 1987 and STJERNQUIST et al. 1987). Much attention has been paid to terminals showing immunoreactivity for neuropeptide Y (NPY). Localization of NPY immunoreactivity shows a certain correspondence to that of adrenergic terminals. Some interest has also been focused on VIP-immunoreactive terminals, the distribution pattern of which shows a certain similarity to that of acetylcholinesterase-positive nerves. Finally, there are a few fibres exhibiting substance P immunoreactivity. These fibres are probably sensory.

The most interesting functional data concern NPY which seems to be stored in large dense cored vesicles in adrenergic terminals (FRIED et al. 1985). It seems to be secreted together with noradrenaline as a cotransmitter when adrenergic nerves are stimulated (LUNDBERG and HÖKFELT 1986; STJERNQUIST et al. 1987; STJÄRNE 1989; ELLIS and BURNSTOCK 1990). In contrast to arterioli, where NPY seems to act as an additive excitatory transmitter, its dominant effect in the guinea pig vas deferens is prejunctional inhibition (autoinhibition) of transmission in the excitatory nerve terminals. In the guinea pig vas deferens NPY is a strong inhibitory modulator of these nerves but a weak postjunctional stimulant.

V. Cotransmission

As a concept the possibility of storage and secretion of more than one transmitter from an autonomic nerve terminal was introduced by BURNSTOCK (1986). The development of the concept was in particular due to two well-known phenomena, i.e., the atropine resistance of part of the neurogenic contractile response of the detrusor and the resistance to α-adrenoceptor blocking agents of part of the neurogenic contractile response of the vas deferens. The early debate about the neurotransmission in the vas deferens has been summarized by SJÖSTRAND (1981). In short: When the nerves of the vas deferens of guinea pig, rat and mouse are stimulated there is a rapid twitch response. Upon prolonged stimulation the organ develops a second slower phase of contraction. Both phases of the contractile response as well as the excitatory junction potentials are blocked by adrenergic neuron blocking agents, but only the second phase of contraction is blocked by α_1-adrenoceptor blocking agents. Originally the resistance to α-adrenoceptor blocking agents of the junction potentials and the twitch response was ascribed to the narrow neuromuscular cleft in the vas deferens making it difficult to reach sufficient concentration of the blocker at postjunctional receptor sites. Another explanation was the introduction of a special junctional adrenoceptor, i.e., the γ-receptor (HIRST and NEILD 1980). This concept has not been generally accepted (see HOLMAN 1987; STJÄRNE 1989). On the other hand, Burnstock's concept that ATP, which is stored together with noradrenaline in the small storage vesicles of adrenergic nerves and together with acetylcholine in the corresponding vesicles of cholinergic nerves, is responsible for most of the twitch response in the vas deferens (and the detrusor) has gradually gained acceptance. The main reason for this is that the stable ATP analogue α,β-methylene ATP blocks the twitch response and the excitatory junction potentials (BURNSTOCK 1986; STJÄRNE 1989).

The relative importance of ATP versus noradrenaline seems, however, to exhibit species differences. Thus, the neurogenic contractile response of human and dog vasa deferentia, which receive few adrenergic terminals and have wide neuroeffector clefts, is completely blocked by α-adrenoceptor blocking agents (ARVER and SJÖSTRAND 1982; WAGNER and SJÖSTRAND 1988). One of the factors that determine the sensitivity of the "adrenergic" neurogenic contractile response seems to be the intimacy of the neuroeffector system. This is suggested by the fact that the neurogenic contractile response of rat and guinea pig vasa deferentia becomes more sensitive to α-adrenoceptor blockade during Wallerian degeneration (HAMMARSTRÖM and SJÖSTRAND 1984a) when the neuroeffector cleft widens (SJÖSTRAND and ANDERSSON-FORSMAN 1991). Furthermore, if the number of adrenergic nerves is reduced in the guinea pig vas deferens the response to field stimulation is blocked by phentolamine (HAMMARSTRÖM and SJÖSTRAND 1984a). It may be that to act as a neurotransmitter ATP demands a narrow neuromuscular junction, while the smaller and more stable noradrenaline molecule is the more important transmitter when the neuromuscular gap is wide and the transmitter has to diffuse through a certain distance to reach the postjunctional receptor. ATP might also in the bovine retractor penis be responsible for part of the excitatory junction potentials which are not completely blocked by phentolamine or prazosin (SAMUELSON et al. 1983; BYRNE and MUIR 1984). In this context one should also be aware of the fact that neurogenic responses may in vivo occasionally be more sensitive to autonomic blocking drugs than in vitro (HAMMARSTRÖM and SJÖSTRAND 1984b).

E. Concluding Remarks

The extent of species differences existing in the distribution and properties of smooth muscle in the external reproductive organs is largely unresolved. This is also true for the innervation of the muscle and the neurotransmission mechanisms of the nerves in question. It is, however, known that the corpus cavernosum of the artiodactyls is, contrary to that of most other mammalian orders, virtually devoid of smooth muscle and autonomic nerves (KLINGE and SJÖSTRAND 1977a). This dissimilarity in angioarchitecture of the penis is one of the factors which contribute to the fact that in the bull and goat the haemodynamics of erection differs from that in other investigated species (BECKETT 1983). As judged by functional studies, the smooth muscle receptors in the external genital organs exhibit species differences that seem to be quite prominent, e.g., in the case of peptide agonists, whereas the differences between the receptors for the classical neurotransmitters seem to be less prominent (KLINGE and SJÖSTRAND 1977a; SJÖSTRAND et al. 1981). According to present knowledge the excitatory innervation is adrenergic in all mammals, and the neuroeffector junction is characterized by high efficiency in several tissues. The nature and function of cholinergic nerves is less well understood. But as far as muscarinic suppression of excitatory adrenergic neurotransmission in concerned, in vitro studies show great similarity between several species (KLINGE and SJÖSTRAND 1977b; HEDLUND et al. 1984; SAENZ DE TEJADA et al. 1989). The species differences in the transmission of impulses from the NANC inhibitory nerves have so far not been widely studied. There is, however, one approach which suggests that here the differences are exceeded in importance by the similarities, i.e., the frequency–response relationships obtained on stimulation of these nerves. In vitro a characteristic common to all species is that marked or maximal relaxation is obtained with low frequencies (KLINGE and SJÖSTRAND 1977a), and the situation seems to be similar in anaesthetized animals (SJÖSTRAND and KLINGE 1979a; VARDI and SIROKY 1993; SJÖSTRAND et al. 1993). Also the experiments performed so far with inhibitors of the nitric oxide synthase have revealed similarities at the neurochemical level. A further characterization of this enzyme from different species and of other factors regulating the release of nitric oxide will probably deepen our insight into the extent of species differences in the inhibitory NANC neurotransmission.

Acknowledgements. The authors gratefully acknowledge the financial support of the Finnish Cultural Foundation and the Swedish Medical Research Council project 14X-07918.

References

Adaikan PG, Karim SMM (1978) Effects of electrical stimulation and drugs on the smooth muscle of the human penis (Abstr 1228). In: Boissier JR, Lechat P (eds) Proceedings of the 7th International Congress of Pharmacology. Pergamon, Oxford, p 474

Adaikan PG, Karim SMM (1981) Adrenoreceptors in the human penis. J Auton Pharmacol 1:199–203

Adaikan PG, Karim SMM, Kottegoda SR, Ratnam SS (1983) Cholinoreceptors in the corpus cavernosum muscle of the human penis. J Auton Pharmacol 3:107–111

Adaikan PG, Kottegoda SR, Ratnam SS (1986) Is vasoactive intestinal polypeptide the principal transmitter involved in human penile erection? J Urol 135:638–640

Adaikan PG, Lau LC, Ng SC, Ratnam SS (1991) Physio-pharmacology of human penile erection – Autonomic/nitrergic neurotransmissions and receptors of the human corpus cavernosum. Asian Pacific J Pharmacol 6:213–217

Alaranta S, Uusitalo H, Klinge E, Palkama A, Sjöstrand NO (1989) Histochemical demonstration of nerve cell bodies in the retractor penis muscle and the penile artery of the bull. Neuroscience 32:823–827

Alaranta S, Klinge E, Pätsi T, Sjöstrand NO (1990) Inhibition of nicotine-induced relaxation of the bovine retractor penis muscle by compounds known to have ganglion-blocking properties. Br J Pharmacol 101:472–476

Alm P, Larsson B, Ekblad E, Sundler F, Andersson K-E (1993) Immunohistochemical localization of peripheral nitric oxide synthase containing nerves using antibodies raised against synthetized C- and N-terminal fragments of a cloned enzyme from rat brain. Acta Physiol Scand 148:421–429

Ambache N, Killick SW (1978) Species differences in postganglionic motor transmission on the retractor penis muscle. Br J Pharmacol 63:25–34

Ambache N, Killick SW, Aboo Zar M (1975) Extraction from ox retractor penis of an inhibitory substance which mimics its atropine-resistant neurogenic relaxation. Br J Pharmacol 54:409–410

Andersson K-E (1993) Pharmacology of lower urinary tract smooth muscles and penile erectile tissues. Pharmacol Rev 45:253–308

Andersson K-E, Holmquist F (1990) Mechanisms for contraction and relaxation of human penile smooth muscle. Int J Impotence Res 2:209–225

Andersson K-E, Hedlund H, Mattiasson A, Sjögren C, Sundler F (1983) Relaxation of isolated human corpus spongiosum induced by vasoactive intestinal polypeptide, substance P, carbachol and electrical field stimulation. World J Urol 1:203–208

Andersson P-O, Bloom SR, Mellander S (1984) Haemodynamics of pelvic nerve induced penile erection in the dog: possible mediation by vasoactive intestinal polypeptide. J Physiol (Lond) 350:209–224

Andersson P-O, Björnberg J, Bloom SR, Mellander S (1987) Vasoactive intestinal polypeptide in relation to penile erection in the cat evoked by pelvic and by hypogastric nerve stimulation. J Urol 138:419–422

Angus JA, Cocks TM (1987) The half-life of endothelium-derived relaxing factor released from bovine aortic endothelial cells in culture. J Physiol (Lond) 388: 71–81
Arver S, Sjöstrand NO (1982) Functions of adrenergic and cholinergic nerves in canine effectors of seminal emission. Acta Physiol Scand 115:67–77
Azadzoi KM, Kim N, Goldstein J, Krane RJ, Saenz de Tejada I (1990) The role of endothelium in the control of corpus cavernosum smooth muscle tone. Int J Impotence Res 2 [Suppl 2]:17–18
Bacq ZM (1935) Recherches sur la pharmacologie et la physiologie du système nerveux autonome, XII. Nature cholinergique et adrénergique des diverses innervations vasomotrices du pénis chez le chien. Arch Int Physiol 40:311–321
Bartfai T, Iverfeldt K, Fisone G (1988) Regulation of the release of coexisting neurotransmitters. Ann Rev Pharmacol Toxicol 28:285–310
Baumgarten HG, Falck B, Lange W (1969) Adrenergic nerves in the corpora cavernosa penis of some mammals. Z Zellforsch 95:58–67
Beckett SD (1983) Circulation to male reproductive organs. In: Shepherd JT, Abboud FM (eds) The cardiovascular system. American Physiological Society, Bethesda, pp 271–283 (Handbook of physiology, vol 3, sect 2)
Bell C, McLean JR (1970) The distribution of cholinergic and adrenergic nerve fibres in the retractor penis and vas deferens of the dog. Z Zellforsch 106:516–522
Benson GS (1983) Penile erection: In search of a neurotransmitter. World J Urol 1:209–212
Benson GS, McConnell J, Lipshultz LJ, Corriere JN (1980) Neuromorphology and neuropharmacology of the human penis. J Clin Invest 65:506–513
Bessou P, Laporte Y (1961) Fibres vasodilatrices cholinergiques innervant le pénis, contenues dans les nerfs hypogastriques, chez le chat. C R Soc Biol (Paris) 155:142–147
Bhargava KN, Prakash R, Bhargava G (1974) Cholinergic and adrenergic innervation of the guinea-pig penis. Mikroskopie 30:262–269
Blanco R, Saenz de Tejada I, Goldstein I, Krane RJ, Wotiz HH, Cohen RA (1988) Cholinergic neurotransmission in human corpus cavernosum. II. Acetylcholine synthesis. Am J Physiol 254:H468–H472
Blanco R, Saenz de Tejada I, Goldstein I, Krane RJ, Wotiz HH, Cohen RA (1990) Dysfunctional penile cholinergic nerves in diabetic impotent men. J Urol 144: 278–280
Bowman A, Drummond AH (1984) Cyclic GMP mediates neurogenic relaxation in the bovine retractor penis muscle. Br J Pharmacol 81:665–674
Bowman A, Drummond AH (1985) The role of cyclic GMP in neurogenic relaxation. In: Kalsner S (ed) Trends in autonomic pharmacology, vol 3. Taylor and Francis, Philadelphia, pp 319–330
Bowman A, Gillespie JS (1981a) An erythrocyte-associated antagonist of inhibitory mechanisms in the bovine retractor penis muscle. Br J Pharmacol 74:181P
Bowman A, Gillespie JS (1981b) Differential blockade of non-adrenergic inhibitory mechanisms in bovine retractor penis and guinea-pig taenia caeci. J Physiol (Lond) 317:92P–93P
Bowman A, Gillespie JS (1982) Block of some non-adrenergic inhibitory responses of smooth muscle by a substance from haemolysed erythrocytes. J Physiol (Lond) 328:11–25
Bowman A, Gillespie JS (1983) Neurogenic vasodilatation in isolated bovine and canine penile arteries. J Physiol (Lond) 341:603–616
Bowman A, McGrath JC (1985) The effect of hypoxia on neuroeffector transmission in the bovine retractor penis and rat anococcygeus muscles. Br J Pharmacol 85:869–875
Bowman A, Gillespie JS, Martin W (1979) The inhibitory material in extracts from the bovine retractor penis muscle is not an adenine nucleotide. Br J Pharmacol 67:327–328

Bowman A, Gillespie JS, Martin W (1981) Actions on the cardiovascular system of an inhibitory material extracted from the bovine retractor penis. Br J Pharmacol 72:365–372
Bowman A, Gillespie JS, Pollock D (1982) Oxyhaemoglobin blocks non-adrenergic non-cholinergic inhibition in the bovine retractor penis muscle. Eur J Pharmacol 85:221–224
Bredt DS, Snyder SH (1990) Isolation of nitric oxide synthetase, a calmodulin-requiring enzyme. Proc Natl Acad Sci USA 87:682–685
Bredt DS, Hwang PM, Snyder SH (1990) Localization of nitric oxide synthase indicating a neural role for nitric oxide. Nature 347:768–770
Bredt DS, Hwang PM, Glatt CE, Lowenstein C, Reed RR, Snyder SH (1991) Cloned and expressed nitric oxide synthase structurally resembles cytochrome P-450 reductase. Nature 351:714–718
Brindley GS (1986) Pilot experiments on the action of drugs injected into the human corpus cavernosum penis. Br J Pharmacol 87:495–500
Brock G, Breza J, Lue TF (1993) Intracavernous sodium nitroprusside: inappropriate impotence treatment. J Urol 150:864–867
Burnett AL, Lowenstein CJ, Bredt DS, Chang TSK, Snyder SH (1992) Nitric oxide: a physiologic mediator of penile erection. Science 257:401–403
Burnett AL, Tillman SL, Chang TSK, Epstein JI, Lowenstein CJ, Bredt DS, Snyder SH, Walsh PC (1993) Immunohistochemical localization of nitric oxide synthase in the autonomic innervation of the human penis. J Urol 150:73–76
Burnstock G (1986) Purines as cotransmitters in adrenergic and cholinergic neurons. Prog Brain Res 68:193–203
Bush PA, Gonzales NE, Ignarro LJ (1992a) Biosynthesis of nitric oxide and citrulline from L-arginine by constitutive nitric oxide synthase present in rabbit corpus cavernosum. Biochem Biophys Res Commun 186:308–314
Bush PA, Aronson WJ, Buga GM, Rajfer J, Ignarro LJ (1992b) Nitric oxide as a potent relaxant of human and rabbit corpus cavernosum. J Urol 147:1650–1655
Byrne NG, Muir TC (1984) Electrical and mechanical responses of the bovine retractor penis to nerve stimulation and drugs. J Auton Pharmacol 4:261–271
Byrne NG, Muir TC (1985) Mechanisms underlying electrical and mechanical responses of the bovine retractor penis to inhibitory nerve stimulation and to an inhibitory extract. Br J Pharmacol 85:149–161
Carati CJ, Goldie RG, Warton A, Henry PJ (1985) Pharmacology of the erectile tissue of the canine penis. Pharmacol Res Commun 17:951–966
Carati CJ, Creed KE, Keogh EJ (1987) Autonomic control of penile erection in the dog. J Physiol (Lond) 384:525–538
Chapelle P-A, Blanquart F, Puech AJ, Held J-P (1983) Treatment of anejaculation in the total paraplegic by subcutaneous injection of physostigmine. Paraplegia 21:30–36
Christ GJ, Maayani S, Valcic M, Melman A (1990) Pharmacological studies of human erectile tissue: characteristics of spontaneous contractions and alterations in α-adrenoceptor responsiveness with age and disease in isolated tissues. Br J Pharmacol 101:375–381
Christ GJ, Schwartz CB, Stone BA, Parker M, Janis M, Gondre M, Valcic M, Melman A (1992) Kinetic characteristics of α_1-adrenergic contractions in human corpus cavernosum smooth muscle. Am J Physiol 263:H15–H19
Connaughton S, Docherty JR (1990) Functional evidence for heterogeneity of peripheral prejunctional alpha$_2$-adrenoceptors. Br J Pharmacol 101:285–290
Dail WG (1987) Autonomic control of penile erectile tissue. Exp Brain Res Ser 16:340–344
Dail WG (1993) Autonomic innervation of male reproductive genitalia. In: Maggi CA (ed) The autonomic nervous system, vol 3. Harwood, London, pp 69–102
Dail WG, Hamill RW (1989) Parasympathetic nerves in penile erectile tissue of the rat contain choline acetyltransferase. Brain Res 487:165–170

Dail WG, Moll MA, Weber K (1983) Localization of vasoactive intestinal polypeptide in penile erectile tissue and in the major pelvic ganglion of the rat. Neuroscience 10:1379–1386
Dail WG, Minorsky N, Moll MA, Manzanares K (1986) The hypogastric nerve pathway to penile erectile tissue: histochemical evidence supporting a vasodilator role. J Auton Nerv Syst 15:341–349
Dail WG, McGuffee L, Minorsky N, Little S (1987) Responses of smooth muscle strips from penile erectile tissue to drugs and transmural nerve stimulation. J Auton Pharmacol 7:287–293
Dail WG, Walton G, Olmsted MP (1989) Penile erection in the rat: Stimulation of the hypogastric nerve elicits increases in penile pressure after chronic interruption of the sacral parasympathetic outflow. J Auton Nerv Syst 28:251–258
Dail WG, Carillo Y, Walton G (1990) Innervation of the anococcygeus muscle of the rat. Cell Tissue Res 259:139–146
Dail WG, Galloway B, Bordegaray J (1993) NADPH diaphorase innervation of the rat anococcygeus and retractor penis muscles. Neurosci Lett 160:17–20
Dalsgaard C-J, Elfvin L-G (1979) Spinal origin of preganglionic fibers projecting onto the superior cervical ganglion and inferior mesenteric ganglion of the guinea pig, as demonstrated by the horseradish peroxidase technique. Brain Res 172:139–143
De Groat WC, Booth AM (1993) Neural control of penile erection. In: Maggi CA (ed) The autonomic nervous system, vol 3. Harwood, London, pp 467–523
De Groat WC, Steers WD (1988) Neuroanatomy and neurophysiology of penile erection. In: Tanagho EA, Lue TF, McClure RD (eds) Contemporary management of impotence and infertility. Williams and Wilkins, Baltimore, pp 3–27
De Groat WC, Steers WD (1990) Autonomic regulation of the urinary bladder and sexual organs. In: Loewy AD, Spyer KM (eds) Central regulation of autonomic functions. Oxford University Press, Oxford, pp 310–333
Dhabuwala CB, Ramakrishna CV, Anderson GF (1985) Beta adrenergic receptors in human cavernous tissue. J Urol 133:721–723
Diederichs W, Stief CG, Lue TF, Tanagho EA (1990) Norepinephrine involvement in penile detumescence. J Urol 143:1264–1266
Domer FR, Wessler G, Brown RL, Charles HC (1978) Involvement of the sympathetic nervous system in the urinary bladder internal sphincter and in penile erection in the anesthetized cat. Invest Urol 15:404–407
Dorr LD, Brody MJ (1967) Hemodynamic mechanisms of erection in the canine penis. Am J Physiol 213:1526–1531
Eadie MJ (1987) Centrally acting anticholinergics. In: Speight TM (ed) Avery's drug treatment: principles and practice of clinical pharmacology and therapeutics. Adis, Auckland, p 1111
Ellis JL, Burnstock G (1990) Neuropeptide Y neuromodulation of sympathetic co-transmission in the guinea pig vas deferens. Br J Pharmacol 100:457–462
Eltze M (1988) Muscarinic M_1- and M_2-receptors mediating opposite effects on neuromuscular transmission in rabbit vas deferens. Eur J Pharmacol 151:205–221
Eränkö O, Klinge E, Sjöstrand NO (1976) Different types of synaptic vesicles in axons of the retractor penis muscle of the bull. Experientia 32:1135–1137
Erina EV (1980) Ganglion-blocking agents in internal medicine. In: Kharkevich DA (ed) Pharmacology of ganglionic transmission. Springer, Berlin Heidelberg New York. pp 417–438 (Handbook of experimental pharmacology, vol 53)
Eva C, Meek JL, Costa E (1985) Vasoactive intestinal peptide which coexists with acetylcholine decreases acetylcholine turnover in mouse salivary glands. J Pharmacol Exp Ther 232:670–674
Fehler E, Burnstock G (1987) Ultrastructural identification of vasoactive intestinal polypeptide and neuropeptide Y-containing nerve fibres in the vas deferens of the guinea-pig. J Auton Nerv Syst 19:235–242
Finberg JPM, Vardi Y (1991) Evidence for a role of EDRF in penile erection. Br J Pharmacol 102:82P

Finberg JPM, Levy S, Vardi Y (1993) Inhibition of nerve stimulation-induced vasodilatation in corpora cavernosa of the pithed rat by blockade of nitric oxide synthase. Br J Pharmacol 108:1038–1042
Fontaine J, Grivegnee AR, Robberecht P (1986a) Evidence against VIP as the inhibitory transmitter in non-adrenergic, non-cholinergic nerves supplying the longitudinal muscle of the mouse colon. Br J Pharmacol 89:599–602
Fontaine J, Schulman CC, Wespes E (1986b) Postjunctional α_1- and α_2-like adrenoceptors in human isolated deep dorsal vein of the penis. Br J Pharmacol 89:493P
Förstermann U, Pollock JS, Schmidt HHHW, Heller M, Murad F (1991) Calmodulin-dependent endothelium-derived relaxing factor/nitric oxide synthase activity is present in the particulate and cytosolic fractions of bovine aortic endothelial cells. Proc Natl Acad Sci USA 88:1788–1792
Fried G, Terenius L, Hökfelt T, Goldstein M (1985) Evidence for differential localization of noradrenaline and neuropeptide Y in neuronal storage vesicles isolated from rat vas deferens. J Neurosci 5:450–458
Fukushi Y, Wakui M (1986a) Possible interaction of cholinergic nerves with two different (pre and post) sites of the neuromuscular junction in guinea-pig vas deferens. J Auton Pharmacol 6:291–297
Fukushi Y, Wakui M (1986b) Pharmacological studies on the role of cholinergic nerves in the neuromuscular transmission in the circular smooth muscle of guinea-pig vas deferens. J Auton Pharmacol 6:299–304
Furchgott RF (1984) The role of endothelium in the responses of vascular smooth muscle to drugs. Annu Rev Pharmacol Toxicol 24:175–197
Furchgott RF (1988) Studies on relaxation of rabbit aorta by sodium nitrite: the basis for the proposal that the acid activated inhibitory factor from bovine retractor penis is inorganic nitrite and the endothelium-derived relaxing factor is nitric oxide. In: Vanhoutte PM, Lausen J (eds) Vascular smooth muscle, peptides, autonomic nerves and endothelium. Raven, New York, pp 401–414 (Vasodilatation, vol 4)
Furchgott RF (1990) Studies on endothelium-dependent vasodilation and the endothelium-derived relaxing factor (The 1989 Ulf von Euler Lecture). Acta Physiol Scand 139:257–270
Furchgott RF, Zawadski JV (1980) The obligatory role of endothelial cells on the relaxation of arterial smooth muscle by acetylcholine. Nature 288:373–376
Gillespie JS, Hunter JC (1982) Subcellular localization of smooth muscle inhibitory factor isolated from the bovine retractor penis muscle. Br J Pharmacol 76:189P
Gillespie JS, Martin W (1978) A smooth muscle inhibitory material extracted from the bovine retractor penis and rat anococcygeus muscles. J Physiol (Lond) 280:45P–46P
Gillespie JS, Martin W (1980) A smooth muscle inhibitory material from the bovine retractor penis and rat anococcygeus muscles. J Physiol (Lond) 309:55–64
Gillespie JS, Xiaorong L (1989) The effect of arginine and L-N-monomethyl arginine on the response of the bovine retractor penis to stimulation of its NANC nerves. Br J Pharmacol 97:453P
Gillespie JS, Hunter JC, Martin W (1981) Some physical and chemical properties of the smooth muscle inhibitory factor in extracts of the bovine retractor penis muscle. J Physiol (Lond) 315:111–125
Gillespie JS, Liu X, Martin W (1989) The effects of L-arginine and N^G-monomethyl L-arginine on the response of the rat anococcygeus muscle to NANC nerve stimulation. Br J Pharmacol 98:1080–1082
Gillespie JS, Liu X, Martin W (1990) The neurotransmitter of the non-adrenergic non-cholinergic inhibitory nerves to smooth muscle of the genital system. In: Moncada S, Higgs EA (eds) Nitric oxide from L-arginine: a bioregulatory system. Excerpta Medica, Amsterdam, pp 147–164
Goldenberg MM (1965) Pharmacologic analysis of atropine-fast nicotine responses of the retractor penis and urinary bladder of the cat. Pharmacologist 7:158

Grieten J, Gerebtzoff M-A (1957) Les cholinestérases dans l'appareil uro-génital. Ann Histochim 2:127–140
Griffith TM, Edwards DH, Lewis MJ, Newby AC, Henderson AH (1984) The nature of endothelium-derived vascular relaxant factor. Nature 308:645–647
Gu J, Polak JM, Probert L, Islam KN, Marangos PJ, Mina S, Adrian TE, McGregor GP, O'Shaughnessy DJ, Bloom SR (1983) Peptidergic innervation of the human male genital tract. J Urol 130:386–391
Gu J, Lazarides M, Pryor JP, Blank MA, Polak JM, Morgan R, Marangos PJ, Bloom SR (1984) Decrease of vasoactive intestinal polypeptide (VIP) in the penises from impotent men. Lancet ii:315–318
Hammarström M, Sjöstrand NO (1984a) Intimacy of the neuroeffector junction and resistance to alpha-adrenoceptor-blockade of the neurogenic contractile response in vasa deferentia from guinea pig and rat. Acta Physiol Scand 122:465–474
Hammarström M, Sjöstrand NO (1984b) Comments on the atropine resistance of the neurogenic contractile response of the rat detrusor muscle. Acta Physiol Scand 122:475–481
Heaton JPW, Morales A, Owen J, Saunders FW, Fenemore J (1990) Topical glyceryltrinitrate causes measurable penile arterial dilation in impotent men. J Urol 143:729–731
Hecker M, Mitchell JA, Harris HJ, Katsura M, Thiemermann C, Vane JR (1990) N^G-monomethyl-L-arginine but not N^ω-nitro-L-arginine is metabolized by endothelial cells to L-citrulline. Eur J Pharmacol 183:648–649
Hedlund H, Andersson K-E (1985a) Comparison of the responses to drugs acting on adrenoreceptors and muscarinic receptors in human isolated corpus cavernosum and cavernous artery. J Auton Pharmacol 5:81–88
Hedlund H, Andersson K-E (1985b) Effects of some peptides on isolated human penile erectile tissue and cavernous artery. Acta Physiol Scand 124:413–419
Hedlund H, Andersson K-E, Mattiasson A (1984) Pre- and postjunctional adreno- and muscarinic receptor functions in the isolated human corpus spongiosum urethrae. J Auton Pharmacol 4:241–249
Henderson VE, Roepke MH (1933) On the mechanism of erection. Am J Physiol 106:441–448
Hibbs JB, Vavrin Z, Taintor RR (1987) L-Arginine is required for expression of the activated macrophage effector mechanism causing selective metabolic inhibition in target cells. J Immunol 138:550–565
Hirst GDS, Neild TO (1980) Some properties of spontaneous excitatory junction potentials recorded from arterioles of guinea-pig. J Physiol (Lond) 303:43–60
Hobbs AJ, Gibson A (1990) L-N^G-nitro-arginine and its methyl ester are potent inhibitors of non-adrenergic, non-cholinergic transmission in the rat anococcygeus. Br J Pharmacol 100:749–752
Holman ME (1987) The gamma controversy. Clin Exp Pharmacol Physiol 14:415–422
Holmquist F, Andersson K-E, Hedlund H (1990a) Actions of endothelin on isolated corpus cavernosum from rabbit and man. Acta Physiol Scand 139:113–122
Holmquist F, Hedlund H, Andersson K-E (1990b) Effects of the α_1-adrenoceptor antagonist R-(−)-YM12617 on isolated human penile erectile tissue and vas deferens. Eur J Pharmacol 186:87–93
Holmquist F, Hedlund H, Andersson K-E (1991a) L-N^G-Nitro-arginine inhibits non-adrenergic, non-cholinergic relaxation of human isolated corpus cavernosum. Acta Physiol Scand 141:441–442
Holmquist F, Stief CG, Jonas U, Andersson K-E (1991b) Effects of the nitric oxide synthase inhibitor N^G-nitro-L-arginine on the erectile response to cavernous nerve stimulation in the rabbit. Acta Physiol Scand 143:299–304
Holmquist F, Hedlund H, Andersson K-E (1992) Characterization of inhibitory neurotransmission in the isolated corpus cavernosum from rabbit and man. J Physiol (Lond) 449:295–311

Hope BT, Michael GJ, Knigge KM, Vincent SR (1991) Neuronal NADPH diaphorase is a nitric oxide synthase. Proc Natl Acad Sci USA 88:2811–2814
Hukovic S, Bubic J (1967) Coeur et vaisseaux sanguins isolés avec leurs nerfs comme moyen de reserche pharmacologique. Pathol Biol (Paris) 15:153–157
Ignarro LJ, Buga GM, Wood KS, Byrns RE, Chaudhuri G (1987) Endothelium-derived relaxing factor produced and released from artery and vein is nitric oxide. Proc Natl Acad Sci USA 84:9265–9269
Ignarro LJ, Bush PA, Buga GM, Wood KS, Fukoto JM, Rajfer J (1990) Nitric oxide and cyclic GMP formation upon electrical field stimulation cause relaxation of corpus cavernosum smooth muscle. Biochem Biophys Res Commun 170:843–850
Iyengar R, Stuehr DJ, Marletta MA (1987) Macrophage synthesis of nitrite, nitrate, and N-nitrosamines: precursors and role of the respiratory burst. Proc Natl Acad Sci USA 84:6369–6373
Jeremy JY, Morgan RJ, Mikhailidis DP, Dandona P (1986) Prostacyclin synthesis by the corpora cavernosa of the human penis: evidence for muscarinic control and pathological implications. Prostaglandins Leukotrienes Med 23:211–216
Jünemann K-P, Lue TF, Luo J-A, Jadallah SA, Nunes LL, Tanagho EA (1987) The role of vasoactive intestinal polypeptide as a neurotransmitter in canine penile erection: a combined in vivo and immunohistochemical study. J Urol 138: 871–877
Kamata K, Sakamoto A, Kasuya Y (1988) Similarities between the relaxations induced by vasoactive intestinal peptide and by stimulation of the non-adrenergic non-cholinergic neurons in the rat stomach. Naunyn Schmiedebergs Arch Pharmacol 338:401–406
Kawanishi Y, Hashine K, Kimura K, Tamura M, Imagawa A (1990) Interaction of VIP and acetylcholine on the relaxation of human corpus cavernosum tissue. Int J Impotence Res 2 [Suppl 1]:9–11
Keast JR (1992) A possible neural source of nitric oxide in the rat penis. Neurosci Lett 143:69–73
Kerr SW, Buchanan LV, Bunting S, Mathews WR (1992) Evidence that S-nitrosothiols are responsible for the smooth muscle relaxing activity of the bovine retractor penis inhibitory factor. J Pharmacol Exp Ther 263:285–292
Kiely EA, Blank MA, Bloom SR, Williams G (1987) Studies on intracavernosal VIP levels during pharmacologically induced penile erections. Br J Urol 59:334–339
Kiely EA, Bloom SR, Williams G (1989) Penile response to intracavernosal vasoactive intestinal polypeptide alone and in combination with other vasoactive agents. Br J Urol 64:191–194
Kim NN, Goldstein I, Krane RJ, Saenz de Tejada I (1990) Neurogenic relaxation of penile smooth muscle. J Urol 143:244A
Kim NN, Azadzoi KM, Goldstein I, Saenz de Tejada I (1991) A nitric oxide-like factor mediates nonadrenergic-noncholinergic neurogenic relaxation of penile corpus cavernosum smooth muscle. J Clin Invest 88:112–118
Kim N, Vardi Y, Padma-Nathan H, Daley J, Goldstein I, Saenz de Tejada I (1993) Oxygen tension regulates the nitric oxide pathway. J Clin Invest 91:437–442
Kimoto Y, Ito Y (1987) Autonomic innervation of the canine penile artery and vein in relation to neural mechanisms involved in erection. Br J Urol 59:463–472
Kimoto Y, Ito Y (1988) Functional innervation patterns in the corpus spongiosum and glans in the dog. Br J Urol 62:597–602
Kimura K, Kawanishi Y, Tamura M, Imagawa A (1989) Assessment of the alpha-adrenergic receptors in isolated human and canine corpus cavernosum tissue. Int J Impotence Res 1:189–195
Kinekawa F, Komori S, Ohashi H (1984) Cholinergic inhibition of adrenergic transmission in the dog retractor penis muscle. Jpn J Pharmacol 34:343–352
Kirkeby HJ, Forman A, Sörensen S, Andersson K-E (1989) Alpha-adrenoceptor function in isolated penile circumflex veins from potent and impotent men. J Urol 142:1369–1371

Kirkeby HJ, Fahrenkrug J, Holmquist F, Ottesen B (1992) Vasoactive intestinal polypeptide (VIP) and peptide histidine methionine (PHM) in human penile corpus cavernosum tissue and circumflex veins: localization and in vitro effects. Eur J Clin Invest 21:24–30
Klinge E (1970) The effect of some substances on the isolated bull retractor penis muscle. Acta Physiol Scand 78:280–288
Klinge E, Penttilä O (1969) Distribution of noradrenaline and acetylcholinesterase in bull and rabbit penile erectile tissue. Ann Med Exp Fenn 47:17–21
Klinge E, Pohto P (1971) Innervation of the bull retractor penis muscle and peripheral autonomic mechamism of erection. Prog Brain Res 34:415–421
Klinge E, Sjöstrand NO (1974) Contraction and relaxation of the retractor penis muscle and the penile artery of the bull. Acta Physiol Scand 93 [Suppl 420]:1–88
Klinge E, Sjöstrand NO (1977a) Comparative study of some isolated mammalian smooth muscle effectors of penile erection. Acta Physiol Scand 100:354–367
Klinge E, Sjöstrand NO (1977b) Suppression of the excitatory adrenergic neurotransmission; a possible role of cholinergic nerves in the retractor penis muscle. Acta Physiol Scand 100:368–376
Klinge E, Sjöstrand NO (1982) Cholinergic suppression of adrenergic neurotransmission in canine corpus cavernosum urethrae. Br J Pharmacol 77:404P
Klinge E, Pohto P, Solatunturi E (1970) Adrenergic innervation and structure of the bull retractor penis muscle. Acta Physiol Scand 78:110–116
Klinge E, Eränkö O, Sjöstrand NO (1978) Cholinergic and adrenergic innervation of the penis artery of the bull: transmitter concentrations and synaptic vesicles. Experientia 34:1624–1626
Klinge E, Alaranta S, Sjöstrand NO (1988) Pharmacological analysis of nicotinic relaxation of the bovine retractor penis muscle. J Pharmacol Exp Ther 245:280–286
Klinge E, Alaranta S, Nissinen R, Sjöstrand NO (1992) Inhibition of nicotine-induced relaxation of the bovine retractor penis muscle by β-adrenoceptor antagonists. Pharmacol Res 25:353–361
Klinge E, Alaranta S, Parkkisenniemi UM, Kostiainen E, Sjöstrand NO (1993) The use of the bovine retractor penis muscle for the assessment of ganglion-blocking activity of neuromuscular blocking and other drugs. J Pharmacol Toxicol Methods 30:197–202
Knispel HH, Goessl C, Beckmann R (1991) Basal and acetylcholine-stimulated nitric oxide formation mediates relaxation of rabbit cavernous smooth muscle. J Urol 146:1429–1433
Knispel HH, Goessl C, Beckmann R (1992) Nitric oxide mediates relaxation in rabbit and human corpus cavernosum smooth muscle. Urol Res 20:253–257
Kostiainen E, Klinge E, Alaranta S, Sjöstrand NO (1991) Inhibition of nicotine-induced relaxation of the bovine retractor penis muscle by nitric oxide synthase inhibitors. Br J Pharmacol 104:396P
Krane RJ, Siroky MB (1981) Neurophysiology of erection. Urol Clin North Am 8:91–102
Langley JN, Anderson HK (1895) The innervation of the pelvic and adjoining viscera. Parts III–V. J Physiol (Lond) 19:85–139
Larsen J-J, Ottesen B, Fahrenkrug J, Fahrenkrug L (1981) Vasoactive intestinal polypeptide (VIP) in the male genitourinary tract. Invest Urol 19:211–213
Leone AM, Palmer RMJ, Knowles RG, Francis PL, Ashton DS, Moncada S (1991) Constitutive and inducible nitric oxide synthases incorporate molecular oxygen into both nitric oxide and citrulline. J Biol Chem 266:23790–23795
Levin RM, Wein AJ (1980) Adrenergic alpha receptors outnumber beta receptors in human penile corpus cavernosum. Invest Urol 18:225–226
Lincoln J, Crowe R, Blacklay PF, Pryor JP, Lumley JSP, Burnstock G (1987) Changes in the vipergic, cholinergic and adrenergic innervation of human penile tissue in diabetic and non-diabetic impotent males. J Urol 137:1053–1059

Liu X, Gillespie JS, Gibson IF, Martin W (1991) Effects of N^G-substituted analogues of L-arginine on NANC relaxation of the rat anococcygeus and bovine retractor penis muscles and the bovine penile artery. Br J Pharmacol 104:53–58
Lundberg JM, Hökfelt T (1986) Multiple co-existence of peptides and classical transmitters in peripheral autonomic and sensory neurons – functional and pharmacological implications. Prog Brain Res 68:241–262
Lundberg JM, Hedlund B, Bartfai T (1982) Vasoactive intestinal polypeptide enhances muscarinic ligand binding in cat submandibular salivary gland. Nature 295:147–149
Marletta MA, Yoon PS, Iyengar R, Leaf CD, Wishnok JS (1988) Macrophage oxidation of L-arginine to nitrite and nitrate: nitric oxide is an intermediate. Biochemistry 27:8706–8711
Martin W, Gillespie JS (1991) L-Arginine-derived nitric oxide: the basis of inhibitory transmission in the anococcygeus and retractor penis muscles. In: Bell C (ed) Novel peripheral neurotransmitters. Pergamon, Oxford, pp 65–79
Martin W, Gillespie JS, Gibson IF (1993) Actions and interactions of N^G-substituted analogues of L-arginine on NANC neurotransmission in the bovine retractor penis and rat anococcygeus muscles. Br J Pharmacol 108:242–247
Martin W, Smith JA, Lewis MJ, Henderson AH (1988) Evidence that inhibitory factor extracted from bovine retractor penis is nitrite, whose acid-activated derivative is stabilized nitric oxide. Br J Pharmacol 93:579–586
Martin W, Gillespie JS, Liu X, Gibson IF (1991) Effects of N^G-substituted analogues of L-arginine on NANC relaxation of the anococcygeus, retractor penis and penile artery. Br J Pharmacol 102:83P
McConnell J, Benson GS (1982) Innervation of human penile blood vessels. Neurol Urodynam 1:199–210
McConnell J, Benson GS, Wood J (1979) Autonomic innervation of the mammalian penis: a histochemical and physiological study. J Neural Transm 45:227–238
Melman A, Henry DP, Felten DL, O'Connor BL (1980) Alteration of the penile corpora in patients with erectile impotence. Invest Urol 17:474–477
Meyhoff HH, Rosenkilde P, Bodker A (1992) Non-invasive management of impotence with transcutaneous nitroglycerin. Br J Urol 69:88–90
Molderings GJ, Göthert M, van Ahlen H, Porst H (1989) Noradrenaline release in human corpus cavernosum and its modulation via presynaptic α_2-adrenoceptors. Fundam Clin Pharmacol 3:497–504
Moore PK, al-Swayeh OA, Chong NWS, Evans RA, Gibson A (1990) L-N^G-Nitro arginine (L-NOARG), a novel, L-arginine-reversible inhibitor of endothelium-dependent vasodilatation in vitro. Br J Pharmacol 99:408–412
Mülsch A, Busse R (1990) N^G-Nitro-L-arginine (N^5-[imino(nitroamino)methyl]-L-ornithine) impairs endothelium-dependent dilations by inhibiting cytosolic nitric oxide synthesis from L-arginine. Naunyn Schmiedebergs Arch Pharmacol 341: 143–147
Muscholl E (1979) Presynaptic muscarine receptors and inhibition of release. In: Paton DM (ed) The release of catecholamines from adrenergic neurons. Pergamon, Oxford, pp 87–109
Myers PR, Minor RL, Guerra R, Bates JN, Harrison DG (1990) Vasorelaxant properties of the endothelium-derived relaxing factor more closely resemble S-nitrosocysteine than nitric oxide. Nature 345:161–163
Ottesen B, Wagner G, Virag R, Fahrenkrug J (1984) Penile erection: possible role for vasoactive intestinal polypeptide as a neurotransmitter. Br Med J 288: 9–11
Ottesen B, Wagner G, Fahrenkrug J (1988) Peptidergic innervation of the sexual organs. In: Sitsen JMA (ed) The pharmacology and endocrinology of sexual function. Elsevier, Amsterdam, pp 66–97 (Handbook of sexology, vol 6)
Palacios M, Knowles RG, Palmer RMJ, Moncada S (1989) Nitric oxide from L-arginine stimulates the soluble guanylate cyclase in adrenal glands. Biochem Biophys Res Commun 165:802–809

Palmer RMJ, Ferrige AG, Moncada S (1987) Nitric oxide release accounts for the biological activity of endothelium-derived relaxing factor. Nature 327:524–526
Palmer RMJ, Rees DD, Ashton DS, Moncada S (1988) L-Arginine is the physiological precursor for the formation of nitric oxide in endothelium-dependent relaxation. Biochem Biophys Res Commun 153:1251–1256
Panula P, Alaranta S, Klinge E, Sjöstrand NO (1991) Neuropeptides in bovine retractor penis muscle and penile artery. Acta Pharm Fenn 100:117
Penttilä O (1966) Acetylcholine, biogenic amines and enzymes involved in their metabolism in penile erectile tissue. Ann Med Exp Fenn 44 [Suppl 9]:1–42
Penttilä O, Vartiainen A (1964) Acetylcholine, histamine, 5-hydroxytryptamine and catecholamine contents of mammalian penile erectile and urethral tissue. Acta Pharmacol (Copenh) 21:145–151
Pickard RS, Powell PH, Zar MA (1991) The effect of inhibitors of nitric oxide biosynthesis and cyclic GMP formation on nerve-evoked relaxation of human cavernosal smooth muscle. Br J Pharmacol 104:755–759
Pickard RS, Powell PH, Zar MA (1993) Evidence against vasoactive intestinal polypeptide as the relaxant neurotransmitter in human cavernosal smooth muscle. Br J Pharmacol 108:497–500
Porst H (1993) Prostaglanding E_1 and the nitric oxide donor linsidomide for erectile failure: a diagnostic comparative study of 40 patients. J Urol 149:1280–1283
Rajfer J, Aronson WJ, Bush PA, Dorey FJ, Ignarro LJ (1992) Nitric oxide as a mediator of relaxation of the corpus cavernosum in response to nonadrenergic, noncholinergic neurotransmission. N Engl J Med 326:90–94
Rees DD, Palmer RMJ, Hodson HF, Moncada S (1989) A specific inhibitor of nitric oxide formation from L-arginine attenuates endothelium-dependent relaxation. Br J Pharmacol 96:418–424
Roy JB, Petrone RL, Said SJ (1990) A clinical trial of vasoactive intestinal peptide to induce penile erection. J Urol 143:302–304
Saenz de Tejada I, Blanco R, Goldstein I, Azadzoi K, de las Morenas A, Krane RJ, Cohen RA (1988) Cholinergic neurotransmission in human corpus cavernosum. I. Responses of isolated tissue. Am J Physiol 254:H459–H467
Saenz de Tejada I, Kim N, Lagan I, Krane RJ, Goldstein I (1989) Regulation of adrenergic activity in penile corpus cavernosum. J Urol 142:1117–1121
Samuelson U, Sjöstrand NO, Klinge E (1983) Correlation between electrical and mechanical activity in myogenic and neurogenic control of the bovine retractor penis muscle. Acta Physiol Scand 119:335–345
Schmalbruch H, Wagner G (1989) Vasoactive intestinal polypeptide (VIP)- and neuropeptide Y (NPY)-containing nerve fibres in the penile cavernous tissue of green monkeys (Cercopithecus aethiops). Cell Tissue Res 256:529–541
Schmidt HHHW, Nau H, Wittfoht W, Gerlach J, Prescher K-E, Klein MM, Niroomand F, Böhme E (1988) Arginine is a physiological precursor of endothelium-derived nitric oxide. Eur J Pharmacol 154:213–216
Schmidt HHHW, Pollock JS, Nakane M, Gorsky LD, Förstermann U, Murad F (1991) Purification of a soluble isoform of guanylyl cyclase-activating-factor synthase. Proc Natl Acad Sci USA 88:365–369
Schoeffter P, Stoclet J-C (1985) Effect of vasoactive intestinal polypeptide (VIP) on cyclic AMP level and relaxation in rat isolated aorta. Eur J Pharmacol 109: 275–279
Segraves RT (1989) Effects of psychotropic drugs on human erection and ejaculation. Arch Gen Psychiatry 46:275–284
Seki N, Suzuki H (1989) Electrical and mechanical activity of rabbit prostate smooth muscles in response to nerve stimulation. J Physiol (Lond) 419:651–663
Sheng H (1991) Effect of LY 83583 on the response to NANC nerve stimulation in rat anococcygeus and bovine retractor penis muscles. Br J Pharmacol 104:137P
Sheng H, Mitchell JA, Nakane M, Warner TD, Pollock JS, Förstermann U, Murad F (1991) Characterisation of NO-synthase from NANC nerve-containing tissues: rat anococcygeus and bovine retractor penis muscles. Br J Pharmacol 104:7P

Sheng H, Schmidt HHHW, Nakane M, Mitchell JA, Pollock JS, Förstermann U, Murad F (1992) Characterization and localization of nitric oxide synthase in non-adrenergic non-cholinergic nerves from bovine retractor penis muscles. Br J Pharmacol 106:768–773
Shirai M, Sasaki K, Rikimaru A (1972) Histochemical investigation on the distribution of adrenergic and cholinergic nerves in human penis. Tohoku J Exp Med 107:403–404
Sjöstrand NO (1961) Effect of some smooth muscle stimulants on the motor response of the isolated guinea pig vas deferens to hypogastric nerve stimulation. Nature 192:1190–1191
Sjöstrand NO (1962) Inhibition by ganglionic blocking agents of the motor response of the isolated guinea-pig vas deferens to hypogastric nerve stimulation. Acta Physiol Scand 54:306–315
Sjöstrand NO (1965) The adrenergic innervation of the vas deferens and the accessory male genital glands. Acta Physiol Scand 65 [Suppl 257]:1–82
Sjöstrand NO (1973a) Effects of acetylcholine and some other smooth muscle stimulants on the electrical and mechanical responses of the guinea-pig vas deferens to nerve stimulation. Acta Physiol Scand 89:1–9
Sjöstrand NO (1973b) Effects of adrenaline, noradrenaline and isoprenaline on the electrical and mechanical responses of the guinea-pig vas deferens to nerve stimulation. Acta Physiol Scand 89:10–18
Sjöstrand NO (1976) Les effecteurs de l'emission du sperme. In: Soulairac A, Gautray JP, Rousseau JP, Cohen J (eds) Système nerveux, activité sexuelle et reproduction. Masson, Paris, pp 263–268
Sjöstrand NO (1981) Smooth muscles of vas deferens and other organs in the male reproductive tract. In: Bülbring E, Brading AF, Jones AW, Tomita T (eds) Smooth muscle: an assessment of current knowledge. Arnold, London, pp 367–376
Sjöstrand NO, Andersson-Forsman C (1991) Ultrastructural changes in early Wallerian degeneration of the neuroeffector system of the rat vas deferens. J Auton Nerv Syst 33:140–141
Sjöstrand NO, Klinge E (1979a) Principal mechanisms controlling penile retraction and protrusion in rabbits. Acta Physiol Scand 106:199–214
Sjöstrand NO, Klinge E (1979b) Changes in penile volume during some cardiovascular reflexes and reactions in rabbit. Acta Physiol Scand 106:327–334
Sjöstrand NO, Klinge E, Himberg J-J (1981) Effects of VIP and other putative neurotransmitters on smooth muscle effectors of penile erection. Acta Physiol Scand 113:403–405
Sjöstrand NO, Eldh J, Alaranta S, Klinge E (1988) Cholinergic mechanisms in human erectile tissue. Acta Physiol Scand 132:19A
Sjöstrand NO, Eldh J, Samuelson UE, Alaranta S, Klinge E (1990) The effects of L-arginine and N^G-monomethyl L-arginine on the inhibitory neurotransmission of the human corpus cavernosum penis. Acta Physiol Scand 140:297–298
Sjöstrand NO, Beckett SD, Klinge E (1993) Nervous regulation of the tone of the retractor penis muscle in the goat. Acta Physiol Scand 147:403–415
Steers WD (1990) Neural control of penile erection. Semin Urol 8:66–79
Steers WD, McConnell J, Benson GS (1984) Anatomical localization and some pharmacological effects of vasoactive intestinal polypeptide in human and monkey corpus cavernosum. J Urol 132:1048–1053
Steers WD, Mallory B, de Groat WC (1988) Electrophysiological study of neural activity in penile nerve of the rat. Am J Physiol 254:R989–R1000
Stief CG, Diederichs W, Benard F, Bosch R, Aboseif S, Lue TF (1989a) Possible role for acetylcholine as a neurotransmitter in canine penile erection. Urol Int 44:357–363
Stief CG, Benard F, Bosch R, Aboseif S, Nunes L, Lue TF, Tanagho EA (1989b) Acetylcholine as a possible neurotransmitter in penile erection. J Urol 141: 1444–1448

Stief CG, Holmquist F, Djamilian M, Krah H, Andersson K-E, Jonas U (1992) Preliminary results with the nitric oxide donor linsidomide chlorhydrate in the treatment of human erectile dysfunction. J Urol 148:1437–1440
Stjärne L (1989) Basic mechanisms and local modulation of nerve impulse-induced secretion of neurotransmitters from individual sympathetic nerve varicosities. Rev Physiol Biochem Pharmacol 112:1–137
Stjernquist M, Owman C, Sjöberg NO, Sundler F (1987) Coexistence and cooperation between neuropeptide Y and norepinephrine in nerve fibers of guinea pig vas deferens and seminal vesicle. Biol Reprod 36:149–155
Thompson JW, Ware MR, Blashfield RK (1990) Psychotropic medication and priapism: a comprehensive review. J Clin Psychiatry 51:430–433
Thüroff JW, Bazeed MA, Schmidt RA, Tanagho EA (1986) Functional pattern of sacral root stimulation. III. Erection. Eur Urol 12:134–140
Traish AM, Carson MP, Kim N, Goldstein I, Saenz de Tejada I (1990) Characterization of muscarinic acetylcholine receptors in human penile corpus cavernosum: studies on whole tissue and cultured endothelium. J Urol 144:1036–1040
Trigo-Rocha F, Hsu GL, Donatucci CF, Lue TF (1993a) The role of cyclic adenosine monophosphate, cyclic guanosine monophoshate, endothelium and nonadrenergic, noncholinergic neurotransmission in canine penile erection. J Urol 149:872–877
Trigo-Rocha F, Aronson WJ, Hohenfellner M, Ignarro LJ, Rajfer J, Lue TF (1993b) Nitric oxide and cGMP: mediators of pelvic nerve-stimulated erection in dogs. Am J Physiol 264:H419–H422
Vardi Y, Siroky MB (1993) Hemodynamics of pelvic nerve induced erection in a canine model. II. Cavernosal inflow and occlusion. J Urol 149:910–914
Virag R, Ottesen B, Fahrenkrug J, Levy C, Wagner G (1982) Vasoactive intestinal polypeptide release during penile erection in man. Lancet ii:1166
Volle RL, Koelle GB (1970) Ganglion stimulating and blocking agents. In: Goodman LS, Gilman A (eds) The pharmacological basis of therapeutics. Macmillan, New York, pp 585–600
Wagner G, Brindley GS (1980) The effect of atropine, α and β blockers on human penile erection. In: Zorgniotti AW, Rossi G (eds) Vasculogenic impotence. Thomas, Springfield, pp 77–81
Wagner G, Sjöstrand NO (1988) Autonomic pharmacology and sexual function. In: Sitsen JMA (ed) The pharmacology and endocrinology of sexual function. Elsevier, Amsterdam, pp 32–43 (Handbook of sexology, vol 6)
Wakui M, Fukushi Y (1986) Dual excitatory actions of acetylcholine at the neuromuscular junction in the guinea-pig vas deferens. Pflügers Arch 406:587–593
Whitelaw GP, Smithwick RH (1951) Some secondary effects of sympathectomy. N Engl J Med 245:121–130
Willis EA, Ottesen B, Wagner G, Sundler F, Fahrenkrug J (1983) Vasoactive intestinal polypeptide (VIP) as a putative neurotransmitter in penile erection. Life Sci 33:383–391
Yui Y, Ohkawa S, Ohnishi K, Hattori R, Aoyama T, Takanishi M, Morishita H, Terao Y, Kawai C (1989) Mechanism for the generation of active smooth muscle inhibitory factor (IF) from bovine retractor penis muscle. Biochem Biophys Res Commun 164:544-549

CHAPTER 15

Urinary Tract

K.E. CREED

A. Introduction

In some lower vertebrates such as teleost fish and frogs, a thin walled bladder, into which urine may be diverted, has developed as a blind sac from the urinary ducts or the cloaca. The urine is normally emptied passively with the faeces but the bladder may be important for salt and water exchange with the blood to maintain water balance. In amniotes a highly distensible urinary bladder has developed from the allantois of the embryo and is present in all adult mammals and a few reptiles. This bladder, which has muscular walls, is drained by a new duct, the urethra, and the urinary ducts (ureters) enter directly into it. It acts as a storage organ that is periodically emptied by muscular contraction and normally no exchange of salts and water occurs across its wall.

The mechanism of micturition was first seriously investigated in the 1800s when the role of the nerves was first recognized (see SHERRINGTON 1892). Since then the physiology and pharmacology has been studied in a number of species and where possible in humans. The walls of the lower urinary tract contain smooth muscle and during micturition the muscle layers of the bladder contract so the vesicular pressure increases and urine passes through the urethra. The activity of the smooth muscle during both storage of urine and expulsion to the exterior is controlled by parasympathetic and sympathetic divisions of the autonomic nervous system. In addition there is an external sphincter of skeletal muscle that receives somatic innervation in the pudendal nerve. This is probably not involved in maintained closure of the urethra but its contraction may cut short flow during voiding.

In young individuals, micturition is a simple spinal reflex triggered by bladder distension or tactile stimulation. Progressively the brain becomes involved so that the pons and medulla facilitate the spinal outflow, and conscious centres in the cortex ensure that micturition only occurs at socially convenient times (BLAIVAS 1982).

In this review the effects of drugs will be considered on the peripheral nerve pathways and on the smooth muscle cells. The detrusor muscle of the bladder and the urethra have been studied most extensively but reference will be made to the ureters, bladder base and central control when appropriate.

B. Species Differences

Experimental work has been carried out in a number of species. Most in vivo research has been on cats and dogs and in vitro work on guinea pigs, rabbits and rats. Other species used include pigs, monkeys and wallabies. It has been found that the basic mechanisms of storage and micturition are the same in all mammals studied but some qualitative differences occur, such as the relative size of the bladder, force of contraction and relative number of cholinergic nerve endings. Another difference is in the pattern and significance of micturition. In many species urine flow is continuous until the bladder is empty. In dogs and pigs, however, urine is expelled in short spurts. At least in dogs, this is probably related to the presence of pheromones secreted into the urine to mark territorial boundaries. In females of a number of species pheromones are present during oestrus and indicate that the female is ready for mating. Mucous glands occur in the ureter of some mammals (e.g, horses) and may reduce the potential damage caused by urine in the lower urinary tract.

C. Anatomy and Physiology

The bladder is classically divided into the trigone and the detrusor muscle. The trigone, which is particularly well developed in humans, is the triangle of thicker tissue between the two ureter openings and the urethra. Functionally it is more appropriate to divide the bladder into the base and the body, the base including the caudoventral part of the detrusor muscle (Elbadawi and Schenk 1968). The smooth muscle layers of the bladder base are continuous with the musculature of the proximal urethra and are arranged in small bundles separated by much connective tissue. The detrusor muscle in the body of the bladder is a thin sheet of relatively large bundles of smooth muscle cells with much less connective tissue (Gosling and Dixon 1975). Bundles are not orientated into distinct layers though there is a tendency for the bundles on the outer surface to be arranged longitudinally between dome and base. The smooth muscle cells are spindle shaped and within a bundle they are closely associated with each other. Nexi or tight junctions are rarely seen though intermediate junctions and desmosomes are present (Elbadawi 1987; Gabella and Uvelius 1990).

The bladder receives urine continuously from the kidneys via the ureters. Waves of contraction, initiated by fluid collecting at the pelvis of the kidney, push the urine down the ureters at 20–30 mm/s. During filling the intravesicular pressure initially shows little increase as the bladder becomes more spherical and the bladder wall stretches as the muscle bundles straighten out. The base pressure is 2–10 mmHg. As the volume increases above a critical level, the pressure increases sharply and superimposed phasic contractions may appear until micturition occurs due to forceful contraction of the detrusor muscle. The pressure may then reach 30–50 mmHg. During filling the pressure in the urethra (especially the proximal urethra close to the bladder) is greater than that in the bladder so that flow is prevented. In the male closure of the proximal urethra also prevents reflux of semen into the bladder during ejaculation. There is no anatomical sphincter at the bladder neck but contraction of smooth muscle throughout the length is partially responsible for closure. In addition, arterial pressure, elastic fibres and folding of the mucosal lining contribute (TULLOCH 1974) and it has been suggested that the position of the trigone and other musculature in the bladder neck may effectively close off the urethra when the detrusor muscle is not contracting. During micturition the trigone is pulled aside and urethral smooth muscle relaxes though this may not be essential so long as bladder pressure exceeds urethral pressure.

D. Innervation and Effects of Drugs

The lower urinary tract is innervated by the hypogastric (sympathetic) and pelvic (parasympathetic) nerves (ELLIOTT 1907). The pelvic nerves leave the spinal cord mostly in sacral roots 1, 2 and 3, which unite to form a single nerve running alongside the rectum. The hypogastric nerve arises from the ventral mesenteric ganglia which in turn receive fibres via the sympathetic chain from lumbar roots 2–5. Some individual and species variation occurs. The hypogastric and pelvic nerves combine to form the pelvic plexus, a diffuse collection of nerve fibres and ganglia on the lateral aspect of the bladder base. From there nerves supply the rectum, bladder, urethra and genital organs. The pelvic plexus also receives input from lumbar branches of the sympathetic chain. These have extensions to the genitalia but do not connect with fibres to the bladder. The plexus spreads to the surface of the bladder as the vesical plexus and ganglion cells have been reported within the walls in several species (DIXON and GOSLING 1987).

I. Hypogastric Nerve

It is considered that the sympathetic nerve activity facilitates storage of urine by constricting the urethra and inhibiting the tone of the detrusor muscle (ELLIOTT 1907). It also reduces urine formation by reducing the glomerular filtration rate and slows propulsion of urine from the kidney to the bladder by decreasing the frequency of contractile waves in the ureter. The hypogastric nerves are predominantly adrenergic, though in some species other nerve types have been identified (TAIRA 1972). The action of noradrenaline (NAdr) can be separated into contraction through α-adrenoceptors, mostly found on the urethra, and inhibition through β-adrenoceptors, mostly on the bladder. Sympathetic nerves in the hypogastric nerve also end on ganglion cells in the pelvic plexus (HAMBERGER and NORBERG

1965) and there is some evidence that their activity inhibits conduction of information along the pelvic nerve (DE GROAT and SAUM 1972).

The role of adrenergic fibres on bladder activity during the storage phase is unclear. Section of the hypogastric nerves increases basal tone and phasic contractions of the bladder suggesting that they contain inhibitory fibres. However, adrenergic nerve fibres are sparse (GOSLING and DIXON 1975) and close arterial injection of noradrenaline or isoprenaline in vivo has only a small effect on bladder pressure (CREED 1979). Noradrenaline decreases spontaneous contractions in vitro and reduces the frequency of spontaneous action potentials recorded with intracellular micro-electrodes in detrusor strips from rabbits and guinea pigs but only at high concentrations (10^{-4} *M* CREED 1971b). The β-adrenoceptors are stimulated by terbutaline and blocked by butoxamine and are therefore classified as β_2-receptors.

The contraction of the urethra, however, is important for continence. It has dense adrenergic innervation (GOSLING and DIXON 1975; ELBADAWI 1987) and stimulation of the hypogastric nerves produces a biphasic response (ELLIOTT 1907; CREED 1979). There is a large increase in pressure which is abolished by α-adrenergic antagonists followed by a small relaxation that is blocked by β-adrenergic antagonists in several species (PERSSON and ANDERSSON 1976). Phenylephrine triggers bursts of action potentials in isolated strips of guinea pig urethra and bladder base (CALLAHAN and CREED 1981). The contraction to noradrenaline is believed to be due to the presence of both α_1- and α_2-adrenoceptors on the smooth muscle cells but the relative abundance varies with the species. In rabbits α_2-receptors dominate so that clonidine is more effective than phenylephrine, and prazosin (an α_1-antagonist) has only minimal effects (ANDERSSON et al. 1984); in humans clonidine has little effect, suggesting that only α_1-adrenoceptors may occur. In all species there are probably α_2-receptors on the nerve endings that restrict further release of noradrenaline.

Little is known concerning the inhibitory action of noradrenaline at the pelvic plexus. Stimulation of hypogastric nerves immediately before pelvic nerve stimulation in guinea pig decreases the electrical response to the pelvic nerve (MCLACHLAN 1977). It is thought that the excitatory response is due to acetylcholine acting on nicotinic receptors and the inhibition to noradrenaline acting on α_2-adrenoceptors, but muscarinic cholinergic receptors have also been recorded in the cat (GALLAGHER et al. 1982; AKASU et al. 1988).

II. Pelvic Nerve

In all species stimulation of the pelvic nerves produces contraction of the bladder resulting in an increase in vesicular pressure and contraction of the urethra (LANGLEY and ANDERSON 1895; ELLIOTT 1907). In healthy human and monkey bladders, the response of the detrusor muscle may be due exclusively to release of acetylcholine since it is blocked by muscarinic antagonists such as atropine and propantheline (SIBLEY 1984). However, other results have suggested that atropine-resistant nerves are also present in humans (COWAN and DANIEL 1983). In other mammals studied the response is partially resistant to atropine (LANGLEY and ANDERSON 1895; URSILLO and CLARK 1956; AMBACHE and ZAR 1970). Records taken from strips of detrusor muscle with the double sucrose gap and micro-electrode techniques clearly indicate that acetylcholine, and the atropine-sensitive component of the response to nerve stimulation, produces a small depolarization with an increase in frequency of action potentials resulting in a sustained increase in tone (CREED et al. 1983). These responses are enhanced by anticholinesterases such as neostigmine. Although cholinergic nerve fibres are spread uniformly through the bladder and urethra, the contractile response of the urethra to pelvic nerve stimulation is always small (CREED and TULLOCH 1978). It is reduced, not abolished, by atropine but not by phentolamine, propanalol, methysergide or quinidine. Acetylcholine increases the frequency of action potentials in guinea pig and rabbit urethra (CALLAHAN and CREED 1981).

The non-cholinergic component in the detrusor muscle produces a large excitatory junction potential (EJP) in the smooth muscle cells after a short latency and with a superimposed spike that produces a large transient contraction (CREED et al. 1983, 1991). The nature of the transmitter has been the subject of much study. It is unlikely to be one of the monoamines since the electrical and mechanical responses to nerve stimulation are unaffected by adrenergic antagonists or by mepyramine or methysergide that block responses to histamine or 5-hydroxytryptamine elsewhere; there is no published evidence for dopamine receptors in the lower urinary tract.

A number of peptides which are now recognized as neurotransmitters in the myenteric plexus of the small intestine have been identified in the lower urinary tract (ALM et al. 1978; GIBBINS et al. 1987). Of these substance P always contracts the detrusor muscle and accelerates spontaneous action potentials recorded from smooth muscle cells (SJÖGREN et al. 1982; MACKENZIE and BURNSTOCK 1984; CALLAHAN and CREED 1986). However, the contraction is slow in onset and the response to nerve stimulation is unaffected by substance P itself or analogues that competitively antagonize the action of substance P elsewhere. Vasoactive intestinal peptide produces relaxation in some species only (SJÖGREN et al. 1985); somatostatin, neurotensin and leuenkephalin may produce small contractile responses and only in some individuals (HUSTED et al. 1981; CALLAHAN and CREED 1986). Other peptides present in the nerves, such as neuropeptide Y and calcitonin gene-related peptide, have not been examined in detail but it seems unlikely that they are responsible for the atropine-resistant contraction as the latency of responses to peptides tends to be longer than that seen in the bladder strips. Until specific antagonists are available, however, peptides cannot be definitely excluded as neurotransmitters.

A strong candidate for the non-cholinergic transmitter in the bladder is adenosine triphosphate (ATP) (BURNSTOCK et al. 1972). Intravenous injections in dogs contract the bladder (CREED and TULLOCH 1978); in vitro ATP produces a rapid, phasic contraction followed by tonic contraction and a transient increase in the frequency of spontaneous action potentials recorded with micro-electrodes from guinea pig strips (CALLAHAN and CREED 1981; FUJII 1988). Furthermore the initial EJPs recorded with the double sucrose gap from strips of rabbit detrusor muscle are selectively blocked by $\alpha\beta$-methylene ATP, a specific competitive antagonist of ATP (FUJII 1988; CREED et al. 1991). The mechanical responses to stimulation of pelvic nerve at low frequencies are particularly sensitive to $\alpha\beta$-methylene ATP whereas high frequency responses are more sensitive to atropine (BRADING and WILLIAMS 1990). However, high concentrations of ATP are necessary to produce responses (close arterial injection of 3–6 mg in dogs; $10^{-5} M$ in vitro) and addition of atropine and $\alpha\beta$-methylene ATP together has a larger effect than the two drugs separately (BRADING 1987; CREED et al. 1991). These results must cast some doubt on the certain identification of ATP as a second excitatory transmitter in the bladder and it is possible that ATP acts to modify release or action of acetylcholine. If ATP is the second transmitter, a third one may also be present in rat and rabbit detrusor muscle, since a residual tetrodotoxin-sensitive response remains when atropine and $\alpha\beta$-methylene ATP are both present (LUHESHI and ZAR 1990; CREED et al. 1991).

In the rabbit both electrical responses recorded with the double sucrose gap are reduced in parallel by hemicholinium or botulinum toxin (CREED et al. 1991). As these two agents selectively act on synthesis and release of acetylcholine, it suggests that the release of the second transmitter is coupled with that of acetylcholine.

Close arterial injection of ATP normally relaxes the urethra of anaesthetized dogs except when high concentrations are used (CREED and TULLOCH 1978). The pressure then closely follows intravesicular pressure, probably because the bladder neck is pulled open. Studies on strips of urethra from guinea pigs confirm that ATP relaxes precontracted smooth muscle and abolishes spike activity (CALLAHAN and CREED 1981). ATP may therefore be released by pelvic nerves in the bladder and urethra during micturition.

III. Modulators

Neurotransmitter agents are released by nerves and react with receptors to produce a response in the target organ. The amplitude of the response, however, may be varied by the presence of other substances. These may affect synthesis, release, action or removal of the transmitter agent.

1. Transmitters

Transmitters themselves produce negative feedback that reduces subsequent release of the transmitter from nerves. In the lower urinary tract it has been shown that α_2-adrenoceptors have this action on release of noradrenaline and acetylcholine so that α_2-antagonists (yohimbine, rauwolscine) increase and agonists (clonidine) decrease the responses to hypogastric and pelvic nerve stimulation (MATTIASSON et al. 1984). Muscarinic receptors have a similar effect on adrenergic nerves in the lower urinary tracts of cats, rabbits and humans (MATTIASSON et al. 1987).

In the myenteric plexus of the small intestine, there is evidence that a single nerve coreleases a traditional transmitter and a peptide (FURNESS et al. 1989). It is postulated that the peptide may be acting to modify the transmitter action. Since many peptides have been identified in nerves to the lower urinary tract it seems probable that they serve a similar function.

An alternative is that a cascade effect exists, so that release of one substance leads to release of another. It has been shown in many other smooth muscle tissues for instance that acetylcholine liberates active agents from endothelium (FURCHGOTT 1983). There are no published records of this in the urinary tract.

2. Hormones

Hormones influence receptors for transmitters. In the lower urinary tract the female hormones in particular are known to modify responses in this way. Progesterone facilitates β-adrenergic responses in rat ureter (RAZ et al. 1972) and also in the uterus of several species. During pregnancy and in the later stages of oestrus and menstrual cycles, when progesterone levels are high, the bladder capacity tends to increase and incontinence may occur. These actions can be explained by the enhancing effect of progesterone on the inhibitory β-adrenoceptors of the bladder and urethra respectively. Oestrogens, on the other hand, facilitate activity at α-adrenoceptors and are used in conjunction with α-agonists (ephedrine, phenylpropanolamine) to treat incontinence in spayed bitches and postmenopausal women (SALMON et al. 1941; OSBORNE et al. 1972). Some of this action may be to restore the integrity of the urethral mucous membrane and smooth muscle (SMITH 1972; RAZ et al. 1973), but oestrogens also increase the sensitivity of the smooth muscle cells to α-agonists in vitro and in vivo in at least some species (CREED 1983; CALLAHAN and CREED 1985). Although the urethra of rabbits shows little change in sensitivity in the presence of oestrogens, an increase in number rather than affinity of receptors has been recorded in the bladder (HODGSON et al. 1978; BATRA and IOSIF 1983; LEVIN et al. 1980).

3. Prostaglandins and Others

At high concentrations prostaglandins (especially the E group) produce contractions of detrusor muscle which are slow (AMBACHE and ZAR 1970).

At concentrations that are too low to have a direct effect, responses to nerve stimulation and to acetylcholine are enhanced (CREED and CALLAHAN 1989). This may be due to a process beyond the receptor (such as Ca availability), but in dog trachea prostaglandins have been shown to modify acetylcholine release from nerves (ITO and TAJIMA 1981). A similar action may occur in the lower urinary tract. As indomethacin is without action in guinea pig in vitro (HILLS 1976; CREED and CALLAHAN 1989), it is unlikely that prostaglandins are released from nerves but may reach the bladder in the circulation or may be released from associated local epithelium. In other species indomethecin relaxes the bladder (HILLS 1976; DEAN and DOWNIE 1978).

It is probable that other naturally occurring hormones and autacoids modify the actions of transmitters. In addition a large number of introduced drugs have this as their prime action or as a significant side effect. Specific antagonists reduce receptor availability; anticholinesterases and local anaesthetics prolong transmitter action by inhibiting removal. Suggested actions of imipramine, which reduces bladder contractility and is used to treat nocturnal enuresis, include blocking cholinergic receptors and preventing re-uptake of noradrenaline.

E. Smooth Muscle

The smooth muscle cells of the urinary tract are joined to each other by low-resistance pathways (nexi) so that they act as a functional syncytium and activity can spread through the muscle layers. In all species studied (including guinea pig, rabbit, dog, rat, wallaby, monkey and human) resting membrane potentials are 30–45 mV and spontaneous action potentials occur. In the detrusor muscle these usually take the form of regular spike-like action potentials at 5–30/min (CREED 1971a); in the urethra action potentials may occur in bursts separated by quiescent periods (guinea pig, CALLAHAN and CREED 1981) or at regular intervals (rabbit, CALLAHAN and CREED 1985); in the ureter bursts of action potentials are usual and at least in the guinea pig recovery of individual spikes within bursts may not occur (KURIYAMA et al. 1967). Despite the sparsity of nexi between cells, action potentials spread through the tissue and they are conducted at a rate of 40–60 mm/s (CREED 1971a).

Contraction of the smooth muscle cell normally follows each action potential and the resting tone of the bladder reflects the frequency of the action potentials. Contraction in response to neurotransmitters may result from an increase in action potential frequency or there may be a more direct action on the contractile mechanism. As with other muscles the contraction involves the formation of cross-bridges between actin and myosin following increase in free intracellular Ca^{2+}. Although it has not been shown specifically for detrusor muscle, it is probable that the Ca^{2+} binds to calmodulin which activates myosin light chain kinase, allowing phosphorylation of myosin and the formation of cross-bridges.

At present few experiments have been carried out on the link between the membrane events and contraction in the urinary tract. In other smooth muscle cells it is believed that agonists may depolarize the membrane so that voltage-sensitive Ca^{2+} channels are opened or they may have a direct effect on Ca^{2+} channels in the membrane (receptor-operated channels) so that Ca^{2+} for contraction enters the cell from the extracellular fluid. Alternatively, a second messenger may be activated, such as inositol trisphosphate (IP_3), which releases Ca^{2+} bound within the cell (HASHIMOTO et al. 1985), or cAMP which modifies some protein kinases necessary for contraction. In the urinary bladder of rabbits and guinea pigs there is some evidence that contraction evoked by ATP is dependent on extracellular Ca^{2+} entering through channels in the membrane whereas acetylcholine can mobilize intracellular Ca^{2+} (MOSTWIN 1985) by increasing the intracellular concentration of IP_3 (IACOVOU et al. 1990; CREED et al. 1991).

Agonist
↓
Hydrolysis of phoshatidylinositol
4,5-bisphosphate (PIP_3) in membrane
↙ ↘
1,2-Diacylglycerol (activates protein kinase C) | Inositol 1,4,5-trisphosphate (releases Ca^{2+})

Theoretically a number of mechanisms could be modified by drugs including membrane excitability, specific ion channels, metabolic pathways controlling Ca^{2+} release within cells and formation of cross-bridges. However, until further research is carried out, the precise mechanisms leading to contraction are not understood and drugs specific to the urinary tract cannot be developed. In practice many drugs have been used empirically to modify activity, the mode of action of which are often not clear. Alternatively, drugs with known actions on other smooth muscles have been tested.

I. Membrane Excitability

Membrane excitability can be reduced by substances which have local anaesthetic actions. Substances such as xylocaine, in vivo, stabilize the membrane and inhibit generation of action potentials. A group of drugs loosely classified as "antispasmodics" probably act at a number of sites including antagonism of muscarinic receptors (FREDERICKS et al. 1978). At least one, dicyclomine, has been reported in addition to have local anaesthetic activity on bladder smooth muscle. Increased excitability is associated with depolarization and this can occur on reduction of Ca^{2+} in the extracellular fluid or on addition of ouabain, which changes the distribution of ions across the resting membrane (Kurihara and Creed 1972).

II. Calcium Channels

Recently attention has been directed to specific ion channels in the membrane and their role in controlling excitability. Tetrodotoxin or replacement of Na^{+} in the perfusing fluid has little action on spontaneous spike activity in the detrusor muscle, suggesting that Na^{2+} entry is not responsible (Creed 1971b), although the plateau phase of the ureter action potential is sensitive to Na^{+} (Kobayashi and Irisawa 1964; Shuba 1981). Reduction of Ca^{2+} or addition of small amounts of other divalent ions such as Mn^{2+} or Ni^{2+} reduce the amplitude and increase the duration of the action potentials (Creed 1971b). It has been confirmed that Ca^{2+} entry is responsible for the upstroke in individual smooth muscle cells with the patch clamp technique in detrusor muscle and in ureter (Klöckner and Isenberg 1985a,b; Lang 1989, 1990). In cells from both tissues, single Ca^{2+} currents were recorded that were voltage dependent and blocked by divalent ions.

The effectiveness of drugs that block Ca^{2+} channels in the membrane varies with the tissue and may reflect the presence of several channels each with different properties. Vascular smooth muscle is relaxed by nifedipine at $10^{-10}\,M$, whereas intestinal smooth muscle requires $10^{-7}\,M$. There are few published reports of the effects of calcium antagonists on the urinary tract. Spontaneous mechanical activity of strips from rabbit and guinea pig detrusor muscle is abolished by nifedipine at $10^{-7}\,M$ (Adamson et al. 1986) and by diltiazem or verapamil at $10^{-5}\,M$ (K.E. Creed, unpublished). The initial non-cholinergic response to nerve stimulation is particularly sensitive to nifedipine, again suggesting that Ca entry across the membrane is important (Bo and Burnstock 1990; Zar et al. 1990).

The Ca^{2+} channel blockers, nifedipine, verapamil and diltiazem, all reduce the amplitude of individual action potentials and increase the duration (Mostwin 1986; K.E. Creed, unpublished). In addition, the pattern of spontaneous activity tends to be changed, particularly by verapamil or the addition of divalent ions (Ni^{2+} or Mn^{2+}). Regular spikes are replaced by bursts of activity, sometimes with incomplete recovery of individual spikes, interspersed with quiescent periods when the resting membrane potential is similar to that in control tissues. Many drugs that block Ca^{2+} channels have been described and this has led to classification of different Ca^{2+} channels. A detailed analysis of the Ca^{2+} currents and the effects on these of drugs and divalent ions has not been reported for the urinary tract. However, it is apparent that Ca^{2+} is important in spike generation and patterns of activity as well as for contraction.

III. Potassium Channels

As in all excitable tissues, K^+ distribution across the membrane is important in establishing the resting membrane potential in smooth muscle cells. The membrane potential is lower than the K equilibrium potential (about 80 mV), suggesting that K^+ is not able to move freely across the membrane and in some cases, including the detrusor of guinea pig, an active Na-K exchange pump may contribute to the K^+ distribution since the membrane is depolarized in the presence of ouabain (KURIHARA and CREED 1972).

Exit of K^+ from cells is responsible for recovery of the action potential (KLÖCKNER and ISENBERG 1985b) and may cause an after-hyperpolarization if K^+ permeability remains high. An increase in permeability could result from the opening of receptor-operated channels by neurotransmitters, or of voltage dependent channels during the upstroke of the action potential or the increased Ca^{2+} within the cell may influence the K^+ channels. Any drug affecting K^+ permeability may therefore modify membrane excitability or the action potential depending on which channels are involved.

K^+ channel blockers, including tetraethylammonium (TEA), procaine, 4-aminopyridine and quinidine, produce contraction of strips of guinea pig detrusor muscle due to depolarization producing increased frequency of action potentials (KURIHARA 1975; FUJII et al. 1990). The duration of each individual action potential (and therefore contraction) is also prolonged by procaine and TEA (KURIHARA 1975; MOSTWIN 1986), and the after-hyperpolarization abolished by procaine. The K^+ current recorded with the patch clamp technique was reduced by TEA and blocked by Cs^+ (KLÖCKNER and ISENBERG 1985a). Conversely, drugs that are considered to open K^+ channels, such as cromakalim, hyperpolarize the membrane and reduce spontaneous mechanical activity in pig, rat and guinea pig detrusor muscle (FOSTER et al. 1989a,b; MALMGREN et al. 1990). Cromakalim and pinacidil also increase efflux of K^+ and sometimes Rb^+ from smooth muscle cells and this is antagonized by TEA and procaine (FOSTER et al. 1989a; FOVAEUS et al. 1989; EDWARDS et al. 1990). Because nerve activity is unaffected, the bladder will still contract in response to nerve stimulation at low concentrations of cromakalim.

IV. Contractile Mechanism

Little work has been carried out on the pathways involved within the cell to produce contraction of urinary tract smooth muscle. Probably the basic mechanisms are similar to those found in other smooth muscles, where myosin phosphorylation is essential for formation of cross-bridges (STULL et al. 1980). This is dependent on Ca^{2+} to activate myosin light chain kinase.

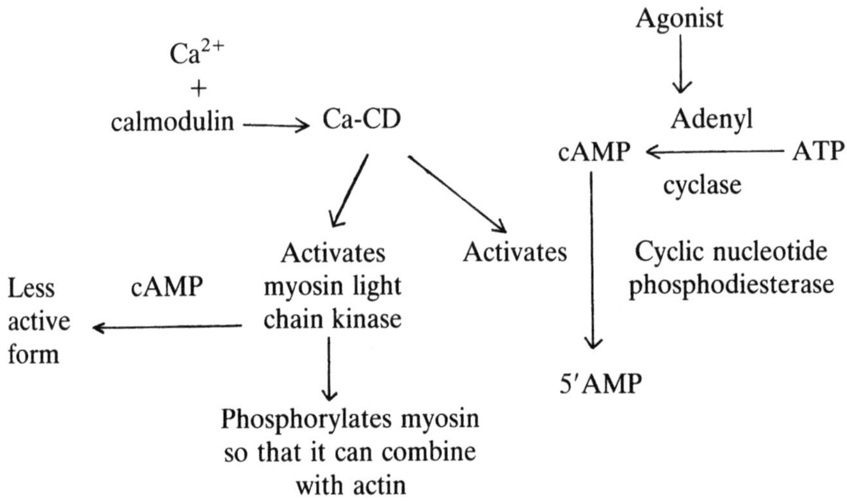

There is evidence that Ca^{2+} may be released from intracellular stores, since IP_3 concentration is increased by acetylcholine (IACOVOU et al. 1990; CREED et al. 1991), and caffeine, which releases bound Ca^{2+} in other smooth muscles, produces contraction of detrusor muscle strips and may cause hyperexcitable bladder. The facilitating action of prostaglandins on detrusor muscle contraction may also be because it makes intracellular Ca^{2+} more available. Papaverine and methylxanthines relax smooth muscle, presumably by inhibiting phosphorylation of myosin and by preventing breakdown of cAMP so that myosin light chain kinase is converted to its less active form. Other drugs, such as the "antispasmodics" and phenothiazines that reduce bladder activity, may also act on the contractile mechanism, but until these are better understood it will not be possible to develop drugs that have selective actions on the urinary tract.

F. Clinical Considerations

The urinary bladder is superficially a simple organ. It stores urine from the kidney until full, then eliminates it via the urethra. Any interference with storage or voiding by drugs, disease or trauma will cause abnormal retention or loss (incontinence) of urine. Such problems occur in the lives of many people at some stage but the treatment may be far from simple.

Incontinence may result from excessive bladder contraction (urge incontinence) or reduced urethral resistance (stress incontinence), whereas retention may be due to inadequate bladder contraction or blockage of the urethra by calculi, enlarged prostate or excessive constriction. Such disorders may be treated with drugs but these commonly produce side effects as few drugs have specific actions on the urinary tract. The drugs most commonly used clinically mimic or antagonize the effects of nerve stimulation, but others have direct effects on the smooth muscle cells.

Drugs to reduce bladder contraction in urge incontinence include anticholinergics (e.g., propantheline or atropine) that also tend to produce dry mouth, tachycardia and disturbances in vision; whereas the heart is more sensitive than the bladder to Ca^{2+} channel antagonists such as nifedipine. Drugs which open K^+ channels (e.g., cromakalim) may prove useful since they reduce spontaneous mechanical activity without inhibiting contraction to nerve stimulation. Drugs which combine anticholinergic action with direct inhibition of the smooth muscle cells (e.g., terodiline and some "antispasmodics") are being tested and may produce the desired effect on the urinary tract with fewer side effects.

For stress incontinence α-adrenergic agonists which produce urethral contraction (e.g., midodrine or phenylpropanolamine) have a greater effect in women when used together with oestrogens. For hypoactive bladder activity, muscarinic agonists (e.g., bethanechol or β-methylcholine chloride) are most commonly used and constricted urethras are relaxed in the short term by α-adrenergic antagonists (e.g., prazosine or phentolamine). A number of drugs normally used for actions on the central nervous system have also been found to have effects on the urinary tract, usually reducing contractions of hyperactive bladder. These include tricyclic antidepressants (e.g., imipramine) and benzodiazepine sedatives (e.g., diazepam). Full reviews of drugs currently used for urological disorders are given elsewhere (ANDERSSON 1988; WEIN 1991).

G. Conclusions

There is much scope for the development of new effective drugs in the urinary tract. Identification and pharmacology of non-cholinergic non-adrenergic transmitters and greater understanding of the biochemical pathways involved in contraction may lead to development of drugs with more selective actions.

References

Adamson GM, Callahan SM, Creed KE (1986) Effects of nifedipine on electrical and mechanical activity of the urinary bladder of the guinea-pig. Proc Aust Physiol Pharmacol Soc 17:136P

Akasu T, Tsurusaki M, Nishimura T, Tokimasu T (1988) Norepinephrine inhibits calcium action potential through α_2-adrenoceptors in rabbit vesical parasympathetic neurons. Neurosci Res 6:186–190

Alm P, Alumets J, Brodin E, Håkanson R, Nilsson G, Sjöberg N-O, Sundler F (1978) Peptidergic (substance P) nerves in the genitourinary tract. Neuroscience 3:419–425

Ambache N, Zar MA (1970) Non-cholinergic transmission by post-ganglionic motor neurones in the mammalian bladder. J Physiol (Lond) 210:761–783

Andersson K-E (1988) Current concepts in the treatment of disorders of micturition. Drugs 35:477–494

Andersson K-E, Larsson B, Sjögren C (1984) Characterization of the alpha-adrenoceptors in the female rabbit urethra. Br J Pharmacol 81:293–300
Batra SC, Iosif CS (1983) Female urethra: a target for estrogen action. J Urol 129:418–420
Blaivas JG (1982) The neurophysiology of micturition: a clinical study of 550 patients. J Urol 127:958–963
Bo X, Burnstock G (1990) The effects of Bay K 8644 and nifedipine on the responses of rat urinary bladder to electrical field stimulation, β,γ-methylene ATP and acetylcholine. Br J Pharmacol 101:494–498
Brading AF (1987) Physiology of bladder smooth muscle. In: Torrens M, Morrison JFB (eds) The physiology of the lower urinary tract. Springer, Berlin Heidelberg New York
Brading AF, Williams JH (1990) Contractile responses of smooth muscle strips from rat and guinea-pig urinary bladder to transmural stimulation: effects of atropine and α,β-methylene ATP. Br J Pharmacol 99:493–498
Burnstock G, Dumsday B, Smythe A (1972) Atropine resistant excitation of the urinary bladder: the possibility of transmission via nerves releasing a purine nucleotide. Br J Pharmacol 44:451–461
Callahan SM, Creed KE (1981) Electrical and mechanical activity of the isolated lower urinary tract of the guinea-pig. Br J Pharmacol 74:353–358
Callahan SM, Creed KE (1985) The effects of oestrogens on spontaneous activity and responses to phenylephrine of the mammalian urethra. J Physiol (Lond) 358:35–46
Callahan SM, Creed KE (1986) Non-cholinergic neurotransmission and the effects of peptides on the urinary bladder of guinea-pigs and rabbits. J Physiol (Lond) 374:103–115
Cowan WD, Daniel EE (1983) Human female bladder and its non-cholinergic contractile function. Can J Physiol Pharmacol 61:1236–1246
Creed KE (1971a) Membrane properties of the smooth muscle membrane of the guinea-pig urinary bladder. Pflugers Arch 326:115–126
Creed KE (1971b) Effects of ions and drugs on the smooth muscle cell membrane of the guinea-pig urinary bladder. Pflugers Arch 326:127–141
Creed KE (1979) The role of the hypogastric nerve in bladder and urethral activity of the dog. Br J Pharmacol 65:367–375
Creed KE (1983) Effects of hormones on urethral sensitivity to phenylephrine in normal and incontinent dogs. Res Vet Sci 34:177–181
Creed KE, Callahan SM (1989) Prostaglandins and neurotransmission at the guinea pig and rabbit urinary bladder. Pflugers Arch 413:299–302
Creed KE, Tulloch AGS (1978) The effect of pelvic nerve stimulation and some drugs on the urethra and bladder of the dog. Br J Urol 50:398–405
Creed KE, Ishikawa S, Ito Y (1983) Electrical and mechanical activity recorded from rabbit urinary bladder in response to nerve stimulation. J Physiol (Lond) 338: 149–164
Creed KE, Ito Y, Katsuyama H (1991) Neurotransmission in the urinary bladder of rabbits and guinea pigs. Am J Physiol 261:C271–C277
Dean DW, Downie JW (1978) Interaction of prostaglandins and adenosine 5′-triphosphate in the non-cholinergic neurotransmission in rabbit detrusor. Prostaglandins 16:245–251
De Groat WC, Saum WR (1972) Sympathetic inhibition of the urinary bladder and of pelvic ganglionic transmission in the cat. J Physiol (Lond) 220:297–314
Dixon JS, Gosling JA (1987) Structure and innervation in the human. In: Torrens M, Morrison JFB (eds) The physiology of the lower urinary tract. Springer, Berlin Heidelbeg New York
Edwards G, Henthorn M, Weston AH (1990) Some in vitro effects of potassium channel openers on rat bladder and portal vein. Eur J Pharmacol 183:2408–2409

Elbadawi A (1987) Comparative neuromorphology in animals. In: Torrens M, Morrison JFB (eds) The physiology of the lower urinary tract. Springer, Berlin Heidelbeg New York
Elbadawi A, Schenk EA (1968) A new theory of the innervation of bladder musculature. I. Morphology of the intrinsic vesical innervation apparatus. J Urol 99: 585–587
Elliott TR (1907) The innervation of the bladder and urethra. J Physiol (Lond) 35:367–445
Foster CD, Fujii K, Kingdon J, Brading AF (1989a) The effect of cromakalim on the smooth muscle of the guinea-pig urinary bladder. Br J Pharmacol 97:281–291
Foster CD, Speakman MJ, Fujii K, Brading AF (1989b) The effects of cromakalim on the detrusor muscle of human and pig urinary bladder. Br J Urol 63:284–294
Fovaeus M, Andersson K-E, Hedlund H (1989) The action of pinacidil in the isolated human bladder. J Urol 141:637–640
Fredericks CM, Green RL, Anderson GF (1978) Comparative in vitro effects of imipramine, oxybutynin and flavoxate on rabbit detrusor. Urology 12:487–491
Fujii K (1988) Evidence for adenosine triphosphate as an excitatory transmitter in guinea-pig, rabbit and pig urinary bladder. J Physiol (Lond) 404:39–52
Fujii K, Foster CD, Brading AF, Parekh AB (1990) Potassium channel blockers and the effects of cromakalim on the smooth muscle of the guinea pig bladder. Br J Pharmacol 99:779–785
Furchgott RF (1983) Role of endothelium in responses of vascular smooth muscle. Circ Res 53:557–573
Furness JB, Morris JL, Gibbins IL, Costa M (1989) Chemical coding of neurones in plurichemical transmission. Annu Rev Pharmacol Toxicol 29:287–306
Gabella G, Uvelius B (1990) Urinary bladder of rat: fine structure of normal and hypertrophic musculature. Cell Tissue Res 262:67–79
Gallagher JP, Griffith WH, Shinnick-Gallagher P (1982) Cholinergic transmission in cat parasympathetic ganglia. J Physiol (Lond) 332: 473–486
Gibbins IL, Furness JB, Costa M (1987) Pathway-specific patterns of the co-existence of substance P, calcitonin gene-related peptide, cholecystokinin and dynorphin in neurones of the dorsal root ganglia of the guinea-pig. Cell Tissue Res 248:417–437
Gosling JA, Dixon JS (1975) The structure and innervation of smooth muscle in the wall of the bladder neck and proximal urethra. Br J Urol 47:549–558
Hamberger B, Norberg K-A (1965) Studies on some systems of adrenergic synaptic terminals in the abdominal ganglia of the cat. Acta Physiol Scand 65:235–242
Hashimoto T, Hirata M, Ito Y (1985) The role of inositol 1,4,5-trisphosphate in the initiation of agonist-induced contractions of dog tracheal smooth muscle. Br J Pharmacol 86:191–199
Hills NH (1976) Prostaglandins and tone in isolated strips of mammalian bladder. Br J Pharmacol 57:464P
Hodgson BJ, Dumas S, Bolling DR, Heesch CM (1978) Effect of estrogen on sensitivity of rabbit bladder and urethra to phenylephrine. Invest Urol 16:67–69
Husted S, Sjögren C, Andersson K-E (1981) Substance P and somatostatin and excitatory neurotransmission in rabbit urinary bladder. Arch Int Pharmacodyn Ther 252:72–85
Iacovou JW, Hill SJ, Birmingham AT (1990) Agonist-induced contraction and accumulation of inositol phosphates in the guinea-pig detrusor: evidence that muscarinic and purinergic receptors raise intracellular calcium by different mechanisms. J Urol 144:775–779
Ito Y, Tajima K (1981) Actions of indomethacin and prostaglandins on neuroeffector transmission in the dog trachea. J Physiol (Lond) 319:379–392
Klöckner U, Isenberg G (1985a) Action potentials and net membrane currents of isolated smooth muscle cells (urinary bladder of the guinea-pig). Pflugers Arch 405:329–339

Klöckner U, Isenberg G (1985b) Calcium currents of cesium loaded isolated smooth muscle cells (urinary bladder of the guinea-pig). Pflugers Arch 405:340–348
Kobayashi M, Irisawa H (1964) Effect of sodium deficiency on the action potential of the smooth muscle cell of ureter. Am J Physiol 206:205–210
Kurihara S (1975) The effect of procaine on the mechanical and electrical activities of the smooth muscle cells of the guinea pig urinary bladder. Jpn J Physiol 25:775–788
Kurihara S, Creed KE (1972) Changes in the membrane potential of the smooth muscle cells of the guinea pig urinary bladder in various environments. Jpn J Physiol 22:667–683
Kuriyama H, Osa T, Toida N (1967) Membrane properties of the smooth muscle of guinea-pig ureter. J Physiol (Lond) 191:225–238
Lang RJ (1989) Identification of the major membrane currents in freshly dispersed single smooth muscle cells of guinea-pig ureter. J Physiol (Lond) 412:375–395
Lang RJ (1990) The whole-cell Ca^{2+} channel current in single smooth muscle cells of the guinea-pig ureter. J Physiol (Lond) 423:453–473
Langley JN, Anderson (1895) The innervation of the pelvic and adjoining viscera. II. The bladder. J Physiol (Lond) 19:71–84
Levin RM, Shofer FS, Wein AJ (1980) Estrogen induced alterations in the autonomic responses of the rabbit urinary bladder. J Pharm Exp Ther 215:614–618
Luheshi G, Zar AB (1990) Purinoceptor desensitization impairs but does not abolish the non-cholinergic motor transmission in rat isolated urinary bladder. Eur J Pharmacol 185:203–208
MacKenzie I, Burnstock G (1984) Neuropeptide action on the guinea-pig bladder; a comparison with the effects of field stimulation and ATP. Eur J Pharmacol 105:85–94
Malmgren A, Andersson K-E, Andersson PO, Fovaeus M, Sjögren C (1990) Effects of cromakalim (BRL 34915) and pinacidil on normal and hypertrophied rat detrusor in vitro. J Urol 143:828–834
Mattiasson A, Andersson K-E, Sjögren C (1984) Adrenoceptors and cholinoceptors controlling noradrenaline release from adrenergic nerves in the urethra of rabbit and man. J Urol 131:1190–1195
Mattiasson A, Andersson K-E, Elbadawi A, Morgan E, Sjögren C (1987) Interaction between adrenergic and cholinergic nerve terminals in the urinary bladder of rabbit, cat and man. J Urol 137:1017–1019
McLachlan EM (1977) Interaction between sympathetic and parasympathetic pathways in the pelvic plexus of the guinea pig. Proc Aust Physiol Pharmacol Soc 8:15P
Mostwin JL (1985) Receptor operated intracellular calcium stores in the smooth muscle of the guinea pig bladder. J Urol 133:900–905
Mostwin JL (1986) The action potential of guinea pig bladder smooth muscle. J Urol 135:1299–1303
Osborne CA, Low DG, Finco DR (1972) Canine and feline urology. Saunders, Philadelphia
Persson CGA, Andersson K-E (1976) Adrenoceptor and cholinoceptor mediated effects in the isolated urethra of cat and guinea-pig. Clin Exp Pharmacol Physiol 3:415–426
Raz S, Zeigler M, Caine M (1972) Hormonal influence on the adrenergic receptors of the ureter. Br J Urol 44:405–410
Raz S, Zeigler M, Caine M (1973) The effects of progesterone on the adrenergic receptors of the urethra. Br J Urol 45:131–135
Salmon UJ, Walter RI, Geist SH (1941) The use of estrogens in the treatment of dysuria and incontinence in postmenopausal women. Am J Obstr Gynecol 42:845–851
Sherrington CS (1892) Notes on the arrangement of some motor fibres in the lumbosacral plexus. J Physiol (Lond) 13:621–772

Shuba MF (1981) Smooth muscle of the ureter: the nature of excitation and the mechanism of action of catecholamines and histamine. In: Bülbring E, Brading AF, Jones AW, Tomita T (eds) Smooth muscle: an assessment of current knowledge. Arnold, London
Sibley GNA (1984) A comparison of spontaneous and nerve-mediated activity in bladder muscle from man, pig and rabbit. J Physiol (Lond) 354:431–443
Sjögren C, Andersson K-E, Husted S (1982) Contractile effects of some polypeptides on the isolated urinary bladder of guinea-pig, rabbit and rat. Acta Pharmacol Toxicol (Copenh) 50:175–184
Sjögren C, Andersson K-E, Mattiasson A (1985) Effects of vasoactive intestinal polypeptide on isolated urethral and urinary bladder smooth muscle from rabbit and man. J Urol 133:136–140
Smith P (1972) Age changes in the female urethra. Br J Urol 44:667–676
Stull JT, Blumenthal DK, Cooke R (1980) Regulation of contraction by myosin phosphorylation – a comparison between smooth and skeletal muscles. Biochem Pharmacol 29:2537–2543
Taira N (1972) The autonomic pharmacology of the bladder. Annu Rev Pharmacol 12:197–208
Tulloch AGS (1974) The vascular contribution to intraurethral pressure. Br J Urol 46:659–664
Ursillo RC, Clark BB (1956) The action of atropine on the urinary bladder of the dog and on the isolated nerve-bladder strip preparation of the rabbit. J Pharmacol Exp Ther 118:338–347
Wein AJ (1991) Practical uropharmacology. Urol Clin North Am 18:269–281
Zar MA, Iravani MM, Luheshi GN (1990) Effect of nifedipine on the contractile responses of the isolated rat bladder. J Urol 143:835–839

Section IV
Vascular System

CHAPTER 16

Molecular Mechanisms of Action of Antihypertensive Agents Ca-Antagonists and K-Channel Openers on Vascular Smooth Muscle

K. KITAMURA and H. KURIYAMA

A. Introduction

The tone of the vascular wall of resistance vessels is one of the most important factors in the control of blood pressure, and is dependent upon the intracellular concentration of Ca ions. Several mechanisms are involved in changing the intracellular Ca concentration of vascular smooth muscle cells. One of the main mechanisms underlying the contraction of smooth muscle cells induced by neurotransmitters, such as noradrenaline and adenosine 5′-triphosphate (ATP), is thought to be the release of Ca ions from their intracellular store site (sarcoplasmic reticulum), which is initiated by the synthesis of inositol 1,4,5-trisphosphate through activation of phospholipase C coupled to the adrenergic and purinergic receptors. However, part of the contraction induced by these stimulants (the slow or sustained component of the contraction) has been suggested to be due to Ca influx through ion channels because removal of extracellular Ca ions inhibits it.

Two types of Ca-permeable ion channel are involved in the stimulant-induced contraction that is sensitive to extracellular Ca ions, namely, the receptor-operated Ca-permeable channel (non-selective cation channel) and the voltage-dependent Ca channel. Reduction of the intracellular Ca ion concentration is caused by activation of Ca-ATPase and the Na-Ca exchange mechanism, which discharges the Ca ion to the extracellular space or to its intracellular store site, although the latter mechanism is thought to have a minor role. Therefore, most antihypertensive agents have been targeted at sites at which Ca mobilization could be directly interfered with. For example, α-blockers prevent activation of the excitatory pathway by reducing the number of sites available for receptor-agonist interactions, whereas Ca-channel blockers (Ca-antagonists) reduce the intracellular Ca concentration by direct inhibition of Ca influx through the voltage-dependent Ca channel (mostly L-type channels). These vasodilators interact directly with Ca mobilization mechanisms in their regulation of the intracellular Ca ion.

On the other hand, the K-channel openers, also vasodilators, are believed to produce vasodilation by indirect inhibition of the voltage-dependent Ca channel, via membrane hyperpolarization. For example, diazoxide, a derivative of thiazide compounds, which are potent diuretic agents, is known to have an antihypertensive action without any diuretic action and has recently been found to be able to open the ATP-sensitive K channel in pancreatic β-cells. Moreover, cromakalim and its isomer, lemakalim, strongly inhibit agonist-induced contractions but not the contraction induced by excess K solution. Although recent reports show that additional mechanisms are also involved in the vasodilating actions of K-channel openers, the main mechanism is thought to involve inhibition of the voltage-dependent Ca channel as a consequence of an induced hyperpolarization of the membrane.

One of the important pharmacological features of the relaxation induced by Ca-antagonists is that some Ca-antagonists, such as dihydropyridine (DHP) derivatives, selectively inhibit contractions in vascular smooth muscle cells rather than those in cardiac cells. One of the possible reasons for such tissue selectivity of Ca-antagonists is that the inhibitory action of the drugs is voltage dependent and the different tissues have different resting membrane potentials. In fact, as the resting membrane potentials of smooth muscle cells were, in general, higher than those in cardiac cells, it could explain why Ca-antagonists inhibit the voltage-dependent Ca channel in smooth muscle cells more potently than those in cardiac muscle cells. However, as little is known about the voltage-dependent Ca channel distributed in smooth muscle cells, we cannot be precise about the mechanism of action of Ca-antagonists in smooth muscle cells.

The application of the patch-clamp technique to studies of the vascular smooth muscle cells has been important in clarifying the biophysical and pharmacological properties of the ion channels present in these cells. We have accumulated knowledge of the properties of the ion channels and of the mechanisms of action of several drugs which modulate the ion channel. In this article, we review the mechanisms of action of Ca-antagonists and K-channel openers on ion channels in smooth muscle cells.

B. Classification of Ca-Antagonists and K-Channel Openers

GODFRAIND et al. (1986) minutely classified Ca-antagonists, including positive modulators of the voltage-dependent Ca channel and intracellular Ca-antagonists. According to GODFRAIND's definition of Ca modulators, "Ca-antagonist" is the general name for drugs which inhibit Ca movements through the plasma membrane or within the cell. Therefore, several drugs which act on the sarcoplasmic reticulum, mitochondria and calmodulin, such as dantrolene, ruthenium red, W-7 and chlorpromazine, could be classified as "Ca-antagonists" in the broad sense. "Ca-antagonists" in the narrow sense which we would like to mention in this article include drugs which might be categorized as "selective Ca-entry blockers". In view of the mechanisms of action of these drugs, "Ca-channel blockers" or "Ca-entry blockers" might be more adequate names for these drugs than "Ca-antagonists". However, since "Ca-antagonist" in the narrow sense is commonly used as the clinical and pharmacological term, in this article we have used the term "Ca-antagonists" in the narrow sense instead of "selective Ca-entry blockers".

Ca-antagonists are classified into several groups depending on their chemical structure. Thus, there are: (a) dihydropyridine (DHP) derivatives (nifedipine, nisoldipine, nitrendipine, manidipine, nimoldipine, nilvadipine, nigludipine, nicardipine, isradipine, amlodipine, benidipine, FRC 8693, etc.); (b) phenylalkylamine derivatives (verapamil, gallopamil, falipamil, etc.); (c) benzothiazepine derivatives (diltiazem, TA3090, etc.); (d) benzothiazine derivative (SD3211); and (e) diphenylpiperazine derivatives (cinnarizine, flunarizine, etc.). These drugs inhibit the voltage-dependent Ca channel selectively, although high concentrations of these drugs are known also to block the voltage-dependent Na and K channels, as well as the Ca channels.

K-channel openers are also classified according to their chemical structure. Thus, there are: (a) benzopyran derivatives (cromakalim, lemakalim, SDZ PCO 400, WAY-120,491, Pr 31-6930, etc.); (b) pyridine or triazine derivatives (minoxidil, SKF11197); (c) thiourea or guanidine derivatives (pinacidil, LY222675, EP-A-0354553, P1060, etc.); (d) pyridine derivatives (nicorandil, KRN 2391); (e) benzothiadiazine derivatives (diazoxide, chlorothiazide, quinethazone); (f) thioformamide derivatives (RP 52891, EP-A-0326297); and (g) dihydropyridine derivative, nigludipine (EDWARDS and WESTON 1990). So far, no evidence has appeared to suggest that each derivative produces a different action on K channels, and the classification derived from their chemical structure does not seem to be important. As several K-channel openers have additional vasodilating actions, the following classification is more useful for K-channel openers at the present state of knowledge: (a) selective K-channel openers (lemakalim); (b) K-channel openers, having Ca-antagonistic actions (cromakalim, minoxidil and nigludipine); (c) K-channel openers, having nitric-compound-like actions (nicorandil); and (d) K-channel openers, interacting with the Ca-release mechanisms (pinacidil).

C. Ion Channels Distributed in Vascular Smooth Muscle Cells

The use of voltage-clamp techniques with whole-cell and patch-membrane methods has enabled various types of ion channels distributed in the membrane of smooth muscle cells to be studied, including Ca, Na, K, Cl and non-selective ion channels.

I. Ca Channels

Since CARBONE and LUX (1984) first reported the presence of two types of voltage-dependent Ca channel in chick dorsal root ganglion (DRG) cells, two types of Ca channel similar to those in DRG cells have been reported in various smooth muscle cells. These two types of Ca channel [transient- (T-) type and long-lasting- (L-) type channels] had characteristics similar to those in cardiac muscle and DRG cells. For example, the T-type channel is inactivated at membrane potentials lower than those for the L-type channel, and is insensitive to Ca-antagonists such as DHP derivatives. Using these criteria, two types of Ca current have been identified from the macroscopic current obtained in the rat mesenteric artery, in cultivated cells from the rat aorta and in the rat portal vein and azygos vein (BEAN et al. 1986; FRIEDMAN et al. 1986; LOIRAND et al. 1986; STUREK and HERMSMEYER 1986). Furthermore, in the guinea pig basilar artery, a third type of voltage-dependent Ca channel, called the "B-type", has been reported and this was less sensitive to nifedipine than the L-type channel (SIMARD 1991). The characteristics of the smooth muscle "B-type" channel were slightly different from those of the "B-type" channel in cardiac cells (ROSENBERG et al. 1988). For example, the "B-type" channel in smooth muscle cells was activated at potentials of, or more positive than, −30 mV and was less sensitive to nifedipine, whereas in cardiac cells this channel was activated at −100 mV and was insensitive to Bay K 8644. Therefore, the cardiac "B-type" channel, but not the smooth muscle "B-type" channel, is activated at the resting membrane potential, suggesting that the cardiac "B-type" channel is a component of the background Ca current. Both "B-type" channels had some properties in common, namely slower rates of rundownn and current decay than those of the L-type channel. However, classification into subtypes from macroscopic currents requires some caution, because elevation of the holding potential inhibits the L-type channel in part, as it does the T-type channel, and a DHP-sensitive Ca channel which had properties similar to the T-type channel was also reported to be present in smooth muscle cells of the guinea pig portal vein (INOUE et al. 1990b).

The clearest criteria for classification of the voltage-dependent Ca channel are obtained from single-channel current recordings. In the majority of smooth muscle tissues examined so far, two types of voltage-dependent Ca channel have been identified which have different values of single-channel conductance. However, in smooth muscle cells obtained from the rabbit small intestine and coronary artery, guinea pig aorta and amphiuma stomach, only one type of DHP derivative-sensitive Ca channel has been detected (CAFFREY et al. 1986; INOUE et al. 1989b; MATSUDA et al. 1990). On the other hand, in the guinea pig taenia coli, three types of Ca channel were reported to be present (YOSHINO et al. 1989). To judge from values of the single-channel conductance, the voltage-dependent Ca channel obtained from various smooth muscle cells can be classified into three groups, namely, those with large (≥20 pS), intermediate (12–15 pS) or small (8 pS) conductances. Two (or possibly only one) of the three types of Ca channel are distributed in each smooth muscle cell. The values of the large, intermediate and small unitary conductances of the voltage-dependent Ca channel were nearly the same as those of the L- (25 pS), N- (neither L- nor T-type; 13 pS) and T-type (8 pS) Ca channels observed in chick DRG cells (NOWYCKY et al. 1985).

The properties of the large and small unitary conductance Ca channels were similar to those of the L- and T-type Ca channels, respectively, in cardiac and neuronal cells, except for the 8 pS Ca channel in the rabbit mesenteric artery which was thought to be a substate of the L-type channel (NELSON and WORLEY 1989). On the other hand, the intermediate unitary conductance Ca channel did not have exactly the same properties as the N-type channel in DRG cells. For example, the threshold potential for channel activation of the intermediate Ca channel in various smooth muscle cells was lower than that for the L-type and N-type channels in chick DRG cells. The properties of the intermediate Ca channel, observed in the rabbit basilar artery and guinea pig vena cava, were very similar to those of the T-type channel observed in different smooth muscle cells. However, it is uncertain whether or not the different values of unitary conductance found in intermediate and small conductance Ca channels in various smooth muscle cells is an adequate criterion for classification. Such differences might be within the margin of error of measurements obtained under the different experimental conditions used by the different researchers, although in guinea pig taenia caeci different unitary conductances for the intermediate and small conductance Ca channels could be observed under exactly the same conditions (YOSHINO et al. 1989).

The intermediate Ca channels recorded in the guinea pig aorta, rabbit mesenteric artery and amphiuma stomach cells were very sensitive to DHP derivatives, although other properties were not systematically investigated. The intermediate Ca channel observed in the rat aorta and amphiuma stomach cells, at least, was not inactivated even when the membrane potential was at a depolarized level for a long time, and the properties of the channel resembled those of the L-type channel (or B-type channel ?) rather than those of the T-type channle. On the other hand, the intermediate channel observed in the guinea pig portal vein had a mixture of the properties of the L- and T-type channels (INOUE et al. 1990b). For example, this channel was inactivated at a lower potential than the L-type channel observed in the same cells, but the rate of the current decay was slower than T-type channel observed in other smooth muscle cells. Furthermore, this intermediate channel was inhibited by DHP derivatives but with a ten times higher concentration than that needed to block the L-type channel. These properties of the intermediate channel in the guinea pig portal vein were well matched with those of the "B-type" channel in the basilar artery of the same species (SIMARD 1991).

As the properties and distribution of the intermediate conductance Ca channel have not yet been fully described, we have no adequate classification for the voltage-dependent Ca channel in smooth muscle cells at present. However, in this article we have conventionally classified the voltage-dependent Ca channel into three groups from the amplitudes of their unitary conductance, that is, into large (L-type), small (T-type) and intermediate conductance channels. It is of interest that the unitary conductance values of the Ca channels depend upon the smooth muscle tissue in which they are found, even if they are classified into the same L-type channel (≥20 pS), and that those of the arterial L-type channel seem to be smaller than those of the visceral and venous L-type channels.

II. Na Channels

Recent experiments have shown that the tetrodotoxin (TTX)-sensitive Na channel is present in the smooth muscle cells of the rabbit pulmonary artery, guinea pig and rat portal veins, rat and human myometria and in the smooth muscle cell-line (DDT_1 MF-2) from the Syrian hamster vas deferens (OKABE et al. 1988, 1990; OKABE and KITAMURA, unpublished observations; MOLLEMAN et al. 1991). OKABE et al. (1988) reported that the K_D value for TTX to block the voltage-dependent Na channel in the rabbit pulmonary artery was 8 nM, which was close to the values for the TTX-sensitive Na channel in neurons or skeletal muscle cells and lower than those for cardiac and denervated skeletal muscle cells (BENDORF et al. 1985; BENOIT et al. 1985; PIDOPLICHKO 1986; REDFERN and THESLEFF 1971). Before these reports, the membrane of smooth muscle cells was thought to have only the TTX-insensitive Na channel, which had been observed in smooth muscle cells of the pregnant rat myometrium and rat azygous vein (AMÉDÉE et al. 1986; STUREK and HERMSMEYER 1986). Unfortunately, we do not know of any role for these TTX-sensitive Na channels, especially in the membrane of the pulmonary artery, because they are fully inactivated at the resting membrane potential of these smooth muscle cells (−60 to −50 mV) and pulmonary arterial cells do not produce action potentials spontaneously or in the presence of low concentrations of agonists.

III. K Channels

Various types of K channels have been classified from the macroscopic and unitary currents in various smooth muscle cells, including cultivated smooth muscle cell lines. However, the relationship between each macroscopic K current and the corresponding single K channel is still unclear. So far, four types of K channel have been classified from the physiological and pharmacological properties of the single-channel current in various smooth muscle cells: namely, the Ca-dependent, the ATP-sensitive, the A-type and the delayed (voltage-dependent) K channels. The first two types are further divided into several subtypes.

1. Ca-Dependent K Channels

The Ca-dependent K channels, from amplitudes and pharmacological profiles, has been subclassified into three subtypes (large, intermediate and small conductance Ca-dependent K channels). The large conductance Ca-dependent K channel, which has been called the "maxi-K" or "big K" channel, is widely and densely distributed in smooth muscle cells (frog and toad stomach, BERGER et al. 1984; rabbit jejunum and guinea pig mesenteric artery, BENHAM et al. 1985b; rabbit portal vein, INOUE et al. 1985; dog trachea, MCCANN and WELSH 1986; toad stomach, SINGER and WALSH 1987; bovine aorta, WILLIAMS et al. 1988; human aorta, BREGESTOVSKI et al. 1988; rabbit trachea, KUME et al. 1990; pregnant rat myometrium, KIHARA et al. 1990). It is very sensitive to tetraethylammonium (TEA), when this agent is applied extracellularly, but relatively resistant to its intracellular application (INOUE et al. 1985). Charybdotoxin and iberiotoxin, components of scorpion venom, such as that of Leirus quinquestriatus hebraeus and that of Buthus tamulus, selectively inhibited the "maxi-K" channels in cultivated bovine aortic smooth muscle cells (GALVEZ et al. 1990). Other K-channel blockers, such as 4-aminopyridine (4-AP) and apamin, did not block the channel. Quinidine, known to be a selective blocker of the Ca-dependent K channel in human red blood cells (BURGESS et al. 1981), also inhibited the "maxi-K" channels in smooth muscle cells of the toad stomach ($\geq 100\,\mu M$; WONG 1989). However, this agent also blocked the voltage-dependent Ca and Ca-independent K currents non-selectively in the guinea pig small intestine and rabbit portal vein at micromolar concentrations (NAKAO et al. 1986; BEECH and BOLTON 1989b). Recently, isoprenaline and cAMP were found to activate the "maxi-K" channel directly through activation of protein kinase A (KUME et al. 1990). WILLIAMS et al. (1988) also reported that GMP and cGMP potentiated the activity of the "maxi-K" channel. Several reports have shown that the "maxi-K" channel can be activated by K-channel openers (see Sect. F).

The small conductance Ca-dependent K channel, found in the rabbit portal vein, was also sensitive to TEA, but at concentrations much higher than those needed to inhibit the "maxi-K" channel (INOUE et al. 1985). However, the properties of this K channel remain obscure. BLATZ and MAGLEBY (1986) found that the small conductance Ca-dependent K channel in skeletal muscle was blocked by apamin. As the unitary conductance of the small conductance Ca-dependent K channel in skeletal muscle was smaller than that of those in the rabbit portal vein (10–14 vs. 92 pS in symmetrical 140 mM K solutions), the two Ca-dependent K channels were different. Apamin did not block the small conductance K channel in the rabbit portal vein (unpublished observations).

The rat portal vein was reported to possess a Ca-dependent K channel with a unitary conductance of 20 pS in symmetrical 140 m*M* K solutions (KAJIOKA et al. 1990) though such a 20-pS K channel was not found in the rabbit portal vein. It was sensitive to millimolar concentrations of TEA and 4-AP, but insensitive to apamin and charybdotoxin. K-channel openers activated this 20-pS K channel in the rat portal vein.

The intermediate conductance Ca-dependent K channel had unique properties in respect of its Ca and voltage dependencies: namely, this K channel was sensitive to extracellular Ca concentration and did not show voltage-dependent activation (INOUE et al. 1986). Millimolar concentrations of TEA inhibited the intermediate Ca-dependent K channel in the rabbit portal vein. A similar extracellular Ca-dependent K channel was also found in the pig coronary artery; however, the unitary conductance was smaller (30 pS) than that in the portal vein (INOUE et al. 1989b). This K channel was inhibited by millimolar concentrations of 4-AP and TEA and potentiated by K-channel openers.

During macroscopic K current recording, membrane depolarization to 0 mV from a relatively less negative holding potential (−60 mV) produced a transient outward K current following the inward Ca current (rabbit intestine, OHYA et al. 1987a; guinea pig intestine, NAKAO et al. 1986; dog trachea, rabbit portal vein, HUME and LEBLANC 1989; MURAKI et al. 1990). The transient K current, abolished in Ca-free solution, was inhibited by elevation of intracellular EGTA concentration, and blocked by low concentrations of TEA (0.1–1 m*M*), charybdotoxin or quinidine, but not by 4-AP. These findings strongly suggest that this current represents a transient activation of the "maxi-K" channel in these smooth muscle cells.

The "maxi-K" channel also contributes to the generation of the spontaneous current oscillation (oscillatory outward current, I_{OO}; spontaneous transient outward currents, STOCs) which has been reported in various smooth muscle cells (BENHAM and BOLTON 1986; OHYA et al. 1987a; HUME and LEBLANC 1989). I_{OO} could be recorded in Ca-free solution for several minutes, and as caffeine, procaine, ryanodine and inositol 1,4,5-trisphosphate ($InsP_3$) modulated its generation this current was thought to be activated by the Ca release from intracellular Ca store sites (BENHAM and BOLTON 1986; OHYA et al. 1987a, 1988; SAKAI et al. 1988; BOLTON and LIM 1989).

The contribution of the intermediate and small conductance Ca-dependent K channels to macroscopic K currents has not yet been clarified, although two extracellular Ca-dependent K channels in the rabbit portal vein and pig coronary artery might be involved in the background current that regulates the resting membrane potential, because these K currents have the property of voltage-independent activation. HUME and LEBLANC (1989) reported that the background current in the rabbit portal vein was inhibited in Ca-free solution, and had a similar TEA sensitivity to that of the intermediate conductance Ca-dependent K channel observed in the same preparations (INOUE et al. 1986). In neuroblastoma cells, the apamin-sensitive (but TEA-resistant) K channel was involved in the generation of the afterhyperpolarization of the action potential (HUGUES et al. 1982). However, in guinea pig intestinal smooth muscle cells, apamin did not inhibit the afterhyperpolarization (NAKAO et al. 1986).

2. ATP-Sensitive K Channels

Since the finding of the ATP-sensitive K channel in cardiac cells by NOMA (1983), the presence of ATP-sensitive K channels has been reported in skeletal muscle, pancreatic β- and nerve cells (see reviews by ASHCROFT 1988; DE WEILLE and LAZDUNSKI 1990). The ATP-sensitive K channel was inhibited by submillimolar concentrations of intracellular ATP, but micromolar concentrations of intracellular ATP or GTP were required to maintain the channel activity. On the other hand, the concentration of Ca ion on either side of the membrane had no influence on the channel activity, indicating that the ATP-sensitive K channel in these tissues should be classified as a Ca-independent K channel.

A similar ATP-sensitive K channel was recently found in vascular smooth muscle cells of the rabbit mesenteric artery and portal vein (STANDEN et al. 1989; KAJIOKA et al. 1991). Millimolar concentrations of TEA and 4-AP, but not of apamin or charybdotoxin, inhibited the ATP-sensitive K channel in smooth muscle cells. The activity of the ATP-sensitive K channel observed in the rabbit mesenteric artery and portal vein was independent of extracellular and intracellular Ca concentrations: however, the Ca-dependent K channels sensitive to K-channel openers recorded in the rat portal vein and pig coronary artery were also classified as ATP-sensitive K channels, because intracellular application of ATP could block the channel activity (KAJIOKA et al. 1991; INOUE et al. 1990a).

Application of glibenclamide, a sulphonylurea, did not depolarize the membrane in the guinea pig and rabbit mesenteric arteries, suggesting no contribution of the ATP-sensitive K channel to the resting membrane potential (NAKASHIMA et al. 1990; NELSON et al. 1990). However, the resting membrane currents at a holding potential of $-40\,\mathrm{mV}$ could be inhibited by glibenclamide in the rat portal vein, indicating that glibenclamide might depolarize the membrane. Therefore, the ATP-sensitive K channel probably participates in the regulation of the resting membrane potential in the rat portal vein. These findings may indicate that the functional role of the ATP-sensitive K channels differs across tissues and species.

We further review the properties of the ATP-sensitive K channel in smooth muscle cells in later sections (Sects. F–H), because this channel is thought to be a target for K-channel openers in smooth muscle cells.

3. A-Type Channels

In the rabbit portal vein and guinea pig ureter, depolarization of the membrane from a lower holding potential ($-80\,\mathrm{mV}$) produced a Ca-independent transient outward current, in addition to the Ca-dependent component

(BEECH and BOLTON 1989a; LANG 1989; IMAIZUMI et al. 1989a). The unitary conductance of the Ca-independent component of the transient outward current was 14 pS. The channel was transiently opened then rapidly inactivated during the depolarization, and also inactivated by elevation of the holding potential (−50 mV; IMAIZUMI et al. 1989a). Compared to nerve cells, smooth muscle cells of the rabbit portal vein and guinea pig ureter had resting membrane potentials that were less negative (between −50 and −60 mV); therefore its functional role in smooth muscle cells may be limited, because a large fraction of the A-type channels would be inactivated at the resting membrane potential. LANG (1989) and IMAIZUMI et al. (1989a) speculated that the A-type channel in the guinea pig ureter participated in the regulation of the action potential, as a counterpart of the Ca current near the resting membrane potential, rather than in the after-hyperpolarization of the spike.

4. Delayed K Channels

A sustained component of the outward current (Ca-independent component; voltage-dependent K current or delayed K current) induced by a membrane depolarization in Ca-free solution could be inhibited by millimolar concentrations of TEA in the rabbit small intestine and rat portal vein (OHYA et al. 1987a; KAJIOKA et al. 1990), and by decimolar concentrations of TEA in the rabbit portal vein and dog trachea (HUME and LEBLANC 1989; BEECH and BOLTON 1989c; MURAKI et al. 1990). However, it was resistant to TEA in the rabbit pulmonary and mesenteric arteries (OKABE et al. 1987a). The delayed K current observed in the rabbit pulmonary artery, portal vein and dog trachea was inhibited by millimolar concentrations of 4-AP. The delayed K channel was also blocked by a number of K-channel blockers, the relative potencies of their inhibitory actions being phencyclidine > quinidine > 4-AP > procaine > TEA (BEECH and BOLTON 1989b). The unitary conductance of this channel was reported to be around 5 pS in the rabbit portal vein (BEECH and BOLTON 1989c).

5. Other K Channels

Although they have not yet been identified from single K-channel currents, other types of K channel have been revealed from macroscopic currents. The muscarinic K current (M current) which was found in toad stomach smooth muscle cells was inhibited by acetylcholine (ACh) or substance P, and activated by isoprenaline or cAMP (SIMS et al. 1985, 1986, 1988). ATP, adrenaline or an unidentified neurotransmitter from non-adrenergic non-cholinergic inhibitory nerves produced a hyperpolarization of the membrane in the guinea pig small intestine which ceased in Ca-free solution and was selectively blocked by apamin (YAMANAKA et al. 1985). As apamin did not depolarize the membrane in smooth muscle cells, the apamin-sensitive K channel does not seem to play a major role in the regulation of the resting membrane potential.

IV. Non-selective Cation Channels

Apart from the selective K channels, several types of non-selective cation channels have also been found in smooth muscle cells. In the rabbit jejunum, rabbit portal vein, guinea pig myometrium and toad stomach, a hyperpolarization-activated channel was present and thought to regulate the activity of spontaneous spike discharges (BENHAM et al. 1987a; COLEMAN and PARKINGTON 1990; HISADA et al. 1991; KAMOUCHI et al. 1991). The unitary conductance of the hyperpolarization-activated channel was 64 pS, and both Na and K ions permeated this non-selective cation channel in toad stomach cells (HISADA et al. 1991). Cs, but not Ba, ions inhibited the hyperpolarization-activated channel (BENHAM et al. 1987b; KAMOUCHI et al. 1991). Other divalent cations, such as Mg or Ca, reduced its unitary conductance (HISADA et al. 1991). In the toad stomach, the hyperpolarization-activated channel appeared identical with the stretch-activated channel in the same cells (HISADA et al. 1991).

Acetylcholine, noradrenaline and ATP also activated the non-selective cation channel in smooth muscle cells of various visceral and vascular tissues, which allowed Na, K, Cs and divalent cations to permeate (ACh, rabbit intestine, BENHAM et al. 1985a; guinea pig intestine, INOUE et al. 1987; ATP, rabbit ear artery, BENHAM and TSIEN 1987; rat myometrium, HONORÉ et al. 1989a; rabbit portal vein, XIONG et al. 1991b; NAd, rabbit ear artery, AMÉDÉE et al. 1990). INOUE et al. (1987) reported that the unitary conductance of the non-selective cation channel activated by ACh in smooth muscle cells of the guinea pig ileum was 20–25 pS under physiological ionic conditions. In the rat portal vein, another non-selective cation channel was found which had a large unitary conductance (200 pS) and was activated by intracellular Ca ions (LOIRAND et al. 1991).

V. Cl Channels

Although intracellular Cl concentration in smooth muscle cells is three to four times as high as predicted from a purely passive distribution, the resting Cl permeability in smooth muscle cells is thought to be very low (P_{Cl}/P_K = 0.04, AICKIN and BRADING 1983; AICKIN and VERMUÉ 1983). Several types of Cl channel have been discovered in various cells; however, only a large conductance Cl channel with a unitary conductance of 340 pS has so far been reported in smooth muscle cells (A7r5 cells; SOEJIMA and KOKUBUN 1988).

The Cl channel current recorded in A7r5 cells was activated at a membrane potential of around 0 mV, rapidly inactivated during depolarization and did not require intracellular Ca ions for its activation, suggesting that it was similar to the maxi-Cl channel observed in the lacrimal gland and skeletal muscle cells and oocytes (see review by FRANCIOLINI and PETRIS 1990). The permeability sequence for the Cl channel in A7r5 cells was similar to that in pulmonary alveolar cells, as was the virtual impermeability of the channel to cations (SCHNEIDER et al. 1985). The presence of a Ca-dependent Cl channel has also been assumed from the increase in Cl conductance on agonist stimulation. In the rat anococcygeus muscle and rat and rabbit portal veins, ACh, NAd, ATP and caffeine all activated the Ca-dependent Cl current (BYRNE and LARGE 1987, 1988; PACAUD et al. 1989; XIONG et al. 1991b; LOIRAND et al. 1991).

An increase in Cl conductance would depolarize the membrane, since the Cl equilibrium potential is around $-20\,\mathrm{mV}$ ($50\,\mathrm{m}M$ intracellular Cl concentration, CASTEELS et al. 1977a; AICKIN and VERMUÉ 1983). However, activation of the Cl channel may not make an important contribution to the NAd-induced depolarization of the membrane, because in the rabbit pulmonary artery depolarization induced by NAd ceased in Na-free solution, suggesting that the non-selective cation channel may play a major role in the depolarization (CASTEELS et al. 1977b; AMÉDÉE et al. 1990). A physiological role for the Cl channel has not yet been established in smooth muscle cells, though contributions to the regulation of cell volume, intracellular pH or resting Cl conductance have been discussed (FRANCIOLINI and PETRIS 1990).

D. Actions of Ca-Antagonists on the Membrane of Smooth Muscle Cells

SUMIMOTO et al. (1988) reported that, in single smooth muscle cells of the pig coronary artery, the K_d value of specifie nifedipine binding was significantly higher than that observed in a microsomal preparation without a change in the number of maximum binding sites (B_{max}) due to a difference in the membrane potential of intact single cells and microsomes. In each preparation, Hill's coefficient was close to 1.0, suggesting a 1:1 binding. PN200-110 binding on polarized and depolarized rat portal veins also showed that the K_d value of PN200-100 binding in depolarized tissue was lower than that in polarized tissue without a change in the B_{max} value (DACQUET et al. 1989). Therefore, depolarization of the membrane might change the binding affinity to DHP derivatives. This evidence favours a voltage-dependent inhibition of Ca-antagonists as observed in electrophysiological experiments. All Ca-antagonists bind to a single polypeptide (α_1-subunit) of the L-type Ca channel in various cells including vascular smooth muscle cells (NAITO et al. 1989); however, DHP and non-DHP derivatives are not bound to the same site on the subunit. Competitive experiments with [^{3}H]-labelled DHP derivatives showed that verapamil and diltiazem replaced the labelled DHP derivative in a non-competitive manner (GODFRAIND et al. 1986; SUMIMOTO et al. 1988). Verapamil and diltiazem reduced the B_{max} value but not the K_d value. On the other hand, flunarizine competitively replaced the DHP derivative from the binding sites without affecting B_{max} (SUMIMOTO et al. 1988). These observations indicate that the DHP-binding site is allosterically linked to non-DHP-binding sites (GODFRAIND et al. 1986).

It was thought that several different actions induced by Ca-antagonists might reflect differences in their binding sites to the Ca channel. One of the differences is that phenylalkylamine, but not DHP derivatives, showed "use-dependent" or "frequency-dependent" inhibition of the voltage-dependent Ca channel, which has been also found in cardiac muscle (EHARA and KAUFMANN 1978; LEE and TSIEN 1983). Such "use-dependent" inhibition of the Ca current by a phenylalkylamine derivative (gallopamil = D600) was also found in rabbit intestinal and vascular smooth muscle cells (TERADA et al. 1987a; HERING et al. 1989). These observations indicate that D600, and possibly other phenylalkylamine derivatives, do not inhibit the Ca channel in the resting state. D600 was also reported to reduce the time constant of current relaxation without a change in the rate of current activation in the guinea pig ventricle and rabbit ear arterial cells using Ba solution, indicating that D600 acts as an open Ca channel blocker (LEE and TSIEN 1983; HERING et al. 1989). However, the open channel blocking action did not participate in the "use-dependent" inhibition induced by D600, because DHP derivatives could also accelerate current relaxation. Another interpretation of the acceleration of the current relaxation by these Ca-antagonists might be that these drugs blocked the inactivated Ca channel rather than the open state during membrane depolarization. There are several pieces of evidence that these Ca-antagonists can bind to the inactivated state with a high affinity (LEE and TSIEN 1983; GODFRAIND et al. 1986; TERADA et al. 1987a; SUMIMOTO et al. 1988; HERING et al. 1989). This effect slows the transition rate from inactivated state to resting state, and, as a consequence, the number of available Ca channels decreases during membrane depolarization. Therefore, both the lack of a blocking action in the resting state and the slow rate of the recovery might be involved in the mechanisms underlying the "use-dependent" or "frequency-dependent" action of phenylalkylamine derivatives.

Block of the Ca channel by DHP derivatives has been characterized as "tonic block" (the Ca current was immediately inhibited to a certain amplitude after administration of the DHP derivatives). As, even at a holding potential of, or more negative than, −100 mV, DHP derivatives were able to inhibit the Ca current, DHP derivatives were thought to block the resting Ca channel. A very clear effect of the DHP derivatives on the L-type Ca channel was the marked shift of the voltage-dependent inhibitory curve of the Ca current to the left (in cardiac cells, SANGUINETTI and KASS 1984; HUME 1985; SANGUINETTI et al. 1986; in smooth muscle cells, BEAN et al. 1986; OKABE et al. 1987b; TERADA et al. 1987a,b; AARONSON et al. 1988; HONORÉ et al. 1989b; NELSON and WORLEY 1989). As voltage-dependent inactivation of the Ca current reflects the ratio between the Ca channels in the resting and inactivated states, the actions of DHP derivatives have been interpreted to reflect high affinity in the inactivated state. Using a technique for rapid drug application, HERING et al. (1988) reported that nifedipine blocked the Ca channel in the resting state, but not in the open or inactivated state, because nifedipine did not change the rate of current relaxation during a depolarization but did inhibit the amplitude of the Ca current evoked by the next depolarization. In this mechanism, the voltage-dependent inhibition of the Ca current by nifedipine could be explained by a voltage-dependent change of the binding affinity to the resting Ca channel. On the other hand, IMAIZUMI et al. (1989b) reported that nifedipine accelerated the Ca current relaxation. This discrepancy might be due to differences in the pulse duration used for producing depolarization. The duration of the current measurement (6s) used by HERING et al. (1988) might be insufficient to adequately reveal the changes in the current decay produced by nifedipine. As D600, which did not produce the resting channel block, also shifted the voltage-dependent inactivation curve, the shift of the voltage-dependent inactivation curve induced by nifedipine may not relate to a channel block in the resting state.

The actions of Ca-antagonists, especially DHP derivatives, on the unitary Ca channel current (unitary Ba current) have also been investigated in several smooth muscle cells (INOUE et al. 1989b; NELSON and WORLEY 1989). In smooth muscle cells of the rabbit ileum longitudinal muscle, INOUE et al. (1989b) reported that nifedipine inhibited the amplitude of the summated single-channel current, due to an increase in the percentage of test puless without channel opening. The same conclusion was also reached for the rabbit mesenteric artery using nisoldipine (NELSON and WORLEY 1989). By contrast, Bay K8644, a Ca-agonist, increased the percentage of test pulese by which channel opening was evoked.

In 1984, HESS et al. proposed a modal model in which the Ca channel possesses fundamentally three different modes (modes 0, 1 and 2), and each mode consists of three channel states (two closed and one open state; the mean open time of mode 2 being much longer than that of mode 1, mode 0 having no open state). Ca-antagonists shift the channel mode to mode 0, and Ca-agonists shift it to mode 2 from mode 1, which is a basal mode for the L-type Ca channel in cardiac cells. In this model, Ca-antagonists and -agonists do not change the kinetics of each state within a mode. KAWASHIMA and OCHI (1988) also demonstrated slow changes in the open probability by analysis of the single-channel current obtained from several thousands of pulses applied repetitively. They found that nitrendipine did not change the averaged mean open time of the channel opening, but reduced the percentage of test pulses which produced channel opening (channel avail- ability). They proposed a three-state model for the slow changes in channel availability (one available and two unavailable states). On the other hand, LACERDA and BROWN (1989) proposed a modified $C_1 \rightarrow C_2 \rightarrow O_1$ model for the action of (−) Bay K8644, because kinetic parameters predicted by the modal model did not fit their observations. They speculated that either closed state binds one DHP molecule in the transition to C_1^* or

C_2^*, then C_2^* binds another DHP molecule with high affinity in the transition to O_1^{**}. They also predicted another low-affinity binding of DHP with O_1 to the third open state (O_1^*) instead of to the inactivated state.

No detailed investigation has been performed of the kinetic properties of the voltage-dependent Ca channel in smooth muscle cells, but INOUE et al. (1989b) observed several interesting behavioural features of the unitary Ca channel current in smooth muscle cells of the rabbit small intestine. They found that two different open states could be classified according to changes in the frequency of stimulation (membrane depolarization). With high frequency ($\geq$0.5 Hz), the open-time histogram was fitted by an exponential, more than half the number of stimulations failed to open the channel (mode 0) and the mean open probability of each trace in which the channel open was seen was less than 0.3 (mode 1). On the other hand, with low frequency ($\leq$0.1 Hz), the mean open-time histogram was fitted by two exponentials (fast and slow components), and the fractional appearances of mode 0 and mode 2 were reduced and increased, respectively. Such frequency-dependent change of the channel mode has not yet been reported in other cells; however, PIETROBON and HESS (1990) recently reported that, in rat ventricle cells, strong depolarizations drove the channel from mode 1 to mode 2 (long openings and high open probability), suggesting that the transition between modes 1 and 2 was voltage dependent. In cardiac cells, cAMP and isoprenaline significantly enhanced the appearance of long openings with high open probability. Furthermore, even in the absence of isoprenaline or strong depolarizations, PIETROBON and HESS (1990) occasionally observed long openings of the L-type channel with high open probability lasting for seconds or minutes. These observations suggest that an unknown third mechanism for channel activation may be at work under drug-free and strong-depolarization-free conditions, in addition to the conformational changes of the Ca channel by membrane depolarization and channel phosphorylation. The frequency-dependent changes of the channel modes observed in ileal smooth muscle cells may be caused by the third or even a fourth mechanism. In each mode, as estimated from the open and closed time histograms, the Ca channel in the ileal smooth muscle had one open and two closed states, as in cardiac cells (HESS et al. 1984).

In rabbit ileal smooth muscle cells, in addition to the shift of the channel mode, nifedipine also modified the channel kinetics within a mode (reduced the mean open time and increased the mean closed time between the bursts in modes 1 and 2). This result differed from those observed in guinea pig ventricular cells using Bay K8644 and nitrendipine, where no change of the mean open and closed times was reported either within a mode (HESS et al. 1984) or in the available state (KAWASHIMA and OCHI 1988). This difference may reflect a difference of the channel kinetics in cardiac and smooth muscle cells. The values of the mean open and closed times of the Ca channel in ileal smooth muscle cells were much slower than those observed in cardiac ventricle cells. If this difference was caused by use of the filter with different cutoff frequencies for smooth muscle and cardiac cells, these results provide us with an another important property of the smooth muscle Ca channel, namely that the Ca channel in smooth muscle is thought to open in bursts, and Ca-antagonists reduce the mean burst duration rather than the mean open time of the channel within a burst.

When the Ca channel of ileal smooth muscle cells was in mode 2, the amplitude of the summated unitary Ca channel currents persisted during the stimulation (150 ms), but nifedipine enhanced the current relaxation by a reduction in the mean open time of the long opening. This indicated that nifedipine acted as an open channel blocker. On the other hand, as nifedipine did not modify the latency of the first channel opening after membrane depolarization, the sequence $C_1 \rightarrow C_2 \rightarrow O$ within each mode may not be modified by the drug. Therefore, INOUE et al. (1989b) concluded that the main mechanism underlying the inhibitory actions of nifedipine on the Ca channel in ileal smooth muscle cells was a reduction of the channel availability, as also reported in cardiac cells (KAWASHIMA and OCHI 1988), as well as a reduction of the rate from O to C_1 and from C_1 to C_2 in each mode. KAWASHIMA and OCHI (1988) and NELSON and WORLEY (1989) also reported that the channel availability decreased in a voltage-dependent manner in cardiac and rabbit mesenteric arterial cells, indicating that an unavailable state, including the prestimulation period, corresponded with the inactivated state, while the prestimulation period followed by the channel-available state (open state) induced by membrane depolarization could be categorized as the resting state. As the Ca-antagonists produce a negative shift of the channel availability curve (KAWASHIMA and OCHI 1988) and elongate the inactivated states, it is concluded that these drugs can bind the Ca channel in the inactivated state with a high affinity rather than in the resting state, and reduce the transition rate from inactivated state to resting state.

E. Molecular Mechanisms Underlying the Actions of Ca-Antagonists on the Ca Channel

The Ca channel in smooth muscle cells that is sensitive to DHP derivatives presumably has nearly the same molecular structure as those in cardiac and skeletal muscle cells, because those Ca channels have similar biophysical, physiological and pharmacological properties. Recent experiments involving cDNA and the expression of mRNA showed that the DHP-sensitive Ca channel in the rat aorta and rabbit airway smooth muscle cells had partly different amino acid sequences from those in cardiac and skeletal muscle cells, indicating that there is tissue specificity of the Ca channel (SLISH et al. 1989; VARADI et al. 1989; KOCH et al. 1989, 1990; BIEL et al. 1990). The DHP-sensitive Ca channels purified from rabbit skeletal muscle were composed of α_1-, α_2-, β-, γ- and δ-subunits (TAKAHASHI et al. 1987; TANABE et al. 1987; CATTERAL et al. 1989). The α_1-subunit, which consists of 1873 amino acids and has a molecular size of 170 kDa, contains the binding site for both DHP and non-DHP type Ca-antagonists, multiple phosphorylation sites and hydrophobic transmembrane segments, and forms a channel pore (STRIESSNIG et al. 1987; TAKAHASHI et al. 1987; TANABE et al. 1987; CATTERAL et al. 1989). On the other hand, the α_1-subunit obtained from airway smooth muscle cells has a larger molecular weight (242.5 kDa) and consists of 2166 amino acids, while the molecular weight of the α_1-subunit of the aortic Ca channel is 243.6 kDa (2169 amino acids). Both smooth muscle α_1-subunits had 65% homology with that of skeletal muscle cells. The aortic α_1-subunit had 93% homology with the cardiac subunit and that from the airway smooth muscle cells had only four small different regions in the amino acid sequence by comparison with the cardiac α_1-subunit, suggesting that both smooth muscle and cardiac Ca channels have the same gene (BIEL et al. 1990; KOCH et al. 1990). From the reconstituted vesicle after purification, a single-channel conductance of 20 pS could be recorded and this was activated by Bay K 8644 and inhibited by D600, suggesting that the purified DHP-binding protein obtained from skeletal muscle cells had the properties of the L-type Ca channel (FLOCKERZI et al. 1986). Injection of mRNA into Xenopus leaves oocytes synthesized the nifedipine-sensitive Ca channel (BIEL et al. 1990). The α_1-subunit contains four internal repeats (domains I, II, III, and IV) and each internal repeat has five hydrophobic segments (S1, S2, S3, S5 and S6) and one positively charged segment (S4), which presumably acts as the voltage sensor of the Ca channel (TANABE et al. 1987; KOCH et al. 1989, 1990; BIEL et al. 1990). The α_1-subunit is the substrate for the cAMP-dependent protein kinase in skeletal, cardiac and smooth muscle cells, but it is uncertain whether or not such cAMP-dependent phosphorylation sites have any function in smooth muscle cells, because cAMP did not potentiate the amplitude of voltage-dependent Ca currents in several smooth muscle preparations (KLÖCKNER and ISENBERG 1985; OHYA et al. 1987c). However, it should be noted that in certain smooth muscle cells, such as the pig coronary artery and A7r5 cell lines from the rat aorta, isoprenaline and forscolin, a direct activator of the adenylate cyclase, augmented the voltage-dependent Ca channels, suggesting that the L-type channel in these smooth muscle cells might have a similar molecular structure to those in the skeletal and cardiac muscle cells (FUKUMITSU et al. 1990; MARKS et al. 1990).

STRIESSNIG et al. (1987, 1990) found that a phenylalkylamine derivative, LU49888, bound the region between the 1349th glutamate and the 1391st tryptophan within the α_1-subunit of skeletal muscle cells and in this region the sequence of amino acids was well preserved in cardiac and smooth muscle cells (MIKAMI et al. 1989; BIEL et al. 1990). This segment of the α_1-subunit contains transmembrane helix S6 of domain IV and seven extracellular and nine intracellular residues. As HESCHELER et al. (1982) reported that phenylalkylamine derivatives could inhibit the L-type Ca channels from the internal surface of the membrane, STRIESSNIG et al. (1990) estimated that LU49888 probably interacts with the intracellular end of helix S6 of domain

IV and the adjacent intracellular nine amino acid residues (the beginning of the long intracellular C-terminal tail). However, in smooth muscle cells of the rabbit portal vein and small intestine, OHYA et al. (1987b) showed that D600 inhibited the voltage-dependent Ca current only on extracellular and not on intracellular application. It is also interesting that the inhibitory actions of D600 and verapamil on the voltage-dependent Ca current in smooth muscle cells are more than 100–1000 times less potent than those in cardiac cells (HESCHELER et al. 1982; LEE and TSIEN 1983; TERADA et al. 1987a; OHYA et al. 1987b; HERING et al. 1989). These results suggest that high-affinity binding sites for phenylalkylamine derivatives identified in the skeletal muscle and possibly cardiac Ca channel may not be present in the smooth muscle Ca channel, although the amino acid sequences of both muscle Ca channels were well preserved. As the amino acid sequence of the phenylalkylamine binding site of the smooth muscle α_1-subunit is identical with that of the cardiac subunit, other segments in the α_1-subunit, which are not preserved in the smooth muscle α_1-subunit (1–16, 322–354, 464–489 and 1302–1330 amino acid residues; BIEL et al. 1990), may also be involved in the high-affinity binding site for phenylalkylamine derivatives which produces the channel block at low concentrations.

On the basis of the results of DHP-binding studies, the binding site for DHP derivatives was thought to be different from the phenylalkylamine or benzothiazepine binding site, but to be close enough to interact. This means, in the case of the cardiac Ca channel, that the DHP receptor is also located in the channel pore facing the cytosol but may not be in the case of smooth muscle cells. When DHP derivatives were applied intracellularly, no inhibition of the voltage-dependent Ca current was induced in cardiac and smooth muscle cells (OHYA et al. 1987b; KASS and ARENA 1989; KASS et al. 1989), suggesting that DHP derivatives only have access to their receptor site from the outer surface. Furthermore, it is well known that DHP derivatives inhibit the single Ca channel current recorded in the cell-attached condition, when drugs are applied in the bath. Therefore, the DHP molecule did not seem to have access to the channel pore directly (through a hydrophilic pathway), but seemed to reach the receptor by lateral diffusion in the membrane (through a hydrophobic pathway). As both neutral and ionized forms of the DHP derivative amlodipine inhibited the Ca current in cardiac cells, KASS and ARENA (1989) speculated that the DHP receptor is buried in the lipid bilayer adjacent to the external end of the channel as proposed by VARDIVIA and CORONADO (1988) using the skeletal muscle Ca channel, incorporated into lipid bilayers. As the pharmacological profiles of the smooth muscle Ca channels response to DHP derivatives were similar to those of cardiac and skeletal muscle Ca channels, we can postulate that the DHP receptor in the smooth muscle Ca channel is also located at a site near the lipid bilayer adjacent to the external region of the channel, and the amino acid residues specific to the smooth muscle α_1-subunit may not interact with the DHP receptor.

F. Heterogeneity of the K Channel Targeted by K-Channel Openers

Many reports have confirmed that the ATP-sensitive K channel is activated by various K-channel openers in cardiac and pancreatic β-cells (ESCANDE et al. 1988, 1989; SANGUINETTI et al. 1988; ARENA and KASS 1989; HIRAOKA and FAN 1989; FAN et al. 1990a; NAKAYAMA et al. 1990). However, the pharmacological profiles of the K channels opened by K-channel openers in various smooth muscle cells are not uniform. For example, the hyperpolarization induced by K-channel openers was Ca independent in the guinea pig mesenteric artery but partly Ca dependent (slow component) in the guinea pig mesenteric vein. Moreover, cromakalim opened a K channel in the guinea pig urinary bladder which had a different permeability to Rb than that observed in the rat portal vein (NAKAO et al. 1988; FOSTER et al. 1989), suggesting that at least two different K channels could be activated by K-channel openers.

Single-channel current recording elicited more than two types of K channel which were activated by K-channel openers. In bovine and rabbit cultured aortic smooth muscle cells, cromakalim activated the large conductance Ca-dependent K channel (200 pS with symmetrical 150 mM K solutions on both sides of the membrane) which was inhibited by scorpion toxin (KUSANO et al. 1987). However, higher concentrations of TEA were required to inhibit this K channel than to block the ordinary "maxi-K" channel. HERMSMEYER (1988) has reported that pinacidil activates the large conductance Ca-dependent K channel, presumably the "maxi-K" channel, in cultured smooth muscle cells of the rat azygos vein. Furthermore, GELBAND et al. (1989), using the microsomal preparations obtained from the rabbit thoracic aorta, have reported that cromakalim increases the open probability of the large conductance Ca-dependent K channel by reducing the closed time (slow component) without a change in the mean open time. They concluded that the "maxi-K" channel is also a target for K-channel openers, as is the ATP-sensitive K channel in smooth muscle cells. On the other hand, in the rat and rabbit portal veins, nicorandil and pinacidil did not modify the activities of the "maxi-K" channel, which was sensitive to low concentrations of TEA ($\leq$1 mM) and to charybdotoxin but insensitive to 4-AP ($\geq$10 mM) and glibenclamide (KAJIOKA et al. 1990, 1991; XIONG et al. 1991). Therefore, they concluded that the target for K-channel openers was not the "maxi-K" channel commonly distributed in these smooth muscle cells.

Although the single-channel conductance and Ca sensitivity of the cromakalim-activated K channel in the rabbit and bovine aorta was quite similar to those of the "maxi-K" channel in various smooth muscle cells, the K-channel opener-sensitive K channel was resistant, but the "maxi-K" channel was very sensitive, to TEA (KUSANO et al. 1987). Therefore, we cannot exclude the possibility that the large conductance Ca-dependent K channel activated by K-channel openers in the rabbit and bovine aorta is a different channel from the "maxi-K" channel, and that there is a subtype of the "maxi-K" channel which is activated by K-channel openers and inhibited by glibenclamide in smooth muscle cells.

STANDEN et al. (1989) found the ATP-sensitive K channel in smooth muscle cells of the rabbit mesenteric artery. This channel was activated by cromakalim and was inhibited by glibenclamide. The single-channel conductance of this ATP-sensitive K channel in the mesenteric artery was relatively large by comparison with those observed in cardiac and pancreatic β-cells. Activation of this arterial ATP-sensitive K channel could be seen in Ca-free solution, suggesting that it was a Ca-insensitive K channel similar to the ATP-sensitive K channel in the cardiac and pancreatic β-cells.

In the rat portal vein, OKABE et al. (1990) reported that cromakalim evoked an outward K current which was blocked by glibenclamide. However, the amplitude of the cromakalim-induced K current in the rat portal vein depended upon the extracellular Ca concentration. The single-channel conductance was very small by comparison with that reported in the rabbit mesenteric artery (10 vs. 80 pS in low-K/high-K conditions). The K channel observed in the rat portal vein was also sensitive to intracellular Ca and ATP concentrations, and inhibited by micromolar concentrations of glibenclamide and millimolar concentrations of both TEA and 4-AP (KAJIOKA et al. 1990). This indicates that K-channel openers activate the ATP-sensitive K channel in the rat portal vein, but that the Ca sensitivity of this K channel differs from that observed in the rabbit mesenteric artery, and also in cardiac and pancreatic β-cells.

Other types of ATP-sensitive K channel have been reported in the pig coronary artery and rabbit portal vein (INOUE et al. 1989a; KAJIOKA et al. 1991). In the pig coronary artery, nicorandil opened a small conductance K channel (30 pS), which was activated by extracellular Ca and inhibited by 5 mM 4-AP. On the other hand, in the rabbit portal vein, KAJIOKA et al. (1991) found that pinacidil activated a novel type of ATP-sensitive K channel (25 pS), sensitive to neither extra- nor intracellular Ca concentration.

From results obtained using whole-cell voltage-clamp techniques, BEECH and BOLTON (1989b) concluded that cromakalim activated the delayed outward current (voltage-dependent K current) in the rabbit portal vein. On the other hand, OKABE et al. (1990) reached the opposite conclusion: that cromakalim inhibited the voltage-dependent K current recorded in the rat portal vein. Although BEECH and BOLTON (1989b) excluded the possible activation of the ATP-sensitive K channel by cromakalim, they also found that glibenclamide completely inhibited the cromakalim-induced K current, but only partly inhibited the delayed K current. Using the same preparation (rabbit portal vein), KAJIOKA et al. (1991) found that pinacidil and cromakalim activated the ATP-sensitive K channel. One of interesting properties of this ATP-sensitive K channel in the rabbit portal vein is that ATP (100 μM) without $MgCl_2$, applied intracellularly, halved the open probability of the ATP-sensitive K channel, but Mg-ATP had no inhibitory effect on the K channel (KAJIOKA et al. 1991). As BEECH and BOLTON (1989b) applied ATP (1 m*M*) intracellularly with 1.2 m*M* Mg in the pipette solution, they probably failed to see the inhibitory action of ATP on the cromakalim-induced K current, because the calculated free ATP concentration might be very low. POST et al. (1991) also excluded a possible involvement of the delayed K current in the current induced by the K-channel openers in the rabbit colon, as the current induced by these drugs was activated without delay.

Therefore, we can conclude that K-channel openers are able to activate at least two types of K channel in various smooth muscle cells, namely, the large conductance Ca-dependent K ("maxi-K") and ATP-sensitive K channels. However, the pharmacological and physiological properties of the "maxi-K" channel and the hyperpolarization induced by the K-channel openers suggest that they are not identical, and several reports have rejected the possibility that the "maxi-K" channel is a target for K-channel openers. Therefore, even though the "maxi-K" channel is also activated by K-channel openers, the "maxi-K" channel may not be a major target channel for these drugs. As the characteristics of the ATP-sensitive K channel differed across tissues and species, more effort must be exerted in this field to clarify the mechanisms of action of K-channel openers.

G. Properties and Gating Mechanisms of the ATP-Sensitive K Channel

Although several ATP-sensitive K channels have been reported to be present in smooth muscle cells, there have been no detailed reports of the biophysical characteristics of the ATP-sensitive K channel. The first report on the ATP-sensitive K channel made by NOMA (1983), who used guinea pig and rabbit ventricular and atrial cells, clearly showed that this channel had an inward rectifying property. This rectification was mainly caused by internal Na and Mg, which plugged the channel from its inner mouth (HORIE et al. 1987). However, another intrinsic factor (a soluble substance of large molecular size) may also contribute to the inward rectification of the ATP-sensitive K channels, because the inward rectifying property was still seen in the patch membrane of permeabilized cells, but not in the inside-out patch membrane (NICHOLS and LEDERER 1990).

In general, the ATP-sensitive K channel in cardiac and pancreatic β-cells (insulin-secreting cultured cells) is said to have one or two open and two closed states (KAKEI et al. 1985; HORIE et al. 1987; SPRUCE et al. 1987; ASHCROFT 1988). However, a more complicated model with one or two open and at least three closed states has been thought to account for the gating kinetics of the ATP-sensitive K channels (DAVIES et al. 1989; FAN et al. 1990a,b). HORIE et al. (1987) demonstrated that intracellular Mg ion blocked the ATP-sensitive K channel with an accompanying reduction in the mean open time and the open and closed time histograms were easily affected by changes in ionic conditions. Furthermore, since activation and inhibition of the ATP-sensitive K channel were recorded in the presence of various nucleotides, such as ATP, ADP, GTP and others, the experimental conditions employed by various investigators appear to modify the channel kinetics.

I. Inhibition by ATP and Its Analogues

The open probability of the ATP-sensitive K channels was reduced by micro- to submillimolar concentrations of intracellular ATP, and application of several-millimolar concentrations of ATP completely blocked the channel opening. Opening of the ATP-sensitive K channel (1 to a few milliseconds) seemed to occur in bursts with a very short shut time in guinea pig ventricular cells and insulin-secreting cell lines (0.3–0.4 ms; FAN et al. 1990a; RIBALET et al. 1989), and possibly in smooth muscle cells of the rabbit mesenteric artery, and rat and rabbit portal veins (NELSON et al. 1990; KAJIOKA et al. 1990, 1991) when the noise was cut at a low frequency ($\leq$300 Hz). The mean open and closed times observed in the rat and rabbit portal veins were larger than those reported in cardiac and pancreatic cells. ATP did not modify the mean open and fast closed times, but reduced the burst duration and increased the longer closed time in guinea pig ventricular cells (FAN et al. 1990a). As the ATP-sensitive K channel showed one open and three closed states, FAN et al. (1990a) proposed the following kinetic scheme: $C_3 \leftrightarrow C_2 \leftrightarrow O \leftrightarrow C_1$, where C_1 is the fast closing state and $O \leftrightarrow C_1$ forms the burst. They estimated that ATP modifies the transitions between C_3, C_2 and O (reduced the forward rate and increased the backward rate). In the rabbit portal vein, the ATP-sensitive K channel had one open and two closed states ($C_2 \leftrightarrow C_1 \leftrightarrow O$), and ATP reduced the mean open time by chopping up the long opening of the ATP-sensitive K channel and increasing the longer closure period (KAJIOKA et al. 1991). However, this does not simply mean that the ATP-sensitive K channel in the two types of muscle cell had different inhibitory mechanisms, as smooth muscle K channels have not shown very fast closed states due to noise reduction. Thus, a kinetic scheme for the ATP-sensitive K channel in the rabbit portal vein may be $C_3 \leftrightarrow C_2 \leftrightarrow O_{BURST}$, and we can suggest that ATP reduced the forward rate to the burst open state and increased the backward rates, as proposed in cardiac cells.

The number of binding sites for ATP molecules was thought to be two in cardiac and smooth muscle cells (FINDLAY 1988; LEDERER and NICHOLS 1989; KAJIOKA et al. 1991) whereas only one inhibitory binding site for ATP was thought to exist in the pancreatic β-cell (KAKEI et al. 1986). ADP also had a weak inhibitory effect with a low Hill coefficient (= 1.0), but AMP had no action on the activity of the ATP-sensitive K channel in cardiac cells (NOMA 1983; LEDERER and NICHOLS 1989). The inhibitory action of ATP on the ATP-sensitive K channel was not modified by the simultaneous presence of Mg ion in heart and rabbit mesenteric arterial muscle cells, but the ATP-sensitive K channel was apparently not blocked by Mg-ATP in pancreatic β-cells or in the smooth muscle cells of the rat and rabbit portal veins (ASHCROFT and KAKEI 1987; STANDEN et al. 1989; LEDERER and NICHOLS 1989; KAJIOKA et al. 1990, 1991). The inhibitory actions of ATP on the ATP-sensitive K channel did not require ATP hydrolysis, for GTP, UTP, CTP and non-hydrolyzable ATP analogues such as AMP-PNP, AMP-PCP and ATPγS also effectively blocked the channel, whereas AMP, GDP and adenosine failed to block it (COOK and HALES 1984; KAKEI et al. 1985; SPRUCE et al. 1985; MISLER et al. 1986; OHNO-SHOSAKU et al. 1987; NOMA and SHIBASAKI 1988; LEDERER and NICHOLS 1989). The order of potency of the inhibitory effects induced by nucleotides in rat ventricular cells was ATP > AMP-PNP > ADP > CTP > GDP = AMP = ITP, and the channel inhibition by ADP or GTP occurred additively with the ATP-induced inhibition, when Mg ion was absent from the cell. When 0.5 mM Mg and 40 μM ATP were present, ADP and GDP restored the channel activity to the same level as that observed in the ATP-free condition (LEDERER and NICHOLS 1989). Therefore, these authors proposed two binding sites for nucleotides (sites 1 and 2). Binding of ATP or AMP-PNP to site 1 increases the affinity of site 2 for the binding of ATP or other nucleotides, and binding to site 2 is essential to block the channel. Mg-ADP and Mg-GDP can compete with ATP at site 1, and reduce ATP binding to site 1 (so that ATP binding to site 2 is then prevented). In the smooth muscle cells of the rabbit portal vein, KAJIOKA et al. (1991) reported that the ATP-sensitive K channel activated by pinacidil was not inhibited by GTP, GTPγS, GDP or GMP, suggesting characteristic differences between the ATP-binding sites in the channels of smooth muscle cells from those in cardiac and pancreatic cells (GTP and GDP had the ability to activate the channel in the absence of Mg as a cofactor).

II. Reactivation by ATP and Its Analogues

In most cells, the activity of the ATP-sensitive K channel was inactivated rapidly ("rundown") after excision of the patch membrane in ATP-free conditions (cardiac cells: TRUBE and HESCHELER 1984; FINDLAY and DUNNE 1986; pancreatic cells: MISLER et al. 1986; OHNO-SHOSAKU et al. 1987; smooth muscle cells: STANDEN et al. 1989; KAJIOKA et al. 1991). However, the channel activity reappeared on application of low concentrations of ATP ($\leq$10 μM) or after removal of high concentrations of ATP ($\geq$1 mM) in the presence of Mg (FINDLAY and DUNNE 1986; MISLER et al. 1986; OHNO-SHOSAKU et al. 1987; RIBALET et al. 1989; TAKANO et al. 1990). Reactivation of the channel by ATP requires the copresence of Mg ions (FINDLAY 1987), and non-hydrolyzable ATP analogues, such as AMP-PNP, AMP-PCP, ATPγA and ADPβS, have little or no ability to reactivate the channel

(DUNNE and PETERSEN 1986a; OHNO-SHOSAKU et al. 1987; TAKANO et al. 1990). These findings strongly indicate that, in cardiac and pancreatic β- cells, ATP-hydrolysis is essential to keep the ATP-sensitive K channel active (the operative state), probably due to phosphorylation of the channel protein, and that in ATP-free conditions the channel enters an inoperative state due to dephosphorylation of the channel.

Other mechanisms may also be involved in channel reactivation in pancreatic β-cells and smooth muscle cells. In pancreatic β-cells, GTP, GTPγS, GDP and GDPβS have been reported to activate the ATP-sensitive K channel in the presence of Mg (DUNNE and PETERSEN 1986b). Neither GTP-binding protein nor phosphorylation of the channel molecule is involved in this activation mechanism, but direct binding to the channel may account for the activation, that is, Mg-guanosine analogue complexes may bind the activating sites of the ATP-sensitive K channel or may compete with ATP at ATP-binding sites (presumably site 1 proposed by LEDERER and NICHOLS 1989).

In the rabbit portal vein, KAJIOKA et al. (1991) found a novel reactivating mechanism for the ATP-sensitive K channel. This channel was easily inactivated by membrane excision and permeabilization (the inoperative state). The channel was not reactivated by application of ATP, ADP or AMP applied intracellularly in the presence of Mg, but it was reactivated by administration of either GDP or GTP, but not of GMP. As GTPγS and GDPβS had no ability to mimic the actions of GDP and GTP had only a weak action by comparison with the GDP-induced reactivation, KAJIOKA et al. (1991) concluded that GTP-binding protein was not involved in the reactivation mechanism of the ATP-sensitive K channel in the rabbit portal vein. Although only hydrolyzable GDP analogues could activate the channel, the mechanism did not require Mg ions, because addition of Mg to GDP or GTP did not modify their actions. Furthermore, they speculated that the binding sites (Hill coefficient is 2) were selective for the diphosphate forms of guanosine, because GDPβS could inhibit, but neither GTP nor $GTP_{\gamma}S$ modified, the channel activity after reactivation by GDP. Therefore, it seems that both steps are essential for activation of the channel by GDP, that is, GDP-binding to the GDP-binding sites and GDP hydrolysis (phosphorylation?).

In guinea pig ventricular cells, TUNG and KURACHI (1991) also found that UDP, IDP, GDP, CDP and ADP had the ability to activate the ATP-sensitive K channel in the inoperative state, but only when Mg ions were present. They estimated that Mg-nucleotide diphosphates (NDPs) may bind at binding sites different from the ATP-induced phosphorylation site, then activate the K channel independently; however, both activating mechanisms possibly work on the same activation gate in the channel. Thus, the reactivating mechanisms of the ATP-sensitive K channel in the rabbit portal vein and guinea pig ventricular cells seemed to be quite similar, except for the requirement for Mg ion. If various ATP-sensitive K channels have this mechanism (GDP or ADP activation), then cellular metabolic conditions which change the cellular concentrations of ADP, ATP, GDP, GTP and other nucleotide may drastically influence the activity of the ATP-sensitive K channel.

H. Mechanisms of Action of K-Channel Openers on the ATP-Sensitive K Channel

Pinacidil did not modify the mean open and fast closed times, but increased the burst duration and reduced the longer closed times in guinea pig ventricular cells (FAN et al. 1990a). In smooth muscle cells [although STANDEN et al. (1989) did not report the values of the mean open and closed times in the presence and absence of cromakalim] cromakalim seems to keep the mean open and closed times constant in the rabbit mesenteric artery. On the other hand, in the rat portal vein, nicorandil and pinacidil both increased the mean open time of the ATP-sensitive (and Ca-dependent) K channel, but, as mentioned before, the very fast component of the closed state (burst state) in the rat portal vein was filtered off (KAJIOKA et al. 1990). Therefore, the results observed in the rat portal vein were not inconsistent with those in guinea pig ventricular cells (FAN et al. 1990a).

The binding sites for K-channel openers are thought to be located at the intracellular face of the cardiac K channel; however, nicorandil was reported to be less effective on the ATP-sensitive K channel in the rat portal vein and pig coronary artery when it was applied intracellularly (HIRAOKA and FAN 1989; KAJIOKA et al. 1990; INOUE et al. 1990a; TUNG and KURACHI 1991). On the other hand, pinacidil or cromakalim could activate the ATP-sensitive K channel from the inside of the membrane in the rat portal veins and pig coronary artery (KAJIOKA et al. 1990, 1991; INOUE et al. 1990a). These results indicate that nicorandil may bind to a different binding site (extracellular face) from those for the other K-channel openers (intracellular face) in the rat portal vein and pig coronary artery. Recently, SHEN et al. (1991) showed that nicorandil activated the cardiac ATP-sensitive K channel from the inside of the membrane, when the channel was activated by NDPs, but not by ATP phosphorylation, while pinacidil and lemakalim activated the cardiac ATP-sensitive K channel when the channel had been activated by either NDPs or ATP phosphorylation. These results indicate that nicorandil may also be able to bind to a site in the channel facing the cytosol, activation of which only interacts with the NDP-binding sites, and other openers may bind to a different binding site(s) located in the same area, which is functionally related to both the NDP-binding and phosphorylation sites. As nicorandil's action in smooth muscle cells is observed in the absence of GDP, it is uncertain whether the insensitivity of the smooth muscle K channel to nicorandil (applied intracellularly) is caused by a lack of the nicorandil receptor at the intracellular face of the channel or by a lack of an activation mechanism on GDP binding. However, if we assume that the channel activation mechanisms are preserved equally well in the outside-out and inside-out conditions, and since activation of the ATP-sensitive K channel by nicorandil has been seen in the outside-out, but not in the inside-out condition, it is likely that the nicorandil receptor is on the extracellular surface of the ATP-sensitive K channel of smooth muscle cells.

Although SHEN et al. (1991) failed to see reactivation of the channel in the inoperative state by low concentrations of pinacidil ($\leq 30\,\mu M$), which produced a marked enhancement of the activity of the channel in the operative state or when activated by NDPs, FAN et al. (1990b) reported that pinacidil ($30\,\mu M$–$1\,mM$) had a reactivating action on the ATP-sensitive K channel in the inoperative state. On the other hand, in the rabbit portal vein, pinacidil ($\leq 0.3\,mM$) did not activate the ATP-sensitive K channel in the inoperative state, and GDP without pinacidil could not activate the channel in the inoperative state, that is, the simultaneous presence of pinacidil and GDP was required to activate the inoperative K channel in the rabbit portal vein (KAJIOKA et al. 1991). These results suggest that pinacidil and channel phosphorylation may interact with a common gate in the cardiac K channel, but in the ATP-sensitive K channel in the rabbit portal vein pinacidil and GDP binding and/or phosphorylation may interact with different gates (open ↔ close gate and operative ↔ inoperative gate).

I. Conclusion

Ca-antagonists interact not only with L-type Ca channels in various smooth muscle cells, but also with several different Ca channels which are not classified as typical L-type channels. In the past decade, the Ca channel current, among other voltage-dependent membrane currents, has been extensively investigated including its inhibition by Ca-antagonists and its regulation by receptor-agonists. However, the characteristics of the Ca channels in smooth muscle cells are not yet fully understood, especially those of the non-L-type channels. In general, the properties of the L-type channel in smooth muscle cells appear quite similar to those in other cells, especially cardiac cells; however, several differences have been reported, and these may be important in assessing the physiological role of the smooth muscle Ca channel. Although the channel structure of the L-type Ca channel in smooth muscle cells is almost the same as that in cardiac cells, such differences in functional properties may indicate that Ca-antagonists and channel molecules interact differently in different muscle cells.

A wider diversity of the target channel for K-channel openers than of those for Ca-antagonists has been reported in smooth muscle cell membranes. In several vascular smooth muscle cells, the presence has been reported of both Ca-dependent and -independent K channels which were activated by K-channel openers and inhibited by glibenclamide and intracellular ATP. These findings strongly suggest that heterogeneity of the K channel is present in smooth muscle cells and causes the different species and tissue responses to drugs. Although such heterogeneity of the ion channel is hindering development in this field, efforts to clarify the individual channels in various smooth muscle cells in various regions and species (including human) will lead to the discovery of new types of drugs which exhibit specific actions in particular cells.

Differences in the ion channels distributed in different tissues also inform us that they will serve different physiological functions. For instance, the physiological roles of the ATP-sensitive K channels seem to differ in the rat and rabbit portal veins and in the pig coronary and rabbit mesenteric arteries. The ATP-sensitive K channels in the rat portal vein and rabbit mesenteric and pig coronary arteries are spontaneously open under resting conditions (without K-channel openers), but the channel activity of the rat portal vein fluctuates during the excitation-relaxation cycle, due to its sensitivity to the intracellular Ca concentration. This channel in the rat portal vein may, at least, participate in the regulation of the duration of the burst discharge of action potentials. As the K channel in the pig coronary artery is activated in the presence of extracellular Ca ions, this channel has an important role in the production of the resting membrane potential. On the other hand, the ATP-sensitive K channel in the rabbit portal vein is closed in the absence of K-channel openers, suggesting no contribution to the resting membrane potential. From this arises another important property of this K channel, that is, endogenous or humoral activating factors may need to be present to activate the ATP-sensitive K channel. Indeed, it has already been reported that calcitonin gene-related peptide, somatostatin and galanin serve as intrinsic activators for the ATP-sensitive K channel in pancreatic β- and smooth muscle cells (De Weille et al. 1988, 1989a, b; Nelson et al. 1990). Further detailed studies are required on the ATP-sensitive K channel that is activated by various K-channel openers, including the role of intrinsic activators to clarify the physiological nature and role of this K channel.

Acknowledgement. The authors are grateful to Dr. R.J. Timms, Birmingham University, for language editing. This work was supported by a Grant-in-Aid for Scientific Research from the Ministry of Science, Education and Culture, Japan.

References

Aaronson PI, Bolton TB, Lang RJ, MacKenzie I (1988) Calcium currents in single isolated muscle cells from the rabbit ear artery in normal-calcium and high-barium solutions. J Physiol (Lond) 405:57–75

Aickin CC, Brading AF (1983) Towards an estimate of chloride permeability in the smooth muscle of guinea-pig vas deferens. J Physiol (Lond) 336:179–197

Aickin CC, Vermué NA (1983) Microelectrode measurement of intracellular chloride activity in the smooth muscle of guinea-pig ureter. Pflugers Arch 336: 179–197

Amédée T, Renaud JF, Jmari K, Lombet A, Mironneau J, Lazdunski M (1986) The presence of Na^+ channels in myometrial smooth muscle cells is revealed by specific neurotoxin. Biophys Biochem Res Commun 137:675–681

Amédée T, Benham CD, Bolton TB, Byrne NG, Large WA (1990) Potassium, chloride and non-selective cation conductances opened by noradrenaline in rabbit ear artery cells. J Physiol (Lond) 423:551–568
Arena JP, Kass RS (1989) Activation of ATP-sensitive K channels in heart cells by pinacidil. Am J Physiol 257:H2092–2096
Ashcroft FM (1988) Adenosine 5′-triphosphate-sensitive potassium channels. Annu Rev Neurosci 11:97–118
Ashcroft FM, Kakei M (1987) Effects of internal Mg^{2+} on ATP-sensitive K-channels in isolated rat pancreatic B-cells. J Physiol (Lond) 390:72P
Bean BP, Sturek M, Puga A, Hermsmeyer K (1986) Calcium channels in muscle cells isolated from rat mesenteric arteries: modulation by dihydropyridine drugs. Circ Res 59:229–235
Beech DJ, Bolton TB (1989a) A voltage-dependent outward current with fast kinetics in single smooth muscle cells isolated from the rabbit portal vein. J Physiol (Lond) 412:397–414
Beech DJ, Bolton TB (1989b) Properties of the cromakalim-induced potassium conductance in smooth muscle cells isolated from the rabbit portal vein. Br J Pharmacol 98:582–588
Beech DJ, Bolton TB (1989c) Two components of potassium current activated by depolarization of single smooth muscle cells from the rabbit portal vein. J Physiol (Lond) 418:293–309
Bendorf K, Boldt W, Nilius B (1985) Sodium current in single myocardial mouse cells. Pflugers Arch 404:190–194
Benham CD, Tsien RW (1987) A novel receptor-operated Ca^{2+}-permeable channel activated by ATP in smooth muscle. Nature 328:275–278
Benham CD, Bolton TB, Lang RJ (1985a) Acetylcholine activates an inward current in single mammalian smooth muscle cells. Nature 316:345–347
Benham CD, Bolton TB, Lang RJ, Tekewaki RJ (1985b) The mechanism of action of Ba^{2+} and TEA on single Ca^{2+}-activated K^+-channels in arterial and intestinal smooth muscle cell membranes. Pflugers Arch 403:120–127
Benham CD, Bolton TB, Denbigh JS, Lang RJ (1987a) Inward rectification in freshly isolated single smooth muscle cells of the rabbit jejunum. J Physiol (Lond) 383:461–476
Benham CD, Hess P, Tsien RW (1987b) Two types of calcium channels in single smooth muscle cells from rabbit ear artery studied with whole-cell and single-channel recordings. Circ Res 61 [Suppl 1]: I-10–I-16
Benham CD, Bolton TB (1986) Spontaneous transient outward currents in single visceral and vascular smooth muscle cells of the rabbit. J Physiol (Lond) 381: 385–406
Benoit E, Corbier A, Dubois J-M (1985) Evidence for two transient sodium currents in the frog node of Ranvier. J Physiol (Lond) 361:339–360
Berger W, Grygorcyk R, Shwarz W (1984) Single K^+ channels in membrane evaginations of smooth muscle cells. Pflugers Arch 402:18–23
Biel M, Ruth P, Bosse E, Hullin R, Stühmer W, Flockerzi V, Hofmann F (1990) Primary structure and functional expression of high voltage activated calcium channel from rabbit lung. FEBS Lett 269:409–412
Blatz AL, Magleby KL (1986) Single apamin-blocked Ca-activated K^+ channels of small conductance in rat cultured skeletal muscle. Nature 323:718–720
Bolton TB, Lim SP (1989) Properties of calcium stores and transient outward currents in single smooth muscle cells of rabbit intestine. J Physiol (Lond) 409:385–401
Bregestovski PD, Printseva OY, Serebryakov V, Stinnakre J, Turmin A, Zamoyski V (1988) Comparison of Ca^{2+}-dependent K^+ channels in the membrane of smooth muscle cells isolated from adult and foetal human aorta. Pflugers Arch 413:8–13

Burgess GM, Claret M, Jenkinson DH (1981) Effects of quinine and apamin on the calcium-dependent potassium permeability of mammalian hepatocytes and red blood cells. J Physiol (Lond) 317:67–90
Byrne NG, Large WA (1987) Actions of noradrenaline on single smooth muscle cells from the rat anococcygeus muscle. J Physiol (Lond) 389:513–525
Byrne NG, Large WA (1988) Membrane ionic mechanisms activated by noradrenaline in cells isolated from the rabbit portal vein. J Physiol (Lond) 404: 557–573
Caffrey JM, Josephson IR, Brown AM (1986) Calcium channels of amphibian stomach and mammalian aorta smooth muscle cells. Biophys J 49:1237–1242
Carbone E, Lux HD (1984) A low voltage-activated fully inactivating Ca channel in vertebrate sensory neurones. Nature 310:501–502
Casteels R, Kitamura K, Kuriyama H, Suzuki H (1977a) The membrane properties of the smooth muscle cells of the rabbit main pulmonary artery. J Physiol (Lond) 271:41–61
Casteels R, Kitamura K, Kuriyama H, Suzuki H (1977b) Excitation-contraction coupling in the smooth muscle cells of the rabbit main pulmonary artery. J Physiol (Lond) 271:63–79
Catteral WA, Seager MJ, Takahashi M, Nunoki K (1989) Molecular properties of voltage-sensitive calcium channels. Adv Exp Med Biol 255:101–109
Coleman HA, Parkington HC (1990) Hyperpolarization activated channels in myometrium: a patch clamp study. Prog Clin Biol Res 327:665–672
Cook DL, Hales CN (1984) Intracellular ATP directly blocks K^+ channels in pancreatic B-cells. Nature 311:271–273
Dacquet C, Loirand G, Rakotoarisoa L, Mironneau C, Mironneau J (1989) (+)-[^{3}H]-PN 200-110 binding to cell membranes and intact strips of portal vein smooth muscle: characterization and modulation by membrane potential and divalent cations. Br J Pharmacol 97:256–262
Davies NW, Spruce AE, Standen NB, Stanfield PR (1989) Multiple blocking mechanisms of ATP-sensitive potassium channels of frog skeletal muscle by tetraethylammonium ions. J Physiol (Lond) 413:31–48
De Weille JR, Lazdunski M (1990) Regulation of the ATP-sensitive potassium channel. In: Narahashi T (ed) Ion channels, vol 2. Plenum, New York, pp 205–222
De Weille JR, Schmid-Antomarchi H, Fosset M, Lazdunski M (1988) ATP-sensitive K^+ channels that are blocked by hypoglycaemia-inducing sulphonylureas in insulin-secreting cells are activated by galanin, a hyperglycaemia-inducing hormone. Proc Natl Acad Sci USA 85:1312–1316
De Weille JR, Schmid-Antomarchi H, Fosset M, Lazdunski M (1989a) Regulation of ATP-sensitive K^+ channels in insulinoma cells: activation by somatostatin and protein kinase C and the role of cAMP. Proc Natl Acad Sci USA 86:2971–2975
De Weille JR, Fosset M, Mourre C, Schmid-Antomarchi H, Bernardi H, Lazdunski M (1989b) Pharmacology and regulation of ATP-sensitive K^+ channels. Pflugers Arch 414 [Suppl 1]:S80–S87
Dunne MJ, Petersen OH (1986a) Intracellular ADP activates K^+ channels that are inhibited by ATP in an insulin-secreting cell line. FEBS Lett 208:59–62
Dunne MJ, Petersen OH (1986b) GTP and GDP activation of K^+ channels that can be inhibited by ATP. Pflugers Arch 407:564–565
Edwards FR, Weston AH (1990) Structure-activity relationships of K^+ channel openers. Trends Pharmacol Sci 11:417–422
Ehara T, Kaufmann R (1978) The voltage- and time-dependent effects of (−)-verapamil on the slow inward current in isolated cat ventricular myocardium. J Pharmacol Exp Ther 207:49–55
Escande D, Thuringer D, Le Guern S, Cavero I (1988) The potassium channel opener cromakalim (BRL 34915) activates ATP-dependent K^+ channels in isolated cardiac myocytes. Biochem Biophys Res Commun 154:620–625

Escande D, Thuringer D, Le Guern S, Courteix J, Laville M, Cavero I (1989) Potassium channel openers act through an activation of ATP-sensitive K^+ channels in guinea-pig cardiac myocytes. Pflugers Arch 414:669–675
Fan Z, Nakayama K, Hiraoka M (1990a) Pinacidil activates the ATP-sensitive K^+ channel in inside-out and cell-attached patch membranes of guinea-pig ventricular myocytes. Pflugers Arch 415:387–394
Fan Z, Nakayama K, Hiraoka M (1990b) Multiple actions of pinacidil on adenosine triphosphate-sensitive potassium channels in guinea-pig ventricular myocytes. J Physiol (Lond) 430:273–295
Findlay I (1987) The effects of magnesium upon adenosine triphosphate-sensitive potassium channels in a rat insulin-secreting cell line. J Physiol (Lond) 391: 611–629
Findlay I (1988) ATP^{4-} and ATP-Mg inhibit the ATP-sensitive K^+ channel of rat ventricular myocytes. Pflugers Arch 412:37–41
Findlay I, Dunn MJ (1986) ATP maintains ATP-inhibited K^+ channels in an operational state. Pflugers Arch 407:238–240
Flockerzi V, Oeken HJ, Hofmann F, Pelzer D, Cavalie A, Trautwein W (1986) Purified dihydropyridine binding site from skeletal muscle T-tubules in a functional calcium channel. Nature 323:66–68
Foster CD, Fujii K, Kingdon J, Brading AF (1989) The effect of cromakalim on the smooth muscle of the guinea-pig urinary bladder. Br J Pharmacol 97:281–291
Franciolini F, Petris A (1990) Chloride channels of biological membranes. Biochim Biophys Acta 1031:247–259
Friedman M, Kurtz GS, Kaczorowski GJ, Katz GM, Reuben JP (1986) Two calcium currents in a smooth muscle cell line. Am J Physiol 250:H669–H703
Fukumitsu T, Hayashi H, Tokuno H, Tomita T (1990) Increase in calcium channel current by β-adrenoceptor agonists in single smooth muscle cells isolated from porcine coronary artery. Br J Pharmacol 100:593–599
Galvez A, Gimenez-Gallego G, Reuben JP, Roy-Contancin L, Feiogenbaum P, Kaczorowski GJ, Garcia ML (1990) Purification and characterization of a unique potent, peptidyl probe for the high conductance calcium-activated potassium channel from venom of the scorpion Buthus tamulus. J Biol Chem 265:11083–11090
Gelband CH, Lodge NJ, Van Breemen C (1989) A Ca^{2+}-activated K^+ channel from rabbit aorta: modulation by cromakalim. Eur J Pharmacol 167:201–210
Godfraind T, Miller R, Wibo M (1986) Calcium antagonism and calcium entry blockade. Pharmacol Rev 38:321–416
Hering S, Bolton TB, Beech DJ, Lim SP (1988) Actions of nifedipine and Bay K8644 is dependent on calcium channel state in single smooth muscle cells from rabbit ear artery. Pflugers Arch 411:590–592
Hering S, Bolton TB, Beech DJ, Lim SP (1989) Mechanism of calcium channel block by D600 in single smooth muscle cells from rabbit ear artery. Circ Res 64: 928–936
Hermsmeyer K (1988) Ion channel effects of pinacidil in vascular muscle. Drugs 36 [Suppl 7]:29–32
Hescheler J, Pelzer D, Trube G, Trautwein W (1982) Does the organic calcium channel blocker D600 act from inside or outside on the cardiac cell membrane? Pflugers Arch 393:287–291
Hess P, Lansman JB, Tsien RW (1984) Different modes of Ca channel gating behaviour favoured by dihydropyridine Ca agonists and antagonists. Nature 311:538–544
Hiraoka M, Fan Z (1989) Activation of the ATP-sensitive K^+ channel by nicorandil (2-nicotinamidoethyl nitrate) in isolated ventricular myocytes. J Pharmacol Exp Ther 250:278–285
Hisada T, Ordway RW, Kirber MT, Singer JJ, Walsh JV Jr (1991) Hyperpolarization-activated cationic channels in smooth muscle cells are stretch sensitive. Pflugers Arch 417:493–499

Honoré E, Martin C, Mironneau C, Mironneau J (1989a) An ATP-sensitive conductance in cultured smooth muscle cells from pregnant rat myometrium. Am J Physiol 257:C297–C305

Honoré E, Amédeé T, Martin C, Dacquet C, Mironneau C, Mironneau J (1989b) Calcium channel current and its sensitivity to (+) isradipine in cultured pregnant rat myometrial cells. An electrophysiological and a binding study. Pflugers Arch 414:477–483

Horie M, Irisawa H, Noma A (1987) Voltage-dependent magnesium block of adenosine-triphosphate-sensitive potassium channel in guinea-pig ventricular cells. J Physiol (Lond) 387:251–272

Hugues M, Romey G, Duval D, Vicent JP, Lazdunski M (1982) Apamin as a selective blocker of the calcium-dependent potassium channel in neuroblastoma cells: voltage-clamp and biochemical characterization of the toxin receptor. Proc Natl Acad Sci USA 79:1308–1312

Hume JR (1985) Comparative interactions of organic Ca^{++} channel antagonists with myocardial Ca^{++} and K^+ channels. J Pharmacol Exp Ther 234:134–140

Hume JR, Leblanc N (1989) Macroscopic K^+ currents in single smooth muscle cells of the rabbit portal vein. J Physiol (Lond) 413:49–73

Imaizumi Y, Muraki K, Watanabe M (1989a) Characteristics of transient outward currents in single smooth muscle cells from the ureter of the guinea-pig. J Physiol (Lond) 427:301–324

Imaizumi Y, Muraki K, Takeda M, Watanabe M (1989b) Measurement and simulation of noninactivating Ca current in smooth muscle cells. Am J Physiol 256:C880–C885

Inoue R, Kitamura K, Kuriyama H (1985) Two Ca-dependent K-channels classified by the application of tetraethylammonium distribute to smooth muscle membranes of the rabbit portal vein. Pflugers Arch 405:173–179

Inoue R, Okabe K, Kitamura K, Kuriyama H (1986) A newly identified Ca^{2+} dependent K^+ channel in the smooth muscle membrane of single cells dispersed from the rabbit portal vein. Pflugers Arch 406:138–143

Inoue R, Kitamura K, Kuriyama H (1987) Acetylcholine activates single sodium channels in smooth muscle cells. Pflugers Arch 410:69–74

Inoue I, Nakaya Y, Nakaya S, Mori H (1989a) Extracellular Ca^{2+}-activated K^+ channel in coronary artery smooth muscle cells and its role in vasodilation. FEBS Lett 255:281–284

Inoue Y, Xiong Z, Kitamura K, Kuriyama H (1989b) Modulation produced by nifedipine of the unitary Ba current of dispersed smooth muscle cells of the rabbit ileum. Pflugers Arch 414:534–542

Inoue I, Nakaya S, Nakaya Y (1990a) An ATP-sensitive K^+ channel activated by extracellular Ca^{2+} and Mg^{2+} in primary cultured arterial smooth muscle cells. J Physiol 430:132

Inoue Y, Oike M, Nakao K, Kitamura K, Kuriyama H (1990b) Endothelin augments unitary calcium channel currents on the smooth muscle cell membrane of guinea-pig portal vein. J Physiol (Lond) 423:171–191

Kajioka S, Oike M, Kitamura K (1990) Nicorandil opens a calcium-dependent potassium channel in smooth muscle cells of the rat portal vein. J Pharmacol Exp Ther 254:905–913

Kajioka S, Kitamura K, Kuriyama H (1991) Guanosine diphosphate activates an adenosine-5′-triphosphate-sensitive K^+ channel in the rabbit portal vein. J Physiol 444:397–418

Kakei M, Noma A, Shibasaki T (1985) Properties of adenosine-triphosphate-regulated potassium channels in guinea-pig ventricular cells. J Physiol (Lond) 363:441–462

Kakei M, Kelly RP, Ashcroft SJH, Ashcroft FM (1986) The ATP-sensitivity of K^+ channels in rat pancreatic β-vells is modulated by ADP. FEBS Lett 208:63–66

Kamouchi M, Xiong Z, Teramoto N, Kajioka S, Okabe K, Kitamura K (1991) Ionic currents involved in vasodilating actions of E4080 a newly synthesized

bradycardia-inducing agent in dispersed smooth muscle cells of the rabbit portal vein. J Pharmacol Exp Ther 259:1396–1403
Kass RS, Arena JP (1989) Influence of pH_o on calcium channel block by amlodipine, a charged dihydropyridine compound. J Gen Physiol 93:1109–1127
Kass RS, Arena JP, Chin S (1989) Cellular electrophysiology of amlodipine: probing the cardiac L-type calcium channel. Am J Cardiol 64:35 I-42
Kawashima Y, Ochi R (1988) Voltage-dependent decrease in the availability of single calcium channels by nitrendipine in guinea-pig ventricular cells. J Physiol 402:219–235
Kihara M, Matsuzawa K, Tokuno H, Tomita T (1990) Effects of calmodulin antagonists on calcium-activated potassium channels in pregnant rat myometrium. Br J Pharmacol 100:353–359
Klöckner U, Isenberg G (1985) Calcium currents of cesium loaded isolated smooth muscle cells (urinary bladder of the guinea-pig). Pflugers Arch 405:340–348
Koch WJ, Hui A, Shull GE, Ellinor P, Scwartz A (1989) Characterization of cDNA clones encoding two putative isoforms of the α_1 subunit of the dihydropyridine-sensitive voltage-dependent calcium channel isolated from rat brain and rat aorta. FEBS Lett 250:386–388
Koch WJ, Ellinor PT, Schwartz A (1990) cDNA cloning of a dihydropyridine-sensitive calcium channel from rat aorta. J Biol Chem 265:17786–17791
Kume H, Takagi K, Satake T, Tokuno H, Tomita T (1990) Effects of intracellular pH on calcium-activated potassium channels in rabbit tracheal smooth muscle. J Physiol (Lond) 424:445–457
Kusano K, Barros F, Katz GM, Roy-Contancin L, Reuben JP (1987) Modulation of K channel activity in aortic smooth muscle by BRL 34915 and scorpion toxin. Biophys J 51:55a
Lacerda AE, Brown AM (1989) Nonmodal gating of cardiac calcium channels as revealed by dihydropyridines. J Gen Physiol 93:1243–1273
Lang RJ (1989) Identification of the major membrane currents in freshly dispersed single smooth muscle cells of guinea-pig ureter. J Physiol (Lond) 412:375–395
Lederer WJ, Nichols CG (1989) Nucleotide modulation of the activity of rat heart ATP-sensitive-K^+ channels in isolated membrane patches. J Physiol (Lond) 419:193–211
Lee KS, Tsien RW (1983) Mechanism of calcium channel blockade by verapamil, D600, diltiazem and nitrendipine in single dialysed heart cells. Nature 302: 790–794
Loirand G, Pacaud P, Mironneau C, Mironneau J (1986) Evidence for two distinct calcium channels in rat vascular smooth muscle cells in short-term primary culture. Pflugers Arch 407:566–568
Loirand G, Pacaud P, Baron A, Mironneau C, Mironneau J (1991) Large conductance calcium-activated non-selective cation channel in smooth muscle cells isolated from rat portal vein. J Physiol (Lond) 437:461–475
Marks TN, Dubyak GR, Jones SW (1990) Calcium currents in the A7r5 smooth muscle-derived cell line. Pflugers Arch 417:433–439
Matsuda JJ, Volk KA, Shibata EF (1990) Calcium currents in isolated rabbit coronary arterial smooth muscle myocytes. J Physiol (Lond) 427:657–680
McCann JD, Welsh MJ (1986) Calcium-activated potassium channels in canine airway smooth muscle. J Physiol (Lond) 372:113–127
Mikami A, Imoto K, Tanabe T, Niidome T, Mori Y, Takeshima H, Narumiya S, Numa S (1989) Primary structure and functional expression of the cardiac dihydropyridine-sensitive calcium channel. Nature 340:230–233
Misler S, Falke LC, Gillkis KD, McDaniel ML (1986) A metabolite-regulated potassium channel in rat pancreatic B-cells. Proc Natl Acad Sci USA 83: 7119–7123
Molleman A, Nelemans A, Van den Akker J, Duin M, den Hertog A (1991) Voltage-dependent sodium and potassium, but no calcium conductances in DDT_1 Mf-2 smooth muscle cells. Pflugers Arch 417:479–484

Muraki K, Imaizumi Y, Kojima T, Kawai T, Watanabe M (1990) Effects of tetraethylammonium and 4-aminopyridine on outward currents and excitability in canine tracheal smooth muscle cells. Br J Pharmacol 100:507–515
Naito K, McKenna E, Schwartz A, Vaghy PL (1989) Photoaffinity labeling of the purified skeletal muscle calcium antagonist receptor by novel benzothiazepine, [^{3}H] azidobutyryl diltiazem. J Biol Chem 264:21211–21214
Nakao K, Inoue R, Yamanaka K, Kiatmura K (1986) Actions of quinidine and apamin on after-hyperpolarization of the spike in circular smooth muscle cells of the guinea-pig ileum. Naunyn Schmiedebergs Arch Pharmacol 334:508–513
Nakao K, Okabe K, Kitamura K, Kuriyama H, Weston AH (1988) Characteristics of cromakalim-induced relaxations in the smooth muscle cells of guinea-pig mesenteric artery and vein. Br J Pharmacol 95:795–804
Nakashima M, Li Y, Seki N, Kuriyama H (1990) Pinacidil inhibits neuromuscular transmission indirectly in the guinea-pig and rabbit mesenteric arteries. Br J Pharmacol 101:581–586
Nakayama K, Fan Z, Marumo F, Hiraoka M (1990) Interrelation between pinacidil and intracellular ATP concentrations on activation of the ATP-sensitive K^+ current in guinea pig ventricular myocytes. Circ Res 67:1124–1133
Nelson MT, Worley JF (1989) Dihydropyridine inhibition of single calcium channels and contraction in rabbit mesenteric artery depends on voltage. J Physiol (Lond) 412:65–91
Nelson MT, Huang Y, Brayden JE, Hesceler J, Standen NB (1990) Arterial dilations in response to calcitonin gene-related peptide involve activation of K^+ channels. Nature 344:770–773
Nichols CG, Lederer WJ (1990) The regulation of ATP-sensitive K^+ channel activity in intact and permeabilized rat ventricular myocytes. J Physiol (Lond) 423: 91–110
Noma A (1983) ATP-regulated K^+ channels in cardiac cells. Nature 305:147–148
Noma A, Shibasaki T (1988) Intracellular ATP and cardiac membrane currents. In: Narahashi T (ed) Ion channels, vol 1. Plenum, New York, pp 183–212
Nowycky MC, Fox AP, Tsien RW (1985) Three types of neuronal calcium channel with different calcium agonist sensitivity. Nature 316:440–443
Ohno-Shosaku T, Zünkler BJ, Trube G (1987) Dual effects of ATP on K^+ currents of mouse pancreatic β-cells. Pflugers Arch 408:133–138
Ohya Y, Kitamura K, Kuriyama H (1987a) Cellular calcium regulates outward currents in rabbit intestinal smooth muscle cell. Am J Physiol 251:C335–C346
Ohya Y, Kitamura K, Kuriyama H (1987b) D600 blocks the Ca^{2+} channel from the outer surface of smooth muscle cell membrane of the rabbit intestine and portal vein. Pflugers Arch 408:80–82
Ohya Y, Kitamura K, Kuriyama H (1987c) Modulation of ionic currents in smooth muscle balls of the rabbit intestine by intracellular perfused ATP and cyclic AMP. Pflugers Arch 408:465–473
Ohya Y, Terada K, Yamaguchi K, Inoue R, Okabe K, Kitamura K, Hirata M, Kuriyama H (1988) Effects of inositol phosphates on the membrane activity of smooth muscle cells of the rabbit portal vein. Pflugers Arch 412:382–389
Okabe K, Kitamura K, Kuriyama H (1987a) Features of 4-aminopyridine sensitive outward current in single smooth muscle cells from the rabbit main pulmonary artery. Pflugers Arch 409:561–568
Okabe K, Terada K, Kitamura K, Kuriyama H (1987b) Selective and long-lasting inhibitory actions of the dihydropyridine derivative, CV-4093, on calcium currents in smooth muscle cells of the rabbit main pulmonary artery. J Pharmacol Exp Ther 243:703–710
Okabe K, Kitamura K, Kuriyama H (1988) The existence of a highly tetrodotoxin sensitive Na channel in freshly dispersed smooth muscle cells of the rabbit main pulmonary artery. Pflugers Arch 411:423–428

Okabe K, Kajioka S, Nakao K, Kitamura K, Kuriyama H, Weston AH (1990) Actions of cromakalim on ionic currents recorded from single smooth muscle cells of the rat portal vein. J Pharmacol Exp Ther 250:832–839
Pacaud P, Loirand G, Mironneau C, Mironneau J (1989) Noradrenaline activates a calcium-activated chloride conductance and increases the voltage-dependent calcium current in cultured single cells of rat portal vein. Br J Pharmacol 97:139–146
Pidoplichko VI (1986) Two different tetrodotoxin-separable inward sodium currents in the membrane of isolated cardiomyocytes. Gen Physiol Biophys 6:593–604
Pietrobon D, Hess P (1990) Novel mechanism of voltage-dependent gating in L-type calcium channels. Nature 346:651–655
Post JM, Stevens RJ, Sanders KM, Hume JR (1991) Effects of cromakalim and lemakalim on slow waves and membrane currents in colonic smooth muscle cells. Am J Physiol 260:C375–C382
Redfern P, Thesleff S (1971) Action potential generation in denervated rat skeletal muscle. II. Action of tetrodotoxin. Acta Physiol Scand 82:70–79
Ribalet B, Ciani S, Eddlestone GT (1989) ATP mediates both activation and inhibition of K (ATP) channel activity via cAMP-dependent protein kinase in insulin-secreting cell lines. J Gen Physiol 94:693–717
Rosenberg RL, Hess P, Tsien RW (1988) L-type channels and calcium-permeable channels open at negative membrane potentials. J Gen Physiol 92:27–54
Sakai T, Terada K, Kitamura K, Kuriyama H (1988) Ryanodine inhibits the Ca-dependent K current after depletion of Ca stored in smooth muscle cells of the rabbit ileal longitudinal muscle. Br J Pharmacol 95:1089–1100
Sanguinetti MC, Kass RS (1984) Voltage-dependent block of calcium channel current in the calf cardiac Purkinje fiber by dihydropyridine calcium channel antagonists. Circ Res 55:336–348
Sanguinetti MC, Krafte DS, Kass RS (1986) Voltage-dependent modulation of Ca channel current in heart cells by Bay K8644. J Gen Physiol 88:369–392
Sanguinetti MC, Scott AL, Zingaro GJ, Siegel PK (1988) BRL 34915 (cromakalim) activates ATP-sensitive K^+ current in cardiac muscle. Proc Natl Acad Sci USA 85:8360–8364
Schneider GT, Cook DI, Gage PW, Young JA (1985) Voltage sensitive, high-conductance chloride channels in the luminal membrane of cultured pulmonary alveolar (type II) cells. Pflugers Arch 404:354–357
Shen WK, Tung RT, Machulda MM, Kurachi Y (1991) Essential role of nucleotide diphosphates in nicorandil-mediated activation of cardiac ATP-sensitive K^+ channel: a comparison with pinacidil and lemakalim. Circ Res 69:1152–1158
Simard JM (1991) Calcium channel currents in isolated smooth muscle cells from the basilar artery of the guinea pig. Pflugers Arch 417:528–536
Sims SM, Singer JJ, Walsh JV Jr (1985) Cholinergic agonists suppress a potassium current in freshly dissociated smooth muscle cells of the toad. J Physiol (Lond) 367:503–529
Sims SM, Walsh JV Jr, Singer JJ (1986) Substance P and acetylcholine both suppress the same K^+ current in dissociated smooth muscle cells. Am J Physiol 251: C580–C587
Sims SM, Singer JJ, Walsh JV Jr (1988) Antagonistic adrenergic-muscarinic regulation of M current in smooth muscle cells. Science 239:190–193
Singer JJ, Walsh JV Jr (1987) Characterization of calcium-activated potassium channels in single smooth muscle cells using the patch-clamp technique. Pflugers Arch 408:98–111
Slish DF, Engle DB, Varadi G, Lotan I, Singer D, Dascal N, Schwartz A (1989) Evidence for the existence of a cardiac specific isoform of the α_1 subunit of the voltage-dependent calcium channel. FEBS Lett 250:509–514
Soejima M, Kokubun S (1988) Single anion-selective channel and its ion selectivity in the vascular smooth muscle cell. Pflugers Arch 411:304–311

Spruce AE, Standen NB, Stanfield PR (1985) Voltage-dependent ATP-sensitive potassium channels of skeletal muscle membrane. Nature 316:736–738
Spruce AE, Standen NB, Stanfield PR (1987) Studies of the unitary properties of adenosine-5′-triphosphate-regulated potassium channels of frog skeletal muscle. J Physiol (Lond) 382:213–236
Standen NB, Quayle JM, Davies NW, Brayden JE, Huang Y, Nelson MT (1989) Hyperpolarizing vasodilators activate ATP-sensitive K^+ channels in arterial smooth muscle. Science 245:177–180
Striessnig J, Knaus HG, Grabner M, Moosburger K, Seits W, Lietz H, Glossmann H (1987) Photoaffinity labelling of the phenylalkylamine receptor of the skeletal muscle transverse tubule calcium channel. FEBS Lett 212:247–253
Striessnig J, Glossmann H, Catteral WA (1990) Identification of a phenylalkylamine binding region within the $\alpha 1$ subunit of skeletal muscle Ca^{2+} channel. Proc Natl Acad Sci USA 87:9108–9112
Sturek M, Hermsmeyer K (1986) Calcium and sodium channels in spontaneously contracting vascular muscle cells. Science 233:475–478
Sumimoto K, Hirata M, Kuriyama H (1988) Characterization of [^{3}H] nifedipine binding to intact vascular smooth muscle cells. Am J Physiol 254:C45–C52
Takahashi M, Seager MJ, Jones JF, Reber BFX, Catteral WA (1987) Subunit structure of dihydropyridine-sensitive calcium channels from skeletal muscle. Proc Natl Acad Sci USA 84:5478–5482
Takano M, Qin D, Noma A (1990) ATP-dependent decay and recovery of K^+ channels in guinea-pig cardiac myocytes. Am J Physiol 258:H45–H50
Tanabe T, Takeshima H, Mikami A, Flockerzi V, Takahashi H, Kangawa K, Kojima M, Matsuo H, Hirose T, Numa S (1987) Primary structure of the receptor for calcium channel blockers from skeletal muscle. Nature 328:313–318
Terada K, Kitamura K, Kuriyama H (1987a) Blocking actions of Ca^{2+} antagonists on the Ca^{2+} channels in the smooth muscle cell membrane of rabbit small intestine. Pflugers Arch 408:552–557
Terada K, Nakao K, Okabe K, Kitamura K, Kuriyama H (1987b) Actions of the 1,4-dihydropyridine derivative, KW-3049, on the smooth muscle membrane of rabbit mesenteric artery. Br J Pharmacol 92:615–625
Trube G, Hescheler J (1984) Inward-rectifying channels in isolated patches of the heart cell membrane: ATP-dependence and comparison with cell-attached patches. Pflugers Arch 401:178–184
Tung RT, Kurachi Y (1991) On the mechanism of nucleotide diphosphate activation of the ATP-sensitive K^+ channel in ventricular cell of guinea-pig. J Physiol 437:239–256
Varadi G, Orlowski J, Scwartz A (1989) Developmental regulation of expression of the $\alpha 1$ and $\alpha 2$ subunit mRNAs of the voltage-dependent calcium channel in a differentiating myogenic cell line. FEBS Lett 250:515–518
Vardivia H, Coronado R (1988) Pharmacological profile of skeletal muscle calcium channels in planar lipid bilayers. Biophys J 53:555a
Williams DL, Katz GM, Roy-Contancin L, Reuben JP (1988) Guanosine 5′-monophosphate modulates gating of high-conductance Ca^{2+}-activated K^+ channels in vascular smooth muscle cells. Proc Natl Acad Sci USA 85:9360–9364
Wong BS (1989) Quinidine blockade of calcium-activated potassium channels in dissociated gastric smooth muscle cells. Pflugers Arch 414:416–422
Xiong Z, Kajioka S, Sakai T, Kitamura K, Kuriyama H (1991a) Pinacidil inhibits the ryanodine-sensitive outward current and glibenclamide antagonizes its action in cells from the rabbit portal vein. Br J Pharmacol 102:788–790
Xiong Z, Kitamura K, Kuriyama H (1991b) ATP activates cationic currents and modulates the calcium current through GTP-binding protein in rabbit portal vein. J Physiol 440:143–165

Yamanaka K, Furukawa K, Kitamura K (1985) The different mechanisms of action of nicorandil and adenosine triphosphate on potassium channels of circular smooth muscle of the guinea-pig small intestine. Naunyn Schmiedebergs Arch Pharmacol 331:96–103

Yoshino M, Someya T, Nishio A, Yazawa K, Usuki T, Yabu H (1989) Multiple types of voltage-dependent Ca channels in mammalian intestinal smooth muscle cells. Pflugers Arch 414:401–409

CHAPTER 17

Drug Effects on the Smooth Muscle of the Coronary System Under Physiological and Pathological Conditions

L. SZEKERES and J.GY. PAPP

A. Introduction

The preceding three sections described the structure, function and responsiveness to endogenous and exogenous substances of smooth muscles in different organ systems, predominantly under normal, physiological conditions. Section IV is devoted to the smooth muscle of the vascular system. The previous chapter dealt with the mode of action of antihypertensive agents protecting against hypertension, a pathological state representing an altered responsiveness of the peripheral vascular bed to drugs.

The present chapter is dedicated to drug effects on the smooth muscle of the coronary system. In view of the great clinical importance of the various pathological manifestations occurring in the coronary system, altered vascular responsiveness due to reduced blood supply will be treated more closely. The pathological and pharmacological redistribution of blood flow within the coronary system will also be discussed. As to the arrangement of the subject matter, the normal function of the coronary system will be treated first, followed by functional changes due to pathological conditions. Since the aim of drug therapy is to combat or alleviate pathological changes, drug effects on the coronary system will be discussed last.

B. Function of the Coronary System

The term "coronary system" is intentionally used, hinting at the complex functions fulfilled by this system. Although there is no fundamental difference between the anatomical structure of the coronary bed and other vascular areas, alone the fact that the coronary system provides for an adequate supply of blood, oxygen and nutrients to the cardiac pump deserves special attention. In addition, more and more evidence is accumulating in favour of the assumption that the coronary smooth muscle, the vascular endothelium and the myocardial cells are closely interrelated from the point of view of regulation of the myocardial blood supply. Therefore, it seems to be more expedient here to relate functional changes and drug effects to this complex system rather than to the coronary smooth muscle alone. On the other hand, it appears to be of importance to consider the anatomical and functional classification, including large and small coronary arteries (100–

300 μm) and arterioles (8–100 μm; TILLMANS and KÜBLER 1984), which constitute the "resistance" vessels; the terminal vascular bed (terminal arterioles, capillaries, venules), the so-called microcirculation; and the venous system, which is made up of "capacitive" vessels.

The myocardium is highly dependent on the rate of blood flow for its supply of oxygen. The heart extracts almost 70% of oxygen from blood in the coronary arteries, and any increase in oxygen supply must therefore be achieved by an increase of coronary blood flow. During strenuous exercise, the coronary blood flow can increase as much as fivefold (POOLE-WILSON 1984).

This subtle adaptation of coronary blood supply to the myocardial oxygen demand is accomplished by several mechanisms, of which metabolic control is certainly predominant. This is shown by the strikingly parallel relationship between changes in myocardial metabolism, as assessed by measurements of myocardial oxygen consumption, and changes in myocardial perfusion (WHITE et al. 1981). In addition, the response is tightly coupled in time. Coronary reactive hyperaemic response following release of a 20-s coronary occlusion can reach an increase in peak blood flow velocity of up to five- to sixfold in man (MARCUS et al. 1981). This is true on a macroscopic scale; however, little is known about the regulation of capillary density, permeability or microcirculatory haemodynamics.

As far the possible mediators between myocardial oxygen consumption and coronary vascular resistance are concerned, potassium (KLINE and MORAD 1978), prostaglandins (MONCADA et al. 1980) and adenosine (BERNE and RUBIO 1979) have been proposed, although convincing evidence has not yet been produced.

Another basic mechanism controlling coronary circulation is the phenomenon of autoregulation, i.e., a change in perfusion pressure is not followed by a parallel shift in flow, because compensatory changes in downstream coronary resistance attenuate alterations in flow due to changes in driving pressure. However, in the heart in situ left ventricular pressure considerably affects myocardial metabolism; thus metabolic regulation will be involved as well. It should be mentioned here that 15% of coronary vascular resistance can be attributed to extravascular compressive forces that appear during cardiac contraction (MARCUS 1984).

The neural regulation of the coronary circulation affects the coronary vessels directly and indirectly by influencing cardiac metabolism secondarily by changes in blood pressure, heart rate and contractility.

The direct effects of sympathetic nerve stimulation produce a moderate coronary constriction. This is followed a few seconds later by the indirect effects of sympathetic stimulation, causing a significant increase in myocardial oxygen consumption (due to the enhanced heart rate, contractility and arterial pressure), which results in a marked coronary dilatation. Among the adrenergic receptors of the coronary vascular bed, only activation of the alpha-1 receptors may cause physiologically meaningful constriction of the coronary vessels. Beta-1 receptor stimulation has insignificant effects and that of beta-2 receptors a moderate dilating action (McRAVEN et al. 1971). Some pathological situations such as coronary stenosis (HEUSCH and DEUSSEN 1983) and hypercholesterolaemia (ROSENDORFF et al. 1981) may enhance the effects of sympathetic nerve stimulation on coronary blood flow. On the other hand, during strenuous exercise alpha-adrenergic stimulation may limit coronary dilatation (MURRAY and VATNER 1979).

The direct effect of vagal stimulation is coronary dilation, which appears only if the marked bradycardia, reducing myocardial oxygen demand, is prevented by pacing (FEIGL 1969). Otherwise the decrease of cardiac metabolism due to parasympathetic stimulation produces indirect coronary constriction.

At present we know very little about the reflex and central control of coronary circulation, and in order to elucidate it more research is needed in this area.

The humoral regulation of coronary circulation is a fairly widely investigated area, and a number of substances circulating in the blood or released from tissues have been found to influence coronary vascular tone. However, according to MARCUS (1982), who reviewed this area, thus far only catecholamines, angiotensin, thyroid hormone and, under certain circumstances, vasopressin (VP) have been found to unequivocally modulate coronary vascular resistance.

A new era in the understanding of humoral regulation of coronary circulation and of the control of vascular tone in general began with the experiments of FURCHGOTT and ZAVADSKI (1980) defining the essential role of the endothelium in mediating relaxation of the vascular smooth muscle to acetylcholine (ACh) in mammalian arteries. Following their observations, a number of substances other than ACh have been shown to mediate endothelium-dependent relaxations, such as adenosine triphosphate (ATP), adenosine diphosphate (ADP), substance P, bradykinin (BK), serotonin (5-HT), VP, angiotensin II (AgII) and histamine. For some of these mediators specific endothelial receptors have been demonstrated by autoradiography (see RALEVICH et al. 1992). Activation of the endothelial receptors by the above-mentioned mediators imitiates a chain of events, releasing endothelium-derived relaxing factor (EDRF), constricting factors (EDCF; VANHOUTTE et al. 1991) or prostaglandins, substances acting on the vascular smooth muscle, thereby affecting the vasomotor tone. These vasoactive factors are released from endothelial cells in response to a variety of physiological stimuli, including stretch, the shear stress of the flowing blood across the cell surface, changes in the partial presure of oxygen, circulating hormones and substances released from blood elements, i.e. aggregating platelets and activated macrophages (see VANHOUTTE 1988).

Evidence indicates that EDRF is, in fact, nitric oxide (NO radical; PALMER et al. 1987), as was originally suggested by FURCHGOTT et al. (1987). It was also shown that L-arginine is the physiological precursor for the formation of nitric oxide in endothelium-dependent relaxation (PALMER et al. 1988).

EDRF (NO) dilatates both large epicardial arteries and also coronary resistance vessels. The vasodilator pathway of EDRF is cyclic guanosine monophosphate (cGMP) dependent. It stimulates guanylate cyclase and increases cGMP levels in smooth muscle. Cyclic GMP-dependent phosphorylation of intracellular key proteins may result in increased extrusion of Ca^{2+}, decreased hydrolysis of phosphatidyl inositol and dephosphorylation of myosin light chain, all of which probably lead to vasodilatation (see BASSENGE and STEWART 1988).

Regulation of coronary collateral flow is dependent on the type of collateral being considered. According to SCHAPER and WUSTEN (1979), they are categorized as native collaterals or developed (mature) collaterals. The native collaterals are present from birth and are thin-walled, small vessels (less than 100 μm) connecting large coronary conduit vessels. Their resistance is 30–40 times that of the minimal resistance of normal coronary vessels. Mature collaterals develop in response to ischaemia; they are thick-walled and large (1–3 mm in diameter). Resistance in active collaterals seems to respond only to changes in extravascular compressive forces, such as alteration in heart rate and arterial pressure. The mature collaterals appear to be responsive to metabolic stimulation; however, it is difficult to determine whether changes in coronary vascular resistance occur as a consequence of dilatation of the collaterals or dilatation of the resistance vessels to the collaterals. Maximal flow in zones served by mature coronary collateral vessels under ideal conditions is 60%–70% of that which can be achieved in the normal coronary vasculature.

C. Changes Due to Pathological Conditions

It was emphasized in the previous section of this chapter that the myocardium is highly dependent on the rate of blood flow for its supply of oxygen and substrates. An increased myocardial demand for blood (arising because of excessive physical activity, enhanced sympathetic tone, increased liberation of catecholamines etc.) will be met under normal conditions by an increased blood supply, since the normal vessels possess a great reserve, enabling them to compensate increased demand by corresponding coronary vasodilatation. If this compensatory function of the coronary vessels is disturbed, ischaemia will arise as a consequence of the imbalance between vascular supply of and myocardial demand for blood. The consequences include angina pectoris, myocardial infarction, acute heart failure, severe arrhythmias and even lethal ventricular fibrillation.

The major cause of failure of coronay circulation to cope with increased demand for blood supply is ischaemic heart disease, which develops on the basis of atherosclerotic degeneration of the coronary vascular bed. As a consequence, sclerotic rigidity of coronary arteries may reduce dilatory reserve and increased demand (by, e.g., physical effort) is not met by increased supply, the typical situation of "effort" angina. In addition, atherosclerotic plaques protruding into the vascular lumen may cause local narrowing and slowing of blood flow. Haemostasis is associated with vortices in the blood flow, promoting the transitory aggregation of platelets and formation of thrombi. This type of aggregation based on transient haemodynamic disturbances may occur according to BORN (1985) in "unstable" angina, explaining why aspirin was very significantly effective in preventing myocardial infarction and death in trials with patients suffering from unstable angina. On the other hand, occlusive coronary thrombosis occurs rapidly, as a consequence of fissure of atheromatous plaques associated with local haemorrhage, which induces platelet aggregation. In such cases aspirin for secondary prevention of myocardial infarction produces no significant benefit.

Although long denied, the concept of vasospasm as a possible cause of imbalance between myocardial supply and demand of blood is gaining ground more and more. The "vasospastic" or Prinzmetal type of angina is based mainly on abnormal coronary artery vasoconstriction (spasm) occurring primarily in atherosclerotic vessels. Vasospasm may play a role in the pathogenesis of unstable angina as well. Elucidation of the importance of endothelium in determining the responses of coronary artery smooth muscle has brought us nearer to understanding the mechanism of this abnormal vasospasm. As already demonstrated by FURCHGOTT and ZAVADSKI (1980), isolated artery preparations (precontracted by addition of noradrenaline, NA) were relaxed by ACh only in the presence of intact endothelium. If the endothelium was removed, then ACh induced a concentration-related contraction. As mentioned in Sect. B, several other "mediators" also require the presence of endothelial cells to produce relaxation. It is obvious that the absence of endothelium in localized segments of coronary arteries (which may be the case in coronary atherosclerosis) could reduce the dilator responses of a variety of endogenous "mediators". It might even convert these responses to constrictor ones. In addition, there is some evidence that coronary artery responses to vasoconstrictor substances might be enhanced in the absence of an intact endothelium (ALLAN et al. 1983). PARRATT (1985) has drawn up a list of possible mediators of coronary spasm. There was evidence for both in vivo coronary vasocontriction and for contraction of coronary arterial smooth muscle in vitro in the case of NA, 5-HT, thromboxane or thromboxanemimetics; whereas in vivo evidence was missing in the case of thrombin and platelet growth factor-2-alpha (PGF-2_{α}), only in vivo evidence was available for leukotrienes C4 and D4, and no evidence for histamine"s constrictory action in vivo, only for that in vitro, was available.

All these might account for localized coronary artery spasm. Since spasm has also been observed in seemingly intact coronary vessels, the possibility of some local defect in the synthesis or release of EDRF or of vasodilator prostaglandins cannot be excluded.

Haemostasis in a certain area due to constriction or obstruction of a major coronary artery supplying this area might influence flow in adjacent coronary arteries and in regions supplied by the latter. Thus short (5-min) occlusion of the left anterior descending coronary artery (LAD) in anaesthetized dogs produced an increase in blood flow in the left circumflex coronary artery (LCX) and the area supplied by this artery (VÉGH et al. 1987). This compensatory increase in LCX blood flow on repeated short LAD occlusion disappeared after "critical" constriction of the LCX. "Critical arterial stenosis" is that degree of constriction beyond which blood flow and pressure distal to the site of stenosis are no longer constant, but decline with progressing stenosis. At this critical degree of constriction, the reactive hyperaemic response to short LCX occlusion disappeared, i.e. maximal hypoxic dilatation of the vascular area perfused by LCX was needed to keep its blood supply just sufficient. After critical LCX constriction, a moderate decline in flow was observed in the area supplied by LAD, especially in the subepicardium. An additional LAD occlusion then resulted in a further decrease in flow, again mainly in the subepicardium. The flow was at similarly low values as during control LAD occlusion. This suggests that following LAD occlusion, subendocardial flow is reduced more than subepicardial flow. The latter probably has a considerable vasodilator reserve, which is exhausted in the presence of a critical LCX constriction. The above changes are in all likelihood due to impairment of the collateral flow from the stenosed LCX to the LAD region. These phenomena were observed in a model more relevant to the clinical situation (in which multivessel artery disease is common) than the usual experimental infarction model with occlusion of one single coronary artery in the otherwise normal myocardium.

D. Effect of Drugs on the Coronary System

The aim of drug therapy is to reverse pathological changes back to the normal state or at least to moderate them. Therefore, discussion of drug effects on the coronary system should be based on disturbances of coronary function and their consequences, which can be influenced by drugs.

Anginal attack is a multifactorial pathological event, originating from a disturbed balance between myocardial blood supply and demand. Ischaemia first appears in the subendocardium, evoking pain by local accumulation of acid metabolites and potassium ions. Another consequence of local ischaemia is a reduced compliance of the ventricular wall, leading to increase of the preload by increased diastolic filling pressure and diastolic wall tension. This might increase extravascular compression of the coronary bed and further reduce coronary flow and myocardial oxygen supply. On the other hand, both ischaemia-induced pain and diminution of cardiac output might stimulate the sympathetic system. Sympathetic stimulation increases afterload by enhancing peripheral resistance, aortic pressure and systolic wall tension.

It also increases heart rate and speed of contractions. All these factors unmistakably enhance myocardial oxygen demand.

In order to break through this vicious circle, pharmacological approaches representing different modes of action should be considered. The coronary system is directly influenced by two groups of drugs: nitrates, predominantly dilatating the smooth muscles of the capacity vessels, i.e. acting on the venous part of the coronary system, and calcium antagonists, which dilate smooth muscles of the coronary resistance vessels, acting on the arterial part of the coronary bed. The third group of antianginal agents, beta blockers, exert their major protective action on extravascular parameters determining myocardial oxygen demand, such as heart rate and speed of contraction.

According to current opinion, the antianginal effect of nitrovasodilators is based on their ability to reduce both preload and afterload. Reduction of preload by dilating the pulmonary vascular bed and the venous "capacity" vessels, thereby reducing the left ventricular blood supply, ventricular volume and left ventricular (end-)diastolic pressure, is considered at present to be the major action of this group of drugs. By reducing intraventricular diastolic pressure, it improves the transmural distribution of myocardial perfusion, because it decreases extrinsic diastolic compression of the subendocardial vessels. This results in an improved perfusion of the ischaemic subendocardium and increased oxygen supply of this latter. Vasodilating nitrates, however, act on coronary arteries as well. They can dilate large coronary arteries and may thus improve via collaterals the blood supply to ischaemic areas in a state of maximal hypoxic dilatation (Winbury et al. 1969). Although they are not considered to influence the small coronary resistance vessels, their influence cannot be denied in the case of coronary arterial spasm. They also act on the peripheral arterial bed and reduce peripheral resistance and blood pressure. This results in a decrease in the afterload and in myocardial oxygen demand.

Recent advances in elucidating the role in vascular physiology of EDRF introduced new aspects for a better understanding of the mode of action of nitrovasodilators. EDRF and atrial natriuretic peptide (ANP) share with the nitrovasodilators the ability to stimulate guanylate cyclase and increase cGMP levels in smooth muscle. How this may lead to vasodilatation was described in Sect. B. There is an important interaction between endogenous activators (such as EDRF or ANP) and exogenous stimulators (such as nitrovasodilatators) of the cGMP dilator system. Enhanced coronary dilating responses to nitroglycerin (NTG) have been shown in coronary artery segments with reduced coronary artery dilatation (Rafflenbeul et al. 1988). This finding may have relevance to the clinical use of nitrates, in that a preferential dilatory effect of NTG on atherosclerotic coronary artery segments (with damaged intimal layer) has been observed. The mechanism of the apparent inhibitory action of endothelium on nitrate-induced dilatation is as yet not known. This phenomenon may also serve as an explanation for the preferential nitrovasodilator actions on veins. It was shown by

SEIDEL and LA ROCHELLE (1987) that continuous basal EDRF release from venous endothelium is relatively small compared with that of arterial endothelium. Thus an endothelial inhibitory action on nitrate-induced dilatation might be less expressed in the venous part of coronary system.

There is evidence that in relatively intact, non-atherosclerotic vessels with only one single site of stenosis, NTG induced a preferential redistribution of coronary blood supply to the area served by the constricted coronary artery. Similar, but less expressed, increase of flow in the stenosed area was observed with the calcium antagonist verapamil, whereas agents possessing potent coronary dilating effect but not showing antianginal action failed to influence blood supply to ischaemic area (SZEKERES 1978). These results were obtained with an animal model of angina pectoris which can be paralleled to anginal attack in young humans in which atherosclerotic changes are not yet prominent. Briefly, LAD was autoperfused and critically constricted in anaesthetized dogs. (For critical constriction, see Sect. C.) At critical constriction, increase in myocardial oxygen demand induced by pacing cannot be compensated for by a corresponding increase of coronary flow in the stenosed area, and therefore ischaemia is produced in the area supplied by the stenosed LAD (SZEKERES et al. 1976). The distribution of coronary flow can be altered by dilating large coronary arteries, a possibility which cannot be excluded even in the case of moderate sclerosis. The importance of this action of NTG has been emphasized by WINBURY et al. (1969). Although arteriolar resistance seems to be fixed by maximal, hypoxic vasodilatation, NTG-induced dilatation of the large coronary arteries may increase collateral flow (FAM and MCGREGOR 1964) or cause redistribution of myocardial blood flow from epicardial to endocardial regions (WINBURY 1971; WINBURY et al. 1971). Either of these effects could explain the beneficial action of NTG.

In the above-described angina model, NTG significantly reduced myocardial ischaemia, indicated by a decrease in the lactate uptake, whereas dipyridamol (DPPP) failed to do this (SZEKERES et al. 1976). This is consistent with the findings of FAM et al. (1966) showing increased oxygen tension in the ischaemic myocardium after NTG and decreased oxygen tension after DPPP, as well as with those of WINBURY et al. (1971) describing a selective increase in endocardial pO_2 after NTG and no increase in either epi- or endocardial pO_2 by DPPP.

These results allow the following conclusions:

1. The antianginal action of coronary dilating drugs seems to be related to their action in improving blood supply to the ischaemic myocardium. Therefore, analysis of the latter is indispensable to define and assess this group of drugs correctly.
2. There is no correlation between action of a coronary dilating drug on the vessels in the normally oxygenated myocardium and that on the coronaries in the ischaemic area. Therefore, analysis of the coronary

action of a drug in the normal heart does not permit us to draw conclusions regarding its possible antianginal effect.

In all these experiments, flow to the area supplied by the critically constricted coronary artery was enhanced by antianginal vasodilators, although the vascular bed of this area was already in a state of maximal metabolic dilatation, and the coronary dilators did not increase but rather depressed the coronary perfusion pressure. Therefore, the question arises as to whether flow can be augmented in the maximally dilated coronary bed (in addition to the unquestionable possibility of a favourable redistribution of flow between ischaemic and non-ischaemic myocardium) by drug-induced further vasodilatation. This requires exclusion of all systemic haemodynamic changes which may affect coronary perfusion pressure or resistance peripheral to the site of stenosis. By using intracoronary injection of drugs into the critically constricted LAD, it was shown that NTG, adenosine and DPPP were all able to augment coronary flow in the vascular area distal to the site of critical stenosis in spite of maximal hypoxic dilatation being present in this area (CSIK et al. 1976).

On the basis of the observations made in the experimental angina model in the dog, the question is raised as to whether relief of angina is entirely dependent on reduction in load. In experiments on the angina model, pacing increased neither venous return nor peripheral resistance. Thus under such conditions these factors did not seem to play an essential role in the antianginal effect of NTG.

Another group of drugs used in angina pectoris are calcium entry blockers, i.e. slow calcium channel inhibitors, commonly called calcium antagonists. The actions of the three most widely used drugs, namely the dihyropyridin-type nifedipine and the cationic amphyphilic-type verapamil (with phenyl alkylamine structure) and diltiazem (with benzothiazepine structure) vary to a great extent. What they have in common is that they all dilate the smooth muscle of the arterial vascular bed. Therefore, they are useful in forms of anginal pectoris based on vasospasm of the coronary arteries. Although the calcium channel blockers, such as nifedipine, produce direct vasodilatation by a mechanism distinct from activation of the guanylate cyclase, i.e. by inhibition of Ca^{2+} entry through slow calcium channels (FLECKENSTEIN 1983), the extent of the direct effect of nifedipine and of other calcium antagonists is of greater magnitude as vessel size decreases and Ca^{2+} sensitivity increases (VATNER and HINTZE 1982). The large initial epicardial artery dilation in response to nifedipine is, according to BASSENGE and STEWART (1988), mostly due to flow-dependent dilatation mediated by EDRF. Nifedipine is a potent coronary and systemic vasodilatator in both isolated arteries and the intact organism. It may also depress myocardial contractility and exert a negative chronotropic effect in vitro. However, in man and in conscious animals, the cardiodepression is compensated for by increased autonomic nervous sytem activity triggered by activation of hypotension-induced baroreflexes (see SZEKERES et al. 1987). On the basis of these results, nifedipine is widely used in therapy of the harmful consequences of ischaemic heart disease, e.g. in vasospastic angina, in classical effort angina and in unstable angina. The antianginal action of nifedipine is a complex phenomenon which may include: (a) dilatation of the large coronary arteries, increasing blood and oxygen supply to the jeopardized myocardium; (b) reduction in myocardial oxygen requirement by a direct negative inotropic effect; (c) reduction in the afterload due to dilatation of peripheral resistance vessels; (d) protection of the ischaemic myocardium from hypoxia by preventing influx of excess calcium, hence preserving mitochondrial integrity (NAYLER et al. 1980). In a model of local myocardial ischaemia (anaesthetized dogs) in which a critical constriction of LCX was combined with sudden occlusion of the LAD, nifedipine increased myocardial blood flow within the stenosed area served by the LCX as well as in the myocardial region supplied by the LAD, mainly in the subepicardium. Accordingly, the drug reduced ischaemic ST-segment elevation only in the epicardium. It is suggested that nifedipine directed flow to the subepicardium of the ischaemic area by improving the collateral circulation. This redistribution of flow resulted in a decrease in the endocardial to epicardial flow ratio (SZEKERES et al. 1987). The nifedipine-induced blood pressure fall in higher doses may initiate a reflex tachycardia which can be compensated for by its combination with an adrenergic beta receptor-blocking drug. This combination is contraindicated when verapamil is used as antianginal agent because of the predominantly negative inotropic effects of this latter drug. In contrast to verapamil, diltiazem is a more potent vasodilator and lacks a significant negative inotropic action. It has been found to be highly effective in the relief of anginal pain related to coronary arterial spasm. In animal experiments, diltiazem has a salutary effect on the ischaemic myocardium both through an increase in collateral flow and by direct action on the myocardium and mitochondrial function of the latter. Diltiazem can even reduce or moderate the harmful effects of reperfusion on the severely ischaemic myocardium (literature see SZEKERES et al. 1985).

In the model of local myocardial ischaemia described above in connection with nifedipine, diltiazem increased myocardial blood flow within the stenosed area supplied by the LCX and produced a marked diminution of the enhanced LVEDP, thus reducing the preload. Increased myocardial blood flow within the stenosed area was associated with enhanced blood flow to the ischaemic myocardium, i.e. diltiazem directed flow to the ischaemic zone by improvement of the collateral circulation (SZEKERES et al. 1985).

The three calcium antagonists with antianginal action discussed above were compared with respect to their coronary dilating, negative inotropic, negative dromotropic and chronotropic effects in the isolated Langendorff heart of guinea pigs (KLAUS 1987). The findings confirmed the in vivo observations, showing that the vasodilating effect is most prominent with nifedipine, possessing at the same time the least negative inotropic action.

Diltiazem exerted a slightly more expressed negative inotropic effect at doses producing similar vasodilatation as nifedipine. The cardiodepressive versus vasodilating effect was most expressed with verapamil. A similar relationship was found regarding the vasodilating versus negative dromotropic effect and negative dromotropic versus negative chronotropic effect.

The third group of agents used in angina pectoris syndrome, i.e. the adrenergic beta receptor blocking agents, do not act directly on the coronary system. They are used mainly for chronic prevention of the anginal attack. This is achieved by their action blocking the myocardial beta-1 receptors in the heart, producing a decline in heart rate and in the speed of contractions. This results in a reduced oxygen demand of the myocardium. In addition, protection is afforded against undue increase of sympathetic drive, which may occur during extreme physical and psychological stress. Since beta blockers also block beta-2 receptors, resulting in vasospasm, their use is contraindicated in vasospastic forms of angina pectoris. In this case, combination with nitrovasodilators or calcium antagonists of the dihydropyridin type can be recommended.

It should also be noted that in addition to the aforementioned typical antianginal drugs, some antiarrhythmic agents also possess spasmolytic vasodilating properties. In strips prepared from the thoracic aorta of rabbits, potential-operated calcium channels were activated by high extracellular potassium concentration (50 mmol/l). The degree of vasospasmolytic activity of the individual drugs was estimated by the inhibition of smooth muscle contracture appearing on depolarization. The order of vasospasmolytic activity on aortic strips was as follows: verapamil > diltiazem > fendiline > prajmaline > quinidine. Futhermore, prajmaline was able to relax the sustained tonic contraction induced by potassium in strips prepared from the LAD and LCX coronary artery as well as from the carotid and the femoral artery of dogs. Accordingly, some antiarrhythmics such as prajmaline and quinidine exert a moderate vasospasmolytic activity which might contribute to combatting dysrhythmias associated with coronary spasm (Horvath et al. 1987).

References

Allan G, Brook CD, Cambridge D, Hladkinsky J (1983) Enhanced responsiveness of vascular smooth muscle to vasoconstrictor agents after removal of endothelial cells. Br J Pharmacol 79:334P

Bassenge E, Stewart DJ (1988) Interdependence of pharmacologically-induced and endothelium-mediated coronary vasodilation in antianginal therapy. Cardiovasc Drugs Ther 2:27–35

Berne RM, Rubio R (1979) Coronary circulation. In: Berne RM (ed) Handbook of physiology, vol 1, section 2. American Physiological Society, Bethesda, pp 873–952

Born GVR (1985) Pathogenetic posibilities. In: Hugenholtz P, Goldman BS (eds) Unstable angina current concepts and management. Schattauer, Stuttgart, pp 13–19

Csik V, Szekeres L, Udvary E (1976) Drug-induced augmentation of coronary flow in vessels with maximum ischemic dilatation. Arch Int Pharmacodyn 224:66–76

Fam WM, McGregor M (1964) Effect of coronary vasodilator drugs on retrograde flow in areas of chronic myocardial ischemia. Circ Res 15:355–365

Fam WM, Nakhjavan FK, Sekely P, McGregor M (1966) The effects of oxygen breathing, nitroglycerin and dipyridamole on oxygen tension in healthy and ischemic areas of the myocardium. In: Proceedings of the international symposium on the cardiovascular and respiratory effects of hypoxia. Karger, New York, pp 375–390

Feigl EO (1969) Parasympathetic control of coronary blood flow in dogs. Circ Res 25:509–519

Fleckenstein A (1983) History of calcium antagonists. Circ Res 52 [Suppl I]:3–16

Furchgott RF, Zavadski JV (1980) The obligatory role of endothelial cells in the relaxation of arterial smooth muscle by acetylcholine. Nature 288:373–376

Furchgott RF, Khan MT, Jothianandan T (1987) Evidence supporting the proposal that endothelium-derived relaxing factor is nitric oxide. Thromb Res 7 [Suppl]:5

Heusch G, Deussen A (1983) The effects of cardiac sympathetic nerve stimulation on perfusion of stenotic coronary arteries in the dog. Circ Res 53:8–15

Horvath A, Papp JGy, Szekeres L (1987) Antiarrhythmic drugs as vaso-spasmolytics. J Mol Cell Cardiol 19 [Suppl III, 109]:34

Klaus W (1987) Differential effects of calcium antagonists on the coronary system. In: Papp JGy (ed) Cardiovascular pharmacology '87. Results, concepts and perspectives. Akademiai Kiado, Budapest, pp 295–303

Kline RP, Morad M (1978) Potassium efflux in heart muscle during activity: extracellular accumulation and its implications. J Physiol (Lond) 280:537–558

Marcus M, Wright C, Doty D, Eastham C, Laughlin D, Krumm P, Fastenow C, Brody M (1981) Measurements of coronary velocity and reactive hyperemia in the coronary circulation of humans. Circ Res 49:877–891

Marcus ML (1982) Humoral control of the coronary circulation. In: Marcus ML (ed) The coronary circulation in health and disease. McGraw-Hill, New York, pp 15–190

Marcus ML (1984) What factors influence coronary flow? In: Hearse DJ, Yellon DM (eds) Therapeutic approaches to myocardial infarct size limitation. Raven, New York, pp 91–107

McRaven DR, Mark AL, Abboud FM, Mayer HE (1971) Responses of coronary vessels to adrenergic stimuli. J Clin Invest 50:773–778

Moncada S, Flower RJ, Vane JE (1980) Prostaglandins, prostacyclin and thromboxane A_2. In: Gilman A (ed) The pharmacological basis of therapeutics. MacMillan, New York, pp 668–681

Murray PA, Vatner SF (1979) Alpha-adrenoceptor attenuation of the coronary vascular response to severe exercise in the conscious dog. Circ Res 45:654–660

Nayler WG, Ferrari R, Williams A (1980) Protective effect of pretreatment with verapamil, nifedipine and propranolol on mitochondrial function in the ischaemic and reperfused myocardium. Am J Cardiol 46:242–247

Palmer RM, Ferrige AG, Moncada S (1987) Nitric oxide release accounts for the biological activity of endothelium-derived relaxing factor. Nature 327:524–526

Palmer RM, Rees DD, Ashton DS, Moncada S (1988) L-Arginine is the physiological precursor for the formation of nitric oxide in endothelium-dependent relaxation. Biochem Biophys Res Commun 153:1251–1256

Parratt JR. (1985) Coronary vascular endothelium, spasm and reperfusion arrhythmias; experimental approaches. In: Hugenholtz P, Goldman BS (eds) Unstable angina current concepts and management. Schattauer, Stuttgart, pp 19–29

Poole-Wilson PA (1984) What causes cell death? In: Hearse DJ, Yellon DM (eds) Therapeutic approaches to myocardial infarct size limitation. Raven, New York, pp 43–61

Rafflenbeul W, Bassenge E, Lichtlen PR (1988) Nitroglycerin-induced dilation is enhanced in coronary artery segments with reduced endothelium-mediated dilation. Z Kardiol 77 [Suppl 1]:73 (abstract 258)
Ralevich V, Lincoln J, Burnstock G (1992) Release of vasoactive substances from endothelial cells. In: Ryan U, Rubanyi GM (eds) Endothelial regulation of vascular tone. Decker, New York, pp 298–328
Rosendorff C, Hoffman JLE, Verrier ED, Rouleau J, Boerboom JE (1981) Cholesterol potentiates the coronary artery response to norepinephrine in anesthetized and conscious dogs. Circ Res 48:320–329
Schaper W, Wusten B (1979) Collateral circulation. In: Schaper W (ed) The pathophysiology of myocardial perfusion. Elsevier/North Holland, Amsterdam
Seidel CL, La Rochelle J (1987) Venous and arterial endothelia; different dilating abilities in dog vessels. Circ Res 60:626–630
Szekeres L (1978) On the mode of action of antianginal drugs. Basic Res Cardiol 73:133–146
Szekeres L, Csik V, Udvary É (1976) Nitroglycerin and dipyridamole on cardiac metabolism and dynamics in a new experimental model of angina pectoris. J Pharmacol Exp Ther 196:15–29
Szekeres L, Udvary É, Végh Á (1985) Importance of myocardial blood flow changes in the protective action of diltiazem in a new model of myocardial ischaemia. Br J Pharmacol 86:341–150
Szekeres L, Udvary É, Végh A (1987) Nifedipine effects in severe myocardial ischaemia in the dog due to left anterior descending coronary occlusion with left circumflex coronary artery constriction. Br J Pharmacol 91:127–137
Tillmanns H, Kübler W (1984) What happens in the microcirculation? In: Hearse DJ, Yellon DM (eds) Therapeutic approaches to myocardial infarct size limitation. Raven, New York, pp 107–125
Vanhoutte PM (1988) Relaxing and contracting factors: biological and clinical research. Humana, Clifton
Vanhoutte PM, Lüscher TF, Gräser T (1991) Endothelium-dependent contractions. Blood Vessels 28:74–83
Vatner SF, Hintze TH (1982) Effects of calcium channel antagonists on large and small coronary arteries in conscious dogs. Circulation 66:579–587
Végh A, Szekeres L, Udvary É (1987) Effect of the blood supply to the normal noninfarcted myocardium on the incidence and severity of early postocclusion arrhythmias in dogs. Basic Res Cardiol 82:159–171
White CV, Kerber RE, Weiss HR, Marcus ML (1981) Effect of atrial fibrillation on wall stress, oxygen consumption and perfusion of the left atrium. Circulation 64 [Suppl 2]:IV-265(225)
Winbury MM (1971) Redistribution of left ventricular blood flow produced by nitroglycerin. An example of integration of macro- and microcirculation. Circ Res 28/29 [Suppl 1]:140–147
Winbury MM, Howe BB, Heffner MA (1969) Effect of nitrates and other coronary dilators on large and small coronary vessels. An hypothesis for the mechanism of action of nitrates. J Pharmacol Exp Ther 168:70–94
Winbury MM, Howe BB, Weiss HR (1971) Effect of nitroglycerin and dipyridamole on epicardial and endocardial oxygen tension – further evidence for redistribution of myocardial blood flow. J Pharmacol Exp Ther 176:184–199

CHAPTER 18

Drugs Affecting the Cerebrovascular Smooth Muscle

M. FUJIWARA and I. MURAMATSU

A. Introduction

The brain tissues demand a sufficient supply of oxygen to maintain their highly integrated functions at a normal level and are more susceptible to circulatory insufficiency than other tissues or organs. Irreversible changes in brain function due to interruption of cerebral circulation lead to a loss of consciousness and finally to a state of brain death. Even a lesser degree of cerebral ischaemia leads to seriously deficient functions of the brain. An adequate continuous blood supply to the brain is essential to ensure the normal function not only of the brain but also of the organism as a whole. In cerebral vessels, autoregulation effectively operates and chemical stimuli produce marked effects. Thus, drugs affecting the cerebrovascular smooth muscle and cerebral circulation are of particular significance.

The action of drugs on the cerebral circulation has been reviewed in detail by SOKOLOFF (1959). The physiology of cerebral circulation including neural, chemical and metabolic regulation has been excellently reviewed by HEISTAD and KONTOS (1984). The obligatory role of endothelial cells in the acetylcholine (ACh)-induced relaxation of arterial smooth muscle (FURCHGOTT and ZAWADZKI 1980) attracted our attention to an endothelium-vascular activity coupling. The newly discovered endothelium-derived relaxing factor (EDRF) was introduced and reviewed by FURCHGOTT (1983) and PEACH et al. (1985). In vivo data on EDRF have been comprehensively reviewed by MARSHALL and KONTOS (1990). An attempt is made in the present review to discuss drug action, with reference to the endothelium and to the unique innervation and receptor mechanisms in the cerebral blood vessels. Data are mainly based on observations in isolated preparations of cerebral blood vessels and to some extent in in vivo studies, which are compared with results in other vascular beds. Agents affecting signal transduction, intracellular second messengers, and regulatory and contractile proteins in the cerebrovascular smooth muscle are not covered.

B. Endothelium-Dependent Vasorelaxation

I. Endothelium-Derived Relaxing Factor

The discovery of endothelium-derived relaxing factor (EDRF) generated by ACh triggered extensive investigation into endothelial control of vascular activity, endothelium-derived vasoactive substances and endothelium-mediated drug actions. In addition to ACh, agents including bradykinin, histamine, ADP, serotonin, substance P and calcium ionophore A23187 also release EDRF (Furchgott et al. 1983; Peach et al. 1985; Angus and Cocks 1989). Evidence has accumulated suggesting that the EDRF is nitric oxide (NO) (Furchgott 1988; Ignarro et al. 1987; Palmer et al. 1987; Moncada et al. 1988). More recently, Palmer et al. (1988) have demonstrated that NO is synthesized from L-arginine in vascular endothelial cells and the production is inhibited by N^G-monomethyl-L-arginine (L-NMMA) (Rees et al. 1989). Kobayashi and Hattori report that N^G-nitro-L-arginine (NO_2Arg) is more potent than L-NMMA (1991).

Although the release of NO was detected by bioassay on rabbit aortic strips (Gryglewski et al. 1986), by chemiluminescence (Palmer et al. 1987) or by mass spectrometry using ^{15}N-labelled L-arginine (Palmer et al. 1988), there are still opinions against the identity of NO. Many investigators suggest that EDRF is a labile nitroso compound which releases NO when it interacts with vascular smooth muscle cells (Palmer et al. 1987; Shikano et al. 1988; Meyers et al. 1990; Ignarro 1990). The most probable candidate for EDRF is now an *S*-nitrosothiol (Ignarro 1990; Meyers et al. 1990; Wei and Kontos 1990). More recently, Vanin has pointed out the very low stability of nitrosothiols in aqueous medium within the natural pH range, and hypothesized that EDRF is a nitrosyl iron complex with low molecular mass thiol ligands, most probably with cysteine (Vanin 1991). Binding of free NO to iron and thiol may be a way to stabilize NO in cells, and to provide both NO transfer in tissues and its release in the active free state.

There may be more than one EDRF (Vanhoutte 1987a). In the rat aorta and main pulmonary artery, ACh releases two different substances, an endothelium-derived relaxing factor (EDRF) and hyperpolarizing factor (EDHF) from the endothelial cells. Neither substance appears to be derived from a pathway dependent on cyclooxygenase (Chen et al. 1988). In the presence of haemoglobin or methylene blue, only the hyperpolarization and a much-reduced relaxation are observed (Chen et al. 1988). The nature of EDHF is still unknown.

Vascular-relaxing action of EDRF is mediated through activation of guanylate cyclase and subsequent increase in the intracellular concentration of cyclic GMP (Furchgott et al. 1984; Holzmann 1982; Ignarro et al. 1984; Rapoport and Murad 1983). Endothelium-dependent relaxations caused by agonists are differently affected by pertussis toxin, and the relaxations caused by α_2-adrenergic or serotonergic receptor stimulation and by aggregated platelets or thrombin, but not bradykinin, ADP or Ca^{2+} ionophore-induced relaxations are mediated by pertussis toxin sensitive G protein (FLAVAHAN and VANHOUTTE 1990). Cyclic GMP activates cyclic GMP-dependent protein kinase, phosphorylates protein and decreases intracellular Ca^{2+} concentration; finally vascular relaxation ensues (MURAD 1986). Reduction of intracellular Ca^{2+} results from activation of the sarcolemmal Ca^{2+} extrusion pump (POPESCU et al. 1985), increase in Ca^{2+} sequestration or binding (LINCOLN 1983) or inhibition of receptor-operated calcium channels (GODFRAIND 1986). Thus, the mechanism of EDRF-induced vasodilation is no different from that of the nitrovasodilator-induced one.

II. EDRF and Cerebral Blood Vessels

Endothelium-dependent vasorelaxation was examined in isolated human basilar artery (HATAKE et al. 1990). Vasorelaxing responses to thrombin reduced with aging, whereas those to bradykinin and Ca^{2+} ionophore did not change significantly (HATAKE et al. 1990). Non-adrenergic, non-cholinergic (NANC) relaxation in response to transmural electrical stimulation was observed in cerebral arteries isolated from various species of animals (DUCKLES et al. 1977; MURAMATSU et al. 1981). A role of NO in neurally induced cerebroarterial relaxation was examined, and it was postulated that NO liberated from vasodilator nerves activated guanylate cyclase in smooth muscle and produced cyclic GMP, resulting in cerebroarterial relaxation (TODA and OKAMURA 1991a). LEE et al. (1982) have shown that cerebral blood vessels receive non-sympathetic vasodilator nerve. On the basis of the fact that transmural nerve stimulation-induced vasodilation of pig cerebral arteries is associated with increases not only in cyclic AMP but also in cyclic GMP content, they have proposed that NO or NO-releasing molecules mediate cerebral neurogenic vasodilation (LEE et al. 1989). They further demonstrated that NANC neurogenic vasodilation in the basilar arteries without endothelial cells of the pig was blocked by NO_2Arg and this blockade was reversed by L-arginine, concluding that NO or an NO-releasing substance may mediate a major component of the NANC neurogenic vasodilation in the pig basilar artery (LEE and SARWINSKI 1991). Toda and Okamura have further proposed that monkey and canine temporal arterial tone is reciprocally regulated by putatively nitroxidergic and adrenergic nerves (TODA and OKAMURA 1991b). The cerebrovascular response to transmural electrical stimulation (DUCKLES et al. 1977; MURAMATSU et al. 1978, 1981; USUI et al. 1985) or nicotine (MURAMATSU et al. 1978) is subjected to species and regional variations. Such variations may be partly due to species differences in involvement of the nitroxidergic nerve in vascular activity. It has already been reported that the response of the rat anococcygeus muscle to NANC nerve stimulation is inhibited by L-NMMA and the inhibition is reversed by L-arginine, suggesting that the transmitter of the nerves involved is NO or a substance releasing it (GILLEPSIE et al. 1989). Neural localizations of NO synthase provide the evidence for an association of NO with neurons (BREDT et al. 1990). Cloning of a cDNA for brain NO synthase reveals the recognition site for NADPH, FAD, flavin nucleotide and calmodulin as well as phosphorylation sites, indicating that the system is regulated by many different factors (BREDT et al. 1991). The amino acid sequence of this enzyme consists of 1429 and the relative molecular mass is 160458. The only known mammalian enzyme with close homology is cytochrome P-450 reductase (BREDT et al. 1991). The NO synthase from macrophage (HIBBS et al. 1987) or polymorphonuclear neutrophils (YUI et al. 1991) may be different from that of vascular endothelium or brain.

The amino acid sequence of inducible NO synthase from mouse macrophages was recently deduced. It contains 1144 amino acids with a predicted molecular size of 130 kDa (LYONS et al. 1992; XIE et al. 1992). It has become apparent that there are at least two types of the NO synthase, constitutive and inducible. Recent advances in physiology, pathophysiology and pharmacology of NO are comprehensively reviewed by MONCADA et al. (1991).

Most studies on production and release of EDRF were performed in isolated large-sized arteries or cultured endothelium removed from such arteries. It is important to confirm the existence of EDRF in vivo. ROSENBLUM et al. (1987) used a helium-neon laser beam either 18 or 36 μm in diameter to illuminate the surface arterioles of the mouse brain following sensitization by intravenous injection of Evans blue dye. Areas 18 or 36 μm in diameter were injured and no longer relaxed to ACh or bradykinin. The same investigators have demonstrated the first in vivo microvascular data showing endothelium dependence of the relaxing response of mice pial arterioles to A23187, which moves calcium into endothelial cells rather than interacting with surface receptors such as ACh or bradykinin (ROSENBLUM and NELSON 1988a). They also show that the endothelium-dependent mechanism for dilation by calcium ionophore is cyclooxygenase dependent, while that for ACh is not (ROSENBLUM et al. 1989). This implies that, in pial arterioles, the EDRF for ACh differs from that for calcium ionophore. It has also been shown that in pial arterioles the EDRF for ACh is different from that for bradykinin. In the arterioles of cats or mice, the EDRF for bradykinin is a free radical, the production of which is prevented by hydroxyl scavengers, while the EDRF for ACh is (at least in cats) aided by radical scavenging (MONCADA et al. 1988; ROSENBLUM 1987; KONTOS et al. 1984, 1990). In conduit arteries, there is evidence that the EDRF is the same for ACh, bradykinin and calcium ionophore (MONCADA et al. 1988); however, there may be other EDRFs released from other large vascular beds and from microvessels. These await further elucidation.

KONTOS et al. (1988) devised an elegant technique for the in vivo assay of EDRF from cerebral microvessels. Anaesthetized cats were implanted with two cranial windows, the assay window subjected to muscarinic blockade with atropine and the donor window superfused with ACh. The superfusate was directed through the assay window with a delay of 6 s. This caused vasodilation equal to that seen in the donor window. Elimination of the vasodilation by lengthening the transit time from the donor to the assay window to longer than 2 min, and inhibition of the dilation by haemoglobin and methylene blue or selective damage to the endothelium were demonstrated. These results are crucial for support of the view that EDRF is released by ACh from cerebral microvessels in vivo.

III. Pathophysiology

Acetylcholine-induced cerebral arteriolar vasodilation is reversed to constriction 30 min after the onset of severe hypertension (WEI et al. 1985). Subsequent treatment with superoxide dismutase (SOD) and catalase partially restores the dilation. The production of free radical from acute hypertension may destroy EDRF and unmask the direct vasoconstrictor effect of ACh on the vascular smooth muscle (WEI et al. 1985). The mechanism of the free radical-mediated damage to cerebral microvessels (KONTOS 1985) and brain parenchyma (SCHMIDLEY 1990) has been well described. The hydroxyl radical also destroys EDRF from ACh in the cerebral microcirculation of the cat in situ (KONTOS et al. 1984; MARSHALL et al. 1988).

MAYHAN et al. (1987) have shown that in situ endothelium-dependent dilation of pial arterioles by ACh or methacholine is absent in stroke-prone spontaneously hypertensive rats (SHRSP), in contrast to normotensive Wistar Kyoto rat (WKY) pial arterioles which dilate in response to ACh and methacholine. Recent studies by KOBAYASHI et al. (1991) in SHRSP and WKY have shown that significant rises in the systolic blood pressure are observed after 1 week of both strains feeding on NO_2Arg. Endothelium-dependent relaxation may have some contribution to blood pressure regulation in the chronic hypertensive state. Captopril, an angiotensin-converting enzyme inhibitor, possesses an endothelium-mediated component of vasodilation, and this dilation may be due to its sulphhydryl group and the ability of the latter to scavenge O_2^-, thereby protecting EDRF (GOLDSCHMIDT and TALLARIDA 1991). Whether endothelium-dependent relaxation reduces in cerebral arteries or pial arterioles of SHRSP following NO_2Arg or L-MMNA feeding is not determined.

Many reports have appeared demonstrating altered endothelium-dependent responses in experimental atherosclerotic animals (HARRISON et al. 1987; BOSSALLER et al. 1987; HEISTAD et al. 1984; YAMAMOTO et al. 1988; VERBEUREN et al. 1990). HEISTAD et al. (1984) found that serotonin dilated both resistance and conduit vessels in the normal and hypercholesterolemic monkey hind limbs in situ and that serotonin constricted the vessels in the atherosclerotic monkey hind limb. Atherosclerosis causes severe alterations in endothelial cells (ROSS 1986) and could have damaged the cells to impair generation or release of EDRF. Ex vivo studies in primates showed that atherosclerosis but not hypercholesterolemia impaired endothelium-dependent dilation by ACh and thrombin (FREIDMAN et al. 1986). Endothelium-dependent vasodilation in response to both ACh and increased blood flow may be lost early in the course of developing atherosclerosis before the appearance of stenosing and occlusive disease in monkeys (MCLENACHAN et al. 1991). Recently, BANK and AYNEDJIAN (1992) demonstrated that, in cholesterol-fed rats, the renal response to ACh infusion was markedly blunted, and suggested that ACh initiated release of vasoconstrictor prostanoids, PGH_2/TXA_2, as well as NO from vascular endothelium. Nisoldipine, a calcium channel blocker, and lovastatin, an HMG CoA reductase inhibitor, attenuate the accumulation of cholesterol and preserve endothelium-dependent relaxation in the aorta of atherosclerotic rabbits (KAPPAGODA et al. 1991; SENARATNE et al. 1991). BOSSALLER et al. (1987) showed that the receptor-mediated dilation to ACh was abolished but that non-receptor-mediated endothelium-dependent dilation to a calcium ionophore A23187 remained intact in atherosclerotic rabbit aorta. On the other hand, HARRISON et al. (1987) showed that there was impairment of both receptor-mediated and non-receptor-mediated dilation in ex vivo atherosclerotic monkey iliac arteries. FÖRSTERMANN et al. (1988) also showed that both receptor- and non-receptor-mediated, endothelium-dependent dilations involving ACh, substance P, bradykinin and A23187 were attenuated in atherosclerotic human coronary arteries. As previously mentioned, ROSENBLUM et al. (1989) have shown that the EDRF for ACh differs from that for calcium ionophore in the cerebral microcirculation. The EDRF from other receptor agonists may also differ from the latter. Arteriosclerotic changes in aorta are produced by feeding the rat a diet containing cholesterol plus vitamin D_2 (KUNITOMO et al. 1981). In such ex vivo arteriosclerotic preparations both ACh and A23187-induced relaxations were attenuated, while aorta from hypercholesterolemic rats fed a diet containing cholesterol without vitamin D_2 responded with relaxation both to either agent as control preparations (KITAGAWA et al. 1992). The Watanabe heritable hyperlipidemic (WHHL) rabbit is the only animal model of human homozygous familial hypercholesterolemia. Endothelium-dependent relaxations to ACh in the thoracic aortas from 2-month-old WHHL rabbits were attenuated in comparison with those from normal Japanese White rabbits, and thereafter the degree of attenuation was progressively promoted with advancing age (6 and 9 months of age), while the endothelium-dependent relaxation of the basilar arteries from WHHL rabbits remained unchanged with age (KITAGAWA et al., unpublished observations). These results are in agreement with the observations on the aortas and basilar arteries from atherosclerotic rabbits fed a cholesterol-rich diet (KANAMARU et al. 1989). There is a report that endothelium-dependent relaxations to ACh in the basilar artery from WHHL rabbits are significantly greater at 6 months of age than corresponding responses at 4 or 12 months of age (STEWART-LEE and BURNSTOCK 1991). The latter authors suggest that lesions formed in the carotid arteries which are known to be susceptible to atherosclerosis have caused a restriction in blood flow in the brain, leading to compensatory vasodilation in the intracranial vessels. Studies on mechanism of the attenuation of ACh-induced endothelium-dependent relaxation of thoracic aortas from WHHL rabbits (10–13 months of age) indicated that not only is the amount of EDRF release by ACh reduced in the presence of atherosclerosis, but the tunica media beneath the atheromatous plaque is also to some extent responsible for the superoxide-induced inactivation of EDRF (TAGAWA et al. 1991). The endothelium-dependent response of cerebrovascular beds in atherosclerotic animals or humans and the nature of impairment of the response require further study.

It is well known that reperfusion after ischaemia produces oxygen radicals which may produce tissue damage (McCORD 1985). These radicals may exaggerate vasoconstriction, spasm and thrombus formation by direct tissue injury and also by destroying EDRF, which plays a protective role in the endothelium in vascular narrowing. MARSHALL et al. (1988) demonstrated that ACh-mediated, endothelium-dependent cerebral vasodilation in the cat was eliminated by topical methylene blue that could generate oxygen radicals. It has been shown that reperfusion after ischaemia results in elimination of ACh-mediated dilation in the cerebral microcirculation of the cat (MAYHAN et al. 1988). Topical application of SOD plus catalase or deferoxamine, a natural product forming iron complexes, preserved the endothelium-dependent dilation after the cerebral ischaemia reperfusion injury (KONTOS 1989). For radical scavenger therapy further studies on the pathophysiological role of the endothelium in this type of cerebrovascular injury are needed.

C. Endothelium-Dependent Vasoconstriction

I. Endothelium-Derived Constricting Factors

Recently much evidence has accumulated indicating that the endothelium can release substances which induce or augment contraction of the underlying smooth muscle (see SÁNCHEZ-FERRER and MARIN 1990; VANHOUTTE et al. 1991). An abrupt decrease in partial pressure of oxygen (hypoxia to anoxia) augments the contractions of isolated canine peripheral, coronal and cerebral arteries (DE MEY and VANHOUTTE 1982; KATUSIC and VANHOUTTE 1986; RUBANYI and VANHOUTTE 1985). Such anoxic or hypoxic facilitation of contraction is abolished or reduced by the removal of the endothelium. In sandwich bioassay preparations of the canine coronary artery, involvement of a diffusible substance in the hypoxic response is demonstrated (RUBANYI and VANHOUTTE 1985). Inhibitors of cyclooxygenase, lipoxygenase or phospholipase A_2 do not prevent the anoxic endothelium-dependent contraction, and a product of the metabolism of arachidonic acid as the mediator is ruled out (RUBANYI and VANHOUTTE 1985). The substance has been termed endothelium-derived constricting factor (EDCF) (VANHOUTTE 1987b). However, the nature of this EDCF is still unknown. The substance that contracts isolated preparations of arterial smooth muscle, probably a peptide, was also termed EDCF by GILLESPIE et al. (1986).

Studies on canine cerebral arteries demonstrated the endothelium-dependent contractions to acetylcholine and arachidonic acid, and thromboxane A_2 (TxA_2) was proposed to be a candidate for the mediator (USUI et al. 1985; SHIRAHASE et al. 1987a,b). The endothelium-dependent contraction insensitive to the inhibition of cyclooxygenase, such as the response to anoxia, was tentatively attributed to the unidentified vasoconstricting substance $EDCF_1$ (VANHOUTTE 1987b; VANHOUTTE and KATUSIC 1988). For the endothelium-dependent contractions sensitive to inhibitors of cyclooxygenase such as responses to arachidonic acid, ACh, A23187, 5-HT or sudden stretch (MILLER and VANHOUTTE 1985; LÜSCHER and VANHOUTTE 1986a,b; KATUSIC et al. 1987, 1988), an unidentified product(s) of cyclooxygenase was held responsible and the contracting factor was tentatively named $EDCF_2$ (VANHOUTTE 1987b; VANHOUTTE and KATUSIC 1988). In recent studies oxygen-derived free radicals contracted the canine basilar artery; SOD prevented the endothelium-dependent contraction evoked by a Ca^{2+} ionophore A23187 in the same blood vessel (KATUSIC and VANHOUTTE 1989). Thus, VANHOUTTE and KATUSIC (1988) have proposed that, in the canine basilar artery, superoxide anion is cyclooxygenase-dependent $EDCF_2$ (KATUSIC and VANHOUTTE 1989). The nature of EDCF in cerebral arteries will be discussed in more detail in the following section.

II. EDCFs and Cerebral Blood Vessels

It has been reported that repetitive applications of ACh to isolated canine basilar artery cause a definite contractile response. Repetitive or time-elapsed applications of ACh may have activated a contractile process, probably cyclooxygenase and TxA_2 synthetase through muscarinic receptors in this preparation, since the response is attenuated by cyclooxygenase inhibitors such as aspirin, TxA_2 synthetase inhibitors such as OKY-046, or atropine (USUI et al. 1983). Earlier studies on the rabbit pulmonary artery have shown that the contractile response to arachidonic acid is mainly due to production of TxA_2 (SALZMAN et al. 1980). ALTIERE et al. (1985) have reported that ACh contracts rabbit intrapulmonary arteries through generation of TxA_2 and that the pulmonary vascular endothelium may contribute to this action of ACh.

A series of experiments performed by FUJIWARA and associates show that ACh-induced contractions in the canine cerebral arteries are endothelium dependent and that the contraction is abolished by atropine, and attenuated, to a greater or lesser extent, by phospholipase A_2 inhibitors such as quinacrine (mepacrine) and manoalid, cyclooxygenase inhibitors such as aspirin and indomethacin, TxA_2 synthetase inhibitors such as OKY-046 and RS-5186, and TxA_2 antagonists such as ONO-3708 and S-1452 (SHIRAHASE et al.

1987a,b; USUI et al. 1983, 1985, 1986). Similar endothelium-dependent contractions are caused by noradrenaline (USUI et al. 1987b), ATP (SHIRAHASE et al. 1988c, 1991a), angiotensins (MANABE et al. 1989), nicotine (SHIRAHASE et al. 1988a), histamine (USUI et al. 1989), arachidonic acid (SHIRAHASE 1987a,b) and A23187 (SHIRAHASE et al. 1988b) in the canine cerebral atreries. The effect of noradrenaline is mediated through α_1-adrenoceptors (USUI et al. 1987b). The purinoceptor involving the endothelium-dependent contraction is P_{2y} (SHIRAHASE et al. 1991a). Angiotensin I is converted to angiotensin II in endothelial cells, then the latter activates the cells to cause endothelium-dependent contraction (MANABE et al. 1989). A role of endothelium-derived prostanoid which activates TXA_2/prostaglandin endoperoxide receptors is proposed in angiotensin-induced vasoconstriction of rat aorta (LIN and NASJLETTI 1991). Nicotine directly activates both nicotinic and muscarinic cholinoceptors in the endothelium and causes contractions. It is unlikely that nicotine acts indirectly via release of ACh (SHIRAHASE et al. 1988a). Histamine causes endothelium-dependent contraction, mediated by H_1-receptors in the canine cerebral artery, whereas the same agent causes endothelium-dependent relaxation, mediated also by H_1-receptors in the monkey artery (USUI et al. 1989). Because ACh causes endothelium-dependent contraction in the monkey cerebral artery, there may be two types of endothelium (EDCF type and EDRF type) or a single type of endothelium with two different signalling processes in this preparation (USUI et al. 1989). The contractions induced by KCl and 5-HT are not affected by denuding the endothelium, indicating unimpairment of smooth muscle cells (USUI et al. 1986). These are schematically shown in Fig. 1. Possible involvement of lipoxygenase pathway in the endothelium-dependent contraction is ruled out because of no alteration of the contraction in the presence of an inhibitor of this enzyme, TMK-777 (SHIRAHASE et al. 1991a). Superoxide dismutase plus catalase also do not affect the endothelium-dependent contraction, at least to 2-methylthio ATP (SHIRAHASE et al. 1991a). On the basis of these results, FUJIWARA et al. (1989a, 1990) proposed that an EDCF or one of the EDCFs in the canine cerebral artery is a cyclooxygenase product of arachidonic acid metabolism, most probably TxA_2, which is produced by activation of phospholipase A_2. In fact, spontaneous release of immunoreactive thromboxane B_2 ($iTxB_2$) from the canine cerebral artery is much higher than that of other vascular beds including the coronary, mesenteric and saphenous arteries. Such release is reduced by indomethacin or OKY-046 or by denuding the endothelium (SHIRAHASE et al. 1987b). PAGANO et al. (1991) suggest that the prostanoid responsible for the vascular smooth muscle-contracting activity of the aortic effluent is a prostaglandin endoperoxide(s) rather than TXA_2. ITO et al. (1991) also suggest that PGH_2 is EDCF in rat aorta. As far as the canine cerebral arteries are concerned, ACh-induced, endothelium-dependent vasocontraction is partially attenunated by TXA_2 synthetase inhibitor, indicating that TXA_2 is involved in this endothelium-dependent vasocontraction (FUJIWARA et al. 1989a, 1990).

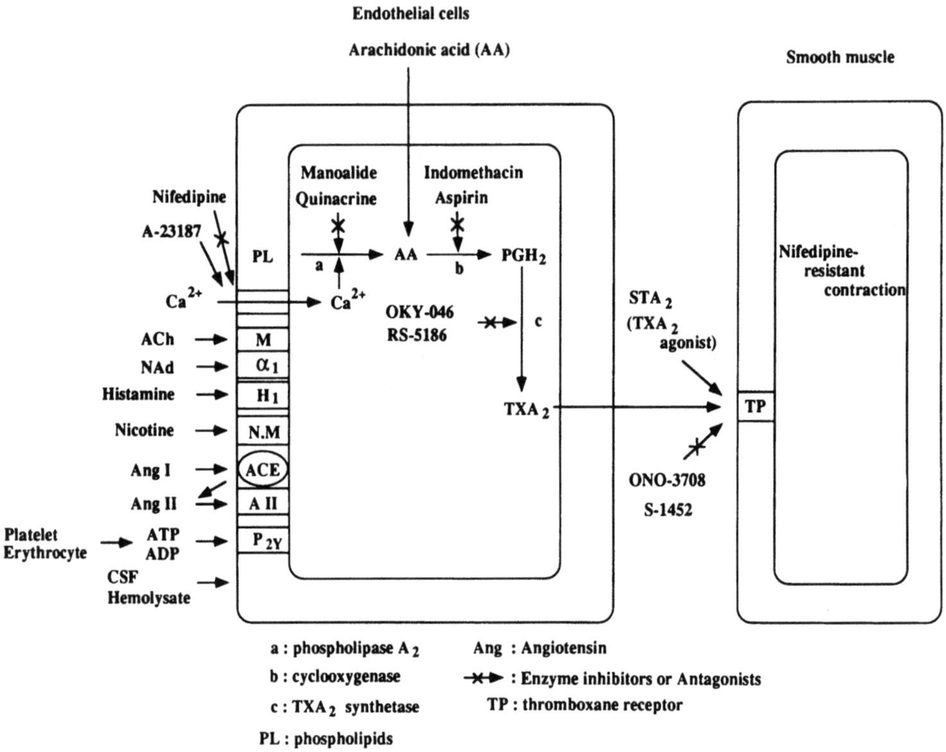

Fig. 1. Endothelium-dependent contraction and possible sites of action of agonists, antagonists and enzyme inhibitors on endothelial and smooth muscle cells in canine cerebral arteries

Demonstration of the presence of TXA_2 synthetase in the same arteries should provide convincing evidence to support our view of the role of TXA_2.

The bovine middle cerebral artery strips respond to the TxA_2-generating system with contraction which is at least three times stronger than that of the coronary artery (ELLIS et al. 1977). The same investigators hypothesized that brain trauma or platelet aggregation in areas of damaged endothelium can release TxA_2 and thus cause constriction of cerebral arteries. ARMSTEAD et al. (1989a) have investigated the effect of ACh on piglet pial arterioles using a closed cranial window, and concluded that prostanoids appear to play a permissive role in ACh-induced pial arteriolar constriction in newborn pigs. TESFAMARIAM et al. (1989) have examined vasoactive prostanoid production caused by ACh in the aorta of rabbits with diabetes mellitus induced by alloxan. Synthesis of TxA_2, measured as $iTxB_2$, is significantly increased in diabetic aortic segments only when the endothelium is present (TESFAMARIAM et al. 1989). The authors suggest that the vasoconstrictor in diabetic aorta is most likely TxA_2 or possibly its precursor, PGH_2. Increased production of vasoconstrictor prostanoids by the endothelium of aorta exposed to elevated glucose may be a consequence of protein kinase C activation (TESFAMARIAM et al. 1991).

In contrast to previous reports (VANHOUTTE 1987b; VANHOUTTE and KATUSIC 1988), RUBANYI (1988b) and SÁNCHEZ-FERRER and MARIN (1990) have proposed at least three distinct EDCFs; $EDCF_1$, a cyclooxygenase metabolite of arachidonic acid; $EDCF_2$, a polypeptide (HICKEY et al. 1985; GILLESPIE et al. 1986; O'BRIEN et al. 1987) that has been isolated and identified as a 21 amino acid peptide named endothelin by YANAGISAWA et al. (1988) and MASAKI (1989); and $EDCF_3$, an unidentified substance released in response to anoxia (RUBANYI and VANHOUTTE 1985) or increases in vascular transmural pressure (RUBANYI 1988a; HARDER et al. 1989). Most of the evidence for $EDCF_1$ in cerebral vessels has been reported in canine basilar arteries (KATUSIC et al. 1987, 1988; SHIRAHASE et al. 1987a,b, 1988a,b,c, 1990, 1991a) and it has also been demonstrated in mouse cerebral arterioles (ROSENBLUM and NELSON 1988b). There is general agreement as to the involvement of a substance originating in the metabolism of arachidonic acid, most likely in the cyclooxygenase pathway; however, the precise chemical nature of $EDCF_1$ is still debated. The $EDCF_1$ has been proposed to be a thromboxane-related compound, probably TxA_2, on the basis of its antagonism by the inhibitors of thromboxane synthetase such as dazoxiben (ALTIERE et al. 1985), OKY-046 or RS-5186 and by the thromboxane antagonists such as ONO-3708 or S-1452 (SHIRAHASE et al. 1987a,b, 1988a–c, 1990). Other investigators reported that inhibition of thromboxane synthetase with imidazole, BW-149H and dazoxiben did not affect endothelium-dependent contractions sensitive to cyclooxygenase inhibitors (MILLER and VANHOUTTE 1985; LÜSCHER and VANHOUTTE 1986a; KATUSIC et al. 1988). The report that in canine basilar arteries TxA_2 seems to be involved in the endothelium-dependent contractions to arachidonic acid and A23187, but not in those to ACh (KATUSIC et al. 1988), suggests the presence of more than one $EDCF_1$. This may explain the fact that, unlike arachidonic acid, ACh did not increase the release of $iTxB_2$ from canine cerebral arteries (SHIRAHASE et al. 1987a,b). Prostaglandins $F_{2\alpha}$ and E_2, closely related compounds (SÁNCHEZ-FERRER and MARIN 1990) or superoxide anion (VANHOUTTE and KATUSIC 1988; KATUSIC and VANHOUTTE 1989) are likely candidates for non-thromboxane $EDCF_1$. Further study is required to clarify these problems.

The endothelium-derived constricting factor(s) counteract the vasorelaxing effect of PGI_2 in the canine basilar artery, but not in the coronary artery (SHIRAHASE et al. 1989). This implies the physiological importance of $EDCF_1$ in maintaining the tone of the cerebral artery. Interactions between EDRF and EDCF in the cerebral artery under physiological and pathophysiological conditions should be the subject of further study. A calcium ionophore, A23187, induces an endothelium-dependent contraction in the canine basilar artery and this contraction is abolished or markedly reduced by lowering the

Ca^{2+} concentration of medium or by nifedipine, whereas the contraction induced by a stable TxA_2 agonist, 9,11-epithio-11,12-methano-TxA_2, is attenuated only slightly (SHIRAHASE et al. 1988b). Thus, Ca^{2+} may play a key role in the production and/or release of $EDCF_1$. Observations with electrical stimulation in feline cerebral arteries imply that Na^+ channels are involved in the release of cyclooxygenase-dependent EDCF (HARDER and MADDEN 1987).

III. Endothelin in Cerebral Blood Vessels

As described in the previous section, endothelin (YANAGISAWA et al. 1988; MASAKI 1989) is termed $EDCF_2$. This peptide was isolated and purified from cultured cells of porcine aortic endothelium. Previous work has indicated that supernatants of cultured bovine endothelial cells had contractile properties on bovine, rat, guinea pig pulmonary (O'BRIEN et al. 1987), pig (HICKEY et al. 1985) and rabbit coronary arteries (GILLESPIE et al. 1986) and that the active principle was protease-sensitive peptidic in nature. In fact, endothelin shows extremely potent, long-lasting vasoconstrictor activity in strips of different regions of arteries from pigs, rats, cats, rabbits, dogs and humans. It also markedly raises arterial blood pressure in anaesthetized rats. Pharmacological analysis of such effects suggests that endothelin acts directly on the smooth muscle cells (YANAGISAWA et al. 1988). Further progress in endothelin research is excellently reviewed by YANAGISAWA and MASAKI (1989a,b) and by SÁNCHEZ-FERRER and MARIN (1990). There are at least three endothelin genes in the human genome and the products of its expression are designated endothelin-1 (ET-1), endothelin-2 (ET-2) and endothelin-3 (ET-3) (INOUE et al. 1989). ET-1 is identical to the originally isolated endothelin from porcine endothelial cells, and only ET-1 is detected in endothelial cells (YANAGISAWA and MASAKI 1989b). The release of endothelin is regulated by the intracellular concentration of Ca^{2+}, involving the calmodulin-dependent pathway and the phospholipase C-mediated phosphoinositide breakdown (EMORI et al. 1989; YANAGISAWA and MASAKI 1989b).

In a pico- or nanomolar range, endothelin (ET-1) induces contractions in a concentration-dependent manner in various regions of arteries from different species including man (see SÁNCHEZ-FERRER and MARIN 1990). Endothelin seems to be more potent as a constrictor on venous than arterial smooth muscle (DENUCCI et al. 1988; MILLER et al. 1989). This peptide also causes contractions in cerebral arteries from various species of animals and man (see SÁNCHEZ-FERRER and MARIN 1990). Vasoconstrictor response of feline cerebral arteries to endothelin is probably due to activation of the influx of Ca^{2+} through L-type Ca^{2+} channels of smooth muscle (SAITO et al. 1989). Maximal contraction of canine and bovine cerebral arteries to endothelin requires the presence of extracellular Ca^{2+} and is independent of the presence of endothelium (SUZUKI et al. 1990b). There are, however, experimental results indicating that endothelin is not directly acting on voltage-operated Ca^{2+} channels in aorta, mesenteric artery and jugular vein (see SÁNCHEZ-FERRER and MARIN 1990). It is interesting that responses of Ca^{2+} channels to pharmacological agents are different between cerebral and peripheral blood vessels, as other characteristics of blood vessels are also subject to regional variations. Reduction in cerebral blood flow to endothelin in vivo has been observed in cats (ROBINSON 1989) and dogs (SUZUKI et al. 1990b). Endothelin acts in feline and canine basilar arteries from the adventitial side, not from the luminal side, to cause vasoconstriction in vivo (MIMA et al. 1989).

Dilation of pial arterioles at a low concentration of endothelin (10^{-10} M) is observed in rats, but the dilation is not observed in the basilar artery (FARACI 1989). In newborn pigs, endothelin can produce either dilation or constriction of cerebral arterioles, depending on concentration (ARMSTEAD et al. 1989b). Furthermore, prostanoids appear to mediate vasodilation induced by the lowest concentration of endothelin and contribute to constriction induced by higher concentration of the same peptide (ARMSTEAD et al. 1989b). Such a vasodilating effect of endothelin is observed in isolated perfused mesentery of the rat, and this effect is ascribed to the release of EDRF (DENUCCI et al. 1988; WARNER et al. 1989). Endothelin is designated to function as a local hormone, released by the endothelial cells to contract the underlying vascular smooth muscle. Fitting with this hypothesis, ET-1 is rapidly removed from the blood. Circulating ET-1 also releases the potent vasodilators prostacyclin and EDRF from endothelial cells, and limits its own vasoconstrictor effects (VANE et al. 1990). The endothelium of porcine cerebral microvessels produces endothelin, and this may regulate the local blood flow within the brain (YOSHIMOTO et al. 1990). The production of endothelin in cerebral microvessel endothelia is dependent on oxygen and carbon dioxide pressure (YOSHIMOTO et al. 1991). Endothelin release in vivo and in vitro is stimulated by endotoxin (SUGIURA et al. 1989).

Recently, two groups reported independently and simultaneously the cloning of cDNA encoding two different endothelin receptors. The deduced amino acid sequence of a bovine endothelin receptor consists of 427 amino acid residues with a relative molecular mass of 48516. The receptor has a transmembrane topology similar to that of other G protein coupled receptors possessing seven membrane-spanning domains with an extracellular N-terminus and a cytoplasmic terminus, and shows specific binding with the highest selectivity to ET-1. This receptor messenger RNA is widely distributed in the central nervous system and peripheral tissues, particularly in the heart and lung (ARAI et al. 1990). The other rat endothelin receptor peptide consists of 441 amino acid residues beginning with a putative 26-residue signal sequence. Thus, the predicted matured peptide consists of 415 amino acid residues with a calculated relative molecular mass of 46901 with seven transmembrane domains, similar to rhodopsin and other G protein coupled receptors. This receptor equally binds all three endothelins. The messenger RNA is detected in many rat tissues including the brain, kidney and lung but not in vascular smooth muscle (SAKURAI et al. 1990). Such a type of receptor with a strong similarity to the rat endothelin receptor was also purified from the bovine lung and its complete nucleotide sequence reported (SAITO et al. 1991). This receptor may be the endothelin receptor on endothelial cells responsible for the release of prostacyclin and EDRF. The former, specific receptor is called ET_A and the latter, non-selective one called ET_B (VANE 1990; SAKURAI et al. 1990). Molecular cloning of ET_A and ET_B receptors was also successfully performed from human placenta cDNA (HOSODA et al. 1991; OGAWA et al. 1991). More recently, it was reported that endothelin-1 caused an endothelium-dependent phasic contraction, in addition to endothelium-independent tonic contraction, in canine basilar artery after more than 5 h of mounting in the organ bath, suggesting the release of TxA_2 from endothelial cells probably through the ET_B receptor (SHIRAHASE et al. 1991b). Cloning of the ET receptors will promote our understanding of the diverse physiological and pathophysiological role of the endothelin family. Selective ET_A antagonists such as BE-18257B, a cyclic pentapeptide isolated from the fermentation products, should provide a valuable tool for the elucidation of the pharmacological and pathophysiological roles of ET-1 (IHARA et al. 1991).

IV. Unidentified EDCF in Cerebral Blood Vessels

In the canine cerebral arteries, hypoxia causes endothelium-dependent contractions (KATUSIC and VANHOUTTE 1986; ELLIOT et al. 1989). Hypoxia-induced contractions of sheep isolated middle cerebral artery are also endothelium dependent (KLAAS and WADSWORTH 1989). These responses are insensitive to cyclooxygenase inhibitors (KATUSIC and VANHOUTTE 1986; KLAAS and WADSWORTH 1989). The possibility that endothelin mediates anoxia-induced, endothelium-dependent response is ruled out (VANHOUTTE et al. 1989). Therefore, the existence of a third EDCF is postulated (RUBANYI 1988b; GREENBERG and DIECKE 1988). In addition to $EDCF_3$, the reduced EDRF is also important in anoxic vasoconstriction, at least in the cerebrovascular bed, as the contraction is attenuated by EDRF inhibitors in the canine (ELLIOT et al. 1989) and sheep cerebral arteries (KLAAS and WADSWORTH 1989). The hypoxia-induced contraction in the canine basilar artery is inhibited by diltiazem and flunarizine (KATUSIC and VANHOUTTE 1986). Voltage-dependent Ca channels may be involved in the contractile response or in the release of unidentified $EDCF_3$ (RUBANYI 1988b).

Transmural pressure-induced, endothelium-dependent contraction in feline cerebral arteries is mediated by a diffusible factor and is not due to inhibition of EDRF (HARDER et al. 1989), in contrast to canine carotid artery, where methylene blue prevents the contraction (RUBANYI 1988a). Endothelium-dependent pressure-induced contraction in isolated canine carotid artery is resistant to the cyclooxygenase blockade with indomethacin (RUBANYI 1988a). In addition, the rapid onset and reversibility of pressure-induced contraction in feline cerebral artery rules out the possibility of endothelin mediation (HARDER et al. 1989). Data in feline cerebral arteries using nordihydroguaiaretic acid (NDGA) suggest that the releasing factor is a product of the metabolism of arachidonic acid via the lipoxygenase or cytochrome P-450 pathways (KAUSER et al. 1989). Furthermore, an apparently similar stimulus such as sudden stretch causes the cyclooxygenase inhibition-sensitive, endothelium-dependent contraction (KATUSIC et al. 1987). Other investigators show the contribution of endothelin in endothelium-dependent vasoconstriction by hypoxia in rat mesenteric artery (RAKUGI et al. 1990). Thus, more studies are needed to characterize the nature of $EDCF_3$.

V. Pathophysiology

Cerebral vasospasm after subarachnoid hemorrhage (SAH) is a major problem requiring treatment by pharmacological agents; however, the pathogenesis is not yet clarified although different hypotheses have been proposed (KASSELL et al. 1985). Cerebrospinal fluid (CSF) collected from patients who had suffered from SAH auses a rapid contraction superimposed by slow, sustained contraction in the canine basilar artery. The response is markedly attenuated by rubbing the luminal side. Data obtained show that CSF with SAH activates the endothelium and releases arachidonic acid metabolites which in turn cause vasoconstriction. Such substance(s) might be derived from blood (USUI et al. 1987a; SHIRAHASE et al. 1990). Other investigators report that segments of canine cerebral arteries exposed to experimental SAH exhibit reduction of endothelium-dependent relaxations, but the endothelium-dependent contractions by various agonists are maintained (KIM et al. 1988).

Endothelin causes a concentration-dependent, long-lasting contraction in canine basilar arteries in vitro and in vivo. Thus it is suggested that this peptide plays an important role in the pathogenesis of vasospasm (IDE et al. 1989; ASANO et al. 1989). The plasma concentration of endothelin in aneurysmal SAH is raised (MASAOKA et al. 1989), and this peptide is found in both plasma and CSF of patients with SAH (FUJIMORI et al. 1990). Interestingly, actinomycin D, a ribonucleic acid synthesis inhibitor, ameliorates cerebral vasospasm after experimental SAH in dogs (SHIGENO et al. 1991). This may lead to the elucidation of the pathogenesis of cerebral vasospasm and also to the availability of a prophylactic adjuvant therapy for patients with SAH (SHIGENO et al. 1991). Endothelin might play an important role in the occurrence of delayed vasospasm of patients with SAH. Nevertheless, other factors could also be involved in the pathogenesis of cerebral vasospasm, and further studies are required to determine whether the cause of the spasm is multiple.

In thoracic aorta from SHR, the endothelium-dependent relaxations evoked by ACh are reduced (LÜSCHER and VANHOUTTE 1986a). ACh causes the simultaneous release of EDRF and EDCF (cyclooxygenase product(s) other than prostacyclin or TxA_2), and the reduced relaxations in the aorta from SHR are not due to a decreased release of EDRF (LÜSCHER and VANHOUTTE 1986a). In the same tissue, oxygen-derived free radicals might be an EDCF and this ultimately causes contraction by stimulation of TxA_2/PGH_2 receptors (AUCH-SCHWELK et al. 1990). The release or production of EDRF and EDCF may be altered at different ages and under different pathological conditions such as hypertension and anoxia.

D. Cerebrovascularly Acting Drugs

I. Adrenoceptor Antagonists and Agonists

The proximal cerebral arteries are mainly innervated by the sympathetic nerves originating from the superior cervical ganglion (TSUKAHARA et al. 1986b) and are responsive to the adrenergic nerve stimulation and adrenergic drugs. Noradrenaline contracts cerebral arterial strips isolated from a variety of species including humans, monkeys, dogs, cats and rabbits (MURAMATSU et al. 1978; SAKAKIBARA et al. 1982; USUI et al. 1985). However, the reactivities to noradrenaline vary between species. For example, compared with human and monkey cerebral arteries the contractile responses of canine cerebral arteries to noradrenaline are smaller and the median effective concentrations are higher. The contractile responses to noradrenaline of human, monkey and rabbit cerebral arteries are antagonized by α_1-adrenoceptor blockers such as prazosin and corynanthine, while those of canine cerebral arteries are insensitive to these drugs and are effectively inhibited by yohimbine. Radioligand binding studies have demonstrated the presence of α_1-adrenoceptors (^{3}H-prazosin-binding sites) and α_2-adrenoceptors (^{3}H-yohimbine-binding sites) in the human and monkey cerebral arteries but only α_2-adrenoceptors in the canine cerebral artery (TSUKAHARA et al. 1985; FUJIWARA et al. 1989b; ALEXANDER and FRIEDMAN 1990). Denervation experiments show that high-affinity α_2-adrenoceptors are located in prejunctional sites and only low-affinity ones on vascular walls of the canine basilar and middle cerebral arteries (TSUKAHARA et al. 1986b). Thus, noradrenaline contraction is mediated through α_1-adrenoceptors in human and monkey cerebral arteries but mediated through low-affinity α_2-adrenoceptors in canine cerebral arteries and the contraction is weak.

In addition to contractile response, MURAMATSU (1991) recently found a relaxing response to noradrenaline in human cerebral arteries (Fig. 2). The relaxation occurred in the preparations (basilar, middle cerebral, posterior cerebral arteries and Willis ring) isolated from three out of five individuals, while the preparations from the remaining two individuals

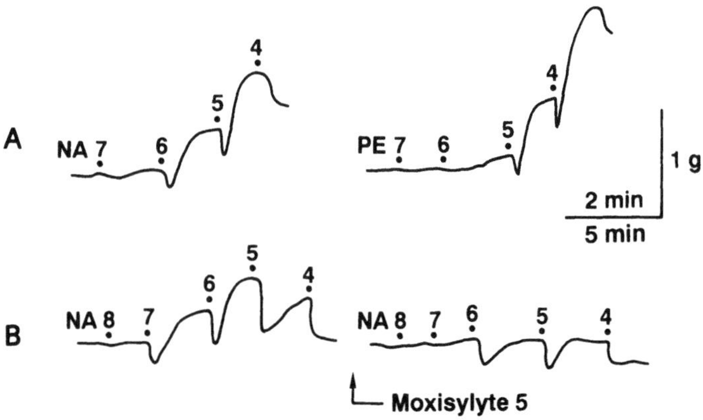

Fig. 2. **A** Effect of noradrenaline (*NA*) and phenylephrine (*PE*) on an isolated human basilar artery. **B** Effects of an α_1-adrenoceptor antagonist, moxisylyte (10^{-5} *M*), on the responses to noradrenaline in an isolated human middle cerebral artery. *Numbers on trace* represent negative logarithmic concentrations of agonist. Noradrenaline or phenylephrine is cumulatively applied (I. Muramatsu, unpublished data)

produced only a contraction in response to noradrenaline. The relaxation was transient and was frequently followed by contraction at high concentrations of noradrenaline. Both relaxing and contractile responses were observed in the presence of propranolol or in endothelium-rubbed preparations. Interestingly, phenylephrine but not clonidine mimicked such dual responses (Fig. 2A), and both the relaxing and contractile responses were inhibited by α_1-adrenoceptor blockers such as prazosin or moxisylyte (Muramatsu 1991). Apparently, moxisylyte inhibits the contractile more than relaxing responses (Fig. 2B). This new finding indicates that, unlike peripheral artery, α_1-adrenoceptors of human cerebral artery mediate the relaxing response, in addition to the contractile response. Such α_1-adrenoceptor-mediated relaxation is well known in the gastrointestinal smooth muscle. The role of sympathetic nerves in regulation of cerebral blood flow has been studied by many workers but their physiological significance is still controversial (Wahl 1985). The α_1-adrenoceptor-mediated dual response and its species, regional and individual variations make our understanding of the physiological role of α_1-adrenoceptors in cerebral blood flow difficult. However, it is interesting to postulate its possible relationship to the autoregulation of cerebral blood flow or occurrence of migraine. Several α-adrenoceptor blockers are now available and clinically used for increasing the cerebral blood flow (Table 1A).

Stimulation of β-adrenoceptors results in an accumulation of cAMP and consequent relaxation of the cerebral blood vessels (Herbst et al. 1979). In human cerebral blood vessels, the presence of β_1- and β_2-adrenoceptors has been demonstrated by radioligand binding experiments (Tsukahara et al.

Table 1. Clinically available drugs which affect cerebral blood flow

Drug	Pharmacological characteristics[a]	Clinical use[a]	Reference
A. α-Adrenoceptor antagonists			
Moxisylyte hydrochloride	α_1-Adrenoceptor selective antagonist CBF increase Inhibition of platelet aggregation Enhancement of mitochondrial function	CBF improvement after infarction and haemorrhage Vertigo	BIRMINGHAM and SZOLCSANYI (1965) MURAMATSU (1991)
Ifenprodil tartrate	α-Adrenoceptor antagonist Increase in vertebral blood flow Inhibition of platelet aggregation *N*-Methyl-D-aspartate receptor antagonist	CBF improvement after infarction and haemorrhage Vertigo Headache	CARRON et al. (1971) MIZUSAWA and SAKAKIBARA (1975) REYNOLDS and MILLER (1989)
Dihydroergotoxin mesylate	α-Adrenoceptor partial agonist Dopamine receptor antagonist 5-HT receptor partial agonist	Symptoms after cerebral infarction and haemorrhage	LOEW and WEIL (1982)
Nicergoline	α_1-Adrenoceptor antagonist CBF increase Inhibition of platelet aggregation	Symptoms after cerebral infarction and haemorrhage	HUGEST et al. (1980) NAGAKAWA et al. (1990)
B. β-Adrenoceptor agonists			
Bamethan sulphate	β-Adrenoceptor agonist CBF increase	Cerebral atherosclerosis	UNNA (1951)
dl-Isoproterenol hydrochloride	β-Adrenoceptor agonist CBF increase	Vertigo	EDVINSSON and OWMAN (1974)
C. β-Adrenoceptor antagonists			
Propranolol hydrochloride	β-Adrenoceptor antagonist 5-HT receptor antagonist	Migraine	DIAMOND and MEDINA (1980) ANDERSSON and PETERSEN (1981)

D. 5-HT receptor antagonists			
Methysergide maleate	5-HT receptor antagonist	Migraine	SALMON (1982)
Dihydroergotamine mesylate	5-HT receptor antagonist α_1-Adrenoceptor partial agonist	Migraine	DIAMOND (1983) FOZARD (1975)
Dimetotiazine mesylate	5-HT receptor antagonist Histamine receptor antagonist	Migraine	KIDO et al. (1968)
Cyproheptadine hydrochloride	5-HT receptor antagonist Histamine receptor antagonist	Migraine	PEROUTKA (1983)
E. Ca antagonists			
Nicardipine hydrochloride	Ca channel antagonist CBF increase	Symptoms after cerebral infarction and haemorrhage	TAKENAKA and HANDA (1979) HADANI et al. (1988)
Flunarizine hydrochloride	Ca channel antagonist	Vertigo Migraine	NAGAO et al. (1986) AGNOLI (1988)
Cinnaridin	Ca channel antagonist CBF increase	Symptoms after cerebral infarction and haemorrhage	HOLMES (1984)
F. Papaverine-like vasorelaxants			
Cinepazide maleate	Inhibition of phosphodiesterase Potentiation of adenosine response CBF increase	Symptoms after cerebral infarction and haemorrhage	MURAMATSU et al. (1984) AKASHI et al. (1979)
Cyclandelate	Vasodilator CBF increase	Symptoms after cerebral infarction and haemorrhage	MIYAZAKI (1971)
Bencyclane fumarate	Vasodilator	Symptoms after cerebral infarction and haemorrhage	KOMLOS and PETÖCZ (1970) KOHLMEYER (1972)
Dilazep dihydrochloride	Vasodilator Potentiation of adenosine response	Vertigo Headache	BUYNISKI et al. (1972) KONDO et al. (1981)

Table 1. Continued

Drug	Pharmacological characteristics[a]	Clinical use[a]	Reference
G. Miscellaneous			
Pentoxifylline	Inhibition of phosphodiesterase	Symptoms after cerebral thrombosis	WARD and CLISSOLD (1987)
Vinpocetine	Inhibition of Ca-dependent phosphodiesterase cGMP increase CBF increase	Symptoms after cerebral infarction and haemorrhage	HAGIWARA et al. (1982)
Propentofylline	Adenosine uptake inhibitor Inhibition of phosphodiesterase	Symptoms after cerebral infarction and haemorrhage	NAGATA et al. (1985) ANDINÉ et al. (1990)
ATP	Purinoceptor agonist	Symptoms after cerebral infarction and haemorrhage Vertigo	MURAMATSU and KIGOSHI (1987) SHIRAHASE et al. (1988c)
Betahistine mesylate	H_1-receptor agonist Relaxation of precapillary vessel	Vertigo	TOMITA et al. (1978) GANELLIN and PARSONS (1982)
Diphenidol hydrochloride	Inhibition of vertebral vasospasm	Vertigo	MATSUOKA et al. (1972)
Ibudilast	Potentiation of PGI_2 actions	Symptoms after cerebral infarction and haemorrhage	ARMSTEAD et al. (1988) HISAYAMA et al. (1989)
Trapidil	CBF increase Antithrombotic effect	Symptoms after cerebral infarction and haemorrhage	SUZUKI et al. (1982)
dl-*α*-Tocopherol nicotinate	Vasodilator Improvement of lipid metabolism	Symptoms after stroke cerebral atheroscrerosis	FONG (1976)
Kallidinogenase	Kinin release	CBF improvement	KAZDA (1975)
Brovincamine fumarate	CBF increase	Symptoms after cerebral infarction and haemorrhage Cerebral atheroscrerosis	KUSHIKU et al. (1985)

[a] Only items for cerebral blood vessels are described.

1986a; ALEXANDER and FRIEDMAN 1990). Some β-adrenoceptor agonists have been clinically used in order to increase the cerebral blood flow (Table 1B).

II. Acetylcholine

Both histochemical and biochemical approaches confirm that ACh is synthesized and stored in the wall of cerebral arteries of every species examined to date (see DUCKLES 1986). Significant choline acetyltransferase activity has been found in cerebral vessels of rabbits, cats and cows (ESTRADA and KRAUSE 1982). Cerebral vessels are also capable of high-affinity choline uptake in rabbits and cats. ACh synthesized by cerebral vessel is released by transmural electrical stimulation (DUCKLES 1982). However, no firm conclusions regarding the physiological function of ACh can be drawn. While relaxing responses to exogenous ACh are abolished when the endothelium is removed, relaxation to transmural nerve stimulation persists and tends to be increased (LEE 1982). Thus, it is unlikely that ACh mediates neurogenic vasodilator responses. The most likely candidate to mediate the vasodilator response may be vasoactive intestinal polypeptide (VIP) (EDVINSSON et al. 1980; DUCKLES and SAID 1982). A role of NO or NO-releasing substance in mediating the neurally induced cerebroarterial vasodilation is mentioned in the section of EDRF and cerebral blood vessels.

Alternative hypotheses for the function of cholinergic nerves have been considered; ACh may act prejunctionally to modulate noradrenaline release. However, the physiological significance of this mechanism is unclear (DUCKLES 1986).

Since the origin of the cholinergic innervation of cerebral arteries has not been well established, it has been difficult to investigate the role of these nerves in vivo. Stimulation of the greater superficial petrosal nerve has been reported to increase cerebral blood flow in dogs (D'ALECY and ROSE 1977). The ACh content in feline cerebral arteries is significantly reduced after removal of the bilateral pterygopalatine ganglia and ciliary ganglia (USUI et al. 1986). Until the role of ACh in control of the cerebral circulation is fully understood, cholinergic or antimuscarinic drugs will not be applicable to cerebrovascular diseases.

III. 5-HT and Its Receptor Antagonists

The presence of serotonin (5-HT) and serotonergic innervation originating from the dorsal or medium raphe nucleus has been demonstrated in the cerebral blood vessels of several animal species including man (GRIFFITH and BURNSTOCK 1983; CHÉDOTAL and HAMEL 1990). Although there was no direct evidence that the amine was released from the serotonergic nerves to produce vascular responses, postjunctional supersensitivity to 5-HT was recently demonstrated in the feline middle cerebral artery after the lesion of the dorsal raphe nucleus (MOREO et al. 1991). 5-HT may also act as a false transmitter in the sympathetic nerves, being taken up into the nerve terminals and released upon nerve stimulation (LEVITT and DUCKLES 1986; SAITO and LEE 1987).

5-Hydroxytryptamine produces contractions in the isolated cerebral arteries of humans, dogs, cats, rabbits, guinea pigs, sheep and rats. The responses in the human and canine cerebral arteries are non-competitively antagonized by ketaneserin (5-HT_2 antagonist) and the residual response is inhibited by spiroxatrine (5-HT_1 antagonist) (NELSON and TAYLOR 1986; TAYLOR et al. 1986; JANSEN et al. 1991a). 5-HT_1 selective agonists sumatriptan (GR 43175) and 8-hydroxy-2-[di-*n*-propylamino]-tetralin (8-OH-DPAT) contract the human and canine cerebral arteries (EDVINSSON and JANSEN 1989; PARSONS et al. 1989). These results suggest that 5-HT_1-like and 5-HT_2 receptors coexist and mediate contractions of cerebrovascular smooth muscle. In contrast, other investigators report that the receptors mediating 5-HT-induced contractions in the canine basilar artery are neither 5-HT_1 nor 5-HT_2 but are similar to those found in the rat stomach fundus (COHEN and COLBERT 1986). In the rat and sheep basilar arteries, 5-HT induces contraction mainly by activation of 5-HT_2 receptors (CHANG et al. 1988; GAW et al. 1990).

5-Hydroxytryptamine also causes relaxation in isolated cerebral arteries. This relaxing response is observed in the human pial and feline middle cerebral arteries pretreated with an alkylating agent, phenoxybenzamine, and is susceptible to blockade by propranolol, suggesting the involvement of β-adrenoceptors (EDVINSSON et al. 1978).

5-Hydroxytryptamine antagonists are well known as effective agents in the prophylaxis of migraine. Methysergide is a mixed $5\text{-HT}_1/5\text{-HT}_2$ antagonist and also elicits a partial agonistic activity at some 5-HT_1 receptors (PEROUTKA 1984; JANSEN et al. 1991b). Dihydroergotamine, dimetotiazine and cyproheptadine show non-selective antagonistic activity to 5-HT receptors in addition to histamine receptors (Table 1D). It is interesting to note that propranolol, a classical β-adrenoceptor antagonist, has a high affinity to 5-HT_{1A} receptor and has been used for the treatment of migrane (DIAMOND and MEDINA 1980; ANDERSSON and PETERSEN 1981; FOZARD 1982) (Table 1C). Recently, selective 5-HT_1 receptor agonists such as sumatriptan have emerged as a novel remedy for acute migraine attacks because they would provide a functional antagonism of the vasodilation and neurogenic inflammation that may occur in dural and cerebral vessels during migrane (BUZZI and MOSKOWITZ 1990; PARSONS 1991).

IV. Neuropeptides

1. Neuropeptide Y

Neuropeptide Y (NPY) in cerebrovascular beds is costored not only with noradrenaline in sympathetic nerves from the superior cervical ganglion but also in a minority with ACh and VIP in parasympathetic nerves originating in the sphenopalatine and otic and internal carotid ganglia (SUZUKI et al. 1990a; TUOR et al. 1990). NPY is a potent contracting peptide in cerebral vessels in vitro and in situ. The contraction is not modified by α- or β-adrenoceptor antagonists or 5-HT blocker, but is strongly attenuated by nifedipine or in a Ca-free medium (EDVINSSON et al. 1987). NPY inhibits adenylate cyclase in feline cerebral blood vessels (FREDHOLM et al. 1985) and slightly depolarizes the membrane in rabbit middle cerebral artery (ABEL and HAN 1989). In peripheral arteries, NPY may potentiate the contractions induced by various agonists (EDVINSSON et al. 1987; OSHITA et al. 1989). However, such a potentiation is not seen in cerebral blood vessels of humans, cats and rats.

In feline spleen neuronally released NPY can elicit α-adrenoceptor blockade-resistant sympathetic contraction (EDVINSSON et al. 1987). However, so far there has been no pharmacological evidence that NPY released from nerve terminals produces cerebrovascular contraction. Recently, VAN RIPER and BEVAN (1991) did an elegant experiment, showing that NPY as well as noradrenaline mediates sympathetic vasoconstriction of the rabbit middle cerebral artery. That is, they found a strong attenuation of the contractile response to electrical field stimulation after the combination of NPY desensitization and α-adrenoceptor blockade, but not after their independent treatment.

2. Vasoactive Intestinal Polypeptide

Vasoactive intestinal polypeptide (VIP) is a relaxing peptide which acts directly on the cerebrovascular smooth muscle (LEE et al. 1984; WHITE 1987). The relaxing effect of VIP is unaffected by cholinoceptor or β-adrenoceptor blockers, indicating the presence of specific VIP receptors. Saturable, specific and reversible binding of $[^{125}I]$VIP is demonstrated in the bovine cerebral artery (SUZUKI et al. 1985). The relaxation occurs in parallel with activation of adenylate cyclase (EDVINSSON et al. 1985; SUZUKI et al. 1988). VIP may stimulate the release of an EDRF which activates adenylate cyclase and relaxes aortic smooth muscle (SATA et al. 1988). Most VIP fibres emanate from the sphenopalatine ganglion, which is parasympathetic and cholinergic in nature (LUNDBERG et al. 1979). Other sources include local microganglia, otic ganglion and cortical VIP neurons (ECKENSTEIN and BAUGHMAN 1984; HARA et al. 1989). VIP release in response to electrical stimulation has been demonstrated in the isolated feline cerebral arteries (BEVAN et al. 1986).

3. Tachykinins

Substance P (SP), neurokinin A (NKA), neurokinin B (NKB) and neuropeptide K (NPK) occur in measurable amounts in cerebral vessels (REGOLI et al. 1984; EDVINSSON et al. 1988). The four tachykinins cause relaxations in the cerebral arteries of various mammals including man and the responses are antagonized by SP antagonists such as spantide (JANSEN et al. 1991a). The relaxation is completely dependent on the presence of intact endothelium, suggesting an involvement of EDRF. The order of potency for tachykinins to produce endothelium-dependent relaxations suggests that the tachykinin receptor involved is of the NK-1 type in the guinea pig basilar artery, while human pial arteries and pig and feline middle cerebral arteries are equipped with a mixture of NK-1 and NK-2 or a non-NK-1 subtype. However, further studies are needed to characterize the tachykinin receptors in cerebral blood vessels. Since tachykinins occur in the sensory nerves originating in the trigeminal ganglia and are distributed in the adventitial layer of cerebral blood vessels, it does not seem logical that the released tachykinins diffuse all the way through the medial muscle coat and elastic layer to reach the endothelial cells (BURNSTOCK 1987). Therefore, there are many problems still to resolve in the elucidation of the physiological significance of tachykinin-induced, endothelium-dependent relaxation.

4. Calcitonin Gene-Related Peptide

Calcitonin gene-related peptide (CGRP) exists in the perivascular SP/NKA fibres in the cerebral circulation (McCULLOCH et al. 1986; UDDMAN et al. 1985). CGRP is a potent dilator of cerebral arteries. Since the relaxing effect of CGRP is independent on the endothelium, it is reasonable to consider that CGRP is a physiological neurovasodilator more than tachykinins of which relaxations are dependent on the endothelium as mentioned above. CGRP activates the adenylate cyclase of cerebral vessels. The effects of CGRP are inhibited by a CGRP fragment CGRP (8–37). Following capsaicin treatment, the majority of perivascular CGRP- and tachykinin-immunoreactive nerve fibres disappear. At present, the fibres seem to be sensory nerves originating in the trigeminal ganglia.

The ubiquitous presence of immunohistochemically visible peptidergic nerves with vasodilatory function is overviewed by OWMAN (1990). In the brain the sensory and parasympathetic pathways for VIP and substance P/CGRP have recently been mapped in detail. These fibre systems may interact with each other and with the sympathetic nervous system. Recent knowledge of the innervation and effects of the dilator neuropeptides in the cerebral circulation is reviewed by EDVINSSON (1991), who documents a linkage of the pathophysiology of SAH and migraine with the release of CGRP. The results obtained by other investigators do not support a substantial role for NPY or VIP as neuroregulators of vascular tone in the pial circulation of the rat, at least under normal conditions (BRAYDEN and CONWAY 1988).

5. Miscellaneous Peptides

The presence of many other neuropeptides in perivascular nerves in cerebral vascular beds has also been demonstrated: bombesin, cholecystokinin, dynorphin B, galanin, neurotensin, somatostatin, vasopressin, etc. However, their physiological roles remain to be resolved (UDDMAN and EDVINSSON 1989).

V. Adenosine, Adenine Nucleotides and Related Drugs

Adenosine, when topically applied to exposed pial arterioles of the cat (BERNE et al. 1974) or infused to the intracarotid artery of the rabbit (HEISTAD et al. 1981), dilates the arterioles or increases the cerebral blood flow. Adenosine and its analogues elicit relaxations of isolated canine cerebral blood vessels mediated through P_1 purinoceptors on the smooth muscle, which are antagonized by methylxanthines (MURAMATSU et al. 1980). The rank order of potency in isolated cerebral arteries of the cat and pig is adenosine 5′-*N*-ethylcarboxamide (NECA) > 2-chloroadenosine ≥ adenosine > *R*-N^6-phenylisopropyladenosine (EDVINSSON and FREDHOLM 1983; MCBEAN et al. 1988). Adenosine and related agonists activate adenylate cyclase in the rat, rabbit, cat and guinea pig cerebral microvessels, resulting in an accumulation of cAMP. The order of potency is the same as that in eliciting relaxations mentioned above (LI and FREDHOLM 1985; SCHUTZ et al. 1982; EDVINSSON and FREDHOLM 1983). These results and binding data with ^{3}H-NECA or 2-chloro-^{3}H-adenosine reveal that the P_1 purinoceptors of cerebral blood vessels are A_2 (adenylate cyclase-stimulating site) rather than A_1 (-inhibitory site) (KALARIA and HARIK 1986, 1988).

A number of compounds that inhibit either adenosine transport or phosphodiesterase including cinepazide, dipyridamole, papaverine and nifedipine potentiate the relaxing response to adenosine in isolated canine cerebral arteries (MURAMATSU et al. 1984) and increase the rat cerebral blood flow (PHILLIS et al. 1984).

Although ATP, ADP and AMP also produce relaxations in isolated canine cerebral vessels, a transient contraction is often observed in responses to ATP and ADP (MURAMATSU et al. 1981; SHIRAHASE et al. 1988c). Part of the transient contraction induced by ATP is endothelium dependent in the canine cerebral artery. In this case, ATP acts on the P_{2Y} purinoceptors on the endothelial cells to release a contractile substance, presumably thromboxane A_2 (SHIRAHASE et al. 1991a). The residual contraction is endothelium independent and is mediated through P_{2X} purinoceptor on the smooth muscle. Such endothelium-independent, P_{2x} purinoceptor-mediated contraction in the canine cerebral arteries is also elicited by electrical transmural stimulation in the canine cerebral arteries (MURAMATSU and KIGOSHI 1987). Since the contractile response to electrical stimulation is completely inhibited not only after P_{2X} purinoceptor desensitization with α,β-methylene ATP but also by guanethidine, MURAMATSU et al. (1981) have proposed that the response is sympathetic in origin but purinergic in nature (MURAMATSU and KIGOSHI 1987). Such sympathetic purinergic responses are found in the peripheral arteries (MURAMATSU 1986; BURNSTOCK 1988; MURAMATSU et al. 1989).

Adenosine is considered to be produced by 5′-nucleotidases at glial foot processes and released into the extracellular space (SCHRADER et al. 1980). The evidence mentioned above suggests that ATP and/or a related substance is released from neuronal tissues and acts as a local vasoactive agent. More recently, a large ATP release from endothelial and vascular smooth muscle cells was demonstrated in peripheral blood vessels (BUXTON et al. 1990; SEDAA et al. 1990). Therefore, it seems that not only adenosine but also ATP or related nucleotides are important in the regulation of cerebral blood flow. ATP has been clinically used in order to dilate vertebral artery.

Various drugs used in cerebrovascular disorders enhance effects of adenosine and/or inhibit phosphodiesterase activity, then accumulate cAMP to reduce cytoplasmic Ca^{2+} concentration. Such drugs and also drugs acting by mechanism other than mentioned are listed in Table 1F,G.

VI. Histamine and Its Antagonists

Studies on feline and canine cerebral artery responses to histamine, both in vitro and in vivo, have established that this amine may exert constriction via H_1-receptors and dilation via H_2-receptors, and possibly in the cat an H_1-mediated dilation (see SERCOMBE and VERRECCHIA 1986). In rabbit cerebral arteries histamine-induced relaxation occurs largely or perhaps wholly via H_2-receptors situated on the endothelium (SERCOMBE and VERRECCHIA 1986). On the other hand, the endothelium-mediated relaxation occurs via H_1-receptors in the rat thoracic aorta (VAN DE VOORDE and LEUSEN 1983), guinea pig pulmonary artery (SATOH and INUI 1984) and rat mesenteric artery (MORITOKI et al. 1986). In monkey basilar artery histamine also causes endothelium-dependent relaxation via H_1-receptors (USUI et al. 1989). There are species differences between H_1- and H_2-receptors mediating endothelium-dependent vasodilation. Nevertheless, betahistine mesylate, a moderately selective H_1-receptor agonist, has had limited clinical use as a vasodilator (TOMITA et al. 1978; GANELLIN and PERSONS 1982) (Table 1G).

VII. Thrombin

Thrombin, well known for its role as a regulator of haemostasis via the proteolytic transformation of fibrinogen to fibrin and via its potent stimulation of platelet aggregation, has been demonstrated to exhibit intrinsic vasoactive properties due to its catalytic activities (WALZ et al. 1986; HAVER and NAMM 1984). Depending on vascular target tissue, thrombin can cause either an endothelium-dependent relaxation (DEMEY et al. 1982; RAPOPORT et al. 1984; WHITE et al. 1984) or an endothelium-independent contraction (HAVER and NAMM 1984; WHITE et al. 1984). In the canine basilar artery, thrombin produces initial relaxation, then contractions which are dependent and independent on the endothelium, respectively (WHITE et al. 1984). The data obtained with chemically modified thrombin analogues indicate clearly that the intrinsic proteolytic activity of the enzyme is required for its vascular actions.

Recently, the proteolytic mechanism whereby thrombin activates its receptor derived from human megakaryocyte-related cells has been clarified by the elegant expression-cloned approach (VU et al. 1991). By acting on a cleavage site found in the primary N-terminal sequence of its receptor, thrombin yields a new N-terminal sequence, $S_{42}FLLRNPNDKYPF_{55}$; it is hypothesized that this newly revealed receptor peptide acts as a "tethered ligand" to activate the cellular receptor. MURAMATSU et al. (1992) recently examined the vascular actions of this receptor peptide and confirmed that the peptide mimics the actions of thrombin in the rat and guinea pig aorta and canine basilar artery.

VIII. Ca Antagonists

As in case of peripheral blood vessels, Ca channels are also present in the cerebral arteries (HARDER 1980; KARASHIMA and KURIYAMA 1981). Therefore, under conditions where the K channels are suppressed by K-channel inhibitors such as tetraethylammonium, cerebral vascular smooth muscle cells are able to produce Ca spikes in response to depolarizing current which are accompanied with a contraction. Both responses are inhibited by inorganic (Mn^{2+}, etc.) or organic Ca^{2+} antagonists (dihydropyridines, diltiazem, verapamil, etc.) (FUJIWARA et al. 1982; NAGAO et al. 1986). The Ca antagonists also inhibit the contractile responses to various agonists (KCl, 5-HT, noradrenaline, NPY, etc.) and relax the basal tension in the cerebral arteries isolated from humans, dogs and rabbits (TOWART 1981; ASANO et al. 1987; BRADT et al. 1981). In general, cerebral arteries are sensitive to Ca antagonists more than peripheral arteries (TOWART 1981; CAUVIN et al. 1983). Cerebral arteries are also more sensitive to a Ca agonist BAY K 8644 in producing a contraction (USKI and ANDERSSON 1985; ASANO et al. 1987). These results suggest that Ca channels in the cerebral arteries may be under states of activation different from those in peripheral arteries. Several agents are clinically used (Table 1E).

There are at least two types of nifedipine-resistant, receptor-operated Ca^{2+} channels (ROC_s) in the rabbit ear artery; nitroglycerin sensitive and nitroglycerin resistant (AKIMOTO et al. 1987). The regulatory mechanism of intracellular free Ca^{2+} in ROC-mediated, TxA_2-induced contraction of the canine basilar artery is different from that of the coronary artery, because in the basilar artery nitroglycerin does not significantly affect the nifedipine-resistant, Ca^{2+}-induced contraction in the presence of a stable TxA_2 analogue, whereas the same agent nearly abolishes such contraction in the coronary artery (USUI et al. 1990). Therefore, drugs that effectively inhibit nitroglycerin-resistant, in addition to nifedipine-resistant, Ca^{2+}-mediated contractions in cerebral arteries should be useful in prevention of cerebral vasospasms if TxA_2 is involved. NP-252, a 1,4-dihydropyridine derivative Ca^{2+} antagonist, has inhibitory effects on ROC_s and/or weak TxA_2 antagonistic effects, in addition to inhibitory effects on VOC_s in cerebral and coronary arteries (AKIMOTO et al. 1991). Management of SAH by calcium antagonists is described by ROBINSON and TEASDALE (1990).

Some Ca antagonists have been demonstrated to have anti-ischaemic actions (ALPS et al. 1988; HADANI et al. 1988; KAMINOW and BEVAN 1991). These effects may be associated with the vasodilator action, inhibition of neurotransmitter release and/or reduction of Ca^{2+} accumulation in ischaemic tissues. HA 1004, a novel intracellular calcium antagonist, increases local cerebral blood flow by increasing collateral blood flow to ischaemic regions in rats (OKADA et al. 1988).

IX. K-Channel Openers

Opening of K channels can also cause relaxation in blood vessels. Cromakalim, pinacidil and nicorandil are categorized as K-channel openers, which hyperpolarize the vascular smooth muscle cells due to an increase in K permeability and then reduce the probability of opening the voltage-dependent Ca channels. ATP-sensitive K channels are a target for the openers, and the effects are blocked by glibenclamide or other hypoglycaemic sulphonylureas (STANDEN et al. 1989). Recently, MASUZAWA et al. (1990) have examined the effects of cromakalim on the canine middle cerebral artery and found that cromakalim can relax the cerebral artery with an increase in ^{86}Rb efflux (used to monitor K permeability), as in the case of peripheral arteries. However, the relaxing effects of cromakalim on the cerebral artery are extremely weak compared with those on the coronary artery used as control, suggesting that there is only sparse distribution of ATP-sensitive K channels in the cerebral artery as compared with the peripheral arteries.

X. Prostanoids and Related Drugs

It has been postulated that because brain arterioles are strongly responsive to prostaglandins and the brain can synthesize prostaglandins from its large endogenous pool of prostaglandin precursors, prostaglandins may be important mediators of changes in cerebral blood flow under normal and abnormal conditions (ELLIS et al. 1979). In fact, products of cyclooxygenase metabolism play important roles in the endothelial regulation of cerebrovascular smooth muscle as already mentioned. There are numerous reports regarding the effects of prostanoids on cerebral vessels and cerebral blood flow (for review, see PICKARD 1981; WHITE and HAGEN 1982). In vitro studies on cerebrovascular smooth muscle effects of prostanoids are reviewed by USKI (1986).

Quinacrine (VALLEE et al. 1979) and manoalide (LOMBARDO and DENNIS 1985) are phospholipase A_2 inhibitors. Aspirin and indomethacin are well known as inhibitors of cyclooxygenase (VANE 1971). Since the discovery of the inhibitory effects of imidazole on human platelet TxA_2 synthetase (NEEDLEMAN et al. 1977; MONCADA et al. 1977), a variety of imidazole derivatives were screened for such activity, including dazoxiben (RANDALL et al. 1981), OKY-046 (IIZUKA et al. 1981), BW149H and RS-5186 (ASAI et al. 1988).

Antagonists for TxA_2 receptors are frequently synthesized and they are extensively reviewed by HALL (1991). They include ONO-3708 (KATSURA et al. 1981; FUJIOKA et al. 1986), AH 23848 (BRITTAIN et al. 1985), SQ 29548 (OGLETREE et al. 1985), S-1452 (NARISADA et al. 1988), AA2414 (SHIRAISHI et al. 1989) and KT2-962 (TOMIYAMA et al. 1990). Ridogrel or R68070 exhibits both TxA_2 synthetase inhibition and TxA_2/prostaglandin endoperoxide receptor blockade combined in one molecule (DECLERCK et al. 1989a,b). S-1452 shows, irrespective of species, a potent and persistent antagonism to TxA_2 actions in various systems including platelets, blood vessels and airway smooth muscles (HANASAKI and ARITA 1988; NARISADA et al. 1988). In addition to its TxA_2 antagonistic activity, AA2414 inhibits 5-lipoxygenase and possesses free-radical scavenging and mast-cell stabilizing

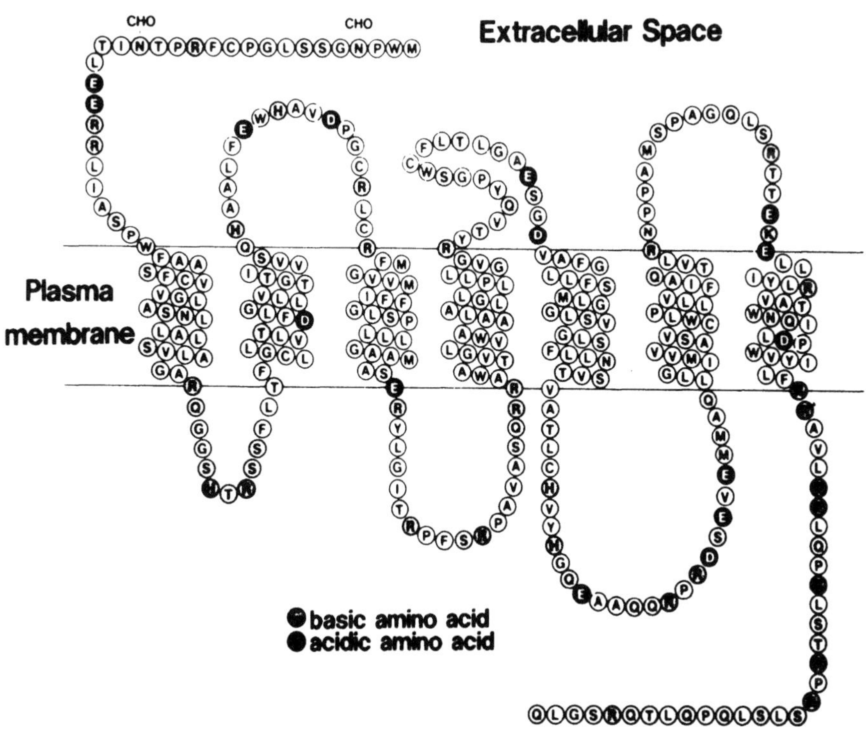

Fig. 3. Amino acid sequence and membrane topology of a human TxA_2 receptor

activity (SHIRAISHI et al. 1989). KT2-962 is a TxA_2/PGH_2 receptor antagonist with a non-prostanoid structure, an azulene derivative (TOMIYAMA et al. 1990). TxA_2 analogues such as U 46,619 (COLEMAN et al. 1981) or STA_2 (KAWAHARA et al. 1983) are used to detect the TxA_2 receptor antagonistic action of compounds. The clinical status of TxA_2 antagonists is listed by HALL (1991).

Using S-1452, a potent and selective antagonist for TxA_2/PGH_2 receptor, human platelet TxA_2 receptor has been purified to apparent homogeneity (USHIKUBI et al. 1989). Recently, HIRATA et al. (1991) obtained a complementary DNA clone encoding this receptor from human placenta and a partial clone from cultured human megakaryocytic leukaemic cells. The placenta cDNA encodes a protein of 343 amino acids with seven putative transmembrane domains (Fig. 3). An identical species of the TxA_2 receptor is present in platelets and vascular tissues. This first report on the molecular structure of an eicosanoid receptor will promote the molecular pharmacology and pathophysiology of the bioactive eicosanoids.

On the basis of our knowledge of EDCF and Ca^{2+} antagonists, there may be several sites of blockade of TxA_2 action in the cerebral blood vessels (Fig. 3). More information regarding the pathophysiological role of TxA_2 in various cerebrovascular disorders will be required to define what type(s) of drugs are suitable for such pathological conditions.

KC-404 or ibudilast (Table 1G) appears to be a selective cerebrovasodilator that potentiates PGI_2- and PGE_2-mediated dilation and has antiplatelet activity (OHASHI et al. 1986), and this agent may be efficacious in the treatment of hypoxic-ischaemic encephalopathy in neonates (ARMSTEDT et al. 1988).

Acknowledgements. We thank Yu Yamaguchi, Satomi Kitagawa and Naomi Aoki for their secretarial assistance. We also thank Professor K. Goto, Institute of Basic Medical Sciences, University of Tsukuba, Japan, Dr. T. Tsukahara, National Cardiovascular Centre, Osaka, Japan, and Dr. N. Suzuki, Mito Red Cross Hospital, Mito, Japan, for their kindness in providing us with valuable and pertinent references. This work was supported in part by a Grant-in-Aid for Scientific Research on Priority Areas: "Vascular Endothelium-Smooth Muscle Coupling", from the Ministry of Education, Science and Culture, Japan, and a grant from the Smoking Research Foundation in Japan.

References

Abel PW, Han CH (1989) Effects of neuropeptide Y on contraction, relaxation, and membrane potential of rabbit cerebral arteries. J Cardiovasc Pharmacol 13:52–63

Agnoli A (1988) The classification of calcium antagonists by the WHO expert committee: relevance in neurology. Cephalalgia 8 [Suppl]:7–10

Akashi A, Hirohashi M, Suzuki I, Shibamura S, Kasahara A (1979) Cardiovascular pharmacology of cinepazide, a new cerebral vasodilator (in Japanese, English abstr). Folia Pharmacol Jpn 75:507–516

Akimoto Y, Kurahashi K, Usui H, Fujiwara M, Shibata S (1987) Nitroglycerin-sensitive and -resistant contraction mediated by receptor-operated calcium channels in rabbit ear artery. Jpn J Pharmacol 44:506–509
Akimoto Y, Kurahashi K, Usui H, Fujiwara M (1991) Vasoinhibitory effect of NP-252, a new dihydropyridine derivative, in canine cerebral artery. Life Sci 48:183–188
Alexander E III, Friedman AH (1990) The identification of adrenergic receptors in human pial membranes. Neurosurgery 27:52–59
Alps BJ, Calder C, Hass WK, Wilson AD (1988) Comparative protective effects of nicardipine, flunarizine, lidoflazine and nimodipine against ischemic injury in the hippocampus of the Mongolian gerbil. Br J Pharmacol 93:877–883
Altiere RJ, Kiritsy-Roy JA, Catravas JD (1985) Acetylcholine-induced contraction in isolated rabbit pulmonary arteries: role of thromboxane A_2. J Pharmacol Exp Ther 236:535–541
Andersson PG, Petersen EN (1981) Propranolol and femoxitine, a 5-HT-uptake inhibitor, in migraine prophylaxis. Acta Neurol Scand 64:280–288
Andiné P, Rudolphi KA, Fredholm BB, Hagberg H (1990) Effect of propentofylline (HWA 285) on extracellular purines and excitatory amino acids in CA1 of rat hippocampus during transient ischemia. Br J Pharmacol 100:814–818
Angus JA, Cocks TM (1989) Endothelium-derived relaxing factor. Pharmacol Ther 41:303–351
Arai H, Hori S, Aramori I, Ohkubo H, Nakanishi S (1990) Cloning and expression of a cDNA encoding an endothelin receptor. Nature 348:730–732
Armstead WM, Mirro R, Leffler CW, Busija DW (1988) The role of prostanoids in the mediation of responses to KC-404, a novel cerebrovasodilator. J Pharmacol Exp Ther 244:138–143
Armstead WM, Mirro R, Busija DW, Leffler CW (1989a) Permissive role of prostanoids in acetylcholine-induced cerebral vasoconstriction. J Pharmacol Exp Ther 251:1012–1019
Armstead WM, Mirro R, Leffler CW, Busija DW (1989b) Influence of endothelin on piglet cerebral microcirculation. Am J Physiol 257:H707–H710
Asai F, Ito T, Ushiyama S, Nagasawa T, Inagaki T, Matsuda K, Oshima T (1988) Anti-platelet effects of RS-5186: a novel thromboxane synthetase inhibitor. Jpn J Pharmacol 46 [Suppl]:279P
Asano M, Aoki K, Suzuki Y, Matsuda T (1987) Effects of Bay K8644 and nifedipine on isolated dog cerebral, coronary and mesenteric arteries. J Pharmacol Exp Ther 647:646–656
Asano T, Ikegaki I, Suzuki Y, Satoh S, Shibuya M (1989) Endothelin and the production of cerebral vasospasm in dogs. Biochem Biophys Res Commun 159:1345–1351
Auch-Schwelk W, Katusic ZS, Vanhoutte PM (1990) Thromboxane A_2 receptor antagonists inhibit endothelium-dependent contractions. Hypertension 15:699–703
Bank N, Aynedjian HS (1992) Role of thromboxane in impaired renal vasodilatation response to acetylcholine in hypercholesterolemic rats. J Clin Invest 89: 1636–1642
Berne RM, Rubio R, Curnich RR (1974) Release of adenosine from ischemic brain. Circ Res 35:262–271
Bevan JA, Buga GM, Moskowitz MA, Said SI (1986) In vitro evidence that vasoactive intestinal peptide is a transmitter of neuro-vasodilation in the head of the cat. Neuroscience 19:597–604
Birmingham AT, Szolcsanyi J (1965) Competitive blockade of adrenergic α-receptors and histamine receptors by thymoxamine. J Pharm Pharmacol 17:449–458
Bossaller C, Habib GB, Yamamoto H, Williams C, Wells S, Henry PD (1987) Impaired muscarinic endothelium-dependent relaxation and cyclic guanosine 5′-monophosphate formation in atherosclerotic human coronary artery and rabbit aorta. J Clin Invest 79:170–174

Bradt L, Andersson KE, Edvinsson L, McKenzie ET, Tamura A, Teasdale G (1981) Effects of extracellular calcium and of calcium antagonists on the contractile response of isolated human pial and mesenteric arteries. J Cereb Blood Flow Metab 1:334–347
Brayden JE, Conway MA (1988) Neuropeptide Y and vasoactive intestinal polypeptide in cerebral arteries of the rat: relationships between innervation pattern and mechanical response. Regul Pept 22:253–265
Bredt DS, Hwang PM, Snyder SH (1990) Localization of nitric oxide synthase indicating a neural role for nitric oxide. Nature 347:768–770
Bredt DS, Hwang PM, Glatt CH, Lowenstein C, Reed RR, Snyder SH (1991) Cloned and expressed nitric oxide synthase structurally resembles cytochrome P-450 reductase. Nature 351:714–718
Brittain RT, Boutal S, Carter MC, Coleman RA, Collington EW, Geisow HP, Hallet P, Hornby EJ, Humphrey PPA, Jack D, Kennedy I, Lumley P, McCabe PJ, Skidmore IF, Thomas M, Wallis CJ (1985) AH 23848: a thromboxane receptor-blocking drug that can clarify the pathophysiologic of thromboxane A_2. Circulation 72:1208–1218
Burnstock G (1987) Mechanisms of interaction of peptide and nonpeptide vascular neurotransmitter systems. J Cardiovasc Parmacol 10:S74–S81
Burnstock G (1988) Sympathetic purinergic transmission in small blood vessels. Trends Pharmacol Sci 9:116–117
Buxton ILO, Walther J, Westfall DP (1990) α-Adenergic receptor-stimulated release of ATP from cardiac endothelial cells in primary culture. Ann NY Acad Sci 603:503–506
Buyniski JP, Losada M, Bierwagen ME, Gardier RW (1972) Cerebral and coronary vascular effects of symmetrical N,N-disubstituted hexahydrodiazepine. J Pharmacol Exp Ther 181:522–528
Buzzi MG, Moskowitz MA (1990) The antimigraine drug, sumatriptan (GR43175), selectively blocks neurogenic plasma extravasation from blood vessels in dura mater. Br J Pharmacol 99:202–206
Carron C, Jullien A, Bucher B (1971) Synthesis and pharmacological properties of series of 2-piperidino alkanol derivatives. Arzneimittel forschung 21:1992–1998
Gauvin C, Loutzenhiser R, Van Breemen C (1983) Mechanism of calcium antagonist -induced vasodilation. Annu Rev Pharmacol Toxicol 23:373–396
Chang J-Y, Hardebo JE, Owman CH (1988) Differential vasomotor action of noradrenaline, serotonin and histamine in isolated basilar artery from rat and guinea-pig. Acta Physiol Scand 132:91–102
Chédotal A, Hamel E (1990) Serotonin-synthesizing nerve fibers in rat and cat cerebral arteries and arterioles: immunohistochemistry of tryptophan-5-hydroxylase. Neurosci Lett 116:269–274
Chen G, Suzuki H, Weston AH (1988) Acetylcholine releases endothelium-derived hyperpolarizing factor and EDRF from rat blood vessels. Br J Pharmacol 95:1165–1174
Cohen ML, Colbert WE (1986) Relationship between receptors mediating serotonin (5-HT) contractions in the canine basilar artery to 5-HT_1, 5-HT_2 and rat stomach fundus 5-HT receptors. J Pharmacol Exp Ther 237:713–718
Coleman RA, Humphrey PPA, Kennedy I, Levy GP, Lumley P (1981) Comparison of the actions of U 46,619, a prostaglandin H_2-analog, with those of prostaglandin H_2 and thromboxane A_2 on some isolated smoothe muscle preparations. Br J Pharmacol 73:773–777
D'Alecy LG, Rose CJ (1977) Parasympathetic cholinergic control of cerebral blood flow in dogs. Circ Res 41:324–331
DeClerck F, Beetens J, de Chaffoy de Courcelles D, Freyne E, Janssen PAJ (1989a) R 68 070: thromboxane A_2 synthetase inhibition and thromboxane A_2/prostaglandin endoperoxide receptor blockade combined in one molecule. I. Biochemical profile in vitro. Thromb Haemost 61:35–42

DeClerck F, Beetens J, Vam de Water A, Vercammen E, Janssen PAJ (1989b) Thromboxane A_2 synthetase inhibition and thromboxane A_2/prostaglandin endoperoxide receptor blockade combined in one molecule: II. Pharmacological effects in vivo and ex vivo. Thromb Haemost 61:43–59
DeMey JG, Vanhoutte PM (1982) Heterogeneous behavior of the canine arterial and venous wall: importance of endothelium. Circ Res 51:439–447
DeMey JG, Claeys M, Vanhoutte PM (1982) Endothelium-dependent inhibitory effects of acetylcholine, adenosine triphosphate, thrombin and arachidonic acid in the canine femoral artery. J Pharmacol Exp Ther 222:166–173
DeNucci A, Thomas R, D'Oreleans-Juste P, Antunes E, Walder C, Warner TD, Vane JR (1988) Pressor effects of circulating endothelium are limited by its removal in the pulmonary circulation and by the release of prostacyclin and endothelium-derived relaxing factor. Proc Natl Acad Sci USA 85:9797–9800
Diamond S (1983) Rational approach to diagnosing and treating headache. III. Therapy. Fam Med Rep 1:39–44
Diamond S, Medina JL (1980) Current thoughts on migraine. Headache 20:208–212
Duckles SP (1982) Choline acetyltransferase in cerebral arteries: modulator of amino acid uptake? J Pharmacol Exp Ther 223:716–720
Duckles SP (1986) Cholinergic innervation of cerebral arteries. In: Owman C, Hardebo JE (eds) Neural regulation of brain circulation. Elsevier, Amsterdam, pp 235–243
Duckles SP, Said SI (1982) Vasoactive intestinal peptide as a neurotransmitter in the cerebral circulation. Eur J Pharmacol 78:371–374
Duckles SP, Lee TJ-F, Bevan JA (1977) Cerebral arterial responses to nerve stimulation in vitro. Species variation in the constrictor and dilator components. In: Owman C, Edvinsson L (eds) Neurogenic control of the brain circulation. Pergamon, Oxford, pp 133–142
Eckenstein F, Baughman SH (1984) Two types of cholinergic innervation in cortex, one colocalized with vasoactive intestinal polypeptide. Nature 309:153–155
Edvinsson L (1991) Innervation and effects of dilatory neuropeptides on cerebral vessels. Blood Vessels 28:35–45
Edvinsson L, Fredholm BB (1983) Characterization of adenosine receptors in isolated cerebral arteries of cat. Br J Pharmacol 80:631–637
Edvinsson L, Jansen I (1989) Characterization of 5-HT receptors mediating contraction of human cerebral, meningeal and temporal arteries: target for GR 43175 in acute treatment of migrane? Cephalalgia 9:39–40
Edvinsson L, Owman CH (1974) Pharmacological characterization of adrenergic alpha and beta receptors mediating vasomotor response of cerebral arteries in vitro. Cire Res 35:835–849
Edvinsson L, Hardebo JE, Owman C (1978) Pharmacological analysis of 5-hydroxytryptamine receptors in isolated intracranial and extracranial vessels of cat and man. Circ Res 42:143–151
Edvinsson L, Fahrenkurg J, Hanko J, Owman C, Sundler F, Uddman R (1980) VIP(vasoactive intestinal polypeptide)-containing nerves of intracranial arteries in mammals. Cell Tissue Res 208:135–142
Edvinsson L, Fredholm BB, Hamel E, Jansen I, Verrecchia C (1985) Perivascular peptides relax cerebral arteries concomitant with stimulation of cyclic adenosine monophosphate accumulation and release of an endothelium-derived relaxing factor in the cat. Neurosci Lett 58:213–217
Edvinsson L, Håkanson R, Wahlestedt C, Uddman R (1987) Effects of neuropeptide Y on the cardiovascular system. Trends Pharmacol Sci 8:231–235
Edvinsson L, Brodin E, Jansen I, Uddman R (1988) Neurokinin A in cerebral vessels: characterization, localization and effects in vitro. Regul Pept 20:181–197
Elliot DA, Ong BY, Bruni JE, Bose D (1989) Role of endothelium in hypoxic contraction of canine basilar artery. Br J Pharmacol 96:949–955

Ellis EF, Nies AS, Oates JA (1977) Cerebral arterial smooth muscle contraction by thromboxane A_2. Stroke 8:480–483
Ellis EF, Wei EP, Kontos HA (1979) Vasodilation of cat cerebral arteries by prostaglandins D_2, E_2, G_2, and I_2. Am J Physiol 273:H381–H385
Emori T, Hirata Y, Ohata K, Shichiri M, Marumo F (1989) Secretory mechanism of immunoreactive endothelin in cultured bovine endothelial cells. Biochem Biophys Res Commun 160:93–100
Estrada C, Krause DN (1982) Muscarinic cholinergic receptor sites in cerebral blood vessels. J Pharmacol Exp Ther 221:85–90
Faraci FM (1989) Effects of endothelin and vasopressin on cerebral blood vessels. Am J Physiol 257:H799–H803
Flavahan NA, Vanhoutte PM (1990) G-proteins and endothelial responses. Blood Vessels 27:218–229
Fong JSC (1976) Alpha-tocopherol; its inhibition on human platelet aggregation. Experientia 32:639–641
Förstermann U, Mügge A, Alheid U, Haverich A, Frölich JC (1988) Selective attenuation of endothelium-mediated vasodilation in atherosclerotic human coronary arteries. Circ Res 62:185–190
Forzard JR (1975) The animal pharmacology of drugs used in the treatment of migraine. J Pharm Pharmacol 27:297–321
Fozard JR (1982) Basic mechanisms of antimigraine drugs. Adv Neurol 33:295–307
Fredholm BB, Jansen I, Edvinsson L (1985) Neuropeptide Y is a potent inhibitor of cyclic AMP accumulation in feline cerebral blood vessels. Acta Physiol Scand 124:467–469
Freidman PC, Mitchell GG, Heistad DD, Armstrong ML, Harrison DG (1986) Atherosclerosis impairs endothelium-dependent vascular relaxation to acetylcholine and thrombin in primates. Circ Res 58:783–789
Fujimori A, Yanagisawa M, Saito A, Goto K, Masaki T, Mima T, Takakura K (1990) Endothelin in plasma and cerebrospinal fluid of patients with subarachnoid haemorrhage. Lancet 336:663
Fujioka M, Nagao T, Kuriyama H (1986) Actions of the novel thromboxane A_2 antagonists, ONO-1270 and ONO-3708, on smooth muscle cells of the guinea-pig basilar artery. Naunyn Schmiedebergs Arch Pharmacol 334:468–474
Fujiwara S, Ito Y, Itoh T, Kuriyama H, Suzuki H (1982) Diltiazem-induced vasodilation of smooth muscle cells of the canine basilar artery. Br J Pharmacol 75: 455–467
Fujiwara M, Shirahase H, Usui H, Manabe K, Kurahashi K (1989a) Endothelium-dependent cerebrovascular contraction. In: Fujiwara M, Narumiya S, Miwa S (eds) Biosignalling in cardiac and vascular systems. Pergamon, Oxford, pp 35–41
Fujiwara M, Tsukahara T, Taniguchi T (1989b) α-Adrenoceptors in human and animal cerebral arteries: alterations after sympathetic denervation and subarachnoid hemorrhage. Trends Pharmacol Sci 10:329–332
Fujiwara M, Usui H, Shirahase H, Kurahashi K, Manabe K, Sawada M (1990) Endothelium-dependent contractile modulation in cerebral artery (abstr). Eur J Pharmacol 183:108–109
Furchgott RF (1983) Role of endothelium in responses of vascular smooth muscle. Circ Res 53:557–573
Furchgott RF (1988) Studies on relaxation of rabbit aorta by sodium nitrite: the basis for the proposal that the acid-activatable inhibitory factor from bovine retractor penis is inorganic nitrite and the endothelium-derived relaxing factor is nitric oxide. In: Vanhoutte PM (ed) Mechanism of vasodilatation, vol 4. Raven, New York, pp 401–414
Furchgott RF, Zawadzki JV (1980) The obligatory role of endothelial cells in the relaxation of arterial smooth muscle by acetylcholine. Nature 288:373–376
Furchgott RF, Cherry PD, Zawadzki JV (1983) Endothelium-dependent relaxation of arteries by acetylcholine, bradykinin and other agents. In: Bevan J, Fujiwara

M, Maxwell RA, Mohri K, Shibata S, Toda N (eds) Vascular neuroeffector mechanisms. 4th International Symposium 1981. Raven, New York, pp 151–157
Furchgott RF, Cherry PD, Zawadzki JV, Joyhianandan D (1984) Endothelial cells as mediators of vasodilation of arteries. J Cardiovasc Pharmacol 2 [Suppl]:S336–S343
Ganellin CR, Parsons ME (1982) Pharmacology of histamine receptors. Wright/PSG, Bristol, Mass
Gaw AJ, Wadsworth RM, Humphrey PPA (1990) Pharmacological characterization of postjunctional 5-HT receptors in cerebral arteries from the sheep. Eur J Pharmacol 197:35–44
Gillespie MN, Owasoyo JO, McMurtry IF, O'Brien RF (1986) Sustained coronary vasoconstriction provoked by a peptidergic substance released from endothelial cells in culture. J Pharmacol Exp Ther 236:339–343
Gillespie JS, Liu X, Martin W (1989) The effects of L-arginine and N^G-monomethyl L-arginine on the response of rat anococcygeus muscle to NANC nerve stimulation. Br J Pharmacol 98:1080–1082
Godfraind T (1986) EDRF and cyclic GMP control gating of receptor-operated calcium channels in vascular smooth muscle. Eur J Pharmacol 126:341–343
Goldschmidt JE, Tallarida RJ (1991) Pharmacological evidence that captopril possesses an endothelium-mediated component of vasodilation: effect of sulfhydryl groups on endothelium-derived relaxing factor. J Pharmacol Exp Ther 257: 1136–1145
Greenberg S, Diecke FPJ (1988) Endothelium-derived and contracting factors: new concepts and new findings. Drug Dev Res 12:131–149
Griffith SG, Burnstock G (1983) Immunohistochemical demonstration of serotonin in nerves supplying human cerebral and mesenteric blood vessels. Lancet 1: 561–562
Gryglewski RJ, Moncada S, Palmer RMJ (1986) Bioassay of prostacyclin and endothelium-derived relaxing factor (EDRF) from porcine aortic endothelial cells. Br J Pharmacol 87:685–694
Hadani M, Young W, Flamm ES (1988) Nicardipine reduces calcium accumulation and electrolyte derangements in regional cerebral ischemia in rats. Stroke 19: 1125–1132
Hagiwara M, Endo T, Hidaka H (1982) Effect of vinpocetine (TCV-3B) on cyclic GMP metabolism (in Japanese, English abstr). Folia Pharmacol Jpn 80:317–323
Hall SE (1991) Thromboxane A_2 receptor antagonists. Med Res Rev 11:503–579
Hanasaki K, Arita H (1988) Characterization of a new compound, S-145, as a specific TXA_2 receptor antagonist in platelets. Thromb Res 50:365–376
Hara H, Jansen I, Ekman R, Hamel E, MacKenzie ET, Uddman R, Edvinsson L (1989) Acetylcholine and vasoactive intestinal peptide in cerebral blood vessels: effect of extirpation of the sphenopalatine ganglion. J Cereb Blood Flow Metab 9:204–211
Harder DR (1980) Comparison of electrical properties of middle cerebral and mesenteric artery in cat. Am J Physiol 239:C23–C26
Harder DR, Madden JA (1987) Electrical stimulation of the endothelial surface of pressurized cat middle cerebral artery results in TTX-sensitive vasoconstrictor. Circ Res 60:831–836
Harder DR, Sánchez-Ferrer CF, Kauser K, Stekiel WJ, Rubanyi GM (1989) Pressure releases a transferable endothelial contractile factor in cat cerebral arteries. Circ Res 65:193–198
Harrison DG, Frediman PC, Armstrong ML, Marcus ML, Heistad DD (1987) Alteration of vascular reactivity in atherosclerosis. Circ Res 61 [Suppl 2]:II-74–II-80
Hatake K, Kakishita E, Wakabayashi, I, Sakiyama N, Hishida S (1990) Effect of aging on endothelium-dependent vascular relaxation of isolated human basilar artery to thrombin and bradykinin. Stroke 21:1039–1043

Haver VM, Namm DH (1984) Characterization of the thrombin-induced contraction of vascular smooth muscle. Blood Vessels 21:53–63
Heistad DD, Kontos HA (1984) Cerebral circulation. In: The cardiovascular system III. Am Physiological Society, pp 137–182 (Handbook Bethesda of physiology)
Heistad DD, Marcus ML, Gourley JK, Busija DW (1981) Effect of adenosine and dipyridamole on cerebral blood flow. Am J Physiol 240:H775–H780
Heistad DD, Armstrong ML, Marcus ML, Piegors DJ, Mark AL (1984) Augmented response to vasoconstrictor stimuli in hypercholesterolemic and atherosclerotic monkeys. Circ Res 54:711–718
Herbst TJ, Raichle ME, Ferrendelli JA (1979) β-Adrenergic regulation of adrenosine 3′,5′-monophosphate concentration in brain microvessels. Science 204:330–332
Hibbs JB Jr, Vavrin Z, Taintor RR (1987) L-arginine is required for expression of the activated selective metabolic inhibition in target cells. J Immunol 138: 550–565
Hickey KA, Rubanyi G, Paul RJ, Highsmith RF (1985) Characterization of a coronary vasoconstrictor produced by cultured endothelial cells. Am J Physiol 248:C550–C556
Hirata M, Hayashi Y, Ushikubi F, Yokota Y, Kageyama R, Nakanishi S, Narumiya S (1991) Cloning and expression of cDNA for a human thromboxane A_2 receptor. Nature 349:617–620
Hisayama T, Takayanagi I, Goromaru N, Okamoto Y (1989) Potentiating effect of 3-isobutyryl-2-isopropylpyrazolo-[1,5a]-pyridine (KC-404), a new cerebral vasodilator with antiplatelet activity, on prostacyclin-induced increase of cyclic AMP content in platelets of rat. Gen Pharmacol 20:183–186
Holmes B (1984) Flunarizine: review of its pharmacodynamic and pharmacokinetic properties and therapeutic use. Drugs 27:6–44
Holzmann S (1982) Endothelium-induced relaxation by acetylcholine associated with larger rises in cGMP in coronary arterial strips. J Cyclic Nucleotide Res 8: 409–419
Hosoda K, Nakao K, Arai H, Suga S, Ogawa Y, Mukoyama M, Shirakami G, Saito Y, Nakanishi S, Imura H (1991) Cloning and expression of human endothelin-1 receptor cDNA. FEBS Lett 287:23–26
Hugest F, Biziere K, Breteau M, Narcisse G (1980) Effects of nicergoline on rat central neurotransmitter receptors: neurochemical profile. J Pharmacol (Paris) 11:257–267
Ide K, Yamakawa K, Nakagomi T, Sasaki T, Saito I, Kurihara H, Yoshizumi M, Yazaki Y, Takakura K (1989) The role of endothelin in the pathogenesis of vasospasm following subarachnoid haemorrhage. Neurol Res 11:101–104
Ignarro LJ (1990) Biosynthesis and metabolism of endothelium-derived nitric oxide. Annu Rev Pharmacol Toxicol 30:535–560
Ignarro LJ, Burke TM, Wood KS, Wolin MS, Kadowitz PJ (1984) Association between cyclic GMP accumulation and acetylcholine-elicited relaxation of bovine intrapulmonary artery. J Pharmacol Exp Ther 228:682–690
Ignarro LJ, Byrns RE, Buga GM, Wood KS (1987) Endothelium-derived relaxing factor from pulmonary artery and vein possesses pharmacological and chemical properties identical to those of nitric oxide radical. Circ Res 61:866–879
Ihara M, Fukuroda T, Saeki T, Nishikibe M, Kojiri K, Suda H, Yano M (1991) An endothelin receptor (ET_A) antagonist isolated from Streptomyces misakiensis. Biochem Biophys Res Commun 178:132–137
Iizuka K, Akahane K, Momose D, Nakazawa M (1981) Highly selective inhibitors of thromboxane synthetase. I. Imidazole derivatives. J Med Chem 24:1139–1148
Inoue A, Yanagisawa M, Kimura S, Kasuya Y, Miyauchi T, Goto K, Masaki T (1989) The human endothelin family: three structurally and pharmacologically distinct isopeptides predicted by three separate genes. Proc Natl Acad Sci USA 86:2863–2867

Ito T, Kato T, Iwama Y, Muramatsu M, Shimizu K, Asano H, Okumura K, Hashimoto H, Satake T (1991) Prostaglandin H_2 as an endothelium-derived contracting factor and its interaction with endothelium-derived nitric oxide. J Hypertens 9:729–736

Jansen I, Alafaci C, McCulloch J, Uddman R, Edvinsson L (1991a) Tachykinins (substance P, neurokinin A, neuropeptide K and neurokinin B) in the cerebral circulation: vasomotor responses in vitro and in situ. J Cereb Blood Flow Metab 11:567–575

Jansen I, Blackburn T, Eriksen K, Edvinsson L (1991b) 5-Hydroxytryptamine antagonistic effect of ICI169,369, ICI170,809 and methysergide in human temporal and cerebral arteries. Pharmacol Toxicol 68:6–13

Kalaria RN, Harik SI (1986) Adenosine receptors of cerebral microvessels and choroid plexus. J Cereb Blood Flow Metab 6:463–470

Kalaria RN, Harik SI (1988) Adenosine receptors and the nucleotide transport in human brain vasculature. J Cereb Blood Flow Metab 8:32–39

Kaminow L, Bevan JA (1991) Clentiazem reduces infarct size in rabbit middle cerebral artery occlusion. Strok 22:242–246

Kanamaru K, Waga S, Tochio H, Nagatani K (1989) The effect of atherosclerosis on endothelium-dependent relaxation in the aorta and intracranial arteries of rabbits. J Neurosurg 70:793–798

Kappagoda CT, Thomson ABR, Senaratne MPJ (1991) Effect of nisoldipine on atherosclerosis in the cholesterol fed rabbit: endothelium dependent relaxation and aortic cholesterol content. Cardiovasc Res 25:270–282

Karashima T, Kuriyama H (1981) Electrical properties of smooth muscle cell membrane and neuromuscular transmission in the guinea-pig basilar artery. Br J Pharmacol 74:495–504

Kassell NF, Sasaki T, Colohan AR, Nazar G (1985) Cerebral vasospasm following aneurysmal subarachnoid hemorrhage. Stroke 16:562–572

Katsura M, Miyamoto T, Hamanaka N, Kondo K, Terada Y, Ohgaki Y, Kawasaki A, Tsuboshima M (1981) In vitro and in vivo effects of new-powerful thromboxane antagonists (3-alkylamino pinane derivatives). In: Samuelsson B, Paoletti R, Ramwell PW (eds) Prostaglandins and leukotrienes in medicine, vol 11. Raven, New York, pp 351–358

Katusic ZS, Vanhoutte PM (1986) Anoxic contractions in isolated canine cerebral arteries: contribution of endothelium-derived factors, metabolites of arachidonic acid, and calcium entry. J Cardiovasc Pharmacol 8 [Suppl 8]:S97–S101

Katusic ZS, Vanhoutte PM (1989) Superoxide anion is an endothelium-derived contracting factor. Am J Physiol 257:H33–H37

Katusic ZS, Shepherd JT, Vanhoutte PM (1987) Endothelium-dependent contraction to stretch in canine basilar arteries. Am J Physiol 252:H671–H673

Katusic ZS, Shepherd JT, Vanhoutte PM (1988) Endothelium-dependent contractions to calcium ionophore A23187, arachidonic acid and acetylcholine in canine basilar arteries. Stroke 19:476–479

Kauser K, Rubanyi GM, Roman RJ, Harder DR (1989) Pressure-induced endothelium-dependent contraction in cat cerebral arteries. In: Seylaz J, MacKenzie ET (eds) Neurotransmisson and cerebrovascular function, vol 1. Elsevier, Amsterdam, pp 71–75

Kawahara Y, Yamanishi J, Furuta Y, Kaibuchi K, Takai Y, Fukuzaki H (1983) Elevation of cytoplasmic free calcium concentration by stable thromboxane A_2 analogue in human platelets. Biochem Biophys Res Commun 117:663–669

Kazda S (1975) Influence of kallikrein on the cerebral flow in the dog. Kininogenases Kallikrein 3:95–100

Kido R, Hirose K, Kojima Y, Eigyo M (1968) Pharmacological studies on dimethylsulfamido-3(dimethylamino-2-propyl)-10-phenotiazine (8599RP) (in Japanese, English abstr). Oyo Yakuri 2:173–179

Kim P, Sundt TM, Vanhoutte PM (1988) Loss of endothelium-dependent relaxations and maintenance of endothelium-dependent contractions in chronic vasospasm following subarachnoid hemorrhage. In: Wilkins RH (ed) Cerebral vasospasm. Raven, New York, pp 145–149

Kitagawa S, Yamaguchi Y, Kunitomo M, Imaizumi N, Fujiwara M (1992) Impairment of endothelium-dependent relaxation in aorta from rats with arteriosclerosis induced by excess vitamin D and a high-cholesterol diet. Jpn J Pharmacol 59:339–347

Klaas M, Wadsworth R (1989) Contraction followed by relaxation in response to hypoxia in the sheep isolated middle cerebral artery. Eur J Pharmacol 168: 187–192

Kobayashi Y, Hattori K (1991) Nitroarginine inhibits endothelium-derived relaxation. Jpn J Pharmacol 52:167–169

Kobayashi Y, Ikeda K, Kakizoe E, Shinozuka K, Nara Y, Yamori Y, Hattori K (1991) Comparison of vasopressor effects of nitro arginine in stroke prone spontaneously hypertensive rats and Wistar-Kyoto rats. Clin Exp Pharmacol Physiol 18:599–604

Kohlmeyer K (1972) Der Einfluß eines neuen Vasodilators (Bencyclan) auf die allgemeine und regionale Hirndurchblutung. Herz Kreislauf 4:196–203

Komlos E, Petöcz LE (1970) Pharmakologische Unterschungen über die Wirkung von *N*-[3-(1-benzylcyclopheptyl-oxy)-propyl]-*N*,*N*-dimethyl-ammonium-hydrogenfumarat. Arzneimittel forschung 20:1338–1357

Kondo S, Kawada M, Sano N (1981) Effects of dilazep on cerebral blood flow under normal conditions and recirculation impairment after cerebral ischemia (in Japanese, English abstr). Folia Pharmacol Jpn 77:205–211

Kontos HA (1985) Oxygen radicals in cerebral vascular injury. Circ Res 57:508–516

Kontos HA (1989) Oxygen radicals in cerebral ischemia. In: Ginsberg MD, Dietruc WD (eds) Cerebrovascular diseases. Raven, New York, pp 365–371

Kontos HA, Wei EP, Povlishock JT, Christman CW (1984) Oxygen radicals mediate the cerebral arteriolar dilation from arachidonate and bradykinin in cars. Circ Res 55:295–303

Kontos HA, Wei EP, Marshall JJ (1988) In vivo assay of endothelium-derived relaxing factor. Am J Physiol 255:H1259–H1262

Kontos HA, Wei EP, Kukreja PC, Ellis EF, Hess ML (1990) Differences in endothelium-dependent cerebral dilation by bradykinin and acetylcholine. Am J Physiol 258:H1261–H1266

Kunitomo M, Kinoshita K, Bando Y (1981) Experimental atherosclerosis in rats fed a vitamin D, cholesterol-rich diet. J Pharmacobio dyn 4:718–723

Kushiku K, Katsuragi T, Mori R, Morishita H, Furukawa T (1985) Cardiovascular effects of brovincamine and possible mechanisms involved. Clin Exp Pharmacol Physiol 12:121–130

Lee TJ-F (1982) Cholinergic mechanism in the large cat cerebral artery. Circ Res 50:870–879

Lee TJ-F, Sarwinski SJ (1991) Nitric oxidergic neurogenic vasodilation in the porcine basilar artery. Blood Vessels 28:407–412

Lee TJ-F, Kinkead LR, Sarwinski S (1982) Norepinephrine and acetylcholine transmitter mechanisms in large cerebral arteries of the pig. J Cereb Blood Flow Metab 2:439–450

Lee TJ-F, Saito A, Berezin I (1984) Vasoactive intestinal polypeptide-like substance: the potential transmitter for cerebral vasodilation. Science 224:898–901

Lee TJ-F, Fang YX, Nickols GA (1989) Cyclic nucleotide and cerebral neurogenic vasodilation. In: Seylaz J, MacKenzie ET (eds) Neurotransmission and cerebrovascular function. Elsevier, Amsterdam, pp 277–280

Levitt B, Duckles SP (1986) Evidence against serotonin as a vasoconstrictor neurotransmitter in the rabbit basilar artery. J Pharmacol Exp Ther 238:880–885

Li Y-O, Fredholm BB (1985) Adenosine analogues stimulate cyclic AMP formation in rabbit cerebral microvessels via adenosine A_2-receptors. Acta Physiol Scand 124:253–259
Lin L, Nasjletti A (1991) Role of endothelium-derived prostanoid in angiotensin-induced vasoconstriction. Hypertension 18:158–164
Lincoln TM (1983) Effects of nitroprusside and 8-bromocyclic GMP on the contractile activity of the rat aorta. J Pharmacol Exp Ther 224:100–107
Loew DM, Weil C (1982) Hydergine in senile mental impairment. Gerontology 28:54–74
Lombardo D, Dennis EA (1985) Cobra venom phospholipase A_2 inhibition by manoalide. J Biol Chem 260:7234–7240
Lundberg J, Hökfelt T, Schultzberg M, Uvnäs-Wallenstein K, Köhler C, Said SI (1979) Occurrence of vasoactive intestinal polypeptide (VIP)-like immunoreactivity in certain cholinergic neurones of the cat: evidence from combined immunohistochemistry and acetylcholinesterase staining. Neuroscience 4:1539–1559
Lüscher TF, Vanhoutte PM (1986a) Endothelium-dependent contractions to acetylcholine in the rat aorta of the spontaneously hypertensive rat. Hypertension 8:344–348
Lüscher TF, Vanhoutte PM (1986b) Endothelium-dependent responses to platelets and serotonin in spontaneously hypertensive rats. Hypertention 8 [Suppl 2]: II55–II60
Lyons CR, Orloff GJ, Cunningham JM (1992) Molecular cloning and functional expression of an inducible nitric oxide synthase from a murine macrophage cell line. J Biol Chem 267:6370–6374
Manabe K, Shirahase H, Usui H, Kurahashi K, Fujiwara M (1989) Endothelium-dependent contractions induced by angiotensin I and angiotensin II in canine cerebral artery. J Pharmacol Exp Ther 251:317–320
Marshall JJ, Kontos HA (1990) Endothelium-derived relaxing factors: a perspective from in vivo data. Hypertension 16:371–386
Marshall JJ, Wei EP, Kontos HA (1988) Independent blockade of cerebral vasodilation from acetylcholine and nitric oxide. Am J Physiol 255:H847–H854
Masaki T (1989) The discovery, the present state, and the future prospects of endothelin. J Cardiovasc Pharmacol 139 [Suppl 5]:S1–S4
Masaoka H, Suzuki R, Hirata Y, Emori T, Marumo F, Hirakawa K (1989) Raised plasma endothelin in aneurysmal subarachnoid hemorrhage. Lancet 2:1402
Masuzawa K, Asano M, Matsuda T, Imaizumi Y, Watanabe M (1990) Possible involvement of ATP-sensitive K^+ channels in the relaxant response of dog middle cerebral artery to cromakalim. J Pharmacol Exp Ther 255:818–825
Matsuoka I, Takaori S, Morimoto M (1972) Effects of diphenidol on the central vestibular and visual systems of cats. Jpn J Pharmacol 22:817–825
Mayhan WG, Faraci FM, Heistad DD (1987) Impairment of endothelium-dependent responses of cerebral arterioles in chronic hypertension. Am J Physiol 253: H1435–H1440
Mayhan WG, Amundsen SM, Faraci FM, Heistad DD (1988) Responses of cerebral arteries after ischemia and reperfusion in cats. Am J Physiol 255:H879–H884
McBean DE, Harper AM, Rudolphi KA (1988) Effects of adenosine and its analogues on porcine basilar arteries: are only A_2 receptors involved? J Cereb Blood Flow Metab 8:40–45
McCord JM (1985) Oxygen-derived free radicals in postischemic tissue injury. N Engl J Med 312:159–163
McCulloch J, Uddman R, Kingman T, Edvinsson L (1986) Calcitonin gene-related peptide: functional role in cerebrovascular regulation. Proc Natl Acad Sci USA 83:5731–5735
McLenachan JM, Williams JK, Fish RD, Ganz P, Selwyn AP (1991) Loss of flow-mediated endothelium-dependent relaxation occurs early in the development of atherosclerosis. Circulation 84:1273–1278

Meyers PR, Minor RL, Guerra R, Bates JN, Harrison DG (1990) Vasorelaxant properties of the endothelium-derived relaxing factor more closely resemble S-nitrosothiol than nitric oxide. Nature 345:161–163
Miller VM, Vanhoutte PM (1985) Endothelium-dependent contractions to arachidonic acid are mediated by products of cyclooxygenase. Am J Physiol 248: H432–H437
Miller VM, Komori K, Burnett JC Jr, Vanhoutte PM (1989) Differential sensitivity to endothelin in canine arteries and veins. Am J Physiol 257:H1127–H1131
Mima T, Yanagisawa M, Shigeno T, Saito A, Goto K, Takakura K, Masaki T (1989) Endothelin acts in feline and canine cerebral arteries from the adventitial side. Stroke 20:1553–1556
Miyazaki M (1971) Effect of cerebral circulatory drugs on cerebral and peripheral circulation, with special reference to aminophylline, papaverine, cyclandelate and isoxsuprine. Jpn Circ J 35:1053–1057
Mizusawa H, Sakakibara E (1975) Effects of ifenprodil tartrate on the cardiovascular system (in Japanese, English abstr). Folia Pharmacol Jpn 71:597–608
Moncada S, Bunting S, Mullane K, Thorogood P, Vane JR, Raz A, Needleman P (1977) Imidazole: a selective inhibitor of thromboxane synthetase. Prostaglandins 13:611–618
Moncada S, Palmer RMJ, Higgs EA (1988) The discovery of nitric oxide as the endogenous nitrovasodilator. Hypertension 12:365–372
Moncada S, Palmer RMJ, Higgs EA (1991) Nitric oxide: physiology, pathophysiology, and pharmacology. Pharmacol Rev 43:109–142
Moreo MJ, Conde MV, Fraile ML, Fernándes-Lomana H, Pablo ALL, Marco EJ (1991) Lesion of the dorsal raphe nucleus induces supersensitivity to serotonin in isolated cat middle cerebral artery. Brain Res 538:324–328
Moritoki H, Hosoki E, Ishida Y (1986) Age-related decrease in endothelium-dependent dilator response to histamine in rat mesenteric artery. Eur J Pharmacol 126:61-67
Murad F (1986) Cyclic guanosine monophosphate as mediator of vasodilation. J Clin Invest 78:1–5
Muramatsu I (1986) Evidence of sympathetic, purinergic transmission in the mesenteric artery of the dog. Br J Pharmacol 87:478–480
Muramatsu I (1991) Selectivity of moxisylyte to α_1-adrenoceptor subtypes (in Japanese, English abstr). J Jpn Coll Angiol 31:767–771
Muramatsu I, Kigoshi S (1987) Purinergic and non-purinergic innervation in the cerebral arteries of the dog. Br J Pharmacol 92:901–908
Muramatsu I, Fujiwara M, Ohsumi Y, Shibata S (1978) Vasoconstrictor and dilator actions of nicotine and electrical transmural stimulation on isolated dog cerebral arteries. Blood Vessels 15:110–118
Muramatsu I, Fujiwara M, Miura A, Shibata S (1980) Reactivity of isolated canine cerebral arteries to adenine nucleotide and adenosine. Pharmacology 21:198–205
Muramatsu I, Fujiwara M, Miura A, Sakakibara Y (1981) Possible involvement of adenine nucleotide in sympathetic neuroeffector mechanisms of dog basilar artery. J Pharmacol Exp Ther 216:401–409
Muramatsu I, Sakakibara Y, Hong S-C, Fujiwara M (1984) Effects of cinepazide on the purinergic responses in the dog cerebral artery. Pharmacology 28:27–33
Muramatsu I, Ohmura T, Oshita M (1989) Comparison between sympathetic adrenergic and purinergic transmission in the dog mesenteric artery. J Physiol (Lond) 411:227–243
Muramatsu I, Laniyonu A, Moore GJ, Hollenberg MD (1992) Vascular actions of thrombin receptor peptide. Can J Physiol Pharmacol 70:996–1003
Nagakawa Y, Akedo Y, Kaku S, Orimo H (1990) Effects of nicergoline on platelet aggregation, plasma viscosity and erythrocyte deformability in geriatric patients with cerebral infarction. Arzneimittel forschung 40:862–864

Nagao T, Suzuki H, Kuriyama H (1986) Effects of flunarizine on electrical and mechanical responses of smooth muscle cells in basilar and ear arteries of the rabbit. Naunyn Schmiedebergs Arch Pharmacol 333:431–438
Nagata K, Ogawa T, Omosu M, Fujimoto K, Hayashi S (1985) In vitro and in vivo inhibitory effects of propentofylline on cyclic AMP phosphodiesterase activity. Arzneimittel forschung 35:1034–1036
Narisada M, Ohtani M, Watanabe F, Uchida K, Arita H, Doteuchi M, Hanasaki K, Kakushi H, Otani K, Hara S (1988) Synthesis and in vitro activity of various derivatives of a novel thromboxane receptor antagonist, (±)-(5Z)-7-[3-endo-[(phenylsulfonyl)amino]bicyclo[2.2.1]hept-2-exo-yl]heptenoic acid. J Med Chem 31:1847–1854
Needleman P, Raz A, Ferrendelli JA, Minkes M (1977) Application of imidazole as a selective inhibitor of thromboxane synthetase in human platelets. Proc Natl Acad Sci USA 74:1716–1720
Nelson DL, Taylor EW (1986) Spiroxatrine: a selective $serotonin_{1A}$ receptor antagonist. Eur J Pharmacol 124:207–208
O'Brien RF, Robbins RJ, McMurtry IF (1987) Endothelial cells in culture produce a vasoconstrictor substance. J Cell Physiol 132:263–270
Ogawa Y, Nakao K, Arai H, Nakagawa O, Hosoda K, Suga S, Nakanishi S, Imura H (1991) Molecular cloning of a non-isopeptide-selective human endothelin receptor. Biochem Biophys Res Commun 178:248–255
Ogletree ML, Harris DN, Greenberg R, Haslanger MF, Nakane M (1985) Pharmacological actions of SQ 29548, a novel selective thromboxane antagonist. J Pharmacol Exp Ther 234:435–441
Ohashi M, Kito J, Nishino M (1986) Mode of cerebral vasodilating action of KC-404 in isolated canine basilar artery. Arch Int Pharmacodyn Ther 280:216–229
Okada T, Shibuya M, Suzuki Y, Ikegaki I, Kageyama N, Asano T, Hidaka H (1988) Effect of the intracellular calcium antagonist HA1004 on cerebral blood flow in rats. Neurol Med Chir (Tokyo) 28:625–630
Oshita M, Kigoshi S, Muramatsu I (1989) Selective potentiation of extracellular Ca^{2+}-dependent contraction by neuropeptide Y in rabbit mesenteric arteries. Gen Pharmacol 20:363–367
Owman C (1990) Peptidergic vasodilator nerves in the peripheral circulation and in the vascular beds of the heart and brain. Blood Vessels 27:73–93
Pagano PJ, Lin L, Sessa WC, Nasjletti A (1991) Arachidonic acid elicits endothelium -dependent release from the rabbit aorta of a constrictor prostanoid resembling prostaglandin endoperoxides. Circ Res 69:396–405
Palmer RMJ, Ferrige AG, Moncada S (1987) Nitric oxide release accounts for the biological activity of endothelium-derived relaxing factor. Nature 327:524–526
Palmer RMJ, Ashton DS, Moncada S (1988) Vascular endothelial cells synthesize nitric oxide from L-arginine. Nature 333:664–666
Parsons AA (1991) 5-TH receptors in human and animal cerebrovasculature. Trends Pharmacol Sci 12:310–315
Parsons AA, Whalley ET, Feniuk W, Connor HE, Humphrey PPA (1989) $5\text{-}HT_1$-like receptors mediate 5-hydroxytryptamine-induced contraction of human isolated basilar artery. Br J Pharmacol 96:434–449
Peach MJ, Loeb AL, Singer HA, Saye JA (1985) Endothelium-derived vascular relaxing factor. Hypertension 7 [Suppl 1]:I-94–I-100
Peroutka SJ (1983) Pharmacology of calcium channel antagonists: novel class of anti-migraine agents? Headache 23:273–283
Peroutka SJ (1984) Vascular serotonin receptors: correlation with $5\text{-}HT_1$ and $5\text{-}HT_2$ binding sites. Biochem Pharmacol 33:2341–2353
Phillis JW, Swanson TH, Barraco RA (1984) Interactions between adenosine and nifedipine in the rat cerebral cortex. Neurochem Int 6:693–699
Pickard J (1981) Role of prostanoids and arachidonic acid derivatives in the coupling of cerebral blood flow to cerebral metabolism. J Cereb Blood Flow Metab 1:361–384

Popescu LM, Panoiu C, Hinescu M, Nutu O (1985) The mechanism of cGMP-induced relaxation in vascular smooth muscle. Eur J Pharmacol 107:393–394
Rakugi H, Tabuchi Y, Nakamaru M, Nagano M, Higashimori K, Mikami H, Ogihara T, Suzuki N (1990) Evidence for endothelin-1 release from resistance vessels of rats in response to hypoxia. Biochem Biophys Res Commun 169: 973–977
Randall MJ, Parry MJ, Hawkeswood E, Cross PE, Dickinson RP (1981) UK 37,248, a novel, selective thromboxane synthetase inhibitor with platelet anti-aggregatory and anti-thrombotic activity. Thromb Res 23:145–162
Rapoport RM, Murad F (1983) Agonist-induced endothelium-dependent relaxation in rat thoracic aorta may be mediated through cGMP. Circ Res 52:352–357
Rapoport RM, Draznin MB, Murad F (1984) Mechanisms of adenosine triphosphate-, thrombin-, and trypsin-induced relaxation of rat thoracic aorta. Circ Res 55:468–479
Rees DD, Palmer RMJ, Hodson HF, Moncada S (1989) A specific inhibitor of nitric oxide formation from L-arginine attenuates endothelium-dependent relaxation. Br J Pharmacol 96:418–424
Regoli D, D'Oleans-Juste P, Escher E, Mizrahi J (1984) Receptors for SP. 1. The pharmacological preparations. Eur J Pharmacol 97:161–170
Reynolds IJ, Miller RJ (1989) Ifenprodil is a novel type of N-methyl-D-aspartate receptor antagonist: interaction with polyamines. Mol Pharmacol 36:758–765
Robinson MJ (1989) The cerebrovascular significance of endothelin. In: Seylaz J, MacKenzie ET (eds) Neurotransmission and cerebrovascular function, vol 1. Elsevier, Amsterdam, pp 301–304
Robinson MJ, Teasdale GM (1990) Calcium antagonists in the management of subarachnoid haemorrhage. Cerebrovasc Brain Metab Rev 2:205–226
Rosenblum WI (1987) Hydroxy radical mediates the endothelium-dependent relaxation produced by bradykinin in mouse cerebral arterioles. Circ Res 61:601–603
Rosenblum WI, Nelson GH (1988a) Endothelium dependence of dilation of pial arterioles in mouse brain by calcium ionophore. Stroke 19:1379–1382
Rosenblum WI, Nelson GH (1988b) Endothelium-dependent constriction demonstrated in vivo in mouse cerebral arterioles. Circ Res 63:837–843
Rosenblum WI, Nelson GH, Povlishock JT (1987) Laser-induced endothelial damage inhibits endothelium-dependent relaxation in the cerebral microcirculation of the mouse. Circ Res 60:169–176
Rosenblum WI, McDonald M, Wormley B (1989) Calcium ionphore and acetylcholine dilate arterioles on the mouse brain by different mechanisms. Stroke 20:1391–1395
Ross R (1986) The pathogenesis of atherosclerosis – an update. N Engl J Med 314:711–718
Rubanyi GM (1988a) Endothelium-dependent pressure-induced contraction of isolated canine carotid arteries. Am J Physiol 255:H783–H788
Rubanyi GM (1988b) Endothelium-derived vasoconstrictor factors. In: Ryan US (ed) Endothelial cell III. CRC, Boca Raton, pp 61–70
Rubanyi GM, Vanhoutte PM (1985) Hypoxia releases a vasoconstrictor substance from the canine vascular endothelium. J Physiol (Lond) 364:45–56
Saito A, Lee T J-F (1987) Serotonin as an alternative transmitter in sympathetic nerves of large cerebral arteries of the rabbit. Circ Res 60:220–228
Saito A, Shiba R, Kimura S, Yanagisawa M, Goto K, Masaki T (1989) Vasoconstrictor response of large cerebral arteries of cats to endothelin, an endothelium-derived vasoactive peptide. Eur J Pharmacol 162:353–358
Saito Y, Mizuno T, Itakura M, Suzuki Y, Ito T, Hagiwara H, Hirose S (1991) Primary structure of bovine endothelin ET_B receptor and identification of signal peptidase and metal proteinase cleavage sites. J Biol Chem 266:23433–23437
Sakakibara Y, Fujiwara M, Muramatsu I (1982) Pharmacological characterization of the alpha adrenoceptors of the dog basilar artery. Naunyn Schmiedebergs Arch Pharmacol 319:1–7

Sakurai T, Yanagisawa M, Takuwa Y, Miyazaki H, Kimura S, Goto K, Masaki T (1990) Cloning of a cDNA encoding a non-isopeptide-selective subtype of the endothelin receptor. Nature 348:732–735

Salmon S (1982) Putative 5-HT central feedback in migraine and cluster headache attacks. Adv Neurol 33:265–274

Salzman PM, Salmon JA, Moncada S (1980) Prostacyclin and thromboxane A_2 synthesis by rabbit pulmonary artery. J Pharmacol Exp Ther 215:240–247

Sánchez-Ferrer CF, Marin J (1990) Endothelium-derived contractile factors. Gen Pharmacol 21:589–603

Sata T, Linden J, Liu L-W, Kubota E, Said SI (1988) Vasoactive intestinal peptide evokes endothelium-dependent relaxation and cyclic AMP accumulation in rat aorta. Peptides 9:853–858

Satoh H, Inui J (1984) Endothelial cell-dependent relaxation and contraction induced by histamine in the isolated guinea-pig pulmonary artery. Eur J Pharmacol 97:321–324

Schmidley JW (1990) Free radicals in central nervous system ischemia. Stroke 21: 1086–1090

Schrader J, Wahl M, Kuschinsky W, Kreutzberg GW (1980) Increase of adenosine content in cerebral cortex of the cat during bicuculline-induced seizure. Pflugers Arch 387:245–251

Schutz W, Steurer G, Tuisl E (1982) Functional identification of adenylate cyclase-coupled adenosine receptors in rat brain microvessels. Eur J Pharmacol 85: 177–184

Sedaa KO, Bjur RA, Shinozuka K, Westfall DP (1990) Nerve and drug-induced release of adenine nucleosides and nucleotides from rabbit aorta. J Pharmacol Exp Ther 252:1060–1067

Senaratne MPJ, Thomson ABR, Kappagoda CT (1991) Lovastatin prevents the impairment of endothelium dependent relaxation and inhibits accumulation of cholesterol in the aorta in experimental atherosclerosis in rabbits. Cardiovasc Res 25:568–578

Sercombe R, Verrecchia C (1986) Role of vascular endothelium in modulating cerebrovascular effects of neurotransmitters. In: Owman C, Hardebo JE (eds) Neural regulation of brain circulation. Elsevier, Amsterdam, pp 27–41

Shigeno T, Mima T, Yanagisawa M, Saito A, Goto K, Yamashita K, Takenouchi T, Matsuura N, Yamasaki Y, Yamada K, Masaki T, Takakura K (1991) Prevention of cerebral vasospasm by actinomycin D. J Neurosurg 74:940–943

Shikano K, Long CJ, Ohlstein EH, Berkowitz BA (1988) Comparative pharmacology of endothelium-derived relaxing factor and nitric oxide. J Pharmacol Exp Ther 247:873–881

Shirahase H, Fujiwara M, Usui H, Kurahashi K (1987a) A possible role of thromboxane A_2 in endothelium in maintaining resting tone and producing contractile responses to acetylcholine and arachidonic acid in canine cerebral arteries. Blood Vessels 24:117–119

Shirahase H, Usui H, Kurahashi K, Fujiwara M, Fukui K (1987b) Possible role of endothelial thromboxane A_2 in the resting tone and contractile response to acetylcholine and arachidonic acid in canine cerebral arteries. J Cardiovasc Pharmacol 10:517–522

Shirahase H, Usui H, Kurahashi K, Fujiwara M, Fukui K (1988a) Endothelium-dependent contraction induced by nicotine in isolated canine basilar artery – possible involvement of a thromboxane A_2 (TXA_2) like substance. Life Sci 42:437–445

Shirahase H, Usui H, Manabe K, Kurahashi K, Fujiwara M (1988b) An endothelium -dependent contraction induced by A-23187, a Ca^{2+} ionophore in canine basilar artery. J Pharmacol Exp Ther 247:701–705

Shirahase H, Usui H, Manabe K, Kurahashi K, Fujiwara M (1988c) Endothelium-dependent contraction and -independent relaxation induced by adenine nucleo-

tides and nucleoside in the canine basilar artery. J Pharmacol Exp Ther 247: 1152–1157
Shirahase H, Usui H, Manabe K, Kurahashi K, Fujiwara M (1989) Vasorelaxing effects of prostaglandin I_2 on the canine basilar and coronary arteries. J Pharmacol Exp Ther 248:769–773
Shirahase H, Usui H, Shimaji H, Kurahashi K, Fujiwara M (1990) Endothelium-dependent contraction induced by platelet-derived substances in canine basilar arteries. J Pharmacol Exp Ther 255:182–186
Shirahase H, Usui H, Shimaji H, Kurahashi K, Fujiwara M (1991a) Endothelium-independent and endothelium-dependent contractions mediated by P_{2X}- and P_{2Y}-purinoceptors in canine basilar arteries. J Pharmacol Exp Ther 256:683–688
Shirahase H, Usui H, Shimaji H, Kurahashi K, Fujiwara M (1991b) Endothelium-independent and -dependent contractions induced by endothelin-1 in canine basilar arteries. Life Sci 49:273–281
Shiraishi M, Kato K, Terao S, Ashida Y, Terashita Z, Kito G (1989) Quinones 4. Novel eicosanoid antagonists: synthesis and pharmacological evaluation. J Med Chem 32:2214–2221
Sokoloff L (1959) The action of drugs on the cerebral circulation. Pharmacol Rev 11:1–85
Standen NB, Quayle JM, Davies NW, Brayden JE, Huang Y, Nelson MT (1989) Hyperpolarizing vasodilators activate ATP-sensitive K^+ channels in arterial smooth muscle. Science 245:177–180
Stewart-Lee AL, Burnstock G (1991) Changes in vasoconstrictor and vasodilator responses of the basilar artery during maturation in the Watanabe heritable hyperlipidemic rabbit differ from those in the New Zealand White rabbit. Arteriosclerosis Thromb 11:1147–1155
Sugiura M, Inagami T, Kon V (1989) Endotoxin stimulates endothelin-release in vivo and in vitro as demonstrated by radioimmunoassay. Biochem Biophys Res Commun 161:1220–1227
Suzuki Y, Yamaguchi K, Shimada S, Kitamura Y, Ohnishi H (1982) Antithrombotic activity and the mechanism of action of trapidil (Roconal). Prostaglandins Leukotrienes Med 9:685–695
Suzuki Y, McMaster D, Huang M, Lederis K, Rorstad OP (1985) Characterization of functional receptors for vasoactive intestinal peptide in bovine cerebral arteries. J Neurochem 45:890–899
Suzuki N, Hardebo JE, Owman C (1988) Origins and pathways of cerebrovascular vasoactive intestinal polypeptide-positive nerves in rat. J Cereb Blood Flow Metab 8:697–712
Suzuki N, Hardebo JE, Kåhrstöm J, Owman C (1990a) Neuropeptide Y co-exists with vasoactive intestinal polypeptide and acetylcholine in the parasympathetic cerebrovascular nerves originating in the sphenopalatione, otic, and internal carotid ganglia of the rat. Neuroscience 36:507–519
Suzuki Y, Satoh S, Ikegaki I, Asano T, Shibuya M, Sugita K, Lederis K, Rorstad O (1990b) Endothelin causes contraction of canine and bovine arterial smooth muscle in vitro and in vivo. Acta Neurochir (Wien) 104:42–47
Tagawa H, Tomoike H, Nakamura M (1991) Putative mechanisms of endothelium-dependent relaxation of the aorta with atheromatous plaque in heritable hyperlipidemic rabbits. Circ Res 68:330–337
Takenaka T, Handa J (1979) Cerebrovascular effects of YC-93, a new vasodilator, in dogs, monkeys and human patients. Int J Clin Pharmacol Biopharm 17:1–11
Taylor EW, Duckles SP, Nelson DL (1986) Dissociation constants of serotonin agonists in the canine basilar artery correlate to K_i values at the $5\text{-}HT_{1A}$ binding site. J Pharmacol Exp Ther 236:118–125
Tesfamariam B, Jakubowski JA, Cohen RA (1989) Contraction of diabetic aorta caused by endothelium-derived PGH_2-TXA_2. Am J Physiol 257:H1327–H1333
Tesfamariam B, Brown ML, Cohen RA (1991) Elevated glucose impairs endothelium-dependent relaxation by activating protein kinase C. J Clin Invest 87:1643–1648

Toda N, Okamura T (1991a) Role of nitric oxide in neurally induced cerebroarterial relaxation. J Pharmacol Exp Ther 258:1027–1032

Toda N, Okamura T (1991b) Reciprocal regulation by putatively nitroxidergic and adrenergic nerves of monkey and dog temporal arterial tone. Am J Physiol 261:H1740–H1745

Tomita M, Gotoh F, Sato T, Amano T, Tanahashi N, Tanaka K, Yamamoto M (1978) Comparative responses of the carotid and vertebral arterial systems of rhesus monkeys to betahistine. Stroke 9:382–387

Tomiyama T, Wakabayashi S, Kosakai K, Yokota M (1990) Azulene derivatives: new non-prostanoid thromboxane A_2 receptor antagonists. J Med Chem 33: 2323–2326

Towart R (1981) The selective inhibition of serotonin-induced contractions of rabbit cerebral vascular smooth muscle by calcium-antagonistic dihydropyridines. Circ Res 48:650–657

Tsukahara T, Taniguchi T, Fujiwara M, Handa H, Nishikawa M (1985) Alterations in alpha adrenergic receptors in human cerebral arteries after subarachnoid hemorrhage. Stroke 16:53–58

Tsukahara T, Taniguchi T, Shimohama S, Fujiwara M, Handa H (1986a) Characterization of beta adrenergic receptors in human cerebral arteries and alteration of the receptors after subarachnoid hemorrhage. Stroke 17:202–207

Tsukahara T, Taniguchi T, Usui H, Miwa S, Shimohara S, Fujiwara M, Handa H (1986b) Sympathetic denervation and alpha adrenoceptors in dog cerebral arteries. Naunyn Schmiedebergs Arch Pharmacol 334:436–443

Tuor UI, Kelly PAT, Edvinsson L, McCulloch J (1990) Neuropeptide Y and the cerebral circulation. J Cereb Blood Flow Metab 19:591–601

Uddman R, Edvinsson L (1989) Neuropeptides in the cerebral circulation. Cerebrovasc Brain Metab Rev 1:230–252

Uddman R, Edvinsson L, Ekman R, Kingman TA, McCulloch J (1985) Innervation of the feline cerebral vasculature by nerve fibers containing calcitonin gene-related peptide: trigeminal origin and co-existence with substance P. Neurosci Lett 62:131–136

Unna K (1951) The pharmacology of new sympathol derivatives. Arch Exp Pathol Pharmakol 213:207–234

Ushikubi F, Nakajima M, Hirata M, Okuma M, Fujiwara M, Narumiya S (1989) Purification of the thromboxane A_2/prostaglandin H_2 receptor from human blood platelets. J Biol Chem 264:16496–16501

Uski TK (1986) Cerebrovascular smooth muscle effects of prostanoids. In: Owman C, Hardebo JE (eds) Neural regulation of brain circulation. Elsevier, Amsterdam, pp 245–260

Uski TK, Andersson KE (1985) Some effects of the calcium promoter BAY K8644 on feline cerebral arteries. Acta Physiol Scand 123:49–53

Usui H, Kurahashi K, Ashida K, Fujiwara M (1983) Acetylcholine-induced contractile response in canine basilar artery with activation of thromboxane A_2 synthesis sequence. IRCS Med Sci 11:418–419

Usui H, Fujiwara M, Tsukahara T, Taniguchi T, Kurahashi K (1985) Differences in contractile responses to electrical stimulation and α-adrenergic binding sites in isolated cerebral arteries of humans, cows, dogs, and monkeys. J Cardiovasc Pharmacol 7 [Suppl 3]:S47–S52

Usui H, Fujiwara M, Tsubomura T, Kurahashi K, Nomura S, Mizuno N (1986) Possible involvement of prostanoids in cholinergic contractile response of canine cerebral artery. In: Owman C, Hardebo JE (eds) Neural regulation of brain circulation. Elsevier, Amsterdam, pp 261–272

Usui H, Fujiwara M, Shirahase H, Kurahashi K, Okamoto S, Miyamoto S (1987a) Endothelium-dependent vasoactivation of the canine cerebral artery by cerebrospinal fluid obtained from the patients who had suffered a subarachnoid hemorrhage. In: Nobin A, Owman C, Arneklo-Nobin B (eds) Neuronal messengers in vascular function. Elsevier, Amsterdam, pp 537–547

Usui H, Kurahashi K, Shirahase H, Fukui K, Fujiwara M (1987b) Endothelium-dependent vasocontraction in response to noradrenaline in the canine cerebral artery. Jpn J Pharmacol 44:228–231
Usui H, Kurahashi K, Shirahase H, Manabe K, Shibata S, Fujiwara M (1989) Endothelium-dependent relaxation in monkey cerebral and contraction in canine cerebral arteries produced by histamine. Med Sci Res 17:1035–1036
Usui H, Akimoto Y, Kurahashi K, Shirahase H, Fujiwara M, Shibata S (1990) Effects of nitroglycerin on stable thromboxane A_2 analogue-induced, nifedipine-resistant contraction in canine basilar artery. Jpn J Pharmacol 54:237–240
Vallee E, Gougat J, Navarro J, Delahayes JF (1979) Anti-inflammatory and platelet antiaggregant activity of phospholipase-A_2 inhibitors. J Pharm Pharmacol 31: 588–592
Van de Voorde J, Leusen I (1983) Role of the endothelium in the vasodilator response of rat thoracic aorta to histamine. Eur J Pharmacol 87:113–120
Van Riper DA, Bevan JA (1991) Evidence that neuropeptide Y and norepinephrine mediate electrical field-stimulated vasoconctriction of rabbit middle cerebral artery. Circ Res 68:568–577
Vane JR (1971) Inhibition of prostaglandin synthesis as a mechanism of action for aspirin-like drugs. Nature [New Biol] 231:232–235
Vane JR (1990) Endothelins come home to roost. Nature 348:207
Vane JR, Anggars E, Botting R (1990) Regulatory functions of the vascular endothelium. N Engl J Med 323:27–36
Vanhoutte PM (1987a) The end of the quest? Nature 327:459–460
Vanhoutte PM (1987b) Endothelium-dependent contractions in arteries and veins. Blood Vessels 24:141–144
Vanhoutte PM, Katusic ZS (1988) Endothelium-derived contracting factor: endothelin and/or superoxide anion? Trends Pharmacol Sci 9:229–230
Vanhoutte PM, Auch-Schwelk W, Boulanger C, Janssen PA, Katusic ZS, Komori K, Miller VM, Schini VB, Vidal M (1989) Does endothelin-1 mediate endothelium-dependent contractions during anoxia? J Cardiovasc Pharmacol 13 [Suppl 5]: S124–S128
Vanhoutte PM, Lüscher TF, Gräser T (1991) Endothelium-dependent contractions. Blood Vessels 28:74–83
Vanin AF (1991) Endothelium-derived relaxing factor is a nitrosyl iron complex with thiol ligands. FEBS Lett 289:1–3
Verbeuren TJ, Jordaens FH, Van Hove CE, Van Hoydonck A-E, Herman AG (1990) Release and vascular activity of endothelium-derived relaxing factor in atherosclerotic rabbit aorta. Eur J Pharmacol 191:173–184
Vu T-KH, Hung DT, Wheton VI, Coughlin SR (1991) Molecular cloning of a functional-thrombin receptor reveals a novel proteolytic mechanism of receptor activation. Cell 64:1057–1068
Wahl M (1985) Local chemical, neural, and humoral regulation of cerebrovascular resistance vessels. J Cardiovasc Pharmacol 7 [Suppl 3]:S36–S46
Walz DA, Anderson GF, Feston JW (1986) Responses of aortic smooth muscle to thrombin and thrombin analogues. Ann NY Acad Sci 485:323–334
Ward A, Clissold SP (1987) Pentoxifylline: a review of its pharmacodynamic and pharmacokinetic properties, and its therapeutic efficacy. Drugs 34:50–97
Warner TD, DeNucci G, Vane JR (1989) Rat endothelin is a vasodilator in the isolated perfused mesentery of the rat. Eur J Pharmacol 159:325–326
Wei EP, Kontos HA (1990) H_2O_2 and endothelium-dependent cerebral arteriolar dilation: implication for the identity of endothelium-derived relaxing factor generated by acetylcholine. Hypertension 16:162–169
Wei EP, Kontos HA, Christman CW, Dewitt DS, Povlishock JT (1985) Superoxide generation and reversal of acetylcholine induced cerebral vasodilation after acute hypertension. Circ Res 57:781–787
White A (1987) Responses of human basilar arteries to vasoactive intestinal polypeptide. Life Sci 41:1155–1163

White RP, Hagen AA (1982) Cerebrovascular actions of prostaglandins. Pharmacol Ther 128:313–331
White RP, Shirasawa Y, Robertson JT (1984) Comparison of responses elicited by alpha-thrombin in isolated canine basilar, coronary, mesenteric, and renal arteries. Blood Vessels 21:12–22
Xie Q-W, Cho HJ, Calaycay J, Mumford RA, Swiderek KM, Lee TD, Ding A, Troso T, Nathan C (1992) Cloning and characterization of inducible nitric oxide synthase from mouse macrophages. Science 256:225–228
Yamamoto H, Bossaller C, Cartwright J Jr, Henry PD (1988) Videomicroscopic demonstration of defective cholinergic arteriolar vasodilation in atherosclerotic rabbit. J Clin Invest 81:1752–1758
Yanagisawa M, Masaki T (1989a) Endothelin, a novel endothelium-derived peptide. Biochem Pharmacol 38:1877–1883
Yanagisawa M, Masaki T (1989b) Molecular biology and biochemistry of the endothelins. Trends Pharmacol Sci 10:374–378
Yanagisawa M, Kurihara H, Kimura S, Tomobe Y, Kobayashi M, Mitsui Y, Yazaki Y, Goto K, Masaki Y (1988) A novel potent vasoconstrictor peptide produced by vascular endothelial cells. Nature 332:411–415
Yoshimoto S, Ishizaki Y, Kurihara H, Sasaki T, Yoshizumi M, Yanagisawa M, Yazaki Y, Masaki T, Takakura K, Murota S (1990) Cerebral microvessel endothelium is producing endothelin. Brain Res 508:283–285
Yoshimoto S, Ishizaki Y, Sasaki T, Murota S (1991) Effect of carbon dioxide and oxygen on endothelin production by cultured porcine cerebral endothelial cells. Stroke 22:378–383
Yui Y, Hattori R, Kosuga K, Eizawa H, Hiki K, Ohkawa S, Ohnishi K, Terao S, Kawai C (1991) Calmodulin-independent nitric oxide synthase from rat polymorphonuclear neutrophils. J Biol Chem 266:3369–3371

CHAPTER 19

Altered Responsiveness of Vascular Smooth Muscle to Drugs in Diabetes

G. Pogátsa

A. Introduction

Cardiovascular alterations are much more frequent and much more severe in diabetic than in metabolically healthy populations (LANGSH et al. 1970). This observation has been made on the basis of the numerous manifestations of vascular diabetic defects throughout the body.

I. Atherosclerosis in Diabetes

Atherosclerosis is the most common and the first detected vascular complication of diabetes. Although the exact cause of the association between diabetes and atherosclerosis remains to be elucidated, disorders of lipid metabolism and hyperinsulinaemia, increased platelet aggregability, increased production of thromboxane B_2 and decreased vascular prostacyclin synthesis have been implicated (for review see STEINER 1981). Diabetic patients with hypertension have shown greater vascular permeability and perivascular fibrosis in the coronary arteries (KUBOTA et al. 1990).

During the past 15 years it has become more and more evident that vascular defects are not only the consequence of an accelerated and more frequent atherosclerosis (HAMBY et al. 1976; DRASH 1976).

II. Microangiopathy in Diabetes

In the early stage of diabetes many vascular alterations are detectable although no signs of atherosclerosis can yet be observed. For example, regional blood flow is lower in many tissues during experimental diabetes (LUCAS and FOY 1977). The microvessel density of various tissues decreases in diabetes (BOHLEN and NIGGL 1979, 1980). Diabetic arterioles exhibit a degeneration of wall structure whereas diabetic capillary basement membrane thickness may increase substantially (SIPERSTEIN et al. 1968; GUNDERSEN et al. 1978). During the autopsy of diabetic subjects immunoglobulin G depositions are frequently discovered in the microvasculature, suggesting that non-enzymatic glycosylation of vascular collagen is an important factor in the pathogenesis of diabetic vascular alterations since immunoglobulins are primarily bound by glycosylated collagens (ZHAO et al. 1990).

These abnormalities in diabetic vessels have been categorized as diabetic microangiopathy, and they may be detected as aneurysms, abnormal vascular low patterns, changes in vessel permeability, local constrictions and alterations in reactivity to various stimuli (PIERCE et al. 1988).

III. Diabetic Vascular Smooth Muscle

Furthermore, the diabetic smooth muscle cell is of a pathological phenotype, different from that of the intima-derived smooth muscle cell, and it plays a major role both in the alterations of the reactivity to various stimuli (PIERCE et al. 1988) and in the formation of intimal thickening through accelerated growth (KAWANO et al. 1990). KASKI et al. (1989) thus found localized patterns of arterial constriction that suggested local smooth muscle hypersensitivity to be a major pathophysiological mechanism, on comparison of angiographic findings during spontaneous and ergonovine-induced coronary artery spasms in diabetes. Moreover, the microvasculature of diabetic subjects is found to have an attenuated vasodilatory response to various agonists (BOHLEN and NIGGL 1979, 1980) and a virtual loss of capacity to control blood flow by autoregulation (HENRIKSEN et al. 1984). Abnormal constriction of arterioles under resting conditions can also be observed in spontaneous and chemically induced diabetes in animals (BOHLEN and NIGGL 1979, 1980). Nevertheless, it is worth pointing out that in the atherosclerotic arteries increased vasoconstrictor responses to noradrenaline, serotonin and histamine have also been reported in various species including man (LÜSCHER 1988a).

B. Altered Vascular Response in Diabetes

I. History

A lack of normal vasodilatory ability was first demonstrated in the arterial bed of the conjunctiva by SZENTIVÁNYI and PÉK (1973) in diabetic subjects. DANDONA et al. (1978) then reported this diminished vasodilatory ability in the cerebral arteries of diabetic patients during carbon monoxide inhalation. At nearly the same time EWALD et al. (1981) also detected a definite decrease in vasodilatory ability in the cutaneous arterial bed, provoking reactive hyperaemia in the forearms of diabetic children. In the coronary and femoral arterial bed a tendency to vasoconstrict was first demonstrated by POGÁTSA (1980) during intra-arterial infusions of catecholamines in alloxan-diabetic dogs. These phenomena are seen as the earliest dysfunctions in diabetic vessels before macroscopic or microscopic lesions are detectable in the arterial wall (PALIK et al. 1982a).

II. Localization

The vasoconstrictor tendency is especially pronounced in the coronary arterial bed, where vasodilatory interventions inversely provoke vasoconstriction. As demonstrated in dogs 3 months after induction of clinically manifest diabetes but without ketosis by 560 μmol/kg alloxan tetrahydrate, both electrical stimulation of the cardiac plexus with increasing frequencies and intracoronary infusion of noradrenaline in increasing doses produce vasodilation in the healthy and vasoconstriction in the diabetic states (PALIK et al. 1982a). Furthermore, reactive hyperaemic responses are considerably longer and flatter in the diabetic coronary arterial bed than in the metabolically healthy ones during a 1-min asphyxia or during a 1-min coronary arterial clamping (Fig. 1) (KOLTAI et al. 1983). Recently, CHILIAN (1990) detected that noradrenaline causes disparate responses in the coronary microvasculature: constriction occurs in vessels greater than 100 μm in diameter, but there is dilation in vessels less than 100 μm in diameter. Since α_1-adrenergic receptors are located in coronary arterioles and arteries while α_2-adrenergic receptors are preferentially located in small coronary arterioles, adrenergic activation can produce dissimilar constrictor effects in the coronary microcirculation. A similar

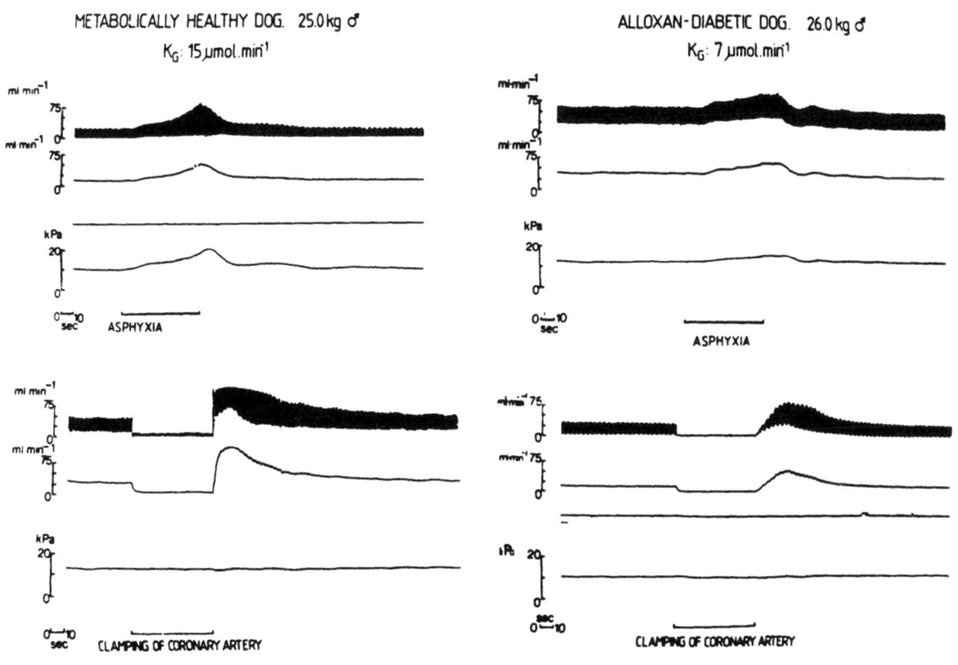

Fig. 1. Results of an experiment showing alterations in characteristics of reactive hyperaemia in coronary arterial bed during 1-min asphyxia (*upper*) or 1-min clamping of coronary artery (*lower*) in metabolically healthy (*left*) and alloxan-diabetic (*right*) dogs. K_G, plasma disappearance rate of glucose

difference between the large and small vessels can also be observed during diabetes in other tissues. For example, in the cremaster muscle the sensitivity of large, so-called first branching order arterioles to noradrenaline did not alter, while that of smaller, so-called second and third branching order arterioles increased initially in diabetes, but returned to normal as the duration of diabetes progressed to the chronic stage (MORFF 1990).

On the other hand, the enhanced vasoconstrictor tendency of the diabetic vascular bed can be demonstrated in various organs. Thus, isolated rings of diabetic carotid artery exhibit an increased response to noradrenaline, methoxamine, phenylephrine and potassium (Agrawal et al. 1987). Vascular sensitivity and reactivity to noradrenaline increased in the hindquarters of rats 1 week after streptozocin treatment and remained unchanged during the observation period (Friedman 1989). In isolated perfused mesenteric vasculature a non-specific decrease in contractile responses to nerve stimulation or vasoactive agonists was observed (Longhurst and Head 1985).

Isolated renal arteries from diabetic humans showed increased constrictor responses to an endoperoxide analogue and decreased dilator responses to isoproterenol (Roth et al. 1982).

Angiotensin induced a similar, dose-dependent increase in conronary blood flow both in the healthy and diabetic states, but it produced a higher elevation in the mean arterial blood pressure in diabetic dogs. The blocakade of vascular α-adrenoceptors with phentolamine diminished coronary blood flow induced by angiotensin only in the healthy condition, but further

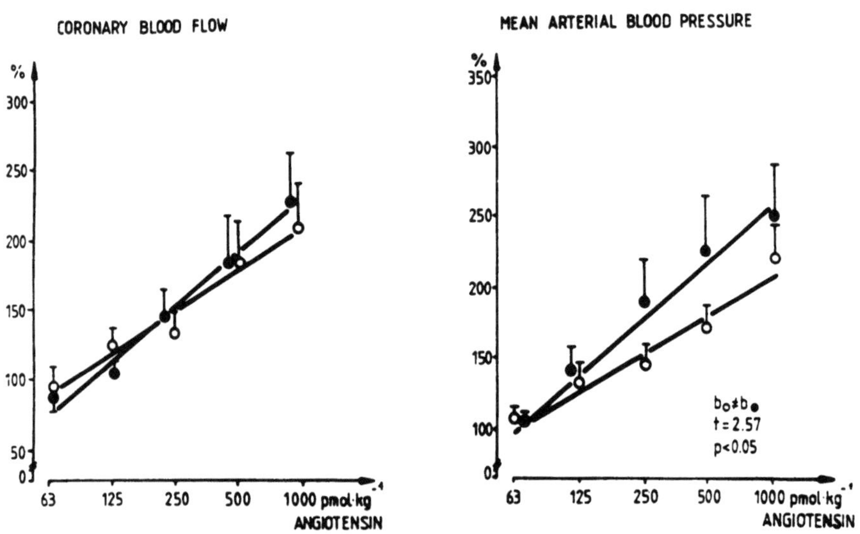

Fig. 2. Effect of phentolamine on angiotensin-induced alterations in coronary blood flow (*right*) and arterial blood pressure (*left*) of healthy (○, $n = 6$) and alloxan-diabetic dogs (●, $n = 6$) (as percentages of basal values)

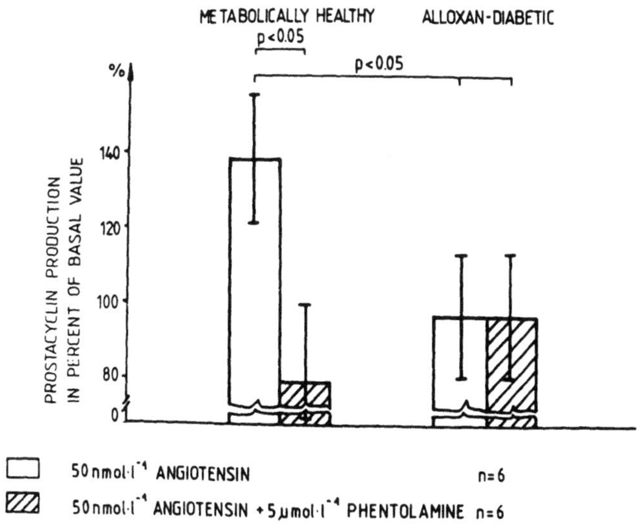

Fig. 3. Effect of phentolamine on angiotensin-induced alterations of prostacyclin production by isolated coronary rings of healthy and alloxan-diabetic dogs (as percentages of basal values)

enhanced the angiotensin-induced elevation of blood pressure in diabetes (Fig. 2). Furthermore, angiotensin potentiated the prostacyclin release only by the healthy vessels, which could be prevented by phentolamine (Fig. 3) (Koltai et al. 1994).

Blood pressure responses to noradrenaline and angiotensin are depressed in short-term diabetes; however, the baroreceptor reflexes are more sensitive to increases in blood pressure. These alterations suggest that there is some type of early phase alteration in the responsiveness of the cardiovascular system to the vasopressor agonists in short-term diabetes (Jackson and Carrier 1983), which may consist of disturbances in sympathetic innervation, membrane permeability, endothelial or ion channel functions and metabolism of catecholamine, calcium and prostaglandin.

Moreover, it cannot be simply assumed that a similar pathoaetiology exists in different tissues of the body during diabetes. Indeed, two independent studies which compared vascular abnormalities in the retina and in the myocardium of diabetic (Waber et al. 1981) and diabetic-hypertensive (Fluckiger et al. 1984) rats found distinct, tissue-specific differences. Retinal capillary basement membrane was thicker in the diabetic rat but coronary vasculature was unaltered. Furthermore, in diabetes, prostacyclin and prostaglandin E_1 lead to contraction in the carotid artery, but unchanged vasodilation in the mesenteric artery (Agrawal and McNeill 1986). Thus, the functional alterations and structural disruptions in the microvasculature which accompany diabetes may depend upon the type of vessel examined. The assumption by many investigators may have been erroneous, therefore,

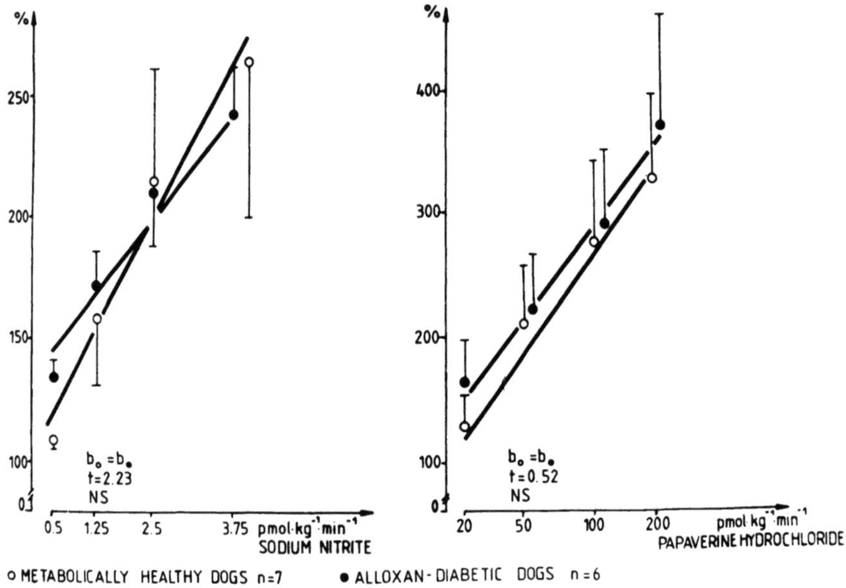

Fig. 4. Alterations in conductivity of coronary arterial bed induced by intracoronary administration of sodium nitrite (*left*) or papaverine hydrochloride (*right*) in healthy (○, $n = 7$) and alloxan-diabetic dogs (●, $n = 6$) (as percentages of basal values)

that when cardiac dysfunction is observed during diabetes similar vasculature disturbances are also present in the heart.

III. Pathomechanism

The question arises, what is the pathomechanism of this phenomenon? First, it is very important to underline that the increased vasoconstrictor tendency of the diabetic coronary arterial bed is not a simple consequence of a decreased vasodilation ability, since the intracoronary administration of sodium nitrite ot papaverine evokes intact and similar vasodilations both in the diabetic and in the healthy coronary arterial beds (Fig. 4) (POGÁTSA 1991b).

1. Adrenergic Receptor

There is no doubt that alterations in adrenergic receptors play an important role in the mechanisms of increased vasoconstictor tendency, since treatment with β-adrenergic antagonist diminishes the differences between the diabetic and healthy coronary arterial beds. However, modification of the adrenergic receptors cannot be the sole cause of these differences, i.e., the simultaneous treatment of α- and β-adrenergic antagonists does not abolish the

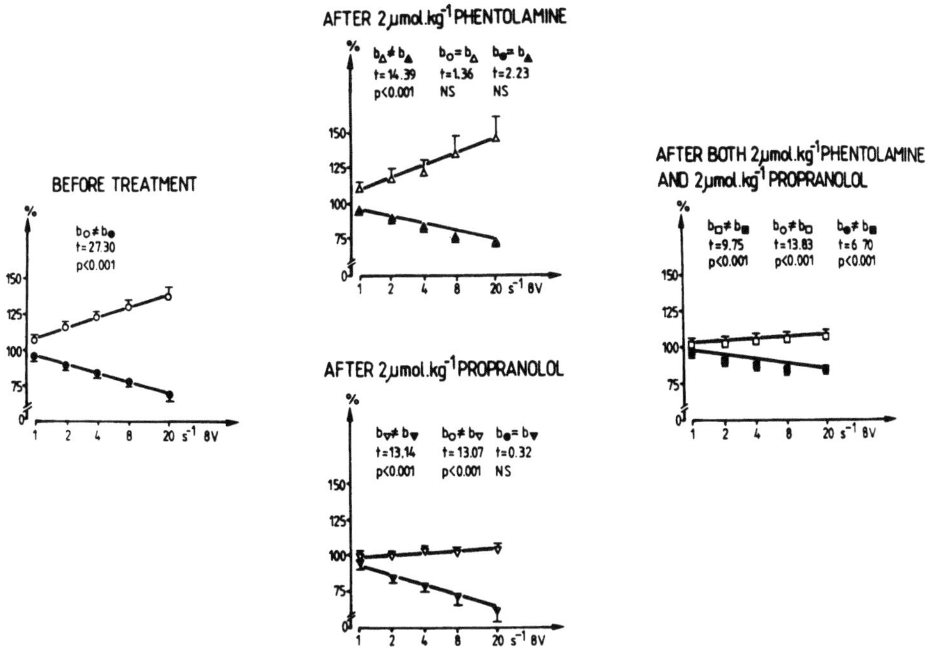

Fig. 5. Effects of phentolamine and propranolol on alterations in conductivity of coronary arterial bed provoked by electrical stimulation of cardiac plexus in healthy (○△▽□, $n = 12$) and alloxan-diabetic dogs (●▲▼■, $n = 12$) (as percentages of basal values)

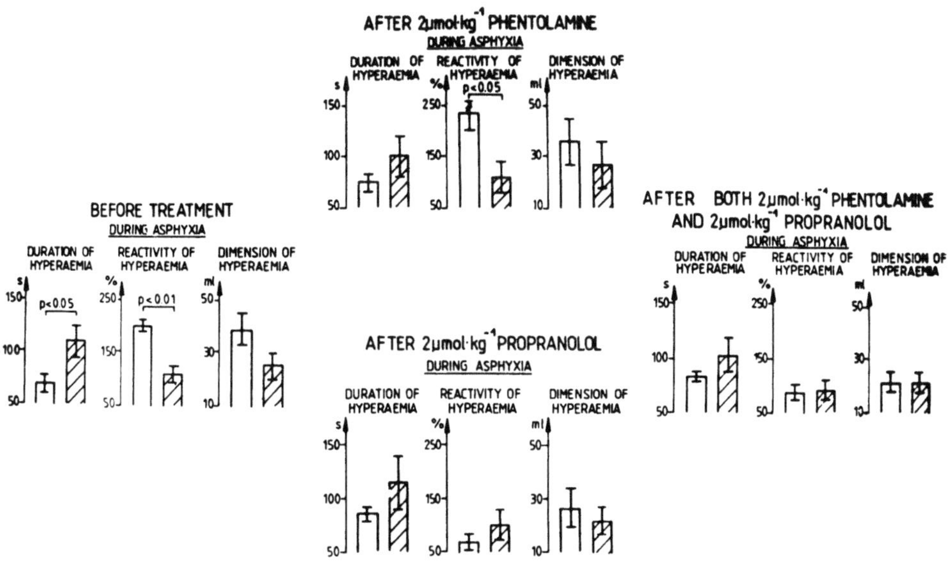

Fig. 6. Effects of phentolamine and propranolol on alterations in characteristics of reactive hyperaemia provoked by 1-min asphyxia in healthy (□, $n = 12$) and alloxan-diabetic dogs (□, $n = 12$)

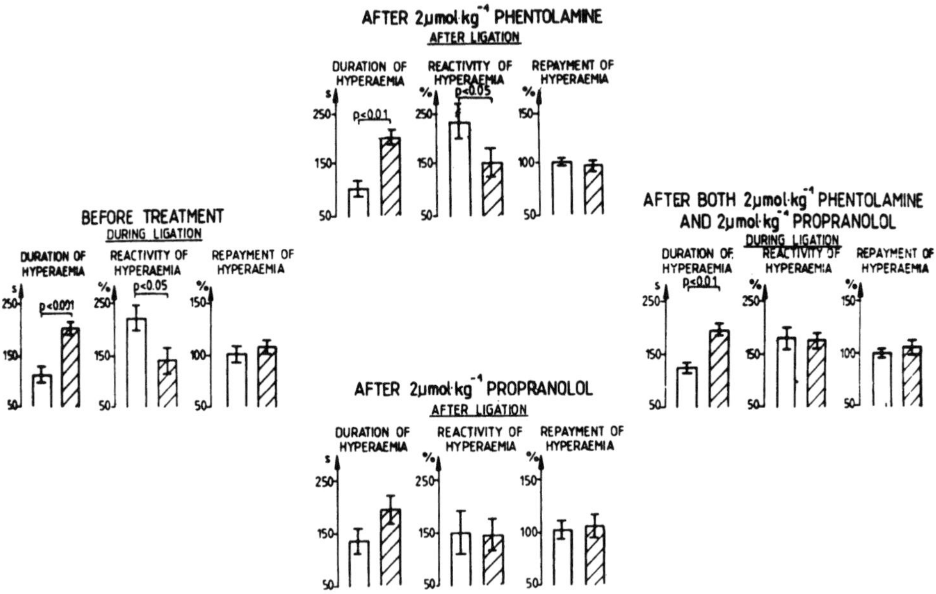

Fig. 7. Effects of phentolamine and propranolol on alterations in characteristics of reactive hyperaemia provoked by 1-min clamping of coronary artery in healthy (□, $n = 12$) and alloxan-diabetic dogs (□, $n = 12$)

differences between the diabetic and healthy coronary arterial beds (Figs. 5–7) (Pogátsa and Koltai 1983).

On the other hand, increased noradrenaline-induced responsiveness of aortae from diabetic rats may be attributed to an increased α_2-adrenoceptor component of the contractile response (Scarborough and Carrier 1983). An increased α_2-adrenoceptor mediated contraction is primarily dependent on extracellular calcium. This suggests that the increased contractile responsiveness of aortae from diabetic rats to α_2-adrenoceptor stimulation drugs may be related to an altered coupling mechanism between the α_2-adrenoceptors and the calcium ion channels or may be due to a change in membrane permeability as a result of the disease process (Scarborough and Carrier 1983). On the other hand, hypo- and hypermagnesaemia altered vascular tone in diabetes. This fact also suggests that calcium distribution, permeability and/or magnesium-calcium exchange sites may be modified in vascular smooth muscle in diabetes (Turlapaty and Altura 1980).

Furthermore, enhanced vascular responsiveness to α-adrenergic agonists in diabetes cannot be attributed to neuronal deterioration, altered post-junctional α-adrenoceptor subtypes, endothelium degeneration or enhanced release of intracellular calcium but is associated with a greater dependency on extracellular calcium. Since, first, transmural vascular nerve stimulation revealed a similar optimal frequency and voltage of stimulation between the non-diabetic and diabetic conditions, diabetic arteries developed greater contractile force than controls. Second, determination of selective α-adrenergic agonist affinities revealed quantitatively similar postjunctional α_1-adrenoceptors in normal and diabetic arteries. Third, disruption of the endothelium did not abolish the enhanced responsiveness of diabetic arteries. In contrast, the increased vascular responsiveness was associated with a greater dependency on extracellular calcium with no change in the response to α-agonist-induced release of calcium from intracellular stores (WHITE and CARRIER 1988).

2. Vasoactive Mediators

Vasoactive mediators that may interact to promote vascular smooth muscle contraction are many and include prostaglandins, platelet-dependent factors, flow-dependent vasomotion and direct modulation of smooth muscle. In diabetes mellitus both the synthesis of prostaglandins and the responsiveness to prostaglandins may be altered.

a) Prostaglandins

α) *Synthesis.* For example, segments from diabetic thoracic and abdominal aorta displayed a marked decrease in prostacyclin production (ROTH et al. 1982). Similarly, prostacyclin production is diminished both in experimentally induced diabetes of the rat and in human type-1 diabetes, and also in atherosclerotic blood vessels (SILBERBAUER et al. 1979). In platelet-rich plasma of diabetic patients, arachidonic acid-stimulated thromboxane B_2 production is positively correlated with fasting plasma glucose levels (HALUSHKA et al. 1981). Furthermore, long-term diabetes leads to diminished prostacyclin release in vessel tissues (JEREMY et al. 1985), and vascular prostacyclin synthesis is also decreased in neonates of diabetic mothers (STURRT et al. 1985). Reduced production of prostaglandins might be responsible for the impaired thromboresistance of diabetic vessels (RÖSEN et al. 1983). Production of prostaglandins is reduced in diabetes presumably due to a reduction in availability of arachidonic acid for conversion to prostaglandins. Moreover, a defect in deacylation-reacylation of arachidonic acid can be demonstrated with an increase in cyclooxygenase activity in the diabetic kidney (SARIBBI et al. 1989).

Furthermore, the pretreatment with cyclooxygenase inhibitors diminishes the differences between the diabetic and healthy coronary arterial beds during the electrical stimulation of cardiac plexus or the intracoronary administration of noradrenaline as well as in the cases of reactive hyperaemic responses evoked by 1-min asphyxia or 1-min clamping of coronary artery (KOLTAI et al. 1985a; POGÁTSA 1991a).

Furthermore, in isolated diabetic arteries it has been observed that basal levels of prostacyclin synthesis by healthy and diabetic vessels are no different. Phenylephrine or noradrenaline potentiated prostacyclin synthesis in healthy subjects, but proved to be ineffective in diabetics. Under hypoxic conditions, prostacyclin production of healthy arteries increased, while that of diabetic ones decreased. In the presence of phentolamine no difference could be detected between the two groups, since it increased in the diabetic arteries (KOLTAI et al. 1988a, 1990). Diabetic coronary and femoral arteries produced more thromboxane than the healthy ones. Phentolamine normalized thromboxane synthesis in the diabetic vessels and enhanced it in the healthy ones. Phenylephrine enhanced prostacyclin production only in the healthy coronaries (KOLTAI et al. 1988b,c).

SHAFFER and MALIK (1982) also demonstrated that activation of cardiac β-adrenoceptors, by stimulation of sympathetic nerves or administration of noradrenaline, enhances prostacyclin synthesis in the heart exclusively through the activation of the cardiac β_1-adrenoceptor. The β-adrenergic relaxation of coronary arteries is not simply related to the total tissue content of cAMP. The enhancement of the relaxing effect of isoproterenol by indomethacin and its inhibition by arachidonic acid evoked similar increases in cAMP. RUBÁNYI et al. (1986) suggested that the enhancement of phosphorylase activity observed in the presence of arachidonic acid should be associated with activation of the type-1 isozyme, and the enhanced relaxation manifested in the presence of indomethacin should be associated with enhanced activation of the type-2 isozyme. Thus, a direct relationship between cAMP, relaxation and phosphorylase can be reconciled with the antiparallel effects of alterations of prostaglandin metabolism by a mechanism in which the effects of cAMP are functionally compartmentalized.

STAM and HÜLSMANN (1977) observed an increased prostacyclin and prostaglandin E_2 synthesis in acutely diabetic hearts after application of arachidonic acid. RÖSEN et al. (1983) also established that in the diabetic heart the enzymatic capacity of the cyclooxygenase reaction, that is to say of prostaglandin synthesis, is not impaired by diabetes but the availability of free arachidonic acid is limited and therefore the basal release of prostacyclin is reduced in the diabetic heart. The regulation of phospholipase activity which mainly controls the amount of arachidonic acid for prostaglandin formation (TOMIYAMA et al. 1976; VOGT 1978) seems, therefore, to represent the key to the understanding of diabetes-induced changes in prostaglandin sythesis. The increased cardiac capacity of the diabetic heart for formation of prostaglandins might represent, therefore, a compensatory mechanism that tries to antagonize the increased thromboxane release seen in the diabetic state. MCNEILL and VADLAMUDI (1982) demonstrated that prostaglandin E_1 activates phosphorylase in diabetic hearts, and supposed that an increased sensitivity of phosphorylase towards cAMP-mediated activation and alteration in cardiac cell compartments caused by diabetes could possibly be the underlying causes for this phosphorylase response. GUDBJARNASON et al. (1987) also confirmed that diabetes is associated with a diminution in fatty acid desaturation in the heart. This alteration of diabetic myocardium might explain the fact that diabetics are more resistant to isoproterenol-induced cardiac necrosis than non-diabetics (EL-HAGE et al. 1985). Furthermore, on the basis of these investigations it can be supposed that similar alterations in prostaglandin synthesis could develop not only in the diabetic myocardium but also in the diabetic vascular smooth muscle.

β) Altered Responsiveness to Prostaglandins. On the other hand, the vasoconstrictor response to arachidonic acid and platelets is enhanced in isolated perfused diabetic rat hearts (REIBEL et al. 1983). The increase in thromboxane B_2 and prostaglandin $F_{2\alpha}$ release occurring under these conditions is comparable in control and diabetic animals, indicating enhanced vascular responsiveness to platelet-derived substances and arachidonic acid products in experimental diabetes of the rat. Coronary vascular response to thromboxane A_2 analogues is enhanced in diabetic animals at high concentrations of the mimetic (ROTH et al.1983). In contrast, unlike that shown in healthy hearts, three different stimuli (i.e., iloprost, arachidonic acid and a thromboxane mimetic U-46619), which were superimposed on ischaemia reperfusion, failed to produce an additional increment in prostacyclin release in postischaemic diabetic hearts, and stimulation of prostacyclin synthesis by coronary vasculature due to ischaemia plus reperfusion per se was not altered in chronic diabetic hearts (PIEPER and GROSS 1990).

Diabetic coronary vasculature also showed an increased responsiveness to leucotrienes C_4 and D_4. Cyclooxygenase inhibitor treatment failed to alter the coronary vascular responses to the leucotrienes (ROTH et al. 1984). It has also been demonstrated that the intracoronary infusion of prostacyclin, in increasing doses, provokes a less effective vasodilation in the diabetic coronary arterial bed than in the healthy ones (KOLTAI et al. 1985a,b, 1986a). Indomethacin (KOLTAI et al. 1985a,b, 1986a) and acetylsalicylic acid pretreatments enhance the effect of prostacyclin in diabetic dogs compared with the diminished reactions of the healthy controls (POGÁTSA 1991). On the other hand, prostaglandin $F_{2\alpha}$ induces an immediate vasoconstriction both in the healthy and in the diabetic coronary arterial bed (KOLTAI and POGÁTSA 1985). Indomethacin (KOLTAI and POGÁTSA 1985) or acetylsalicylic acid pretreatment diminishes this immediate vasoconstrictor effect only in healthy dogs, demonstrating a considerable difference between healthy and diabetic dogs in the case of indomethacin pretreatment. In diabetic coronary vasculature, REIBEL et al. (1983) also demonstrated an increased reactivity to vasoconstrictor prostaglandins which could be partially reversed by insulin treatment. In the healthy state this immediate vasoconstrictor effect of prostaglandin $F_{2\alpha}$ is followed several minutes later by vasodilation (KOLTAI and POGÁTSA 1985). Therefore, the drug could be administered only in bolus form. Indomethacin (KOLTAI and POGÁTSA 1985) or acetylsalicyclic acid pretreatment abolishes this delayed vasodilation. It is well known from the literature (DEDECKERE and HOOR 1980) that prostaglandin $F_{2\alpha}$ may enhance the synthesis of endogenous prostacyclin. Hence, it seems reasonable to assume that the absence of vasodilation in the diabetic state and its disappearance after cyclooxygenase blockade can be explained by the fact that the supply of arachidonic acid is diminished in diabetes and its metabolism is blocked by cyclooxygenase inhibition. However, with the blockade of cyclooxygenase these differences ceased only in the cases of reactive hyperaemia and not in the cases of electrical stimulation of the cardiac plexus or intracoronary administration of noradrenaline.

These facts indicate that prostaglandin metabolism also plays a role in the mechanism of the increased vasoconstrictor tendency of diabetic coronary vessels, and, in addition, confirm the hypothesis of Colwell et al. (1975), who were the first to suggest that the development of vascular disturbances and the increased platelet aggregability are a consequence of disturbed prostacyclin formation and thromboxane synthesis in diabetic patients. However, the alterations in the metabolism of prostaglandins during diabetes could not fully explain the pathomechanism of the increased vasoconstrictor tendency of the diabetic coronary vessels, and play only a minor role in this process. However, their role becomes significant when cyclooxygenase inhibitors are administered to diabetics, for example in rheumatism.

b) Adenosine

Because it is well known that adenosine is the most important vasodilator metabolite, the alterations of coronary arterial conductivity were also investigated during the administration of this substance. The vasodilation evoked by increasing doses of intracoronary adenosine infusion is considerably smaller in diabetic than in healthy dogs (Koltai et al. 1984; Pogátsa et al. 1986a). Indomethacin (Pogátsa et al. 1986a) and acetylsalicylic acid pretreatments normalize the reduced coronary arterial response in diabetic dogs. Similarly, 12 weeks after the induction of diabetes, the diabetic rat aortas exhibited less relaxation in response to theophylline when they were previously maximally contracted with phenylephrine ($10^{-4}\,M$). This diabetes-induced decrease in relaxation of the phenylephrine contracture was not reversed by insulin treatment (Pfaffman et al. 1983). Therefore, it can be concluded that adenosine-induced vasodilation is also diminished in diabetes.

c) Endothelin

Other potentially important influences of vasoactive mediators which may interact to promote vascular smooth muscle contraction include endothelial-dependent and endothelial-independent compounds, as well as modulations of endothelial functions by the atherosclerotic process. For example, endothelium-dependent relaxation by acetylcholine is severely impaired in the diabetic rat aorta (Pieper and Gross 1988). Moreover, the relaxation to acetylcholine is expressed much more strongly in diabetic than in non-diabetic coronary arterial strips (Gebremedhin et al. 1988). The mechanical removal of the endothelial layer in coronary arterial strips enhances the maximum contractile force generated by diabetic vessels whereas it does not influence that of non-diabetic arteries (GEBREMEDHIN et al. 1988). In free radical-exposed vessels, endothelium-dependent relaxation by acetylcholine is reduced by half in non-diabetic vessels and totally abolished in diabetic vessels. Nevertheless, diabetic vessels could still be fully relaxed by nitroglycerin or papaverine (PIEPER and GROSS 1988). The augmentation of the adrenergic responses in diabetic carotid arteries upon electrical stimulation and to exogenously added noradrenaline is dependent on the endothelium while endothelium-dependent relaxation to acetylcholine is unaltered in the carotid artery of the rabbit (COHEN and ZIRNAY 1986). Finally, in diabetic mesenteric microvessels histamine and bradykinin – which are potent permeability-increasing factors – may antagonize the vasoconstrictor response to noradrenaline through an action on endothelial cells with increased vascular permeability and changes in composition of extracellular fluid, since reactive processes of endothelial cells to permeability factors are affected by diabetes mellitus. However, the response of microvessels to acetylcholine and papaverine – which are devoid of permeability-increasing properties – is not influenced by diabetes (FORTES et al. 1983).

All these results refer to the assumption that the increased vasoconstrictor tendency in diabetes mellitus is a consequence of different processes, involving modification of adrenergic receptors, alterations in the metabolism of prostaglandins and injury of the endothelial layer in the coronary arterial bed.

d) Insulin

The question also arises whether the above-described alterations of the diabetic vascular smooth muscle are really consequences of diabetes only, in which case insulin treatment is required to prevent or normalize them. Therefore, the effect of insulin treatment was also investigated. It was demonstrated that effective insulin treatment could prevent and normalize the diabetic reactions induced by electrical stimulation of the cardiac plexus (POGÁTSA et al. 1986b, 1988b). This is also true with intracoronary (POGÁTSA et al. 1986b, 1988b) or intrafemoral (POGÁTSA et al. 1988a) noradrenaline infusion, while insulin treatment does not influence the decreased vasodilation induced by intracoronary (POGÁTSA et al. 1986b, 1988b) or intrafemoral (POGÁTSA et al. 1988a) prostacyclin infusion or the increased vasoconstriction induced by intracoronary administration of prostaglandin $F_{2\alpha}$ in the diabetic coronaries (POGÁTSA et al. 1986b, 1988b). Therefore, only the altered reactions evoked by sympathetic stimuli could be prevented and normalized by insulin treatment in both diabetic coronary and femoral arterial beds. Reactions to prostaglandins are not influenced. These results suggest that in diabetic coronary and femoral arteries the prostaglandins play an important role in the development of the lack of appropriate vasodilatory ability. The carbohydrate metabolic disorder is supposed to be indirectly related to the diabetic alterations of vascular lipid metabolism. Furthermore, a decrease in mean postocclusive reactive hyperaemia was observed in diabetic children which improved during the first few months of insulin treatment (Kobbah et al. 1985), but was related to the degree of long-term diabetic control (Lorini et al. 1987).

On the other hand, insulin may evoke vasodilation. The vasodilatory effects of insulin have been related to the activation of β-adrenergic receptors and at higher dosages to some unknown mechanism in the human forearm (Creager et al. 1985). In this context, it is of interest that the relaxations induced by insulin in isolated arterial rings, including human arteries, are endothelium dependent (Thom et al. 1987). Changes in the circulating levels of (i.e., hypo- or hyperinsulinaemia) and/or in the endothelial responsiveness to this hormone could therefore modulate vascular tone in diabetes and contribute to abnormal vascular reactivity in this disease (Lüscher 1988b).

IV. Clinical Consequences

Finally, the question arises whether these observations have practical clinical consequences. It is a well-known fact that diabetes can evoke numerous alterations in the cardiovascular system. Heart disease is therefore a major cause of death in diabetic patients (Garcia et al. 1969). Consequently, one of the most important diabetic alterations may be the increased vasoconstrictor tendency, especially in the coronary arterial system, because it counteracts the increase in collateral myocardial blood flow, and it may thereby cause severe alterations during myocardial hypoxia or ischaemia, contributing to the severe outcome of diabetic cardiac complications (Crall and Roberts 1978). The abnormalities in the coronary arterioles may lead to reperfusion damage in the myocardium and have been implicated as primary causes of cardiac contractile dysfunction and failure during diabetes mellitus. Maximal constriction, which might be needed to obstruct a relatively normal artery, is not necessary to obstruct flow in a severely stenosed vessel (Lambert and Pepine 1990). Furthermore, it has been demonstrated that myocardial tissue blood flow may decrease in the diabetic state during hypoxia (Fig. 8), while it may increase in the metabolically healthy state when the sympathetic tone is increased or 60 nmol/kg noradrenaline is infused intravenously over 10 min, during a similarly long inhalation of carbon monxide mixed with air in 5–7 vol%, producing a carbon monoxide haemoglobin level of 40%–60%, in alloxan-diabetic dogs, 3 months after induction of clinically manifest diabetes but without ketosis (Koltai et al. 1986b). Therefore, the size of the myocardial infarction is larger in such alloxan-diabetic dogs than in metabolically healthy dogs, 48 h after ligation of the left anterior descending coronary artery just below its first major oblique branch (Fig. 9) (Palik et al. 1982b). The possibility that this complication may be a contributing factor in heart dysfunction during diabetes mellitus requires much further research. Unfortunately, most of the evidence

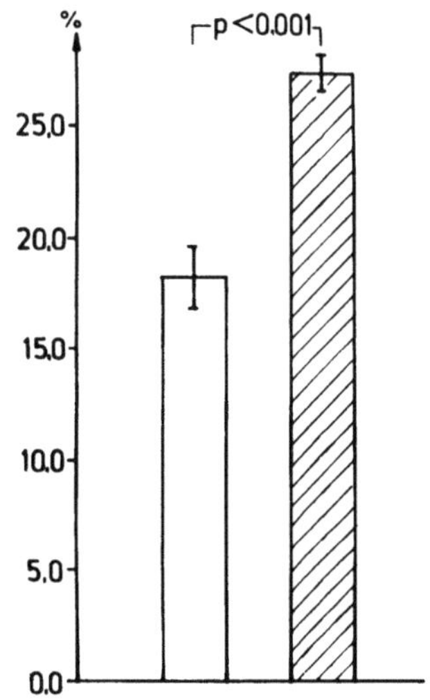

Fig. 8. Alteration in conductivity of myocardial microcirculation during carbon monoxide intoxication (10 min long inhalation of carbon monoxide mixed with air in 5–7 vol%, producing a carbon monoxide-haemoglobin level of 40%–60%) in healthy (□, $n = 6$) and alloxan-diabetic dogs (▨, $n = 6$) (as percentages of basal values)

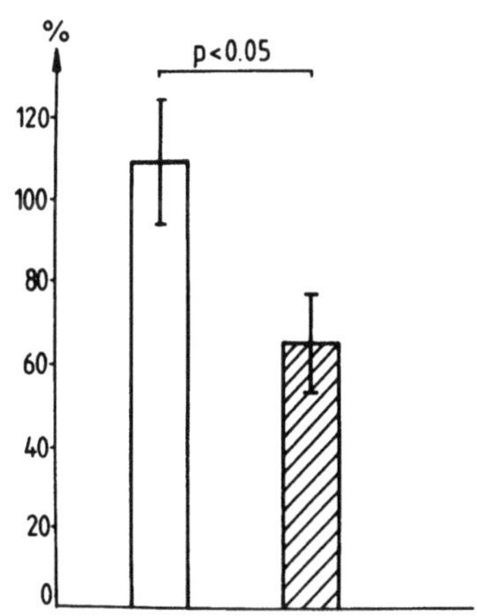

Fig. 9. Alterations in size of necrotic area related to left ventricular myocardium 48 h after ligation of left anterior descending coronary artery just below its first major oblique branch in metabolically healthy (□, $n = 6$) and alloxan-diabetic dogs (▨, $n = 6$)

reported from humans to support an involvement of the microvasculature in the depression of cardiac performance during diabetes is largely conjectural. Many studies have reported abnormalities in heart function, or electrocardiographic recordings from diabetic patients and, after observing no evidence of coronary stenosis or ischaemic heart disease, have concluded the cause of this dysfunction to be microvascular in origin (Rubler et al. 1978; Sanderson et al. 1978; Shah 1980; Kannel et al. 1974). However, no direct evidence supports such a conclusion in these studies (Pierce et al. 1988). It is assumed, therefore, that there is a coexistence of different diabetic alterations in the diabetic vasculature. Nevertheless, the altered responsiveness of vascular smooth muscle to drugs is a major contributor to the increased cardiovascular mortality rate in diabetic conditions.

C. Summary

During the past 15 years it has become evident that an increased vasoconstrictor tendency exists in diabetes in the arterial beds of different organs. This vasoconstrictor tendency is most strongly expressed in the coronary arteries, where sympathetic stimulations exert vasodilation in the healthy and vasoconstriction in the diabetic arterial beds. Consequently, myocardial tissue blood flow decreases during hypoxia in the diabetic, whereas it increases in the non-diabetic state. Furthermore, infarcted myocardial zone is extended in diabetic dogs compared with non-diabetic ones 48 h after ligation of the left anterior descending coronary artery just below its first major oblique branch. Alterations in the endothelium and in the adrenergic vascular and prostaglandin mechanisms are involved in the pathomechanisms of this phenomenon, since pretreatment with adrenergic antagonists or with cyclooxygenase inhibitors considerably diminishes the increased vasoconstrictor tendency in the diabetic state. Moreover, after mechanical removal of the endothelial layer in the coronary arterial strips the maximum force generated by diabetic vessels exceeded the maximum contraction produced by non-diabetic arteries, and the doseresponse curve of diabetic arteries to phenylephrine is steeper than it is in non-diabetic strips. All these phenomena appear in such an early state when macro- and microscopic alterations cannot be observed in the vessels and sodium nitrite or papaverine evokes similar vasodilation in the diabetic and non-diabetic coronary arterial beds.

References

Agrawal DK, McNeill JH (1986) Effect of prostaglandins E_1 and I_1 in vascular smooth muscle of alloxan-diabetic rabbits. Fed Proc 45:424

Agrawal DK, Bhimji S, McNeil JH (1987) Effect of chronic experimental diabetes on vascular smooth muscle function in rabbit carotid artery. J Cardiovasc Pharmacol 9:584–593

Bohlen HG, Niggl BA (1979) Arteriolar anatomical and functional abnormalities in juvenile mice with genetic or streptozotocin-induced diabetes mellitus. Circ Res 45:390–396

Bohlen HG, Niggl BA (1980) Early arteriolar disturbances following streptozotocin-induced diabetes mellitus in adult mice. Microvasc Res 20:19–29

Chilian WM (1990) Adrenergic vasomotion in the coronary microcirculation. Basic Res Cardiol 85 [Suppl 1]:111–120

Cohen RA, Zitnay KM (1986) Augmented adrenergic responses of diabetic carotid arteries are dependent on the endothelium. (Abstr). Circulation 74 [Suppl 2]: 413

Colwrll JH, Chambers A, Laimins M (1975) Inhibition of labile aggregation-stimulating substance (LASS) and platelet aggregation in diabetes mellitus. Diabetes 24:684–687

Crall FM, Roberts WC (1978) The extramural and intramural coronary arteries in juvenile diabetes mellitus. Am J Med 64:221–230

Creager MA, Liand CS, Coffman JD (1985) Beta-adrenergic-mediated vasodilator response to insulin in the human forearm. J Pharmacol Exp Ther 235:709–714

Dandona P, James IM, Newbury RA, Woollard ML, Beckett AG (1978) Cerebral blood flow in diabetes mellitus: evidence of abnormal cerebrovascular reactivity. Br Med J 2:325–326

DeDeckere EAM, Hoor TF (1980) $PGF_{2\alpha}$ stimulates release of PGE_2 and PGI_2 in the isolated perfused rat heart. Adv Prostaglandin Thromboxane Res 7:658–665

Drash AL (1976) Hyperlipidemia and the control of diabetes mellitus. Am J Dis Child 130:1057–1058

El-Hage AN, Herman EH, Jordan AW, Ferrans VJ (1985) Influence of the diabetic state on isoproterenol-induced cardiac necrosis. J Mol Cell Cardiol 17:361–369

Ewald U, Tuemo T, Rooth G (1981) Early reduction of vascular reactivity in diabetic children detected by transcutaneous oxygen electrode. Lancet 8213: 1287–1288

Fluckiger W, Perrin IV, Rossi GL (1984) Morphometric studies on retinal micro-angiopathy and myocardiopathy in hypertensive rats (SHR) with induced diabetes. Virchows Arch [B] 47:79–94

Fortes ZB, Leme JG, Scivoletto R (1983) Vascular reactivity in diabetes mellitus: role of the endothelial cell. Br J Pharmacol 79:771–781

Friedman JJ (1989) Vascular sensitivity and reactivity to norepinephrine in diabetes mellitus. Am J Physiol 256:H1134–H1138

Garcia MJ, McNamara PM, Gordon T, Kannell WB (1969) Morbidity and mortality in diabetics in the Framingham population. Diabetes 23:537–546

Gebremedhin D, Koltai MZ, Pogátsa G, Magyar K, Hadházy P (1988) Influence of experimental diabetes on the mechanical responses of canine coronary arteries: role of endothelium. Cardiovasc Res 22:537–544

Gudbjarnason S, El-Hage AN, Whitehurst VE, Simental F, Balázs T (1987) Reduced arachidonic acid levels in major phospholipids of heart muscle in the diabetic rat. J Mol Cell Cardiol 19:1141–1146

Gundersen HJG, Osterby R, Lundbeak K (1978) The basement membrane controversy. Diabetologia 15:361–363

Halushka PV, Mayfield R, Wohltmann HJ, Rogers RC, Goldberg AK, McCoy SA, Loadholt CB, Colwell JA (1981) Increased platelet arachidonic acid metabolism in diabetes mellitus. Diabetes 30 [Suppl 2]:44–48

Hamby RI, Sherman L, Mehta J, Aintablian A (1976) Reappraisal of the role of the diabetic state in coronary artery disease. Chest 70:251–257

Henriksen O, Kastrup J, Parving HH, Lassen NA (1984) Loss of autoregulation of blood flow in subcutaneous tissue in juvenile diabetes. J Cardiovasc Pharmacol 6:S666–S670

Jackson CV, Carrier GO (1983) Influence of short-term experimental diabetes on blood pressure and heart rate in response to norepinephrine and angiotensin II in the conscious rat. J Cardiovasc Pharmacol 5:260–265

Jeremy JY, Thompson CS, Mikhailidis DP, Dandona P (1985) Experimental diabetes mellitus inhibits prostacyclin synthesis by the rat penis: pathological implications. Diabetologia 28:365–368

Kannel WB, Hjortland M, Castelli WP (1974) Role of diabetes in congestive heart failure: the Framingham study. Am J Cardiol 34:29–34

Kaski JC, Maseri A, Vejar M, Crea F, Hackett D (1989) Spontaneous coronary artery spasm in variant angina is caused by a local hyperreactivity to a generalized constrictor stimulus. J Am Coll Cardiol 14:1456–1463
Kawano M, Kanzaki T, Koshikawa T, Morisaki N, Saito Y, Yoshida SH (1990) Pathologic phenotype of aortic smooth muscle cell causes diabetic macroangiopathy (Abstr)? Arteriosclerosis 10:841a
Kobbah M, Ewald U, Tuvemo T (1985) Vascular reactivity during the first year of diabetes in children. Acta Paediatr Scand Suppl 320:56–63
Koltai MZ, Pogátsa G (1985) Die Interaktion der Prostaglandinen und Adenosin in der Regulation der Koronardurchblutung (Abstr). Z Kardiol 74 [Suppl 5]: 911
Koltai MZ, Jermendy G, Kiss V, Wagner M, Pogátsa G (1984) The effect of sympathetic stimulation and adenosine on coronary circulation and heart function in diabetes mellitus. Acta Physiol Hung 63:119–125
Koltai MZ, Wagner M, Pogátsa G (1983) Altered hyperemic response of the coronary arterial bed in alloxan-diabetes. Experientia 39:738–740
Koltai MZ, Hadházy P, Malomvölgyi B, Pogátsa G (1985a) The role of prostaglandins in the altered coronary reactivity of alloxan-diabetic dogs (Abstr). G Arteriosclerosi 1 [Suppl 1]:119
Koltai MZ, Hadházy P, Pogátsa G (1985b) Effects of prostaglandins on coronary arteries (Abstr). J Mol Cell Cardiol 17 [Suppl 3]:120
Koltai MZ, Hadházy P, Malomvölgyi B, Kiss V, Pogátsa G (1986a) Effect of prostacyclin on the coronary, femoral and coeliac arterial bed in diabetes mellitus. Adv Pharmacol Res Pract 3:377–382
Koltai MZ, Wagner M, Balogh I, Kiss V, Köszeghy A, Pogátsa G (1986b) Effect of acute hypoxia on cardiac function in alloxan-diabetic dogs. Basic Res Cardiol 81:92–100
Koltai MZ, Kösen P, Hadházy P, Ballagi-Pordány G, Köszeghy A, Pogátsa G (1988a) Effects of hypoxia and adrenergic stimulation induced alterations in PGI_2 synthesis by diabetic coronary arteries. J Diabetes Complic 1:5–7
Koltai MZ, Rösen P, Hadházy P, Ballagi-Pordány G, Köszeghy A, Pogátsa G (1988b) Relationship between vascular adrenergic receptors and prostaglandin biosyntheses in canine diabetic coronary arteries. Diabetologia 31:681–686
Koltai MZ, Rösen P, Hadházy P, Ballagi-Pordány G. Aranyi Z, Pogátsa G (1988c) Hypoxia-induced alterations of prostaglandin synthesis mediated by α-adrenoceptors in canine coronary arteries (Abstr). J Mol Cell Cardiol 20 [Suppl 5]:611
Koltai MZ, Rösen P, Ballagi-Pordány G, Hadházy P, Pogátsa G (1990) Increased vasoconstrictor response to norepinephrine in femoral vascular bed of diabetic dogs. Is thromboxane A_2 involved? Cardiovasc Res 24:707–710
Koltai MZ, Rösen P, Hadházy P, Ballagi-Pordány G, Aranyi Z, Pogátsa G (1994) The role of vascular adrenergic mechanism in the haemodynamic and prostacyclin stimulating effects of angiotensin in diabetic dogs. Circ Res (in press)
Kubota I, Fukuhara T, Kinoshita M (1990) Permeability of small coronary arteries and myocardial injury in hypertensive diabetic rats. Int J Cardiol 29:349–355
Lambert CR, Pepine CJ (1990) Coronary artery spasm: American view. Coronary Artery Dis 1:654–659
Langsh HG, Nowak W, Mohnike A (1970) Diabetes mellitus: 10. Makroangiopathie und Neuropathie bei Diabetes mellitus. Z Arztl Fortbild 64:867–871
Longhurst PA, Head RJ (1985) Responses of the isolated perfused mesenteric vasculature from diabetic rats: the significance of appropriate control tissues. J Pharmacol Exp Ther 235:45–49
Lorini R, Chirico G, Larizza D, Cortona L, Rondini G, Severi F (1987) Vascular reactivity in diabetic children. Acta Paediatr Scand 76:151–152
Lucas PD, Foy JM (1977) Effects of experimental diabetes and genetic obesity on regional blood flow in the rat. Diabetes 26:786–792

Lüscher TF (1988a) Atherosclerosis: vascular responsiveness of atherosclerotic blood vessels. In: Lüscher TF (ed) Endothelial vasoactive substances and cardiovascular disease. Karger, Basel, pp 83–85
Lüscher TF (1988b) Diabetic vascular disease. In: Lüscher TF (ed) Endothelial vasoactive substances and cardiovascular disease. Karger, Basel, p 111
McNeill JH, Vadlamudi RVSV (1982) Effects of acute and chronic experimental diabetes on rat cardiac cyclic AMP and phosphorylase-A levels. Fed Proc 41:1358
Morff RJ (1990) Microvascular reactivity to norepinephrine at different arteriolar levels and durations of streptozocin-induced diabetes. Diabetes 39:354–360
Palik I, Koltai MZ, Wagner M, Kolonics I, Pogátsa G (1982a) Altered adrenergic responses of the coronary arterial bed in alloxan-diabetic dogs. Experientia 38:934–935
Palik I, Koltai MZ, Wagner M, Kolonics I, Pogátsa G (1982b) Effects of coronary occlusion and norepinephrine on the myocardium of alloxen-diabetic dogs. Basic Res Cardiol 77:499–505
Pfaffman MA, Dudley P, Prater A (1983) Relationship between untreated and insulin-treated diabetes and vascular relaxation. Arch Int Pharmacodyn Ther 266:131–143
Pieper GM, Gross GJ (1988) Oxygen free radicals abolish endothelium-dependent relaxation in diabetic rat aorta. Am J Physiol 255:H825–H833
Pieper GM, Gross GJ (1990) Differential response of postischemic diabetic myocardium to a thromboxane-mimetic. Eicosanoids 3:127–133
Pierce GN, Beamish RE, Dhalla NS (1988) Dysfunction of the cardiovascular system during diabetes: etiology of heart failure during diabetes: microvascular lesions in the heart. In: Pierce GN, Beamish RE, Dhalla NS (eds) Heart dysfunction in diabetes. CRC, Boca Raton, pp 63–67
Pogátsa G (1980) Altered adrenergic response of the coronary and femoral arterial bed in alloxan-diabetic dogs. Adv Physiol Sci 27:213–226
Pogátsa G (1991a) Effect of prostaglandins on the diabetic heart and coronary circulation. In: Nagano M, Dhalla NJ (eds) The diabetic heart. Raven, New York, pp 45–58
Pogátsa G (1991b) The role of diabetic vascular alterations in the development of myocardial ischaemia. Bratisl Lek Listy 92:24–33
Pogátsa G, Koltai MZ (1983) Altered vascular reactivity of coronary and femoral arterial beds in alloxan-diabetic dogs. Proc Int Union Physiol Sci 15:493
Pogátsa G, Koltai MZ, Hadházy P, Köszeghy A, Ballagi-Pordány G (1986a) Interaction between prostanoids and vasodilating endogenous agents in the altered reactivity in diabetes mellitus. Proc Int Union Physiol Sci 16:595
Pogátsa G, Koltai MZ, Köszeghy A, Ballagi-Pordány G (1986b) Effects of insulin treatment and indomethacin on altered vascular reactivity in diabetes mellitus (Abstr). J Mol Cell Cardiol 18 [Suppl 2]:98
Pogátsa G, Koltai MZ, Hadházy P, Köszeghy A, Ballagi-Pordány G (1988a) Insulin induced reversibility of altered responsiveness in femoral arterial bed of diabetic dogs. Diabetes Res 9:41–45
Pogátsa G, Koltai MZ, Ballagi-Pordány G (1988b) Effect of insulin treatment on the altered coronary vascular reactions in diabetes (Abstr). J Mol Cell Cardiol 20 [Suppl 5]:60
Reibel DK, Roth DM, Lefer BL, Lefer AM (1983) Hyperreactivity of coronary vasculature in platelet-perfused hearts from diabetic rats. Am J Physiol 245: H640–H645
Rösen P, Senger W, Freuerstein J, Grote H, Reinauer H, Schrör K (1983) Influence of streptozotocin diabetes on myocardial lipids and prostaglandin release by the rat heart. Biochem Med 30:19–33
Roth DM, Reibel DK, Lefer AM (1982) Altered vascular reactivity and prostacyclin generation in diabetic rats. Fed Proc 41:856

Roth DM, Reibel DK, Lefer AM (1983) Vascular responsiveness and eicosanoid production in diabetic rats. Diabetologia 24:372–376
Roth DM, Reibel DK, Lefer AM (1984) Altered coronary vascular responsiveness to leucotrienes in alloxan-diabetic rats. Circ Res 54:388–395
Rubányi G, Galvas P, DiSalvo J, Paul RJ (1986) Eicosanoid metabolism and β-adrenergic mechanisms in coronary arterial smooth muscle: potential compartmentation of cAMP. Am J Physiol 250:C406–C412
Rubler S, Sajadi MRM, Araoye MA, Holford FD (1978) Noninvasive estimation of myocardial performance in patents with diabetes. Effect of alcohol administration. Diabetes 27:127–134
Sanderson JE, Brown DJ, Rivellese A, Kohner E (1978) Diabetic cardiomyopathy? An echocardiography study of young diabetics. Br Med J 1:404–407
Sarubbi D, McGriff JC, Quilley J (1989) Renal vascular responses and eicosanoid release in diabetic rats. Am J Physiol 257:F762–F768
Scarborough NL, Carrier GO (1983) Increased α_2-adrenoreceptor mediated vascular contraction in diabetic rats. J Auton Pharmacol 3:177–183
Shaffer JE, Malik KU (1982) Activation of cardiac β-adrenoceptors enhances the output of prostaglandins in the rabbit heart. Fed Proc 41:1766
Shah S (1980) Cardiomyopathy in diabetes mellitus. Angiology 31:502–504
Silberbauer K, Schernthaner G, Sizinger H, Clopath P, Piza-Katzer H, Winer M (1979) Diminished prostacyclin generation in human and experimentally induced (streptozotocin, alloxan) diabetes mellitus. Thromb Hemost 42:334
Siperstein MD, Unger RH, Madison LL (1968) Studies of muscle capillary membrane in normal subjects, diabetic and prediabetic patients. J Clin Invest 47:1973–1999
Stam H, Hülsmann WC (1977) Effect of fasting and streptozotocin-diabetes on the coronary flow in isolated rat hearts: a possible role of endogenous catecholamines and prostaglandins. Basic Res Cardiol 72:365–375
Steiner G (1981) Diabetes and atherosclerosis. Diabetes 30 [Suppl 2]:1–7
Stuart MJ, Setty Y, Sunderji S, Boone S, Ganley C (1985) Abnormalities in vascular arachidonic acid metabolism in the infant of the diabetic mother (Abstr). Pediatr Res 19:321A
Szentiványi M, Pék L (1973) Characteristic changes of vascular adrenergic reactions in diabetes mellitus. Nature [New Biol] 243:276–277
Thom SA, Hughes AD, Martin G, Sever PS (1987) Endothelium-dependent relaxation in isolated human arteries and veins. Clin Sci 73:547–552
Tomiyama M, Minagawa A, Suzuki T, Munahata J, Iwasaki T, Tanaka K, Nakajima S, Takayama K (1976) Muscle metabolism during exercise. Jpn Diabetic Soc 19:130–140
Turlapaty PDMV, Altura BM (1980) Magnesium ions and contractions of alloxan-diabetic vascular muscle. Artery 6:375–384
Vogt W (1978) Role of phospholipase A_2 in prostaglandin formation. Adv Prostaglandin Thromboxane Res 3:89–95
Waber S, Meister V, Rossi GL, Mordasini C, Reisen WF (1981) Studies on retinal microangiopathy and coronary macroangiopathy in rats with streptozotocin-induced diabetes. Virchows Arch [B] 37:1–10
White RE, Carrier GO (1988) Enhanced vascular-adrenergic neuroeffector system in diabetes: importance of calcium. Am J Physiol 255:H1036–H1042
Zhao JB, Mikata A, Azuma K (1990) Immunoglobulin deposits in diabetic microangiopathy. Observations in autopsy materials. Acta Pathol Jpn 40:729–734

Subject Index

MIX
Papier aus verantwortungsvollen Quellen
Paper from responsible sources
FSC® C105338

FSC
www.fsc.org

If you have any concerns about our products,
you can contact us on
ProductSafety@springernature.com

In case Publisher is established outside the EU,
the EU authorized representative is:
Springer Nature Customer Service Center GmbH
Europaplatz 3, 69115 Heidelberg, Germany

Printed by Libri Plureos GmbH
in Hamburg, Germany